T0391746

Meteoroids

This definitive guide provides advanced students and researchers with a detailed yet accessible overview of all of the central topics of meteor science. Leading figures from the field summarise their active research on themes ranging from the physical composition of meteoroids to the most recent optical and radar observations and ongoing theoretical developments. Crucial practical issues are also considered, such as the risk posed by meteoroids – to spacecraft, and on the ground – and future avenues of research are explored. Taking advantage of the latest dynamical models, insights are offered into meteor flight phenomena and the evolution of meteoroid streams and complexes, as well as descriptions of the in-depth laboratory analysis of recovered material. The rapid rate of progress in twenty-first-century research makes this volume essential reading for anyone who wishes to understand how recent developments broaden our understanding of meteors, meteoroids and their origins.

GALINA O. RYABOVA is Principal Researcher at Tomsk State University. Her research interests include the mathematical modelling of meteoroid streams and interrelations in the asteroid–comet–meteoroid complex. She has served on the Organizing Committee of the International Astronomical Union's Commission on "Meteors, Meteorites and Interplanetary Dust", and is a council member of the International Meteor Organization (IMO).

DAVID J. ASHER is Visiting Research Fellow at Armagh Observatory and Planetarium, whose interests include solar system dynamics. His work with Robert H. McNaught brought the Leonid meteor storms of 1999–2002 to public attention worldwide. He has served on the Organizing Committee of the IAU's "Meteors, Meteorites and Interplanetary Dust" Commission and, through his long involvement with the IMO, maintains extensive links with the active amateur meteor community in many countries.

MARGARET D. CAMPBELL-BROWN is Associate Professor at the University of Western Ontario and Vice-President of the IAU's Commission on "Meteors, Meteorites and Interplanetary Dust". Her research focuses on the origin of sporadic meteoroids and their physical and chemical properties, and her telescopic tracking system for meteors has produced unprecedented data on the fragmentation and dynamics of small meteoroids.

Cambridge Planetary Science

Series Editors:

Fran Bagenal, David Jewitt, Carl Murray, Jim Bell, Ralph Lorenz, Francis Nimmo, Sara Russell

Books in the Series:

1. Jupiter: The Planet, Satellites and Magnetosphere[†]
   Edited by Bagenal, Dowling and McKinnon
   978-0-521-03545-3

2. Meteorites: A Petrologic, Chemical and Isotopic Synthesis[†]
   Hutchison
   978-0-521-03539-2

3. The Origin of Chondrules and Chondrites[†]
   Sears
   978-1-107-40285-0

4. Planetary Rings[†]
   Esposito
   978-1-107-40247-8

5. The Geology of Mars: Evidence from Earth-Based Analogs[†]
   Edited by Chapman
   978-0-521-20659-4

6. The Surface of Mars[†]
   Carr
   978-0-521-87201-0

7. Volcanism on Io: A Comparison with Earth[†]
   Davies
   978-0-521-85003-2

8. Mars: An Introduction to its Interior, Surface and Atmosphere[†]
   Barlow
   978-0-521-85226-5

9. The Martian Surface: Composition, Mineralogy and Physical Properties
   Edited by Bell
   978-0-521-86698-9

10. Planetary Crusts: Their Composition, Origin and Evolution[†]
    Taylor and McLennan
    978-0-521-14201-4

11. Planetary Tectonics[†]
    Edited by Watters and Schultz
    978-0-521-74992-3

12. Protoplanetary Dust: Astrophysical and Cosmochemical Perspectives[†]
    Edited by Apai and Lauretta
    978-0-521-51772-0

13. Planetary Surface Processes
    Melosh
    978-0-521-51418-7

14. Titan: Interior, Surface, Atmosphere and Space Environment
    Edited by Müller-Wodarg, Griffith, Lellouch and Cravens
    978-0-521-19992-6

15. Planetary Rings: A Post-Equinox View (Second edition)
    Esposito
    978-1-107-02882-1

16. Planetesimals: Early Differentiation and Consequences for Planets
    Edited by Elkins-Tanton and Weiss
    978-1-107-11848-5

17. Asteroids: Astronomical and Geological Bodies
    Burbine
    978-1-107-09684-4

18. The Atmosphere and Climate of Mars
    Edited by Haberle, Clancy, Forget, Smith and Zurek
    978-1-107-01618-7

19. Planetary Ring Systems
    Edited by Tiscareno and Murray
    978-1-107-11382-4

20. Saturn in the 21st Century
    Edited by Baines, Flasar, Krupp and Stallard
    978-1-107-10677-2

21. Mercury: The View after MESSENGER
    Edited by Solomon, Nittler and Anderson
    978-1-107-15445-2

22. Chondrules: Records of Protoplanetary Disk Processes
    Edited by Russell, Connolly Jr. and Krot
    978-1-108-41801-0

23. Spectroscopy and Photochemistry of Planetary Atmospheres and Ionospheres
    Krasnopolsky
    978-1-107-14526-9

24. Remote Compositional Analysis: Techniques for Understanding Spectroscopy, Mineralogy, and Geochemistry of Planetary Surfaces
    Edited by Bishop, Bell III and Moersch
    978-1-107-18620-0

25. Meteoroids: Sources of Meteors on Earth and Beyond
    Edited by Ryabova, Asher and Campbell-Brown
    978-1-108-42671-8

† Reissued as a paperback

METEOROIDS
Sources of Meteors on Earth and Beyond

Edited by
GALINA O. RYABOVA
Tomsk State University

DAVID J. ASHER
Armagh Observatory and Planetarium

MARGARET D. CAMPBELL-BROWN
University of Western Ontario

CAMBRIDGE
UNIVERSITY PRESS

University Printing House, Cambridge CB2 8BS, United Kingdom

One Liberty Plaza, 20th Floor, New York, NY 10006, USA

477 Williamstown Road, Port Melbourne, VIC 3207, Australia

314–321, 3rd Floor, Plot 3, Splendor Forum, Jasola District Centre, New Delhi – 110025, India

79 Anson Road, #06–04/06, Singapore 079906

Cambridge University Press is part of the University of Cambridge.

It furthers the University's mission by disseminating knowledge in the pursuit of education, learning, and research at the highest international levels of excellence.

www.cambridge.org
Information on this title: www.cambridge.org/9781108426718
DOI: 10.1017/9781108606462

© Cambridge University Press 2019

This publication is in copyright. Subject to statutory exception and to the provisions of relevant collective licensing agreements, no reproduction of any part may take place without the written permission of Cambridge University Press.

First published 2019

Printed the United Kingdom by TJ International Ltd. Padstow Cornwall

A catalogue record for this publication is available from the British Library.

Library of Congress Cataloging-in-Publication Data
Names: Ryabova, Galina O., 1955- editor. | Asher, David J., 1966- editor. | Campbell-Brown, Margaret D., 1976- editor.
Title: Meteoroids : sources of meteors on Earth and beyond / edited by Galina O. Ryabova
(Tomsk State University, Russian Federation), David J. Asher (Armagh Observatory and Planetarium),
Margaret D. Campbell-Brown (University of Western Ontario).
Description: Cambridge ; New York, NY : Cambridge University Press, 2019. |
Includes bibliographical references and index.
Identifiers: LCCN 2019004230 | ISBN 9781108426718 (hardback : alk. paper)
Subjects: LCSH: Meteoroids.
Classification: LCC QB738 .M485 2019 | DDC 523.5/1-dc23
LC record available at https://lccn.loc.gov/2019004230

ISBN 978-1-108-42671-8 Hardback

Cambridge University Press has no responsibility for the persistence or accuracy of URLs for external or third-party internet websites referred to in this publication and does not guarantee that any content on such websites is, or will remain, accurate or appropriate.

Contents

List of Contributors	page *ix*
Preface	*xiii*
Acknowledgements	*xiv*
Introduction: Meteor Astronomy in the Twenty-First Century DAVID ASHER, MARGARET CAMPBELL-BROWN AND GALINA RYABOVA	1

Part I Meteor Physics … 7

1. Modelling the Entry of Meteoroids … 9
   OLGA POPOVA, JIŘÍ BOROVIČKA, AND MARGARET D. CAMPBELL-BROWN

2. Physical and Chemical Properties of Meteoroids … 37
   JIŘÍ BOROVIČKA, ROBERT J. MACKE, MARGARET D. CAMPBELL-BROWN,
   ANNY-CHANTAL LEVASSEUR-REGOURD,
   FRANS J. M. RIETMEIJER, AND TOMÁŠ KOHOUT

Part II Meteor Observations on the Earth … 63

3. Radar Observations of Meteors … 65
   JOHAN KERO, MARGARET D. CAMPBELL-BROWN, GUNTER STOBER,
   JORGE LUIS CHAU, JOHN DAVID MATHEWS, AND ASTA PELLINEN-WANNBERG

4. Meteors and Meteor Showers as Observed by Optical Techniques … 90
   PAVEL KOTEN, JÜRGEN RENDTEL, LUKÁŠ SHRBENÝ, PETER GURAL, JIŘÍ BOROVIČKA,
   AND PAVEL KOZAK

Part III Meteors on the Moon and Planets … 117

5. Extra-Terrestrial Meteors … 119
   APOSTOLOS CHRISTOU, JÉRÉMIE VAUBAILLON, PAUL WITHERS, RICARDO HUESO,
   AND ROSEMARY KILLEN

6. Impact Flashes of Meteoroids on the Moon … 136
   JOSÉ M. MADIEDO, JOSÉ L. ORTIZ, MASAHISA YANAGISAWA, JESÚS ACEITUNO
   AND FRANCISCO ACEITUNO

	Part IV	**Interrelations**	159
	7	From Parent Body to Meteor Shower: The Dynamics of Meteoroid Streams Jérémie Vaubaillon, Luboš Neslušan, Aswin Sekhar, Regina Rudawska, and Galina O. Ryabova	161
	8	Asteroid–Meteoroid Complexes Toshihiro Kasuga and David Jewitt	187
	9	Minor Meteor Showers and the Sporadic Background Iwan P. Williams, Tadeusz J. Jopek, Regina Rudawska, Juraj Tóth, and Leonard Kornoš	210
	10	Interstellar Meteoroids Mária Hajduková Jr., Veerle Sterken and Paul Wiegert	235
	Part V	**Hazard**	253
	11	The Meteoroid Impact Hazard for Spacecraft Gerhard Drolshagen and Althea V. Moorhead	255
	12	Impact Hazard of Large Meteoroids and Small Asteroids Vladimir Svetsov, Valery Shuvalov, Gareth Collins, and Olga Popova	275
	Index		299

A color plate section can be found between pages 274 and 275

Contributors

FRANCISCO ACEITUNO
Sierra Nevada Observatory
Institute of Astrophysics of Andalusia
Spanish Council for Scientific Research
Spain

JESÚS ACEITUNO
Spanish-German Astronomical Centre
Calar Alto
Almería
Spain

DAVID J. ASHER
Armagh Observatory and Planetarium
Armagh, BT61 9DG
Northern Ireland
United Kingdom

JIŘÍ BOROVIČKA
Department of Interplanetary Matter
Astronomical Institute of the Czech Academy of Sciences
Ondřejov
Czech Republic

MARGARET D. CAMPBELL-BROWN
Department of Physics and Astronomy
University of Western Ontario
London, ON
Canada

JORGE LUIS CHAU
Department Radar Remote Sensing
Leibniz-Institute of Atmospheric Physics
Kühlungsborn
Germany

APOSTOLOS CHRISTOU
Armagh Observatory and Planetarium
Armagh, BT61 9DG
Northern Ireland
United Kingdom

GARETH COLLINS
Department of Earth Science and Engineering
Imperial College London
London
United Kingdom

GERHARD DROLSHAGEN
Division for Medical Radiation Physics and Space Environment
Fak. VI Medicine and Health Sciences
Carl von Ossietzky University Oldenburg
Germany

PETER GURAL
Gural Software Development and Analysis LLC
Sterling, VA
USA

MÁRIA HAJDUKOVÁ JR.
Astronomical Institute of the Slovak Academy of Sciences
Bratislava
Slovakia

RICARDO HUESO
Applied Physics I
Bilbao School of Engineers
University of the Basque Country
Bilbao
Spain

DAVID JEWITT
Earth, Planetary and Space Sciences / Physics and Astronomy
University of California, Los Angeles
Los Angeles, CA
USA

TADEUSZ J. JOPEK
Institute Astronomical Observatory
Faculty of Physics
Adam Mickiewicz University
Poznań
Poland

TOSHIHIRO KASUGA
Public Relations Center
National Astronomical Observatory of Japan
Tokyo
Japan
and
Department of Physics
Kyoto Sangyo University
Kyoto
Japan

JOHAN KERO
Solar Terrestrial and Atmospheric Research Programme
Swedish Institute of Space Physics
Kiruna
Sweden

ROSEMARY KILLEN
NASA Goddard Space Flight Center
Planetary Magnetospheres
Greenbelt, MD
USA

TOMÁŠ KOHOUT
Department of Geosciences and Geology
University of Helsinki
Helsinki
Finland
and
Institute of Geology of the Czech Academy of Sciences
Prague
Czech Republic

LEONARD KORNOŠ
Faculty of Mathematics, Physics and Informatics
Comenius University
Bratislava
Slovakia

PAVEL KOTEN
Department of Interplanetary Matter
Astronomical Institute of the Czech Academy of Sciences
Ondřejov
Czech Republic

PAVLO KOZAK
Astronomical Observatory
Taras Shevchenko National University of Kyiv
Kyiv
Ukraine

ANNY-CHANTAL LEVASSEUR-REGOURD
LATMOS
Sorbonne Université, CNRS, IPSL, UVSQ
Paris
France

ROBERT J. MACKE
Vatican Observatory
Vatican City-State

JOSÉ M. MADIEDO
Faculty of Experimental Sciences
University of Huelva
Huelva
Spain

JOHN DAVID MATHEWS
Department of Electrical Engineering
The Pennsylvania State University
University Park, PA
USA

ALTHEA MOORHEAD
NASA Meteoroid Environment Office
Marshall Space Flight Center, EV44
Huntsville, AL
USA

LUBOŠ NESLUŠAN
Astronomical Institute of the Slovak Academy of Sciences
Tatranská Lomnica
Slovakia

JOSÉ L. ORTIZ
Institute of Astrophysics of Andalusia,
Spanish Council for Scientific Research
Granada
Spain

ASTA PELLINEN-WANNBERG
Department of Physics
Umeå University
Umeå
Sweden

OLGA POPOVA
Institute of Dynamics of Geospheres
Russian Academy of Sciences
Moscow
Russia

JÜRGEN RENDTEL
Leibniz-Institute for Astrophysics Potsdam (AIP)
Research Branch Cosmic Magnetic Fields
Potsdam
Germany
and
International Meteor Organization
Potsdam
Germany

FRANS J. M. RIETMEIJER
Department of Earth and Planetary Sciences
University of New Mexico
Albuquerque, NM
USA

REGINA RUDAWSKA
Scientific Support Office
European Space Research and Technology Centre
   (ESA/ESTEC)
AZ Noordwijk
The Netherlands

GALINA O. RYABOVA
Research Institute of Applied Mathematics and Mechanics
Tomsk State University
Tomsk
Russia

ASWIN SEKHAR
Centre for Earth Evolution and Dynamics
Faculty of Mathematics and Natural Sciences
University of Oslo
Norway

LUKÁŠ SHRBENÝ
Department of Interplanetary Matter
Astronomical Institute of the Czech Academy of Sciences
Ondřejov
Czech Republic

VALERY SHUVALOV
Institute of Dynamics of Geospheres
Russian Academy of Sciences
Moscow
Russia

VEERLE STERKEN
Institute of Applied Physics (IAP) and
    Astronomical Institute University of Bern (AIUB)
Faculty of Science, Department of Physics and Astronomy
Bern
Switzerland

GUNTER STOBER
Department Radar Remote Sensing
Leibniz-Institute of Atmospheric Physics
Kühlungsborn
Germany

VLADIMIR SVETSOV
Institute of Dynamics of Geospheres
Russian Academy of Sciences
Moscow
Russia

JURAJ TÓTH
Faculty of Mathematics, Physics and Informatics
Comenius University
Bratislava
Slovakia

JÉRÉMIE VAUBAILLON
Institute for Celestial Mechanics
    and Calculation of Ephemerides
Paris Observatory
Paris
France

PAUL WIEGERT
Department of Physics and Astronomy
University of Western Ontario
London ON
Canada

IWAN P. WILLIAMS
School of Physics and Astronomy
Queen Mary University of London
London E1 4NS
United Kingdom

PAUL WITHERS
Astronomy Department
Boston University
Boston, MA
USA

MASAHISA YANAGISAWA
University of Electro-Communications
Department of Engineering Science
Tokyo
Japan

Preface

This Meteoroids-book project came about as an initiative of IAU Commission F1 (the "Meteors, Meteorites and Interplanetary Dust" commission of the International Astronomical Union). Among existing research-level publications covering topics related to meteors and meteoroids was "Meteors in the Earth's Atmosphere" edited by Edmond Murad and Iwan P. Williams, published in 2002 by Cambridge University Press (ISBN 0521804310). This book reviewed the state of knowledge at the time: it focused mostly on the observations of meteors in the Earth's atmosphere and the near-Earth environment, and therefore additional important topics such as space dust measurements and evolution of meteoroid streams were included. The 2006 book "Meteor Showers and their Parent Comets" by Peter Jenniskens (Cambridge University Press 2006, ISBN 978-0521853491) has established itself as an extremely widely used book in our meteor research community, focusing on the characteristics of all known showers, meteor shower and outburst predictions, and interrelationships with comets and asteroids. The focus of "Interplanetary Dust" edited by Eberhard Grün, Bo Å. S. Gustafson, Stanley F. Dermott and Hugo Fechtig (Springer 2001, ISBN 3540420673) was primarily interplanetary, circumplanetary and interstellar dust, with one chapter focusing on the Leonid meteor shower. Both technical and scientific progress in meteor science have rapidly increased, and we realized the need had arisen in recent years for a new reference book reporting the current knowledge.

For this reason, during the "Meteoroids 2016" conference in Noordwijk, the Organizing Committee (OC) of IAU C.F1 established an editorial board for this book. Chapter proposals were assessed, and the range of chapters chosen, by the Commission F1 OC. One requirement was that the topic of each chapter be sufficiently general, with, for example, no chapter focusing exclusively on just a single meteoroid stream. The second requirement – that topics discussed in the earlier published works described previously could be considered if they needed an essential update – aimed to complement these important existing books. And the third very significant requirement, although not an absolute one, was that institutions on a chapter's authors should cover at least two countries, with all authors on a chapter if possible from different institutions. Our aim was to strengthen the existing international and inter-institutional relations in meteor science and establish new links.

This new volume is intended for interested researchers, including undergraduate and PhD students, with attention given to clarity of the text for non-experts. So while it is written for experts in what is a rather specialist field, there will be useful information for meteor astronomers at a range of levels. The book contains twelve chapters, written by leading scientists, that review the major aspects of meteor astronomy, summarizing the progress and current state of the art of the field. A feature common to every chapter is to finish with a future work section, thus establishing a framework for the directions in which the subject will move in the next decade and beyond.

Acknowledgements

The project, which we called "Meteoroids-book" in our IAU Commission F1, lasted three years, and in addition to the chapter authors many people made significant contributions. We appreciate the Cambridge University Press Referees' and Series Editors' careful consideration of the book proposal and their thoughtful and constructive comments. The CUP staff were very helpful and we are particularly grateful to our Senior Commissioning Editor Vince Higgs for all his advice and support, and to Sarah Lambert for her extensive help as the project progressed. Finding the ideal title for the volume was proving problematic until Jiří Borovička, President of IAU C.F1 at that time, came up with the excellent idea for the title you can now see on our nice cover. Our scientific referees put a lot of work into their careful reviews, and in addition to those fifteen of our colleagues who preferred to make this valuable contribution to our project anonymously it is a pleasure to thank George Flynn, Tadeusz Jopek, Donovan Mathias, Petr Pokorný, Duncan Steel, Giovanni Valsecchi, Joel Younger, and especially Peter Brown, who reviewed several chapters. The SAO/NASA Astrophysics Data System (ADS), now so widely used, has been of great value in a volume intended to serve as a reference work and with such numerous citations to the literature. And finally we would like to thank our TeXpert Regina Rudawska for her technical assistance with LaTeX and Overleaf: she made the bumpy process of the book consolidation much easier than we would have found it.

Introduction: Meteor Astronomy in the Twenty-First Century

David Asher, Margaret Campbell-Brown and Galina Ryabova

We can only speculate how long ago a human first consciously noticed a shooting star in the sky. Writings go back thousands of years. The Egyptian hieratic papyrus of the Hermitage museum in St. Petersburg (archive number 1115) dates from between the twentieth and seventeenth centuries BC and mentions a falling star in the "Tale of the Shipwrecked Sailor" (Astapovich, 1958). Aristotle hypothesised on the nature of meteors, though the correct scientific basis and the connection with meteorites had to wait until Ernst Chladni's work two hundred years ago: see Beech (1995), and chapter 3 of Littmann (1998). For a long time there had been civilisations around the world that kept careful and extensive records that we can now recognise relate to meteor outbursts, meteor showers or fireballs. Notable among such ancient records are those from China, Japan and Korea (Imoto and Hasegawa, 1958; Zhuang, 1977; Ahn, 2005). Many records include exact dates, of immense value in modern studies to test our ideas of how processes in space have operated over millennia. More about the history of meteor observations and meteor work can be found in Williams and Murad (2002), chapter 1 of Jenniskens (2006) and references therein. Humans have seen meteors for millennia; since Chladni it has been known that underlying the meteor phenomenon is the existence of solid objects in space, which we call meteoroids.

The fundamental scientific study is concerned with meteoroids; meteors are the light, ionisation, sound and other phenomena produced when meteoroids collide with a planetary atmosphere. Recently, the International Astronomical Union updated the definitions of meteor and meteoroid, motivated by the blurring of the line between asteroids and meteoroids. When the previous definitions were put in place in 1961, a meteoroid was defined as "a solid object moving in interplanetary space, of a size considerably smaller than an asteroid and considerably larger than an atom or molecule" (Millman, 1961). This worked well until improvements in asteroid searches began to find many objects smaller than 100 m, some as small as a few metres (Beech and Steel, 1995). The Chelyabinsk impact was caused by an object 19 m in diameter; according to the old definitions it could not be called an asteroid because it was not observed in interstellar space, while the smaller, ≈ 3 m 2008 TC$_3$, which struck the Earth one day after it was discovered (Jenniskens et al., 2009), was an asteroid.

In order to resolve this, Commission F1 of the IAU proposed to establish a size threshold to divide asteroids from meteoroids (Borovička, 2016). There is no natural size limit, as the population is continuous from small to large objects, so an arbitrary limit of 1 metre was chosen; objects larger than this are asteroids (or comets, if they show activity). At the smaller end, a division was introduced between meteoroids and interplanetary dust; in this case, the natural division is the size at which a particle is too small to produce light and ionisation when it strikes a planetary atmosphere. This limit depends on the speed of the object, but an arbitrary limit of 30 µm was chosen as being characteristic. Remarks to the definitions include some elasticity, so that any object that causes meteor phenomena may be called a meteoroid; the Chelyabinsk impactor was both an asteroid and a meteoroid. Also, any natural object observed in space, even if below the 1-metre threshold, may be called an asteroid.

Although the science is of *meteoroids*, with meteors a manifestation thereof, the term 'meteor science' is often used to encompass the study of meteoroids. Meteor science continues to be studied for scientific and practical reasons. The practical includes the development of the ability to mitigate effects of impacting meteoroids when potentially harmful: the hazard on Earth and to spacecraft must be understood over a wide size range of impactors. Scientifically, the motivation is to understand ongoing processes in nature: how comets and asteroids evolve, or what happens to their debris, in space and in planetary atmospheres. As in other sciences, observation or experiment combine with theory to elucidate what processes really occur. Models fits data if given processes operate. Theory suggests that various forces in principle could act on particles moving in the Solar System (e.g., radiative, electrostatic, relativistic). If the forces are not directly observable, a model should predict an observable consequence. For example, if the radiative Poynting–Robertson effect influences the dynamical evolution of small grains (meteoroids) in space, this can help to explain observations at the Earth relating to the Geminid stream (Jakubík and Neslušan, 2015) or at Mercury relating to the Taurids (Christou et al., 2015). Despite the fact that cometary dust trails have been observed, the dust particle concentrations in space cannot be detected by modern instruments, as a rule. Therefore studies of the dynamics and structure of meteoroid streams, and particularly the orbital resonances that operate, are important (Soja et al., 2011; Kortenkamp, 2013). A good model of the composition and structure of the meteoroids themselves can predict how they will interact with the atmosphere, including the early release of volatiles like sodium (Vojáček et al., 2019). The interaction strongly depends on fragmentation, which affects both the meteor light and the dynamics of the meteoroid (Ceplecha and Revelle, 2005; Borovička et al., 2007).

The twenty-first century brings great opportunities to advance meteor science. Modern astronomy has been characterised by each new generation of telescopes seeing fainter and with better resolution, allowing discoveries that drive the theoretical understanding. In the study of meteors, this new century has seen both increased observational precision and huge increases in the size and availability of databases. This has improved the reliability with which orbits are computed and streams are identified, and the details of physical processes during the meteor flight, such as fragmentation. Computer speed, including access to supercomputers where necessary, leads to more elaborate models better able to match observations. The space age has brought space telescopes and missions that provide *in situ* data. The missions can be to the parent bodies that are the source of meteoroids, or to other planets where the effects of the meteoroids impacting the atmosphere or surface are detectable by various means. The spacecraft themselves can carry dust detection instruments (examples in Grün et al., 2002). We develop these points in the following text.

Observations in Earth's Atmosphere and Elsewhere

Theory is often data driven: observation and datasets are the basis for our science. The ionisation trails of meteors scatter radiation very efficiently, and can be used to calculate shower and sporadic meteor activity. The ionised region around the head of the meteor is smaller, but high-power, large-aperture radars are capable of tracking these faint objects and have opened up the very smallest meteor-producing cosmic particles for study (see Chapter 3, Kero et al., 2019). The light from meteors can be used to characterise nighttime meteor sources, and can provide strong constraints to ablation modelling, particularly when spectral data are gathered (more details in Chapter 4, Koten et al., 2019). Lidars can be used to study the deposition of meteoric material in the atmosphere, and large meteoroids produce shockwaves detectable at the surface through infrasound and seismic observations. All observing techniques have different strengths and weaknesses, so multi-technique observations have great value to cross-calibrate measurements and to provide a wealth of information about single events, particularly meteorite-dropping bolides.

Away from Earth, naturally, much less has been observed, but this is changing. The meteor phenomenon is in principle observable on other planets with atmospheres. Moreover, if the meteors are not observed, their aftereffects can be: ablating meteoroids cause layers of metal atoms and ions to be deposited in planetary atmospheres (Chapter 5, Christou et al., 2019). Airless bodies provide a different environment for impacting meteoroids. It is possible to detect both the impact phenomenon itself, in the form of impact flashes on hitting the surface (Chapter 6, Madiedo et al., 2019), and the aftereffects, one source of neutrals in the exospheres coming from impact vaporisation when the surface is bombarded (Christou et al., 2019, see Section 5.7). Of course impact flashes on Jupiter (Christou et al., 2019, see Section 5.6) are a meteor (superbolide) phenomenon, not surface impacts.

Until recently, the lunar dust cloud had only been expected to exist. The Lunar Atmosphere and Dust Environment Explorer (LADEE) mission discovered it and mapped its density distribution. Moreover, several meteoroid streams were detected, and the Geminid stream radiant was determined (Szalay and Horányi, 2016). It has been realised that "the Moon can be used as an enormous meteoroid detector" (Szalay, 2017).

Detection of meteoroid influx to the Earth and other worlds can be viewed from complementary perspectives: a means to study the meteoroids or an effect on the target bodies themselves. Internal sources of oxygen are apparently inadequate to explain levels of oxygen-bearing species on Titan and the giant planets (Plane et al., 2018) and delivery by meteoroids provides a possible external source. Infalling matter may have had an important role in delivering organics to the early Earth (chapter 34 of Jenniskens, 2006).

The properties of asteroid (3200) Phaethon, including its small orbit and small perihelion distance, as well as its association with one of the most prominent annual meteor showers, the Geminids, have prompted much debate about its nature (chapter 22 of Jenniskens, 2006). The study of Phaethon is an excellent example of the value of space-based telescopes, with the mass loss observed by STEREO (Chapter 8, Kasuga and Jewitt, 2019, see Section 8.3.1). Moreover we expect to learn about the dust from Phaethon *in situ* with JAXA's DESTINY+ mission. Another example of STEREO's results was the discovery of the existence of a dust ring at the orbit of Venus (Jones et al., 2013). Such a resonance ring is known to exist around the Earth's orbit. The observations of the venusian dust ring should lead to improved understanding of the factors influencing its formation.

That will not be the first mission to a parent body of meteoroids and dust; plenty of others have given valuable scientific results (Chapter 2, Borovička et al., 2019, see Section 2.2.4), including ESA's recent Rosetta mission to comet 67P/Churyumov-Gerasimenko. This comet's perihelion distance is presently above 1.2 AU so that although its activity releases meteoroids, it is not apparently a source of meteors on Earth. However, being a typical Jupiter family comet, its orbit can change substantially within decades or centuries (Królikowska, 2003) and the unstable nature of its meteoroid stream is confirmed by dynamical modelling (Soja et al., 2015).

Meteoroid Ablation Modelling and Stream Modelling

Major meteor showers have always been relatively easy to identify in datasets around the time of their maxima, when stream meteoroids greatly outnumber sporadics. Minor streams and showers (Chapter 9, Williams et al., 2019) are now increasingly identified as datasets grow and powerful search techniques are developed. Meteor velocity measurements are one of the best examples of the importance of observational precision enhancing theoretical understanding. Critical in evaluating the existence of hyperbolic meteoroids with an interstellar origin (Chapter 10, Hajduková et al., 2019), sufficient velocity accuracy will also help to characterise, for example, trail structure within meteoroid streams (Chapter 7, Vaubaillon et al., 2019, see Section 7.7). Our mathematical models need more precise data than has been available to date, so the parameters in the models are often poorly constrained. However, the advent of high-precision data allows the range of errors to be narrowed (e.g. Abedin et al. 2015). Modern computer power not only enables the processing of massive observational datasets; it can greatly improve the reliability of physical models. To model the meteoroidal influx to the Earth or other planets (Wiegert et al., 2009; Pokorný et al., 2017), the evolution of particles is modelled beginning with their creation from the parent body populations, through streams and later dispersion into the sporadic background.

Interpretation of observational data is based strongly on physical models. The interaction between meteoroids and the atmosphere depends on many factors, including the size and speed of the meteoroid and the height at which ablation takes place. Small meteoroids may interact with individual air molecules, while larger, more deeply penetrating meteoroids form a shock front that changes the rate of energy and momentum transfer (Chapter 1, Popova et al., 2019). Fragmentation is a particularly difficult issue to handle, which can affect conclusions about the density and structure of meteoroids (Chapter 2, Borovička et al., 2019). High-precision observations are a significant advantage in constraining the many free parameters in ablation models, including deceleration and fragmentation, and telescopic tracking systems may be particularly useful in this context (Campbell-Brown et al., 2013).

Practical Applications

Although the nineteenth century brought the realisation that the Earth's passage through cometary debris streams causes annual meteor showers, it is essentially at the start of the twenty-first century that outburst forecasts have become routine. The reliability of forecasts depends on a number of factors, often relating to the parent comet (Vaubaillon, 2017). Providing a forecast for human observers to view one of the sky's great displays can be regarded as a practical application of meteor astronomy. A high level meteor storm is not only a visual spectacle to be witnessed just a few times in a lifetime by travelling to the right point on Earth; in the space age it can very briefly increase the hazard to spacecraft by orders of magnitude (Ma et al., 2007). But such a storm is of short duration, with the Earth traversing the really dense region of space perhaps in under an hour. Overall, the risk to spacecraft is dominated by the sporadic background (Chapter 11, Drolshagen and Moorhead, 2019, see Section 11.5.2.4). The hazard is not restricted to spacecraft orbiting Earth; the 2014 approach of Comet C/2013 A1 (Siding Spring) to Mars showed the importance of computing dust and meteoroid impact risks to spacecraft elsewhere (Moorhead et al., 2014).

A detailed understanding of the inner Solar System meteoroid population has become essential to modern society. Models are now available which quantitatively map the temporal and spatial density variations of meteoroids in interplanetary space (see Chapter 11, Drolshagen and Moorhead, 2019, and Chapter 7, Vaubaillon et al., 2019, Section 7.4.8).

The discovery of lunar water (Pieters et al., 2009) has renewed interest in colonising the Moon. In this regard estimation of the meteoroid flux to the lunar surface becomes a practical task. Until LADEE, it was monitored by observations of visual light flashes from large meteoroids with masses > 1 kg (Suggs et al., 2014). Our understanding of meteoroid and dust fluxes on airless bodies has improved in the last two decades (see Szalay et al., 2018, and references therein).

Incidentally, spacecraft themselves yield extensive data on occasions when they re-enter Earth's atmosphere (Yamamoto et al., 2011) behaving like large fireballs and detectable via the various phenomena associated with such events. There is, moreover, the advantage of known entry parameters.

The other category of hazard is at larger sizes. Near-Earth asteroid surveys beginning in the 1970s and continuing today have catalogued potential impactors, gradually extending the size limit downwards as advancing survey capabilities scan more sky to deeper magnitudes. The more likely occurrences remain those at intermediate sizes (Tunguska, Chelyabinsk) and even if the individual events are unforeseen, research gives an understanding of their frequency and effects (Chapter 12, Svetsov et al., 2019).

Meteor Showers and Meteor Nomenclature

With the most famous showers, there is never ambiguity as to what is being referred to, e.g., the Perseids are the debris of comet 109P/Swift-Tuttle. One theme of this book is that well-defined or well-structured streams eventually disperse into the sporadic background. Inevitably some showers are hard to detect or define. In such cases it may be unclear whether or not different authors are writing about the same shower. The common use of a shower or stream list as standards would help to promote clarity in the literature. Over the years some lists of showers or streams have become widely used, e.g., the list of Cook (1973) and the International Meteor Organization's "Working list of meteor showers" (Rendtel, 2014).

In addition to its efforts to formalise what we mean by a meteor shower and a meteoroid stream (Chapter 9, Williams et al., 2019, see Section 9.2), the IAU, via its Meteor Data Center (MDC), now maintains a shower list (see Section 9.4.3) which can serve as a standard in shower nomenclature. Provisional names are assigned to newly discovered showers. The IAU's Commission F1 operates a Working Group on Meteor Shower Nomenclature, which recommends the showers that are well enough established to be officially approved by the IAU. There are presently 112 established showers (see Tables 9.2, 9.3 and 9.4). As has been traditional (though not quite universal, which is one of the problems when no standard exists), the shower name relates to the part of the sky from which the shower's meteors appear to radiate (Jenniskens, 2008), the radiant being a well-defined direction for meteoroids arriving on similar orbital paths. As well as the name, the MDC ensures that every shower has a unique three-letter code (relating to the name as far as the uniqueness constraint allows), and a unique number which can exceed three digits, as at the time of writing, the number in the list of provisional and established showers has just passed a thousand. The reader will see the IAU number and/or three-letter code widely used in the chapters in this book. For example, the Phoenicids are PHO/254 or #254. As noted by Jenniskens (2008), this system is certainly clearer than the same shower being variably described as the Draconids, γ-Draconids, October Draconids, Giacobinids or Giacobini-Zinnerids.

Meteoroids: Sources of Meteors on Earth and Beyond

The subsequent chapters present twelve aspects of meteoroid research, concentrating on recent and current developments. A chosen feature of the book is that every chapter concludes with a section envisaging the most important future research directions. We can foresee that future advances will come from space missions, or ground-based observational surveys, or outburst predictions, that we know are going to be undertaken. Other advances will be serendipitous: Asteroid 2008 TC$_3$ and the Almahata Sitta meteorite notwithstanding (Chapter 4, Koten et al., 2019, see Section 4.3.3.2), we do not know when the next spectacular bolide or meteorite fall will be. Meteor science will move in exciting directions in the twenty-first century.

References

Abedin, A., Spurný, P., Wiegert, P. et al. 2015. On the age and formation mechanism of the core of the Quadrantid meteoroid stream. *Icarus*, **261**, 100–117.

Ahn, S.-H. 2005. Meteoric activities during the 11th century. *Monthly Notices of the Royal Astronomical Society*, **358**, 1105–1115.

Astapovich, I. S. 1958. *Meteor Phenomena in the Earth Atmosphere*. Moscow, USSR: Fizmatgiz. In Russian.

Beech, M. 1995. The makings of meteor astronomy: Part IX. *WGN, Journal of the International Meteor Organization*, **23**, 48–50.

Beech, M., and Steel, D. 1995. On the definition of the term Meteoroid. *Quarterly Journal of the Royal Astronomical Society*, **36**, 281–284.

Borovička, J. 2016. About the definition of meteoroid, asteroid, and related terms. *WGN, Journal of the International Meteor Organization*, **44**, 31–34.

Borovička, J., Spurný, P., and Koten, P. 2007. Atmospheric deceleration and light curves of Draconid meteors and implications for the structure of cometary dust. *Astronomy & Astrophysics*, **473**, 661–672.

Borovička, J., Macke, R. J., Campbell-Brown, M. D. et al. 2019. Physical and chemical properties of meteoroids. In: G. O. Ryabova, D. J. Asher, and M. D. Campbell-Brown (eds), *Meteoroids: Sources of Meteors on Earth and Beyond*. Cambridge, UK: Cambridge University Press, pp. 37–62.

Campbell-Brown, M. D., Borovička, J., Brown, P. G., and Stokan, E. 2013. High-resolution modelling of meteoroid ablation. *Astronomy & Astrophysics*, **557**, A41.

Ceplecha, Z., and Revelle, D. O. 2005. Fragmentation model of meteoroid motion, mass loss, and radiation in the atmosphere. *Meteoritics and Planetary Science*, **40**, 35–54.

Christou, A., Vaubaillon, J., Withers, P., Hueso, R., and Killen, R. 2019. Extra-terrestrial meteors. In: G. O. Ryabova, D. J. Asher, and M. D. Campbell-Brown (eds), *Meteoroids: Sources of Meteors on Earth and Beyond*. Cambridge, UK: Cambridge University Press, pp. 119–135.

Christou, A. A., Killen, R. M., and Burger, M. H. 2015. The meteoroid stream of comet Encke at Mercury: Implications for MErcury Surface, Space ENvironment, GEochemistry, and Ranging observations of the exosphere. *Geophysical Research Letters*, **42**, 7311–7318.

Cook, A. F. 1973. A working list of meteor streams. In: Hemenway, C. L., Millman, P. M., and Cook, A. F. (eds), *Evolutionary and Physical Properties of Meteoroids, Proceedings of the International Astronomical Union's Colloquium #13, Held at the State University of New York, Albany, NY, on June 14–17, 1971*. *NASA SP-319*. Washington, D.C.: National Aeronautics and Space Administration, pp. 183–191.

Drolshagen, G., and Moorhead, A. V. 2019. The meteoroid impact hazard for spacecraft. In: G. O. Ryabova, D. J. Asher, and M. D. Campbell-Brown (eds), *Meteoroids: Sources of Meteors on Earth and Beyond*. Cambridge, UK: Cambridge University Press, pp. 255–274.

Grün, E., Dikarev, V., Krüger, H., and Landgraf, M. 2002. Space dust measurements. In: Murad, E. and Williams, I. P. (eds), *Meteors in the Earth's Atmosphere*. Cambridge, UK: Cambridge University Press, pp. 35–75.

Hajduková, M., Sterken, V., and Wiegert, P. 2019. Interstellar meteoroids. In: G. O. Ryabova, D. J. Asher, and M. D. Campbell-Brown (eds), *Meteoroids: Sources of Meteors on Earth and Beyond*. Cambridge, UK: Cambridge University Press, pp. 235–252.

Imoto, S., and Hasegawa, I. 1958. Historical records of meteor showers in China, Korea, and Japan. *Smithsonian Contributions to Astrophysics*, **2**, 131–144.

Jakubík, M., and Neslušan, L. 2015. Meteor complex of asteroid 3200 Phaethon: Its features derived from theory and updated meteor data bases. *Monthly Notices of the Royal Astronomical Society*, **453**, 1186–1200.

Jenniskens, P. 2006. *Meteor Showers and their Parent Comets*. Cambridge, UK: Cambridge University Press.

Jenniskens, P. 2008. The IAU meteor shower nomenclature rules. *Earth Moon and Planets*, **102**, 5–9.

Jenniskens, P., Shaddad, M. H., Numan, D. et al. 2009. The impact and recovery of asteroid 2008 TC$_3$. *Nature*, **458**, 485–488.

Jones, M. H., Bewsher, D., and Brown, D. S. 2013. Imaging of a circumsolar dust ring near the orbit of Venus. *Science*, **342**, 960–963.

Kasuga, T., and Jewitt, D. 2019. Asteroid–meteoroid complexes. In: G. O. Ryabova, D. J. Asher, and M. D. Campbell-Brown (eds), *Meteoroids: Sources of Meteors on Earth and Beyond*. Cambridge, UK: Cambridge University Press, pp. 187–209.

Kero, J., Campbell-Brown, M., Stober, G. et al. 2019. Radar observations of meteors. In: G. O. Ryabova, D. J. Asher, and M. D. Campbell-Brown (eds), *Meteoroids: Sources of Meteors on Earth and Beyond*. Cambridge, UK: Cambridge University Press, pp. 65–89.

Kortenkamp, S. J. 2013. Trapping and dynamical evolution of interplanetary dust particles in Earth's quasi-satellite resonance. *Icarus*, **226**, 1550–1558.

Koten, P., Rendtel, J., Shrbený, L. et al. 2019. Meteors and meteor showers as observed by optical techniques. In: G. O. Ryabova, D. J. Asher, and M. D. Campbell-Brown (eds), *Meteoroids: Sources of Meteors on Earth and Beyond*. Cambridge, UK: Cambridge University Press, pp. 90–115.

Królikowska, M. 2003. 67P/Churyumov-Gerasimenko – Potential target for the Rosetta mission. *Acta Astronomica*, **53**, 195–209.

Littmann, M. 1998. *The Heavens on Fire*. Cambridge, UK: Cambridge University Press.

Ma, Y., Xu, P., and Li, G. 2007. Redefinition of the meteor storm from the point of view of spaceflight security. *Advances in Space Research*, **39**, 616–618.

Madiedo, J. M., Ortiz, J. L., Yanagisawa, M., Aceituno, J., and Aceituno, F. 2019. Impact flashes of meteoroids on the Moon. In: G. O. Ryabova, D. J. Asher, and M. D. Campbell-Brown (eds), *Meteoroids: Sources of Meteors on Earth and Beyond*. Cambridge, UK: Cambridge University Press, pp. 136–158.

Millman, P. M. 1961. Meteor News. *Journal of the Royal Astronomical Society of Canada*, **55**, 265–267.

Moorhead, A. V., Wiegert, P. A., and Cooke, W. J. 2014. The meteoroid fluence at Mars due to Comet C/2013 A1 (Siding Spring). *Icarus*, **231**, 13–21.

Pieters, C. M., Goswami, J. N., Clark, R. N. et al. 2009. Character and spatial distribution of OH/H_2O on the surface of the Moon seen by M^3 on Chandrayaan-1. *Science*, **326**, 568–572.

Plane, J. M. C., Flynn, G. J., Määttänen, A. et al. 2018. Impacts of cosmic dust on planetary atmospheres and surfaces. *Space Science Reviews*, **214**, 23.

Pokorný, P., Sarantos, M., and Janches, D. 2017. Reconciling the dawn-dusk asymmetry in Mercury's exosphere with the micrometeoroid impact directionality. *The Astrophysical Journal*, **842**, L17.

Popova, O., Borovička, J., and Campbell-Brown, M. 2019. Modelling the entry of meteoroids. In: G. O. Ryabova, D. J. Asher, and M. D. Campbell-Brown (eds), *Meteoroids: Sources of Meteors on Earth and Beyond*. Cambridge, UK: Cambridge University Press, pp. 9–36.

Rendtel, J. (ed.) 2014. *Meteor Shower Workbook 2014*. Hove, BE: International Meteor Organization.

Soja, R. H., Baggaley, W. J., Brown, P., and Hamilton, D. P. 2011. Dynamical resonant structures in meteoroid stream orbits. *Monthly Notices of the Royal Astronomical Society* **414**, 1059–1076.

Soja, R. H., Sommer, M., Herzog, J. et al. 2015. Characteristics of the dust trail of 67P/Churyumov-Gerasimenko: An application of the IMEX model. *Astronomy & Astrophysics*, **583**, A18.

Suggs, R. M., Moser, D. E., Cooke, W. J., and Suggs, R. J. 2014. The flux of kilogram-sized meteoroids from lunar impact monitoring. *Icarus*, **238**, 23–36.

Svetsov, V., Shuvalov, V., Collins, G., and Popova, O. 2019. Impact hazard of large meteoroids and small asteroids. In: G. O. Ryabova, D. J. Asher, and M. D. Campbell-Brown (eds), *Meteoroids: Sources of Meteors on Earth and Beyond*. Cambridge, UK: Cambridge University Press, pp. 275–298.

Szalay, J. R. 2017. Lunar dust exosphere. In: Cudnik, B. (ed), *Encyclopedia of Lunar Science*. Cham: Springer International Publishing, pp. 1–4.

Szalay, J. R., and Horányi, M. 2016. Detecting meteoroid streams with an in-situ dust detector above an airless body. *Icarus*, **275**, 221–231.

Szalay, J. R., Poppe, A. R., Agarwal, J. et al. 2018. Dust phenomena relating to airless bodies. *Space Science Reviews*, **214**, 98.

Vaubaillon, J. 2017. A confidence index for forecasting of meteor showers. *Planetary and Space Science*, **143**, 78–82.

Vaubaillon, J., Neslušan, L., Sekhar, A., Rudawska, R., and Ryabova, G. 2019. From parent body to meteor shower: the dynamics of meteoroid streams. In: G. O. Ryabova, D. J. Asher, and M. D. Campbell-Brown (eds), *Meteoroids: Sources of Meteors on Earth and Beyond*. Cambridge, UK: Cambridge University Press, pp. 161–186.

Vojáček, V., Borovička, J., Koten, P., Spurný, P., and Štork, R. 2019. Properties of small meteoroids studied by meteor video observations. *Astronomy & Astrophysics*, **621**, A28.

Wiegert, P., Vaubaillon, J., and Campbell-Brown, M. 2009. A dynamical model of the sporadic meteoroid complex. *Icarus*, **201**, 295–310.

Williams, I. P., and Murad, E. 2002. Introduction. In: Murad, E., and Williams, I. P. (eds), *Meteors in the Earth's Atmosphere*. Cambridge, UK: Cambridge University Press, pp. 1–11.

Williams, I. P., Jopek, T. J., Rudawska, R., Tóth, J., and Kornoš, L. 2019. Minor meteor showers and the sporadic background. In: G. O. Ryabova, D. J. Asher, and M. D. Campbell-Brown (eds), *Meteoroids: Sources of Meteors on Earth and Beyond*. Cambridge, UK: Cambridge University Press, pp. 210–234.

Yamamoto, M.-Y., Ishihara, Y., Hiramatsu, Y. et al. 2011. Detection of acoustic/infrasonic/seismic waves generated by hypersonic re-entry of the HAYABUSA capsule and fragmented parts of the spacecraft. *Publications of the Astronomical Society of Japan*, **63**, 971–978.

Zhuang, T.-s. 1977. Ancient Chinese records of meteor showers. *Chinese Astronomy*, **1**, 197–220.

Part I

Meteor Physics

1
Modelling the Entry of Meteoroids

Olga Popova, Jiří Borovička, and Margaret D. Campbell-Brown

1.1 Introduction

1.1.1 Ablation and Fragmentation

Meteoroids represent the material remaining from the formation of the solar system and carry unique information on the earliest forming solids. They enable the study of the structure and composition of these small-scale solids, which form the seeds of planets. Meteoroids are most easily studied using the atmosphere as a detector, as they produce light, ionisation and sonic waves during interaction with the atmosphere. Under the right conditions, the passage of the meteoroid through the atmosphere may result in a meteorite falling to Earth or even the formation of an impact crater (Svetsov et al., 2019), but in most cases only the luminosity and ionisation are available for analysis. Therefore, meteoroid properties (physical, chemical and all other possible properties) need to be determined through observations. The most obvious way to evaluate meteoroid properties based on observational data is to apply a model to fit the data.

The mass loss process is called *ablation*. Ionised and luminous areas, metal layers and smoke dust appear in the atmosphere due to ablation. Meteoric material is involved in atmospheric chemistry (Dressler and Murad, 2001; Plane et al., 2018, and references therein). Meteor spectra confirm the presence of Fe, Si, Mg, H, Na, Ca, Ni, Mn, Cr, Al, Ti, FeO, AlO, MgO, OH due to ablation (Ceplecha et al., 1998). Meteors are sometimes considered as a source of organic material deposited into the atmosphere during ablation (Jenniskens et al., 2000b).

Ablation is dependent on the meteoroid size and mass, the entry velocity, the altitude of flight and the meteoroid properties. The ablation rate determines the deposition of mass, and influences the momentum and energy release into the atmosphere. Meteor radiation and ionisation, which allow us to observe meteor phenomena, are controlled by the ablation rate:

$$I = -\tau \cdot \frac{dE_k}{dt} = -\tau \cdot \left(\frac{V^2}{2}\frac{dM}{dt} + MV\frac{dV}{dt}\right);$$
$$q = -\frac{\beta}{\mu V}\frac{dM}{dt} \quad (1.1)$$

here E_k, V and M are the meteoroid kinetic energy, velocity and mass, τ is the luminous efficiency and β is the ionising efficiency, q is the linear electron concentration, μ is the average mass of an ablated meteoroid atom and I is the intensity of radiation.

The classical models of meteor ablation use conservation of energy and momentum to determine the light and/or ionisation produced by a meteoroid as a function of time. Different physical processes causing mass loss are taken into consideration in the dependence on entry conditions. Meteoroids lose mass mostly as a result of vapour production. The strongly temperature-related meteoroid mass loss processes are generally called thermal ablation and other processes are usually excluded from consideration except for special cases when they are important.

The total size range of meteoroids entering the Earth's atmosphere is very large. Their sizes range from about micron dust to 10-km impactors. For some of these objects ablation in the atmosphere doesn't play a role. Analysis based on meteor physics equations similar to that introduced in the 1960s (see e.g. Popova (2004) and references therein), shows that for an impact speed of 40 km s^{-1}, stony objects ($\leq 10^{-6}$ m) decelerate before being substantially heated, and the heating of objects with radius $R \leq 10^{-4}$ m is limited by thermal re-radiation. The recent analysis of micrometeoroids collected on arctic surface snow point to mechanical destruction and weathering (Duprat et al., 2007). Rietmeijer (2002) points out that even certain classes of interplanetary dust particles that are collected almost intact in the Earth's atmosphere show traces of flash heating to 300–1000°C. For a large-scale impact it should be mentioned that cosmic objects larger than about one hundred meters will lose only a small part of their mass and energy in the atmosphere (at average impact velocity and entry angle). So, ablation is most important for cosmic objects roughly in the range of 10^{-4}–100 m. Note that the largest annual event appears to have initial kinetic energy of about $E_k \sim 5$–10 kt TNT (Nemtchinov et al., 1997; Brown et al., 2002b) (i.e. mass $M \sim 150$–300 t and diameter ~ 4–5 m).

In addition to the basic equations of ablation for a single particle, fragmentation must be taken into account. Evidence for fragmentation may be direct (observed breakups in flight, meteorite strewn and crater fields) or inferred from multiple differences between theoretical predictions and observations. For example, numerous studies of the light curves of faint meteors (e.g. Jacchia, 1955; Jones and Hawkes, 1975; Fleming et al., 1993; Murray et al., 1999; Koten et al., 2004, and others) have shown that light curve shapes are extremely variable, and do not match the light curve predicted by the single-body model. Irregular ionisation profiles, the scatter of underdense decay times in radar meteors and short trails, flashes and flares on light curves, luminosity in shutter breaks, the difference between dynamic and photometric masses and other data may be explained by fragmentation. So modelling of meteoroid entry includes fragmentation models.

Different fragmentation mechanisms are considered when interpreting observations and modelling. Disruption due to heating or due to aerodynamical loading are the main ones. Several types of fragmentation, which include decay of a meteoroid into few non-fragmenting pieces; progressive disintegration into successively disintegrating fragments; quasi-continuous fragmentation (a gradual release of the smallest fragments from the body and their subsequent evaporation) and simultaneous ejection of large number of small particles (giving rise to meteor flares), are usually considered. Various combinations of these types may be observed and included in modelling for the same body.

The ablation and fragmentation proceed at the altitudes of optical and radio meteor observations, i.e. mainly at 130–20 km. High-altitude meteors were registered as high as 200 km (see Section 1.2.1). For a specific object the ablation and fragmentation altitudes are dependent on its size (larger bodies penetrate deeper), on entry velocity (the higher the velocity, the higher the aerodynamical loading and the higher the incoming energy flux) and on the meteoroid origin, composition and structure (cometary material is deposited higher than asteroidal matter: Rietmeijer, 2000).

Small and large meteoroids in the atmosphere are observed by different methods, have different ablation altitudes, and their interaction with the atmosphere occurs in different flow regimes (see Subsection 1.1.2). The physical conditions during meteoroid entry change considerably as a function of altitude and different processes are responsible for ablation at different stages of meteoroid flight. Different models are used to describe the entry of small and large meteoroids. Many models aim to reproduce meteoroid behavior in the atmosphere (deceleration and/or light curves) in different flow conditions. Other models are trying to describe the physical conditions that occur around the meteor body. The main goal of this chapter is to describe the current state of modelling of different scale meteor phenomena, to review current entry models, and to discuss their boundaries and limitations.

Following an introduction of ablation theory and description of the flow regimes and their boundaries (Subsections 1.1.2–1.1.4), the chapter is broken up into three sections covering different regimes of meteoroid-atmosphere interaction: free molecular flow (Section 1.2), the transition regime (Section 1.3), and continuous flow (Section 1.4). Subsections are devoted to the main issues of each flow regime. Sputtering (Subsection 1.2.1), luminous and ionisation efficiencies (Subsection 1.2.2), and head echoes and ionisation radius (Subsection 1.2.3) are included in the free molecular flow section. Fragmentation of small meteoroids as well as the parameters of the luminous area are discussed in Subsections 1.2.4 and 1.2.5. The formation of the screening vapour cloud around the meteoroid and current modelling efforts in the transition regime (including heat transfer coefficient estimates, description of the conditions in the luminous area, etc.) are presented in Section 1.3. In the continuous flow section, the ablation coefficient and luminous efficiency are discussed in Subsections 1.4.1 and 1.4.2. Modelling of spectra, fragmentation models and a short description of hydrodynamical modelling are described in Subsections 1.4.3–1.4.5. The subsections for different regimes are not the same, since different emphasis is placed on different issues in each regime. The ionisation efficiency has been studied for free molecular flow since radars observe faint meteors, and there are no detailed studies of ionisation in bolides. In contrast, spectral modelling has been done primarily for bolides, since these spectra are rich and thermal equilibrium conditions can be assumed. Concluding remarks are presented in Section 1.5.

1.1.2 Different Regimes of Meteoroid Interaction with the Atmosphere

The physical conditions during meteoroid entry change considerably as a function of altitude; in the range of ablation altitudes, the atmospheric density varies by orders of magnitude (from 10^{-10} kg m^{-3} at 200 km altitude down to 1 kg m^{-3} at the ground). Corresponding momentum and energy fluxes are equal to $\rho_a V^2$ and $\rho_a V^3/2$ in the absence of meteoroid surface shielding, where ρ_a is the atmosphere density at the altitude of flight. For a meteor radiating between 130 and 80 km altitude the fluxes increase more than 3000 times from the beginning height to the end. This large variation in the fluxes leads to the fact that the conditions of meteoroid-atmosphere interaction change along the trajectory.

The local flow regime around the falling body determines the heat transfer and mass loss processes. Two limiting cases are evident in the meteoroid interaction with the atmosphere. If the meteoroid is small enough, or the altitude of flight is large enough, the mean free path of the air molecules is larger than the meteoroid size. The flow can be considered to be individual particles moving in straight lines, and the meteoroid is effectively under particle bombardment, which causes the meteoroid heating and an appearance of evaporated atoms/molecules with thermal velocities (Figure 1.1). The appropriate gas dynamic regime is determined by the magnitude of a Knudsen number, which represents the ratio of the molecule mean free path l to a characteristic body dimension R: $Kn = l/R$. The free-molecular flow corresponds to $Kn > 10$, where interparticle collisions are negligible. As the atmosphere density increases, the mean free path decreases. When the Knudsen number becomes small compared to unity, of the order of $Kn \leq 0.1$, the medium can be treated as a continuous one and described in terms of the macroscopic variables: velocity, density, pressure and temperature. The reduction of the free path length with a decrease of flight altitude leads first to the formation of a viscous layer around the body and then to the formation of a shock wave in front of it. A large meteoroid at a relatively low altitude (where the shock wave is formed) is satisfactorily described by hydrodynamic models.

However, the Knudsen number for undisturbed air is insufficient to describe air-meteoroid interaction because the presence of evaporated molecules affects the flow.

Bronshten (1983) suggested using a modified Knudsen number Kn_r: $Kn_r = (V_r/V)Kn$, which takes into account the increase in the concentration of evaporated molecules near the meteoroid due to the difference between the thermal velocity V_r and the meteoroid velocity V. This correction shifts the boundaries between the flow regimes upward in height. Since the difference between the thermal velocity and the velocity of the meteoroid affects the physics of the interaction process, these boundaries depend on both the size and the speed of the meteoroid. In the process of modelling the transition flow

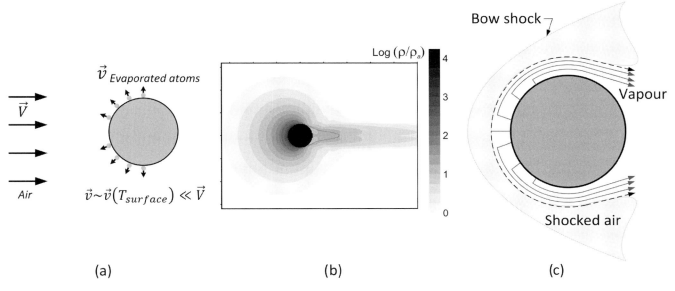

Figure 1.1. Schematic picture of the interaction of a meteoroid with the atmosphere in different flow regimes; (a) free molecular regime, where air particles reach the meteoroid surface, causing the appearance of evaporated atoms; (b) in the transition regime a vapour cloud is formed in front of the meteoroid, density distribution according the air-beam model (see Section 1.3) is shown; (c) a shock wave is formed in the continuous flow conditions.

conditions, it is necessary to take into account the shielding of the meteoroid surface by evaporated material and the subsequent reduction of the heat transfer coefficient (Popova, 2004 and references therein).

There are independent estimates of the boundary between the free-molecule and transition regimes (Popova et al., 2000, 2001; Stokan and Campbell-Brown, 2015). The high velocity of fast meteors produces a high evaporation rate and the vapour pressure exceeds aerodynamical loading (Popova et al., 2000). A vapour cloud is formed around the body and screens the surface from direct impacts of incoming air molecules. The corresponding upward shift in the flow regime boundaries is consistent with the estimate from the modified Knudsen number Kn_r (Popova et al., 2000, 2001; Popova, 2004). For example, for the most part in Leonids (larger than about 10^{-3} m) the interaction takes place in the transition regime from free-molecule flow to continuous due to their high entry velocity (Figure 1.2).

Stokan and Campbell-Brown (2015) compared the number density of atmospheric and evaporated particles at first collision. They found that for a representative meteoroid travelling at 40 km s^{-1} with a radius of 10^{-3} m (corresponding to an approximate mass of 10^{-5} kg with a density of 1000 kg m^{-3}), the shielding of the meteoroid surface should be taken into account below about 105 km altitude, which is in agreement with other estimates (in Figure 1.2 these estimates are recalculated for 70 km s^{-1} velocity).

Thus, the ratio of the meteoroid size to the atmospheric mean free path at different altitudes, corrected for the meteoroid velocity, determines the mode of interaction with the atmosphere and the character of the ablation (Bronshten, 1983; Popova, 2004). Large meteoroids lose most of their mass in the continuum flow regime, whereas small meteoroids interact with the atmosphere mainly in the free molecular flow or transition flow

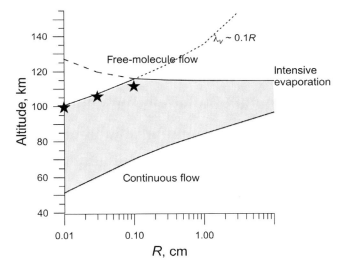

Figure 1.2. Boundaries of different flow regimes taking into account the presence of an ablation vapour cloud in front of a meteor. The intensive evaporation altitude and the line where the free path length in vapour λ_v is ten times smaller than meteoroid size R, are shown (Popova et al., 2000). The estimates correspond to 70 km s^{-1} entry velocity and cometary meteoroid composition. Stars mark the altitudes where the ratio of evaporated to atmospheric particles is significant according to the estimates by Stokan and Campbell-Brown (2015). Grey area corresponds to transition regime.

regimes. Very roughly, the boundary may be estimated as $\sim$1 cm (Borovička, 2005a). Large meteoroids spend a longer time in the atmosphere before they disintegrate and give rise to more phenomena; they can be significantly decelerated, produce more complex light curves, and can drop meteorites in some cases.

1.1.3 Main Equations Used in Modelling

The physical conditions during the meteoroid entry change considerably as a function of altitude and different processes may be responsible for ablation at different stages of meteoroid flight through the atmosphere. Therefore, we must emphasise that it is incorrect to apply the approximations/equations obtained for one regime to another.

The thermal energy received by the meteoroid from the impinging air molecules is balanced by radiative loss, temperature increase, melting, phase transitions, and by vapourisation of the meteoric constituents (Levin, 1956; Jones and Kaiser, 1966; Lebedinets et al., 1973). The standard heat balance equation for a spherical particle may be written as follows:

$$\pi R^2 \Lambda \frac{\rho_a V^3}{2} = 4\pi R^2 \varepsilon \sigma_b (T_s^4 - T_0^4)$$
$$+ \frac{4}{3}\pi R^3 \rho_m c \frac{dT_s}{dt} - Q\frac{dM}{dt} \quad (1.2)$$

The energy flux received from the impacting air molecules ($1/2 \cdot \Lambda \rho_a V^3$) is used for thermal radiation cooling, meteoroid heating and ablation. Here Λ is the accommodation coefficient (or heat transfer coefficient), which determines the fraction of incoming energy flux reaching the meteoroid surface. The heat transfer coefficient Λ is often denoted as C_h, especially in papers devoted to modelling in the continuous flow conditions. If there is no shielding in free molecular flow, the Λ value is equal to unity. The first term on the right-hand side of the equation is the radiation loss, where ε is the emissivity coefficient, σ_b is the Stefan-Boltzmann constant, and T_s and T_0 are the temperature of the particle surface and the atmospheric environment, respectively. The second term is the heat consumed to increase the temperature of the particle (c is the bulk specific heat, ρ_m is the particle density). The last term is the heat consumed in the transfer of particle mass into the gas phase, where Q is the ablation heat including all the energy needed to melt and/or vapourise meteoroid material.

A meteoroid, even one that is only 1 mm in size, will not heat uniformly. To determine the temperature at the surface, the thermal conduction equation may be solved in the meteoroid interior simultaneously with the modelling of the entry (Čapek and Borovička, 2017), or a simplification can be used. It is assumed that a shell of the meteoroid, the thickness of which is determined by the material parameters, heats uniformly, while the interior remains cool (Love and Brownlee, 1991; Campbell-Brown and Koschny, 2004; McAuliffe and Christou, 2006).

Mass is considered to be lost through sublimation and evaporation, vapour thus being the final stage of majority of the ablated material. It is often assumed that sublimation begins as soon as the meteoroid temperature starts to rise (Lebedinec and Šušková, 1968; Lebedinets et al., 1973; Love and Brownlee, 1991; Moses, 1992; Adolfsson et al., 1996; Campbell-Brown and Koschny, 2004; Rogers et al., 2005, and many others) with the temperature dependent mass loss rate being modelled using the Knudsen–Langmuir formula (Bronshten, 1983):

$$dM/dt = -4\pi R^2 p_v(T_s)\sqrt{\frac{\mu}{2\pi k_b T_s}},$$
$$\log_{10} p_v = A_v - B_v/T_s. \quad (1.3)$$

Here k_b is the Boltzmann constant, p_v is the saturated vapour pressure and A_v and B_v are empirically or theoretically determined constants for specific substance. The influence of external gas pressure on the evaporation is neglected, i.e. this approach is fully justified in the frame of free molecular flow and can't be applied in the continuous flow conditions. In the transition regime the application of (Equation 1.3) is limited by increasing counterpressure.

The energy equation (Equation 1.2) is widely used in numerous papers devoted to the entry of small meteoroids. Its right-hand side is modified by different authors depending on the purpose of the study and the size of meteoroids under consideration. For example, the absorption of solar radiation may be included (Moses, 1992; McAuliffe and Christou, 2006), and the atmosphere radiation may be excluded (Love and Brownlee, 1991; Moses, 1992, and others). In the case of small particles, along with thermal cooling, it is necessary to take into account the energy and the mass lost to sputtering. Neither process is significant in the case of large meteoroids.

McNeil et al. (1998) calculated vapour pressures of the various melt constituents and introduced the concept of differential ablation. They assumed sequential release of different compounds according to their volatility. A current example of an ablation model which also predicts the injection rates of individual elements is the Chemical Ablation MODel (CABMOD) (Vondrak et al., 2008). Genge (2017) incorporated partial melting behaviour of particles to study micrometeorite formation.

In addition, other energy and mass losses are considered and corresponding terms are included in Equations (1.2) and (1.3). At high temperatures, meteoroids may lose mass through spraying of the melted layer on the surface (Bronshten, 1983; Campbell-Brown and Koschny, 2004; Briani et al., 2013; Čapek and Borovička, 2017). The deeper the meteoroid penetrates into the atmosphere, the larger the received energy flux will be. The altitude at which the received energy flux exceeds the energy losses from meteoroid heating and thermal radiative cooling may be called the height of intensive evaporation. For porous bodies with $R \sim 0.1$–10 cm this altitude is about 110–130 km (Lebedinets, 1980; Bronshten, 1983). Below this altitude the incoming energy contributes mainly to ablation; heat conduction and the thermal radiation can be excluded from consideration (Lebedinets, 1980; Bronshten, 1983). Equation (1.2) is transformed into the following mass loss equation:

$$\frac{dM}{dt} = -\Lambda \cdot \frac{\pi R^2 \rho_a V^3}{2Q} \quad (1.4)$$

which is correct after the beginning of intensive evaporation. The conditions in the body itself usually are not of interest, it is assumed that the meteoroid surface temperature remains at the melting/boiling value. Ablation modelling of bolides and photographic meteors is usually restricted to this equation; the stage of meteoroid heating is not included into the modelling (Ceplecha et al., 1998). An additional mechanism was included in Equation (1.4) by Borovička et al. (2007). The authors suggested that small fragments can be detached from the meteoroid, producing additional mass losses, and called this process erosion.

The dominant role of thermal ablation in meteoroid mass loss has been questioned by some authors. Spurný and Ceplecha (2008) proposed triboelectric charging as the most important energy transfer inside a meteoroid able to explain millisecond flares and some other observed phenomena.

A full system of equations describing meteoroid entry also includes the rate of penetration of the meteoroid into the atmosphere. The atmospheric height (H) varies with time according to the relationship

$$\frac{dH}{dt} = -V \cdot \cos\theta, \quad (1.5)$$

where θ is the zenith angle of meteoroid entry. Conservation of linear momentum leads to the drag equation which specifies the deceleration of the meteoroid:

$$\frac{d\vec{V}}{dt} = -\pi R^2 \cdot \Gamma \cdot \frac{\rho_a V^2}{M} \cdot \frac{\vec{V}}{|V|} + \vec{g}. \quad (1.6)$$

Here Γ is the drag coefficient, which is a dimensionless quantity expressing the efficiency of the momentum transfer. The drag coefficient is often denoted as C_d, and Equation (1.6) may include an additional factor of 1/2. The second (gravitational) term in Equation (1.6) is much smaller than the first one in the range of meteoroid entry velocities. This term is important when the object is essentially decelerated, as well as the variation of the entry angle (for example, for meteorite-producing meteoroids):

$$\frac{d\theta}{dt} = -g \cdot \sin\theta / V. \quad (1.7)$$

Neglecting the gravitational term, the meteoroid's deceleration may be written as:

$$\frac{dV}{dt} = -\frac{\Gamma A \rho_a V^2}{M^{1/3} \rho_m^{2/3}}. \quad (1.8)$$

Here ρ_m is the meteoroid density, and A is the shape factor, which is equal to the cross-sectional area S of the meteoroid multiplied by $(\rho_m/M)^{2/3}$ (this factor relates cross section to volume, i.e. it has geometric meaning and for spherical objects it is equal to 1.21). Usually it is assumed that the meteoroid shape is unchanged during ablation (Bronshten, 1983). The drag coefficient Γ can reach a value of 2. The horizontal and vertical position of the meteoroid can be updated using this equation along with a spherical Earth.

The differential equations of meteoroid motion and ablation (Equations (1.4) and (1.8)) are often transformed to connect the mass change and velocity (Bronshten, 1983; Ceplecha et al., 1998) through the ablation coefficient σ, which allows us to determine meteoroid mass in dependence on velocity under the condition of constancy of the coefficients in the equations (see equation 1.15 and Bronshten (1983)). The ablation coefficient is:

$$\sigma = \frac{\Lambda}{2Q\Gamma}. \quad (1.9)$$

A solution to the equations of meteoroid motion and ablation is also used in the form of $l_{tr} = l_{tr}(t)$, i.e., the distance, l_{tr}, travelled by the body in its trajectory as a function of time (Pecina and Ceplecha, 1983, 1984; Ceplecha et al., 1993, 1998). The distance l_{tr} is directly observed. In this case there are two independent parameters, i.e. the ablation coefficient and the shape-density coefficient, which is:

$$K = \Gamma A \rho_m^{-2/3}. \quad (1.10)$$

To calculate a meteor's position and brightness (or electron line density) as a function of time a set of Equations (1.6) and (1.4) with (1.1) and (1.5), or a set of Equations (1.6), (1.2), (1.3) and (1.1), (1.5) are used.

The parameters that determine the action of the atmosphere on the meteoroid, i.e. drag Γ and heat transfer coefficient Λ, are dependent on the modelled flow regime and in principle vary along the trajectory (Bronshten, 1983; Popova, 2004).

Full-scale hydrodynamic modelling (i.e., solution of the Euler equations for mass, momentum and energy conservation (Landau and Lifshitz, 1987)) with radiative transfer is used to include the simulation of atmospheric shock waves, their interaction with the surface and other entry-related events (see a review by Artemieva and Shuvalov, 2016).

1.1.4 Influence of Model Assumptions

Properties of meteoroids (size/mass, density, structure, etc.) cannot be obtained directly from observational data (trajectory, light curve, ionisation curve, spectra, strewn fields). Theoretical models are used to extract the meteoroid properties from observations, and certainly, the result depends on the assumed model and on the coefficients used in the model, as well on assumed material properties.

Light curves and deceleration profiles were produced for a number of small meteoroids ($< 10^{-4}$ kg), yielding values for initial meteoroid mass and density (Stokan and Campbell-Brown, 2015). The comparison of observed and modelled light curves for one of these meteors is shown in Figure 1.3. The modelled light curve labelled (A) was obtained assuming the mass loss is proportional to the kinetic energy imparted to the meteoroid, i.e. Equation (1.4). The initial value of the mass corresponds to an altitude of 98 km. This meteoroid is small, the energy losses on its heating and the thermal radiation affect the light curve. Application of the model Equations (1.2) and (1.3), starting from an altitude of about 150 km, demonstrated that the

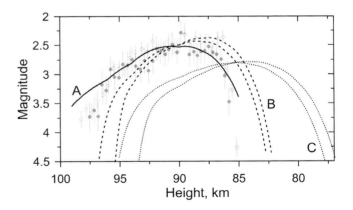

Figure 1.3. Comparison of light curves obtained under different assumptions for the same initial values of mass and density. The observed light curve from two stations is shown by the grey points with error bars. The solid curve (A) corresponds to the mass loss Equation (1.4); the dashed curves (B) correspond to Equation (1.2) with two different vapourisation rates (Equation 1.3); the dotted curves (C) are calculated by Equations (1.2) and (1.3), taking into account vapour screening of the meteoroid surface following (Nelson et al., 2005).

model light curves (labeled B) will be a worse match for the observational data. The ablation rate and corresponding light curve (B) are dependent on the assumed relation for vapour pressure; results for two pressure dependencies are shown. This particular meteor has an entry velocity of about 30 km s^{-1}, and screening of the meteoroid surface may play a role. Light curves that are obtained with a heat transfer coefficient adjusted for screening (C) are shifted to lower altitudes. In all the considered modelling cases, the assumed luminous efficiency was the same.

It should be noted here that if the aerodynamical loading is comparable with vapour pressure it alters the ablation rate and the ablation equation (1.4) should be modified (see for example Baldwin and Allen 1968; Baldwin and Sheaffer 1971).

When speaking about the properties of meteoroids and their action on the atmosphere, it should be remembered that there is uncertainty in the determined parameters that depends strongly on the chosen ablation model and is caused by the use of poorly known coefficients under various assumptions. Additionally, the meteor considered was supposed to be non-fragmenting, and fragmentation brings even more complexities.

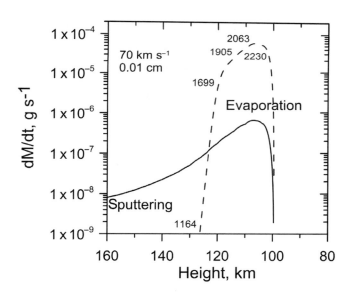

Figure 1.4. Comparison of ablation rate due to evaporation and due to sputtering for a stony meteoroid 0.01 cm in radius entering the atmosphere at 70 km s^{-1}. The expression for vapour pressure is taken from Opik (1958), and the sputtering yield from Popova et al. (2007); numbers correspond to the temperature of the meteoroid (in K).

1.2 Free Molecular Flow Regime

1.2.1 Non-Thermal Mass Loss

High-altitude meteors, i.e. those observed beyond classical ablation heights (with beginning heights > 130 km altitude), constitute a small fraction of the total number of observed meteors. They are often observed for meteoroids with high entry velocity. Koten et al. (2006) found that, of 164 shower meteors observed, about 5% were high-altitude meteors with beginning heights between roughly 130 and 150 km. Optical meteors have been observed as high as 199 km (Spurný et al., 2000a, 2000b). Radar meteor echoes have been observed at 130–200 km (see compilation by Mann et al., 2011; Gao and Mathews, 2015). During the 1998 Leonids, Brosch et al. (2001) monitored echoes from meteors at altitudes of up to 400 km. They have interpreted the origin of these trails as sputtering of meteoroids by atmospheric particles at these altitudes.

Several processes may occur in a solid whose surface is under bombardment by energetic particles. If the surface atom receives energy greater than the surface binding energy (which is close to the sublimation heat and is about 1–10 eV), it may be ejected from the surface. This process is called physical sputtering. In this case the meteoroid particles leave the surface as a result of receiving momentum from the collision cascade induced by the incident particles, which may be ions, neutral atoms or molecules. If the energetic incoming particles react chemically with the surface atoms and change their ejection behaviour, the process is called chemical sputtering.

The energies of air particles are high enough at meteor velocities to cause physical sputtering (at $V \sim 70$ km s^{-1} the energy of air particles is about 400–800 eV). Threshold projectile kinetic energy should be several (4–8) times higher than the surface binding energy to sputter a surface particle (Behrisch and Eckstein, 2007), since not all impinging energy is transferred to one particle.

The action of air particles on the meteoroid leads to both heating of the meteoroid and sputtering of the meteoroid surface. Physical sputtering doesn't depend on temperature to a great extent and may cause mass losses in the upper part of the meteor trajectory before the meteoroid is strongly heated (Figure 1.4). Sputtering was considered as a mass loss mechanism (may be called non-thermal ablation) by a number of authors under different assumptions (Levin, 1956; Opik, 1958; Lebedinets et al., 1973; Lebedinets, 1980) or as a reason for additional meteoroid deceleration (Coulson, 2002).

Sputtering is quantified by the sputtering yield, Y, the number of atoms removed per incident particle. The yield depends on the projectile energy, physical and chemical properties of the solid target, and on the type of sputtered atom. Several sputtering considerations (Hill et al., 2005; Rogers et al., 2005; Kero, 2008) have been guided by the approach to sputtering of astrophysical materials developed by Tielens et al. (1994). According to Hill et al. (2005), Y is no smaller than about 0.3–0.7 for a velocity of 70 km s^{-1}. Vinkovic (2007) assumed a sputtering efficiency $Y \sim 1$ for projectile energies of hundreds of eV (Field et al., 1997). Popova et al. (2007) considered the air-meteoroid interaction with a Monte Carlo-type physical model, which allowed them to describe the sputtering of the meteoroid surface. It was shown that sputtering occurs at meteor velocities exceeding 30 km s^{-1}, and sputtering yield increased with meteoroid velocity (up to $Y \sim 2$ for chondritic material). Fast particles (sputtered and reflected) can carry out up to $10 - 15\%$ of the incoming energy. Values of energy loss included in sputtering estimates are the main source of uncertainties (about an order of magnitude) in the sputtering characteristics (May et al., 2000).

The sputtering is included in the mass loss and heat flux equations as follows:

$$\left(\frac{dM}{dt}\right)_{sput} = -Y \cdot S \cdot \rho_a V \cdot \mu/\mu_a \quad (1.11)$$

$$\left(\frac{dE}{dt}\right)_{sput} = -\left(\frac{dM}{dt}\right) \cdot Q_s$$

or $\quad (1.12)$

$$\left(\frac{dE}{dt}\right)_{sput} = \Lambda_{sput} \cdot S \cdot \rho_a \cdot V^3/2$$

where S is the sputtered surface, Q_s and Λ_{sput} are the specific heat of sputtering and the fraction of incoming heat flux which is lost through sputtering. It is often assumed that $Q_s = Q$ (heat of ablation) (Rogers et al., 2005; Vondrak et al., 2008), i.e. that sputtered particles are ejected with thermal velocities similar to evaporated particles. If the ejection velocity is larger, the value Q_s should be determined in some other way. The second possibility to include energy losses from sputtering is to determine the fraction Λ_{sput} of incoming heat flux, which contributes to sputtering (Lebedinets et al., 1973). Popova et al. (2007) estimated that this fraction can reach up to about 10%, which includes the energy carried out by sputtered and reflected particles. Sputtering may affect the drag coefficient due to momentum transfer (Levin, 1956; Bronshten, 1983) just like a vapour cloud formation (see Section 1.3) but this effect needs additional study. This has been attempted (see Coulson, 2002), but the found effect looks overestimated. In general, the values Y, Q_s and Λ_{sput} are poorly known.

The fraction of mass sputtered depends on assumptions about sputtering parameters (Y and Q_s or Λ_{sput}) and on the initial size of the meteoroid and its velocity. In contrast to the recent suggestion by Rogers et al. (2005) and Hill et al. (2005), sputtering is not the main mass loss process in most cases. Kero (2008) reconsidered the mass loss using a numerical model of sputtering and thermal ablation similar to that used by Rogers et al. and Hill et al. He suggested that the atmospheric densities in the Rogers et al. (2005) paper are a factor of 100 too high and that this led the authors to overestimate the mass loss caused by sputtering (note that the atmospheric densities are not explicitly given in the Rogers et al. (2005) paper). According to Vondrak et al. (2008) the effect of sputtering is small (< 10% of mass) for meteoroids with velocities up to 45 km s^{-1} and heavier than 10^{-11}g; the fastest particles up to masses of 10^{-14}g are reduced to approximately half of their initial mass by sputtering.

Popova et al. (2007) have suggested that the oxygen luminosity observed at high altitudes can be generated by the interaction of the sputtered atoms and ions with the background ionosphere (several percent of sputtered species are ionised). The sputtered atoms and ions typically have kinetic energies more than 1000 times the energy of surrounding ionospheric particles. When thermalising through collisions they generate luminosity. Formation of oxygen triplet radiation at 777 nm, which occurs due to interaction of sputtered particles with the surrounding atmosphere, was considered. Model light curves produce a dependence of brightness on air density which is similar to observed light curves (Popova et al., 2007). The formation of a luminous area at high altitudes due to interaction of sputtered atoms with the atmosphere was considered by Vinkovic (2007). He found that the initial particle energy is redistributed by secondary collisions and is enough to form a large disturbed area around the meteoroid. The shape of the disturbed area and its changes with altitude coincided with observations. Rosenberg (2008) suggests that fast ions sputtered from a meteoroid might generate waves in the background plasma possibly detectable by radar scattering.

The sputtering models used in the studies mentioned previously have limitations. For meteoroids with entry velocity slower than about 40 km s^{-1}, energies of impinging atmospheric species are close to the threshold energies for sputtering and those thresholds are in most cases determined with great uncertainty. The sputtering yield at sub-keV energies can be influenced by many processes. The precise composition and structure of meteoroids, as well the state of their surfaces, are unknown. Atmospheric species may react with the meteoroid surface.

1.2.2 Luminous and Ionisation Efficiencies

When comparing models to observations, either the luminous intensity or the electron density is usually used to determine how well the model fits the observations, in addition to the observed deceleration. In order to simulate the light or ionisation produced by the modelled meteoroid, the efficiency with which light or electrons are produced must be estimated. There is substantial uncertainty in both, particularly the luminous efficiency.

The luminous efficiency is determined as the fraction of a meteoroid's kinetic energy converted to light in a particular bandpass (Equation (1.1)). The luminous efficiency depends on the speed of the meteor, because the likelihood of exciting electrons which will then radiate optically depends on the collision energies involved. It depends strongly on the composition of the meteoroid: most faint meteors emit in a few bright metal lines, particularly sodium, magnesium and iron (Borovička et al., 2005), so the presence or absence of one of these metals affects the brightness. It also depends on the spectral sensitivity of the detector, since certain spectral lines will only contribute if the detector is sensitive to the blue or the red end of the spectrum. The luminous efficiency is normally calculated experimentally in the bandpass of a specific optical instrument, and may be converted to panchromatic (light emitted in all visible wavelengths) or bolometric (electromagnetic radiation at all wavelengths). Finally, the luminous efficiency may depend on the mass of the meteoroid and the height at which it ablates, since this affects the interaction physics between the meteoroid and the atmosphere. Many faint meteors produce a significant amount of light in iron lines, so the luminous efficiency of iron is often used to calculated the total luminous efficiency; it is worth remembering that the importance of iron in the spectra of faint meteors varies from dominating emission to being absent (Borovička et al., 2005), and therefore the trend of luminous efficiency on speed does not necessarily follow that of iron.

Several groups have attempted to calculate the luminous efficiency theoretically, based on the probability of atomic collisions producing optical emission. Opik (1955) found a luminous efficiency which increased linearly with speed and was on the

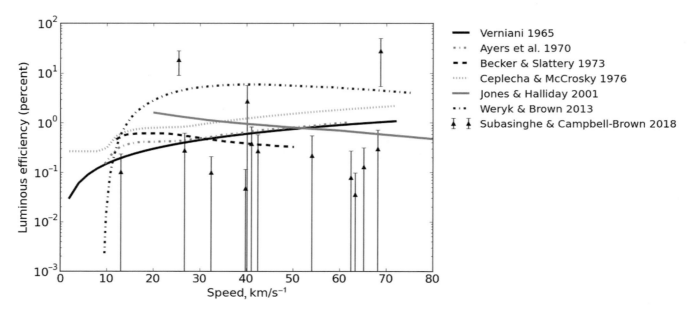

Figure 1.5. Luminous efficiency meteors as a function of speed for selected studies.

order of a few tenths of a percent; he had very little experimental data on which to base his calculations. Jones and Halliday (2001), using lab measurements from Früchtenicht and Becker (1973) as constraints, calculated an excitation probability for iron, though they pointed out the lack of data on scattering cross sections for elastic and inelastic scattering at meteoroid speeds. The excitation probability leads to a luminous efficiency in the photographic bandpass which decreases with speed from about 1.5 to 0.5% for speeds from 20 to 70 km s^{-1}: they did not calculate it for lower speeds. The results are shown in Figure 1.5.

Optical observations of meteors can be used to measure luminous efficiency, provided the mass of a meteoroid can be determined independent of the light. The mass can be determined from the measured deceleration of the meteoroid, if a density is assumed. If the meteoroid fragments in flight, this dynamic mass gives only the mass of the largest (leading) fragment, so luminous efficiencies calculated in this way will be too large if fragmentation is present and not taken into account. Verniani (1965) used 413 precisely reduced Super-Schmidt meteors to find the dependence of luminous efficiency on speed. He recognised that these meteors were fragmenting, and applied a correction, though it is worth noting that the correction relied on an assumed luminous efficiency. Like Öpik (1955), Verniani found that luminous efficiency was directly [linearly] dependent on speed. He then obtained an absolute calibration for this line using a single meteor; one asteroidal object ablating below 65 km with deceleration measured at only six points, four of which were used by him. He found a luminous efficiency that varied from approximately 0.2 to 1% over the full range of meteor speeds (shown in Figure 1.5). In spite of the problems with this study, Verniani's result is widely used, for example by Ceplecha and McCrosky (1976), who used artificial meteors at low speeds but the Verniani speed dependence at higher speed. Other groups have found luminous efficiencies for individual fireballs, including meteorite-dropping events (ReVelle, 1979;

Halliday et al., 1981; Ceplecha and Revelle, 2005), most of which had speeds between 10 and 20 km s^{-1}. The measured values ranged from about 0.1 to 6%. Here it should be mentioned that meteorite-dropping fireballs radiate mainly in continuous flow conditions, in which the shock wave is formed and significantly affects the radiation produced. In order to address the problems of unknown mass, shape and composition, Ayers et al. (1970) used data from artificial meteors to calculate the luminous efficiency of iron. They fired artificial meteors into the atmosphere at speeds between 8 and 16 km s^{-1}, and found that the luminous efficiency increased steeply with speed from 1 to 2.7% for pure iron particles; Ceplecha and McCrosky (1976) used this to calculate the luminous efficiency of chondritic material with approximately 28% iron. Ayers et al. (1970) used the results of lab experiments to infer that the luminous efficiency levels off at higher speeds; Figure 1.5 shows their results for stone, along with Ceplecha and McCrosky's adjusted result.

More recently, Subasinghe and Campbell-Brown (2018) used high-resolution observations of faint meteors made with the Canadian Automated Meteor Observatory (CAMO) tracking system to find meteors which did not fragment, or which had a clear fragment which itself did not fragment further. Of the fifteen meteors they used, thirteen were leading fragments and two were apparent single-body meteors; the speeds of the meteors ranged from 13 to 68 km s^{-1}. Twelve of the meteors had luminous efficiencies between 0.05 and 0.5%; the remaining three were between 2 and 30%, but were either single-body meteors (which may in fact have had unresolved fragments) or had large uncertainties. There was no trend in luminous efficiency with speed, as shown in Figure 1.5. These low luminous efficiencies are interesting, but since they come from leading fragments, their light production may not be representative of the general population.

In order to reach higher speeds than possible with artificial meteors but to have controlled conditions, experiments have been done firing micron-sized particles into low-pressure gas chambers to simulate ablation. The light produced can be measured and compared to the known mass of the particle. Becker and Friichtenicht (1971), Becker and Slattery (1973) and Friichtenicht and Becker (1973) used an apparatus of this kind to measure the luminous efficiency of iron, copper, aluminum and silicon for speeds between 11 and 47 km s^{-1}. For each element, they found a peak efficiency at some speed between 14 and 20 km s^{-1}, after which the efficiency decreased with speed. The luminous efficiency of iron in the photographic bandpass was larger than those of any of the other elements by nearly an order of magnitude, which is why later works using their results have used only the luminous efficiency of iron. The luminous efficiency peaked at just over 1% for pure iron particles; the results obtained by Becker and Slattery (1973) are shown in Figure 1.5. The particles used in these experiments are much smaller than optical meteoroids (by three orders of magnitude) and the gas pressure corresponds to a height of around 50 km, much higher than the pressures encountered by faint meteors near 100 km. It is not possible to simulate millimetre-sized particles in the appropriate pressure range, since an evacuated chamber would need to be kilometres in length to contain the entire ablation process. Recently, Thomas et al. (2016) have constructed a new ablation simulator based on the work of Friichtenicht et al. (1968), but it has not yet been used to calculate luminous efficiencies.

If the ionisation efficiency is known (or better known than the luminous efficiency), then simultaneous observations with radar and optical techniques allow the luminous efficiency to be calculated. Saidov and Simek (1989) used a combination of theoretical estimates and simultaneous radar and optical observations to show that their ionisation probabilities and luminous efficiencies were compatible, and found luminous efficiencies around 0.5%, with a slight decreasing trend with speed for meteors faster than 20 km s^{-1}. Weryk and Brown (2013) assumed that the theoretical work of Jones (1997) on the ionisation probability was correct, and used this to calculate the luminous efficiency of over a hundred simultaneous optical and radar meteors; the trend with speed is shown in Figure 1.5. They found that the ratio of ionisation probability to luminous efficiency increased with speed, and led to a luminous efficiency which increased rapidly with speed to a maximum of around 5% bolometric near 30 km s^{-1}, and fell slightly with speed at higher speeds. This luminous efficiency is nearly an order of magnitude higher than other studies at low speeds.

The disagreement among all the previously mentioned studies, both with respect to the magnitude of the luminous efficiency and the dependence with speed, indicate that more work must be done in this area.

The ionisation probability, β, is the chance that an evaporated meteoroid atom will produce a free electron. It is often referred to as the ionisation coefficient, since it is possible for β to be greater than one, in cases where more than one electron is freed by a single meteoric atom (for example, if an ionised meteoritic atom has also ionised an air molecule). Meteor radars measure the scattering cross section of the ionised region around the head of a meteor or the ionised trail; in either case this can be converted to an electron line density, or the number of electrons produced by the meteor per unit length (Equation (1.1)).

The ionisation coefficient was computed theoretically by Massey and Sida (1955), but there were problems with their treatment, including the lack of a cutoff at low speeds. Each atom requires a minimum amount of energy to ionise, and no ionisation should be seen at energies below this value. Experimental work by Slattery and Friichtenicht (1967), using the ablation chamber described in the luminous efficiency section, found the ionisation probability for iron particles with speeds between 20 and 40 km s^{-1}, increasing with speed and peaking close to unity. Bronshten (1983) used the experimental work of Slattery and Friichtenicht for iron and the theoretical work of Massey and Sida for other meteoric elements, and found $\beta \propto V^{3.42}$, which has the same problem at low speed as Massey and Sida.

A significant improvement was made by Jones (1997), who derived the primary ionisation probability (which ignores multiple electrons produced by a single meteor atom) and calibrated it with the results of Slattery and Friichtenicht (1967). For faint meteors with speeds lower than 35 km s^{-1}, he found

$$\beta \approx 9.4 \times 10^{-6}(V-10)^2 V^{0.8} \tag{1.13}$$

so that no ionisation is produced for speeds lower than 10 km s^{-1}. For brighter and faster meteors, the assumptions that there is no secondary ionisation and no dissociative recombination may not be valid.

Recently, Thomas et al. (2016) and DeLuca et al. (2018) have redone the lab measurements of iron particles in air, and found that they matched the Jones data, provided those were recalibrated: they also used the more general form derived by Jones for the probability of an electron being produced in the first collision:

$$\beta_0 = \frac{c(V-V_0)^n v^{0.8}}{1 + c(V-V_0)^n V^{0.8}} \tag{1.14}$$

This can be integrated to find the probability of ionisation over all collisions required to thermalise an atom. They found that a reasonable fit was achieved with $n = 1.6$ and $c = 1.60 \times 10^{-4} \pm 0.03 \times 10^{-4}$, in units of s$^{2.4}$ km$^{-2.4}$, $V_0 \sim 8.91$ km s^{-1}.

1.2.3 Head Echo and Ionisation Radius Modelling

The scattering of radio waves from ionised meteor plasma is a complex process that depends on the distribution and density of free electrons. In order to understand how radiation is reflected and how received power relates to the meteoroid mass that produced the ionisation, modelling of meteor plasma is useful.

Transverse scatter radars (described in more detail in Kero et al., 2019) scatter radiation from sections of a meteor's ionised trail roughly a kilometre long, depending on the wavelength of the radar; these echoes occur when the trail is at right angles to the radar beam (specular reflection). The trail forms with an initial radius in a short time as the fast-moving meteoroid atoms thermalise, and then it diffuses. Underdense trails from smaller meteoroids scatter radiation from the entire depth of the trail, while overdense trails are radiatively thick and effectively scatter radiation from the surface of a cylinder. Underdense trails can suffer attenuation when the trail radius is of the order of the wavelength of the radar: this produces a height ceiling effect since trails that form higher in the atmosphere (at lower atmospheric density) have a larger initial radius. The amount of attenuation depends both on the size of the trail and on the density profile of electrons in the trail.

Manning (1958) used a kinetic model to examine the dependence of initial radius with the neutral mean free path in the atmosphere, assuming the ions had a Gaussian distribution in the trail. He found that the initial radius was approximately 14 mean free paths, using scattering cross sections appropriate to thermal speeds. Massey and Sida (1955) point out that these cross sections are not appropriate to atoms at meteoric speeds, and Jones (1995) redid the analysis using more appropriate scattering cross sections. He found that the trails did not have a Gaussian density distribution, but rather had a compact core in a more diffuse distribution.

The time it takes meteor trails to diffuse into the background is sometimes used (in radars without interferometers) to infer the meteor height, since the diffusion timescale depends on the mean free path. The most basic theory predicts that at altitudes above 95 km, diffusion perpendicular to the magnetic field lines is inhibited due to gyration of electrons around the magnetic field: the trail is thus elliptical or even ribbon-like (Ceplecha et al., 1998). Dyrud et al. (2001) modelled the diffusion of meteor trails and found that instabilities at heights around and above 100 km tend to even out the diffusion of trails, preventing the formation of strongly elongated cross sections. An analytic 2D model was developed and tested by Dimant and Oppenheim (2006a, 2006b), assuming a Gaussian distribution of electron density in the trail, and looked specifically at the electromagnetic fields in the plasma, which are important to diffusion and the development of field-aligned irregularities, regions where density variations in the electron density can scatter radiation non-specularly from meteor trails. A full 3D model of trails including winds and random magnetic field orientations with respect to the trail was done by Oppenheim and Dimant (2015); they point out that including atmospheric inhomogeneities and larger meteoroids in the future will help the understanding of non-specular trails in particular.

It is important to understand the formation of the ionised head around the meteoroid for head echo measurements, mainly made by high power, large aperture (HPLA) radars (described in more detail in Kero et al., 2019). The relationship between the electron line density produced – HPLA radars also use electron line density, even though they measure the moving cloud of free electrons surrounding the meteoroid – and the mass of the meteor can be calculated with the ionisation coefficient (described in Section 1.2.2), but the relationship between the measured scattering cross section and the electron density is much more difficult to determine. Close et al. (2004) assumed a Gaussian, spherically symmetric electron distribution in the meteor head, and modelled the scattering of radio waves from the plasma at different frequencies; their results were consistent with measurements made mainly at two frequencies with the ALTAIR radar. The same model was used to calculate meteoroid masses in Close et al. (2005).

A 2D model of meteor head ionisation (which also included the formation of the trail and field-aligned irregularities) was used by Dyrud et al. (2007) to reproduce the main features of head echoes. A full three-dimensional model, including the Earth's magnetic field, was done by Marshall and Close (2015) and found the "overdense area" of meteor heads as a function of frequency. The densest core of the head may be radiatively opaque, depending on the density and the wavelength of the radar, and larger overdense areas produce stronger echoes. This model was used with three-frequency head echo data to test different radial density profiles: the best fit was not Gaussian but r^{-2}, r being the distance from the centre of the ionisation (Marshall et al., 2017). They note that it is difficult to characterise the peak ionisation at the centre, since this radiatively thick part of the core is not penetrated by the radar beam; in addition, more frequencies would better constrain the shape. In contrast, a kinetic model by Dimant and Oppenheim (2017b,a) found that the electron density should fall as $1/r$ around the head of the meteoroid, changing to r^{-2} in the trail some distance behind the meteoroid.

1.2.4 Fragmentation Models

Meteoroid ablation models start with single body theory (Section 1.1.3). The complication is that faint meteoroids fragment, and the way in which they fragment is not always obvious from observations (see Borovička et al., 2019, section 3). The fact that faint meteors fragment was inferred by Jacchia (1955) from the observation that meteors recorded with the Super-Schmidt cameras were shorter than expected, with symmetric light curves. The classical single-body equations given above predict light curves which peak toward the end. Opik (1958) coined the term "dustball" to describe a meteoroid that consists of refractory grains lightly bound by a connective "glue"; this model was quantified by Hawkes and Jones (1975).

Babadzhanov (2002) summarised the types of fragmentation a meteoroid may undergo as it ablates, pointing out that a single meteor may undergo several different fragmentation processes. It may break into comparably-sized, non-fragmenting pieces (sometimes called gross fragmentation); it may break into pieces that subsequently fragment themselves (progressive fragmentation); it may lose small fragments from the surface while retaining a large portion of its original mass (quasi-continuous fragmentation); or it may produce sudden bursts of fragments, causing a flare in the meteor's light curve. He used a model of quasi-continuous fragmentation to obtain masses and densities of a set of Super-Schmidt meteors. Bellot Rubio et al. (2002) subsequently used single-body theory to model about three quarters of the same set of meteors; since each meteor had only a handful of measured points, the constraints were not strong.

A dustball model in which meteoroids fragment into fundamental grains when they reach a threshold temperature (thermal disruption) was used by Campbell-Brown and Koschny (2004) to model Leonid meteors. The grain sizes in the model may be set to a power law or Gaussian distribution, or the distribution may be manually set. Meteors were started at 180 km and equations (1.8), (1.3), and (1.2) were solved for each fragment. Borovička et al. (2007) used a different implementation of the dustball model to match Draconid meteor observations; this version started the meteors at the initial observed height and a reasonable ablation temperature, and allowed them to shed fragments in a quasi-continuous way, with the duration and mass of the fragmentation varied to match the observations. Both of these models were tested on high-resolution observations from the Canadian Automated Meteor Observatory (CAMO) tracking system (Campbell-Brown et al., 2013), with each model predicting the morphology of the meteor in the narrow field without prior knowledge. Neither model did a satisfactory job, emphasising the need for more constraints on observations in order to determine the fragmentation mode of a meteoroid. A particular meteor's light curve and deceleration data may be matched by several very different fragmentation mechanisms, producing densities and masses which may vary by factors of several.

Bellot-Rubio's assertion that many faint meteors are adequately described without fragmentation is undermined by CAMO's tracking system; Subasinghe et al. (2016) showed that 90% of meteors observed with the high-resolution camera showed significant fragmentation, either in the form of distinct fragments or long bright wakes produced by unresolved fragments. There is evidence that the remaining 10% of meteors with short observed wakes may also be fragmenting; Campbell-Brown (2017) attempted to model one of these meteors with a single-body model that used multiple compositions and allowed for changing meteoroid density during ablation, and determined that no single-body model could explain the short light curve. A model involving quasi-continuous fragmentation was more successful at matching the light curve while not exceeding the wake observed on the high-resolution camera. Most faint meteors fragment, and it is possible that all faint meteors do.

1.2.5 Shape and Size of Luminous Area in Meteors as an Indication of Fragmentation

Cook et al. (1962) described measurements of the optical width of meteor trails observed during the Geminid shower of 1957 ($V \sim 35$ km s^{-1}). Nine meteor trails were measured. The meteor brightness was about +4 to +9 magnitude. The average width was about 1–3 m, though two showed widths of 6 m, probably due to fragmentation.

Usual video observations are generally integrated over a relatively long exposure time (~ 0.04 s) that corresponds to a meteoroid path about 2 km along the trajectory (for Leonid meteors). The same is true of photographic observations. So, in most observations, meteor radiation is averaged over a large distance along the trajectory, exceeding the estimated sizes of meteor wake (see the following paragraphs).

Observations with short exposure times are useful to study the luminous area. A number of meteors were recorded by Babadzanov and Kramer (1968) with exposures of $\sim 10^{-4}$ s. They found different types of images (from "dotted" images to distinctly visible disc and a long tail). Fast meteors mainly reveal a head and tail with average lengths of about 50–150 m (up to 400 m). The wake of a bright meteor (brighter than -7^{mag} during a flare) was about 50–150 m at 85–80 km altitude and reached a maximal value (~ 300 m) at the moment of the flare (80 km altitude). Babadzanov and Kramer (1968) modelled the intensity distribution in the wakes assuming these wakes are formed due to detachment of tiny particles, but did not explain the observations. They suggested that radiative recombination is responsible for wake formation.

Several papers dealing with instantaneous photographs were published by Babadzhanov's group. The width of the luminous meteor area wasn't given directly there. Novikov et al. (1996) tried to reproduce meteor coma recorded on instantaneous photographs considering one kind of fragmentation – husking (i.e. destruction by the detachment of a large number of small debris particles from the surface of the parent meteor body, which do not further fragment). They stated that the results obtained agreed with the observed sizes, which are about 4–20 m in width. The meteor width was obtained under the assumption that all fragmented particles possess an initial velocity of detachment of about 100 km s^{-1}. The nature of this velocity is not explained. Meteor wakes in nine faint sporadic meteors were also investigated by Fisher et al. (2000) (with exposures $\sim 4 \times 10^{-4}$ s). They suggested that wake formation could be explained by a grain model (Hawkes and Jones, 1975) due to grain separation and further ablation of these grains. Four meteors were found to have statistically significant wake. For one meteor the wake reached 1500–2000 m in length, and the other three cases were restricted to maximum dimensions of less than 800 m. Combining these observations with data from from earlier wake studies (a total of 52 meteors, 3 wakes) they concluded that fewer than about 12% of faint meteors exhibit detectable wake. There are no data on meteor velocities. The brightness of this data sample also was not given and may be attributed by analogy with other data sets of this Canadian meteor team as +5 to +2^{mag} (Murray et al., 1999) or +8 to +2^{mag} as extremes (Campbell et al., 1999). They estimated meteoroid masses as 10^{-5} to 10^{-7} kg for Leonid campaign observations ($R \sim 0.2 - 0.03$ cm).

Additionally, three sporadic meteors demonstrated transverse separation, although no detailed description of these events was given by Fisher et al. (2000). Precursor UV photoionisation was suggested as an explanation for a kilometre-sized halo around a Leonid meteor between 104 and 110 km altitude (Stenbaek-Nielsen and Jenniskens, 2004), but this was dismissed as ineffective.

The width of the luminous meteor region was studied with the high-resolution tracking camera of the CAMO system. Stokan et al. (2013) studied 30 meteors of around +3 magnitude, and found that widths below 100 km were typically on the order of 20 m, while at higher altitudes the widths were up to 100 m, with one meteor showing a width of 170 m. They showed that the use of a scattering cross section appropriate to meteor speeds was necessary in order for a simple collisional model to explain the observed widths. They point out that, in addition to the overall sensitivity of the system, the spectral sensitivity may play a major role in determining the measured width. Since each subsequent collision suffered by a meteoroid atom as it thermalises is at lower energies and further from the meteoroid, the energy of the emitted photons may also decrease with distance, so that red-sensitive image intensifiers would see wider trails than blue-sensitive photographic film.

Stokan and Campbell-Brown (2014) found nine meteors among the high-resolution CAMO tracking data that showed significant transverse separation of fragments, the fragments moving apart at speeds of about 100 m s^{-1}. Simple models of meteoroid rotation and charging showed that these factors alone were unlikely to explain the high transverse speeds, while at the high altitudes observed, lift was insignificant.

Explosive devolatilisation was proposed as a cause of the high transverse speeds.

Stokan and Campbell-Brown (2015) chose nine high-resolution CAMO meteors with short wakes (where short refers to wakes less than 100 m), and modelled them with a simple hard-sphere collisional model. The luminous area was approximated by the region in which collisions exceeded the excitation potential of meteoroid atoms: when this was weighted by the energy of the collisions, a better match to the observed luminosity was obtained. The widths of the meteors matched well without any need for laterally spreading fragments, but the modelled wakes were too short, which the authors took to imply that even these short wakes must be a result of unresolved fragmentation.

1.3 Transition Regime

In the transition regime, the surface of a meteoroid is shielded from the impinging atmosphere particles by a vapour layer, which is formed due to intensive evaporation, but the shock wave is not yet formed. Given a meteoroid size, the regime boundaries are determined by the entry velocity and material properties (Figure 1.2). The detailed observations that were collected during the international scientific study of the Leonids in the beginning of the 2000s have provided a wealth of data on fast meteors, with an entry velocity of ~ 71 km s^{-1} and a wide range of meteor sizes (see the book edited by Jenniskens et al., 2000a). These data gave rise to the development of models which, in the first approximation, could explain the observations. The formation of a vapour cloud around fast meteor bodies was considered in the context of the air beam model (Popova et al., 2000) and by Monte Carlo simulation (Boyd, 2000). The Monte Carlo modelling allows meteoroid screening to be considered (Nelson et al., 2005).

An analogy between meteoroid evaporation under impacts of air molecules and a high-energy ion beam acting on a target in a vacuum allows the impinging air molecules to be considered as a particle beam with an energy flux $\rho_a V^3/2$, whereas the vapour formed may be described using gasdynamics. The energy transfer when air particles penetrate the layer of evaporated molecules may be described similarly to radiation transfer, assuming some effective absorption coefficient. Consideration of particle dynamics by Monte Carlo simulation showed that the energy transfer takes place in about $n \sim 10$ collisions (Popova et al., 2000, 2001). The mass effective absorption coefficient for energy transfer may be written as $\kappa_\lambda = n l_{av} \rho_v \sim 10^8 - 10^6$ cm^2 g^{-1}, where l_{av} is free path length of air in the vapour and ρ_v is vapour density.

The absorption of incoming energy takes place in a layer with mass $m = 1/\kappa_\lambda$. This mass being evaporated, the vapour that is formed screens the surface and decreases the evaporation rate. When the formed vapour pressure exceeds that of the air ($p_v \gg p_a$), the vapour expands in the gasdynamic regime as if into vacuum. The outer layers of vapour attain supersonic velocity and become transparent for impinging air molecules. Absorption occurs mainly in the subsonic part of the flow. For lower vapour densities, screening decreases and the ablation rate and vapour density increase. For larger vapour densities, screening increases and vapour production and density decrease. In this self-regulating screening regime (Nemtchinov, 1967; Maljavina and Nemtchinov, 1972) the density of the vapour in the sonic cross section is conditioned by the meteoroid size and the effective absorption coefficient κ_λ.

Vapour parameters in the whole flow area may be found directly by hydrodynamic numerical simulation or by solving the set of ordinary differential equations (Nemtchinov, 1967; Popova et al., 2000). Vapour parameters depend on meteoroid size, velocity, altitude of flight and composition. Near the meteoroid surface, the vapour has the lowest temperature, close to that of evaporation. Vapour temperature increases with body size R and with velocity V (in accordance with high temperature vapour spectral component behavior (Borovička, 1994)). Vapour temperature also increases with decreasing altitude, so spectra should be altitude dependent (this agrees with observations by Abe et al., 2000). The density of vapour is about 10^{-6} g cm^{-3}–10^{-9} g cm^{-3} and rapidly decreases with distance from the particle (Popova et al., 2000, 2001). The vapour radiation decreases the maximal temperature, and at low altitudes its action increases vapour density and cloud size. Calculated vapour spectra show general agreement with observations (Popova et al., 2000, 2001)

The formation of a vapour cloud was confirmed by 2D gasdynamical numerical modelling based on the SOVA multi-material hydrocode (Shuvalov, 1999; Popova et al., 2000, 2001). The mass and momentum fluxes were assumed to be absorbed in the same way as the energy flux, and with the same value of absorption coefficient. The impulse absorption acts as a counterpressure. The total size of the dense vapour cloud is about 5–10 times the body size, beyond which its density falls below the ambient vapour density (Figure 1.6).

This air-beam model allows a qualitative picture and the main flow characteristics to be estimated. Treating the air-meteoroid interaction in the transition regime in a framework of an air-beam model takes into account evaporation, the heating of the forming vapour, screening of the meteoroid surface, and radiation from the heated vapour (Popova et al., 2000, 2001).This gasdynamic model cannot be applied to a nonablating body, but it may be used for an ablating one. A model needs to be developed to determine more precisely the energy and momentum transfer from air to vapour, air-vapour mixing, and to include air radiation and non-elastic collisions.

The expanding vapour cloud, the velocity of which is defined both by the meteoroid velocity and the vapour expansion velocity, can be considered as a flux of high velocity particles in the undisturbed atmosphere. The interaction of this flux with the ambient air may be described (as a first approximation) in the frame of a vapour beam model (the same as the air-beam model). A portion of the vapour cloud with density exceeding the density of the ambient air (which looks like a sphere about 12 cm of radius for a 1-cm-radius meteoroid at an altitude of 115 km at velocity 72 km s^{-1}) creates a disturbed area (Figure 1.6). The resulting flow looks like a typical gasdynamical flow around a body in the lower atmosphere, but this is not in fact the case. A strong shock wave is not formed; however, there is a heated layer in front of the meteoroid. Density variations as well as the heating in the leading part of the wake are considerably lower than the shock parameters resulting from a Hugoniot adiabat.

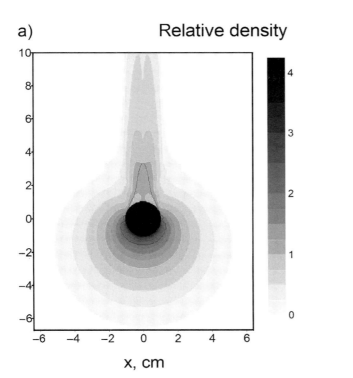
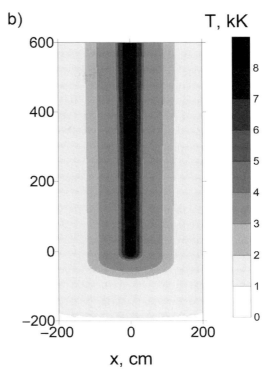

Figure 1.6. (a) Distribution of relative density (i.e., a ratio of local density to normal ambient air density in log scale, i.e. $\log(\rho/\rho_a)$) around a 1 cm body at 100 km altitude; energy, momentum, and mass absorption are taken into account. (b) Temperature distribution in the trail formed by a 1-cm-radius meteoroid at an altitude of 115 km.

The width of a rarefied hot (with temperatures exceeding 5 kK) region in the far wake equals about 1 m for the 1-cm-radius meteoroid. These values are still lower than the observed width of meteor trails, although they considerably exceed the meteoroid size and the size of vapour cloud. The results can be recalculated to other altitudes of flight and meteoroid sizes: if the vapour is assumed to be an ideal gas, then the flow around the flying body depends only on one dimensionless parameter (Popova et al., 2000).

The two-dimensional flow field about a typical Leonid meteoroid entering the Earth's atmosphere was simulated by the Direct Simulation Monte Carlo technique (Boyd, 2000). A spherical meteoroid of diameter 1 cm, entering the Earth's atmosphere at an altitude of 95 km at a speed of 72 km s^{-1} was treated. The physical modelling for this initial study of meteoroid flows is kept to a relatively simple level. Air is represented by three chemical species (N_2, O_2 and O). The vapour ablated from the meteoroid surface is assumed to be a single chemical species. Thermal relaxation is allowed between the translational, rotational and vibrational energy modes. The Variable Hard Sphere collision model is employed (Bird, 1994).

A cloud of meteoroid vapour of very high density formed in front of the body. The peak density of the vapour is more than two orders of magnitude higher than the peak density of the air species. Behind the meteoroid, the air species' densities are almost identical to the case without ablation, and the density of the ablated material slowly decays. The temperatures rapidly decay immediately behind the meteoroid and then slowly decrease with distance. The temperatures far behind the meteoroid are significantly higher than those computed for the case with no ablation. The meteoroid ablation leads to a large region of high-temperature air in the wake of the meteoroid, with values of several thousand degrees.

Due to the build-up of ablated vapour around the meteoroid, there was a sufficient number of intermolecular collisions to maintain the gas in the wake in a state of thermal equilibrium. One of the most important predictions from the computations is the physical size of the meteoroid trail. The influence of ablation on the formation of a long hot meteor wake is critical; the size of the hot meteor wake significantly increases if ablation is taken into account. The trail diameter and length are predicted to be about 5–6 m and 20–40 m respectively.

There is a need to include dissociation and ionisation reactions of the air species colliding at very high energy with ablated meteoroid vapour as well as radiative losses. In addition, there is the need to improve the meteoroid ablation model to include more microscopic phenomena. Nevertheless, the modelling confirmed the formation of the vapour cloud around the body and that the meteor wake is the main source of radiation.

Nelson et al. (2005) conducted Direct Simulation Monte Carlo investigations and determined the heat transfer coefficient Λ for a number of velocities ($V=30, 50, 70$ km s^{-1}), two meteoroid sizes ($R = 1, 8$ cm), and wide range of Knudsen parameters ($Kn \sim 0.44 - 175$). Their modelling confirmed

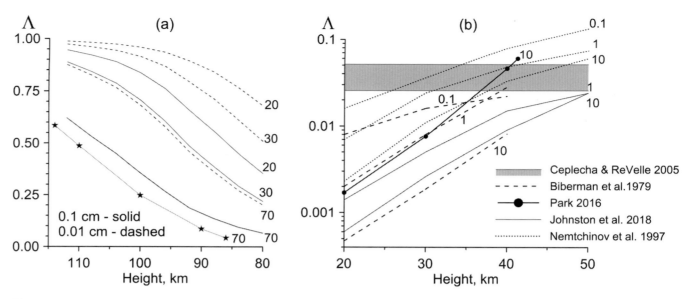

Figure 1.7. Ablation model parameters for transition and continuous flow regimes. (a) Dependence of heat transfer coefficient on altitude in the transition regime for two different sizes (velocities V are marked at the curves, km s^{-1}) according to Nelson et al., 2005, starred curve corresponds to air-beam model (Popova et al., 2000); (b) Heat transfer coefficient in continuous regime according to Johnston et al. (2018); Nemtchinov et al. (1997); Ceplecha and Revelle (2005); Park (2016); Biberman et al. (1980) (size in m is marked at the curves, $V=20$ km s^{-1}).

the formation of a cloud of ablation products with a density exceeding the density of the surrounding atmosphere by several orders of magnitude. The efficiency of heat transfer essentially decreases in the presence of ablated material in comparison with the free-molecule value ($\Lambda = 1$). They also suggested extrapolation formulae outside their range and confirmed the applicability of the extrapolation. The values of heat transfer calculated based on their results are given in Figure 1.7a. The heat transfer coefficient decreases with decreasing altitude and increasing velocity, in agreement with the estimates described previoiusly. For a 0.1 cm body at 90 km altitude, the heat transfer coefficient decreases by a factor of 5 compared to the free molecule value for Leonid meteor velocities. The effect of mass transfer on the drag coefficient Γ is opposite to that on heat transfer, i.e. with increasing blowing ratio (i.e. increasing ablation) the heat transfer decreases and the drag increases.

The models described in this section confirmed that a vapour cloud is formed around a meteoroid in transition flow conditions, and its formation changes the efficiency of energy transfer from the impacting air constituents to the meteoroid. Models predict the size of the disturbed area around the meteoroid and estimate the thermodynamical parameters there. The results may be used for improving small meteoroid modelling and for the determination of the corresponding meteoroid properties.

1.4 Continuous Flow Regime

1.4.1 Ablation Coefficient

The ablation rates of the meteoroids interacting with Earth's atmosphere at middle altitudes have been studied by many investigators. Ablation may occur due to both convective and radiative heating, and different estimates of the convective and radiative parts of the heat transfer coefficient Λ are used by different authors (Allen et al., 1963; ReVelle, 1979; Baldwin and Sheaffer, 1971, and others). For large meteoroids, meteoroids of diameter 0.1–1 m and larger (the precise boundary, size as well as Λ itself, are dependent on altitude of flight, velocity and material properties), ablation occurs mostly due to radiative heating and convective heating is negligible. The radiative heating rate increases roughly linearly with body radius, whereas the convective heating rate decreases as the inverse square root of the radius (Baldwin and Sheaffer, 1971; Park, 2014). In addition, the evaporated material blocks convective heat flux from reaching the ablating surface. Both the shock wave radiation and vapour radiation contribute to the ablation.

The value of the coefficient Λ may be estimated both from observations and from modelling results. Applying the meteor physics equations to observational data collected by bolide observational networks permits a determination of the ablation coefficient σ, which is connected to Λ by Equation (1.9) (i.e. $\sigma = \Lambda/(2Q\Gamma)$), and the shape-density coefficient K average along trajectory (Pecina and Ceplecha, 1983; Ceplecha et al., 1998, 1993; Bellot Rubio et al., 2002) or varying along it (Ceplecha and Revelle, 2005).

Assuming all coefficients in the governing equations do not vary along the meteoroid trajectory and no fragmentation occurs, the mass of a meteoroid M at some velocity V may be written as follows

$$\frac{M}{M_0} = \exp\left(\frac{\sigma}{2} \cdot (V^2 - V_0^2)\right) \quad (1.15)$$

where V_0 and M_0 are the entry velocity and initial mass (Levin, 1956; Bronshten, 1983; Ceplecha et al., 1998). The luminous trajectory ends at velocities of roughly 4 km s^{-1}, i.e. a meteoroid entering at 20 km s^{-1} will retain about 7% of its initial mass at the end of its luminous track if $\sigma \sim 0.014$ s^2 km^{-2} (i.e. $\Lambda \sim 0.09$), or about $10^{-9} M_0$ for a larger value of $\sigma \sim 0.1$ s^2 km^{-2} ($\Lambda \sim 0.64$).

Different groups of meteoroids were recognised by Ceplecha and McCrosky (1976) on the basis of their ability to penetrate into the Earth's atmosphere. The existence of the groups was also confirmed from the distribution of ablation coefficients (Ceplecha et al., 1993, 1998). A typical value of the ablation coefficient is 0.014 s^2 km^{-2} for group I (ordinary chondrites) and 0.042 s^2 km^{-2} for group II (carbonaceous chondrites). For fragile bolides (type III, attributed to a cometary origin) the ablation coefficient is estimated as $\sim$0.1–0.2 s^2 km^{-2}. The ablation coefficient determined in this way is affected by fragmentation.

The entry of several specific bolides was reconsidered with the fragmentation (FM) model by (Ceplecha and Revelle 2005). This model has a large number of free parameters that need to be estimated during the initial stage of modelling and are re-calculated during the solution process. The solution itself is a search for the best fit both to the light curve and trajectory data (height as function of time). This model may be used for cases where there is high-precision observational data. This method allows the so-called intrinsic ablation coefficient be determined, which is supposed to be free from fragmentation influence. The intrinsic ablation coefficient of these bolides was low, mostly in the range from 0.004 to 0.008 s^2 km^{-2} (that corresponds to about $\Lambda \sim 0.024$–0.048), and appeared to not depend on the bolide type, which was unexpected. The authors claimed that the apparent ablation coefficients reflect the process of fragmentation, whereas the bolide types indicate the severity of the fragmentation process Ceplecha and Revelle (2005). It should be noted here that the values of the intrinsic coefficients obtained are dependent on assumed fragmentation modes.

Theoretical estimates of Λ and σ have been presented by different researchers. Originally, the suggested dependences were an extrapolation of the available results on heat fluxes far beyond the range of parameters in which they were obtained, and have different restrictions. More often the influence of radiation was only included for the optically thin approximation or/and screening by the evaporation products wasn't taken into account. Biberman et al. (1978, 1979, 1980); Golub' et al. (1996, 1997); Nemtchinov et al. (1997); Park (2014, 2016); Johnston et al. (2018) considered the influence of radiation and ablation on the heating of the meteoroid more accurately, although these models are also not free from restrictions.

Biberman et al. (1978, 1979, 1980) considered the 1D radiative-convective transfer in the bow shock layer for a wide range of parameters, taking into account blowing. The body nose radius was varied from 0.1 to 10 m, the velocity from 10 to 50 km s^{-1}, and the altitude above Earth's surface from 0 to 70 km. The flow structure and heat transfer in the shock layer was considered in the vicinity of the stagnation point in the inviscid approximation. When determining optical properties, only the continuous part of the real absorption coefficient was taken into account. The thermodynamical properties of equilibrium air (Kuznetsov, 1965) and the optical properties of high temperature air (Avilova et al., 1970) were used. A heat-shielding material consisting of light atoms was considered to be the ablating surface (Biberman et al., 1978). The values of Λ obtained were mainly in the range 0.001 − 0.028 for $H \sim 20 - 40$ km, $R \sim 0.1$-1 m, $V \sim 10$-20 km s^{-1} (Figure 1.7b).

In the ablating piston model of Golub' et al. (1996, 1997), radiation transfer in the air and in the meteoritic vapour was taken into account for 0.1–10 m radius meteoroids with velocities in the range of 10–30 km s^{-1}, at 20–50 km altitude (Nemtchinov et al., 1997). The model is based on an analogy between the one-dimensional nonstationary motion of a cylindrical piston in the air and the two-dimensional quasistationary flow around the body (Chernyi, 1959; Hayes and Probstein, 1959). Detailed tables of the spectral opacities of vapours of irons and chondrites (Kosarev et al., 1996; Kosarev, 1999) were used. The same optical and thermodynamical properties for air were used as in Biberman et al. (1980). The ablation process was assumed to occur uniformly over the surface of the meteoroid in accordance with the value of the radiation flux. Melting and shedding of the melted layer were also incorporated into the model. Values of Λ were mainly in the range 0.002–0.08 for $H \sim 20$–40 km, $R \sim 0.1$–1 m, $V \sim 10$–20 km s^{-1} and the corresponding σ is 0.007–0.08 s^{2}km^{-2} (Nemtchinov et al., 1997), not far from the intrinsic ablation coefficient values defined by Ceplecha and Revelle (2005)(Figure 1.7b).

Park (2014) applied an inviscid stagnation line analysis and determined ablation rates taking into account radiative transfer in the approximation of radiative heat diffusion with the mean absorption Rosseland coefficients, which were found previously (Park, 2013). The calculated ablation rates appeared to be smaller than ones determined by Golub' et al. (1996), although it should be noted here that only data on the ablation of iron meteoroids were given in Golub' et al. (1996), whereas data on H-chondrite ablation were presented in Nemtchinov et al. (1997), and the ablation rates for H-chondrites are smaller than those for irons. In Park (2016) radiation intensity was calculated line-by-line. The calculated heating/ablation rates decrease with nose radius, peak at altitudes between 30 and 40 km, and do not always increase with velocity. They can be higher or lower than the values calculated using the Rosseland approximation, and unexpectedly do not asymptotically approach the Rosseland values at high densities. According to Johnston et al. (2018) the corresponding values of the heat transfer coefficient are about $\Lambda \sim 0.002 - 0.05$ for a 10 m-sized object at $V \sim 20$ km s^{-1}.

The flow in front of the meteoroid was determined by a Navier-Stokes solver in Johnston et al. (2018). The radiation coupling assumes one-dimensional radiative transport along body-normal rays through the shock layer. Detailed air thermochemistry was included, and 26 additional species (in fractions corresponding to LL-chondrites) were added to the flowfield to account for meteor ablation products, as these have a significant impact on the shock layer radiation. A number of cases were considered, i.e. velocities ranging from 14 to 20 km s^{-1}, altitudes ranging from 20 to 50 km, and nose radii ranging from 1 to 100 m. The heat transfer coefficients from these simulations are below 0.045 for the range of cases. The comparison of Λ values ($R \sim 10$ m, $V \sim 20$ km s^{-1}, $H \sim 20$–50 km) demonstrated agreement between their results and those of Biberman; the Park and Nemtchinov values, however, are more than a factor of two larger than the results by Johnston et al. (2018)(Figure 1.7b).

The blowing of the melt was included in ablation losses by Golub' et al. (1996), and for iron meteoroids this blowing is responsible for as little as 4% of the mass loss rate (10 m body at 40 km altitude, 20 km s^{-1}) and as much as 100% for a 0.1 m meteoroid at the same altitude with 10 km s^{-1} velocity. Chen (2016) considered a thermal ablation model for silicates, which includes mass losses through the balance between evaporation and condensation, and through the moving molten layer driven by the surface shear force and pressure gradient. He showed that the mass loss through the moving molten layer is negligibly small for heat-flux conditions at around 1 MW/cm^2, which approximately corresponds to altitudes below 45 km for 20 km s^{-1} entry velocity. Additionally, for conditions with a heat flux around the order of 1 kW/cm^2, the surface recession is mostly driven by the moving molten layer (at roughly 90 km for a 20 km s^{-1} entry).

Modern modelling efforts result in low values of the ablation and heat transfer coefficients due to effective surface shielding by massive ablation. These results are in agreement with analysis of precise observational data, assuming some fragmentation model. At the same time, the ablation coefficient averaged over the trajectory is larger, as some kind of fragmentation accompanies every meteoroid entry even if there are no obvious manifestations of disruption. Applying a value of $\Lambda \sim 0.1$ or similar (Bland and Artemieva, 2006, and others) in different studies, which aimed to determine the energy release and/or to reproduce the dynamics and light curves of various bolides, means implicit inclusion of continuous fragmentation. If the low values of ablation coefficient are used, then fragmentation in some form must be included to reproduce observational data.

1.4.2 Luminous Efficiency

The luminous efficiency values determined for small meteors (see Section 1.2.2) cannot be directly applied to large fireballs or bolides, which are interacting with the atmosphere in continuous flow. The luminous efficiency is expected to be dependent on velocity, size (or mass), composition and altitude of flight (Nemtchinov et al., 1997). Even a simple extrapolation of the luminous efficiency derived for large Prairie Network and European Network bolides (Revelle and Rajan, 1979; Revelle, 1980) to larger satellite-observed bolides by US Government Sensors (Tagliaferri et al., 1994; Nemtchinov et al., 1997; Brown et al., 2016) may be unsuitable because the masses of the meteoroids detected from the satellites might be higher by two or three orders of magnitude. An increase in the geometrical and optical thicknesses of the shock-heated air and vapour layers may change the radiation fluxes and the values of luminous efficiency.

The luminous efficiency can either be extracted from the analysis of observational data (Ceplecha et al., 1996; Ceplecha, 1996; Halliday et al., 1981; Ceplecha and Revelle, 2005) or be obtained in the course of radiative hydrodynamic numerical simulations (Golub' et al., 1996, 1997; Nemtchinov et al., 1997).

The instantaneous luminous efficiency was derived from the analysis of observational data on meteoroids of about 0.01 – 1 m in size taking into account their disruption (Ceplecha and Revelle, 2005). This efficiency was found to depend on velocity and mass. The velocity-dependence from Pecina and Ceplecha (1983), which was also used by Revelle and Ceplecha (2001), proved to work well in actual fireball modelling by the semi-empirical model (Borovička et al., 2013a), at least for low velocities ($V < 20$ km s^{-1}). The mass-dependency was modified by Revelle and Ceplecha (2001). The increase around masses of 1 kg was kept but the difference in τ between small and large masses was found to be much smaller (factor of $\sim$2) than the factor of 10 proposed by Revelle and Ceplecha (2001). This does not mean that luminous efficiencies of small meteoroids high in the atmosphere are not small. The recommended luminous efficiency τ (in percent) for fireball modelling is then:

$$\ln \tau = 0.567 - 10.307 \ln V + 9.781 (\ln V)^2 - 3.0414 (\ln V)^3 \\ + 0.3213 (\ln V)^4 + 0.347 \tanh(0.38 \ln M) \quad (1.16)$$
$$\text{for } V < 25.372$$
$$\ln \tau = -1.4286 + \ln V + 0.347 \tanh(0.38 \ln M) \quad (1.17)$$
$$\text{for } V \geq 25.372,$$

where velocity, V, is in km s^{-1} and mass, M, in kg (ln is natural logarithm and tanh is hyperbolic tangent) (Borovička et al., 2013a). These equations give $\tau = 5\%$ at 15 km s^{-1} for large masses and 2.5% for small masses (Figure 1.8). The increase for higher velocities is nearly linear (strictly linear above 25.4 km s^{-1}), the decrease for lower velocities is steeper. For converting the radiated energy into absolute magnitudes, the relation of 1500 W corresponding to zero magnitude is used under the assumption of an effective temperature in the radiative area of about 4500 K (Ceplecha et al., 1998). The corresponding magnitude is usually suggested to be the visible (panchromatic) or bolometric one, the irradiated energy corresponds to the whole spectral range.

The brightness of meteors is measured in magnitudes by comparisons with stars, and it is measured only in a given limited spectral range, which for network bolides is mostly in the panchromatic passband from about 3600 to 6600 A. Different assumptions of the spectrum of the meteor result in different conversion factors between the physical units and magnitudes. Halliday et al. (1981) derived an average efficiency of about 4–8% for radiation energy in the panchromatic passband for the Innisfree meteorite, which would result in an efficiency 2–3 times larger for the whole spectrum, that is larger than Equation (1.16) predicts. At the same, time Ceplecha and Revelle (2005) obtained values smaller than about 1.2% for the whole spectrum for Innisfree.

For satellite network bolides a common assumption is that the source radiation follows a 6000 K blackbody and, in this case, Ceplecha et al. (1998) suggested 1100 W per zero magnitude for the whole spectrum, whereas a larger value of about 3200 W was applied by others (Tagliaferri et al., 1994; Brown et al., 1996). The relation between satellite-recorded power and magnitude quoted by Brown et al. (2013) corresponds to a conversion factor of about 2100 W. Therefore the uncertainty in radiation intensity may reach 2–3 times in physical units (or up to 1 magnitude).

This conversion (from magnitude in the limited observational passband to the energy irradiated in the whole spectrum) implicitly assumes that the ratio of the fraction of the energy radiated in the observational band and beyond does not differ from this

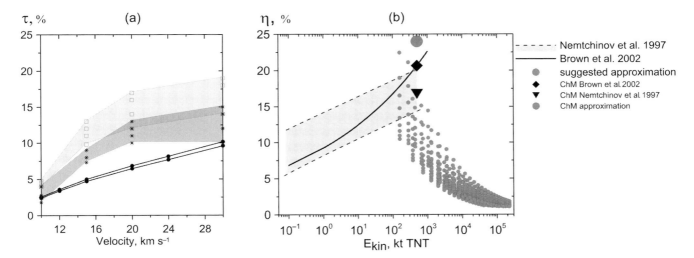

Figure 1.8. Ablation model parameters for continuous flow regime. (a) Luminous efficiency τ of meteors in the whole spectral range. Black pointed curves correspond to Equations 1.16 and 1.17 for two different meteoroid masses of about 23 kg (lower curve) and 23000 kg (upper curve) (i.e. 14 cm and 1.4 m in radius objects with density 2000 kg m^{-3}). Polygons with symbols (23 kg – stars; 23000 kg – squares) correspond to the ablating model results (Nemtchinov et al., 1997) for the same masses. This model predicts the luminous efficiency dependence on the altitude of flight, rows of symbols correspond to the different altitudes of flight (top row – 20 km, bottom – 50 km, data for 30 and 40 km are in between). (b) Integral luminous efficiency η for 20–120 m asteroidal impactors (grey circles Popova et al., 2017) $\theta \sim 25 - 65°$ and $V \sim 15 - 25$ km s^{-1} correspond to the most probable values, observed for 1–10 m impactors (Brown et al., 2016). Filled area and thin black line demonstrate efficiency for meters-sized chondritic bodies according to Nemtchinov et al. (1997) and Brown et al. (2002a) correspondingly.

ratio for the spectrum of a black body with a certain temperature (Ceplecha et al., 1998). Such a ratio may not be satisfied in the case of the real and/or modelled radiation field, which also increases the uncertainty.

The ablating piston model currently provides the best known theoretical estimates of luminous efficiency for ordinary chondrites and irons of about 0.1–10 m in size at middle altitudes (Golub' et al., 1996, 1997; Nemtchinov et al., 1997). The model demonstrates that the nearest region of wake (at a distance of several diameters of the body) is mainly responsible for the emission in the panchromatic wavelength range; the UV emission mainly originates from the bolide's head; whereas the IR radiation is produced in the far wake. Total radiation losses for large (>0.1 m) iron and H-chondrites do not differ substantially (no more than by a factor of two or three) from that of a pure silica body or from values obtained in the framework of an air-radiation dominated model. This is explained by the fact that the radiation of the vapour in the ablating piston model is substantially governed by the radiation emitted by the shock-heated air, absorbed and re-emitted by the vapour.

These theoretical luminous efficiencies grow with velocity, body size and atmospheric density. The luminous efficiency in the panchromatic passband increases from 0.7 to 5% for a 14 cm ordinary chondritic object at 15 km s^{-1} velocity with altitude decrease from 50 to 20 km (Golub' et al., 1997). However, it should be remembered that the spectrum differs from the black body one and changes with altitude, velocity and size of the object. So, luminous efficiencies in different wavelength ranges (i.e. for different types of sensors) also differ. The efficiencies for the same object within the sensitivity region of the photoelectric sensors of the satellite observational system (1–3 eV) are about 4–8% (Golub' et al., 1997). The energy losses in the whole spectrum exceed the output in the panchromatic wavelength range by roughly a factor of 2-3. Bolometric efficiencies for a 14 cm object at 15 km s^{-1} vary from 7 to 9% depending on altitude, and exceed 10% for 20 km s^{-1} (Figure 1.8a) (Nemtchinov et al., 1997). The theoretical luminous efficiencies demonstrate greater dependence on size (mass) than Equations (1.16) and (1.17) and dependence on altitude, which is not included in Equations (1.16) and (1.17) (Figure 1.8a).

The integral luminous efficiency η connects the total light production and initial kinetic energy of the meteoroid $\eta = E_r/E_k$; it can be useful in characterising the entry of large bodies, and is useful if the instantaneous luminous efficiency changes with altitude and deceleration. For 0.1–10 m scale bodies the integral luminous efficiency η was determined based on the results of radiation hydrodynamic simulations (Nemtchinov et al., 1997). Multiple additional parameters such as material strength, fragmentation model, the heat transfer coefficient and entry angle play some role in defining η within this model framework. The empirical method relates E_r to the total initial energy estimates generated independently by airwave measurements, and several fireballs for which meteorites have been found and which had instrumental data (photographic, video) providing well-calibrated initial energies (Brown et al., 2002b). The dependences obtained suggest that η should be about 5–15% for bodies of chondritic composition with energies between 0.1 and 50 kt TNT with the efficiency increasing as energy increases (Figure 1.8b) (Nemtchinov et al., 1997; Brown et al., 2002b). Recently these values of integral luminous efficiency allowed the initial kinetic energy of the 500 kt Chelyabinsk meteoroid to be determined based on the irradiated energy and light curve; these estimates were in agreement with values of kinetic energy determined by other methods (Brown et al., 2013; Popova et al., 2013). The integral luminous efficiency for the Chelyabinsk event was estimated to be 14–17% (Popova et al., 2013) or 17% (Brown et al., 2013). Consideration of the Chelyabinsk entry in the frame of a quasi-liquid model (QL model), in which the entering body is treated as a strengthless quasi-liquid object, and its deformation and the flow are described by the hydrodynamic equations, results in a slightly larger value of 18.4% (Shuvalov et al., 2017).

The trend of increasing luminous efficiency with impactor energy should stop at some initial kinetic energy, as large impactors reach the surface with negligible energy losses and produce explosive craters. Recent modelling of the entry of large chondritic and cometary objects (30–300 m) in the frame of a quasi-liquid model (see Svetsov et al., 2019, for more detail) allows η values to be determined (Svetsov and Shuvalov, 2017) and a scaling relation to be suggested (Popova et al., 2017). The efficiency η decreases with energy for large objects, unlike the dependence for small bodies (Figure 1.8b). It can be assumed that this decrease is connected with an increase of the optical thickness of the emitting region, which leads to radiation losses mainly from its surface.

Roughly, the uncertainties mentioned in this section mean that the luminous efficiency for asteroidal meteoroids 0.1–1 m in size is known within a factor about 2–3, and may be worse for smaller bodies.

1.4.3 Modelling Spectral Radiation

Important information about the composition of meteoroids and about the ablation process and physical conditions in the radiating plasma can be obtained from meteor spectra. Existing models of spectral radiation are either fully theoretical or simplified, aimed at the interpretation of observed spectra.

Theoretical spectra of bolides produced by large meteoroids (20 cm–20 m) at heights between 20 and 60 km were presented by Golub' et al. (1997). The spectra are based on the one-dimensional model of meteoroid ablation and radiation of Golub' et al. (1996), which computes temperature profiles of vapours and air and assumes local thermal equilibrium (except that electron temperature was different from the temperature of atoms and ions). The resulting spectra consist of a continuum and line radiation, while the importance of continuum increases with increasing meteoroid size and velocity and with decreasing height in the atmosphere. The spectra computed with Golub's model were compared with the observed spectra of the Benešov bolide in Borovička et al. (1998b). The observed spectrum showed lower continuum than the theoretical spectrum. The difference may be partly due to the fact that the meteoroid was fragmenting during flight and the radiating volume was more complicated than assumed in the model.

Theoretical spectra were also recently computed by Park (2018) using the opacities of Park (2013) and ablation model of Park (2014). When comparing the excitation temperatures from the computed and observed line intensities (for Benešov), the theory gave much higher values (>10000 K vs. ~ 5000 K). The reason is probably that the theory considered vapour radiation only from the region in front of the meteoroid, between its surface (the wall) and the shock wave, and not the larger (and cooler) regions around and behind the meteoroid. On the other hand, Park (2018) accounted for meteoroid fragmentation into smaller pieces.

A simple model for the analysis of observed spectra was developed by Borovička (1993) on the basis of previous work by Ceplecha. The whole radiating gas is assumed to be in thermal equilibrium at a temperature T. The principles are the same as in the subsequently developed Laser Induced Breakdown Spectroscopy (LIBS), used to analyse the chemical composition of various materials, including meteorites (De Giacomo et al., 2007). The computed and observed intensities of atomic spectral lines are compared and model parameters are adjusted to obtain the best match. In addition to temperature, the input parameters are the concentrations of individual chemical species, n_i, and the size of the radiating volume. In fact, only the column densities, i.e. concentrations integrated along the line of sight across the radiating volume, $N_i = \int n_i \, ds$, and the cross section of the radiating volume, P, are important for the emerging spectrum. The emerging radiation intensity at frequency ν is

$$I_\nu = B_\nu(T)\,(1 - e^{-\tau_\nu}), \qquad (1.18)$$

where $B_\nu(T)$ is the Planck function for temperature T and τ_ν is the optical thickness of the radiating gas along the line of sight at frequency ν. In case of optically thin radiation ($\tau_\nu \ll 1$), $I_\nu = B_\nu \tau_\nu$. In the limiting case of very large optical thickness at all wavelengths (for very large bodies), the spectrum converges to a black body spectrum and information about chemical composition is lost.

If the spectrum is dominated by emission lines, radiative transfer can be considered only in spectral lines. The optical thickness of a spectral line is proportional to the atom column density, level population, and transition probability. Since thermal equilibrium is assumed, the population of atomic levels is given by the Boltzmann distribution. We then have

$$\tau_\nu \sim N_i \, gf \, e^{-E_1/k_b T} \, \Phi(\nu), \qquad (1.19)$$

where g is statistical weight, f is oscillator strength, and E_1 is the excitation potential of the lower level of the transition (these parameters can be found in databases of atomic lines), and $\Phi(\nu)$ is the line profile, which can be assumed to follow the Voigt function. The line profile is important only for lines which are optically thick (in the center). The damping constant, γ, influencing the line profile is then another free parameter of the model (see Borovička, 1993). Since line profiles are not resolved in observed meteor spectra, only the total line intensities,

$$I = \int I_\nu \, d\nu, \qquad (1.20)$$

are important. While intensities of optically thin lines are proportional to their optical thickness at the center wavelength, intensities of optically thick lines increase only slowly with increasing optical thickness (e.g. with increasing column density) since the line core is saturated and only the line wings become brighter. This can be seen on the so-called emission curve of growth (Ceplecha, 1964). It is sometimes assumed in spectral analyses that all lines are optically thin, and this may lead to erroneous results for bright meteors.

All atoms, ions, and possibly also molecules with identified lines in the spectrum are included in the calculations.

For molecules, it is better to include individual rotational lines in calculations even if they are not resolved. Atoms and ions are taken as independent species in this phase. The ratio of concentrations of atoms and ions of the same element could in principle be used to compute electron density from the Saha equation and to evaluate ionisation degrees of all elements. However, as discussed by Borovička (1994), there are two spectral components in bright meteors. Most lines of neutral atoms belong to the main component of $T \sim 5000$ K, while most lines of singly ionised atoms belong to the second component of $T \sim 10000$ K (the lines of Ca II are bright in both components). In result, electron density must be evaluated separately for each component using the calculated P, N_i, and T and assuming some geometry of the radiating volume. For a possible approach see Borovička (1993).

The results of meteor spectral analyses concerning chemical composition of meteoroids are discussed in Borovička et al. (2019). Meteor spectra also provide insight into the ablation process. One important effect is incomplete evaporation. The abundances of refractory elements (Al, Ca, Ti) in the radiating gas were found to vary along the meteor trajectory by Borovička (1993). They remained underabundant (in comparison with meteorite composition) along the whole trajectory but increased in meteor flares and toward the meteor end. The latter tendency was very pronounced in the deeply penetrating Benešov bolide (Borovička and Spurný, 1996). It is obvious that refractory elements are not vapourised from the meteoroid surface. They are released in solid or liquid phase and are vapourised incompletely or too late, i.e. just before leaving the hot radiating region. The vapourisation of chondritic melt was computed theoretically using the MAGMA code by Schaefer and Fegley (2004). Indeed, unless the temperature was very high (6,000 K), most of the Al and Ca vapourised within the last 5% of the material in their simulations.

The work of Schaefer and Fegley (2004) was used by Borovička (2005b) in the study of the Senohraby fireball spectrum. No pronounced spectral change occurred during the broad fireball maximum (-10 mag) at a height of 46.5 km. The volume and mass of the radiating region reached their maxima here. The effective diameter of the vapour cloud was about 12 m and the temperature was near 4,000 K. The temperature increased rapidly after a major fragmentation event (identified independently by Spurný and Ceplecha, 2004) at a height of 42 km, reaching 4700 K. The volume and mass of the radiating gas decreased after the fragmentation while electron density and abundances of refractory elements increased. The fraction of vapourised material was estimated at 90–95% both before and after the fragmentation but the vapourisation temperature increased from 2400 K to 3200 K after the fragmentation. The fragmentation therefore temporarily disturbed the shielding of the meteoroid, allowing the heat transfer to increase.

Berezhnoy and Borovička (2010) modelled the formation of molecules in the vapour cloud. For typical fireball temperatures of 4000–5000 K, most elements are expected to be in the form of atoms and ions. Notable exceptions are Si and C, which are expected to be mainly in the form of SiO and CO. When the plasma cools, more molecules will be formed in the wakes and afterglows. The abundances of metal oxides (FeO, MgO, AlO etc.) are highest at temperatures of 2000–2500 K. Borovička and Berezhnoy (2016) analysed molecular radiation in the Benešov bolide. Molecular bands were indeed particularly bright in the wake and in the radiating cloud left behind the bolide. FeO was present from the highest altitudes and was probably ablated directly in molecular form at high altitudes. CaO was first detected just below 50 km and its intensity, relatively to FeO, strongly increased towards lower altitudes. AlO behaved as FeO rather than CaO at lower altitudes. MgO was observed only in the radiating cloud. The CaO/FeO and AlO/FeO behaviour was consistent with wake temperatures of about 3,500 K.

1.4.4 Fragmentation Models

When comparing modelling efforts and observations, it becomes obvious that fragmentation has a much stronger effect than deceleration and ablation. Meteor models typically differ in how they handle fragmentation. The hydrodynamical models that consider all effects, i.e., ablation, radiation and fragmentation, are sparse, and have so far only been applied to several large objects (mainly $\geq 10^8$ kg or $\geq \sim 30$ m; see Svetsov et al., 1995), or used to predict asteroid hazard consequences (see Svetsov et al., 2019, and Section 1.4.5). These models cannot predict details of the fragmentation process, such as the simultaneous production of dust and fragments. The behaviour of smaller meteoroids is typically described by standard equations (see Section 1.1.3) supplemented by a fragmentation model.

It is usually assumed that the destruction of a meteoric body begins at the moment when the aerodynamic pressure in the vicinity of a stagnation point becomes equal to some constant describing the strength of meteoroid material. There are different approaches to the choice of characteristic strength in the breakup criterion (see Svetsov et al., 1995, for some discussion); compressive, tensile, or shear strengths may be utilised by different authors, depending on the assumed mechanism of breakup. During the atmospheric loading, the internal stresses in the meteoroid are proportional to the ram pressure (Tsvetkov and Skripnik, 1991), and so the stagnation pressure at breakup may be taken as the breakup criterion and measure of the meteoroid strength in the disruption.

The natural fracture distribution in asteroids results in a low strength at entry, from 0.1 to approximately 1 MPa on first breakup, and a maximal strength on breakup of $1 - 10$ MPa, compared to the average measured tensile strength of the similar meteorite classes, which is about 30 MPa (Popova et al., 2011). According to the statistical strength theory (Weibull, 1951) and direct observations on natural rocks (e.g. Hartmann, 1969), the strength of a body in nature tends to decrease as body size increases. The effective strength is usually expressed as $\sigma_{str} = \sigma_s (M_s/M)^\alpha$, where σ_{str} and M are the effective strength and mass of the larger body, σ_s and M_s are those of a small specimen, and α is a scaling factor. A value of exponent $\alpha \sim 0.25$ is most often used for the exponent in modelling (Tsvetkov and Skripnik, 1991; Svetsov et al., 1995), whereas fireball data have produced values between 0.1 and 0.75 (Popova et al., 2011). The strength scaling law is commonly assumed in meteoroid fragmentation theories (e.g. Baldwin and Sheaffer, 1971; Nemtchinov and Popova, 1997; Borovička et al., 1998b; Bland and Artemieva, 2006). Bland and Artemieva (2006) suggested the use of a small variation in strength (about 10% around predicted values), but there could be much more significant deviations (Popova, 2011; Popova et al., 2011).

1.4.4.1 Pancake Models

When the ram pressure exceeds the assumed apparent strength of a meteoroid, the liquid-like or pancake model assumes that the meteoroid is disrupted into a swarm of small bodies, which continue their flight as a single mass with an increasing, pancake-like, cross section (Figure 1.9) (Grigorian, 1979; Hills and Goda, 1993; Chyba, 1993). The smallest fragments can be easily evaporated and fill the volume between larger pieces. If the time between fragmentations is smaller than the time for fragment separation, all the fragments move as a unit, and a swarm of fragments and vapour penetrates deeper, being deformed by the aerodynamical loading like a drop of liquid. As a whole, this process may be described in the frame of a single-body model. This fragmentation was initially suggested for the disruption of relatively large bodies.

Hills and Goda (1993) defined the spreading rate by equating the kinetic energy of the expanding fragments to the work done by the air to increase the area of the cloud of fragments, which resulted in the following relation for the radius increase rate U: $U = K_1 V \sqrt{\frac{\rho_a}{\rho_m}}$ where K_1 is a constant. It was suggested that meteoroid spreading will cease when the ram pressure in front of the meteoroid drops below the strength of the meteoroid. In applications of this model to smaller objects, other constraints on maximal spreading are often assumed (a choice of limiting maximal value). In Chyba (1993) and Collins et al. (2005) the spreading rate is defined by a force balance on the walls of the disrupted meteoroid, which is approximated as an incompressible, strengthless cylinder, deformed by an average differential stress proportional to the ram pressure. Avramenko et al. (2014) derived an alternative spreading rate using dimensional analysis and a chain reaction analogy for the cascading disruption of meteoroid fragments.

The coefficient K_1 was estimated to be in the range 0.3–1 based on fitting of the observational data (Borovička et al., 1998a; Popova, 2011). The light intensity rises more sharply if a larger value of K_1 is used. A decrease of K_1 decreases maximum intensity altitudes and increases the distance between the altitude of breakup and the altitude of maximum intensity. The total irradiated energy also slightly decreases. In an unconstrained pancake model, the meteoroid is allowed to expand without bound, while in a constrained model, the growing radius is limited to a multiple of the initial diameter of the meteoroid (the so-called pancake factor; see Collins et al., 2017).

When describing specific events, this model needs to be adjusted by constraints (Popova, 2011). An application of the pancake model with different additional settings allowed the altitudes of airbursts and energy deposition caused by the impacts of large meteoroids to be determined (Chyba et al., 1993; Hills and Goda, 1993; Borovička et al., 1998a; Collins et al., 2005, and others). Bright flares in the light curves of large bolides are typically explained as fragmentation. For a number of satellite-observed bolides, these light curves were successfully reproduced in the frame of a pancake model (Svetsov et al., 1995; Nemtchinov et al., 1997; Popova and Nemchinov, 2008).

Several clouds may be formed in successive breakup stages of the meteoroid. A few dust clouds were formed in the disruption of the 2005 bolide (Klekociuk et al., 2005), the Marshall Island bolide of 1 February 1994 (McCord et al., 1995), the Almahata Sitta bolide and the Chelyabinsk meteoroid. In these cases, the light curve may be reproduced under the assumption that initially formed few pieces, which are fragmented further into two or three pancake-like clouds of fragments. The shape of the

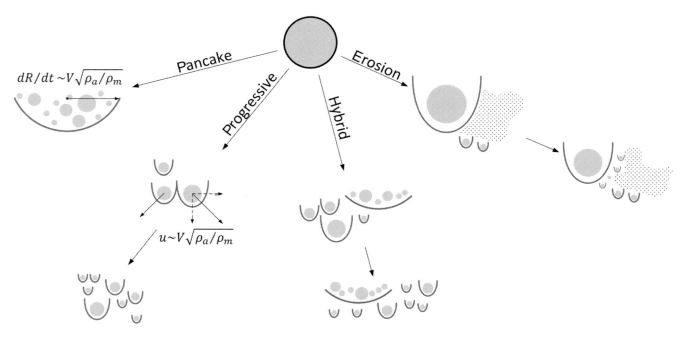

Figure 1.9. Schematic picture of different fragmentation scenarios: (a) Pancake; (b) Progressive fragmentation; (c) Hybrid; (d) Erosion.

curve provides some constraints on the amount of mass under fragmentation at each breakup stage.

The pancake model does not estimate the size and numbers of fragments formed, so there is no information about the numbers and sizes of meteorites and/or the strewn field.

1.4.4.2 Progressive Fragmentation Models

A progressive fragmentation model suggests that meteoroids are disrupted into fragments which continue their flight as independent bodies and may be disrupted further (Figure 1.9). Similar models have been suggested in numerous papers, beginning with Levin (1956). At the initial stage, fragments interact through their shock waves, which results in the appearance of lateral velocity (Passey and Melosh, 1980): $U = K_2 V \sqrt{\frac{\rho_a}{\rho_m}}$ where K_2 is determined by the interaction of fragments' shock waves, and is about 0.45 for two identical fragments (Artemieva and Shuvalov, 1996; Bland and Artemieva, 2006). Larger values of lateral velocity were found in actual bolides (Borovička and Kalenda, 2003; Borovička et al., 2013b). Azovskii and Shuvalov (2002) demonstrated that this coefficient may grow due to the influence of fragment shapes. The progressive fragmentation approach allows a mass-velocity distribution of fragments on the surface to be defined with their corresponding craters or meteorite strewn field (Passey and Melosh, 1980; Nemtchinov and Popova, 1997; Bland and Artemieva, 2006; Popova et al., 2013; Brown et al., 2013).

The possibility of describing the fate of individual fragments is the main advantage of the progressive fragmentation type models, and it is extremely important. The number of fragments changes over the process of the disruption from 1 (a parent body) to an arbitrarily large value, depending on the assumed properties of the meteoroid. This type of model usually incorporates the strength scaling law mentioned in Section 1.4.4, and different assumptions about the distribution of the mass of formed fragments. Bland and Artemieva (2006) suggested that each fragmentation of a single body results in two fragments with smaller mass and usually higher strength. Each fragment is subject to additional fragmentation later if the dynamic loading exceeds the updated fragment strength.

The mass distribution of fragmented rocks is often described by a power law (Hartmann, 1969; Fujiwara et al., 1989). The power law distribution has also been used in the description of meteorite finds (Jenniskens et al., 1994; Hildebrand et al., 2006) and in the modelling of meteoroid entry (Nemtchinov and Popova, 1997; Borovička et al., 1998a; Popova et al., 2003). Disruption into two fragments leads to a light curve with a single peak. The formation of a large number of fragments may result in the appearance of flashes on the light curve, and it slightly shifts the light curve to higher altitude.

The unmodified progressive fragmentation model may be applied if the number of fragments is not extremely large and they can each be considered independent. The application of this model to the entry of Chelyabinsk resulted in too many pieces and too much mass in meteorites, and in other differences with observations. The huge number of fragments formed at altitudes of 50–20 km during a short time suggests that the fragments cannot be considered to be independent (Popova et al., 2013).

1.4.4.3 Hybrid Fragmentation Models

The complete picture of fragmented-body motion is comparatively complicated, as both the pancake and progressive fragmentation scenarios are simultaneously realised in real events at different stages of a meteoroid's passage through the atmosphere. More complicated semi-analytical models that take into account splitting into separate fragments, treated as a set of liquid-like or solid particles, have been developed and applied for the modelling of ground-based and satellite-observed bolides (Figure 1.9). This hybrid model was used to model the Benešov and the Marshall Island bolides (Nemtchinov and Popova, 1997; Borovička et al., 1998a). More elaborate versions were used later (Ceplecha and Revelle, 2005; Popova, 2011; Brown et al., 2013; Popova et al., 2013; Bronikowska et al., 2017; Register et al., 2017; Wheeler et al., 2017).

Popova et al. (2013) used a hybrid combination of pancake and progressive fragmentation approaches (called hybrid model) in order to describe the light curve and meteorite-strewn field of the Chelyabinsk meteoroid. Wheeler et al. (2017) presented a hybrid model modification (called a fragment-cloud model (FCM)), and perfomed a sensitivity study to demonstrate how variations in the model's parameters affect the energy deposition results.

1.4.4.4 Semi-Empirical Fragmentation Model

The semi-empirical fragmentation model was developed to fit the observed light curves and dynamical data of bright meteors (fireballs and superbolides) with the aim of determining the meteoroid's entry mass and fragmentation processes during atmospheric penetration. It was first used to study the Košice superbolide and meteorite fall (Borovička et al., 2013a). The principle of the model is simple. For each individual fragment, the basic single body equations for drag, mass loss and radiation are used (see Section 1.1.3). All fragments are supposed to be independent and the fireball light curve is constructed by a summation of the contributions of all individual fragments at any given time.

The trajectory and entry velocity of the modelled fireball is considered to be known. Modelling starts at a chosen point along the trajectory. The input parameters are the meteoroid mass (M), density (ρ_m), shape (A), drag (Γ) and ablation (σ) coefficients. In fact, only $K = \Gamma A \rho_m^{-2/3}$, M and σ are relevant. The equations of drag and mass loss are not integrated numerically; the integral solution of Pecina and Ceplecha (1983, 1984) is used instead. The computation is therefore fast but the coefficients K and σ must be kept constant, at least until the next fragmentation event. To compute luminosity, both the mass loss and deceleration terms are considered (with equal luminous efficiency). Since the model uses a steady-state ablation approach (i.e. does not compute preheating and start of ablation), the luminosity near the fireball beginning can be overestimated.

The main purpose of the model is to reveal fragmentation points. This is done manually by the trial-and-error method. Reliable light curves with high temporal resolution from radiometers (i.e. photomultipliers) are the most suitable for modelling. A flare or sudden increase in light curve slope is an indication of fragmentation. The model allows three kinds of objects to be formed in any fragmentation (Figure 1.9):

Individual fragments. Each fragment is characterised by its mass and parameters K and σ. It is also possible to create several identical fragments. In that case the computation is performed only once and the luminosity produced is multiplied by the number of fragments.

Dust. Dust is a large number of fragments of a certain size range. All fragments are assumed to have the same K and σ. Further parameters are the upper and lower mass limits for fragments, M_u and M_l, and the mass distribution index, s. A power-law mass distribution is assumed, i.e. $n(M) \sim M^{-s}$, where $n(M)$ is the number of fragments of mass M. To simplify the computation, fragments are sorted into mass bins. Computation is performed only once for each mass bin and the luminosity produced is multiplied by the number of fragments in that bin. The number of bins per order of magnitude in mass, z, can be chosen (usually $z \leq 10$). If we define the mass sorting parameter $p = 10^{-1/z}$, the actual number of mass bins is $k = 1 + \log(M_l/M_u)/\log p$. The number of fragments of the largest mass M_u is

$$n_u = \frac{M_f}{k M_u}, \quad \text{for} \quad s = 2 \quad (1.21)$$

$$n_u = \frac{M_f}{M_u} \cdot \frac{1-p^{(2-s)}}{1-p^{k(2-s)}}, \quad \text{for} \quad s \neq 2, \quad (1.22)$$

where M_f is the total mass of the dust. The masses and numbers of fragments in the i-th bin are $M_i = M_u p^i$ and $n_i = n_u (M_u/M_i)^{(s-1)}$, respectively ($i = 0, 1, \ldots, k-1$).

Eroding fragments. Individual fragments and dust released at once cannot explain all features on fireball light curves. The released dust produces flares with a steep onset, while light curves often contain smooth humps. They can be explained by the concept of eroding fragments. The concept was first developed for faint meteors, where the whole meteor could be fitted by one erosion event (Borovička et al., 2007)[1]. Erosion is in fact a quasi-continuous fragmentation – dust is released gradually from the meteoroid. The rate of erosion is described by the erosion coefficient, η_{er}, which is analogous to the ablation coefficient, σ. The total mass loss of the eroding fragment is therefore $dM/dt = K(\sigma + \eta_{er}) M^{2/3} \rho_a V^3$. In each time step, only the ablated part of the mass immediately contributes to the radiation. The eroded part is distributed into fragments in the same way as is done for dust. The eroded fragments continue their flight and ablate and radiate independently of the eroding fragment. In summary, the parameters of an eroding fragment are the same as for individual fragments and dust combined, plus the erosion coefficient.

Figure 1.10 illustrates features on a simulated light curve which can be produced by various parameters of fragmentation. By combining more fragmentation events, the main characteristics of observed fireball light curves can be fitted. To restrict the freedom of too many free parameters, it is advisable to keep some parameters the same or in a narrow range for all (or most) fragments. Namely, the meteoroid density is kept at a value corresponding to the known or supposed material type (e.g. meteorite), ΓA is ~ 0.7, $s = 2$, and $\sigma \sim 0.005$ s^2 km^{-2}, which

---

[1] There is a misprint in Equation 15 of that paper. The correct relation is given in Equation 1.22 in this chapter.

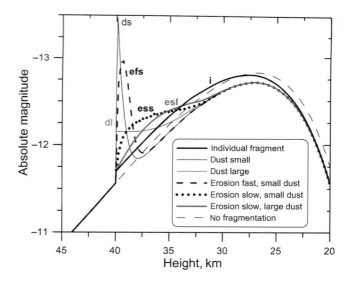

Figure 1.10. Modeled light curves using various fragmentation modes of the semi-empirical model. In all cases, 10 kg is lost at a height of 40 km from a 100 kg meteoroid moving at 15 km s^{-1}. Dust parameters are either $M_u = 10^{-3}$ kg, $M_l = 10^{-5}$ kg (small dust case) or $M_u = 0.1$ kg, $M_l = 0.01$ kg (large dust case). In both cases, $s = 2$ and $z = 10$. Erosion is either fast with $\eta_{er} = 1$ s^2 km^{-2} or slow with $\eta_{er} = 0.1$ s^2 km^{-2}. For all fragments and dust, $\sigma = 0.005$ s^2 km^{-2}, $\Gamma A = 0.7$, and $\rho_m = 3400$ kg m^{-3}. Trajectory slope is 45°. Luminous efficiency was computed from Equation 1.16.

is the intrinsic value of the ablation coefficient corresponding to "pure ablation" without fragmentation (Ceplecha and Revelle, 2005). The free parameters are the positions of fragmentations, numbers and masses of fragments, mass limits of the dust, and the erosion coefficients.

The semi-empirical model was used to study the fragmentation during several meteorite falls and in other fireballs (e.g. Borovička et al., 2013a, 2015, 2017, see also Chapter 2, Borovicka et al., 2019). The positions of important fragmentations are usually obvious from the light curve. The derived fragmentation strengths of the meteoroids are therefore reliable. The exact fragmentation sequence and the masses of fragments are more difficult to find. The masses can be better restricted if there are also dynamical data (deceleration). From photographic observations, dynamics are usually only available for the most massive (leading) fragment. High resolution video data, where fragments can be tracked individually, would be ideal to study details of fragmentation. Occasional video data show that fragmentation can be complex, with fragments exchanging the leading position (Borovička and Kalenda, 2003).

The semi-empirical model is somewhat similar to the FM model of Ceplecha and Revelle (2005). The FM model, however, uses the concept of a main meteoroid, from which big or small fragments are separated. The shape-density coefficient, K, of the main meteoroid is allowed to vary significantly during the flight. The duration of flares is taken from observations and can reach several seconds. The concept of eroding fragments offers a physical explanation of these long "flares", though other explanations (such as expanding debris clouds) are also possible. The FM model further assumes changes of luminous efficiency with height, normalised to fireball maximum. The velocity- and mass-dependencies of τ are almost identical in both models.

1.4.5 Hydrodynamical Modelling

Two- and three-dimensional models are able to describe directly the flow around the meteoroid. Great progress in the development of 2D-3D models occurred due to the impact of comet Shoemaker-Levy 9 (see review by Artemieva and Shuvalov, 2016). Published models can explain deformation, deceleration and light impulse from bolides (Ahrens et al., 1994; Boslough et al., 1994; Svetsov et al., 1995, and others).

The quasi-liquid model (QL) (the crushed body can be treated as a strengthless object, which behaves as a liquid) was fruitfully used in numerical hydrodynamic simulations of the impacts of asteroids into the Venusian atmosphere (Korycansky and Zahnle, 2003), the impacts of the fragments of the comet Shoemaker-Levy 9 into the Jovian atmosphere (Crawford et al., 1994; Mac Low and Zahnle, 1994; Shuvalov et al., 1997; Pond et al., 2012), the airburst of the Tunguska cosmic object (Shuvalov and Artemieva, 2002) and in simulations of the consequences of the entry of asteroids up to 300 m in diameter into the Earth's atmosphere (Shuvalov et al., 2013, 2016).

During its entry the meteoroid is deformed, flattened and fragmented into a nonuniform debris jet. The near-wake and the debris jet itself consist of highly mixed vapour and shock-compressed air. At the second stage, the elongated debris jet is decelerated and most of the energy is released. Hot air and vapour accelerate upward and create a ballistic plume. Although formally the QL model is valid for meteoroids larger than about 40 m (Svetsov et al., 1995) it was successfully used to the model of the Chelyabinsk meteoroid (Shuvalov et al., 2017).

Hydrodynamic codes based on the QL model can be used for simulations of all stages of the impact: the flight through the atmosphere, formation of a crater, and the interaction of crater ejecta with the atmosphere. The main restriction of the hydrodynamic approach to airbursts is the quasi-liquid assumption, which does not take into account strength and the formation of separated solid fragments, which can continue their flight independently after meteoroid disruption and reach the ground.

Robertson and Mathias (2017) applied different yield curves to model asteroid disruption. Strong asteroids break up and create peak energy deposition close to the altitude at which ram dynamic pressure equals the material strength. Weak asteroids break up at high altitude but travel as a unit until the pressure at lower altitude is high enough to disperse the rubble. So different yield assumptions change the breakup mechanics but have a limited effect on the burst altitude and energy deposition.

Hydrodynamical modelling also makes it possible to consider the consequences of large meteoroid impacts, including simulations of atmospheric shock waves carried out with energy deposition along the trajectory determined either from the light curve (Brown et al., 2013; Popova et al., 2013; Aftosmis et al., 2016; Nemec et al., 2017), from a simple entry model (Avramenko et al., 2014) or from the result of QL-model entry (Shuvalov et al., 2013, 2016, 2017). The radiation field on the Earth's surface was treated with hydrodynamical entry modelling (Shuvalov et al., 2017; Svetsov and Shuvalov, 2017).

In addition, luminous efficiencies and ablation coefficients obtained in hydrodynamical modelling are included in semi-analytical entry models.

1.5 Concluding Remarks and Future Issues

Due to the variety of physical conditions that occur during the entry of meteoroids into the atmosphere, modelling efforts require the use of a wide range of different models – from sputtering of the surface under direct impacts of particles up to high-fidelity gasdynamical modelling. When starting to analyse observational data or to describe any physical phenomenon connected with meteors and bolides, one needs to understand clearly in what conditions this phenomenon is occurring and what physical processes need to be taken into account. It should be also remembered that all models reviewed in this chapter have their own uncertainties and limitations.

The local flow regime around the falling body determines the heat transfer and mass loss processes. Ablation depends on the size/mass, velocity, altitude of flight and meteoroid properties. In the free molecule regime, the ablation rate can be computed relatively easily, but many details are still unclear, such as the determination of luminous/ionising efficiencies, the role of differential ablation, state of ablated material, and spectra formation. In addition, the processes of fragmentation undergone by the meteoroid, which is closely connected with the question of meteoroid structure, need to be clarified. Meteoroid modelling in the transition regime has made evident progress during recent years. DSMC and hybrid modelling are good tools here. However, only preliminary estimates of the parameters in the ablated material have been done to the present. Despite great progress during recent years, there is no totally self-consistent 2D-3D model, including radiation and ablation for the whole range of sizes/velocities/altitudes of flight and meteoroid material in the continuous regime. Detailed comparison of observational and model results has been done for a limited number of events, due both to incomplete observational data and modelling problems. Advances in computing power, coupled with high-resolution optical and radar observations of meteoroid ablation, should be able to address some of the outstanding issues in the near future.

Acknowledgments

JB was supported by grant no. 16-00761S from GA ČR. OP was supported by the project of the Russian Academy of Sciences no. 0146-2017-0003.

References

Abe, S., Yano, H., Ebizuka, N. and Watanabe, J.-I. 2000. First results of high-definition TV spectroscopic observations of the 1999 Leonid meteor shower. *Earth, Moon, and Planets*, **82**, 369–377.

Adolfsson, L. G., Gustafson, B. Å. S. and Murray, C. D. 1996. The Martian atmosphere as a meteoroid detector. *Icarus*, **119**, 144–152.

Aftosmis, M. J., Nemec, M., Mathias, D. L. and Berger, M. J. 2016. Numerical simulation of bolide entry with ground footprint prediction. In: *54th AIAA Aerospace Sciences Meeting, AIAA-0998*.

Ahrens, T. J., Takata, T., O'Keefe, J. D. and Orton, G. S. 1994. Impact of comet Shoemaker-Levy 9 on Jupiter. *Geophysical Research Letters*, **21**, 1087–1090.

Allen, H. J., Seiff, A. and Winovich, W. 1963. Aerodynamic heating of conical entry vehicles at speeds in excess of Earth parabolic speed. *NASA Technical Report NASA TR R-185*.

Artemieva, N. A., and Shuvalov, V. V. 1996. Interaction of shock waves during the passage of a disrupted meteoroid through the atmosphere. *Shock Waves*, **5**, 359–367.

Artemieva, N. A., and Shuvalov, V. V. 2016. From Tunguska to Chelyabinsk via Jupiter. *Annual Review of Earth and Planetary Sciences*, **44**, 37–56.

Avilova, I. V., Biberman, L. M., Vorob'ev, V. S. et al. 1970. *Optical Properties of Heated Air*. Moscow: Nauka. In Russian.

Avramenko, M. I., Glazyrin, I. V., Ionov, G. V. and Karpeev, A. V. 2014. Simulation of the airwave caused by the Chelyabinsk superbolide. *Journal of Geophysical Research: Atmospheres*, **119**, 7035–7050.

Ayers, W. G., McCrosky, R. E. and Shao, C. -Y. 1970. Photographic observations of 10 artificial meteors. *SAO Special Report*, **317**.

Azovskii, A. N., and Shuvalov, V. V. 2002. The influence of geometric shape on the motion of fragments of a dissociated meteoroid. *Solar System Research*, **36**, 334–340.

Babadzhanov, P. B., and Kramer, E. N. 1968. Some results of investigations of instantaneous meteor photographs. In: L. Kresak, and P. Millman (eds.), *Physics and Dynamics of Meteors*. Dordrecht: D. Reidel, pp. 128–142.

Babadzhanov, P. B. 2002. Fragmentation and densities of meteoroids. *Astronomy & Astrophysics*, **384**, 317–321.

Baldwin, B., and Allen, H. J. 1968. A method for computing luminous efficiencies from meteor data. *NASA Technical Report NASA-TN-D-4808*.

Baldwin, B., and Sheaffer, Y. 1971. Ablation and breakup of large meteoroids during atmospheric entry. *Journal of Geophysical Research*, **76**, 4653–4668.

Becker, D. G., and Friichtenicht, J. F. 1971. Measurement and interpretation of the luminous efficencies of iron and copper simulated micrometeors. *The Astrophysical Journal*, **166**, 699–716.

Becker, J. C., and Slattery, D. G. 1973. Luminous efficiency measurements for silicon and aluminum simulated micrometeors. *The Astrophysical Journal*, **186**, 1127–1140.

Behrisch, R., and Eckstein, W. 2007. Introduction and overview. In: R. Berisch, and W. Eckstein (eds.), *Sputtering by Particle Bombardment*. Berlin: Springer, pp. 1–20.

Bellot Rubio, L. R., Martnez González, M. J., Ruiz Herrera, L. et al. 2002. Modeling the photometric and dynamical behavior of Super-Schmidt meteors in the Earth's atmosphere. *Astronomy & Astrophysics*, **389**, 680–691.

Berezhnoy, A. A., and Borovička, J. 2010. Formation of molecules in bright meteors. *Icarus*, **210**, 150–157.

Biberman, L. M., Bronin, S. Ya., Brykin, M. V. and Mnatsakanian, A. Kh. 1978. Influence of gaseous heat-shield destruction products on the heat transfer in the neighborhood of the stagnation point on a blunt body. *Mekhanika Zhidkosti i Gaza*, **13**, 129–136.

Biberman, L. M., Bronin, S. Ya. and Brykin, M. V. 1979. Heat transfer in hypersonic flow in conditions of strong radiation-convective interaction. *Thermal Physics of High Temperatures*, **17**, 84–91.

Biberman, L. M., Bronin, S. Ya. and Brykin, M. V. 1980. Moving of a blunt body through the dense atmosphere under conditions of severe aerodynamic heating and ablation. *Acta Astronautica*, **7**, 53–65.

Bird, G. A. 1994. *Molecular Gas Dynamics and the Direct Simulation of Gas Flows*. Oxford: Clarendon Press.

Bland, Ph. A., and Artemieva, N. A. 2006. The rate of small impacts on Earth. *Meteoritics & Planetary Science*, **41**, 607–631.

Borovička, J. 1993. A fireball spectrum analysis. *Astronomy & Astrophysics*, **279**, 627–645.

Borovička, J. 1994. Two components in meteor spectra. *Planetary and Space Science*, **42**, 145–150.

Borovička, J. 2005a. Physical and chemical properties of meteoroids as deduced from observations. In: D. Lazzaro, S. Ferraz-Mello, and J. Fernandez, (eds.), *Proceedings of the 229th Symposium of the International Astronomical Union*. Cambridge: Cambridge University Press, pp. 249–271.

Borovička, J. 2005b. Spectral investigation of two asteroidal fireballs. *Earth, Moon, and Planets*, **95**, 279–293.

Borovička, J., and Berezhnoy, A. A. 2016. Radiation of molecules in Benešov bolide spectra. *Icarus*, **278**, 248–265.

Borovička, J., and Kalenda, P. 2003. The Morávka meteorite fall: 4. Meteoroid dynamics and fragmentation in the atmosphere. *Meteoritics & Planetary Science*, **38**, 1023–1043.

Borovička, J., and Spurný, P. 1996. Radiation study of two very bright terrestrial bolides and an application to the comet S-L 9 collision with Jupiter. *Icarus*, **121**, 484–510.

Borovička, J., Popova, O. P., Nemtchinov, I. V., Spurnỳ, P. and Ceplecha, Z. 1998a. Bolides produced by impacts of large meteoroids into the Earth's atmosphere: Comparison of theory with observations. I. Beneov bolide dynamics and fragmentation. *Astronomy & Astrophysics*, **334**, 713–728.

Borovička, J., Popova, O. P., Golub', A. P., Kosarev, I. B. and Nemtchinov, I. V. 1998b. Bolides produced by impacts of large meteoroids into the Earth's atmosphere: Comparison of theory with observations. II. Beneov bolide spectra. *Astronomy & Astrophysics*, **337**, 591–602.

Borovička, J., Koten, P., Spurný, P., Boček, J. and Štork, R. 2005. A survey of meteor spectra and orbits: Evidence for three populations of Na-free meteoroids. *Icarus*, **174**, 15–30.

Borovička, J., Spurnỳ, P. and Koten, P. 2007. Atmospheric deceleration and light curves of Draconid meteors and implications for the structure of cometary dust. *Astronomy & Astrophysics*, **473**, 661–672.

Borovička, J., Tóth, J., Igaz, A. et al. 2013a. The Košice meteorite fall: Atmospheric trajectory, fragmentation, and orbit. *Meteoritics & Planetary Science*, **48**, 1757–1779.

Borovička, J., Spurný, P., Brown, P. et al. 2013b. The trajectory, structure and origin of the Chelyabinsk asteroidal impactor. *Nature*, **503**, 235–237.

Borovička, J., Spurnỳ, P., Šegon, D. et al. 2015. The instrumentally recorded fall of the Križevci meteorite, Croatia, February 4, 2011. *Meteoritics & Planetary Science*, **50**, 1244–1259.

Borovička, J., Spurnỳ, P., Grigore, V. I. and Svoreň, J. 2017. The January 7, 2015, superbolide over Romania and structural diversity of meter-sized asteroids. *Planetary and Space Science*, **143**, 147–158.

Borovička, J., Macke, R. J., Campbell-Brown, M. D. et al. 2019. Physical and chemical properties of meteoroids. In: G. O. Ryabova, D. J. Asher, and M. D. Campbell-Brown (eds.), *Meteoroids: Sources of Meteors on Earth and Beyond*. Cambridge: Cambridge University Press, pp. 37–61.

Boslough, M. B., Crawford, D. A., Robinson, A. C. and Trucano, T. G. 1994. Mass and penetration depth of Shoemaker-Levy 9 fragments from time-resolved photometry. *Geophysical Research Letters*, **21**, 1555–1558.

Boyd, I.D. 2000. Computation of atmospheric entry flow about a Leonid Meteoroid. *Earth, Moon, and Planets*, **82**, 93–108.

Briani, G., Pace, E., Shore, S. N. et al. 2013. Simulations of micrometeoroid interactions with the Earth atmosphere. *Astronomy & Astrophysics*, **552**, A53.

Bronikowska, M., Artemieva, N. A. and Wünnemann, K. 2017. Reconstruction of the Morasko meteoroid impact – Insight from numerical modeling. *Meteoritics & Planetary Science*, **52**, 1704–1721.

Bronshten, V.A. 1983. *Physics of Meteoric Phenomena*. Dordrecht: D. Reidel Publishing Co.

Brosch, N., Schijvarg, L. S., Podolak, M., and Rosenkrantz, M. R. 2001. Meteor observations from Israel. In: B. Warmbein (ed.), *Proceedings of the Meteoroids 2001 Conference*. Noordwijk: ESA Publications Division, pp. 165-173.

Brown, P., Wiegert, P., Clark, D. and Tagliaferri, E. 2016. Orbital and physical characteristics of meter-scale impactors from airburst observations. *Icarus*, **266**, 96–111.

Brown, P. G., Hildebrand, A. R., Green, D. W. E. et al. 1996. The fall of the St-Robert meteorite. *Meteoritics & Planetary Science*, **31**, 502–517.

Brown, P. G., ReVelle, D. O., Tagliaferri, E. and Hildebrand, A. R. 2002a. An entry model for the Tagish Lake fireball using seismic, satellite and infrasound records. *Meteoritics & Planetary Science*, **37**, 661–675.

Brown, P. G., Spalding, R. E., ReVelle, D. O., Tagliaferri, E. and Worden, S. P. 2002b. The flux of small near-Earth objects colliding with the Earth. *Nature*, **420**, 294–296.

Brown, P. G., Assink, J. D., Astiz, L. et al. 2013. A 500-kiloton airburst over Chelyabinsk and an enhanced hazard from small impactors. *Nature*, **503**, 238–241.

Campbell, M. D., Hawkes, R. L. and Babcock, D. D. 1999. Light curves of shower meteors: Implications for physical structure. In: W. J. Baggaley and V. Porubcan, (eds.), *Proceedings of the Meteoroids 1998 Conference*. Bratislava: Astronomical Institute of Slovak Academy of Sciences, pp. 363–366.

Campbell-Brown, M. 2017. Modelling a short-wake meteor as a single or fragmenting body. *Planetary and Space Science*, **143**, 34–39.

Campbell-Brown, M. D., and Koschny, D. 2004. Model of the ablation of faint meteors. *Astronomy & Astrophysics*, **418**, 751–758.

Campbell-Brown, M. D., Borovička, J., Brown, P. G. and Stokan, E. 2013. High-resolution modelling of meteoroid ablation. *Astronom.Astrophys.*, **557**, A41.

Čapek, D., and Borovička, J. 2017. Ablation of small iron meteoroids– First results. *Planetary and Space Science*, **143**, 159–163.

Ceplecha, Z. 1964. Study of a bright meteor flare by means of emission curve of growth. *Bulletin of the Astronomical Institutes of Czechoslovakia*, **15**, 102–112.

Ceplecha, Z. 1996. Luminous efficiency based on photographic observations of the Lost City fireball and implications for the influx of interplanetary bodies onto Earth. *Astronomy & Astrophysics*, **311**, 329–332.

Ceplecha, Z., and McCrosky, R. E. 1976. Fireball end heights: A diagnostic for the structure of meteoric material. *Journal of Geophysical Research*, **81**, 6257–6275.

Ceplecha, Z., and ReVelle, D. O. 2005. Fragmentation model of meteoroid motion, mass loss, and radiation in the atmosphere. *Meteoritics & Planetary Science*, **40**, 35–54.

Ceplecha, Z., Spurný, P., Borovička, J. and Keclková, J. 1993. Atmospheric fragmentation of meteoroids. *Astronomy & Astrophysics*, **279**, 615–626.

Ceplecha, Z., Spalding, R. E., Jacobs, C. F. and Tagliaferri, E. 1996. Luminous efficiencies of bolides. In: *Proceedings SPIE 2813, Characteristics and Consequences of Orbital Debris and Natural Space Impactors*. Bellingham, WA: SPIE, pp. 46–56.

Ceplecha, Z., Borovička, J., Elford, W. G. et al. 1998. Meteor phenomena and bodies. *Space Science Reviews*, **84**, 327–471.

Chen, Y. 2016. Thermal ablation modeling for silicate materials. In: *AIAA Paper 2016-1514*.

Chernyi, G. G. 1959. *Techemya gaza s bolshoi sverklzzvukovoi skorost'yu (Gas Rows with a High Supersonic Speed)*. Moscow: Fizmatgiz, (in Russian).

Chyba, C. F. 1993. Explosions of small Spacewatch objects in the Earth's atmosphere. *Nature*, **363**, 701–703.

Chyba, C. F., Thomas, P. J. and Zahnle, K. J. 1993. The 1908 Tunguska explosion: Atmospheric disruption of a stony asteroid. *Nature*, **361**, 40–44.

Close, S., Oppenheim, M., Hunt, S. and Coster, A. 2004. A technique for calculating meteor plasma density and meteoroid mass from radar head echo scattering. *Icarus*, **168**, 43–52.

Close, S., Oppenheim, M., Durand, D., and Dyrud, L. 2005. A new method for determining meteoroid mass from head echo data. *Journal of Geophysical Research: Space Physics*, **110**, A09308.

Collins, G. S., Melosh, H. J. and Marcus, R. A. 2005. Earth impact effects program: A web-based computer program for calculating the regional environmental consequences of a meteoroid impact on Earth. *Meteoritics & Planetary Science*, **40**, 817–840.

Collins, G. S., Lynch, E., McAdam, R. and Davison, T. M. 2017. A numerical assessment of simple airblast models of impact airbursts. *Meteoritics & Planetary Science*, **52**, 1542–1560.

Cook, A. F., Hawkins, G. S. and Stienon, F. M. 1962. Meteor trail widths. *The Astronomical Journal*, **67**, 158–162.

Coulson, S. G. 2002. Resistance of motion to a small, hypervelocity sphere, sputtering through a gas. *Monthly Notices of the Royal Astronomical Society*, **332**, 741–744.

Crawford, D. A., Boslough, M. B., Trucano, T. G. and Robinson, A. C. 1994. The impact of comet Shoemaker-Levy 9 on Jupiter. *Shock Waves*, **4**(1), 47–50.

De Giacomo, A., Dell'Aglio, M., de Pascale, O., Longo, S. and Capitelli, M. 2007. Laser induced breakdown spectroscopy on meteorites. *Spectrochimica Acta Part B*, **62**, 1606–1611.

DeLuca, M., Munsat, T., Thomas, E. and Sternovsky, Z. 2018. The ionization efficiency of aluminum and iron at meteoric velocities. *Planetary and Space Science*, **156**, 111–116.

Dimant, Y. S., and Oppenheim, M. M. 2006a. Meteor trail diffusion and fields: 1. Simulations. *Journal of Geophysical Research: Space Physics*, **111**, A12312.

Dimant, Y. S., and Oppenheim, M. M. 2006b. Meteor trail diffusion and fields: 2. Analytical theory. *Journal of Geophysical Research: Space Physics*, **111**, A12313.

Dimant, Y. S., and Oppenheim, M. M. 2017a. Formation of plasma around a small meteoroid: 1. Kinetic theory. *Journal of Geophysical Research: Space Physics*, **122**, 4669–4696.

Dimant, Y. S., and Oppenheim, M. M. 2017b. Formation of plasma around a small meteoroid: 2. Implications for radar head echo. *Journal of Geophysical Research: Space Physics*, **122**, 4697–4711.

Dressler, R. A., and Murad, E. 2001. The gas-phase chemical dynamics associated with meteors. In: R. Dressler (ed), *Chemical Dynamics in Extreme Environments*. Singapore: World Scientific, pp. 268–348.

Duprat, J., Engrand, C., Maurette, M. et al. 2007. Micrometeorites from Central Antarctic snow: The CONCORDIA collection. *Advances in Space Research*, **39**, 605–611.

Dyrud, L. P., Oppenheim, M. M. and vom Endt, A. F. 2001. The anomalous diffusion of meteor trails. *Geophysical Research Letters*, **28**, 2775–2778.

Dyrud, L. P., Kudeki, E. and Oppenheim, M. 2007. Modeling long duration meteor trails. *Journal of Geophysical Research: Space Physics*, **112**, A12307.

Field, D., May, P. W., Pineau des Forets, G. and Flower, D. R. 1997. Sputtering of the refractory cores of interstellar grains. *Monthly Notices of the Royal Astronomical Society*, **285**, 839–846.

Fisher, A. A., Hawkes, R. L., Murray, I. S., Campbell, M. D. and LeBlanc, A. G. 2000. Are meteoroids really dustballs? *Planetary and Space Science*, **48**, 911–920.

Fleming, D. E. B., Hawkes, R. L. and Jones, J. 1993. Light curves of faint television meteors. In: J. Stohl and I. P. Williams (eds), *Meteoroids and their Parent Bodies*. Bratislava: Astronomical Institute SAS, pp. 261–264.

Friichtenicht, J. F., and Becker, D. G. 1973. Determination of meteor parameters using laboratory simulation techniques. *NASA Special Publication*, **319**, 53.

Friichtenicht, J. F., Slattery, J. C. and Tagliaferri, E. 1968. A laboratory measurement of meteor luminous efficiency. *The Astrophysical Journal*, **151**, 747–758.

Fujiwara, A., Cerroni, P., Davis, D. R. et al. 1989. Experiments and scaling laws for catastrophic collisions. In: R. Binzel, T. Gehrels, and M. Matthews (eds), *Asteroids II*. Tucson: University of Arizona Press, pp. 240–265.

Gao, B., and Mathews, J. D. 2015. High-altitude meteors and meteoroid fragmentation observed at the Jicamarca Radio Observatory. *Monthly Notices of the Royal Astronomical Society*, **446**, 3404–3415.

Genge, M. J. 2017. The entry heating and abundances of basaltic micrometeorites. *Meteoritics & Planetary Science*, **52**, 1000–1013.

Golub', A. P., Kosarev, I. B., Nemchinov, I. V. and Shuvalov, V. V. 1996. Emission and ablation of a large meteoroid in the course of Its motion through the Earth's atmosphere. *Solar System Research*, **30**, 183–197.

Golub', A. P., Kosarev, I. B., Nemtchinov, I. V. and Popova, O. P. 1997. Emission Spectra of Bright Bolides. *Solar System Research*, **31**, 85–98.

Grigorian, S. S. 1979. Motion and destruction of meteors in planetary atmospheres. *Cosmic Research*, **17**, 724–740.

Halliday, I., Griffin, A. A. and Blackwell, A. T. 1981. The Innisfree meteorite fall: - A photographic analysis of fragmentation, dynamics and luminosity. *Meteoritics*, **16**, 153–170.

Hartmann, W. K. 1969. Terrestrial, lunar, and interplanetary rock fragmentation. *Icarus*, **10**, 201–213.

Hawkes, R. L., and Jones, J. 1975. A quantitative model for the ablation of dustball meteors. *Monthly Notices of the Royal Astronomical Society*, **173**, 339–356.

Hayes, W. D., and Probstein, R. F. 1959. *Hypersonic Flow Theory*. New York and London: Academic Press.

Hildebrand, A. R., McCausland, P. J. A., Brown, P. G. et al. 2006. The fall and recovery of the Tagish Lake meteorite. *Meteoritics & Planetary Science*, **41**, 407–431.

Hill, K. A., Rogers, L. A., and Hawkes, R. L. 2005. High geocentric velocity meteor ablation. *Astronomy & Astrophysics*, **444**, 615–624.

Hills, J. G., and Goda, M. P. 1993. The fragmentation of small asteroids in the atmosphere. *The Astronomical Journal*, **105**, 1114–1144.

Jacchia, L. G. 1955. The physical theory of meteors. VIII. Fragmentation as cause of the faintmeteor anomaly. *The Astrophysical Journal*, **121**, 521–527.

Jenniskens, P., Betlem, H., Betlem, J. et al. 1994. The Mbale meteorite shower. *Meteoritics*, **29**, 246–254.

Jenniskens, P., Rietmeijer, F., Brosch, N., and Fonda, M. (eds). 2000a. *Leonid Storm Research*. Dordrecht: Springer Netherlands.

Jenniskens, P., Wilson, M. A., Packan, D. et al. 2000b. Meteors: A delivery mechanism of organic matter to the early Earth. *Earth Moon and Planets*, **82**, 57–70.

Johnston, C. O., Stern, E. C., and Wheeler, L. F. 2018. Radiative heating of large meteoroids during atmospheric entry. *Icarus*, **309**, 25–44.

Jones, J., and Hawkes, R. L. 1975. Television observations of faint meteors. II - Light curves. *Monthly Notices of the Royal Astronomical Society*, **171**, 159–169.

Jones, J., and Kaiser, T. R. 1966. The effects of thermal radiation, conduction and metoriod heat capacity on meteoric ablation. *Monthly Notices of the Royal Astronomical Society*, **133**, 411–420.

Jones, W. 1995. Theory of the initial radius of meteor trains. *Monthly Notices of the Royal Astronomical Society*, **275**, 812–818.

Jones, W. 1997. Theoretical and observational determinations of the ionization coefficient of meteors. *Monthly Notices of the Royal Astronomical Society*, **288**, 995–1003.

Jones, W., and Halliday, I. 2001. Effects of excitation and ionization in meteor trains. *Monthly Notices of the Royal Astronomical Society*, **320**, 417–423.

Kero, J. 2008. High resolution meteor exploration with tristatic radar methods. Ph.D. thesis.

Kero, J., Campbell-Brown, M., Stober, G. et al. 2019. Radar Observations of Meteors. In: G. O. Ryabova, D. J. Asher, and M. D. Campbell-Brown (eds), *Meteoroids: Sources of Meteors on Earth and Beyond*. Cambridge, UK: Cambridge University Press, pp. 65–89.

Klekociuk, A. R., Brown, P. G., Pack, D. W. et al. 2005. Meteoritic dust from the atmospheric disintegration of a large meteoroid. *Nature*, **436**(7054), 1132–1135.

Korycansky, D. G., and Zahnle, K. J. 2003. High-resolution simulations of the impacts of asteroids into the venusian atmosphere III: Further 3D models. *Icarus*, **161**, 244–261.

Kosarev, I. B. 1999. Calculation of thermodynamic and optical properties of the vapors of cosmic bodies entering the earths atmosphere. *Journal of Engineering Physics and Thermophysics*, **72**, 1030–1038.

Kosarev, I. B., Loseva, T. V., and Nemchinov, I. V. 1996. Vapor optical properties and ablation of large chondrite and ice bodies in the Earth's atmosphere. *Solar System Research*, **30**, 265–278.

Koten, P., Borovička, J., Spurný, P., Betlem, H. and Evans, S. 2004. Atmospheric trajectories and light curves of shower meteors. *Astronomy & Astrophysics*, **428**, 683–690.

Koten, P., Spurný, P., Borovička, J. et al. 2006. The beginning heights and light curves of high-altitude meteors. *Meteoritics & Planetary Science*, **41**, 1305–1320.

Kuznetsov, N. M. 1965. *Thermodynamic Functions and Impact Adiabates for Air under High Temperatures*. Moscow: Mashinostroenie. in Russian.

Landau, L. D., and Lifshitz, E. M. 1987. *Fluid Mechanics*. Oxford: Pergamon.

Lebedinec, V. N., and Šuškova, V. B. 1968. Evaporation and deceleration of small meteoroids. In: L. Kresak, and P. Millman (eds), *Physics and Dynamics of Meteors*. Dordrecht: D. Reidel, pp. 193–204.

Lebedinets, V. N. 1980. *Dust in the Upper Atmosphere and in Space Meteors*. Leningrad: Gidrometeoizdat, 248p. In Russian.

Lebedinets, V. N., Manochina, A. V. and Shushkova, V. B. 1973. Interaction of the lower thermosphere with the solid component of the interplanetary medium. *Planetary and Space Science*, **21**, 1317–1332.

Levin, B. I. 1956. *Fizicheskaia teoriia meteorov i meteornoye veschestvo v Solnechnoi sisteme*. Moscow: Izdatel'stvo Akademii Nauk SSSR. In Russian. Exists also in German translation: *Physikalische Theorie der Meteore und die meteoritische Substanz im Sonnensystem*. Berlin: Akademie-Verlag (1961).

Love, S. G., and Brownlee, D. E. 1991. Heating and thermal transformation of micrometeroids entering the Earth's Atmosphere. *Icarus*, **89**, 26–43.

Mac Low, M., and Zahnle, K. 1994. Explosion of comet Shoemaker-Levy 9 on entry into the Jovian atmosphere. *The Astrophysical Journal*, **434**, L33–L36.

Maljavina, T. B., and Nemtchinov, I. V. 1972. Parameters of stationar vapor jet with radial symmetry, heated by laser radiation. *Prikladnaya mekhanika i technicheskaya fizika* (Applied mechanics and technical physics). (In Russian), **5**, 59–75.

Mann, I., Pellinen-Wannberg, A., Murad, E. et al. 2011. Dusty plasma effects in near earth space and interplanetary medium. *Space Science Reviews*, **161**, 1–47.

Manning, L. A. 1958. The initial radius of meteoric ionization trails. *Journal of Geophysical Research*, **63**, 181–196.

Marshall, R. A., and Close, S. 2015. An FDTD model of scattering from meteor head plasma. *Journal of Geophysical Research: Space Physics*, **120**, 5931–5942.

Marshall, R. A., Brown, P. and Close, S. 2017. Plasma distributions in meteor head echoes and implications for radar cross section interpretation. *Planetary and Space Science*, **143**, 203–208.

Massey, H. S. W., and Sida, D. W. 1955. XXIII. Collision processes in meteor trails. *The London, Edinburgh, and Dublin Philosophical Magazine and Journal of Science*, **46**, 190–198.

May, P. W., Pineau des Forts, G., Flower, D. R. et al. 2000. Sputtering of grains in C-type shocks. *Monthly Notices of the Royal Astronomical Society*, **318**, 809–816.

McAuliffe, J. P., and Christou, A. A. 2006. Modelling meteor ablation in the venusian atmosphere. *Icarus*, **180**, 8–22.

McCord, T. B., Morris, J., Persing, D. et al. 1995. Detection of a meteoroid entry into the Earth's atmosphere on February 1, 1994. *Journal of Geophysical Research*, **100**, 3245–3249.

McNeil, W. J., Lai, Sh. T. and Murad, E. 1998. Differential ablation of cosmic dust and implications for the relative abundances of atmospheric metals. *Journal of Geophysical Research*, **103**, 10899–10912.

Moses, J. I. 1992. Meteoroid ablation in Neptune's atmosphere. *Icarus*, **99**, 368–383.

Murray, I. S., Hawkes, R. L., and Jenniskens, P. 1999. Airborne intensified charge-coupled device observations of the 1998 Leonid shower. *Meteoritics & Planetary Science*, **34**, 949–958.

Nelson, D. A., Baker, R. L., and Yee, P. P. 2005. Heat and mass transfer in hypervelocity rarefied flow. Technical report ATR2006(9368)-1. The Aerospace Corporation.

Nemec, M., Aftosmis, M. J., and Brown, P. G. 2017. Numerical prediction of meteoric infrasound signatures. *Planetary and Space Science*, **140**, 11–20.

Nemtchinov, I. V. 1967. Stationary regime of motion of substance vapor heated by radiation under the conditions of lateral spreading. *Prikladnaya Matematika i Mekhanika (Applied Mathematics and Mechanics)*, **31**, 300–319, (in Russian).

Nemtchinov, I. V., and Popova, O. P. 1997. An analysis of the 1947 Sikhote-Alin Event and a comparison with the Phenomenon of February 1, 1994. *Solar System Research*, **31**, 408–420.

Nemtchinov, I. V., Svetsov, V. V., Kosarev, I. B. et al. 1997. Assessment of kinetic energy of meteoroids detected by satellite-based light sensors. *Icarus*, **130**, 259–274.

Novikov, G. G., Pecina, P. and Blokhin, A. V. 1996. Fragmentation of meteoroids and the structure of meteor coma. *Astronomy & Astrophysics*, **312**, 1012–1016.

Opik, E. J. 1955. Meteors and the upper atmosphere. *Irish Astronomical Journal*, **3**, 165–181.

Opik, E. J. 1958. *Physics of Meteor Flight in the Atmosphere*. New York: Interscience Publishers, 1958.

Oppenheim, M. M., and Dimant, Y. S. 2015. First 3-D simulations of meteor plasma dynamics and turbulence. *Geophysical Research Letters*, **42**, 681–687.

Park, C. 2013. Rosseland mean opacities of air and H-chondrite vapor in meteor entry problems. *Journal of Quantitative Spectroscopy and Radiative Transfer*, **127**, 158–164.

Park, C. 2014. Stagnation region radiative ablation of asteroidal meteoroids in the Rosseland Approximation. *Journal of Thermophysics and Heat Transfer*, **28**, 598–607.

Park, C. 2016. Inviscid-flow approximation of radiative ablation of asteroidal meteoroids by line-by-line method. In: *AIAA paper 2016-0506*.

Park, C. 2018. Radiation phenomenon for large meteoroids. *Astronomy & Astrophysics*, accepted.

Passey, Q. R., and Melosh, H. J. 1980. Effects of atmospheric breakup on crater field formation. *Icarus*, **42**, 211–233.

Pecina, P., and Ceplecha, Z. 1983. New aspects in single-body meteor physics. *Bulletin of the Astronomical Institutes of Czechoslovakia*, **34**, 102–121.

Pecina, P., and Ceplecha, Z. 1984. Importance of atmospheric models for interpretation of photographic fireball data. *Bulletin of the Astronomical Institutes of Czechoslovakia*, **35**, 120–123.

Plane, J. M. C., Flynn, G. J., Määttänen, A. et al. 2018. Impacts of cosmic dust on planetary atmospheres and surfaces. *Space Science Reviews*, **214**, #23.

Pond, J. W. T., Palotai, C., Gabriel, T.s et al. 2012. Numerical modeling of the 2009 Impact Event on Jupiter. *The Astrophysical Journal*, **745**, 113–121.

Popova, O. 2004. Meteoroid ablation models. *Earth, Moon, and Planets*, **95**, 303–319.

Popova, O. 2011. Passage of bolides through the atmosphere. In: *Meteoroids: The Smallest Solar System Bodies*. NASA/CP-2011-216469, pages 232–242.

Popova, O., Nemtchinov, I. and Hartmann, W. K. 2003. Bolides in the present and past Martian atmosphere and effects on cratering processes. *Meteoritics & Planetary Science*, **38**(6), 905–925.

Popova, O., Strelkov, A. and Sidneva, S. 2007. Sputtering of fast meteoroids surface. *Advances in Space Research*, **39**, 567–573.

Popova, O., Borovička, J., Hartmann, W. K. et al. 2011. Very low strengths of interplanetary meteoroids and small asteroids. *Meteoritics & Planetary Science*, **46**, 1525–1550.

Popova, O. P., Glazachev, D. O., Podobnaya, E. D., Svetsov, V. V. and Shuvalov, V. V. 2017. Radiation and ablation of large meteoroids decelerated in the Earth's atmosphere. European Planetary Science Congress, **11**, EPSC2017–812.

Popova, O. P., and Nemchinov, I. V. 2008. Bolides in the Earth atmosphere. In: V. V. Adushkin, and I.V. Nemchinov (eds), *Catastrophic Events Caused by Cosmic Objects*. Berlin: Springer, pp. 131–162.

Popova, O. P., Sidneva, S. N., Shuvalov, V. V. and Strelkov, A. S. 2000. Screening of meteoroids by ablation vapor in high-velocity meteors. *Earth, Moon, and Planets*, **82-83**, 109–128.

Popova, O. P., Sidneva, S. N., Strelkov, A. S. and Shuvalov, V. V. 2001. Formation of disturbed area around fast meteor body. In: B. Warmbein (ed), *Proceedings of the Meteoroids 2001 Conference*. Noordwijk: pp. 237–245.

Popova, O. P., Jenniskens, P., Emel'yanenko, V. et al. 2013. Chelyabinsk Airburst, damage assessment, meteorite recovery, and characterization. *Science*, **342**, 1069–1073.

Register, P. J., Mathias, D. L. and Wheeler, L. F. 2017. Asteroid fragmentation approaches for modeling atmospheric energy deposition. *Icarus*, **284**, 157–166.

ReVelle, D. O., and Ceplecha, Z. 2001. Bolide physical theory with application to PN and EN fireballs. In: B. Warmbein (ed), *Proceedings of the Meteoroids 2001 Conference*. Noordwijk: ESA Publications Division, pp. 507–512.

ReVelle, D. O., and Rajan, R. S. 1979. On the luminous efficiency of meteoritic fireballs. *Journal of Geophysical Research*, **84**, 6255–6262.

ReVelle, D. O. 1979. A quasi-simple ablation model for large meteorite entry - Theory vs observations. *Journal of Atmospheric and Solar-Terrestrial Physics*, **41**, 453–473.

ReVelle, D. O. 1980. A predictive macroscopic integral radiation efficiency model. *Journal of Geophysical Research*, **85**, 1803–1808.

Rietmeijer, F. J. M. 2002. Collected extraterrestrial materials: Interplanetary dust particles micrometeorites, meteorites, and meteoric dust. In: E. Murad and I. P. Williams (eds), *Meteors in the Earth's Atmosphere: Meteoroids and Cosmic Dust and their Interactions with the Earth's Upper Atmosphere*. Cambridge, UK: Cambridge University Press, pp. 215–245.

Rietmeijer, F. J. M. 2000. Interrelationships among meteoric metals, meteors, interplanetary dust, micrometeorites, and meteorites. *Meteoritics & Planetary Science*, **35**, 1025–1041.

Robertson, D. K., and Mathias, D. L. 2017. Effect of yield curves and porous crush on hydrocode simulations of asteroid airburst. *Journal of Geophysical Research*, **122**, 599–613.

Rogers, L. A., Hill, K. A. and Hawkes, R. L. 2005. Mass loss due to sputtering and thermal processes in meteoroid ablation. *Planetary and Space Science*, **53**, 1341–1354.

Rosenberg, M. 2008. On the possibility of a lower-hybrid instability driven by fast ions sputtered from a meteoroid. *Planetary and Space Science*, **56**, 1190–1193.

Saidov, K. H., and Simek, M. 1989. Luminous efficiency coefficient from simultaneous meteor observations. *Bulletin of the Astronomical Institutes of Czechoslovakia*, **40**, 330–332.

Schaefer, L. and Fegley, B. 2004. Application of an equilibrium vaporization model to the ablation of chondritic and achondritic meteoroids. *Earth Moon and Planets*, **95**, 413–423.

Shuvalov, V., Svetsov, V., Popova, O. and Glazachev, D. 2017. Numerical model of the Chelyabinsk meteoroid as a strengthless object. *Planetary and Space Science*, **147**, 38–47.

Shuvalov, V. V., Artem'eva, N. A., Kosarev, I. B., Nemtchinov, I. V. and Trubetskaya, I. A. 1997. Numerical simulation of the bolide phase of the impact of Comet Shoemaker-Levy 9 fragments on Jupiter. *Solar System Research*, **31**, 393–400.

Shuvalov, V. V. 1999. Multi-dimensional hydrodynamic code SOVA for interfacial flows: Application to the thermal layer effect. *Shock Waves*, **9**, 381–390.

Shuvalov, V. V., and Artemieva, N. A. 2002. Numerical modeling of Tunguska-like impacts. *Planetary and Space Science*, **50**, 181–192.

Shuvalov, V. V., Svettsov, V. V. and Trubetskaya, I. A. 2013. An estimate for the size of the area of damage on the Earths surface after impacts of 10–300-m asteroids. *Solar System Research*, **47**, 260–267.

Shuvalov, V. V., Popova, O. P., Svettsov, V. V., Trubetskaya, I. A. and Glazachev, D. O. 2016. Determination of the height of the meteoric explosion. *Solar System Research*, **50**, 1–12.

Slattery, J. C., and Friichtenicht, J. F. 1967. Ionization probability of iron particles at meteoric velocities. *The Astrophysical Journal*, **147**, 235–244.

Spurný, P., and Ceplecha, Z. 2004. Fragmentation model analysis of EN270200 fireball. *Earth, Moon, and Planets*, **95**, 477–487.

Spurný, P., and Ceplecha, Z. 2008. Is electric charge separation the main process for kinetic energy transformation into the meteor phenomenon? *Astronomy & Astrophysics*, **489**, 449–454.

Spurný, P., Betlem, H., van't Leven, J. and Jenniskens, P. 2000a. Atmospheric behavior and extreme beginning heights of the 13 brightest photographic Leonids from the ground-based expedition to China. *Meteoritics & Planetary Science*, **35**, 243–249.

Spurný, P., Betlem, H., Jobse, K., Koten, P., and van't Leven, J. 2000b. New type of radiation of bright Leonid meteors above 130 km. *Meteoritics & Planetary Science*, **35**, 1109–1115.

Stenbaek-Nielsen, H. C., and Jenniskens, P. 2004. A "shocking" Leonid meteor at 1000 fps. *Advances in Space Research*, **33**, 1459–1465.

Stokan, E., and Campbell-Brown, M. D. 2014. Transverse motion of fragmenting faint meteors observed with the Canadian Automated Meteor Observatory. *Icarus*, **232**, 1–12.

Stokan, E., and Campbell-Brown, M. D. 2015. A particle-based model for ablation and wake formation in faint meteors. *Monthly Notices of the Royal Astronomical Society*, **447**, 1580–1597.

Stokan, E., Campbell-Brown, M. D., Brown, P. G. et al. 2013. Optical trail widths of faint meteors observed with the Canadian Automated Meteor Observatory. *Monthly Notices of the Royal Astronomical Society*, **433**, 962–975.

Subasinghe, D., and Campbell-Brown, M. 2018. Luminous efficiency estimates of meteors-II. Application to Canadian Automated Meteor Observatory meteor events. *The Astronomical Journal*, **155**, 88.

Subasinghe, D., Campbell-Brown, M. D. and Stokan, E. 2016. Physical characteristics of faint meteors by light curve and high-resolution observations, and the implications for parent bodies. *Monthly Notices of the Royal Astronomical Society*, **457**, 1289–1298.

Svetsov, V., and Shuvalov, V. 2017. Thermal radiation from large bolides and impact plumes. *European Planetary Science Congress*, **11**, EPSC2017–26.

Svetsov, V., Shuvalov, V., Collins, G. and Popova, O. 2019. Impact hazard of large meteoroids and small asteroids. In: G. O. Ryabova, D. J. Asher, and M. D. Campbell-Brown (eds), *Meteoroids: Sources of Meteors on Earth and Beyond*. Cambridge, UK: Cambridge University Press, pp. 275–298.

Svetsov, V. V., Nemtchinov, I. V. and Teterev, A. V. 1995. Disintegration of large meteoroids in Earth's atmosphere: Theoretical models. *Icarus*, **116**, 131–153.

Tagliaferri, E., Spalding, R., Jacobs, C., Worden, S. P. and Erlich, A. 1994. *Detection of Meteoroid Impacts by Optical Sensors in Earth Orbit*. Tucson: University of Arizona Press, pp. 199–220.

Thomas, E., Horányi, M., Janches, D. et al. 2016. Measurements of the ionization coefficient of simulated iron micrometeorites. *Geophysical Research Letters*, **43**, 3645–3652.

Tielens, A. G. G. M., McKee, C. F., Seab, C. G. and Hollenbach, D. J. 1994. The physics of grain-grain collisions and gas-grain sputtering in interstellar shocks. *The Astrophysical Journal*, **431**, 321–340.

Tsvetkov, V. I., and Skripnik, A. Y. 1991. Atmospheric Fragmentation of meteorites according to the Strength Theory. *Solar System Research*, **25**, 273–280.

Verniani, F. 1965. On the luminous efficiency of meteors. *Smithsonian Contributions to Astrophysics*, **8**, 141–171.

Vinkovic, D. 2007. Thermalization of sputtered particles as the source of diffuse radiation from high altitude meteors. *Advances in Space Research*, **39**, 574–582.

Vondrak, T., Plane, J. M. C., Broadley, S., and Janches, D. 2008. A chemical model of meteoric ablation. *Atmospheric Chemistry & Physics Discussions*, **8**, 14557–14606.

Weibull, W. 1951. A statistical distribution function of wide applicability. *Journal of Applied Mechanics*, **10**, 140–147.

Weryk, R. J., and Brown, P. G. 2013. Simultaneous radar and video meteors II: Photometry and ionisation. *Planetary and Space Science*, **81**, 32–47.

Wheeler, L. F., Register, P. J. and Mathias, D. L. 2017. A fragment-cloud model for asteroid breakup and atmospheric energy deposition. *Icarus*, **295**, 149–169.

2

Physical and Chemical Properties of Meteoroids

Jiří Borovička, Robert J. Macke, Margaret D. Campbell-Brown, Anny-Chantal Levasseur-Regourd,
Frans J. M. Rietmeijer, and Tomáš Kohout

2.1 Introduction

Almost all meteoroids are debris from asteroids and comets. Determining the physical properties of meteoroids from the interpretation of meteor observations can therefore provide some clues to the properties of asteroidal and cometary materials on size scales from tenths of millimeters to tens of meters and thus on their formation and evolution in the solar system. Since meteor observations provide pre-entry orbits, studied meteoroids can be linked to their source regions in the solar system or, in cases of meteor showers, directly to their parent bodies. The physical properties and internal structure of asteroids and comets are of great interest for various reasons, both scientific and practical, but are poorly known.

One quantity that gives insight into the internal structure of large bodies is bulk density. For asteroids, and even more for cometary nuclei, bulk density is difficult to measure because one needs to derive both the volume and the mass of the body, neither of which is straightforward. The most reliable values are known for bodies visited by spacecraft, such as (1) Ceres and (4) Vesta from the Dawn mission, (21) Lutetia and comet 67P/Churyumov-Gerasimenko from the Rosetta mission, (253) Mathilde and (433) Eros from the NEAR Shoemaker mission, or (25143) Itokawa from the Hayabusa mission.

Asteroid densities can be compared with densities of meteorites. With the exception of lunar and Martian meteorites, most meteorites are believed to come from asteroids, though some researchers argue that some carbonaceous chondrites may originate from comets (e.g. Gounelle et al., 2006). Carbonaceous chondrites can have densities as low as 1,600 kg m^{-3}; other stony meteorites fall in the range 2,800–3,800 kg m^{-3}; and metallic (iron) meteorites up to 7,900 kg m^{-3} (for more details see Section 2.2.1 in this chapter).

Asteroids are assigned to their meteorite analogs using reflectance spectroscopy. In general, S-type asteroids have the same compositions as ordinary chondrites. This has been confirmed by samples returned from Itokawa by the Hayabusa mission (Nakamura et al., 2011). C-type asteroids are linked to carbonaceous chondrites, and M-type asteroids are possibly related to iron meteorites. Note that the whole classification is more complex (see e.g. DeMeo et al., 2015).

The bulk densities of large differentiated asteroids generally correspond to their types: 2,162 kg m^{-3} for C-type Ceres (Park et al., 2016), and 3,456 kg m^{-3} for stony (V-type) Vesta (Russell et al., 2012). Smaller asteroids mostly have lower densities: Eros (S-type, size 34 km) has a density of 2,670 kg m^{-3} (Veverka et al., 2000); Itokawa (S-type, 0.5 km) 1,950 kg m^{-3} (Abe et al., 2006); Mathilde (C-type, 66 km) 1,300 kg m^{-3} (Veverka et al., 1997). The relatively large (121 km) stony asteroid Lutetia has a high density of 3400 kg m^{-3} (Pätzold et al., 2011).

The fact that some asteroids have lower densities than meteorites can have two explanations: either some asteroids are fully or partly composed of low-density materials, which are too weak to survive passage through Earth's atmosphere and are therefore not represented in meteorite collections, or asteroids contain voids, i.e. empty space, within their volume. The voids may range in size from microscopic (referred to as microporosity) to macroscopic (macroporosity), with void sizes ranging from less than a micrometer to tens of meters. It is thought that many medium-sized asteroids are so-called rubble piles, i.e. aggregates of smaller blocks held together only by their mutual gravity (Pravec and Harris, 2000). Rubble piles can be produced by re-accumulation of debris after catastrophic asteroid collisions. Large-scale voids then arise quite naturally.

Porosity, in percent, is defined as $p = 100(1 - \rho_b/\rho_m)$, where ρ_b is the bulk density and ρ_m is material density. An extensive discussion of the densities and porosities of asteroids is given in Carry (2012).

Comet nuclei contain substantial amounts of ices and their densities could therefore be expected to be lower than those of asteroids. Preliminary estimations from non-gravitational forces together with local estimations for flown-by nuclei, have suggested that their densities are always below 1,000 kg m^{-3} (e.g. Levasseur-Regourd et al., 2009; Thomas et al., 2013), meaning that the porosities of the refractory components and of the ices within cometary nuclei must be large. The most precise value is known for 67P/Churyumov-Gerasimenko: about 530 kg m^{-3} (Jorda et al., 2016; Pätzold et al., 2016). Comets are diverse objects with variable dust-to-ice ratios in their released material and various properties of the released dust. Comet nuclei visited by spacecraft have different shapes, surface morphologies, and activity. Dust samples returned from the coma of comet 81P/Wild 2 by the Stardust mission also contained high-temperature minerals (Zolensky et al., 2006). The comet which has been studied in the most detail is 67P/Churyumov-Gerasimenko, thanks to twenty-six months of data from the Rosetta rendezvous mission (A'Hearn, 2017; Barucci and Fulchignoni, 2017). Dust particles in its coma were shown to present a large amount of high-molecular-weight organics (e.g. Bardyn et al., 2017), and to consist of porous aggregates with a hierarchical structure (e.g. Mannel et al., 2016). Other samples of dust particles of presumably mainly cometary origin are found in some interplanetary dust particles (IDPs) collected in Earth's atmosphere, and some micrometeorites.

Meteor observations sample all types of solar system materials, which can evolve into Earth-intersecting orbits. They can in principle answer many questions: Which asteroids are the parent bodies of individual meteorite types? Are meteorites representative samples of meteorite-producing meteoroids? Are some meter-sized bodies rubble piles? Are some meteoroids and small asteroids mixed from different types of materials? Are there asteroidal materials that are too weak to produce meteorites? What fraction of bodies on asteroidal orbits are in fact remnants of comets? Are those asteroids that have formed meteoroid streams (such as the Geminids and Quadrantids) normal asteroids, dormant comets, or another type of transitional object? What is the structure of cometary material (dust) on different scales? What are the sources of IDPs and micrometeorites? Are there differences in the composition and structure of dust between Jupiter-family and Halley-type comets? Can comets deliver meteorites? Were chondrules incorporated into cometary material? How is asteroidal and cometary material affected by space weathering and by close approaches to the Sun?

In this chapter, we will first review the known properties of solar system materials studied in laboratories (meteorites, IDPs, micrometeorites, and cometary and asteroidal samples returned by space missions) or in situ by spacecraft. After that, the methods of inferring meteoroid properties from meteor observations will be discussed. Meteoroid properties will be compared with those of the laboratory materials. Small (smaller than about a centimeter) and big meteoroids will be discussed separately, in part because of different observational techniques used, but mainly because of differences in interaction with the atmosphere between millimeter-sized and decimeter-sized (or larger) bodies. The current state of knowledge in light of the above questions will be presented. Finally, the remaining open questions and potential ways to solve them will be discussed.

2.2 Materials Studied in the Laboratory

2.2.1 Meteorites

The term "meteorite" refers to any non-terrestrial rock that has fallen to the surface of the Earth. Meteorites fall everywhere on the planet, with roughly homogeneous distribution, though some places on Earth are far better for preservation and recovery of meteorites. Dry deserts with few terrestrial rocks are ideal recovery locations. Examples of places where disproportionate numbers of meteorites have been found include Antarctica, Saharan Africa, and the arid parts of the central and southwestern USA.

Most meteorites are asteroidal in origin, though some come from the Moon, Mars, and possibly comets. Meteorites can be collected and studied in the laboratory, and are the best material available for understanding the composition, structure, and physical properties of meteoroids and asteroid parent bodies. Nevertheless, it is important to acknowledge that meteorites represent a biased population of asteroidal material, limited to those objects dynamically favorable for reaching Earth and surviving entry into the atmosphere and reaching the ground more or less intact. Thus, near-Earth asteroids that encounter

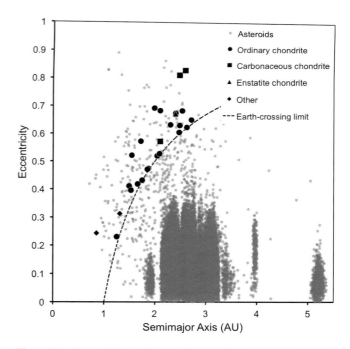

Figure 2.1. Eccentricity and semimajor axes of the pre-atmospheric orbits of meteorite-producing fireballs. Gray dots represent asteroids with known orbits. Fireball data from Meier (2017) and sources therein.

Earth near their perihelion or aphelion would be more heavily sampled than those from further out in the asteroid belt. Comet nuclei and those few Trans-Neptunian Objects that get tossed into the inner solar system would tend to reach the Earth with a high relative velocity. This reduces the likelihood of intact remnants surviving to be collected as meteorites. There are only a handful of meteorites for which we know the pre-entry orbit, though due to the growth of fireball observation networks and security camera systems, the population of fireballs with known orbits has been increasing in recent years. Only twenty-six such events with recovered meteorites have been published by mid 2017 (see summaries in Borovička et al., 2015a; Meier, 2017; plus Bland et al., 2016). All of these have semi-major axes of about 2.7 AU or less (Figure 2.1), and have relative velocities at atmospheric entry of less than 29 km s^{-1}, with an average of about 18 km s^{-1}. While orbits of asteroids can evolve by various dynamical mechanisms from the outer main belt to Earth-crossing orbits (as may be the case with the Tagish Lake meteorite), this is still a very small percentage of asteroids and it is possible that some populations of asteroids are not represented at all among known meteorites.

Strength and other physical properties are also biasing factors; our collections may be biased toward stronger specimens with lower microporosity than perhaps are representative of asteroids of the same type. This will be discussed in greater detail shortly.

2.2.1.1 Meteorite Classification

Due to the diversity of sources, there are many classes and types of meteorites. Most basically, all meteorites can be divided into stony, stony-iron, and iron meteorites. The iron meteorites originated from the cores of differentiated parent bodies that had subsequently shattered during the early history of the solar system. As the name suggests, they are predominantly composed of iron in the form of metallic Fe-Ni alloys, with nickel content typically between 5.7 and 16 wt% (Hutchison, 2004). They may contain sulfide inclusions, sometimes several centimeters across. They are further subdivided into four types (I to IV) based on Ga and Ge concentrations (highest in I, lowest in IV; cf. Wasson and Kimberlin, 1967). They have been spectroscopically linked to M-type asteroids (Chapman and Salisbury, 1973), though likely mixed with silicates at least near the surface (e.g. Hardersen et al., 2011).

Stony-irons contain roughly even amounts of silicate and metal, with two primary groups being mesosiderites (a roughly even mix of rock and metal) and pallasites, which are composed of large olivine crystals embedded in a Fe-Ni metallic matrix. Pallasites are generally thought to have formed near the core-mantle boundary of differentiated parent bodies, though there are competing formation theories based on impact mixing (e.g. Yang et al., 2010) or partial melting of a chondritic parent body (e.g. Boesenberg et al., 2012).

Of the stony meteorites, some (achondrites) are rocky silicates that originated from the crusts of differentiated parent bodies. Some of these have been determined to be lunar or Martian in origin. Howardites, eucrites, and diogenites (HEDs) have been linked to the asteroid (4) Vesta (McCord et al., 1970; Consolmagno and Drake, 1977; Ruzicka et al., 1997; McSween et al., 2011). Most stony meteorites are chondrites; that is, meteorites of primitive composition with elemental abundances (except for volatiles) that are close to solar composition. While they may have experienced significant thermal or aqueous alteration in their histories, they still generally preserve the basic structural arrangement of matrix and inclusions that they acquired at the time of parent body formation. They are generally characterized by the presence of chondrules, or resolidified silicate melt droplets up to several millimeters in diameter, embedded in a matrix of small (~ 1 μm) silicate grains, metals, and other inclusions. (This is not universal for all chondrites; the CI type carbonaceous chondrites contain almost no chondrules and are entirely matrix.) Chondrites originated on undifferentiated parent bodies, or possibly undifferentiated portions of partially differentiated parent bodies (see Elkins-Tanton et al., 2011). Chondrites are further subdivided into three classes: carbonaceous chondrites, ordinary chondrites, and enstatite chondrites.

Carbonaceous chondrites are subdivided into several groups (CI, CM, CK, CO, CV, CR, CB, CH, and some ungrouped specimens) that vary in composition and matrix abundance. CI are the most chemically primitive, with overall elemental composition closest to solar abundances except for the most volatile elements. All carbonaceous chondrites are characterized by refractory lithophile (i.e. rock-forming) element abundances (normalized to Mg and CI composition) greater than or equal to 1 (Figure 2.2), and are not necessarily high in carbon despite their dark, sooty appearance. They are spectroscopically linked to low-albedo C-type asteroids (Chapman et al., 1975), with possible connections to X- or K-types (Burbine et al., 2002), and a handful of friable, ungrouped carbonaceous chondrites such as Tagish Lake may be linked to D- or P- type asteroids in the outer belt (Hildebrand et al., 2006). Several classes of carbonaceous

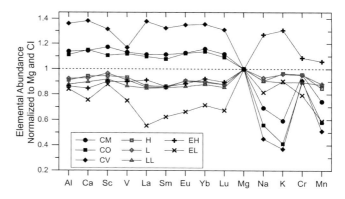

Figure 2.2. Lithophile abundances/Mg for chondrites, normalized to CI. Data from Wasson and Kallemeyn (1988); Anders and Grevesse (1989). The elements to the left of Mg are refractory, those to the right are volatile.

chondrites contain refractory calcium- and aluminum-rich inclusions (CAIs) that are thought to be among the earliest substances to have solidified out of the early solar nebula (cf. Allègre et al., 1995; Russell et al., 2006).

Ordinary chondrites have refractory lithophile / Mg abundances (normalized to CI) lower than 1. They are by far the most abundant of meteorite falls. Ordinary chondrites are further subdivided according to iron and metal content: H – high Fe (~ 8 vol%) (Righter et al., 2006) and Fe^0/FeO of ~ 0.58 (Weisberg et al., 2006); L – low Fe (~ 3 vol%), with Fe^0/FeO of ~ 0.29; and LL – low Fe (~ 1.5 vol%), low metal, with Fe^0/FeO of ~ 0.11. Here Fe^0 is metallic iron.

Enstatite chondrites also contain abundant FeO-free enstatite, and are not common among meteorite falls. Like ordinary chondrites, they have also been subdivided into high-Fe (EH) and low-Fe (EL), but although the two groups are distinct in chemistry, mineralogy, and texture, the high-Fe versus low-Fe distinction has been called into question in recent years (Rochette et al., 2008; Macke et al., 2010).

A final major grouping of meteorites is primitive achondrites. These are compositionally similar to chondrites, but have been thoroughly reprocessed. They may have formed in melted zones of only partially differentiated parent bodies. Examples include ureilites, which are compositionally related to CV chondrites (cf. Goodrich et al., 2007), as well as acapulcoites, and lodranites. Atomic abundances of major elements in various types of chondrites and achondrites are given in Table 2.1.

Among meteorite falls (those specimens observed to fall and recovered shortly thereafter), ordinary chondrites are by far the most abundant, constituting about 81% of all falls. Carbonaceous chondrites make up about 4% of falls, as do iron meteorites. Enstatite chondrites are about 2%. Achondrites make up about 7%, and everything else together constitutes about 2% (Table 2.2). Antarctic finds, which preserve a record of falls over many thousands of years and which are far more abundant than observed falls, exhibit a similar abundance pattern.

2.2.1.2 Meteorite Formation and Structure

Though valuable information may be obtained from all meteorites, chondritic meteorites have best preserved a record of

Table 2.1. *Atomic abundances (relative to Mg) of common elements for chondrites and achondrites*

	Na	Ca	Cr	Mn	Fe	Al	Si	Ti	Ni
CI	0.053	0.06	0.013	0.009	0.82	0.080	0.94	0.002	0.046
CM	0.037	0.07	0.012	0.006	0.78	0.091	0.95	0.003	0.042
CO	0.030	0.07	0.011	0.005	0.74	0.089	0.95	0.003	0.040
CV	0.024	0.08	0.012	0.004	0.71	0.109	0.93	0.003	0.038
H	0.048	0.05	0.012	0.007	0.85	0.073	1.04	0.002	0.047
L	0.050	0.05	0.012	0.008	0.63	0.074	1.07	0.002	0.033
LL	0.048	0.05	0.011	0.008	0.53	0.070	1.07	0.002	0.028
EH	0.068	0.05	0.014	0.009	1.19	0.069	1.36	0.002	0.068
EL	0.043	0.04	0.010	0.005	0.68	0.067	1.14	0.002	0.038
Eucrite	0.053	1.01	0.011	0.041	1.34	0.564	3.00	0.029	
Howardite	0.016	0.24	0.009	0.018	0.64	0.170	1.46	0.007	
Diogenite	0.001	0.05	0.007	0.011	0.36	0.019	0.94	0.001	
Angrite	0.003	1.61	0.003	0.013	1.13	0.408	2.18	0.059	
Acapulcoite	0.030	0.05	0.010	0.009	1.43	0.030	0.70	0.002	0.125
Lodranite	0.005	0.01	0.001	0.005	2.34	0.002	0.41	0.000	0.096
Winonaite	0.040	0.05	0.002	0.002	2.09	0.040	0.72	0.002	0.166
Brachinite	0.023	0.06	0.005	0.007	0.57	0.025	0.65	0.002	0.016
Ureilite	0.001	0.01	0.003	0.006	0.35	0.002	0.48	0.001	0.005
Aubrite	0.002	0.02	0.000	0.000	0.04	0.002	0.65	0.000	0.001

Chondrite data from Wasson and Kallemeyn (1988). Achondrite abundances based on Hutchison (2004) and sources therein.

Table 2.2. *Relevant properties by meteorite type*

Meteorite Type		Abundance (Falls)[a]	Petrologic Type	Matrix[b]	Grain Density[c] (mean) kg m^{-3}	Bulk Density[c] (mean) kg m^{-3}	Porosity[c] (mean)	Strength[d,e] Compres. MPa	Tensile MPa	Cosmic Ray Exp. Age peaks[f] Ma
Ordin. Chond.	H	33%	3–6	10–15%	3,700	3,350	9.5%	76–327	22–42	6–8; 33
	L	38%	3–6	10–15%	3,600	3,300	8%	20–1100	2–62	28; 40
	LL	10%	3–6	10–15%	3,400	3,200	9.5%	64–160	22*	15
Carb. Chond.	CI	< 1%	1	99+%	2,400	1,600	35%	–	0.7–2.8	<2
	CM	1%	2	70%	2,900	2,200	25%	50–82	2.0–8.8	<2
	CO	1%	3	34%	3,500	3,100 2,700[g]	19% 20%[g]	–	31*	4–40
	CV	1%	3	40%	3,500	3,400[h]	4%[h]	22–58	28*	3–23
	CK	< 1%	3–6	40%	3,500	3,000	18%	–	–	9–37
Other CC		1%	–	–	–	–	–	–	–	–
Enstat. Chond.	EH	1%	4–5[i]	2–15%	3,700	3,600	3%	–	–	25–40
	EL	1%	6[i]	2–15%	3,700	3,600	2%	–	–	25–40
Achond.	HED	6%	–	–	3,300	2,900	11%	–	–	19–25; 33–39
	Ure	< 1%	–	–	3,400	3,200	4%	–	56*	<35
Other Ach.		1%	–	–	–	–	–	–	–	–
Irons		4%	–	–	≤ 7,900	–	–	–	–	(III) 650 (IVA) 255 217

[a]Meteoritical Bulletin Database; [b]Weisberg et al. (2006); [c]Macke (2010); [d]Flynn et al. (2018), and sources therein; [e]Slyuta (2017), and sources therein; [f]Eugster et al. (2006), and sources therein; [g]oxidized CV subgroup; [h]reduced CV subgroup; [i]in most cases; *value based on only one data point.

their formation and alteration history. Therefore, this discussion will be limited to chondrites unless otherwise stated. It is still a mystery how chondrites formed; in particular how they lithified from an original accumulation of chondrules, dust grains, metal grains, and other inclusions that formed in the early stages of the solar system. Based on degree of equilibration of chondrules and other factors, chondrites are assigned a petrologic type indicating the degree of thermal or aqueous alteration they experienced, with type 3 having minimal alteration, types 1 and 2 primarily aqueously altered, and types 4–6 (and 7) are thermally altered, the degree of thermal alteration increasing from 4 through to 7 (cf. Van Schmus and Wood, 1967). Type 3 carbonaceous chondrites experienced maximum temperatures of less than 900 K, in some cases lower than 600 K (Schwinger et al., 2016). Chondrites of types 1–3 contain within their matrix a few presolar grains, identified by isotopic abundances several standard deviations from solar. These are small grains (up to a few microns) that were preserved intact from the time of their formation in astrophysical environments (supernova envelopes, AGB star atmospheres, etc.). These include SiC, TiC, graphite, nanodiamonds, oxides, silicate grains, and more (e.g. Zinner, 1996; Mostefaoui and Hoppe, 2004; Nittler et al., 2008). It is a matter of ongoing study as to how parent bodies lithified while preserving all of these signatures, though shock compaction may have played a significant role (Weidenschilling and Cuzzi, 2006).

Beyond thermal and aqueous alteration, meteorites also exhibit histories of break up and recombination. Several ordinary chondrites are brecciated; they broke up and were relithified by shock lithification. These are characterized by large angular chondritic inclusions. There is also evidence for asteroidal break-up and re-accumulation. The asteroid 2008 TC_3, for instance, was composed of rocks of several different meteoritic types. Whole stones from its resultant meteorite, Almahata Sitta, are either ureilites, enstatite chondrites, ordinary chondrites, or carbonaceous chondrites (Shaddad et al., 2010), which would each have formed separately and then been recombined in a new object.

In addition to annealing, shock has considerable effects on internal structure. Three effects are particularly notable: it compresses open pore space (see Section 2.2.1.3), it creates cracks throughout the rock, and it liquefies metallic inclusions, causing metal transport. The last effect manifests in metallic veins filling shock cracks and in the formation of nanometer-scale metal droplets. These in turn shorten the average light-scattering distance, causing a severe reduction in albedo (shock blackening). The shock stage of meteorites is classified on scale from S1 (unshocked) to S6 (very strongly shocked) or S7 (whole rock melted) (Stöffler et al., 1991, 2018).

Since cosmic rays do not generally penetrate deeper than a few meters and most of the mass of a typical asteroid is interior to that depth, the time from the final break-up of the meteorite precursor (parent body or subsequent asteroid) can be estimated by determining the cosmic ray exposure age (CRE), assuming cosmic ray flux has been constant over time. High-energy cosmic rays can induce nuclear reactions (see Eugster et al., 2006), and the abundance of the products of these reactions relative to their production rates provides the exposure age. CRE ages are typically a few million to tens of millions of years, though iron meteorites typically exhibit CRE ages of hundreds of millions of years, up to a maximum of about 1.5 Ga. A majority of H chondrites have CRE ages of between 6 and 8 million years, indicating a catastrophic event in the history of the H-chondrite precursor object at that time (Marti and Graf, 1992). Other CRE peaks are given in Table 2.2.

2.2.1.3 Physical Properties

A full treatment of all meteorite physical properties would fill a chapter on its own. Properties that have been studied include density, porosity, magnetic susceptibility, compressive and tensile strength, heat capacity, thermal conductivity, Young's modulus, sound speed, and others. Several articles give a fuller treatment of some or all of these properties (e.g. Consolmagno et al., 2008; Macke, 2010; Slyuta, 2017; Flynn et al., 2018). Here, we highlight bulk and grain density, porosity, and compressive and tensile strength since these properties are most relevant for comparison with meteor data.

Meteorites exhibit a range of densities and porosities based on their composition and structure. Ordinary chondrite falls are on average 8–10% porous, with grain densities ranging from 3,400 to 4,000 kg m^{-3} (increasing from LL to L to H), and bulk densities ranging from 2,800 to 3,800 kg m^{-3} (Macke, 2010), carbonaceous chondrites vary by type, but (except for CR and CH) have lower densities than ordinary chondrites (Table 2.2), and average porosities ranging from 18–35%, with CI among the most porous group. Enstatite chondrites have greater metal content than ordinary chondrites, so have higher densities, but are much less porous on average, only about 2–3%. Lightly shocked chondrites (shock stage S1, <5 GPa) tend to be more porous than more shocked ones, up to twice as porous. The average porosity of S1 carbonaceous chondrites is about 21%, while more strongly shocked specimens are generally less than 10% porous (Macke, 2010). The nature of pore space also varies; pore space in lightly shocked chondrites is primarily in the form of intergranular space within the matrix, while more shocked specimens are more tightly compact with most pore space accounted for by shock-induced cracks (Consolmagno et al., 2008; Friedrich et al., 2017). Exposure to the terrestrial environment results in oxidation of metallic iron into goethite and other low-density products, which expand to fill in available pore space, so densities and porosities of weathered ordinary and enstatite chondrite finds are significantly reduced from those of their counterparts among observed falls. Density and porosity of iron meteorite falls have not been extensively studied, but densities vary based on abundance of inclusions and max out at around 7,900 kg m^{-3}, the density of solid iron-nickel alloy (Macke, 2010). Porosities are in general not more than a few percent.

There have not been extensive studies of the compressive strengths of meteorites, and even fewer studies of tensile strengths, because their measurement usually requires stressing the meteorite to the breaking point (though some have attempted to study compressive strength nondestructively by measuring sound velocities through specimens). Of those meteorites measured destructively, chondrite stones typically have compressive strengths on the order of several tens to a few hundred MPa, with tensile strength an order of magnitude less.

Carbonaceous chondrite (CM and CV) compressive strengths typically lie within 30–50 MPa, with tensile strengths from 2–30 MPa. Some very friable carbonaceous chondrites are considerably weaker; for instance, the best measurement for the compressive strength of the primitive ungrouped carbonaceous chondrite Tagish Lake is only 0.7 MPa (Brown et al., 2002), about the same as its tensile strength (Slyuta, 2017). Data are lacking for compressive strengths of CI chondrites, but the few measurements of their tensile strengths range from 0.7 to 2.8 MPa (Slyuta, 2017), an order of magnitude less than that of their CM counterparts. Ordinary chondrites tend to be much stronger, but exhibit a much greater range of compressive strengths, mostly between 60–460 MPa, with an average of 190 MPa. Tensile strengths for this class mostly lie between 16 and 40 MPa, averaging 29 MPa, with no obvious difference between H, L, and LL subclasses. Meteorite strength does exhibit an inverse correlation with porosity. A study of intact stones of the Košice meteorite found greater variation and higher average porosity among stones with small mass than among those greater than a few tens of grams (Kohout et al., 2014a), indicating that more porous portions of meteorites are weaker and will tend to break up into smaller pieces. Those stones that survive intact are themselves much stronger than their precursor meteoroids. This fact will be discussed in more detail in Section 2.3.4.

2.2.2 Interplanetary Dust Particles

Interplanetary Dust Particles (IDPs) are tiny meteoric matter that was decelerated in Earth's upper atmosphere and survived with the least possible thermally-induced chemical, physical, and/or textural modifications. Once IDPs reach rest velocity in the upper stratosphere, they begin settling to lower altitudes, thus no longer carrying a record of their orbital histories. While falling in the stratosphere, they can be collected on inertial-impact, flat-plate collectors deployed by the NASA Cosmic Dust Program indiscriminate of particle sizes, from ~ 2 μm porous chondritic IDPs to compact, ~ 25 μm particles. This collection program originated at the NASA Ames Research Center in the mid-1970s with the first U2 aircraft IDP collections (Brownlee et al., 1976) prior to transfer in May 1981 to the NASA Johnson Space Center, where it is a still-ongoing program. The emphasis has been on the mostly ~ 10 μm chondritic porous (CP) IDPs (Figure 2.3, left).

In addition to CP IDPs, there are non-chondritic IDPs of similar sizes (Table 2.3). When used with some discretion, a 10-micron limit includes almost all CP IDPs and non-aggregate IDPs (Rietmeijer, 2002). Larger particles, called Cluster IDPs, break apart on impact with the collector.

A small number of chondritic IDPs are hydrated and have developed a characteristic smooth surface (Figure 2.3, center). These hydrated chondritic IDPs are dominated by serpentine and montmorillonite layer silicate minerals (Schramm et al., 1989), but they are still poorly characterized. The CP IDPs are most likely comet dust (Brownlee, 2014), but hydrated chondritic IDPs might have asteroid links.

The CP IDPs are fine-grained aggregates of numerous submicron grains with an average composition that fortuitously matches the CI composition. Most minerals are anhydrous but very small amounts of hydrous minerals (serpentine; montmorillonite) can be present. CP IDPs consist of numerous GEMS (glass with embedded metal and sulfides) spheres ~ 45 to ~ 100 nm in size. A recent paradigm shift in cosmochemistry now holds that GEMS formed in the early solar system, probably via vapor phase condensation, and only

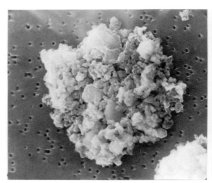 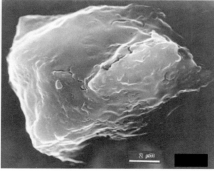 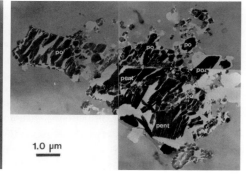

Figure 2.3. **(left)** Scanning electron microscope image of a CP IDP that is an aggregate of numerous small grains. The CP IDPs have no preferred shapes or sizes. They might be fragments of larger aggregates that broke apart along the weakest structural link in the fine-grained matrix either on atmospheric entry or on impact onto the flat-plate collectors (Rietmeijer, 2002). It is quite remarkable that these CP IDPs are not much larger than 10 μm. This particle is ~ 10 μm in size (courtesy the National Aeronautics and Space Administration; Particle W7029B13 (NASA number S-82-27575)

(center) Scanning electron microscope image of a hydrated chondritic IDP. Their typically smooth surface shows dehydration cracks that probably formed during atmospheric entry heating (courtesy the National Aeronautics and Space Administration; Particle W7017B12 (NASA number S-81-39963)

(right) Transmission electron microscope image of sulfide IDP that is a mixture of pyrrhotite (po) and pentlandite (pent). Flash-heating during atmospheric entry caused rapid thermal expansion of this particle followed by rapid quenching in the upper atmosphere. This process rendered the sulfide particle brittle which caused sample loss during thin-section preparation (IDP particle L2005B22, reproduced from Rietmeijer (2004), courtesy *Meteoritics & Planetary Science*).

Table 2.3. *Classification of interplanetary dust particles*

Type	Subtype	Composition
IDPs	Chondritic porous IDPs mostly < 10 μm	GEMS < 100 nm micron Fe,Ni-sulfide; Mg,Fe-silicate; Ca,Al-silicate; Mg,Ca-silicate amorphous and equilibrated ferromagnesiosilica grains
	Non-chondritic IDPs < 10 μm	Fe(Ni)-sulfide grains (pyrrhotite, pentlandite); Mg,Fe-silicate grains (olivine; pyroxene); Ca,Al silicate grains (plagioclase); refractory HT mineral aggregate grains
Cluster IDPs		$\sim$ 10 μm IDPs $\sim$ 10 μm to 25 μm silicate and Fe(Ni)-sulfide particles

1–6% of GEMS might be circumstellar grains (Keller and Messenger, 2011). They are the dominant grains in the matrix of all aggregate IDPs. The GEMS matrix can be host to a variety of small grains, viz. (1) Mg-Fe silicates, mostly olivine and pyroxene, (2) low-Ni and Ni-free pyrrhotite (FeS) and (3) fewer Mg-Ca-Al silicates, e.g. plagioclase ($CaAl_2Si_2O_8$), diopside ($MgCaSi_2O_6$) and Mg-wollastonite, (4) (rare) SiO_2 crystals, and (5) compound ferromagnesiosilica grains of Mg,Fe-olivine, Mg,Fe-pyroxene crystals and amorphous silica (Rietmeijer, 1998) (Table 2.3).

Non-chondritic IDPs are $\sim$10 μm (1) Mg-rich Mg,Fe-olivine, (2) sulfides (Figure 2.3, right) with evidence of atmospheric flash-heating, and (3) aggregates of refractory high-temperature minerals, e.g. hibonite, perovskite and melilite-group minerals (Table 2.3). These non-chondritic particles often have small GEMS clusters, or melted GEMS clusters, adhered to their surface. It appears that CP IDPs and these non-aggregate IDPs might be "fragments" from larger aggregate Cluster IDPs that include much larger, $\sim$10 μm to 25 μm, Mg-pyroxene, plagioclase, Fe(Ni)-sulfide and mixed silicate-sulfide grains (Table 2.3). These larger cluster particles survived atmospheric entry with their pre-entry properties (almost) intact. On the collector surface they form small piles of associated grains.

The CP IDPs contain rare elemental carbon particles, including diamond, but they contain only very small amounts of surviving organic (CHON) material. Cluster IDPs are rich in surviving CHON materials, probably because their low atmospheric entry velocities allowed all but their most volatile CHON material to survive.

Chondritic IDPs have a wide range of porosities (Rietmeijer, 1998). There are particles with zero porosity, particles with low to moderate porosities (2–50%, mean 12%) and rare particles with extreme porosities (70–80% and >90%). From the earliest days of dust collections, the open aggregate texture of CP IDPs was thought to be pore space that was originally filled by water-ice when resident on a comet. The main source of these IDPs is probably the interplanetary dust cloud (also called zodiacal cloud), fed by $\sim$10 μm dust ejected from Jupiter family comets. If so, IDPs might resemble some of the particles from comet 81P/Wild 2 collected by the NASA Stardust mission.

All decelerating IDPs experience flash-heating to temperatures ranging between $\sim$300°C and $\sim$1,000°C depending on their entry velocity and mass. Flash-heating causes thermally-induced mineral changes, and even complete melting. In CP IDPs, flash-heating causes the formation of a thin Fe-oxide rim on individual mineral grains. In non-chondritic sulfide IDPs, flash-heating results in the formation of a continuous Fe-oxide rim on pyrrhotite, $Fe_{(1-x)}S$ ($x = 0$ to 0.2) on top of a vesicular zone due to sulfur loss, internal crystallographic disorder and (rare) precipitation of pure sulfur spheres. Non-chondritic IDPs such as Mg-rich olivine and Ca,Mg-silicate IDPs suffer partial melting causing surface-melt ablation to rounded particles with internal Fe-oxide precipitates and (rare) vesicle formation in individual minerals, and even "boiling" leaving extreme vesicles in the surviving IDP (Rietmeijer, 1998, 2002).

2.2.3 Micrometeorites

Micrometeorites (MMs) are $\sim$ 10–100 μm cosmic particles that survive atmospheric entry and settle down to Earth's surface. This distinguishes them from IDPs, which are collected high in the stratosphere. The typical collection areas of MMs are places where they tend to concentrate, such as ice sheets in arctic regions, desert environments, or sediments. Based on their physical appearance and degree of heating experienced during atmospheric entry, we distinguish three basic groups of MMs: melted, partially melted, and unmelted ones. Each group is further divided into classes and types as proposed by Genge et al. (2008).

Unmelted MMs represent material that reaches the atmosphere at entry velocities slow enough to prevent substantial melting, and thus these are almost identical to IDPs. Thermal entry effects are limited to some loss of volatiles and dehydration. Sometimes, a magnetite rim surrounding the particle is developed due to entry heating (Toppani et al., 2001; Toppani and Libourel, 2003) and is equivalent to the fusion crust found in meteorites. Particles without magnetite rims are possibly the least altered MMs. Based on their internal texture and composition they can be divided into coarse-grained (Cg), fine-grained (Fg), refractory, or ultracarbonaceous MMs. The typical size of unmelted MMs is of the order tens of microns.

Fine-grained micrometeorites (Figure 2.4, left) are usually composed of hydrated silicates and thus have affinities to

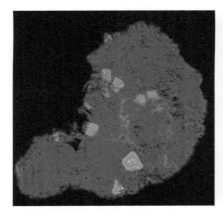 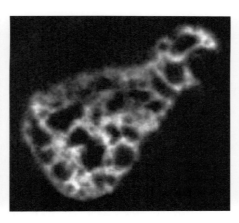 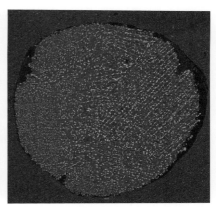

Figure 2.4. Examples of micrometeorites collected from ice in Novaya Zemlya, Russia. Catalogue numbers correspond to the database in Kohout et al. (2014b).
(left) 210 μm-sized unmelted micrometeorite no. 12 with hydrated silicates and iron sulfide inclusions (bright)
(center) 230 μm-sized partly melted scoriaceous micrometeorite no. 11 with abundant vesicles
(right) 390 μm-sized entirely melted cosmic spherule no. 2 showing barred olivine structure with lamellas of olivine and magnetite-rich (white) melt. Micrometeorites no. 2 and 12 are back-scattered electron microscope images, no. 11 is X-ray microtomography image.

carbonaceous chondrites and are of asteroid origin. Coarse-grained micrometeorites are, on the other hand, composed mostly of dry silicates with large crystals readily distinguishable in microscope sections or X-ray microtomography scans; their origin is linked to ordinary chondrites and their parent bodies. They are most likely fragments of chondrules. Apart from coarse silicate mineral assemblages, particles composed of single crystals also exist. Some particles contain zones of both hydrated Fg and dry Cg material and thus, parent bodies containing both types of these silicates may exist. In addition to chondritic unmelted MMs, CgMMs with achondritic compositions are also found.

Refractory MMs are usually silicate FgMMs containing inclusions of refractory minerals similar to CAIs found in chondrites. Some particles are dominated by silicates while some may be dominated by refractory minerals. Both unaltered and altered CAI-like assemblages are found.

Ultracarbonaceous MMs (also UCAMMs, or Ultracarbonaceous Antarctica Micrometeorites) are a rare class of unmelted micrometeorites composed of amorphous carbon with embedded dry silicates, pyrrhotite, or kamacite. The composition of silicates may vary within one particle and thus is unequilibrated. The content of carbon is sometimes higher than in CI chondrites. Presolar grains are also often present. These properties are similar to the cometary dust composition observed on Stardust and Rosetta samples. The exact origin of the Ultracarbonaceous MMs cannot be determined. Nesvorný et al. (2010, 2011), however, suggest on the basis of interplanetary dust cloud modeling that the majority of MMs are of cometary (Jupiter family) origin.

At higher entry velocities friction with air causes substantial heating accompanied by loss of volatiles, dehydration, and silicate recrystallization of the MMs forming Scoriaceous (Sc) MMs (Figure 2.4, center). Typically relict original or recrystallized silicate grains such as olivine and pyroxenes exist surrounded by vesicles and melt fractions. The formation of vesicles is related to thermal decomposition of hydrated silicates, carbonaceous matter, or sulfides. Metal distribution seems to be mostly unaffected at this stage. The formation of vesicles gives the ScMMs substantial porosity sometimes exceeding 50%. Some ScMMs show at least the partial presence of a magnetite rim and thus, their original shape may be at least partly preserved. The possible changes in shape of ScMMs are rather the result of plastic deformation than re-melting. The typical size range of ScMMs is similar to unmelted MMs.

Once the melting has affected most of the MM material a droplet-shaped melted particle is formed. The melted MMs are also referred to as Cosmic Spherules (CSs). Unlike unmelted and partially melted MMs the original shape of CSs is not preserved and is rather fully the result of atmospheric entry and associated melting. CSs are devolatilized due to their substantial heating.

Cosmic spherules show a wide range of compositions and textures. Their size is typically hundreds of microns. S-type CSs are formed predominantly by dry silicates, G-type by magnetite dendrites in silicate glass, and I-type are dominated by magnetite and wüstite with frequent large vesicles in their center indicative of rapid cooling. The texture of S-type CSs can range from apparently featureless glassy and cryptocrystalline to barred olivine (Figure 2.4, right) with lamellas of olivine and magnetite-bearing glass or porphyritic with large skeletal silicate crystals within glass. Sometimes, a substantial portion of relict unmelted original crystals can be found, and such CSs are referred to as coarse-grained CSs. Some S-type spherules contain one or two large iron metal or iron-oxide beads close to their perimeter. Sometimes the bead is lost during atmospheric entry and a large void is found instead.

The density and porosity values of MMs were compiled from data by Genge et al. (2008), Taylor et al. (2011), Kohout et al. (2014b), and Genge (2017). The porosity of unmelted micrometeorites varies considerably between 0 and 50%, with some examples such as ultracarbonaceous MMs having values over 50%. Scoriaceous MMs have higher porosity in general due to vesicle formation, with a typical range of 16–70% due to volatile evaporation and partial melting (scoriaceous phase). The density of unmelted or partly melted MMs is hard to measure due to their small size. However, grain densities can be determined based on mineralogy, with approximate values between 1,500 and 3,500 kg m^{-3}. Completely melted CSs are characterized by loss of porosity (1%) due to re-melting. Density data on several examples measured indicate almost identical bulk and grain densities of 3,200 kg m^{-3} and 3,300 kg m^{-3}, respectively. However, some glassy CSs were reported to contain a significant amount of spherical vesicles forming porosity up to 50%.

There is a clear trend in the evolution of internal structure and porosity in MMs with their size and entry conditions. Smaller MMs have a higher chance of surviving atmospheric entry intact at lower entry velocities or shallower trajectories than larger MMs. This can explain why observed unmelted or ScMMs are typically an order of magnitude smaller (~10 μm) than melted CSs (~100 μm). Initial porosity of unmelted MMs shows a wide scatter based on their type and origin. The volatile-rich or hydrated MMs tend to develop large vesicles due to heat during atmospheric entry, forming highly porous ScMMs. The presence of volatiles and hydrated phases creating vesicles during entry heating has been proposed by Genge (2017) to create a parachute effect increasing the chance of survival of hydrated MMs by a factor of two compared to dry MMs.

Larger particles entering at faster velocities or on steeper trajectories experience higher heating and melt completely, which often results in a loss of volatiles and collapse of porosity. In this scenario, larger particles have a higher chance of survival, as the smaller ones may entirely evaporate. This can possibly explain a larger size of CSs compared to unmelted or ScMMs.

2.2.4 Cometary and Asteroidal Dust

Dust particles and small meteoroids orbiting the Sun form the interplanetary dust cloud. The interplanetary dust cloud has a very low spatial density. As suggested by the distribution of the zodiacal light, i.e. the solar light scattered by the particles that build the interplanetary dust cloud, the spatial density of the cloud increases towards the Sun and its near-ecliptic surface of symmetry. Since the dust particles may be blown away by solar radiation pressure or may spiral towards the Sun under the Poynting–Robertson effect, sources are needed to replenish the cloud. In the inner solar system, the main sources are likely to be the dust released by cometary nuclei and the dust originating from asteroids (for reviews on interplanetary dust see e.g. Grün et al., 2001).

Various approaches, corresponding to analyses of the zodiacal light (Levasseur-Regourd, 1998; Levasseur-Regourd et al., 1999a; Lasue et al., 2007, 2015), to a dynamical model of dust in the inner solar system (Nesvorný et al., 2010, 2011), and to an improved model of the zodiacal cloud infrared emission (Rowan-Robinson and May, 2013), independently indicate that the contribution of cometary dust to the interplanetary dust cloud is above 70% in Earth's vicinity. In the remainder of Section 2.2.4, we summarize progress in the understanding of dust particles over the last decades, first in comets and secondly in asteroids (see e.g. Levasseur-Regourd et al., 2018; Szalay et al., 2018, for corresponding reviews).

2.2.4.1 Cometary Dust, First Approaches

Flyby missions to comet 1P/Halley by Vega 1, Vega 2 and Giotto in March 1986, several weeks after its perihelion passage, had already given clues to the specificity of cometary dust. The presence of dust particles with different physical properties, likely to consist of low-density aggregates (about 100 kg m^{-3}) with very low geometric albedos (ranging between 3 and 10%), was inferred (Levasseur-Regourd et al., 1999b; Fulle et al., 2000). The presence of organic compounds was immediately suspected (Kissel et al., 1986) in the dust, although the relative velocities during the flybys were very high (about 70 km s^{-1}). The dust was found to be dominated by Mg-Fe-Si grains with Fe/(Fe+Mg) compositions ranging from zero to one, with no preferred compositions plus a few FeS-dominated grains (Jessberger et al., 1988; Fomenkova et al., 1992). The dust particles with masses above about 10^{-16} kg, were found to be less rich in very low Fe/(Fe+Mg) ratios than the lighter ones, below 5×10^{-19} kg (Mukhin et al., 1991). These (Mg,Fe)-silicates had an average Fe/(Fe+Mg) = 0.35 (atomic %) composition. Metallic Fe, iron-oxide and Fe-sulfide nanoparticles were reported (Fomenkova et al., 1992). Collectively these nanoparticles caused an average nearly CI-chondritic bulk nucleus composition, although the comet Mg/Fe ratio is slightly higher than the CI ratio (see Table 2.4). The light particles included a considerable number of Mg-only particles that could be MgO, Mg(OH)$_2$ or MgCaO$_3$ (Mukhin et al., 1991; Fomenkova et al., 1992). It could explain the non-CI Mg/Fe ratio.

The NASA Stardust mission has confirmed, by a flyby of comet 81P/Wild 2 in 2004, the presence of organic compounds (Brownlee et al., 2004). Moreover, samples collected by its return capsule have provided unique results about dust released in the coma. Although the dust particles were collected at high velocity (6.1 km s^{-1}), possibly inducing thermal alteration, analyses of impacts on aluminum foils and tracks in aerogel cells established that the dust particles were more or less cohesive and rich in complex organics, possibly including glycine (Hörz et al., 2006; Burchell et al., 2008; Matrajt et al., 2008; Elsila et al., 2009). Refractory particles including chondrule-like igneous particles were detected as well (Zolensky et al., 2006; Gainsforth et al., 2015). The particles were chemically heterogeneous up to the largest grains (Flynn et al., 2006).

Table 2.4. *Atomic abundances (relative to Mg) of common elements for dust of two comets measured in situ*

	C	N	O	Na	Ca	Cr	Mn	Fe	Al	Si	Ti	Ni
1P/Halley	8.1	0.42	8.9	0.10	0.063	0.009	0.005	0.52	0.068	1.85	0.004	0.041
67P/Churyumov-Gerasimenko	48	1.7	48	0.70	0.048	0.023	0.038	2.5	0.15	8.8		

1P data data from Jessberger et al. (1988). 67P data from Bardyn et al. (2017).

Their mean composition is consistent with CI composition, but individual particles show highly variable deviations from the CI composition. The minerals found in comet 81P/Wild 2 (Fe,Ni-sulfides, Mg,Fe,Ca-silicates, Ca,Al,K-feldspars and even highly refractory CAI-inclusions) are typically found in the matrices of unequilibrated ordinary chondrite meteorites (Zolensky et al., 2008; Joswiak et al., 2009). This leaves the intriguing notion that should the Wild 2 dust become fully hydrated, the result would be a CI body.

Meanwhile, remote telescopic observations of the targets of cometary missions and of bright comets have taken place. Infrared spectroscopy has provided clues to the presence of crystalline silicates and of micron-sized particles or of larger aggregates of sub-micron-sized grains, as reviewed in Hanner and Bradley (2004). In the visible domain, polarimetric observations have given clues to variability in the properties (e.g. composition, size, porosity) of dust ejected within a given comet (Levasseur-Regourd, 1999; Hadamcik and Levasseur-Regourd, 2003). Experimental simulations have provided satisfactory matches for mixtures of porous aggregates of submicron-sized MgSiO, FeSiO, and C grains and compact Mg-silicates (Hadamcik et al., 2007), and have also given clues to the properties and evolution of interplanetary dust particles (Hadamcik et al., 2018). Meanwhile, numerical simulations have suggested the presence of mixtures of compact particles and aggregates, both consisting of minerals and more absorbing organics (Lasue and Levasseur-Regourd, 2006; Levasseur-Regourd et al., 2008; Kolokolova and Kimura, 2010; Kiselev et al., 2015). Furthermore, sky surveys of the infrared thermal emission have not only contributed to point out similarities between the properties of cometary and interplanetary dust particles (Levasseur-Regourd et al., 2007), but have also allowed the discovery of cometary dust trails (e.g. Sykes and Walker, 1992; Reach et al., 2007; Stevenson et al., 2014); such trails, observed in the vicinity of the orbits of short-period comets, are likely to be built of rather large particles of low geometric albedo (e.g. Ishiguro et al., 2002; Agarwal et al., 2010; Arendt, 2014). They probably represent very young meteoroid streams evolved from quite large dust particles present in jets near the surface of active nuclei.

2.2.4.2 Cometary Dust, Present Understanding

More recently, the ESA Rosetta rendezvous mission spent twenty-six months surveying the nucleus and the coma of comet 67P/Churyumov-Gerasimenko (hereafter 67P/C-G), operating on a wide range of distances to the Sun (encompassing its perihelion passage in August 2015) and distances to the nucleus, until its final landing in September 2016. Three instruments were specifically devoted to studies of dust particles. COSIMA, a dust mass spectrometer (to which an optical microscope was associated) analyzed particles in the 10 to 1000 μm range; MIDAS, the first atomic force microscope flown in space, built 3D images of dust in the 10s nm-to-10s μm size range; GIADA measured optical cross-sections, speeds, momentum and flux of dust particles. In addition, the OSIRIS cameras provided images of large dust particles in the coma (e.g. Glassmeier et al., 2007).

Results on the physical properties reveal that dust particles are essentially aggregates of grains, with morphologies ranging from very fluffy fractal particles to very compact ones, and high porosities, from at least 50% to much higher values (Langevin et al., 2016; Mannel et al., 2016; Fulle and Blum, 2017). They present a hierarchical structure on a range of sizes covering over 3 orders of magnitude (Bentley et al., 2016).

Results on the composition of dust particles from the 67P/C-G mission provide evidence for the prominence of organic matter in the dust ejected by the nucleus, with carbon present in the form of macromolecules that bear mass spectral similarities with the insoluble organic matter in carbonaceous meteorites (Fray et al., 2016). This macromolecular component for dust particles in the coma corresponds to an abundance of roughly 45% in mass (Bardyn et al., 2017) and possibly 70% in volume (Levasseur-Regourd et al., 2018). The atomic abundances are, except for the carbon overabundance, within a factor of 3 of CI composition (Bardyn et al., 2017). However, the Mg/Fe and Ca/Fe ratios show opposite deviations from CI than in 1P/Halley (Table 2.4; see also Figure 2.8). The Na/Fe and Si/Fe ratios are higher than CI in both comets (although the error bars are large). Besides, a few new organic molecules were detected from dust particles encountered by the Philae lander (Goesmann et al., 2015; Wright et al., 2015). Both the carbonaceous composition and the porous structure of dust particles are consistent with values inferred for the interior of the nucleus by bi-static radar probing (Herique et al., 2016).

2.2.4.3 Asteroidal Dust

Asteroids present a wide variety in their sizes and compositions, as revealed by remote observations including spectroscopic observations. As discovered by several asteroidal flybys and rendezvous missions since 1991, their surfaces are covered with granular and unconsolidated materials called regoliths, resulting from thermal shock, micrometeorite impacts, and other space weathering processes (e.g. Barucci et al., 2011). These dust particles are progressively released from low-gravity small asteroids; they are also injected into the interplanetary dust cloud, together with bigger debris, after collisions between asteroids.

In September 2005, the JAXA Hayabusa spacecraft visited the silicaceous S-type asteroid (25143) Itokawa, and found it to be mostly covered by boulders. A sample of less than 1 g of small dust particles, present on a smooth region covered by tiny particles (MUSES-C Regio), was collected in a recovery capsule that returned to Earth in June 2010. Typical particles present a composition similar to LL ordinary chondrites and show evidence of long-term thermal metamorphism (Nakamura et al., 2011). With sizes ranging from a few μm to several hundred μm, these small particles have probably been formed by successive impacts and moved by shockwaves to build up the smooth terrain on which they were collected (Tsuchiyama et al., 2011).

More information is expected from the JAXA Hayabusa2 mission to C-type asteroid (162173) Ryugu and from the NASA OSIRIS-REx mission to B-type asteroid (101955) Bennu (Lauretta et al., 2015). Both asteroids are low-albedo carbonaceous objects, belonging to the Apollo group. The approaches to the asteroids began in 2018. Samples are to be collected and brought to Earth for analysis in 2020 (Hayabusa2) and 2023 (OSIRIS-REx). While it is likely, from comparisons between densities of cometary nuclei and of asteroids, that dust particles and debris of asteroidal origin are more compact than dust particles within dust streams originating from comets, further analysis of samples from asteroids should provide a better understanding of the diversity in composition and structure of their dust particles.

Although space missions to comets and asteroids have increased our knowledge of these objects, the huge diversity of these small bodies may preclude drawing final conclusions. Indeed, Interplanetary Dust Particles and micrometeorites retain uncertainties about their origin. It is nevertheless of major interest that some fraction of CP IDPs are porous aggregates of anhydrous minerals with large amounts of organics and that UCAMMs present a major fraction of carbon consisting of polyaromatic organic matter (e.g. Flynn et al., 2003; Dobrică et al., 2012; Flynn et al., 2013). While their porous structures have possibly favored their survival in atmospheric entry (Dobrică et al., 2010; Levasseur-Regourd and Lasue, 2011), their structures and compositions compare well with the properties of cometary dust, as revealed by the Rosetta mission.

2.3 Meteoroid Properties From Meteor Observations

Meteor observations sample all types of materials encountering the Earth and provide indirect clues about their structure and composition. In this section we review the current status in this field.

2.3.1 Meteor Data

The most relevant data for studies of meteoroid properties are obtained by optical techniques of meteor observations, i.e. either videos or photographs, possibly combined with radiometers. More information about optical techniques can be found in Chapter 4, Koten et al. (2019), see Section 4.2. Obtained data include the meteor trajectory (which is usually assumed to be straight and can be therefore defined by its beginning and end points), the position of the meteor in the trajectory as a function of time, and meteor brightness as a function of time (the light curve). Meteor velocity and deceleration as a function of time or height are computed by fitting the measured positions. The heliocentric orbit of the meteoroid is computed from the velocity vector at the trajectory beginning, before any significant deceleration due to atmospheric drag occurs.

Inferring meteoroid physical properties from meteor data is not straightforward. It relies on modeling the meteoroid interaction with the atmosphere and comparing the resulting output with the observed quantities. Various models of meteoroid entry were discussed in Popova et al. (2019), see Chapter 1. The known trajectory and entry velocity are input parameters for the models. The compared quantities include the meteor light curve (i.e. radiation) and deceleration (i.e. dynamics). As shown in Popova et al. (2019), Sections 1.2.5 and 1.4.4, the main factor influencing meteor flight is in many cases meteoroid fragmentation. It is therefore an advantage if the observational technique also provides the physical appearance of the meteor at various times, e.g. the presence of wake and positions of separated fragments, if there are any (i.e. meteor morphology).

If morphology is not available, indirect evidence of fragmentation can be used. Types of such evidence include increase of deceleration, caused by the sudden decrease of mass of the meteoroid, meteor flares produced by sudden release of small fragments, or other irregularities of the light curve. To quantify the fragmentation effects from this evidence, meteor models must be used.

Very small meteoroids are better observed by radar systems, in particular the High Power Large Aperture (HPLA) radar. More information about radar techniques of meteor observations can be found in Kero et al. (2019) (see Chapter 3). HPLA radar systems directly measure meteor distance and radial (Doppler) velocity. Full three-dimensional trajectories and velocities can be obtained if these data are combined with interferometry. The analogy to meteor brightness is radar cross section. It is proportional to the number of free electrons in the meteor head.

Additional information about meteoroid interaction with the atmosphere can be obtained from temporal evolution of meteor spectra. Spectroscopy also provides information about meteoroid composition. For a general description of meteor spectra and information on currently running spectroscopic projects see Koten et al. (2019), i.e., Section 4.2.2.

2.3.2 Physical Properties of Small Meteoroids

2.3.2.1 Direct Observations of Meteoroid Fragmentation

Both radar and optical observations can be used to study the way in which small meteoroids fragment. Variations in the strength of an echo from a radar may be caused by returns from two or more individual fragments separated by a significant fraction of the radar's wavelength and moving relative to one another. Oscillations consistent with returns from multiple fragments have been observed on several radar systems (e.g. Campbell-Brown and Close, 2007; Mathews et al., 2010), but the most convincing observations are from the multistatic EISCAT HPLA radar. Kero et al. (2008) used three receivers separated by hundreds of kilometers to observe head echoes from very different angles. A radar echo that alternately increases and decreases may be caused by alternating constructive and destructive interference between fragments, but such alternating echoes could also be caused by instrumental effects or a rapidly changing ablation rate. The tristatic observations showed different rates of beating at the different receiver locations, consistent with the different relative speed of fragments along the different lines of sight. Single-station radar observations are not able to rule out other causes of signal oscillation. For trail echoes, Fresnel holography (Elford, 2004) can be used to look for enhancements in the electron density at certain points along the line of ionization, which may represent individual fragments, but the enhancements are barely above the noise level.

Fragments can be observed distinctly in optical systems, but only if the separation of the fragments is larger than the resolution of the system. For example, the CILBO automated meteor observatory on the Canary Islands (Koschny et al., 2013) has a pixel scale of 2.3′ per pixel (30 degrees and 768 pixels across), giving a spatial resolution of about 75 m per pixel (at a range of 110 km). The influx system of the Canadian Automated Meteor Observatory (CAMO) has a field of view 20 degrees across and a frame is 1400 pixels, giving a pixel scale of 0.86′ per pixel; at a typical range for this system (105 km), the spatial resolution is 26 m per pixel (Weryk et al., 2013). Only a small number of meteors in standard faint meteor cameras show measurable fragmentation, usually in the form of luminous wake rather than resolved fragments (Fisher et al., 2000).

Telescopic observations typically do not give better resolution in meteor observations, because the large angular speed of meteors smears the image by many degrees even in a short video frame. A small telescope has been successfully used to observe faint meteors at meter-scale resolution, using a pair of mirrors to direct the light from the meteor into the telescope, minimizing the motion of the meteor in the field of view. The CAMO tracking system uses wide-field (30 degrees) cameras to detect meteors, and tracks them with cameras mounted on a telescope. The narrow-field cameras have fields of view of about 1.5 degrees, and can typically resolve details as small as 4 meters (Weryk et al., 2013). This system allows the fragmentation behavior of meteoroids to be characterized for a large number of observations.

Only about 5% of meteors observed in the CAMO tracking system show a group of discrete fragments, but 85% show long wakes consistent with a large number of unresolved grains (Subasinghe et al., 2016). The remaining 10% of meteors show short wakes or no wakes, but may still be fragmenting: Campbell-Brown (2017) attempted to fit one of these meteors with a single-body model, and found that a small amount of continuous fragmentation was necessary to explain the light curve, while not producing more wake than observed. Fragments are normally observed to spread along the line of motion of the meteoroid, because of their differing deceleration rates, but a small number of meteoroids show lateral speeds of tens or hundreds of m s^{-1}, most likely because of explosive devolatilization (Stokan and Campbell-Brown, 2014).

2.3.2.2 Indirect Evidence of Physical Properties

Even without direct measurements of fragmentation, the behavior of a meteor as it ablates can be inferred from its light curve. According to the classical theory of meteor ablation, a homogeneous, single-body meteoroid that ablates self-similarly (without changing shape) will have a peak toward the end of the light curve. The F-parameter is used as a measure of the skewness of a light curve: it is defined as the distance (or time) from the beginning of light production to the maximum light divided by the distance (or time) from the beginning to the end of light production. A classical, single-body light curve has an F-parameter of about 0.7 (e.g Beech, 2009).

If an object fragments, individual fragments may produce classical light curves, but the sum of the curves will not be classical. A meteoroid consisting of many small grains held loosely together, which disrupts before ablating, will produce a symmetric light curve (Hawkes and Jones, 1975): this is the dustball model. The light curve is symmetric because small grains produce all their light in the early part of the trajectory, while the largest grains peak toward the end of the curve after the smallest grains have vanished. In general, faint meteors have symmetric light curves, first noticed by Jacchia (1955) in the Super-Schmidt camera meteor data. The F-parameters of a number of meteor showers have been studied by Koten et al. (2004, 2015). Most showers have a wide spread in individual F-parameters, but the average parameter is between 0.50 and 0.59 for the Taurids, Perseids, Leonids, Orionids, and Geminids. The Draconids (October Draconids, IAU #009) had, on average, light curves with the earliest peaks, with a mode of the F-parameter distribution of 0.35. Draconid meteoroids are known to be among the most fragile observed, which implies that early peaked light curves indicate more fragile material. Faint meteors have, on average, symmetric light curves, but larger meteoroids (Perseids larger than 10^{-5} kg, for example) have mostly late peaked light curves (Koten et al., 2004). This was interpreted (for example, by Borovička, 2006) as an indication that small meteoroids have completely disrupted before the onset of ablation, while larger meteoroids continue to fragment during ablation.

The average F-parameter may help to distinguish fragile meteoroids from strong ones, but in general light curve shape is a poor predictor of the fragmentation behavior for faint meteors. Subasinghe et al. (2016) expected to find that meteors with long wakes in the narrow field CAMO system (hundreds of meters long) would have symmetric light curves, consistent with crumbling objects, while objects showing little fragmentation would have late-peaked light curves. While the crumbling meteors did indeed, on average, have F-parameters near 0.5 on average, so did meteors with short wakes.

Since most faint meteors are fragile and fragment into smaller pieces, the height to which they penetrate in the atmosphere may not be a useful measure of their strength or structure. The beginning height, however, gives some indication of the energy required for the meteoroid to begin ablating intensively; this was first investigated by Ceplecha (1967). He defined the k_B parameter to characterize the amount of kinetic energy liberated through interaction with the atmosphere required to begin ablation of the meteoroid:

$$k_B = \log \rho_B + \frac{5}{2} \log v_\infty - \frac{1}{2} \log \cos z_R. \tag{2.1}$$

Here ρ_B is the atmospheric density at the beginning of the observed trail (in g cm^{-3}), v_∞ is the pre-deceleration speed (cm s^{-1}), and z_R is the zenith angle. A histogram of this parameter produced two peaks in Super-Schmidt camera data, which Ceplecha called A and C, while the region between the peaks was called group B. Later, using small-camera and intensified video meteors, he refined the classification, given in the following text (Ceplecha, 1988). The measured beginning height depends on the sensitivity of the system, so the same meteor will have a different k_B parameter when observed on different systems. The group limits given in the following text are for Super-Schmidt cameras: the less-sensitive small-camera systems should subtract 0.30 from their calculated k_B parameters, while more sensitive intensified video systems should add 0.15.

- Asteroidal Meteors: $8.00 \leq k_B$; ordinary chondrites
- Group A: $7.30 \leq k_B < 8.00$; carbonaceous chondrites (comets or asteroids)
- Group B: $7.10 \leq k_B < 7.30$; Dense cometary material
- Group C: $6.60 \leq k_B < 7.10$; Regular cometary material
- Group D: $k_B < 6.60$; Soft cometary material

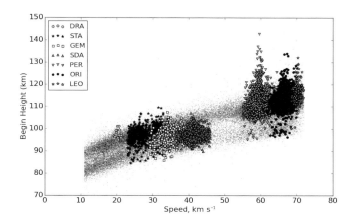

Figure 2.5. Plot of begin height against speed for meteors observed with the CAMS system, with shower meteors marked.

Meteors of group B have perihelion distances $q \leq 0.30$ AU.

The effect of the zenith angle on the k_B parameter is small compared to the initial speed, so the effect on the k_B parameter can be seen in a plot of begin height against initial speed. Meteors generally fall in two groups, corresponding to the two peaks in the k_B histogram, which on the begin height – speed plot are two curved lines. The upper line, at higher begin heights, consists of meteors from Ceplecha's C groups, and the lower group is the meteors in Group A, which are presumably stronger or more refractory. Figure 2.5 shows a begin height – speed plot for CAMS (Cameras for Allsky Meteor Surveillance) data (Jenniskens et al., 2016a, one of the largest datasets of faint meteors), with showers marked. Most Halley-type comet showers, like the Perseids, Orionids and Leonids, are members of Group C (a few, the highest in the plot, are group D), as are Southern Taurids from comet 2P/Encke. The Draconids, with a Jupiter family comet parent, belong to Group D. The more refractory Geminids fall between the two main sporadic groups, in Ceplecha's Group B, and the same is true for Southern δ Aquariids. Both of the latter showers have low perihelion distances.

Jenniskens et al. (2016b) points out that Ceplecha's k_B parameter assumes that the temperature at the surface of the meteoroid depends on $v_\infty^{2.5}$, while the change in begin height with speed seems to follow v_∞^2. They formulate a k_c parameter:

$$k_c = H_b + (2.86 - 2.00 \log v_\infty)/0.0612, \qquad (2.2)$$

where H_b is the begin height in km, and v_∞ is the speed in km s^{-1}. This parameter produces a cleaner separation of the A and C groups in the begin height – speed plot, particularly at lower speeds.

The k_B or k_c parameters are a more reliable measure of the friability of meteors than the light curve shape. For example, Campbell-Brown (2015) found a class of slow, faint meteors that have very low beginning heights, but that also have very small F-parameters, and therefore early peaks; the light curve shapes imply that these are fragile, but the low beginning heights imply the reverse. Borovička et al. (2005) had previously observed a number of meteors with these characteristics with a spectral system; iron was the only meteoroid line in their spectra. Čapek and Borovička (2017) explained the early peaks as the melting and oxidation of the iron combined with spraying of iron droplets.

Inferring the bulk density of meteoroids would be a better way to examine their structure, since we expect most primitive meteoroids have similar grain densities and the bulk density will therefore be a function of their porosity. In principle, the density can be determined by modeling the light curve and deceleration of the meteoroid, and varying the density until a good match is obtained, the most important variable being the ratio of surface area to mass, which determines the deceleration. The difficulty with this approach is that fragmentation also changes the surface area, so the way in which the meteoroid is assumed to fragment will have a strong effect on the density obtained. For example, Babazhanov (2002) and Bellot Rubio et al. (2002) both modeled an identical set of Super-Schmidt photographic meteors to find their densities. Bellot-Rubio et al. were able to fit nearly three quarters of the meteors assuming the meteoroid did not fragment at all, while Babadzhanov assumed all of the meteoroids underwent quasi-continuous fragmentation. In general, the densities found with the single-body model were lower than those found assuming fragmentation was occurring, as one would expect.

Absolute densities, in the absence of detailed information about fragmentation, are problematic, but relative densities of shower meteors are likely useful, providing there are not significant differences in the fragmentation behavior between different types of meteoroids. Subasinghe et al. (2016) found that there was no significant difference in the morphology of meteoroids in different classes of cometary and asteroidal orbits in the CAMO high-resolution observations, so this assumption seems reasonable. Both Babazhanov (2002) and Bellot Rubio et al. (2002) found that Geminid meteors had the highest densities, and that Perseid densities were much lower.

Another potential problem with density measurements is the fact that most modeling is done by trial and error: when a good match to the observations is achieved, the modelling stops. There may, however, be many combinations of parameters which will match the data. To address this, Kikwaya et al. (2011) modeled over 100 CAMO influx meteors with hundreds of thousands of combinations of parameters for each meteor. They found a best-fit density, but also a range of possible densities for each one. They found that meteors with long-period, high-inclination orbits had densities of around 1,000 kg m^{-3}, while meteors in both asteroidal and Jupiter family comet orbits had much higher densities of around 4,000 kg m^{-3}. This work, however, was done before high-resolution meteor observations were possible, so the uncertainty due to fragmentation remains.

Borovička et al. (2007) inferred densities of about 300 kg m^{-3} for four 5–10 mm-sized Draconid meteoroids. The meteoroids were found to be consistent with porous aggregates of grains of sizes 20–100 μm, which were gradually released during the first halves of the meteor trajectories; the precise size of the grains is likely model dependent. Borovička et al. (2014) studied eight larger Draconid meteoroids and confirmed their porous nature but also found significant differences in the fragmentation behavior and thus the internal structure of meteoroids within the same stream. Some of the meteors showed smooth and flat light curves, large decelerations, long wakes, and early release of sodium. All these are indications of complete and quick disintegration into small grains. Other meteors showed smaller decelerations and wakes, presence of Na along the whole trajectory, and terminal flares. The analysis revealed that bulk densities were not significantly different from the former case and grain release started at similar heights, but it was slower and part of the meteoroids disrupted abruptly at the end.

2.3.3 Composition of Small Meteoroids

For studies of meteoroid composition, spectral lines of meteoric origin are relevant. Unfortunately, lines of only a few chemical elements can be typically measured in spectra of ordinary meteors produced by small meteoroids. The reason is partly natural and partly instrumental. Under the physical conditions of the radiating plasma from small meteoroids (temperature, column density), there are indeed only a few dominating lines in the visual region (450–650 nm). These are the Na I doublet at 589 nm (multiplet 1 according to Moore, 1945) and the Mg I triplet at 518 nm (multiplet 2), which are supplemented by somewhat fainter but more numerous lines of Fe I at 527–545 nm (multiplet 15). There are other bright lines in the blue part of the spectrum below 450 nm, but their study is hampered by low resolution and often also by low sensitivity in the blue region of video spectrographs used for observations of ordinary meteors. Most lines in the blue region belong to Fe. The Ca I line at 423 nm or Ca II lines at 393 and 397 nm can be measurable, while Cr I (425–430 nm), Mn I (403 nm), and Mg I (383 nm) are mixed with Fe. The infrared part above 650 nm contains mostly lines of atmospheric origin (O, N, and N_2 bands). Although oxygen is present in meteoroids as well, a comparison of intensities of O and N lines shows that the O/N ratio corresponds to atmospheric composition and the observed oxygen is therefore dominantly atmospheric. Meteoric lines which may be observable in infrared include another Na I doublet at 819 nm and a K I line at 770 nm (the other K I line at 776 nm overlaps with atmospheric absorption by O_2).

The paucity of spectral lines prevents the computation of the temperature and composition of radiating plasma around small meteoroids directly. Nevertheless, the intensities of observable lines reflect to some extent the abundances of corresponding elements. Borovička et al. (2005) computed theoretical line intensities of Mg I (multiplet 2), Na I (1) and Fe I (15) for a reasonable range of temperatures (3500–5500 K), sizes and densities of a radiating plasma cloud corresponding to meteors of normal brightness (so that the plasma is not optically too thick, as is the case in fireballs) and composition of CI chondrites. The resulting multiplet intensity ratios (sum of all lines of the given multiplet) were plotted in a Mg-Na-Fe ternary diagram. The Na/Mg intensity ratio proved to depend strongly on temperature because the Na line has a much lower excitation potential than the Mg line, and because Na has a low ionization potential, and is therefore highly ionized at higher temperatures. By plotting the observed Na/Mg intensity ratio as a function of meteor velocity, it was found that the ratio decreases with velocity up to about 40 km s^{-1} and then remains constant. The temperature dependence is therefore reflected in the velocity dependence.

When theoretical intensity ratios for CI composition are plotted in the ternary diagram, they form a relatively well-defined band. Meteoroids with chondritic composition should therefore

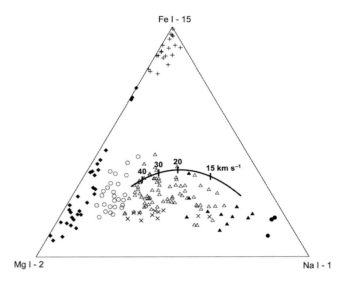

Figure 2.6. Ternary diagram of intensity ratios of selected multiplets of Mg, Na, and Fe in meteor spectra. The curve with velocity marks is the chondritic line, where spectra of meteoroids with chondritic composition should be spread according to their entry velocities. The symbols show actual meteor measurements from Borovička et al. (2005) and Vojáček et al. (2015). Meteors were classified as Normal (empty triangles), Na poor (empty circles), Na free (filled diamonds), Na enhanced (filled triangles), Na rich (filled circles), Fe poor (crosses), and Irons (plus signs).

scatter around the center of the band, which can be called the chondritic line. The position along the line will depend on entry velocity. The chondritic line is plotted in Figure 2.6 together with actual data from 179 observed meteors taken from Borovička et al. (2005) and Vojáček et al. (2015). Line intensities were integrated along meteor trajectories. We can see that many meteors occupy other regions of the diagram than those expected for chondritic composition. Small meteoroids, especially sporadic ones, are chemically surprisingly diverse, even if we take into account only three elements. The most notable groups are Irons, Na free, Na rich, and Mainstream meteoroids; these groups are described in the following paragraphs.

2.3.3.1 Irons

The spectra are dominated by Fe lines. Na is absent and Mg is absent or faint – it is difficult to say because some Fe lines overlap with the Mg line at 518 nm. Meteors classified as Irons also have other common characteristics. They begin lower than other meteors of the same velocity. The light curve is often, though not always, strongly asymmetric with a rapid rise and slower decline after maximum. The trajectories are typically quite short.

Iron meteoroids are almost certainly identical with the population of small refractory meteoroids detected by Campbell-Brown (2015). One might associate them with iron meteorites. However, here we deal with the population of very small meteoroids. They represent a significant fraction of sporadic meteors fainter than magnitude +3 (Campbell-Brown, 2015) but their proportion decreases rapidly with increasing meteor brightness.

While Irons form more than 10% in our sample in Figure 2.6, they are completely absent in a sample of 35 spectra of meteors brighter than zero magnitude (Rudawska et al., 2016). Iron meteorites, or at least the large ones, are therefore likely produced by a different population of large iron meteoroids. We suspect that small iron meteoroids may be impact melt products of asteroid collisions while large iron meteorites are fragments of differentiated asteroids, although some small iron meteorites (of type IAB and ungrouped) could be impact melts as well (Wasson, 2011).

2.3.3.2 Na-Free Meteoroids

Another fraction of more than 10% of sporadic meteors is formed by Na-free meteoroids. The spectra contain both Mg and Fe, but the Na line is missing. These meteoroids are found on two types of orbits. The first group have orbits with perihelion distances smaller than 0.2 AU. The loss of volatile sodium can be ascribed to thermal desorption in the vicinity of the Sun. The process was studied by Čapek and Borovička (2009) and was found to work on timescales of thousands of years provided that millimeter-sized meteoroids have some porosity. The nonporous parts (grains) have to be smaller than ~ 100 μm.

The second group of Na-free meteoroids resides in cometary orbits (Halley type) with high inclinations and perihelia far from the Sun. It was proposed (Borovička et al., 2005) that they are part of the cometary crust and the loss of sodium was induced by cosmic ray exposure in the Oort cloud for billions of years. However, so far this process has not been studied in detail.

Similarly to Irons, Na-free meteors exhibit lower beginning heights than average meteors. This fact has been interpreted as evidence of their higher mechanical strength.

2.3.3.3 Mainstream Meteoroids

In contrast to Irons, Na-free meteoroids do not form an isolated group in the ternary diagram. There is a smooth transition from Na-free to Na-poor and to Normal meteoroids. Normal meteoroids have nearly chondritic composition, though it seems that on average they have a somewhat lower Fe content. Meteoroids with very low Fe line intensity were classified as Fe-poor, but the boundary is somewhat arbitrary. According to their orbits and high beginning heights, which suggests fragility, Fe-poor meteoroids are likely of cometary origin. This is in accordance with Mg-rich cometary dust such as is seen in comet 1P/Halley.

There are meteoroids with a somewhat brighter Na line than expected, and they were classified as Na-enhanced. Not only must the position in the ternary diagram be taken into account, but also the entry speed. Again, the boundary with Normal meteoroids is somewhat arbitrary.

Together, the Normal, Na-poor, Na-enhanced, and Fe-poor meteoroids are called Mainstream meteoroids. Their composition does not differ much from chondritic composition.

2.3.3.4 Na-Rich Meteoroids

The spectra of Na-rich meteoroids are dominated by the Na line. They are relatively rare; there are only three cases in the sample in Figure 2.6. Unfortunately, the orbits of two of them are unknown; the third one had a Jupiter family orbit. Borovička et al. (2008) studied in detail another meteor, which was classified only as Na-enhanced, but also had a Jupiter family orbit and low inclination. The K line was observed in the spectrum. Mg and especially Fe were found to be depleted (relatively to CI composition) in comparison with both alkali metals. One may speculate that that meteoroid and perhaps all Na-rich meteoroids are somewhat related to the salts, which were detected to be present on Ceres (De Sanctis et al., 2016), Io (Lellouch et al., 2003), or Enceladus (Postberg et al., 2011).

2.3.3.5 The Classification of Shower Meteoroids

Most meteoroids from major showers have mainstream composition. The exceptions are the Southern δ Aquariids and Geminids (Vojáček et al., 2015). The Southern δ Aquariids were found to be Na-free, which is not surprising owing to their low perihelion distance ($q = 0.07$ AU). Sodium was, nevertheless, found in two bright Southern δ Aquariid meteors (Rudawska et al., 2016). Similarly, Geminids ($q = 0.14$ AU) have a wide range of Na content from Na-free to Normal with a trend of Na increasing with meteor brightness (Borovička, 2010). Borovička et al. (2010) concluded that Na content depends on pore sizes in Geminid meteoroids.

Quadrantids are mostly Na-poor (Koten et al., 2006; Borovička et al., 2010; Madiedo et al., 2016) despite the fact that the perihelion distance is currently large. It was, nevertheless, low 1500 years ago (Kasuga and Jewitt, 2019, see Section 8.3.2). The depletion of Na suggests that Quadrantid material was exposed to solar heat at that time and not hidden deep inside the parent body. Since the Quadrantid stream is believed to be much younger than 1500 years (Abedin et al., 2015), it probably means that the material that now forms the stream was originally located near the surface of the parent asteroid (196256) 2003 EH_1.

Some meteoroids of Halley-type streams, such as the Perseids and Leonids, were classified as Fe-poor, which corresponds with their cometary origin. Draconids were found to have normal compositions despite differences in their physical structure (Borovička et al., 2014). Also, Taurids have mostly normal composition; nevertheless, the content of both Na and Fe varies widely from meteor to meteor (Matlovič et al., 2017).

2.3.4 Physical Properties of Large Meteoroids

The fact that there are large meteoroids with very different physical properties became obvious from large differences in penetration depths of fireballs into the atmosphere. On the basis of fireball end heights, Ceplecha and McCrosky (1976) defined the P_E parameter, which is still used today:

$$P_E = \log \rho_E - 0.42 \log m_\infty + 1.49 \log v_\infty \\ - 1.29 \log \cos z_R, \qquad (2.3)$$

where ρ_E is the density of the atmosphere at the end of the fireball luminous trajectory in g cm^{-3}; v_∞ is the fireball entry speed in km s^{-1}; z_R is the zenith distance of the radiant (i.e. the

Table 2.5. *Classification of fireballs.*

Type	I	II	IIIA	IIIB
P_E	< -4.60	< -5.25	< -5.70	≥ -5.70
A_L	> 5.36	> 4.13	> 3.18	≤ 3.18
σ	< 0.025	< 0.075	< 0.15	≥ 0.15
Group	ast?	A	C	D
ρ_m	3,500	2,000	750	270

For definitions of P_E and A_L see Equations (2.3) and (2.5), respectively; σ is the apparent ablation coefficient in s^2 km^{-2} or kg MJ^{-1}; ρ_m is assigned typical bulk density in kg m^{-3}.

angle between the trajectory and vertical direction); and m_∞ is the initial mass of the meteoroid determined from the light curve:

$$m_\infty = 2 \int \frac{I\,dt}{\tau v^2} + m_E. \qquad (2.4)$$

Here $I = 10^{-0.4M}$ is fireball luminosity (M is fireball absolute magnitude); τ is luminous efficiency, which is a function of velocity, v; t is time; and m_E is terminal mass (usually negligible). Note that the original luminous efficiency from Ceplecha and McCrosky (1976) must be used when computing P_E, although it is now known to be underestimated (and produces therefore unrealistically high m_∞). To avoid problems with luminous efficiency, Ceplecha (1988) gives also a nearly equivalent parameter A_L:

$$A_L = 2\log(\rho_E / \cos z_R) + 5 \log v_\infty - 0.83 \log(S), \qquad (2.5)$$

where S is the integral of luminosity along the trajectory, $S = \int I\,dt$, and ρ_E is to be substituted in g m^{-3} (!).

Fireballs were classified into four types defined by Ceplecha and McCrosky (1976): I, II, IIIA, and IIIB. The values of the criteria are given in Table 2.5. Type I was identified with stony material (ordinary chondrites in particular), type II with carbonaceous chondrites, type IIIA is called regular cometary material, and type IIIB soft cometary material. The corresponding classification of small meteoroids (Section 2.3.2.2) and estimated typical bulk densities of each type are given as well (but most 'ast' meteoroids may in fact be Irons). There is no direct correspondence between physical classification and orbit. Type IIIB fireballs can be encountered on asteroidal orbits.

We caution that the classification scheme in Table 2.5 was developed for 'ordinary' fireballs of magnitudes between about −5 and −15. Misleading results can be obtained when applied to much brighter events (superbolides). They must be analyzed individually. Note also that iron meteoroids do not fill well into the scheme – they usually fall into the IIIA category.

The penetration depth obviously depends on the ablation rate. Fireball classification can therefore also be done, even more rigorously, on the basis of the *apparent* ablation coefficient, σ, computed from the dynamics of the fireball (for values of σ see Table 2.5). The disadvantages are that good dynamical data must be available for the whole trajectory, the fireball must show appreciable deceleration (which is seldom the case for fireballs disappearing high in the atmosphere), and the computation is much more complex (see e.g. Ceplecha et al., 1993).

More detailed analysis of fireball data (dynamics, light curves) revealed that differences in penetration ability cannot be explained simply by differences in ablation coefficients and bulk densities of meteoroids. The main factor is atmospheric fragmentation. Ceplecha and ReVelle (2005) found that in the absence of fragmentation, the *intrinsic* ablation coefficient (due solely to melting and evaporation) is nearly the same for all fireball types and is as low as 0.005 s^2 km^{-2}. Fragmentation is therefore the main mechanism of mass loss for all fireball types. An excellent example of severe atmospheric fragmentation in an ordinary chondrite is the Morávka meteorite fall, where individual fragments could be studied on a casual video record (Borovička and Kalenda, 2003).

In contrast to small cometary meteoroids, where fragmentation is probably a thermal process, large meteoroids fragment mechanically under the action of pressure differences on their front and back side. Without going into detail of this complicated process, meteoroid strength can be roughly defined by the dynamic pressure acting at the moment of fragmentation, given by $p = \rho v^2$, where ρ is atmospheric density and v is meteoroid velocity at the fragmentation point (the drag coefficient should be also included in the equation but since it is nearly the same for all fireballs and of the order of unity, it is usually ignored). Determining the strength of a meteoroid and its parts requires revealing fragmentation points along the fireball trajectory using fireball dynamics, radiation, or morphology. For details of fragmentation modeling, see Popova et al. (2019), i.e. Section 1.4.4.

The results of fragmentation modeling of seven observed falls of ordinary chondrites, one fall of a carbonaceous chondrite, two cometary fireballs and one unusual fireball are compared in Figure 2.7. There is an interesting pattern of two stage fragmentation for ordinary chondrites. The first disruptions usually occur at pressures 0.04–0.12 MPa. After a quiet period, second stage disruptions start at 1–3 MPa. Fragmentation can then continue at higher pressures or even after the pressure has started to decrease because of deceleration (see e.g. Borovička and Kalenda, 2003). These results also confirm the earlier results obtained using a more heterogeneous set of data for meteorite falls by Popova et al. (2011), namely that the bulk strength of meteoroids is much lower than the strength of the meteorites they produce (see Table 2.2 for meteorite strengths) and that the meteoroid strengths do not depend on meteoroid mass, which is at odds with the widely-used Weibull theory (Weibull, 1951).

The reason for the low strengths of stony meteoroids is probably internal cracks acquired during collisions in interplanetary space. Fireball data suggest that the strengths of cracks are not random but cumulate around two values, ∼0.08 MPa and ∼2 MPa. Only Žďár nad Sázavou disrupted at only 0.016 MPa, which can be possibly ascribed to low petrologic type and high microporosity of that body. In rare cases, such as the Carancas crater forming event (e.g. Brown et al., 2008; Tancredi et al., 2009), cracks can be absent. The Carancas H4-5 meteoroid, with an estimated initial mass of several tons, survived dynamic pressures of at least 20 MPa without fragmentation (Borovička and Spurný, 2008). The largest well-documented impacting body, Chelyabinsk (LL5, 10,000 tons, 19 m size; Brown et al., 2013a; Popova et al., 2013), started to disrupt at 0.7 MPa, severe disintegration occurred at 1–5 MPa, and only a small part of the original body had a strength of 15–18 MPa (Borovička et al.,

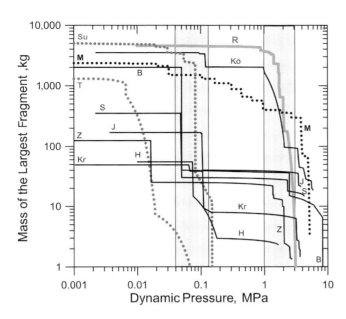

Figure 2.7. Results of fragmentation modeling of selected fireballs. The mass of the largest surviving fragment is plotted as a function of increasing dynamic pressure. Ordinary chondrite meteorite falls are plotted with dark solid lines: B = Benešov (LL3.5+H5), H = Hradec Králové (LL5), J = Jesenice (L6), Ko = Košice (H5), Kr = Križevci (H6), S = Stubenberg (LL6), Z = Žďár nad Sázavou (L3). The others are two cometary superbolides, Su = Šumava and T = Taurid EN311015, the carbonaceous chondrite meteorite fall M = Maribo (CM2), and R = the unusual Romanian superbolide of 2015 January 7. The ranges of pressures, where ordinary chondrites usually fragment, are grayed. Model results are from Borovička (2016) for B, Spurný et al. (2010) (slightly revised) for J, Borovička et al. (2013a) for Ko, Borovička et al. (2015b) for Kr, Borovička and Spurný (1996) for Su, and Borovička et al. (2017) for R, T, and M. Models for H, S, and Z have not been published, yet.

2013b). There are other examples of meteoroids which started to fragment only at pressures ≥ 1 MPa (Borovička and Spurný, 2008; Popova et al., 2011).

The question of whether some meter-sized meteoroids can be gravitational aggregates (rubble piles) with nearly zero strength was discussed by Borovička (2016). The conclusion was negative for Benešov and Chelyabinsk. In principle, however, a disruption into several similarly sized fragments under low pressures at high altitudes would be difficult to recognize from fireball data.

The carbonaceous Maribo fragmented all the way along its trajectory, but part of it survived pressures of up to ∼5 MPa (Figure 2.7). This confirms the conclusion of Popova et al. (2011) about Tagish Lake that carbonaceous chondrites reach the same strength as ordinary chondrites despite essential difference in composition. The strength of carbonaceous bodies is, however, governed by their primordial structure rather than cracks, since the fragmentation strength approaches the tensile strength of carbonaceous (CI, CM) meteorites (Slyuta, 2017).

The two large cometary bodies (type IIIB) in Figure 2.7 were both destroyed completely at pressures of about 0.1 MPa. These were likely low-density, high-porosity bodies which disintegrated into mm-sized dust at atmospheric entry. For the Taurid, the disintegration was 99% finished at 0.02 MPa. Smaller members of the same stream were, nevertheless, classified as type II, and some even as type I (Spurný et al., 2017). The Taurid material (from comet 2P/Encke) therefore contains more compact parts on the centimeter scale, though dynamic pressures they encountered did not exceed several tenths of MPa. Borovička and Jenniskens (2000) observed a Leonid meteoroid (from comet 55P/Tempel-Tuttle) that disintegrated at ∼0.1 MPa but one mm-sized fragment emerged and survived up to 2 MPa. Structural inhomogeneity of cometary material was confirmed by Kokhirova and Borovička (2011), who reported one type I fireball among ten Leonids. Some cometary meteoroids spontaneously disrupt under zero pressure in interplanetary space (Watanabe et al., 2003; Koten et al., 2017). In contrast was the compact sporadic Karlštejn type I fireball on a cometary orbit (Spurný and Borovička, 1999a), which was Na-free and probably represented part of a cometary crust.

Geminids behave differently than cometary showers in the sense that deep penetration is not an exception and large meteoroids are more compact and probably more dense than small ones (Borovička et al., 2010). Even meteorite dropping from a Geminid is possible (Spurný and Borovička, 2013; Madiedo et al., 2013, although the latter data seem to be not very reliable). Although possible meteorite-dropping Taurids have been reported as well (Brown et al., 2013b; Madiedo et al., 2014), in the case of Taurids there is a danger of contamination by sporadic fireballs.

An interesting case was the 2015 Romanian superbolide (Borovička et al., 2017), which survived almost intact until 1 MPa and then disintegrated rapidly, so that no macroscopic fragments larger than a few grams remained. It seems that 1 MPa was the intrinsic material strength. This is almost two orders of magnitude lower strength than for uncracked ordinary chondrites and almost two orders of magnitude higher strength than for similarly sized Taurid material. Since its orbit was asteroidal, it seems that the Romanian bolide represents a new type of weak but structurally homogeneous asteroidal material, which is not represented in meteorite collections.

2.3.5 Composition of Large Meteoroids

In this section we will review the results of spectroscopy of bright fireballs obtained with high resolution spectrographs providing multi-line spectra. The identification of major spectral lines has been known for a long time (Millman, 1935). Such spectra allow us to compute also abundances of various elements in the radiating gas at various points along the fireball trajectory. The method was described e.g. in Borovička (1993). The principles are the same as in the subsequently invented Laser Induced Breakdown Spectroscopy (LIBS), which among many other applications was used to measure the chemical composition of meteorites (De Giacomo et al., 2007). In both cases the excitation temperature and elemental abundances are determined from the intensities of emission lines in the spectrum. While in the case of LIBS the sample is ablated by a laser beam, in the case of fireballs the meteoroid is ablated by high velocity interaction with the atmosphere. This is a source of potential complications since the conditions change along the trajectory as the meteoroid penetrates to deeper atmospheric layers, and also from fireball to fireball depending on entry velocity, trajectory slope and meteoroid size.

Borovička (1994a) showed that spectra of fireballs contain two components. The main component is always present and is produced at excitation temperatures typically between 3,500–5,500 K. It contains mostly lines of neutral metals such as Fe I, Mg I, Na I, Ca I, Cr I, Mn I. In good spectra faint lines of Ti I, Ni I, Co I, or Li I can also be present. Si I and Al I have one or two lines, respectively, which can be relatively bright but are located close to other lines, and good spectral resolution is therefore needed to measure them. A K I line is present in the near infrared. Because of the low ionization potential of Ca and low excitation potential of Ca II lines, Ca II lines are also present in the main component. The second component is produced at excitation temperatures of about 10,000 K and is much stronger in fast fireballs than in slow fireballs (and is absent in faint meteors). It contains lines of Ca II (which are present in both components), Mg II, Si II, Fe II, and sometimes neutral H I. The second component is probably connected with the meteor shock waves. It is not quite clear to which components the atmospheric lines (N I and O I) belong. They need high excitation but are present also in faint (fast) meteors. Important elements that cannot be observed in ground-based meteor spectra are carbon and sulfur. Their strongest lines lie in the ultraviolet.

As already shown by Borovička (1993), relative abundances of elements vary along the fireball trajectory. An obvious effect is incomplete evaporation of refractory elements (Al, Ca, Ti). It follows from the spectra that these elements are ablated mostly in solid or liquid form (as small particles or droplets) and their abundances in the meteoroid cannot be determined directly from the abundances in the radiating gas. The evaporation is more complete in deep atmospheric layers (Borovička and Spurný, 1996) or just after meteor flares (Borovička, 2005). The abundances of other elements (Cr, Na) were found to vary as well, though to a lesser extent, so there may be other unresolved effects which influence gas composition. Other difficulties may arise from the overlapping of different spectral components, including the wake and afterglow radiation, which is not in thermal equilibrium (Borovička and Jenniskens, 2000).

Figure 2.8 compiles some abundances derived from fireball data. Abundances are presented as deviations from CI composition, normalized to Fe. Shower meteoroids are compared with two comets with in situ measured dust compositions, and sporadic meteoroids are compared with LL and EH chondrites. Perseids and Leonids, two showers on Halley-type orbits, show hints of enhancements of Na, Mg, and Si and depletion of Cr. The same trend, except for Cr depletion, is visible for 1P/Halley. Comet 67P/C-G does show depletion in Cr but is even more depleted in Mg, which is not observed in any fireball. The depletion of refractory Al and Ca in fireballs is caused by incomplete evaporation and only by chance nearly corresponds to 67P/C-G. The Al and Ca abundances in Figure 2.8 were derived from the main spectral component. The lines of ionized Ca are very bright in Perseid and Leonid fireballs and, in fact, suggest Ca overabundance (Borovička, 2004). It seems therefore

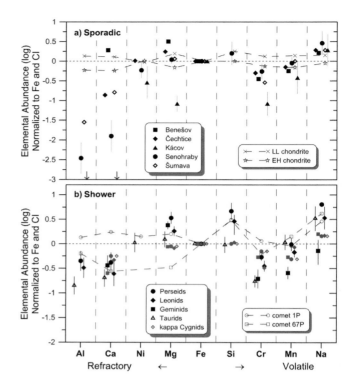

Figure 2.8. Relative abundances of nine chemical elements as derived from the spectra of selected fireballs. Abundances are plotted as deviations (in logarithmic scale) from CI composition, normalized to iron. In panel **a)** data for sporadic fireballs Benešov (average for heights 25–21 km), Čechtice (45 km), Kácov (average of whole trajectory), Senohraby (42 km), and Šumava (62 km) are given. The data are taken form Borovička and Spurný (1996) and Borovička (2005). Values for LL and EL chondrites (Table 2.1) are plotted for comparison. In panel **b)** data for fireballs from five meteor showers are given. Data for black symbols were taken from Borovička (2007) and references therein; data for smaller gray symbols are from Trigo-Rodriguez et al. (2003). In some cases data for several fireballs of the same shower were averaged. In situ measured dust compositions of two comets (Table 2.4) are plotted for comparison.

that Perseids and Leonids deviate from CI composition in a similar manner as 1P dust. The data must be, however, taken with caution since Trigo-Rodriguez et al. (2003) obtained nearly CI ratios of Mg, Fe, and Si, and Jenniskens (2007) obtained Mg/Fe ratio < 1 for two Leonids. It is not clear if differences between individual meteoroids or differences in approaches of individual authors are responsible for these different results.

The differences in Na content between individual Geminids are surely real and can be explained by a different degree of thermal desorption of Na in the vicinity of the Sun (see also Section 2.3.3). The same effect may affect the moderately volatile Mn. The Mg/Fe ratio in Geminids seems to be similar to that in Leonids and Perseids, although the orbit and parent body of the Geminids is asteroidal. Taurids, on the other hand, have nearly CI composition, except for an enhancement in Na and depletion of Cr. It is, however, not clear if Cr depletion (observed to some degree in all fireballs) is intrinsic to the meteoroid, since Cr variations were observed along meteor trajectories (Borovička,

1993, 2005). An effect of incomplete evaporation may be at work.

Sporadic fireballs in Figure 2.8 include the extremely bright Šumava fireball of fragile nature (type IIIB), which may be related to the Taurid complex. The abundances indeed show the same trend as the Taurids. Other fireballs were asteroidal; Benešov dropped LL and H chondrites (Spurný et al., 2014). The high abundance of Mg computed from the Benešov spectrum suggest that the LL composition probably prevailed in the meteoroid. Note that abundances were determined below heights of 25 km, where complete evaporation was reached, and no Ca depletion was observed in this part of Benešov's spectrum. Other studied fireballs were not luminous down to these low heights. According to the Mg abundance, Čechtice may be a LL chondrite as well. Senohraby is closer to CI or H composition (see Table 2.1) in this respect; Na abundance seems, however, to be relatively high.

In the Mg-Fe-Na classification of small meteoroids (Section 2.3.3), all fireballs mentioned previously would fall into the mainstream group, either Normal or with a tendency to be Fe-poor or Na-enhanced. In general, compositions far from chondritic are rare among fireball spectra. This is consistent with the prevalence of chondrites among meteorite falls (Table 2.2) and with nearly chondritic composition of cometary dust. As expected, iron meteoroids are occasionally observed (Halliday, 1960; Ceplecha, 1966). Among 70 fireball spectra recorded by the European Fireball Network in 2017, two showed iron compositions (unpublished data). An anomaly is the Kácov fireball, which shows strong depletion in Mg and Ca (Figure 2.8; the measured value of log[Ca/Fe] = −5.4 is off the scale). Such an Mg-poor composition has not been observed among small meteoroids so far and does not correspond to any known meteorite type. The possibility that the results are influenced by the low entry speed of the fireball (13.9 km s^{-1}) must, however, be verified. Other non-chondritic fireballs include the Na-free Karlštejn fireball on a retrograde orbit (Spurný and Borovička, 1999b) and a Na-poor probable diogenite in 3:1 resonance with Jupiter (Borovička, 1994b).

The examples of Benešov and especially Almahata Sitta (ureilite+EH+EL+H+L+LL+CB+R; Shaddad et al., 2010; Bischoff et al., 2010) meteorite falls showed us that one meteoroid can in fact be a mixture of more meteorite types. Another case was the Galim meteorite fall that contained LL and EH specimens (Rubin, 1997). If the bulk compositions of all types are similar or one type forms just a minority, we cannot expect to reveal such cases from fireball spectra. Meteorite recoveries and their analyses in the laboratory are needed. However, for a long time the paradigm has been that all meteorites from one fall must be of the same type, so only one or a few meteorites were analyzed from each fall. If meteorites of different types were found within the same strewn field, they were assumed to originate from different falls. The examples which can be found in the Meteoritical Bulletin Database[1] include Gao-Guenie + Gao-Guenie (b) (H+CR), Markovka + Markovka (b) (H+L), or La Ciénega + La Ciénega (b) (H+L). It is possible, though not certain, that we are dealing here with heterogeneous meteoroids or that there were other mixed falls, which have not been recognized. The production of heterogeneous meteoroids by asteroid collision and debris re-accumulation was discussed in connection with Almahata Sitta by Horstmann and Bischoff (2014) and Goodrich et al. (2015).

2.4 Summary

We have described the current knowledge of interplanetary materials as studied by different techniques. Figure 2.9 summarizes the known materials on a size – density chart. The best studied materials are meteorites, which represent sizeable samples suitable for detailed analysis by laboratory techniques. They provide information about various types of asteroids. The information is, however, not complete. Not only are there dormant comets among asteroids as evidenced by IIIB fireballs on asteroidal orbits, but there are also objects, such as the Romanian meteoroid, with material properties neither cometary nor of any known meteorite type. Material properties of Geminids from asteroid (3200) Phaethon, though still not well understood, also lie somewhere between typical asteroids and typical comets. In addition, parent meteoroids differ from the meteorites they produce – at least in the presence of cracks. Fireball data suggest that there are two types of cracks in stony meteorites: weak ones with strengths 0.04–0.12 MPa, which are not always present, and common ones with strengths 1–3 MPa, which are only rarely absent. On the other hand, there is no evidence for almost strengthless rubble piles among meter-sized asteroidal meteoroids. Some meteoroids are re-accumulated debris of various meteorite types but even those have some strength, comparable to the strength of weak cracks. We did not consider friction but it is unlikely that friction between heterogeneous parts with random shapes could hold the body together until pressures of tens of kPa.

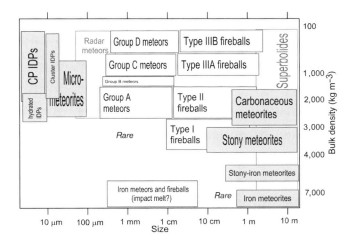

Figure 2.9. Schematic representation of sizes and densities of different types of materials studied in laboratories (gray boxes) and during atmospheric penetration. Size refers to pre-atmospheric size. Meteor detection size limits are only rough because in practice they strongly depend on entry speed.

---

[1] www.lpi.usra.edu/meteor/metbull.php

Typical cometary material has high microporosity on all size scales, from microns to meters. It seems that dust particles are aggregates presenting a hierarchical structure. This structure is, however, very variable, even within one comet, especially on smaller scales up to centimeters with some almost compact aggregates, and highly porous ones with a fractal dimension below 2, which gives constraints on the formation of comets in the early solar system. The properties of cometary dust may have allowed the delivery through the interplanetary cloud of a significant amount of organic-rich particles on the Earth's surface during the Late Heavy Bombardment, as derived from studies of IDPs collected in the stratosphere (Flynn et al., 2004) and from later studies, with emphasis on the composition and physical properties of cometary dust particles provided by the Rosetta mission (Levasseur-Regourd et al., 2018). The question if comets contain CI chondrites is still open, though it looks unlikely, since confirmed cometary materials are anhydrous. But chondrule-like objects may be present.

Interplanetary dust and small meteoroids up to cm-size are dominated by cometary and carbonaceous bodies. Ordinary chondritic material is rare at these sizes, at least in the vicinity of Earth. This statement is based both on orbital and physical studies but we have to warn that deriving physical properties of small meteoroids such as density is still a challenge. It is still to be understood why there is a large fraction of iron bodies among small meteoroids on asteroidal orbits. Independent interplanetary cloud models predict that the maximum mass influx is from small ($< 100 \mu$m) particles originating from Jupiter family comets (e.g. Nesvorný et al., 2010, 2011). Because of their small size and low entry velocity, they are difficult to detect (and thus confirm) even by HPLA radars (Janches et al., 2017).

For compositional studies, the CI composition is a benchmark. Individual types of chondritic meteorites differ a little from that composition, most notably in the Mg/Fe ratio. There are also indications that dust in Halley-type comets has a higher Mg/Fe ratio than CI, while Jupiter family comets and comet Encke have nearly CI composition. Large meteoroids coming from differentiated asteroids naturally have non-chondritic compositions, but they form a minority of bodies. Small meteoroids are chemically more diverse, partly because they are subject to secondary effects such as formation of impact melt iron bodies or loss of volatiles (in particular Na) during close approaches to the Sun or by long-term irradiation by cosmic rays. There are also indications that some small meteoroids are salty objects from the outer asteroid belt or from satellites of major planets.

2.5 Future Work

There is still a lot to understand from existing data and to do in the coming years. Some hypotheses from the previous section need to be verified: impact melts, salty meteoroids, cosmic-ray irradiation effects, differences in cometary Mg/Fe ratios, etc. Atmospheric fragmentation of small meteoroids is still not well understood and the structure of Geminids and Quadrantids remains poorly known. But the perspectives are promising. CCD cameras with high resolution provide better precision in observations than in the past. Tracking systems such as CAMO are particularly valuable for fragmentation studies. Radiometers provide precise light curves of fireballs. Spectroscopy will benefit from DSLR cameras with high resolution and a bit depth of at least 14 bits.

Much attention needs to be devoted to improving data reduction methods. The best possible precision of each individual meteor is needed to reliably study meteoroid properties. Meteor ablation and fragmentation models can be improved using simultaneous data on meteor dynamics, radiation, and morphology. Spectroscopic analysis also needs improvements in order to reveal not only major but also minor differences in meteoroid compositions.

Synthesis with other techniques will continue. It would be useful if IDP and micrometeorite studies were devoted to a wide range of particles, not concentrated mostly on chondritic ones. Meteorites have already been much studied, but there is a deficiency of data on their mechanical strengths. It would be also helpful to characterize as many meteorites as possible from each multiple fall, at least by approximate techniques such as magnetic susceptibility. Laboratory experiments and numerical simulations of meteoroid ablation can help to better understand meteor physics. Other processes relevant to meteoroids, such as cosmic ray dust irradiation, could be also studied in the laboratory (the emphasis so far has been on ices). Finally, ongoing and future missions to asteroids and comets will provide important information about specific objects.

Acknowledgments

JB was supported by grant no. 16-00761S from GA ČR. ACLR acknowledges support from Centre National d'Etudes Spatiales (CNES) in the realization of instruments devoted to space exploration of comets and in the initial part of their scientific analysis. FJMR was supported by NASA LARS grant NNX14AF21G. TK was supported by Academy of Finland project no. 257487 and institutional support RVO 67985831 of the Institute of Geology of the CAS.

References

Abe, S., Mukai, T., Hirata, N. et al. (2006). Mass and local topography measurements of Itokawa by Hayabusa. *Science*, **312**, 1344–1349.

Abedin, A., Spurný, P., Wiegert, P. et al. (2015). On the age and formation mechanism of the core of the Quadrantid meteoroid stream. *Icarus*, **261**, 100–117.

Agarwal, J., Müller, M., Reach, W. et al. (2010). The dust trail of Comet 67P/Churyumov-Gerasimenko between 2004 and 2006. *Icarus*, **207**, 992–1012.

A'Hearn, M.F. (2017) Comets: looking ahead. *Philosophical Transactions of the Royal Society of London, Series A*, **375**, 20160261.

Allègre, C. J., Manhès, G., and Gopel, C. (1995). The age of the Earth. *Geochimica et Cosmochimica Acta*, **59**, 1445–1456.

Anders, E., and Grevesse, N. (1989). Abundances of the elements: Meteoritic and solar. *Geochimica et Cosmochimica Acta*, **53**, 197–214.

Arendt, R. G. (2014). DIRBE comet trails. *The Astronomical Journal*, **148**, 135.

Babadzhanov, P. B. (2002). Fragmentation and densities of meteoroids. *Astronomy and Astrophysics*, **384**, 317–321.

Bardyn, A., Baklouti, D., Cottin, H. et al. (2017). Carbon-rich dust in comet 67P/Churyumov-Gerasimenko measured by COSIMA/Rosetta. *Monthly Notices of the Royal Astronomical Society*, **469**, S712–S722.

Barucci, M. A., Dotto, E., and Levasseur-Regourd, A. C. (2011). Space missions to small bodies: asteroids and cometary nuclei. *The Astronomy and Astrophysics Review*, **19**, 48.

Barucci, M. A., and Fulchignoni, M. (2017). Major achievements of the Rosetta mission in connection with the origin of the solar system. *The Astronomy and Astrophysics Review*, **25**, 3.

Beech, M. (2009). On the shape of meteor light curves, and a fully analytic solution to the equations of meteoroid ablation. *Monthly Notices of the Royal Astronomical Society*, **397**, 2081–2086.

Bellot Rubio, L. R., Martínez González, M. J., Ruiz Herrera, L. et al. (2002). Modeling the photometric and dynamical behavior of Super-Schmidt meteors in the Earth's atmosphere. *Astronomy and Astrophysics*, **389**, 680–691.

Bentley, M. S., Schmied, R., Mannel, T. et al. (2016). Aggregate dust particles at comet 67P/Churyumov-Gerasimenko. *Nature*, **537**, 73–75.

Bischoff, A., Horstmann, M., Pack, A., Laubenstein, M., and Haberer, S. (2010). Asteroid 2008 TC_3—Almahata Sitta: A spectacular breccia containing many different ureilitic and chondritic lithologies. *Meteoritics and Planetary Science*, **45**, 1638–1656.

Bland, P. A., Towner, M. C., Sansom, E. K. et al. (2016). Fall and recovery of the Murrili meteorite, and an update on the Desert Fireball Network. *79th Annual Meeting of the Meteoritical Society*, abstract #6265.

Boesenberg, J. S., Delaney, J. S., and Hewins, R. H. (2012). A petrological and chemical reexamination of Main Group pallasite formation. *Geochimica et Cosmochimica Acta*, **89**, 134–158.

Borovička, J. (1993). A fireball spectrum analysis. *Astronomy and Astrophysics*, **279**, 627–645.

Borovička, J. (1994a). Two components in meteor spectra. *Planetary and Space Science*, **42**, 145–150.

Borovička, J. (1994b). Meteor spectra – possible link between meteorite classes and asteroid families. In: Y. Kozai, R. P. Binzel, T. Hirayama, eds., *75 Years of Hirayama Asteroid Families: The Role of Collisions in the Solar System History*, Astronomical Society of the Pacific Conference Series, **63**, 186–191.

Borovička, J. (2004). Elemental abundances in Leonid and Perseid meteoroids. *Earth Moon and Planets*, **95**, 245–253.

Borovička, J. (2005). Spectral investigation of two asteroidal fireballs. *Earth Moon and Planets*, **97**, 279–293.

Borovička, J. (2006). Physical and chemical properties of meteoroids as deduced from observations. Pages 249–271 of: Lazzaro, D., Ferraz-Mello, S., and Fernández, J. A. (eds), *Asteroids, Comets, Meteors. Proceedings of the 229th Symposium of the International Astronomical Union held in Búzios, Rio de Janeiro, Brasil, 7–12 August 2005*. Cambridge, UK: Cambridge University Press.

Borovička, J. (2007). Properties of meteoroids from different classes of parent bodies. Pages 107–120 of: Milani, A., Valsecchi, G. B., and Vokrouhlicky, D. (eds), *Near Earth Objects, our Celestial Neighbors: Opportunity and Risk. Proceedings of the 236th Symposium of the International Astronomical Union held in Prague, Czech Republic, 14–18 August 2006*. Cambridge, UK: Cambridge University Press.

Borovička, J. (2010). Spectroscopic Analysis of Geminid Meteors. Pages 42–51 of: Rendtel, J. and Vaubaillon, J. (eds), *Proceedings of the 26th International Meteor Conference held in Bareges, France, 7–10 June, 2007*. International Meteor Organization.

Borovička, J. (2016). Are some meteoroids rubble piles?. Pages 80–85 of: *Asteroids: New observations, New Models. Proceedings of the 318th Symposium of the International Astronomical Union*. Cambridge, UK: Cambridge University Press.

Borovička, J., and Jenniskens, P. (2000). Time resolved spectroscopy of a Leonid fireball afterglow. *Earth Moon and Planets*, **82**, 399–428.

Borovička, J., and Kalenda, P. (2003). The Morávka meteorite fall: 4. Meteoroid dynamics and fragmentation in the atmosphere. *Meteoritics and Planetary Science*, **38**, 1023–1043.

Borovička, J., and Spurný, P. (1996). Radiation study of two very bright terrestrial bolides and an application to the comet S-L 9 collision with Jupiter. *Icarus*, **121**, 484–510.

Borovička, J., and Spurný, P. (2008). The Carancas meteorite impact – Encounter with a monolithic meteoroid. *Astronomy and Astrophysics*, **485**, L1–L4.

Borovička, J., Koten, P., Spurný, P., Boček, J., and Štork, R. (2005). A survey of meteor spectra and orbits: evidence for three populations of Na-free meteoroids. *Icarus*, **174**, 15–30.

Borovička, J., Spurný, P., and Koten, P. (2007). Atmospheric deceleration and light curves of Draconid meteors and implications for the structure of cometary dust. *Astronomy and Astrophysics*, **473**, 661–672.

Borovička, J., Koten, P., Spurný, P., and Štork, R. (2008). Analysis of a low density meteoroid with enhanced sodium. *Earth Moon and Planets*, **102**, 485–493.

Borovička, J., Koten, P., Spurný, P., et al. (2010). Material properties of transition objects 3200 Phaethon and 2003 EH_1. Pages 218–222 of Fernandez, J. A., Lazzaro, D., Prialnik, D., and Schulz, R. (eds), *Icy Bodies of the Solar System, Proceedings of the 263rd Symposium of the International Astronomical Union*. Cambridge, UK: Cambridge University Press.

Borovička, J., Tóth, J., Igaz, A. et al. (2013a). The Košice meteorite fall: Atmospheric trajectory, fragmentation, and orbit. *Meteoritics and Planetary Science*, **48**, 1757–1779.

Borovička, J., Spurný, P., Brown, P. et al. (2013b). The trajectory, structure and origin of the Chelyabinsk asteroidal impactor. *Nature*, **503**, 235–237.

Borovička, J., Koten, P., Shrbený, L., Štork, R., and Hornoch, K. (2014). Spectral, photometric, and dynamic analysis of eight Draconid meteors. *Earth Moon and Planets*, **113**, 15–31.

Borovička, J., Spurný, P., and Brown, P. (2015a). Small near-Earth asteroids as a source of meteorites. In: P. Michel, F. E. DeMeo, and W. F. Bottke, eds., *Asteroids IV*. Tucson: University of Arizona Press, pp. 257–280.

Borovička, J., Spurný, P., Šegon, D. et al. (2015b). The instrumentally recorded fall of the Križevci meteorite, Croatia, February 4, 2011. *Meteoritics and Planetary Science*, **50**, 1244–1259.

Borovička, J., Spurný, P., Grigore, V. I., and Svoreň, J. (2017). The January 7, 2015, superbolide over Romania and structural diversity of meter-sized asteroids. *Planetary and Space Science*, **143**, 147–158.

Brown P. G., ReVelle D. O., Tagliaferri E., and Hildebrand A. R. (2002). An entry model for the Tagish Lake fireball using seismic, satellite and infrasound records. *Meteoritics and Planetary Science*, **37**, 661–675.

Brown, P., ReVelle, D. O., Silber, E. A. et al. (2008). Analysis of a crater-forming meteorite impact in Peru. *Journal of Geophysical Research (Planets)*, **113**, E09007.

Brown, P. G., Assink, J. D., Astiz, L. et al. (2013a). A 500-kiloton airburst over Chelyabinsk and an enhanced hazard from small impactors. *Nature*, **503**, 238–241.

Brown, P., Marchenko, V., Moser, D. E., Weryk, R., and Cooke, W. (2013b). Meteorites from meteor showers: A case study of the Taurids. *Meteoritics and Planetary Science*, **48**, 270–288.

Brownlee, D. E. (1985). Cosmic dust: Collection and research. *Annual Review of Earth and Planetary Sciences*, **13**, 147–173.

Brownlee, D. E. (2014). Comets. In: H. D. Holland and K. K. Turekian, eds., *Treatise on Geochemistry* (2nd edition), **2**, 355–362. available at: https://doi.org/10.1016/B978-0-08-095975-7.00128-5

Brownlee, D. E., Tomandl, D. A., and Hodge, P. W. (1976). Extraterrestrial particles in the stratosphere. In: H. Elsasser and H. Fechtig, eds., *Interplanetary Dust and the Zodiacal Light*. New York: Springer-Verlag, pp. 279–284.

Brownlee, D. E., Horz, F., Newburn, R. L. et al. (2004). Surface of young Jupiter family comet 81 P/Wild 2: View from the Stardust spacecraft. *Science*, **304**, 1764–1769.

Burbine, T. H., McCoy, T. J., Meibom, A., Gladman, B., and Keil, K. (2002). Meteoritic Parent Bodies: Their Number and Identification. In: W. Bottke, A. Cellino, P. Paolicci, and R. Binzel, eds., *Asteroids III*. Tucson: University of Arizona Press, pp. 653–667.

Burchell, M. J., Fairey, S. A. J., Wozniakiewicz, P. et al. (2008). Characteristics of cometary dust tracks in Stardust aerogel and laboratory calibrations. *Meteoritics and Planetary Science*, **43**, 23–40.

Campbell-Brown, M. (2015). A population of small refractory meteoroids in asteroidal orbits. *Planetary and Space Science*, **118**, 8–13.

Campbell-Brown, M. (2017). Modelling a short-wake meteor as a single or fragmenting body. *Planetary and Space Science*, **143**, 34–39.

Campbell-Brown, M. D., and Close, S. (2007). Meteoroid structure from radar head echoes. *Monthly Notices of the Royal Astronomical Society*, **382**, 1309–1316.

Čapek, D., and Borovička, J. (2009). Quantitative model of the release of sodium from meteoroids in the vicinity of the Sun: Application to Geminids. *Icarus*, **202**, 361–370.

Čapek, D., and Borovička, J. (2017). Ablation of small iron meteoroids–First results. *Planetary and Space Science*, **143**, 159–163.

Carry, B. (2012). Density of asteroids. *Planetary and Space Science*, **73**, 98–118.

Ceplecha, Z. (1966). Complete data on iron meteoroid (Meteor 36221). *Bulletin of the Astronomical Institutes of Czechoslovakia*, **17**, 195–206.

Ceplecha, Z. (1967). Classification of meteor orbits. *Smithsonian Contributions to Astrophysics*, **11**, 35–60.

Ceplecha, Z. (1988). Earth's influx of different populations of sporadic meteoroids from photographic and television data. *Bulletin of the Astronomical Institutes of Czechoslovakia*, **39**, 221–236.

Ceplecha, Z., and McCrosky, R. E. (1976). Fireball end heights - A diagnostic for the structure of meteoric material. *Journal of Geophysical Research*, **81**, 6257–6275.

Ceplecha, Z., and ReVelle, D. O. (2005). Fragmentation model of meteoroid motion, mass loss, and radiation in the atmosphere. *Meteoritics and Planetary Science*, **40**, 35–50.

Ceplecha, Z., Spurný, P., Borovička, J., and Keclíkova, J. (1993). Atmospheric fragmentation of meteoroids. *Astronomy and Astrophysics*, **279**, 615–626.

Chapman, C. R., Morrison, D., and Zellner, B. (1975). Surface properties of asteroids: A synthesis of polarimetry, radiometry and spectrometry. *Icarus*, **25**, 104–130.

Chapman, C. R., and Salisbury, J. W. (1973). Comparisons of meteorite and asteroid spectral reflectivities. *Icarus*, **19**, 507–522.

Consolmagno, G. J., and Drake, M. J. (1977). Composition and evolution of the eucrite parent body: Evidence from rare earth elements. *Geochimica et Cosmochimica Acta*, **41**, 1271–1282.

Consolmagno, G. J., Britt, D. T., and Macke, R. J. (2008). The significance of meteorite density and porosity. *Chemie der Erde*, **68**, 1–29.

Cotto-Figueroa, D., Asphaug, E., Garvie, L. A. J. et al. (2016). Scale-dependent measurements of meteorite strength: Implications for asteroid fragmentation. *Icarus*, **277**, 73–77.

De Giacomo, A., Dell'Aglio, M., de Pascale, O., Longo, S., and Capitelli, M. (2007). Laser induced breakdown spectroscopy on meteorites. *Spectrochimica Acta Part B*, **62**, 1606–1611.

DeMeo, F. E., Alexander, C. M. O., Walsh, K. J., Chapman, C. R., and Binzel, R. P. (2015). The compositional structure of the asteroid belt. In: P. Michel, F. E. DeMeo, and W. F. Bottke, eds., *Asteroids IV*. Tucson: University of Arizona Press, pp. 13–41.

De Sanctis, M. C., Raponi, A., Ammannito, E. et al. (2016). Bright carbonate deposits as evidence of aqueous alteration on (1) Ceres. *Nature*, **536**, 54–57.

Dobrică, E., Engrand, C., Duprat, J., and Gounelle, M. (2010). A statistical overview of Concordia Antarctic micrometeorites, *Meteoritics and Planetary Science*. **45** Suppl., A46.

Dobrică, E., Engrand, C., Leroux, H., Rouzaud, J. N., and Duprat, J. (2012). Transmission electron microscopy of Concordia ultracarbonaceous Antarctic micrometeorites (UCAMMs): Mineralogical properties. *Geochimica et Cosmochimica Acta*, **76**, 68–82.

Elford, W. G. (2004). Radar observations of meteor trails, and their interpretation using Fresnel holography: A new tool in meteor science. *Atmospheric Chemistry and Physics*, **4**, 911–921.

Elkins-Tanton, L. T., Weiss, B. P., and Zuber, M. T. (2011). Chondrites as samples of differentiated planetesimals. *Earth and Planetary Science Letters*, **305**, 1–10.

Elsila, J. E., Glavin, D. P., and Dworkin, J. P. (2009). Cometary glycine detected in samples returned by Stardust. *Meteoritics and Planetary Science*, **44**, 1323–1330.

Eugster, O., Herzog, G. F., Marti, K., and Caffee, M. W. (2006). Irradiation records, cosmic-ray exposure ages, and transfer times in meteorites. In: D. S. Lauretta and H. Y. McSween Jr., eds., *Meteorites and the Early Solar System II*. Tucson: University of Arizona Press, pp. 829–852.

Fisher, A. A., Hawkes, R. L., Murray, I. S., Campbell, M. D., and LeBlanc, A. G. (2000). Are meteoroids really dustballs?. *Planetary and Space Science*, **48**, 911–920.

Flynn, G. J., Keller, L. P., Feser, M., Wirick, S., and Jacobsen, C. (2003). The origin of organic matter in the solar system: Evidence from the interplanetary dust particles. *Geochimica et Cosmochimica Acta*, **67**, 4791–4806.

Flynn, G. J., Keller, L. P., Jacobsen, C., and Wirick, S. (2004). An assessment of the amount and types of organic matter contributed to the Earth by interplanetary dust. *Advances in Space Research*, **33**, 57–66.

Flynn, G. J., Bleuet, P., Borg, J., et al. (2006). Elemental compositions of Comet 81P/Wild 2 samples collected by stardust. *Science*, **314**, 1731–1735.

Flynn, G. J., Wirick, S., and Keller, L. P. (2013). Organic grain coatings in primitive interplanetary dust particles: Implications for grain sticking in the Solar Nebula. *Earth, Planets and Space*, **65**, 1159–1166.

Flynn, G. J., Consolmagno, G. J., Brown, P., and Macke, R. J. (2018). Physical properties of the stone meteorites: Implications for the properties of their parent bodies. *Chemie der Erde*, **78**, 269–298.

Fomenkova, M. N., Kerridge, J. F., Marti, K., and McFadden, L.-A. (1992). Compositional trends in rock-forming elements of Comet Halley dust. *Science*, **258**, 266–269.

Fray, N., Bardyn, A., Cottin, H. et al. (2016). High-molecular-weight organic matter in the particles of comet 67P/Churyumov-Gerasimenko. *Nature*, **538**, 72–74.

Friedrich, J. M., Ruzicka, A., Macke, R. J., et al. (2017). Relationships among physical properties as indicators of high temperature deformation or post-shock thermal annealing in ordinary chondrites. *Geochimica et Cosmochimica Acta*, **203**, 157–174.

Fulle, M., Levasseur-Regourd, A. C., McBride, N., and Hadamcik, E. (2000). In-situ dust measurements from within the coma of 1P/Halley: First order approximation with a dust dynamical model. *The Astronomical Journal*, **119**, 1968–1977.

Fulle, M., and Blum, J. (2017). Fractal dust constrains the collisional history of comets. *Monthly Notices of the Royal Astronomical Society*, **469**, S39–S44.

Gainsforth, Z., Butterworth, A. L., Stodolna, J. et al. (2015). Constraints on the formation environment of two chondrule-like igneous particles from comet 81P/Wild 2. *Meteoritics and Planetary Science*, **50**, 976–1004.

Genge, M. J. (2017). An increased abundance of micrometeorites on Earth owing to vesicular parachutes. *Geophysical Research Letters*, **44**, 1679–1686.

Genge, M. J., Engrand, C., Gounelle, M., and Taylor, S. (2008). The classification of micrometeorites. *Meteoritics and Planetary Science*, **43**, 497–515.

Glassmeier, K. H., Boehnhardt, H., Koschny, D., Kührt, E., and Richter, I. (2007). The Rosetta mission: Flying towards the origin of the Solar System. *Space Science Reviews*, **128**, 1–21.

Goesmann, F., Rosenbauer, H., Bredehft, J. H. et al. (2015). Organic compounds on comet 67P/Churyumov-Gerasimenko revealed by COSAC mass spectrometry. *Science*, **349** (6247), aab0689.

Goodrich, C. A., Van Orman, J. A., and Wilson, L. (2007). Fractional melting and smelting on the ureilite parent body. *Geochimica et Cosmochimica Acta*, **71**, 2876–2985.

Goodrich, C. A., Hartmann, W. K., O'Brien, D. P. et al. (2015). Origin and history of ureilitic material in the solar system: The view from asteroid 2008 TC$_3$ and the Almahata Sitta meteorite. *Meteoritics and Planetary Science*, **50**, 782–809.

Gounelle, M., Spurný, P., and Bland, P. A. (2006). The orbit and atmospheric trajectory of the Orgueil meteorite from historical records. *Meteoritics and Planetary Science*, **41**, 135–150.

Grün, E., Gustafson, B. A. S., Dermott, S., and Fechtig, H., eds. (2001). *Interplanetary Dust*, Berlin: Springer.

Hadamcik, E., and Levasseur-Regourd, A. C. (2003). Imaging polarimetry of cometary dust: Different comets and phase angles, *Journal of Quantitative Spectroscopy and Radiative Transfer*, **79–80**, 661–678.

Hadamcik, E., Renard, J. B., Rietmeijer, F. J. M. et al. (2007). Light scattering by fluffy MgFeSiO and C mixtures as cometary analogs (PROGRA2 experiment). *Icarus*, **190**, 660–671.

Hadamcik, E., Lasue, J. B., Levasseur-Regourd, A. C., and Renard, J. B. (2018). Analogues of interplanetary dust particles to interpret the zodiacal light polarization. *Planetary and Space Science*, in press. https://doi.org/10.1016/j.pss.2018.04.022

Halliday, I. (1960). The spectrum of an asteroidal meteor fragment. *The Astrophysical Journal*, **132**, 482–485.

Hanner, M. S., and Bradley, J. P. (2004). Composition and mineralogy of cometary dust. In: M. C. Festou, H. U. Keller, and H. A. Weaver, eds., *Comets II*. Tucson: University of Arizona Press, pp. 555–564.

Hardersen, P. S., Cloutis, E. A., Reddy, V., Mothé-Diniz, T., and Emery, J. P. (2011). The M-/X-asteroid menagerie: Results of an NIR spectral survey of 45 main-belt asteroids. *Meteoritics and Planetary Science*, **46**, 1910–1938.

Hawkes, R. L., and Jones, J. (1975). A quantitative model for the ablation of dustball meteors. *Monthly Notices of the Royal Astronomical Society*, **173**, 339–356.

Herique, A., Kofman, W., Beck, P. et al. (2016). Cosmochemical implications of CONSERT permittivity characterization of 67P/CG. *Monthly Notices of the Royal Astronomical Society*, **462**, S516–S532.

Hildebrand, A. R., McCausland, P. J. A., Brown, P. G. et al. (2006). The fall and recovery of the Tagish Lake meteorite. *Meteoritics and Planetary Science*, **41**, 407–431.

Horstmann, M., and Bischoff, A. (2014). The Almahata Sitta polymict breccia and the late accretion of asteroid 2008 TC3. *Chemie der Erde*, **74**, 149–183.

Hörz, F., Bastien, R., Borg, J. et al. (2006). Impact features on Stardust: Implications for comet 81P/Wild 2 dust. *Science*, **314**, 1716–1719.

Hutchison R. (2004). *Meteorites: A Petrologic, Chemical and Isotopic Synthesis*. Cambridge, UK: Cambridge University Press.

Ishiguro, M., Watanabe, J., Usui, F. et al. (2002). First detection of an optical dust trail along the orbit of 22P/Kopff. *The Astrophysical Journal*, **572**, L117–L120.

Jacchia, L. G. (1955). The physical theory of meteors. VIII. Fragmentation as cause of the faintmeteor anomaly. *The Astrophysical Journal*, **121**, 521–527.

Janches, D., Swarnalingam, N., Carrillo-Sanchez, J. D. et al. (2017). Radar detectability studies of slow and small Zodiacal Dust Cloud particles. III. The role of sodium and the head echo size on the probability of detection. *The Astrophysical Journal*, **843**, 1.

Jenniskens, P. (2007). Quantitative meteor spectroscopy: Elemental abundances. *Advances in Space Research*, **39**, 491–512.

Jenniskens, P., Nénon, Q., Albers, J., et al. (2016a). The established meteor showers as observed by CAMS. *Icarus*, **266**, 331–354.

Jenniskens, P., Nénon, Q., Gural, P. S., et al. (2016b). CAMS newly detected meteor showers and the sporadic background. *Icarus*, **266**, 384–409.

Jessberger, E. K., Christoforidis, A., and Kissel, J. (1988). Aspects of the major element composition of Halley's dust. *Nature*, **332**, 691–695.

Jorda, L., Gaskell, R., Capanna, C. et al. (2016). The global shape, density and rotation of Comet 67P/Churyumov-Gerasimenko from preperihelion Rosetta/OSIRIS observations. *Icarus*, **277**, 257–278.

Joswiak, D. J., Brownlee, D. E., Matrajt, G., Westphal, A. J., and Snead, C. J. (2009). Kosmochloric Ca-rich pyroxenes and FeO-rich olivines (Kool grains) and associated phases in Stardust tracks and chondritic porous interplanetary dust particles: Possible precursors to FeO-rich type II chondrules in ordinary chondrites. *Meteoritics and Planetary Science*, **44**, 1561–1588.

Kasuga, T. and Jewitt, D. (2019). Asteroid–meteoroid complexes. In: G. O. Ryabova, D. J. Asher, and M. D. Campbell-Brown, eds., *Meteoroids: Sources of Meteors on Earth and Beyond*. Cambridge, UK: Cambridge University Press, pp. 187–209.

Keller, L. P., and Messenger, S. (2011). On the origins of GEMS grains. *Geochimica et Cosmochimica Acta*, **75**, 5336–5365.

Kero, J., Szasz, C., Pellinen-Wannberg, A. et al. (2008). Three-dimensional radar observation of a submillimeter meteoroid fragmentation. *Geophysical Research Letters*, **35**, L04101.

Kero, J., Campbell-Brown, M., Stober, G. et al. (2019). Radar Observations of Meteors. In: G. O. Ryabova, D. J. Asher, and M. D. Campbell-Brown, eds., *Meteoroids: Sources of Meteors on Earth and Beyond*. Cambridge, UK: Cambridge University Press, pp. 65–89.

Kikwaya, J.-B., Campbell-Brown, M., and Brown, P. G. (2011). Bulk density of small meteoroids. *Astronomy and Astrophysics*, **530**, A113

Kiselev, N., Rosenbush, V., Levasseur-Regourd, A. C., and Kolokolova, L. (2015). Comets. In: L. Kolokolova, J. Hough, and A. C. Levasseur-Regourd, eds., *Polarimetry of Stars and Planetary Systems*. Cambridge, UK: Cambridge University Press, pp. 379–404.

Kissel, J., Sagdeev, R.Z., Bertaux, J.L. et al. (1986). Composition of comet Halley dust particles from VEGA observations. *Nature*, **321**, 280–282.

Kohout T., Havrila K., Tóth, J. et al. (2014a). Density, porosity and magnetic susceptibility of the Košice meteorite shower and homogeneity of its parent meteoroid. *Planetary and Space Science*, **93**, 96–100.

Kohout, T., Kallonen, A., Suuronen J.-P. et al. (2014b). Density, porosity, mineralogy, and internal structure of cosmic dust and alteration of its properties during high velocity atmospheric entry. *Meteoritics and Planetary Science*, **49**, 1157–1170.

Kokhirova, G. I., and Borovička, J. (2011). Observations of the 2009 Leonid activity by the Tajikistan fireball network. *Astronomy and Astrophysics*, **533**, A115.

Kolokolova, L. and Kimura, H. (2010). Comet dust as a mixture of aggregates and solid particles: model consistent with ground-based and space-mission results. *Earth, Planets and Space*, **62**, 17–21.

Koschny, D., Bettonvil, F., Licandro, J. et al. (2013). A double-station meteor camera set-up in the Canary Islands - CILBO. *Geoscientific Instrumentation, Methods and Data Systems*, **2**, 339–348.

Koten, P., Borovička, J., Spurný, P., Betlem, H., and Evans, S. (2004). Atmospheric trajectories and light curves of shower meteors. *Astronomy and Astrophysics*, **428**, 683–690

Koten, P., Borovička, J., Spurný, P. et al. (2006). Double station and spectroscopic observations of the Quadrantid meteor shower and the implications for its parent body. *Monthly Notices of the Royal Astronomical Society*, **366**, 1367–1372.

Koten, P., Vaubaillon, J., Margonis, A. et al. (2015). Double station observation of Draconid meteor outburst from two moving aircraft. *Planetary and Space Science*, **118**, 112–119.

Koten, P., Čapek, D., Spurný, P. et al. (2017). September epsilon Perseid cluster as a result of orbital fragmentation. *Astronomy and Astrophysics*, **600**, A74.

Koten, P., Rendtel, J., Shrbený, L. et al. (2019). Meteors and Meteor Showers as Observed by Optical Techniques. In: G. O. Ryabova, D. J. Asher, and M. D. Campbell-Brown, eds., *Meteoroids: Sources of Meteors on Earth and Beyond*. Cambridge, UK: Cambridge University Press, pp. 90–115.

Langevin, Y., Hilchenbach, M., Ligier, N. et al. (2016). Typology of dust particles collected by the COSIMA mass spectrometer in the inner coma of 67P/Churyumov-Gerasimenko. *Icarus*, **271**, 76–97.

Lasue, J., and Levasseur-Regourd, A. C. (2006). Porous aggregates of irregular sub-micron sized grains to reproduce cometary dust light scattering observations. *Journal of Quantitative Spectroscopy and Radiative Transfer*, **100**, 220–236.

Lasue, J., Levasseur-Regourd, A. C., Fray, N., and Cottin, H. (2007). Inferring the interplanetary dust properties from remote observations and simulations. *Astronomy and Astrophysics*, **473**, 641–649.

Lasue, J., Levasseur-Regourd, A. C., and Lazarian, A. (2015). Interplanetary dust. In: L. Kolokolova, J. Hough, and A. C. Levasseur-Regourd, eds., *Polarimetry of Stars and Planetary Systems*. Cambridge, UK: Cambridge University Press, pp. 419–436.

Lauretta, D. S., Bartels, A. E., Barucci, M. A. et al. (2015). The OSIRIS-REx target asteroid (101955) Bennu: Constraints on its physical, geological, and dynamical nature from astronomical observations. *Meteoritics and Planetary Science*, **50**, 834–849.

Lellouch, E., Paubert, G., Moses, J. I., Schneider, N. M., and Strobel, D. F. (2003). Volcanically emitted sodium chloride as a source for Io's neutral clouds and plasma torus. *Nature*, **421**, 45–47.

Levasseur-Regourd, A. C. (1998). Zodiacal light, certitudes and questions. *Earth, Planets and Space*, **50**, 607–610.

Levasseur-Regourd, A. C. (1999). Polarization of light scattered by cometary dust particles: Observations and tentative interpretations. *Space Science Reviews*, **90**, 163–168.

Levasseur-Regourd, A. C., and Lasue, J. (2011). Inferring sources in the interplanetary dust cloud, from observations and simulations of zodiacal light and thermal emission. In: *Meteoroids: The Smallest Solar System Bodies* (NASA/CP-2011-216469), pp. 66–75.

Levasseur-Regourd, A. C., Cabane, M., and Haudebourg, V. (1999a). Observational evidence for the scattering properties of interplanetary and cometary dust clouds: An update. *Journal of Quantitative Spectroscopy and Radiative Transfer*, **63**, 631–641.

Levasseur-Regourd, A. C., McBride, N., Hadamcik, E., and Fulle, M. (1999b). Similarities between in situ measurements of local dust scattering and dust flux impact data within the coma of 1P/Halley. *Astronomy and Astrophysics*, **348**, 636–641.

Levasseur-Regourd, A. C., Mukai, T., Lasue, J., and Okada, Y. (2007). Physical properties of cometary and interplanetary dust. *Planetary and Space Science*, **55**, 1010–1020.

Levasseur-Regourd, A. C., Zolensky, M., and Lasue, J. (2008). Dust in cometary comae: present understanding of the structure and composition of dust particles. *Planetary and Space Science*, **56**, 1719–1724.

Levasseur-Regourd, A. C., Hadamcik, E., Desvoivres, E., and Lasue, J. (2009). Probing the internal structure of the nucleus of comets. *Planetary and Space Science*, **57**, 221–228.

Levasseur-Regourd, A. C., Agarwal, J., Cottin, H. et al. (2018). Cometary Dust. *Space Science Reviews*, **214**, 64 (56pp.)

Macke, R. J. (2010). *Survey of Meteorite Physical Properties: Density, Porosity and Magnetic Susceptibility.* Ph.D. dissertation. University of Central Florida, Orlando, FL, USA.

Macke, R. J., Consolmagno, G. J., Britt, D. T. and Hutson, M. L. (2010). Enstatite chondrite density, magnetic susceptibility and porosity. *Meteoritics and Planetary Science*, **45**, 1513–1526.

Madiedo, J. M., Trigo-Rodríguez, J. M., Castro-Tirado, A. J., Ortiz, J. L., and Cabrera-Caño, J. (2013). The Geminid meteoroid stream as a potential meteorite dropper: A case study. *Monthly Notices of the Royal Astronomical Society*, **436**, 2818–2823.

Madiedo, J. M., Ortiz, J. L., Trigo-Rodríguez, J. M. et al. (2014). Analysis of bright Taurid fireballs and their ability to produce meteorites. *Icarus*, **231**, 356–364.

Madiedo, J. M., Espartero, F., Trigo-Rodríguez, J. M. et al. (2016). Observations of the Quadrantid meteor shower from 2008 to 2012: Orbits and emission spectra. *Icarus*, **275**, 193–202.

Mannel, T., Bentley, M.S., Schmied, R. et al. (2016). Fractal cometary dust - A window into the early Solar system. *Monthly Notices of the Royal Astronomical Society*, **462**, S304–S311.

Marti, K., and Graf, T. (1992). Cosmic-ray exposure history of ordinary chondrites. *Annual Review of Earth and Planetary Sciences*, **20**, 221–243.

Mathews, J. D., Briczinski, S. J., Malhotra, A., and Cross, J. (2010). Extensive meteoroid fragmentation in V/UHF radar meteor observations at Arecibo Observatory. *Geophysical Research Letters*, **37**, L04103.

Matlovič, P., Tóth, J., Rudawska, R., and Kornoš, L. (2017). Spectra and physical properties of Taurid meteoroids. *Planetary and Space Science*, **143**, 104–115.

Matrajt, G., Ito, M., Wirick, S. et al. (2008). Carbon investigation of two Stardust particles: A TEM, NanoSIMS, and XANES study. *Meteoritics and Planetary Science*, **43**, 315–334.

McCord, T. B., Adams, J. B., and Johnson, T. V. (1970). Asteroid Vesta: Spectral reflectivity and compositional implications. *Science*, **168**, 1445–1447.

McSween, H. Y. Jr., Mittlefehldt, D. W., Beck, A. W., Mayne, R. G., and McCoy, T. J. (2011). HED Meteorites and Their Relationship to the Geology of Vesta and the Dawn Mission. *Space Science Reviews*, **163**, 141–174.

Meier, M. M. M. (2017). Meteoriteorbits.info - Tracking all known meteorites with photographic orbits. *Lunar and Planetary Science XLVIII*, abstract #1178.

Millman, P. M. (1935). An analysis of meteor spectra: Second paper. *Annals of Harvard College Observatory*, **82**, 149–177.

Mostefaoui, S., and Hoppe, P. (2004). Discovery of abundant in situ silicate and spinel grains from red giant stars in a primitive meteorite. *The Astrophysical Journal*, **613**, L149–L152.

Moore, C. E. (1945). A multiplet table of astrophysical interest. *Contributions from the Princeton University Observatory*, **20**, 1–110.

Mukhin, L., Dolnikov, G., Evlanov, E., et al. (1991). Re-evaluation of the chemistry of dust grains in the coma of Comet Halley. *Nature*, **350**, 480–481.

Nakamura, T., Noguchi, T., Tanaka, M., et al. (2011). Itokawa dust particles: A direct link between S-Type asteroids and ordinary chondrites. *Science*, **333**, 1113–1116.

Nesvorný, D., Jenniskens, P., Levison, H. F., et al. (2010). Cometary origin of the zodiacal cloud and carbonaceous micrometeorites: Implications for hot debris disks. *The Astrophysical Journal*, **713**, 816–836.

Nesvorný, D., Janches, D., Vokrouhlický, D. et al. (2011). Dynamical model for the zodiacal cloud and sporadic meteors. *The Astrophysical Journal*, **743**, 129.

Nittler, L. R., Alexander, C. M. O., Gallino, R. et al. (2008). Aluminum-, calcium- and titanium-rich oxide stardust in ordinary chondrite meteorites. *The Astrophysical Journal*, **682**, 1450–1478.

Park, R. S., Konopliv, A. S., Bills, B. G. et al. (2016). A partially differentiated interior for (1) Ceres deduced from its gravity field and shape. *Nature*, **537**, 515–517.

Pätzold, M., Andert, T. P., Asmar, S. W. et al. (2011). Asteroid 21 Lutetia: Low mass, high density. *Science*, **334**, 491–492.

Pätzold, M., Andert, T., Hahn, M. et al. (2016). A homogeneous nucleus for comet 67P/Churyumov-Gerasimenko from its gravity field. *Nature*, **530**, 63–65.

Popova, O., Borovička, J., Hartmann, W. K. et al. (2011). Very low strengths of interplanetary meteoroids and small asteroids. *Meteoritics and Planetary Science*, **46**, 1525–1550.

Popova, O. P., Jenniskens, P., Emel'yanenko, V. et al. (2013). Chelyabinsk airburst, damage assessment, meteorite recovery, and characterization, *Science*, **342**, 1069–1073.

Popova, O., Borovička, J., and Campbell-Brown, M. (2019). Modelling the Entry of Meteoroids. In: G. O. Ryabova, D. J. Asher, and M. D. Campbell-Brown, eds., *Meteoroids: Sources of Meteors on Earth and Beyond*. Cambridge, UK: Cambridge University Press, pp. 9–36.

Postberg, F., Schmidt, J., Hillier, J., Kempf, S., and Srama, R. (2011). A salt-water reservoir as the source of a compositionally stratified plume on Enceladus. *Nature*, **474**, 620–622.

Pravec, P., and Harris, A. W. (2000). Fast and slow rotation of asteroids. *Icarus*, **148**, 12–20.

Reach, W. T., Kelley, M. S., and Sykes, M. V. (2007). A survey of debris trails from short period comets. *Icarus*, **191**, 298–322.

Rietmeijer, F. J. M. (1998). Interplanetary Dust Particles. In: J. J. Papike (ed), *Planetary Materials. Reviews in Mineralogy*, **36** (Chantilly, VA: Mineralogical Society of America), pp. 2-1–2-95.

Rietmeijer, F. J. M. (2002). The earliest chemical dust evolution in the solar nebula. *Chemie der Erde, Geochemistry*, **62**, 1–45.

Rietmeijer, F. J. M. (2004). Dynamic pyrometamorphism during atmospheric entry of large (~ 10 micron) pyrrhotite fragments from cluster IDPs. *Meteoritics and Planetary Science* **39**, 1869–1887.

Righter, K., Drake, M. J., and Scott, E. (2006). Compositional Relationships Between Meteorites and Terrestrial Planets. In: D. S. Lauretta and H. Y. McSween Jr., eds., *Meteorites and the Early Solar System II*. Tucson: University of Arizona Press, pp. 803–828.

Rochette, P., Gattacceca, J., Bonal, L. et al. (2008). Magnetic classification of stony meteorites: 2. Non-ordinary chondrites. *Meteoritics and Planetary Science*, **43**, 959–980.

Rowan-Robinson, M., and May, B. (2013). An improved model for the infrared emission from the zodiacal dust cloud: Cometary, asteroidal and interstellar dust. *Monthly Notices of the Royal Astronomical Society*, **429**, 2894–2902.

Rubin, A. E. (1997). The Galim LL/EH polymict breccia: Evidence for impact-induced exchange between reduced and oxidized meteoritic material. *Meteoritics and Planetary Science*, **32**, 489–492.

Rudawska, R., Tóth, J., Kalmančok, D., Zigo, P., and Matlovič, P. (2016). Meteor spectra from AMOS video system. *Planetary and Space Science*, **123**, 25–32.

Russell, S. S., Hartmann, L., Cuzzi, J. et al. (2006). Timescales of the Solar Protoplanetary Disk. In: D. S. Lauretta and H. Y. McSween Jr., eds., *Meteorites and the Early Solar System II*. Tucson: University of Arizona Press, pp. 233–251.

Russell, C. T., Raymond, C. A., Coradini, A. et al. (2012). Dawn at Vesta: Testing the protoplanetary paradigm. *Science*, **336**, 684–686.

Ruzicka, A., Snyder, G. A., and Taylor, L. A. (1997). Vesta as the howardite, eucrite and diogenite parent body: Implications for the size of a core and for large-scale differentiation. *Meteoritics and Planetary Science*, **32**, 825–840.

Schramm, L. S., Brownlee, D. E., and Wheelock, M. M. (1989). Major element composition of stratospheric micrometeorites. *Meteoritics* **24**, 99–112.

Schwinger, S., Dohmen, R., and Schertl, H.-P. (2016). A combined diffusion and thermal modeling approach to determine peak temperatures of thermal metamorphism experienced by meteorites. *Geochimica et Cosmochimica Acta*, **191**, 255–276.

Shaddad, M. H., Jenniskens, P., Numan, D. et al. (2010). The recovery of asteroid 2008 TC3. *Meteoritics and Planetary Science*, **45**, 1557–1589.

Slyuta, E. N. (2017). Physical and mechanical properties of stony meteorites. *Solar System Research*, **51**, 64–85. (Originally published in Russian in *Astronomicheskii Vestnik*, **51**, 72–95.)

Spurný, P., and Borovička, J. (1999a). EN010697 Karlštejn: the first type I fireball on retrograde orbit. In: W. J. Baggaley and V. Porubcan, eds., *Meteoroids 1998*. Bratislava: Astronomical Institute of Slovak Academy of Sciences, pp. 143–148.

Spurný, P., and Borovička, J. (1999b). Detection of a high density meteoroid on cometary orbit. In: J. Svoreň, E. M. Pittich, and H. Rickman, eds., *Evolution and Source Regions of Asteroids and Comets*. pp. 163–168.

Spurný, P., and Borovička, J. (2013). Meteorite dropping Geminid recorded, *Meteoroids 2013 Conference*, abstract #061. available at: (http://pallas.astro.amu.edu.pl/Meteoroids2013/main_content/data/abstracts.pdf)

Spurný, P., Borovička, J., Kac, J. et al. (2010). Analysis of instrumental observations of the Jesenice meteorite fall on April 9, 2009. *Meteoritics and Planetary Science*, **45**, 1392–1407.

Spurný, P., Haloda, J., Borovička, J., Shrbený, L., and Halodová, P. (2014). Reanalysis of the Benešov bolide and recovery of polymict breccia meteorites – old mystery solved after 20 years. *Astronomy and Astrophysics*, **570**, A39.

Spurný, P., Borovička, J., Mucke, H., and Svoreň, J. (2017). Discovery of a new branch of the Taurid meteoroid stream as a real source of potentially hazardous bodies. *Astronomy and Astrophysics*, **605**, A68.

Stevenson, R., Bauer, J. M., Kramer, E. A. et al. (2014). Lingering grains of truth around comet 17P/Holmes. *The Astrophysical Journal*, **787**, 116.

Stöffler, D., Keil, K., and Scott, E. R. D. (1991). Shock metamorphism of ordinary chondrites. *Geochimica et Cosmochimica Acta*, **55**, 3845–3867.

Stöffler, D., Hamann, C., and Metzler, K. (2018). Shock metamorphism of planetary silicate rocks and sediments: Proposal for an updated classification system. *Meteoritics and Planetary Science*, **53**, 5–49.

Stokan, E., and Campbell-Brown, M. D. (2014). Transverse motion of fragmenting faint meteors observed with the Canadian Automated Meteor Observatory. *Icarus*, **232**, 1–12.

Subasinghe, D., Campbell-Brown, M. D., and Stokan, E. (2016). Physical characteristics of faint meteors by light curve and high-resolution observations, and the implications for parent bodies. *Monthly Notices of the Royal Astronomical Society*, **457**, 1289–1298.

Sykes, M. V. and Walker, R. G. (1992). Cometary dust trails. I - Survey. *Icarus*, **95**, 180–210.

Szalay, J. R., Poppe, A. R., Agarwal, J. et al. (2018). Dust phenomena relating to airless bodies. *Space Science Reviews*, **214**, 98 (47pp.)

Tancredi, G., Ishitsuka, J., Schultz, P. H. et al. (2009). A meteorite crater on Earth formed on September 15, 2007: The Carancas hypervelocity impact. *Meteoritics and Planetary Science*, **44**, 1967–1984.

Taylor, S., Jones, K. W., Herzog, G. F. and Hornig, C. E. (2011). Tomography: A window on the role of sulfur in the structure of micrometeorites. *Meteoritics and Planetary Science*, **46**, 1498–1509.

Thomas, P. C., A Hearn, M. F., Veverka, J. et al., (2013). Shape, density, and geology of the nucleus of Comet 103P/Hartley 2. *Icarus*, **222**, 550–558.

Toppani, A., Libourel, G., Engrand, C., and Maurette, M. (2001). Experimental simulation of atmospheric entry of micrometeorites. *Meteoritics and Planetary Science*, **36**, 1377–1396.

Toppani, A., and Libourel, G. (2003). Factors controlling compositions of cosmic spinels: Application to atmospheric entry conditions of meteoritic materials. *Geochimica et Cosmochimica Acta*, **67**, 4621–4638.

Trigo-Rodriguez, J. M., Llorca, J., Borovička, J., and Fabregat, J. (2003). Chemical abundances determined from meteor spectra: I Ratios of the main chemical elements. *Meteoritics and Planetary Science*, **38**, 1283–1294.

Tsuchiyama, A., Uesugi, M., Matsushima, T. et al. (2011). Three-dimensional structure of Hayabusa samples: Origin and evolution of Itokawa regolith. *Science*, **333**, 1125–1128.

Van Schmus, W. R., and Wood, J. A. (1967). A chemical-petrologic classification for the chondritic meteorites. *Geochimica et Cosmochimica Acta*, **31**, 747–765.

Veverka, J., Thomas, P., Harch, A. et al. (1997). NEAR's Flyby of 253 Mathilde: Images of a C asteroid. *Science*, **278**, 2109–2114.

Veverka, J., Robinson, M., Thomas, P. et al. (2000). NEAR at Eros: Imaging and spectral results. *Science*, **289**, 2088–2097.

Vojáček, V., Borovička, J., Koten, P., Spurný, P., and Štork, R. (2015) Catalogue of representative meteor spectra. *Astronomy and Astrophysics*, **580**, A67.

Wasson, J. T. (2011). Relationship between iron-meteorite composition and size: Compositional distribution of irons from North Africa. *Geochimica et Cosmochimica Acta*, **75**, 1757–1772.

Wasson, J. T., and Kallemeyn, G. W. (1988). Compositions of Chondrites. *Philosophical Transactions of the Royal Society of London, Series A*, **325**, 535–544.

Wasson, J. T., and Kimberlin, J. (1967). The chemical classification of iron meteorites-II. Irons and pallasites with germanium concentrations between 8 and 100 ppm. *Geochimica et Cosmochimica Acta*, **31**, 2065–2093.

Watanabe, J.-I., Tabe, I., Hasegawa, H. et al. (2003). Meteoroid clusters in Leonids: Evidence of fragmentation in space. *Publications of the Astronomical Society of Japan*, **55**, L23–L26.

Weibull, W. A. (1951). A statistical distribution function of wide applicability. *Journal of Applied Mechanics*, **10**, 140–147.

Weidenschilling, S. J. and Cuzzi, J. N. (2006). Accretion Dynamics and Timescales: Relation to Chondrites. In: D. S. Lauretta and H. Y. McSween Jr., eds., *Meteorites and the Early Solar System II*. Tucson: University of Arizona Press, pp. 473–485.

Weisberg, M. K., McCoy, T. J., and Krot, A. N. (2006). Systematics and Evaluation of Meteorite Classification. In: D. S. Lauretta and H. Y. McSween Jr., eds., *Meteorites and the Early Solar System II*. Tucson: University of Arizona Press, pp. 19–52.

Weryk, R. J., Campbell-Brown, M. D., Wiegert, P. A. et al. (2013). The Canadian Automated Meteor Observatory (CAMO): System overview. *Icarus*, **225**, 614–622.

Wright, I.P., Sheridan, S., Barber, S. J. et al. (2015). CHO-bearing organic compounds at the surface of 67P/Churyumov-Gerasimenko revealed by Ptolemy. *Science*, **349** (6247), aab0673.

Yang, J., Goldstein, J. I., and Scott, E. R. D. (2010). Main-group pallasites: Thermal history, relationship to IIIAB irons, and origin. *Geochimica et Cosmochimica Acta*, **74**, 4471–4492.

Zinner E. (1996). Presolar material in meteorites: An overview. Pages 3–26 of: Bernatowicz, T. J. and Zinner, E. (eds), *Astrophysical Implications of the Laboratory Study of Astrophysical Materials, AIP Conference Proceedings* vol. **402**. Melville, NY: AIP Publishing.

Zolensky, M. E., Zega, T. J., Yano, H. et al. (2006). Mineralogy and petrology of Comet 81P/Wild 2 nucleus samples. *Science*, **314**, 1735–1739.

Zolensky, M., Nakamura-Messenger, K., Rietmeijer, F. et al. (2008). Comparing Wild 2 particles to chondrites and IDPs. *Meteoritics and Planetary Science*, **43**, 261–272.

Part II

Meteor Observations on the Earth

3

Radar Observations of Meteors

Johan Kero, Margaret D. Campbell-Brown, Gunter Stober, Jorge Luis Chau,
John David Mathews, and Asta Pellinen-Wannberg

3.1 Introduction

In addition to light, a meteoroid entering the atmosphere produces a transient plasma that expands to a trail due to ambipolar diffusion and drifts with the local wind. Radar (RAdio Detection And Ranging) offers complementary ways to optical recording of meteors beyond the limitations posed by daytime sky brightness and cloud coverage. Radar also enables detection of very small-mass meteoroids, i.e. faint meteors which could, in principle, only be optically detected using large-scale telescopes and under optimal sky conditions.

The radar operating frequencies used to study meteors span the frequency range from MHz to GHz. Radar echoes are caused by radio waves scattered from the ionized plasma created by meteoroid-atmosphere collisions. Head echoes are scattered from the dense ionized region near the meteoroid, while trail echoes come from the trail of electrons left behind the meteoroid. Since meteor plasmas are embedded in the background ionosphere, many aspects must be taken into account when studying the scattering of radio waves. The characters of the echoes from the head and the trail are very dissimilar and they are optimally monitored with two different types of radar.

As the name describes, specular meteor radars (SMRs) observe meteors with meteor trails oriented such that they provide specular reflection towards the radar. For the common case of a monostatic radar system where the transmitter and receiver are collocated, the trail must be oriented perpendicular to the radar beam. The ionization in the meteor trail exceeds the background ionospheric levels locally along the trail and, for faint meteors, diffuses away on the order of seconds. The process is observable with a standard, low-power (kW) Yagi antenna with a wide radiation pattern. Statistically, many meteors fulfill the specular direction requirement. The most commonly used frequency range for this method is 15–40 MHz. The lower limit lies just above the maximum ionospheric plasma frequency, and the upper limit is imposed by the radio meteor height ceiling effect.

High-Power Large-Aperture (HPLA) radars are sensitive to head echoes from meteors traveling in any direction through the measurement volume formed by the radar beam. As the name implies, the transmitted power of these radars is in the MW class. The large antenna aperture forms a radar beam with a narrow main lobe of only a few degrees or less, corresponding to a lobe a few kilometers broad at meteor heights. The high power and narrow beam gives a power density several orders of magnitude higher than specular radars. Typical meteor detection rates for HPLA radars are therefore very high in spite of the small collecting area. Operating frequencies for HPLA radar systems range from 50 MHz to beyond 1 GHz.

Figure 3.1 shows two examples of HPLA radar events containing a head echo, a specular (transverse) trail echo, and non-specular trail (or range-spread trail) echoes caused by a meteoroid atmospheric entry over (a) Jicamarca at a tropical latitude and another one over (b) the MAARSY radar north of the Arctic circle. Such plots are often referred to as range-time-intensity diagrams.

In both plots, the head echo starts in the upper left corner and is depicted as a thin line of strong intensity showing how the range from the radar to the meteor head echo target decreased as the meteoroid penetrated deeper into the atmosphere.

The meteoroid trajectories in these examples were aligned such that their trails of meteor plasma at time ~ 3 (a) and ~ 10.5 (b) were perpendicular to the direction towards the transmitter. This particular geometry leads to strong specular (transverse) trail echoes. In the figure, these specular echoes are depicted by the horizontal lines of strong signal intensity, i.e. constant range as a function of time.

Finally, Figure 3.1 also contains non-specular (also called range-spread) trail echoes from regions located further away from the radar than the head and the specular trail echo at any particular instant of time and are spread in range.

In this chapter we review the scattering mechanisms and radar techniques used for observations of head echoes, specular trail echoes, and non-specular trail echoes. We also explain how routine meteor trail echo observations are used for determining the wind and temperature in the mesosphere and lower thermosphere. The chapter concludes by discussing suggested future work to improve our current understanding of radar meteors and enhance the contributions to meteor science.

3.2 Meteor Head Echoes

Faint and transient echoes from the ionization in the immediate vicinity of approaching meteors were observed and characterized for the first time by Hey et al. (1947). These appeared in addition to the stronger and more enduring broadside reflections from meteor trails. McKinley and Millman (1949) introduced the concept of head echo to describe the transient echoes from the target moving with the geocentric velocity of a meteoroid.

Jones and Webster (1991) reported about 700 head echoes during different meteor showers from 25 years of observations with SMRs. Their thorough analysis on visual meteors and radar

66 Kero et al.

Table 3.1. *Definitions of abbreviations*

FDTD	Finite-Difference Time Domain
GPS	Global Positioning System
HPLA	High-Power Large-Aperture
MLT	Mesosphere and Lower Thermosphere
PRF	Pulse Repetition Frequency
RCS	Radar Cross Section
SMR	Specular Meteor Radar
SNR	Signal-to-Noise Ratio
UHF	Ultra High Frequency
VHF	Very High Frequency

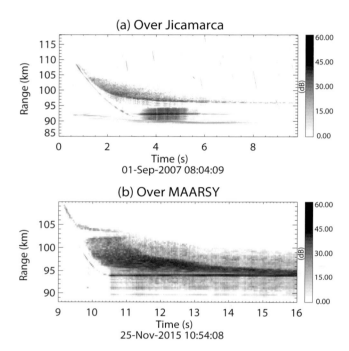

Figure 3.1. Radar observations of the three types of meteor echoes from the same meteoroid, i.e., meteor-head, specular, and non-specular (or range-spread) trail echoes, over (a) Jicamarca, and (b) MAARSY. The figure vertical axis is range, horizontal axis time, and color depicts echo intensity.

head echoes showed that the appearance of the latter seemed to be completely stochastic. They concluded that a prerequisite for a proper theory of head echo scattering is a detailed knowledge of the ionized coma in the vicinity of a single meteoric grain. Steel and Elford (1991), also operating specular radars, deduced that further experimental work needed to employ some completely different techniques from those used in SMRs.

The first efforts to observe meteors with HPLA radars were made in the 1960s at the Royal Radar Establishment in Malvern (Greenhow and Watkins, 1964) and with the Millstone Hill radar at Haystack (Evans, 1965). To increase the observational volume of the narrow beams, the radars were tilted to low elevations. Evans (1966) also pointed the radar towards several low-altitude shower radiants and recorded velocities and decelerations for down-the-beam meteors. These are probably the first reported HPLA radar head echo observations, even though Evans did not mention the concept as such. In the conclusions, Evans points out the complexity of trying to determine accurate mass values for these meteoroids and mentions fragmentation as one of the problems.

Most scientific HPLA radars have been developed in order to apply the incoherent scatter radar technique to study the ionosphere and various phenomena such as aurora. The target is in this case the beam-filling plasma, which, depending on conditions, may require minutes of time-integration to increase the signal above the background noise. This differs fundamentally from the coherent reflections from the transient, strongly Doppler-shifted and compact meteor targets flashing through the radar beam in a fraction of a second.

After Evans (1966), the first new meteor head echoes using HPLA radars were observed with the EISCAT Ultra High Frequency (UHF) radar in December 1990 (Pellinen-Wannberg and Wannberg, 1994). The aim was to study how meteor showers influence sporadic E-layers and their ion content. For this purpose broad filters were employed, which, surprisingly, let the strongly Doppler-shifted meteor head echoes appear in the data.

During the 1990s, signal processing and data storage capabilities progressed rapidly and soon made millisecond time resolution of meteor parameters possible. This opened the observational windows to the ionized coma in the vicinity of a single meteoric grain. Most HPLA radars around the world started to run meteor head echo observations in preparation for the expected Leonid storms when Earth was passing through dust trails of the parent comet, 55P/Tempel-Tuttle, close to its 33-year perihelion. It was a great disappointment that the storm did not cause any significantly increased radar head echo activity even though the sky was full of beautiful visual Leonids (Janches et al., 2000b; Close et al., 2002a; Chau and Woodman, 2004). The immediate conclusion was that these radars are so powerful and sensitive that they see deep into the faint sporadic background, while the number of shower meteors in this size-regime was comparably small.

The HPLA radar systems used for head echo observations have diverse system characteristics in terms of operating frequency, dish or phased array antenna, aperture size etc., as summarized in Table 3.2. The broad range of system characteristics, as well as their locations at different longitudes and latitudes on Earth, make the radar data complementary in several respects. Methods of head echo analysis have been developed more or less independently at several of the facilities, and with emphasis on different aspects of meteor science and/or radio science issues (e.g. Pellinen-Wannberg and Wannberg, 1994; Mathews et al., 1997; Close et al., 2000; Sato et al., 2000; Chau and Woodman, 2004; Mathews et al., 2008; Malhotra and Mathews, 2011; Schult et al., 2013).

3.2.1 Head Echo Scattering – Theory and Observations

In classical meteor radar terminology, meteor plasmas are classified as underdense or overdense, depending on their volume density of electrons. If the density is sufficiently low, radio waves penetrate the trail and scattering occurs independently from the free electrons in the plasma (e.g. Lovell and Clegg, 1948).

Table 3.2. *Characteristics of HPLA radars essential for meteor observations. Dish antenna radars are monostatic (Section 3.2.2.1) unless consisting of several separated receiver stations (Section 3.2.2.2) while phased arrays can be used for interferometric observations (Section 3.2.2.3).*

Radar	Geographical Location	Frequency [MHz]	Antenna and Aperture	Peak Power and Max Duty Cycle
ALTAIR	Kwajalein Atoll Marshall Islands	160 422	Parabolic dish: 1 660 m^2	6MW, 5%
AMISR	Alaska, USA Resolute Bay, Canada	440	Phased array: 715 m^2	2MW, 10%
Arecibo	Puerto Rico	430	Spherical dish: 73 000 m^2	2MW, 6%
EISCAT UHF	Northern Scandinavia	930	Parabolic dish: 800 m^2	2MW, 12%
EISCAT VHF	Northern Scandinavia	224	Parabolic cylinder dish: 4 800 m^2	1.6MW, 12%
EISCAT Svalbard Radar: ESR	Spitsbergen	500	Parabolic dishes: 800 m^2, 1 400 m^2	1MW, 12%
EISCAT 3D	Northern Scandinavia	233	3–5 phased arrays: 3–5 × 3 850 m^2	5–10MW, 25%
Jicamarca	Peru	49.9	Phased array: 85 000 m^2	1.5MW, 6%
MAARSY	Norway	53.5	Phased array: 6 300 m^2	0.8MW, 5%
Millstone Hill	Massachusetts USA	440	Parabolic dishes: 1 660, 3 525 m^2	2.5MW, 6%
MU	Shikaragi Japan	46.5	Phased array: 8 300 m^2	1MW, 5%
PANSY	Showa Station Antarctica	47	Phased array: 18 000	0.5MW, 5%
Sondrestrøm	Greenland	1 290	Parabolic dish: 800	3MW, 3%

If, however, the density is large enough for secondary scattering from electron to electron become important, the electrons no longer behave as independent scatterers. In this case the incident wave cannot penetrate the plasma freely. The number of electrons per m^3 (N) causes the dielectric constant (κ) of an ionized gas to differ from the vacuum value of unity according to

$$\kappa = 1 - \frac{N\lambda^2}{\pi}r_e \simeq 1 - 81\frac{N}{f^2}, \qquad (3.1)$$

where λ is the wavelength of the incident radiation (m), $r_e \approx 2.8 \times 10^{-15}$ m is the classical electron radius and f is the frequency (Hz) corresponding to the radar wavelength. If $\kappa < 0$ throughout an appreciable volume of the ionized gas, the plasma is classified as overdense. This means that the radio waves are reflected from the plasma as they would be from a perfect conductor. However, the transition from underdense to overdense scattering is not expected to be distinct.

3.2.1.1 Critical Frequency, Plasma Frequency and Resonances

The critical frequency obtained from Equation (3.1) when $\kappa = 0$ refers originally to the plane wave with the lowest frequency (f) that can penetrate a deep homogeneous layer of plasma with density N (Herlofson, 1951). The so-defined critical frequency happens to be equal to the plasma frequency of an infinite layer of plasma. It is important to note, however, that the plasma frequency depends on the geometrical shape of the system containing the plasma and not only the charge density. The plasma frequency of a cylindrical shape is determined by $\kappa = -1$; the electric field outside a spherical plasma is even greater and the plasma frequency is determined by $\kappa = -2$ (Herlofson, 1951). Hence the plasma frequency of a spherical shape is reduced to one third of the infinite layer value.

Herlofson (1951) showed that plasma resonance effects are likely to occur for meteor trails if the probing radar frequency is close to the plasma frequency of its charge distribution.

More recently, Dyrud et al. (2008a) investigated the effect of plasma resonances in numerical simulations of meteor head echoes. For radar frequencies near the plasma frequency, large fluctuations occurred in their simulated radar cross section (RCS) as a function of frequency. This could explain why data sets from different radars are difficult to incorporate into one single consistent scattering model.

3.2.1.2 Radar Cross Section

The RCS of meteor head echo targets can be evaluated from signal-to-noise ratio (SNR) by rewriting the classical radar equation (e.g. Skolnik, 1962) as

$$\text{RCS} = \frac{(4\pi)^3 \, P_r \, R^4}{G_r(\theta, \phi) \, G_t(\theta, \phi) \, \lambda^2 \, P_t}, \quad (3.2)$$

where
- P_r = received power,
- R = target range,
- G_t = transmitter antenna gain,
- G_r = receiver antenna gain,
- θ = azimuth of target (positive east of north),
- ϕ = elevation of target,
- λ = radar wavelength, and
- P_t = transmitted power.

The received power is given by

$$P_r = \text{SNR} \times T_{noise} \, k_B \, b_w, \quad (3.3)$$

where SNR is the signal-to-noise ratio, T_{noise} is the equivalent noise temperature, k_B is the Stefan-Boltzmann constant, and b_w is the receiver bandwidth. $T_{noise} = T_{sys} + T_{cosmic}$ is the sum of the system noise (T_{sys}) and the cosmic background radio noise that depends on frequency and pointing direction with respect to radio sources on the celestial sphere.

One of the approaches to find meteoroid mass from head echo observations is to calculate the ratio of meteoroid mass to cross-sectional area, i.e. the ballistic parameter, from the measured deceleration and velocity (e.g. Janches et al., 2000b). This is referred to as the dynamical mass and is highly sensitive to correctly estimated deceleration uncertainties. Estimating the uncertainty of observed deceleration is possible (although difficult) when using interferometric and multi-station head echo radar data, while it is impossible if a monostatic radar without interferometry was used (Kero et al., 2008a). Also, if fragmentation occurs, the dynamical mass will typically represent that of the dominating fragment and not the mass of the meteoroid before it broke into parts.

Another approach is to convert the RCS directly to mass using a scattering model (e.g. Close et al., 2005). Scattering model mass determination does not rely on deceleration, but instead involves converting the RCS measurements to plasma density, and is complicated by still unresolved assumptions about the plasma distribution near the meteoroid as further outlined later in this section. Another issue is the ionization coefficient, which needs to be taken into account (e.g. Williams et al., 2017) and strongly depends on velocity as inferred from laboratory-based ablation experiments (Thomas et al., 2016).

Combining the previous two methods, a third method is to use the measured velocity, altitude, and RCS time profile data of each event for comparison with an ablation model by adjusting the input parameters of test particles propagated down from the top of the atmosphere to the observation altitude (e.g. Kero, 2008; Dyrud and Janches, 2008; Janches et al., 2009; Fucetola, 2012). One of the advantages of this approach is the possibility of estimating the atmospheric entry mass in addition to the instantaneous mass at the time of detection, by modeling the whole mass loss process, part of which may have taken place before detection. As a matter of fact, Dyrud and Janches (2008) found that the original mass above the atmosphere is about one to two orders of magnitude larger than that estimated at the time of observation and up to three orders of magnitude larger than the masses estimated using dynamical methods alone.

Janches et al. (2009) demonstrated an additional advantage of the approach by investigating features in the time profiles of meteor head echoes by comparing them to the output of a differential ablation model, where different chemical constituents are free to ablate at different altitudes along the meteoroid trajectory.

Dynamical modeling to estimate fireball mass from optical data have made use of extended Kalman filters (Sansom et al., 2015). This approach provides a rigorous way of propagating uncertainties in trajectory states, which would be useful also for radar head echo observations.

Fragmentation is an issue which is generally not fully taken into account in any of the three previously mentioned approaches due to the current difficulties in constraining the role of fragmentation in observations. This needs to be properly considered before the observations can be fully used to constrain the meteoroid mass flux (Plane, 2012).

Mathews et al. (1997) and Mathews (2004) interpret the Arecibo 430 MHz UHF radar meteor head echo observations using an underdense coherent scattering mechanism. Here, each electron scatters independently and hence the radar target can be considered as a coherent ensemble of n electrons with a total backscattering cross-section of

$$\sigma_{BS} = 4\pi \, n^2 \, r_e^2, \quad (3.4)$$

where r_e is the classical electron radius defined in Section 3.2.1. Mathews (2004) considers the ensemble of n electrons located within a quarter-wavelength diameter volume from the meteoroid to be contributing to coherent scattering, and, for underdense scattering to be valid, assumes that this distribution is relatively thin.

Wannberg et al. (1996) suggested that EISCAT Very High Frequency (VHF) and UHF observations could be accounted for by assuming an overdense scattering mechanism. The radar target was considered to be a perfectly conducting sphere of plasma with a radius equal to the radius of the isocontour surface where the plasma density drops below critical. The volume around the meteoroid in which this condition is fulfilled was estimated to be a few centimeters wide for submillimeter particles. This is in the Rayleigh scattering regime as discussed further in the next few paragraphs.

Mathews et al. (1997) pointed out a possible flaw in the overdense scattering assumption for the head echo case. The small size of the target means that an electromagnetic wave incident on a plasma should always penetrate it to some depth. The range of this evanescent wave (the skin depth) into a uniform plasma of critical density is proportional to the wavelength.

It was therefore argued that overdense scattering cannot occur for targets that are smaller than the radar wavelength. While this statement is certainly true for uniform density targets, it is not clear what the plasma density profile in the immediate vicinity of a meteoroid looks like. If the density increases monotonically near the meteoroid, the evanescent wave may fade out completely even if the target is small compared to the wavelength.

Marshall and Close (2015) developed a fully three-dimensional Finite-Difference Time Domain (FDTD) physics-based meteor plasma scattering model in order to describe the relationship between a meteor plasma distribution and its RCS. Previous FDTD modeling work reported by Dyrud et al. (2008a,b) was effectively made two-dimensional by choosing a perfect electric conductor boundary in the top and bottom of the box and it did not take into account the background of the Earth's magnetic field.

According to the head echo simulations by Marshall and Close (2015) the overdense assumption – i.e. the cross-sectional area of the meteor plasma where the critical frequency exceeds the radar frequency – is a good approximation of the RCS. The simulation result holds for meteor plasmas that are small as well as large compared to the radar wavelength.

If a head echo target is considered to be a perfectly conducting sphere with radius $r_a \ll \lambda$, the value of its RCS will follow from Rayleigh scattering theory, where RCS is a function of r_a^6. Indeed, this is also what is found in the FDTD model results by Marshall and Close (2015). Furthermore, if the circumference of the sphere is equal to or larger than the radar wavelength, the FDTD simulations indicate that RCS converges to the cross-sectional area of the sphere (πr_a^2) as in the optical regime without Mie resonances that would be expected from scattering off a "hard" target. This is not surprising, as all illuminated electrons should contribute to the head echo, including those outside the overdense region and in the forming trail.

One of the remaining open questions is the actual plasma distributions near the meteoroid. Recent development of kinetic theory for meteor head echoes by Dimant and Oppenheim (2017a,b) suggests that the plasma density within less than a collisional mean free path from the meteoroid surface should drop as l^{-1} where l is the distance from the meteoroid center. Further out, at distances much longer than one mean free path behind the meteoroid, Dimant and Oppenheim (2017b) find that the density decreases as l^{-2}.

Multi-frequency radar observations of the same head echo provide a way to develop and compare different theories and assumptions. Marshall et al. (2017) report a total of sixty-three meteor head echoes observed simultaneously on all three frequencies by the CMOR system (further described in Section 3.3), a data set which was used for constraining the plasma distribution function. Further verification is still needed in terms of simultaneous multi-frequency radar observations spanning a larger range of RCS.

The few multi-frequency HPLA radar head echo data sets available to date are from ALTAIR (Close et al., 2002b), EISCAT (Westman et al., 2004) and Arecibo (Mathews et al., 2010). Both the EISCAT and Arecibo data set were monostatic without interferometry, which means that the antenna gain cannot be taken into account in order to calculate meaningful multi-frequency RCS measurement values to compare with simulations. ALTAIR used two differently polarized feed horns to receive left-circular and right-circular polarized return signals separately, and four additional left-circular receive horns to determine the angle of arrival of the radar return to within a fraction of the beam width (Close et al., 2002b). This multi-frequency data has been used to develop analytical theories and methods to determine meteoroid masses from RCS measurements, as presented in e.g. Close et al. (2004, 2005, 2007, 2012). Head echoes recorded simultaneously with different operating frequencies were found to exhibit larger RCS at the lower operating frequency.

An important limitation pointed out by Marshall et al. (2017) is that the simple relationship between meteor plasma distribution and RCS used as a basis for current simulations (e.g. Marshall et al., 2017; Dimant and Oppenheim, 2017b) is valid only for single body meteoroids. If, or when, a meteoroid breaks apart into fragments, the RCS will be complicated by the new arrangement of mass and the plasma distribution will no longer be spherical. Interference from multiple scattering returns due to two or more fragments will yield RCS measurements that are not straightforward to interpret and model.

Briczinski et al. (2009); Mathews et al. (2010); Malhotra and Mathews (2011); Zhu et al. (2016) find signatures in the vast majority of head echo events observed using HPLA radar systems such as Arecibo, Jicamarca and AMISR that they interpret as potential fragmentation events. These radar systems have different operating frequencies and diverse system characteristics (cf. Table 3.2). Given that these systems are monostatic (with or without interferometry), other effects causing similar signatures in the SNR/RCS time profiles such as differential ablation (Janches et al., 2009) cannot be exclusively differentiated from fragmentation (Kero et al., 2008c). Also, it should be noted that other studies such as that by Sparks et al. (2009), using similar data as Malhotra and Mathews (2011), did not report fragmentation signatures in the vast majority of the echoes. It is clear that fuller meteor physics studies will become possible only when the radio science aspects of this issue are sorted out.

3.2.1.3 High-Altitude Meteors

High-altitude optical meteors have been unambiguously observed (e.g. Fujiwara et al., 1998; Spurný et al., 2000). Analysis of a spectrum from an unusually high Perseid fireball, with a beginning height at 170 km, showed that it contained emissions from atmospheric constituents only while above 130 km altitude (Spurný et al., 2014). Below 110 km altitude, the spectrum was similar to other Perseid fireballs with spectral lines corresponding to the ablation of meteoric constituents. This agrees with the review by Koten et al. (2006) who found that sputtering could explain high-altitude luminosity, while ablation dominates below about 130 km. Koten et al. (2006) found a clear positive relation between observed beginning height and estimated photometric mass for the majority of meteors, also in line with sputtering being the cause of their high-altitude appearance. However, some less bright meteors were also found to begin significantly higher than comparable meteors of similar mass usually do. Koten et al. (2006) suggest high-altitude fragmentation to be the cause, which would increase the total atmospheric cross section of the fragments as compared to a single body, and thereby enable detection.

High-altitude radar meteors have also been reported (Brosch et al., 2013; Gao and Mathews, 2015) but the evidence is not compelling and they remain a source of controversy in the community. Vierinen et al. (2014) showed that high-altitude EISCAT VHF detections reported by Brosch et al. (2013) almost certainly were caused by meteors observed in side lobes. Gao and Mathews (2015) report high-altitude head echoes from observations with the interferometric Jicamarca radar. However, the limited number of baselines in the interferometer layout does not allow them to be distinguished from ambiguous detections of meteors at large range but normal altitude in antenna side lobes.

Vondrak et al. (2008) showed through ablation and sputtering modeling that the sputtering rate at 150 km altitude is 7 orders of magnitude lower than the peak ablation rate in the meteor zone at around 85 km altitude. To be detectable as a head echo target at 150 km altitude due to sputtering therefore requires a large meteoroid. From a statistical point of view, the flux of such large meteoroids is too low to explain the high-altitude radar meteors reported by Brosch et al. (2013) and Gao and Mathews (2015), which favors the explanation that these are actually ambiguous side lobe detections of meteors at lower altitude than reported (Vierinen et al., 2014). Future multi-station radar observations are needed in order to confirm, or deny, the existence of high-altitude radar meteors.

3.2.2 Head Echo Observation Techniques

Meteor head echo experiments have been implemented at all radar facilities listed in Table 3.2. The experiments have in common that they have used transmitted radar pulses of typical lengths from a few to a few hundred microseconds, with a pulse repetition frequency accommodated to utilize the maximum radar system duty cycle. The duty cycle is usually of the order of 5–10%. The radar transmitter power, duty cycle and antenna aperture determine the system sensitivity, but the radar frequency is also important. As was discussed in Section 3.2.1, when the same meteor is observed using more than one radar frequency, its RCS is generally found to have a larger value at the lower radar frequency (Close et al., 2004).

The interpulse period is used for reception of echoes with the range r being determined from the time after transmission t via the relation $r = ct/2$, where c is the speed of light. Head echo analysis starts with scanning the data either in the power domain or the frequency domain, to find occurrence times of meteor events, and then determining the time profiles of the basic parameters: range, line-of-sight velocity, and signal-to-noise ratio (SNR).

3.2.2.1 Monostatic Radar Measurements

Achieving a high accuracy in the radial velocity (the component along the line-of-sight from the radar) from a monostatic radar measurement is straightforward using the Doppler shift and, if more than a single data point is available, pulse-to-pulse phase correlation, as detailed by e.g. Kero et al. (2012a). However, a monostatic radar without interferometry only provides radial velocity; it cannot measure the velocity of the meteoroid along its trajectory, nor the meteor radiant position in the sky, without assumptions on the angle between the meteoroid trajectory and the radar beam. One way to estimate the meteoroid trajectory angle with respect to the beam axis is comparison of the antenna gain pattern with the detected SNR time profile (e.g. Dyrud and Janches, 2008).

Monostatic measurements cannot be used for meteoroid orbit calculations unless it can be assumed that the angle between the meteoroid trajectory/radiant location and the beam pointing direction is arbitrarily small. However, a monostatic radar with a steerable dish can, to a limited extent, be used to distinguish and study meteors from meteor showers by pointing the radar straight at the shower radiant. In fact, this is what was done in some of the first head echo observations by Evans (1966), who pointed the Millstone Hill radar towards several different major meteor shower radiants during their expected maximum meteor activity. Still, a strict filtering has to be applied to the data in order to avoid contamination by detections of sporadic meteors at arbitrary angles to the beam.

The detection rate of shower meteors in head echo data is generally only a few percent of the total sporadic rate due to the very low limiting mass of the observation method and the excess of low-mass sporadic meteors compared to meteor showers (e.g. Kero et al., 2012c). Only during meteor storm activity levels might the rate of shower meteor head echoes possibly outnumber the sporadic head echo rate.

Monostatic measurements are of limited use for studying the orbits of meteoroids as the angle of the trajectory and the beam can only be estimated statistically. Janches et al. (2000a) assumed down-the-beam trajectories (i.e. no tangential velocity) to calculate geocentric meteoroid velocities from the measured vertical (radial) meteoroid velocity component observed with the vertically pointed monostatic Arecibo radar. The measured radial component is a fraction of the meteoroid velocity and its magnitude depends on the angle to the beam. Janches et al. (2000a) interpreted head echoes with small radial velocity components as near-antapex micrometeors with atmospheric speeds below Earth's escape velocity (11.2 km s^{-1}). Such small radial velocity observations more likely correspond to meteors with significant tangential velocity components.

Results from orbit calculations assuming down-the-beam geometries (Janches et al., 2000a, 2001; Meisel et al., 2002) should be interpreted with care, as the radiant direction and geocentric velocity could be far from the assumptions depending on the tangential velocity component. This issue is discussed in Chapter 10 (Hajdukova et al., 2019) in the context of interpreting the reliability of scientific reports claiming meteor observations of interstellar particles.

There is no way of knowing where in the beam pattern a meteoroid trajectory was from head echo data without interferometry, multiple receiver stations or simultaneous optical data. However, that information is needed to convert the measured signal to RCS. If the ionization profile of a meteor is smooth and long enough to pass several side lobes, one approach to try to compensate for the beam pattern is to fit a theoretical antenna beam pattern profile to the measured time profile of the SNR. Janches et al. (2004) used this technique on the 250 strongest Arecibo radar events, which due to their strong SNR are likely to have passed through the main lobe. Most of the events appeared to agree well with the assumption of less than 15° deviation from vertical trajectories. However, these results cannot be generalized to infer trajectories of weaker and/or shorter duration events without clear antenna main beam and side lobe patterns in their SNR time profiles.

Michell et al. (2019) used the Arecibo radar together with an optical imager and found that five out of nineteen simultaneously detected meteors were in the far side lobes of the radar, beyond the main-beam and first side lobe. This indicates that a considerable fraction of the events detected by the UHF radar could be in far side lobes and therefore have significantly underestimated RCS, given the smaller backscattered powers received in the side lobes. Further, deceleration measurements from monostatic observations cannot be used in ballistic equations to estimate meteoroid mass unless the meteoroid is traveling down the beam as was assumed by Janches et al. (2000b). A non-zero tangential component will give rise to an apparent deceleration for a vertically pointed radar and thus an underestimation of the meteoroid mass (Kero et al., 2008a).

Janches et al. (2014b, 2015, 2017) provide good examples of how to utilize monostatic radar observations. Here, models of the zodiacal dust cloud are used to generate radial velocity distributions, which are then compared with the observed radial velocity distribution. Efforts are made to explain the lack of observed small, low-velocity meteoroids that, according to the model, should be the dominant portion of the meteoric mass flux into Earth's atmosphere. In addition, Janches et al. (2017) modeled the head echo aspect sensitivity and found an angular dependence that becomes stronger as the particle mass decreases. Specifically, the angular effect would be particularly strong for the smallest particles currently only detectable by the Arecibo radar, with the conclusion that these smallest meteoroids are expected to produce detectable ionization only if entering the atmosphere on near-vertical trajectories. This would resolve the assumption of down-the-beam geometry for the smallest and slowest meteors. However, using a monostatic radar without additional information such as optical observations (Michell et al., 2019), there is no unambiguous way of distinguishing a slow meteoroid traveling down-the-beam from a faster meteoroid with a significant tangential velocity component. Similarly, there is no unambiguous way of distinguishing a small-RCS meteor in the main lobe from a large-RCS meteor in a far-side lobe.

3.2.2.2 Multi-Station Radar Measurements

An advantage of multi-station (sometimes called multistatic) head echo observations over monostatic observations is the possibility of measuring more than one radial component of the meteoroid velocity vector (Kero et al., 2008a). Three independent measured components enable the calculation of an accurate meteoroid velocity and trajectory direction, thus also the radiant, from a single radar pulse measurement. A time profile of several received pulses gives an accurate velocity profile useful for deceleration determination.

HPLA radars all have main lobe beam widths that are narrow compared to the full extent of typical meteor ionization profiles. Therefore, the observed initial meteor altitude as well as final altitude may be limited by the beam width rather than the onset and end of the ionization process itself. The finite beam width has to be taken into account when statistics of initial and final altitudes are investigated.

High-resolution time profiles enable the measured parameters of range/height, velocity and RCS to be used to study not only meteoroid orbits (Szasz et al., 2008) but also meteoroid-atmosphere interaction processes such as ablation and fragmentation (Kero et al., 2008c). It is also possible to estimate meteoroid masses using the dynamical method, a scattering theory to convert RCS to electron densities, or by comparison of the measured profile to the output of an ablation model (Kero, 2008). Monostatic data can be used for comparison with ablation modeling when proper constraints are taken into account, as was done by e.g. Dyrud and Janches (2008); Janches et al. (2009) using Arecibo.

The tristatic EISCAT radar system offers the unique ability to observe the same meteor from three directions simultaneously. When the tristatic UHF system was used to investigate the aspect angle dependency of meteor head echoes, Kero et al. (2008b) found a very weak aspect sensitivity at the operating frequency of 930 MHz. Meteors were detected at virtually all possible aspect angles, all the way out to 130° from the direction of meteoroid propagation, limited by the antenna pointing directions. This means meteoroids on all atmospheric flight trajectories radiating from above the local horizon give rise to observable head echoes if they enter a radar beam, independent of the direction of the radar beam.

The RCS at the UHF frequency was found to be close to isotropic in the whole range of measurable meteor trajectory alignments occurring during the experiments. This means that the head echo target, at least at the UHF frequency, is consistent with an essentially spherical shape, as was first reported by Close et al. (2002a) who investigated the polarization of meteor head echoes in ALTAIR data. A tristatic polarization study further concluded that the average polarization of UHF head echoes is independent of meteor direction of arrival and echo strength (Wannberg et al., 2011). In observations conducted at the lower 224 MHz VHF operating frequency, using both a vertically oriented beam and a beam tilted north, Vierinen et al. (2014) found enhanced RCS for meteor trajectories near perpendicular to the beam, suggesting a possible contribution from trail electrons to the head echo. The lack of enhanced RCS at the 930 MHz UHF operating frequency for trajectories aligned perpendicular to the beam as compared to the simultaneously measured bistatic RCS at two different non-perpendicular observing geometries shows that the initial trail radius must be large compared to the UHF radar wavelength $\lambda_{UHF} \sim 0.33$ m, and that contributions from trail electrons at that wavelength therefore are negligible.

It should be noted that the aspect sensitivity of meteor head echoes from very small meteoroids according to modeling studies by Janches et al. (2014b, 2015, 2017) differs from the isotropic head echo targets reported from observations with ALTAIR (Close et al., 2002a) and EISCAT UHF (Kero et al., 2008b). Janches et al. (2017) show that this effect should be most significant for the smallest and slowest particles that are at the limit of detection using Arecibo, and negligible for the meteors observed at other radars.

Multi-station observations are especially useful for studying meteoroid fragmentation. Kero et al. (2008c) provided the first strong observational evidence of a submillimeter-sized meteoroid breaking apart into two distinct fragments, and demonstrated the need for having multiple receiver stations to discriminate the signature of fragmentation from other fluctuations in the meteor RCS, as discussed in Chapter 2, Borovička et al. (2019).

3.2.2.3 Interferometric Radar Measurements

A monostatic interferometric radar does not enable an instantaneous measurement of the meteoroid velocity vector, but gives an opportunity to locate the meteor target for each received radar pulse in addition to measuring the radial velocity component. A high-resolution time profile of range, azimuth and elevation enables the change in position of the target to be used for determining the trajectory alignment with respect to the radar beam and converting the radial velocity component to meteoroid velocity. Details for how the pulse-to-pulse position is combined with the simultaneously measured radial velocity are given by e.g. Chau and Woodman (2004); Kero et al. (2012a); Schult et al. (2013).

This conversion enables the radiant to be located and the meteoroid velocity to be determined. The quality depends strongly on the trajectory estimation from the target's transverse motion. Meteoroid radiants can be determined typically within less than $1°$, but careful error estimations have to be performed to determine the uncertainties in the derived parameters (Kero et al., 2012a). Interferometric observations have been key to revealing information on the low-mass component of meteor showers, as well as the sporadic complex, as described in Section 3.2.3.

3.2.3 Meteor Showers and Sporadic Meteor Sources

Interferometric head echo observations by Chau et al. (2007) using the Jicamarca radar provided the first head echo data set where the six sporadic source regions, previously found in other orbital surveys (Jones and Brown, 1993), were recognized. Moreover, being sensitive to a lower mass, the Jicamarca observations revealed new narrow-width source regions collocated with the previously known Apex sources, characterized by a higher mean velocity and therefore observed at higher altitudes than the traditional Apex source meteors. Due to the initial radius effect (cf. Section 3.3.3) this high-velocity apex population is not easily observed by SMRs due to their inefficiency at detecting trails formed by very fast meteoroids.

Modeling work by Wiegert et al. (2009) revealed that retrograde Halley-family comets, and in particular 55P/Tempel-Tuttle, could be the dominant source of these central enhancements. Further, Nesvorný et al. (2011b) studied the fate of small debris particles produced by Oort Cloud Comets. They found that intermediate particle sizes (100 μm) have the highest Earth-impact probability and are expected to produce meteors with radiants within the apex source, also in agreement with the head echo observations.

Kero et al. (2012c) used the interferometric Shigaraki MU radar to distinguish all six sporadic source regions, as well as the ring depleted of meteor radiants at a radius of about $55°$ from the apex first reported from CMOR SMR measurements (Campbell-Brown, 2008).

At the latitude of Shigaraki ($35°$N), the south toroidal meteor source is always below $10°$ elevation and visible above the local horizon only around the autumnal equinox and for a short period of each diurnal cycle. Surprisingly, the south toroidal source is clearly discernable in the MU radar data and its determined orbital properties agree well with the orbital properties found for the north toroidal source (Kero et al., 2012c). When the very disparate equivalent observation times of their respective location on the celestial sphere are taken into account, their relative strengths are also equal. This clearly shows that a narrow-beam interferometric radar pointed towards zenith is also useful for detecting low-elevation meteors.

The MAARSY radar located near Andenes, Norway at high northern latitude ($69°$N) is a modern version of the MU radar with similar characteristics (Schult et al., 2013). A notable difference is that MAARSY is operated in a continuous mode with a time-sliced experiment schedule. This enables meteor head echo observations with a minimum coverage generally exceeding 50%. In contrast, the operation schedule of the MU radar, similar to all other HPLA radars in Table 3.2, only enables head echo observations on a campaign basis.

Schult et al. (2017) report the first continuous head echo survey at northern latitudes longer than two years. The data set contains 900 000 meteors, which is almost an order of magnitude more than the only currently comparable head echo data set, collected at the MU radar with measurement campaigns covering typically one diurnal cycle per month during 2009–2010 (Kero et al., 2012c), and five times more than the Jicamarca data set obtained in less than ninety hours of observations spread over five years (Chau et al., 2007).

The observation programs with Jicamarca, MU and MAARSY have enabled direct measurements of the strengths of the sporadic source regions at the meteoroid mass range accessible with HPLA radars, similar to the observation programs carried out with transverse meteor radar systems further described in Section 3.3. Prior to this, model assumptions were the only way to utilize monostatic HPLA radar diurnal detection rates to infer properties of the meteoroid input function of small particles (e.g. Janches et al., 2006).

The first definitive meteor shower detection using an interferometric HPLA radar was reported by Chau and Galindo (2008) using the Jicamarca radar to study the η-Aquariids and the Perseids. In addition to the Orionids, the MU radar has been used to study the October Draconids (Kero et al., 2012b; Fujiwara et al., 2016), and MU and MAARSY have been used to study the Geminids (Kero et al., 2013; Schult et al., 2013). Recently, Schult et al. (2018) identified thirty-three of the established IAU catalog meteor showers in MAARSY data using a wavelet shower search algorithm. They found that $\sim 1\%$ of all measured head echoes at an estimating limiting mass of 10^{-8} kg were associated with meteor showers.

The tristatic radar studies described in Section 3.2.2.2 showed that the head echo aspect sensitivity, at least at the 930 MHz operating frequency, is near to isotropic and not a reason why meteors with radiants near the local horizon should be less detectable than meteors with radiants located high in the sky. There are, however, other mechanisms giving rise to filtering effects. One of them is a purely geometric effect: the flux of meteors from a given radiant will, if the meteor zone is regarded as a thin layer (with respect to Earth's radius), depend on the radiant altitude angle a as $\sin(a)$. In addition, the ablation profile of the meteor depends on the atmospheric density gradient along the meteoroid trajectory (Janches et al., 2017). Finally, the observation volume of the radar is a beam (with side lobes) and the collecting area (probability of detection) depends on both the angle the trajectory makes with the beam and the maximum RCS of the meteor along its trajectory.

One way to take into account how good detection algorithms implemented on the radar data are at providing detections as a function of radiant elevation and RCS/SNR is to use a probability-of-detection approach similar to that of visual meteors (Zvolankova, 1983). This has to be done on each radar system independently due to their different characteristics and detection algorithms. As an example, when implemented on MU radar data, Kero et al. (2011) found for head echoes from the Orionid meteor shower a radiant altitude dependence $\sim \sin^{1.5}(a)$. When correcting the detection rate of Orionids in the MU radar data for radiant elevation dependence, the activity could be tracked as long as the shower radiant was more than $10°$ above the local horizon. The maximum detection rate of Orionids reached 50 ± 7 per hour when the radiant culminated at $71°$ above the horizon. This rate corresponds to $\sim 7\%$ of the rate of sporadic meteors, which peaked at about 700 ± 26 per hour in the morning hours when the apex source region culminated. Taking into account the estimated collecting area of MU as a function of RCS, varying from about 1 to 1000 km^2 within the RCS span (-40 dBsm < RCS < 20 dBsm) of the detected meteors, Kero et al. (2011) converted the detection rate to a zenithal equivalent flux of Orionids of about ~ 1 km^{-2}h^{-1}. This is the first reported attempt to convert interferometric HPLA radar shower detections to meteoroid flux. As the RCS of meteor head echoes span several orders of magnitude, it is expressed in units of decibel-relative-to-a-square-metre (dBsm), where 0 dBsm is equivalent to the return from a 1 m^2 sphere according to $\log_{10}(1 \text{ m}^2) = 0$ dBsm.

3.3 Specular Meteor Radar Systems and Results

Meteoroids in the millimeter size range produce ionized trails – tapered cylinders, the radius of which depends on height – many kilometers long. The process of scattering radiation off this trail is described in detail in McKinley (1961), Ceplecha et al. (1998) and Baggaley (2002). An underdense trail has a sufficiently low electron density that scattering occurs from all parts of the trail; an overdense trail is radiatively thick, so that scattering occurs essentially at the surface of the cylinder; this is analogous to the under- and overdense head echo cases. When the trail forms at right angles to a radar beam, the radar can receive a strong echo even at low transmitter power because of the length of the Fresnel scattering region. The amplitude and phase of an underdense echo changes in a predictable way as the meteor moves through Fresnel zones. As the meteor moves, the distance between the forming trail and the radar changes, so that echoes from electrons in the scattering regions of the trail are alternately in and out of phase. The first Fresnel zone, responsible for most of the scattered power, lies either side of the t_0 point, which is

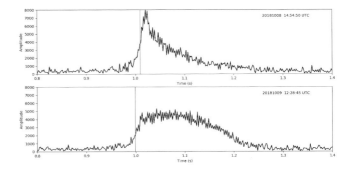

Figure 3.2. (Top) A time-amplitude plot of an underdense echo observed with the Canadian Meteor Orbit Radar (CMOR). (Bottom) A transition/overdense echo from CMOR. The t_0 point is marked on both echoes with a vertical line.

the point of closest approach to the radar and also the specular point; this zone is the region where echoes are in phase with the radiation scattered from the t_0 point. This zone has a length of $\sqrt{2\lambda R_0}$, approximately 1.4 km if the wavelength is 10 m and the range 100 km. Initially, the phase of the received signal oscillates while the amplitude steadily grows; the peak amplitude is reached as the meteor covers the first Fresnel zone; and then the amplitude oscillates as the meteor passes through subsequent Fresnel zones. At the same time, the trail of the underdense echo diffuses outward, causing the amplitude of the signal to decay. Figure 3.2 shows time-amplitude plots for an underdense and overdense echo.

The most fundamental measured quantities for specular scattering are the amplitude and phase of the echo, and its range from the radar; these only require a single transmitter and receiver. For some radars the height may be inferred from the echo ambipolar diffusion time, though this method has considerable uncertainty (see Section 3.4.4). Most astronomical meteor radars use multiple receiving antennas arranged in an interferometer to determine the position of the echo in the sky; a commonly used arrangement is that described in Jones et al. (1998), which uses five receiving antennas spaced by 2.0 and 2.5 wavelengths. With the range, this allows the height of the meteor to be determined with good precision, subject to the uncertainty in the range and ambiguity in the echo direction.

3.3.1 Speed Measurements

The speed of the meteoroid may be deduced from the Fresnel oscillations in amplitude as the meteor traverses the Fresnel zones (Ellyett and Davies, 1948), but because of noise and meteoroid fragmentation (Elford and Campbell, 2001) this method only works for about 10% of meteors. Cervera et al. (1997) used instead the phase of the echo before the meteoroid reaches the t_0 point, called the pre-t_0 method; the phase is not strongly affected by multiple fragments and can be measured well even when the amplitude of the echo is close to the noise level. It works especially well for fast meteors, where the amplitude oscillations may be too rapid to be adequately sampled at the receiver. The phase must be corrected for the drift of the trail in the wind, and it must be unwrapped, since it typically undergoes a number of 2π cycles before reaching the t_0 point. This method was automated by Hocking (2000), who also used the amplitude oscillations in the post-t_0 portion of the echo to constrain the speeds to approximately 1%; it works on approximately 10% of meteor trail echoes. Elford (2004) used a full Fresnel transform to fit both the amplitude and phase of each echo; the meteor is modeled at various speeds, with the speed interval being narrowed successively until the best match is found. This method works on a larger number of echoes than the simpler Fresnel fit; it also produces not only the speed of the meteor, but also the distribution of ionization in the trail. The basic Fresnel method assumes the meteor is a uniform cylinder that lengthens with time, whereas the Fresnel transform models the meteor as a moving ionization distribution, which then provides an image of the meteor, allowing wake and sometimes even individual fragments to be identified, though the modeling process is not necessarily unique. The method was implemented by Holdsworth et al. (2007), who were able to measure speeds for 70% of meteors on a radar with a low pulse repetition frequency (PRF), effectively less than 200 Hz.

Multiple receivers can also be used to find both the speed and the trajectory of a meteor, allowing the orbit to be computed. The Harvard Radio Meteor Project (HRMP) used six outlying receivers separated from the main one by a few kilometers, but did not use them to measure speeds; instead they matched the Fresnel oscillations to find the best speed on each receiver, and used the time differences among the receivers to infer the radiant (Hawkins, 1963). The spacing of the receivers means that each receiver has a different t_0 point, and speed measurements at these echo points allowed decelerations to be measured over the span of the meteor trail. The Advanced Meteor Orbit Radar (AMOR) used a hybrid approach to interferometry, with a narrow vertical fan beam that restricted observations to a narrow range of azimuths, and three receivers to measure the elevation of the echo. In addition, two more receivers were located ≈ 8 km from the main site, forming a right-angled triangle with the main site; time delays on echoes received at the three sites allowed both the speed and trajectory to be determined, without the need for Fresnel analysis, though if Fresnel oscillations were visible, decelerations were also calculated (Baggaley et al., 1994b). The Canadian Meteor Orbit Radar (CMOR) (Jones et al., 2005; Ye et al., 2013) detects meteors over the whole sky, using interferometers for echo location, and has outlying stations (initially two, later upgraded to five) to calculate time-of-flight speeds and trajectories. The additional stations mean a larger number of meteors have trajectories with the necessary geometry to produce the minimum of three echoes at different sites, and for ideal geometry all six stations produce very precise speeds. The Southern Argentina Agile Meteor Radar (SAAMER) Orbital System (Janches et al., 2013) has two outlying stations to calculate orbits. The time-of-flight method does not require a clean Fresnel signal, so it can be performed on fragmenting meteors, but it is slightly less precise, in the range of 3–5% uncertainty.

One difficulty with measuring meteor speeds with SMRs is correcting for meteor deceleration in the atmosphere. Because the amplitude of the signal is attenuated if the radial width of an underdense trail is large compared to the wavelength of the radar, meteors occurring high in the atmosphere are not detected (for typical frequencies, meteors are mainly seen between 80 and 100 km; more details are in Section 3.3.3). Meteors that are detected have therefore already been slowed by interactions with the atmosphere. Brown et al. (2004) developed an empirical correction using shower meteors observed at different heights to obtain deceleration as a function of meteoroid speed and observed height.

3.3.2 Meteor Shower Detections

It is not strictly necessary to measure the trajectories of individual meteors in order to determine the activity of meteors from a particular radiant. Jones and Jones (2006) developed a statistical method (based on Jones, 1977) to find concentrations of radiants in the sky. The basic idea is that meteor trail echoes will occur along a great circle at 90° to the radiant, in order to satisfy the specular condition. One can therefore scan the sky, and for each radiant, count the number of echoes that lie on a band with some small width, centered on the great circle of that radiant. Obviously the bands for different radiants overlap, and therefore sporadic radiants will contaminate shower radiants. Jones determined a weighting function (with a maximum at the center of the band) that falls to zero and then becomes negative at a width exceeding the expected width of the source: radiants far from the trial radiant will have both positive and negative weights and therefore do not contribute to the activity, while meteors belonging to the trial radiant will all be counted. This method is able to locate the radiants of showers with activity significantly above the local sporadic activity to within a degree in single-station CMOR observations, and give an indication of the shower's activity, though sporadic meteors with the same radiant and different speeds will contaminate the measurement. The method was used by Younger et al. (2009) to search for meteor showers in the southern hemisphere using single-station radar data; they found thirty-seven showers, including nine new ones; Janches et al. (2013) used the SAAMER radar to identify thirty-two meteor showers, including two newly identified ones.

If the orbits of meteors have been measured, a search for associated orbits may be performed to find meteor showers. It is not enough to look for clusters of radiants, since sporadic meteors may produce random groupings and contaminate real meteor showers; for more details, see Chapter 9 (Williams et al., 2019). Sekanina (1970a,b) used the Southworth and Hawkins D (D_{SH}) criterion to search for meteors belonging to the major showers. The D criterion is a measure of the difference between orbits: a small value indicates possible association, but the exact value that should be used varies by shower, since some are more dispersed than others. Sekanina gradually increased the D cutoff value, and found using modeling that the number of orbits associated with a shower should plateau around the appropriate cutoff, then increase again as more sporadic meteors are associated with the shower. He found that values of D_{SH} between 0.18 and 0.4 were appropriate to the 11 major showers he studied. Sekanina then used the method on nearly 20,000 HRMP meteors and found 72 streams beyond the 11 used to test the method (Sekanina, 1973).

If the number of orbits is very large, a wavelet search may be done to find showers. Galligan and Baggaley (2002a,b) applied this technique to AMOR radar data, and found six meteor showers; because the AMOR radar was very sensitive, it detected mainly sporadic meteors in evolved orbits. In the wavelet method, a kernel with a certain width in radiant and speed is convolved with the speeds and radiants in the set of meteor orbits: a local maximum in the wavelet coefficient identifies a shower. This method was used on CMOR data (Brown et al., 2008); the initial survey using a 2D radiant wavelet discovered twelve unrecognized showers, and found that the activity of some showers (like the Taurids) was much longer than previously reported. An updated survey (Brown et al., 2010) using a full 3D (radiant plus speed) wavelet found 117 showers, including 62 new showers. The Geminids and Quadrantids were also found to have activity periods much longer than originally thought. One million orbits from SAAMER were used for a southern hemisphere 3D wavelet survey of meteor showers (Pokorný et al., 2017), and found fifty-eight showers, including thirty-four new showers. Most of these new showers were components of the south toroidal source or in the vicinity of that least studied sporadic source, and can be used in the future to constrain the origin of the toroidal complex, as northern hemisphere data was used in Pokorný et al. (2014).

3.3.3 Meteor Fluxes and Mass Distributions

The number of meteors incident on the Earth in a certain mass range is interesting for a number of reasons. It helps to narrow down the global mass influx of meteoroids, a subject with much uncertainty (Plane, 2012). The change in shower flux with time can help to constrain meteoroid models, and estimate the hazard to spacecraft.

The number of meteors per unit area per unit time which strike the Earth depends on the limiting mass. It is therefore important to calculate the limiting mass of a radar as a function of speed, and to have an estimate of the mass distribution of the population being studied in order to compare fluxes from different studies. The limiting mass of a radar can be determined theoretically using the radar parameters to find the limiting received power which will be detected. The power received from a trail as it forms is a sum of the contributions from each electron, taking into account their relative phase – a more complete treatment can be found in Ceplecha et al. (1998) or McKinley (1961). Given that the transmitter power is P_T, the gain of the transmitter and receiver antennas are G_T and G_R, the wavelength of the radar is λ, r_e is the scattering radius of an electron, q is the line density of electrons in the meteor trail, and R_0 is the range from the radar to the t_0 point, the received power is:

$$P_R = P_T G_T G_R \lambda^3 q^2 r_e^2 \left(C^2 + S^2\right) / \left(64\pi^2 R_0^3\right) \quad (3.5)$$

with all quantities in SI units. Here the integral along the trail is taken into account by the term $\left(C^2 + S^2\right)$, where C and S are the Fresnel integrals, $C = \int \cos(\pi x^2/2) dx$ and $S = \int \sin(\pi x^2/2) dx$, where x is the Fresnel parameter, which takes into account the range, wavelength and distance along the trail. This formulation assumes that the decay time of the echo is much larger than the time for the meteor to traverse the Fresnel zones, otherwise the power will be reduced by destructive interference in the diffusing trail. Once the whole trail has formed, and assuming it does not begin or end close to the t_0 point, so that there are a number of Fresnel zones to each side, the Fresnel integrals sum to approximately 2, and the total received power will be:

$$P_R = 2.51 \times 10^{-32} P_T G_T G_R \lambda^3 q^2 / R_0^3. \quad (3.6)$$

For a given threshold received power, the minimum detectable line density can be found using Equation (3.6). The instantaneous mass loss can be found from the line density using the ionization efficiency, β:

$$q = -\frac{\beta}{\mu v} \frac{dm}{dt} \quad (3.7)$$

where μ is the average atomic mass of the meteoroid and v is the speed. The ionization efficiency is very uncertain: see Chapter 1, Popova et al. (2019), for a detailed discussion. In practice, the effective limiting radar magnitude may be higher than the theoretical value, depending on the efficiency of the software used to detect echoes.

The mass distribution of meteor showers is normally assumed to be a power law, such that the number of meteoroids with masses between m and $m + dm$ is:

$$dN \propto m^{-s} dm \quad (3.8)$$

The cumulative number of meteoroids with masses greater than m is proportional to m^{1-s}. A mass index s of 2 indicates that there is equal mass in small particles and large particles; a value less than 2 implies an excess of mass in large particles, and a value greater than 2 implies that there is more mass in small particles. One complication with radar is that only a small fraction of the trail contributes most of the scattered power, so the electron line density which is calculated does not correspond to the total mass, or even the peak ionization of the meteor. Kaiser (1960) showed that even with this random sampling of meteor trails, the distribution of amplitudes has the same slope as the distribution of masses, and that therefore amplitudes can be used to calculate the mass distribution index.

The mass distribution index of sporadic meteors has been measured with AMOR (Baggaley, 1999; Galligan and Baggaley, 2004), and was found to vary over the year from 1.8 to 2.1, with an average value of 2.03. Blaauw et al. (2011b), using CMOR, found an average sporadic value of 2.17, with variation between 2.0 and 2.3; it is not clear if the difference is due to imprecise measurements, or simply differences in the population at different masses. A slightly lower average value of 2.10 was found for sporadic meteors by Pokorný and Brown (2016), also using CMOR. Shower meteor mass distribution indices are normally lower than those of sporadic meteors; Blaauw et al. (2011a) found indices between 1.6 and 1.8 at the peak times of six major meteor showers with CMOR data.

Once the limiting mass and mass distribution are known, the flux can be calculated. There are four main biases which reduce the amplitude of underdense radar echoes and therefore reduce the number of echoes seen by the radar: Faraday rotation, the initial radius effect, the PRF effect, and the finite velocity effect. These are described in detail in Galligan and Baggaley (2004), who used them to correct the orbital distribution of faint meteors at the Earth; each is described later in this section. More details can also be found in Ceplecha et al. (1998).

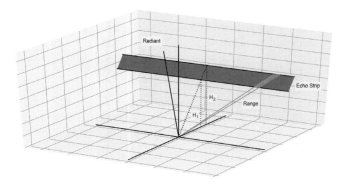

Figure 3.3. Geometry of the strip of sky in which a specular radar receives echoes from a given radiant.

The collecting area of a single-station specular radar for a particular radiant is a strip on a great circle at right angles to the radiant, covering the range of heights at which meteors from the radiant ablate (which is higher for faster meteors and lower for slower ones). Kaiser (1960) developed this technique: one integrates along the echo strip for a meteor shower, taking into account the changing sensitivity of the radar along the echo strip due to range, transmitter and receiver gain, and uses the mass distribution index of the shower to adjust the area to a constant limiting mass. The geometry for the integration is shown in Figure 3.3.

This method was improved with the addition of corrections for the initial radius effect and the finite velocity effect (described later in this section) by Belkovich and Tohktasev (1974a,b), who used it to calculate the flux of Quadrantids. An improved version of this method was used by Ryabova (2008, 2009) to calculate the flux of the Geminids, using numerical integration instead of analytic approximations for the radar sensitivity and the meteor layer thickness.

The method of Kaiser was also improved by Campbell-Brown and Jones (2006), using a more realistic height distribution for faint meteors from video data than that used by Kaiser. They also included a correction for Faraday rotation at this stage, since the attenuation from this effect depends strongly on the direction in the sky. Faraday rotation affects the polarization of the radar beam as it travels through the ionosphere along Earth's magnetic field. For each point on the echo line, the attenuation is $\cos(2\Omega)$, where the Faraday rotation angle is integrated along the path from the radar to the echo line (e.g. Ceplecha et al., 1998):

$$\Omega = \frac{2.36 \times 10^4}{f^2} \int \overline{B} \cos \chi \, N(s) \, ds \qquad (3.9)$$

Here $\overline{B}$ is the local geomagnetic field, χ is the angle between the radar beam path and the magnetic field, N is the electron density, and f is the frequency of the radar. The rotation of the beam is the same on the return trip, and in the same direction, so the total angle by which the polarization is rotated is 2Ω. Faraday rotation affects lower frequency radars more strongly, but because of the height ceiling effect, only the D region of the atmosphere contributes significantly. This means that Faraday rotation is normally only significant for a few hours shortly after local noon, when the ionosphere has its maximum vertical extent; even then, the direction of the radar beam must be close to the direction of the local magnetic field.

The most important correction to the flux is the initial radius correction, which takes into account echoes missed because of destructive interference in wide underdense trails. This effect is more severe for fast meteors, since they ablate high in the atmosphere. At higher altitudes, the lower atmospheric density allows the meteor trail to spread significantly in the time it takes to cross the first Fresnel zone, so that the maximum amplitude reached is much smaller than for a meteor ablating at lower altitude. Radars operating at higher frequencies, with wavelengths of a few meters, are most severely affected, since the beams will suffer more destructive interference. Radars operating at wavelengths of tens of meters are affected much less, but terrestrial interference tends to be worse at those frequencies. The most recent measurements of initial radius using radar were done by Jones et al. (2005), who compared the amplitude of echoes on two of the three CMOR frequencies. They obtained effective radii of order 1 m, and corresponding corrections to the observed rate of meteors varying from 10% at low speeds to more than a factor of two at high speeds. Their results are in broad agreement with other observations of initial radius at heights between 80 and 100 km and with modeling efforts (see Chapter 1, Popova et al. (2019), for a more detailed review). Jones et al. (2005) also emphasized the role of fragmentation in initial radius: in the absence of fragmentation, all trails should have similar electron density profiles and be of similar sizes at the same height. Because fragments can easily spread laterally by a few meters, they can significantly increase the radius of the trail; this produces a scatter in the observed initial radius attenuation with height. At the lowest heights, less scatter is observed, which the authors attribute to the complete ablation of most fragments at the end of the meteor's trail, leaving only the largest.

The PRF effect biases radar observations against meteors with very short durations. In general, a radar must receive a number of pulses from a meteor in order to confirm detection, often 4 to 6. If the meteor has decayed below the noise level in fewer pulses than this, the meteor will not be recorded. Radars with higher frequency (shorter wavelengths) and lower PRFs will be affected more strongly. Galligan and Baggaley (2004) determined that, for AMOR, the PRF effect is only significant at heights above 100 km, where the initial radius effect has already reduced the number of observed echoes significantly.

The finite velocity effect is an additional attenuation for slow meteors. It takes into account the fact that slow meteors take more time to cross the first Fresnel zone, so that more diffusion has taken place when the echo reaches its maximum amplitude compared to fast meteors at the same altitude. Because these slow meteors generally ablate at lower heights, the effect is generally small compared to the initial radius effect.

Once the collecting area for a given radiant has been calculated and corrected for observing biases, the number of echoes on the echo line (as found using the method described in Section 3.3.2) can be used to calculate the flux. Because single-station radars cannot distinguish between shower meteors and sporadics from the same radiant, the sporadic background must be subtracted. Campbell-Brown (2004) used the CMOR flux from the Daytime Arietid radiant before and after the activity period of the shower to estimate the baseline helion sporadic flux, and subtracted this from the calculated Daytime Arietid flux. This method has also been used to calculate fluxes from CMOR observations of the 2005 and 2012 October Draconid outbursts (Campbell-Brown and Jones, 2006; Ye et al., 2014), the η-Aquariid shower (Campbell-Brown and Brown, 2015), and the Camelopardalid outburst in 2014 (Campbell-Brown et al., 2016).

The collecting area of an orbital radar is more complicated, because in order to determine a meteoroid's orbit, in addition to satisfying the specular condition at the main site, meteors must have the required geometry to reflect radiation to at least two remote receivers, constraining further the number of meteors observed. This problem has not, as yet, been solved, and therefore shower activities determined with this method are generally given as wavelet coefficients rather than fluxes (e.g. Brown et al., 2010). Solving this issue would essentially remove the problem of sporadic contamination of shower fluxes.

The activity of meteors from the sporadic sources may also be measured with SMRs. The first analysis, including daytime radiants, of sporadic activity over the Northern hemisphere was performed with HRMP data, and confirmed the existence of the six sporadic sources (Sekanina and Southworth, 1975). The debiasing of this orbital sample was improved by Taylor (1995) and Taylor and Elford (1998), correcting an error in the velocity debiasing used in the original study. A decade of AMOR data was used to redo the orbital distribution (Galligan and Baggaley, 2004); compared to this study, CMOR sees fewer high-speed meteors (Campbell-Brown, 2008), even after debiasing. This may be due to differences in the sporadic populations at the different masses being observed. Sporadic orbital distributions from SAAMER, however, agree more closely with CMOR results in spite of the former's limiting mass being closer to AMOR's than CMOR's (Janches et al., 2015).

Annual variations in the activity of the sporadic sources have been measured by Campbell-Brown and Jones (2006) and Campbell-Brown and Wiegert (2009). The activity of each of the sources varies by a factor of about 30% over the year, and in a consistent way from year to year. Unlike the helion, antihelion, and apex sources, the radiant distribution of the north toroidal source also varies through the year, consistent with it being a series of extended showers rather than a uniform background of evolved meteoroids. The orbits and activity of the sources have been used to constrain meteoroid models (e.g. Wiegert et al., 2009; Pokorný et al., 2014; Nesvorný et al., 2010, 2011a,b).

3.4 Atmospheric Dynamics Using Specular Meteor Radar

Meteor trail plasma is due to collisions with atmospheric constituents immediately decelerated and moves/drifts with the ambient atmosphere. Thus, the drift velocity of the meteor trail is a measure of neutral winds at the altitude of the observation. Based on the interferometric analysis it is possible to measure for each detected meteor with a sufficiently high SNR the position relative to the radar location and a Doppler shift corresponding to the radial motion of the drifting meteor trail.

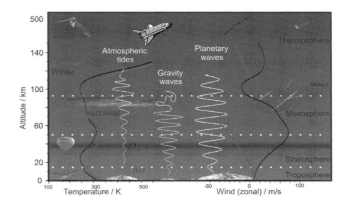

Figure 3.4. Schematic view of the vertical structure of the atmosphere at mid- and high latitudes. (A black and white version of this figure will appear in some formats. For the colour version, please refer to the plate section.)

Specular meteor radar (SMR) observations are valuable sources of information to investigate atmospheric dynamics in the Mesosphere/Lower Thermosphere (MLT), an atmospheric region characterized by a huge variability covering time scales from several years down to minutes. A part of this variability is driven by waves, ranging from planetary waves and atmospheric tides to the smaller scale gravity waves. Planetary waves result from Earth's rotation and atmospheric tides are global-scale oscillations excited by the day-night temperature difference, the Moon's gravity etc. Atmospheric gravity waves form e.g. when air flows over a mountain range and buoyancy pushes air up. Figure 3.4 shows the vertical structure of the atmosphere with a temperature and wind profile and the typical waves and their sources.

During the past decade SMRs have become a widespread scientific instrument around the world. There are radars deployed from Antarctica (Holdsworth et al., 2008) to the Arctic Svalbard (Hall et al., 2002) covering almost all latitudes from the South to the North. Most of these radars are nearly continuously in operation and provide observations of meteor detections, winds, or the mesopause region temperature estimated from the ambipolar diffusing trails. Over the years the system performance of SMRs has evolved as more powerful transmitters became available, leading to a significant enhancement in the number of meteor detections, and hence to much better measurements of the obtained MLT parameters. The increased measurement statistics also led to an enhanced altitude coverage of the obtained winds and to derive the gravity wave momentum flux (Hocking, 2005).

3.4.1 Deriving Mean Horizontal Winds

Most SMRs operate with commercial software packages for detection and initial identification of specular meteors, and also provide a module for obtaining horizontal winds after an internal quality check. Detection algorithms and analysis techniques for ATRAD systems are described by Holdsworth et al. (2004) and for SKiYMET systems by Hocking et al. (2001).

The wind estimation from SMRs uses a procedure similar to that used with Doppler beam swinging radars. As meteors occur randomly in space and time, the data is binned in altitude and time. The wind for each bin is then estimated using the radial wind equation:

$$v_{rad} = u \cos\phi \sin\theta + v \sin\phi \sin\theta + w \cos\theta \qquad (3.10)$$

here u, v, w are the 3D wind components, v_{rad} is the observed radial drift velocity for each meteor, θ and ϕ are the off-zenith and azimuth angle, respectively, measured relative to the radar. The azimuth angle ϕ is given here in the mathematical convention with reference to East and counterclockwise rotation. By binning the meteors in time and altitude it is possible to derive instantaneous winds for each bin. This procedure is described in more detail in Hocking et al. (2001) and is often referred to as the all-sky fit. The first SMRs were able to measure winds in the altitude range between 82–97 km using a 6 kW transmitter. Later systems employing more powerful transmitters could provide mean winds at altitudes between 75–110 km. Typically the resulting winds have a temporal resolution of 1 hour and use a vertical resolution of 2–3 km.

Because of the continuous and autonomous operation of SMRs, they provide valuable information about the mean horizontal winds (e.g., Mitchell et al., 2002; Kumar et al., 2007; Jacobi et al., 2007; Das et al., 2010). The wind analysis is usually done applying a least squares fit, assuming that the vertical wind is negligible ($w \approx 0$); given the large observation volume of up to 400 km in diameter this assumption seems to be well justified. By fitting the radial wind equation in plane geometry it is possible to obtain mean horizontal winds at the MLT, an altitude region that is not easily accessible by other remote sensing techniques.

Over the past decade many scientists have contributed with studies investigating all types of waves from SMR wind measurements. Sandford et al. (2006) showed a spectrum indicating the presence of all of these different waves from SMR wind observations. Figure 3.5 shows a wavelet spectra of more than a year showing the seasonal behavior of planetary waves, tides and gravity waves at a northern hemisphere mid-latitude station. The dominating wave at this latitude through the year is the semi-diurnal tide. As SMRs are so widespread around the world, and the wind measurements are known to be of reasonable quality, it is suitable to validate/compare them with global assimilated models such as the Navy Global Environmental Model (McCormack et al., 2017).

SMR observations of mean winds have provided a significant contribution to the understanding of MLT dynamics and variability covering nearly all scales of waves, and to the study of vertical coupling processes like sudden stratospheric warmings (e.g., Matthias et al., 2012; de Wit et al., 2014). Due to their continuous measurements, it is also possible to investigate atmospheric tides (e.g., Deepa et al., 2006; Conte et al., 2017). Depending on the latitude of the instrument either the diurnal or semi-diurnal tide can be a dominating feature. In particular, at mid-latitudes the semi-diurnal tide can reach up to 100 m s^{-1} amplitude. It is also possible to combine several meteor radars to perform a tidal Hough mode decomposition for a more sophisticated analysis (Yu et al., 2015). Other studies apply phase difference

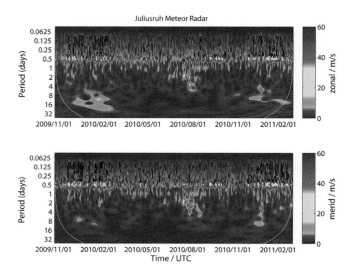

Figure 3.5. Wavelet spectra of zonal and meridional wind measurements for a mid-latitude station in Juliusruh. (A black and white version of this figure will appear in some formats. For the colour version, please refer to the plate section.)

techniques (He et al., 2018) to obtain the wave number, which is not accessible otherwise from local ground-based observations.

Although the procedure outlined in the preceding paragraphs, assuming a plane geometry and a negligible vertical velocity, already provides reasonable results for the horizontal wind, a more sophisticated wind analysis could further improve the quality of the deduced winds and may even provide a chance to drop the assumption of $w = 0$. Egito et al. (2016) tried to infer mean vertical winds applying the standard analysis technique, which resulted in rather high mean vertical motions of several m s^{-1} over five days. It should be noted that the vertical wind estimates are extremely susceptible to contamination by horizontal components even for small errors in estimated zenith angle.

A critical aspect in obtaining winds is the vertical shear of the wind field and the precision of the altitude determination of each meteor. Hocking et al. (2001) uses a mean Earth radius to compute the height above ground for each detected meteor, i.e., spherical approximation, which is valid for relatively small volumes. The height estimation and wind derivation can be further improved using a full Earth geometry and transforming the observed azimuth and off-zenith ϕ and θ into local East-North-Up coordinates using the geodetic longitude and latitude for each meteor. Although the corrections are rather small when compared to the local spherical geometry (typically less than 4° in azimuth and off-zenith angle and 400 m in altitude), they significantly improve the estimated vertical velocities and reduce the error in the altitude computation for each meteor.

Depending on the scientific needs, more complex wind retrieval algorithms using full non-linear error propagation and a regularization to account for the irregular spatial and temporal sampling can be applied (Stober et al., 2017). Such algorithms solve the radial wind equation accounting for the statistical uncertainties of the radial velocity, the zonal, meridional and vertical winds and the uncertainty of the azimuth and off-zenith angle. As the statistical errors for the azimuth and off-zenith angles are not available for all data sets, the normal retrieval uses the values given in Jones et al. (1998). The first iteration is based on the standard least square solution to provide the initial weights and then the algorithm solves the radial wind equation, updating the weights and errors with each iteration. This new retrieval also includes a regularization in space and time. Therefore the vertical and temporal derivative for each wind field component and grid point is computed. This allows an additional weight to be given to each meteor depending on its true occurrence time and altitude compared to the bin center reference time and altitude. The result is a smoother wind in all three vector components, but more importantly, a much more realistic estimate of error in the retrieved quantities.

3.4.2 Multistatic Specular Meteor Radar Measurements

Multi-static meteor observations are well-established for astronomical studies (e.g., Baggaley et al., 1994a; Webster et al., 2004); see Section 3.3. Recently the potential of multi-static meteor measurements to investigate atmospheric dynamics was demonstrated in Stober and Chau (2015) by using a standard SMR and installing a passive receive-only station 118 km away. This concept is called MMARIA (Multi-static Multi-frequency Agile Radar for Investigation of the Atmosphere). The multi-static geometry is beneficial to study the spatial structure in the wind field, and to apply spatial wind retrievals to access higher-order kinematic properties in the wind.

This opens new possibilities to study gravity wave dynamics in the MLT and to observe inhomogeneities in the wind field. A simple but sufficient approximation is the so-called Volume Velocity Processing (Waldteufel and Corbin, 1979). Chau et al. (2017) applied this approach combining the Andenes and Tromsø SMR in northern Norway. The basic idea is to express the zonal and meridional wind component including horizontal gradients:

$$u = u_0 + \frac{\partial u}{\partial x} \times (x - x_0) + \frac{\partial u}{\partial y} \times (y - y_0)$$
$$v = v_0 + \frac{\partial v}{\partial x} \times (x - x_0) + \frac{\partial v}{\partial y} \times (y - y_0). \quad (3.11)$$

The coordinates (x_0, y_0) are a freely selectable reference, which could be the center position between two multi-static stations, (x, y) measures the distance of each meteor to the reference coordinates and u_0, v_0 represents a mean zonal and meridional wind. This approach allows a first order approximation of the wind field in terms of a horizontal divergence to be derived, along with relative vorticity and a stretching and shearing deformation.

As demonstrated in Stober and Chau (2015); Chau et al. (2017) with two systems, it is possible to build more extended SMR networks consisting of several multi-static links or several SMRs with overlapping observation volumes. The first MMARIA network has been installed in Germany and makes use of two standard SMRs in Juliusruh and Collm and three passive links. The passive stations consist of a receiving array in the Jones configuration to measure the angle of arrival for each detected meteor. A schematic of a multi-static forward scatter

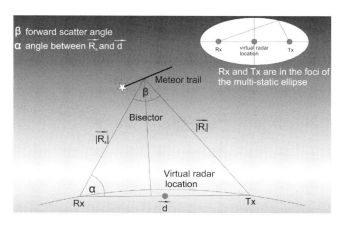

Figure 3.6. Scheme of multi-static meteor detection using one transmitter and a receiver station with interferometry to measure the angle of arrival α. The distance d between the stations is known.

link is shown in Figure 3.6. As the receive station uses an array similar to the standard SMR, the angle α is known.

Further, the radars need to be synchronized with a 1 pulse per second Global Positioning System (GPS) disciplined rubidium clock to ensure sufficient stability in timing and frequency. Due to the synchronization of the radars, the total distance $|\vec{R_i}| + |\vec{R_s}|$ of the transmitted and scattered radio wave is measured. The distance between the stations $\vec{d}$ is taken from the GPS coordinates. As this configuration allows all relevant quantities to be measured, it is possible to determine all triangle parameters directly:

$$|\vec{R_s}| = \frac{R_n^2 - |\vec{d}|^2}{2 \times (R_n - |\vec{d}| \cos(\alpha))}. \quad (3.12)$$

R_n is given for pulsed systems by the employed PRF:

$$R_n = |r| + n \times R_0. \quad (3.13)$$

The r is the measured ambiguous range and R_0 is defined as $R_0 = c/\text{PRF}$ with c being the speed of light. The integer n can take values from $.. -1, 0, 1, 2..$ depending on the PRF. A more detailed description of the procedure is outlined in Stober and Chau (2015).

At present the MMARIA network in Germany has three permanent passive links available: Kuehlungsborn-Juliusruh, Kuehlungsborn-Collm and Juliusruh-Collm. On a campaign basis the network is complemented by making use of two continuous wave transmitters in Luebs and Schwerin, which are received in Kuehlungsborn as well (Vierinen et al., 2016). In the future, the SMR network in Germany is going to be extended by adding further stations around the existing infrastructure to further improve the coverage.

This type of experimental setup is suitable to retrieve arbitrary wind fields by making use of more complicated mathematical algorithms. Although SMRs observe thousands of meteors each day, it is still an ill-posed mathematical problem as the number of unknowns can exceed the number of observations. However, with proper setting of the horizontal resolution and sufficient temporal integration it should be possible to retrieve the horizontal structure of the wind field.

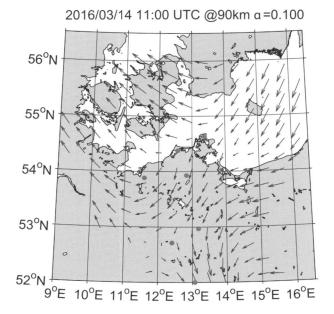

Figure 3.7. Obtained horizontally resolved wind field using the MMARIA Germany meteor radar network.

Figure 3.7 shows an example of a derived wind field making use of the MMARIA network described at the beginning of this section. Preliminary results and specific details of the method are given by Stober et al. (2018).

3.4.3 Estimation of Gravity Wave Momentum Flux

Gravity waves are of particular interest for the MLT as they are expected to contribute significantly to the energy budget in this region (Fritts and Alexander, 2003; Becker, 2012). Gravity waves transport energy and momentum from their source region through the atmosphere until they dissipate. Primary gravity waves excited in the troposphere or lower stratosphere (orographic or from transient sources) propagate upward and transfer their energy and momentum as body forces to the mean flow, which may give rise to secondary waves. To investigate this gravity wave – mean flow interaction in the MLT, Hocking (2005) suggested a straightforward method to estimate the momentum transfer flux. The basic idea is to decompose the radial wind velocities into a mean flow $(\bar{u}, \bar{v}, \bar{w})$ and a gravity wave distortion (u', v', w') given by:

$$\begin{aligned} u &= \bar{u} + u' \\ v &= \bar{v} + v' \\ w &= \bar{w} + w'. \end{aligned} \quad (3.14)$$

Using this ansatz in the radial velocity equation leads to 6 terms describing the squared velocity fluctuations (u'^2, v'^2, w'^2) and the zonal and meridional momentum fluxes $u'w'$ and $v'w'$ and to the term $u'v'$.

However, the technique requires a suitable decomposition of the radial velocities in order to obtain the gravity wave-induced momentum flux. There have been many studies investigating the optimal method to decompose the winds to infer reliable radial velocity fluctuation v'_{rad} (e.g., Antonita et al., 2008; Fritts et al., 2010; Vincent et al., 2010; Placke et al., 2011; Fritts et al., 2012a; Andrioli et al., 2013). One result of all these studies was that the number of meteors required must be much larger than for a successful wind fit. Thus, high power systems like SAAMER in Argentina employing a 60 kW transmitter and using beam forming on transmission (Fritts et al., 2010) are preferable for these studies. Subsequently, similar systems have been deployed in Trondheim (de Wit et al., 2014) and the Drake passage in Antarctica (Fritts et al., 2012b). In spite of providing estimates that are subject to greater uncertainties, Fritts et al. (2012a) showed that conventional radars having lower power and a single transmission antenna are still able to give reasonable gravity wave momentum flux estimates at the altitudes having the highest meteor counts.

Recently, de Wit et al. (2017) provided observational evidence suggesting that secondary gravity waves may be generated in the austral winter polar stratosphere and observed in the MLT. This is an example of how SMRs can provide measurements of fundamental processes that were only predicted by models (Becker and Vadas, 2018).

3.4.4 Temperature Derivation from Ambipolar Diffusion

Over the years, two methods have evolved to derive the atmospheric temperature in the meteor zone: the temperature gradient method proposed in Hocking (1999) and the pressure model approach presented in Nielsen et al. (2001); Hall et al. (2004). The temperature gradient model outlined in Hocking et al. (1997); Hocking (1999) assumes a hydrostatic dependency of the pressure with altitude (barometric height formula of pressure $p = p_0 \exp(mg/RT \times z)$) and takes the vertical derivative to get rid of some constants:

$$T_m = s \left(2 \frac{dT}{dz} + \frac{m\,g}{R} \right). \quad (3.15)$$

Here s is the slope of the vertical profile of the decay time, m is the mean mass of an atmospheric molecule, g is the acceleration due to Earth gravity at the mean altitude, R is the specific gas constant of the atmosphere and dT/dz is the temperature gradient. The temperature so obtained is referenced to the mean altitude of all observed meteors entering the analysis, which is in most cases around 90 km altitude. However, the gradient method requires an *a priori* knowledge or guess of the seasonal dependent vertical temperature gradient, which makes it a semi-empirical method or a model/empirical hybrid.

The pressure model method uses the ambipolar diffusion equation:

$$D_a = k_{amb} \frac{T^2}{p}. \quad (3.16)$$

Here D_a is the ambipolar diffusion constant determined from the decay time, k_{amb} is the zero field mobility constant (Hocking, 1999) and T and p are the atmospheric temperature and pressure, respectively.

Both methods have their advantages and disadvantages as they depend on empirical models of the vertical temperature gradient or the pressure. Typically, satellite observations are used to provide complementary information (e.g., Stober et al., 2008;

Dyrland et al., 2010; Yi et al., 2016, and many other studies). Due to the long and continuous operation of some meteor radars it has even been possible to obtain some temperature trends in the MLT at polar latitudes (Hall et al., 2012). However, Younger et al. (2014) showed that the decay times of underdense meteor trails are reduced in the lower portions of the meteor region due to plasma neutralization initiated by the attachment of positive trail ions to neutral atmospheric molecules. Decreased echo decay times cause meteor radars to produce erroneously high estimates of the ambipolar diffusion coefficient at heights below 90 km, which affects the temperature estimation there. Also, there are still many open questions related to the ambipolar diffusion of meteor plasma. As seen from non-specular trails, the presence of charged particles or dust can alter the diffusion significantly, enhancing or shortening the lifetime of the trail. How good is the assumption of ambipolar diffusion in the case of strong geomagnetic activity due to electron or proton precipitation? These questions remain open and require more observations similar to the ECOMA project (Rapp et al., 2012).

3.4.5 Neutral Air Density Variations

In the last decade the idea to estimate neutral air density variations from SMR observations has been revived. This topic is of great interest to meteor studies, since models of meteoroid ablation must use an atmospheric density profile. As outlined in Section 3.4.4 on temperature estimation from ambipolar diffusing trails, there is also a connection to neutral air density using the ideal gas equation and replacing pressure by density in Equation (3.16). Takahashi et al. (2002) conducted SMR observations and used temperatures from OH-airglow to estimate the neutral air density.

Observations of MLT atmospheric densities are sparse and come mainly from in-situ measurements from sounding rockets (Luebken et al., 1994). However, very often empirical models like the NRLMSIS-00 (Picone et al., 2002) provide at least some value of the atmospheric density. These models are most likely sufficient to represent a certain seasonal behavior, but lack the short-term variability induced by atmospheric waves.

The biggest challenge in measuring the density is presented by the precision of the altitude measurement for each technique. For example, assuming a scale height of 5 km and an altitude resolution of 1 km leads automatically to an error of approximately 20% in the derived densities. This is of particular importance for intercomparisons between different instruments. Satellites can also measure the atmospheric density in the MLT. At present the SABER instrument (Remsberg et al., 2008; Rezac et al., 2015) on board the TIMED satellite and MLS (Microwave Limb Sounder) (Livesey et al., 2006; Schwartz et al., 2008) on board the Aura space vessel are used to infer atmospheric densities in the MLT. However, the measurement errors of those satellite observations are also in the range of 15–20% for the neutral air density at the MLT, and sometimes even larger.

Another technique that appears to be more promising to obtain atmospheric density from SMR observations is related to the height of the meteor layer itself (Jacobi et al., 2007; Lima et al., 2015). These techniques assume that the observed height of the meteor layer corresponds to a constant atmospheric density surface. Younger et al. (2015) argued that the inflection point of the vertical diffusion profile corresponds to a constant density surface, and using MLS satellite temperatures and geopotential heights obtained an absolute measurement of neutral air density under the assumption that the MLS instrument derives a 'true' value.

There is another possible way to obtain atmospheric density variations from SMRs using meteor ablation modeling. For the sporadic meteor population most of the meteoroid parameters are well known, which enables the simulation of meteors of a given mass, entry angle and velocity assuming a mean population corresponding to SMR observations. As a result, any change of the meteor ablation altitude can be used to infer a change of the atmospheric density. Stober et al. (2012) applied this method to measurements of three meteor radars at mid- and high latitudes and showed that the atmospheric densities obtained showed similar wave features as seen in the wind and MLT temperatures. The ablation model used for this study was described in more detail in Schult et al. (2015).

The potential of SMRs to measure changes in the neutral air density either directly or indirectly by observing variations of the altitude of the meteor layer is important to investigate solar cycle responses of the MLT. There are many studies indicating that solar cycle 23 was exceptional with the lowest solar activity in the years 2008–2010 (Jacobi et al., 2011; Stober et al., 2014; Lima et al., 2015; Liu et al., 2017). All of these studies indicated that the MLT showed some solar cycle response. Early work on the effect of solar variability on meteor radar density estimates includes Lindblad (1976); Ellyett and Kennewell (1980).

3.5 Non-Specular Trail Echoes

As mentioned in Section 3.1, there are three main classes of meteor echoes: (a) Head echoes coming from plasma as it forms in front of the meteoroid, (b) specular trail echoes that are due to Fresnel scattering when the radar points perpendicular to the meteor trajectory, and (c) non-specular trail echoes, also known as range-spread trail echoes. Figure 3.1 shows events with all three types of echoes from one meteoroid observed at equatorial latitudes and one meteoroid at polar latitudes with the Jicamarca and the MAARSY HPLA radars, respectively.

The typical characteristics of non-specular echoes are: (a) they are spread in range, (b) they typically last longer than most specular meteor trail echoes, (c) they are mainly observed at VHF frequencies, and (d) they are observed when the radar points non-perpendicular to the meteor trail, i.e., the specular condition is not required. In contrast to the head-echoes and specular (transverse) echoes, the scattering mechanism of non-specular meteor echoes is not as well understood.

Currently there are two main mechanisms explaining non-specular echoes: one related to field-aligned plasma irregularities, where the geometry with respect to the geomagnetic field ($\overline{B}$) is a key parameter (e.g. Chapin and Kudeki, 1994; Dyrud et al., 2007; Yee and Close, 2013; Oppenheim and Dimant, 2015), and one that invokes the presence of charged meteoric dust that reduces the diffusion, i.e., increases the life time, of atmospheric/ionospheric irregularities (e.g., Kelley, 2004). The great majority of non-specular echoes have been associated with the former mechanism. Close et al. (2008) showed that the frequency of non-specular echoes falls off rapidly ($\sim$3 dB per degree) as a radar moves its pointing direction away from perpendicular to the geomagnetic field ($\overline{B}$).

Non-specular echoes associated with field-aligned irregularities have been studied the most, mainly for two reasons: (1) they occur more often, a few hundred per hour at low latitudes, compared to a few per week at high latitudes, and (2) there are more VHF HPLA radars at mid- and low latitudes where perpendicular-to-$\overline{B}$ observations are possible, than at high latitudes where such a condition is not satisfied with the main radar beams.

Observations supporting the importance of charged meteoric dust are very few. Kelley et al. (1998) observed a long duration non-specular meteor trail with the Poker Flat radar in Alaska. Astonishingly, this event was also observed with in-situ instrumentation on board a rocket. Using the SOUSY Svalbard radar, Röttger (2000) reported non-specular meteor echoes that resemble the low-latitude observations. At the time, it was assumed that the echoes came down the beam, due to their strength. No physical process was postulated, but those echoes definitely did not come from field-aligned irregularities. Close et al. (2011) have suggested that, in order to explain the observed features in duration, polarization and pointing angle of the long-duration non-specular meteor trail observed with the ALTAIR radar, both mechanisms are needed.

More recently, Chau et al. (2014) provided additional evidence that non-specular meteor echoes can come from non-field aligned irregularities. They postulated that the echoes exist because of turbulence in charged dust arising from the meteor. The non-specular meteor echoes, as seen from their range-time features in MAARSY data, resemble those frequently observed at low latitudes. However, those low-latitude measurements were for radars pointing close to perpendicular to $\overline{B}$. Using the ALTAIR radar, Close et al. (2008) showed that the signal strength falls off at approximately 3 dB per degree away from perpendicular to $\overline{B}$. If the 3 dB per degree reduction of signals extends to signals this far off from $\overline{B}$, we would expect at least a 250 dB signal reduction, far too much to allow MAARSY to make such a long-duration observation. Therefore, field-aligned plasma irregularities cannot explain non-specular meteor echo observations at high-latitudes.

Kelley (2004) suggested that high Schmidt numbers due to the presence of charged meteor dust could explain non-specular meteor echoes in a manner similar to the explanation of polar mesospheric summer echoes (PMSE). The latter are radar echoes occurring over a broad wavelength range ($\sim$20 cm–100 m) close to the summer mesopause (e.g., Rapp and Lübken, 2004).

Kelley et al. (1998) suggested that a train of charged dust of $\sim$20–50 nm size behind the meteor can produce a high Schmidt number and therefore reduce the electron diffusion rate. These suggestions were supported by in-situ rocket measurements. Later, Kelley et al. (2013) demonstrated that the trails are severely convectively unstable, therefore the ablated charged meteor dust particles immersed in turbulent flow can create a Bragg-scattering similar to PMSE.

The duration of the echoes could range from a few milliseconds to a few minutes. In the case of field-aligned echoes the duration is associated with the angle with respect to $\overline{B}$,
the background atmospheric conditions (e.g., day or night), the electric field, and the velocity and geometry of the meteor (e.g., Chapin and Kudeki, 1994; Dyrud et al., 2007; Malhotra et al., 2007; Sugar et al., 2010; Oppenheim and Dimant, 2015). For example, Yee and Close (2013) have shown that the ambipolar diffusion coefficient is not sufficient, and external electric fields and anomalous cross-field diffusion need to be added. From the high latitude observations it is clear that by adding the effects of meteoric charged dust, the size and composition of the meteoroid plays a significant role, particularly in those events lasting a few seconds or more.

But how long can the charged dust survive? Positively charged dust will be neutralized by attachment of surrounding electrons. Negatively charged dust will be neutralized by an attachment of surrounding ions and by photodetachment during the day. The duration is limited by the time constant of electron attachment, which is inversely proportional to the square of dust radii. For example, for $\sim$30 nm dust grains this time will be $\sim$1 s. For the nighttime conditions recombination of negatively charged dust will be $\sim$237 times longer than the recombination time of positively charged dust. Therefore the time constant will be determined by the diffusion time. For daytime conditions, the lifetime of charged dust depends on the optical properties of dust material, for example the time constant for a 30 nm diameter Fe_2O_3 particle is $\sim 10^{-4}$ s and $\sim$1 s for SiO_2 (e.g., Rapp, 2009; Megner and Gumbel, 2009; Plane et al., 2015). Therefore, for the same meteoroid characteristics (composition, size, velocity, etc.) we expect the associated trail echoes to last longer during the night than during the day.

As in the case of the previous meteor echoes, non-specular meteors are not unique to HPLA radar systems. They are so strong that all conventional meteor systems should be able to detect them. Currently, standard analysis programs on these specular meteor systems do not search for long-lasting meteor trail echoes. For example, during the Leonid meteor shower events of 2001 and 2002, using the Andenes specular meteor system, a few tens of these events were observed during a three-hour period around the peak of the Leonid meteor shower.

Besides understanding these echoes, they have already been used to study the MLT horizontal wind profiles at low latitudes (e.g. Oppenheim et al., 2009; Li et al., 2012; Oppenheim et al., 2014; Li et al., 2014). Each profile can be obtained with relatively high temporal resolution ranging from a few seconds to minutes and high altitudinal resolution (a few hundreds of meters), allowing the identification of dominant wind shears in this region. In addition to the horizontal wind profiles, such echoes might also be used to investigate atmospheric turbulence and meteor dust, if their understanding is improved.

3.6 Future Work

For transverse scatter radars, Fresnel transforms are poised to greatly improve the measured speeds and orbits of meteors, once an automated algorithm has been implemented. Radars using multiple stations will have much better deceleration measurements, providing better constraints to meteoroid properties. Fresnel transforms will also provide information on fragmentation of meteoroids from local peaks in the measured electron density along the trail. Simultaneous radar and high-resolution optical observations will be of particular importance in refining this technique.

Rigorous collecting areas for orbital radars will allow shower fluxes to be better determined without the sporadic contamination present in single-station fluxes. Since many shower radiants are within sporadic source regions, complete elimination of sporadic contamination is impossible, but minimization of sporadic contamination will be further aided by a better characterization of the sporadic flux. Fluxes will be measurable for minor showers with maxima close to the rate of sporadics, and the time interval over which fluxes can be determined for major showers will be increased. Southern hemisphere observations will better constrain models of shower complexes and the sporadic environment.

In order to determine the stability of sporadic fluxes from year to year, the effects of the solar cycle on meteor detection rates must be well understood. In addition to changing the altitudes of meteors (see Section 3.4.5), the density profile of the atmosphere changes, which makes individual meteor echoes brighter or fainter and can alter the rate of detection.

For head echo meteor studies, simultaneous radar and optical observations may provide constraints on some of the free parameters needed to calibrate the radar scattering and photometric mass scales. Recently, Brown et al. (2017) presented a comparison of 105 double-station optical and MAARSY head echo observations and reviewed previously published comparisons at other radars.

Jenniskens et al. (2018) found in a Southern hemisphere survey that video cameras and radar often see different showers and sometimes measure different semi-major axis distributions for the same meteoroid stream. Can we constrain the age or degree of evolution of meteor showers by combined radar-optical observations? Are there high-inclination showers in both hemispheres that could be related? Can we constrain the age of these showers depending on Northern-Southern counterpart existence?

The possibility of combining several meteor radars with overlapping observation volumes opens new possibilities to access mesospheric dynamics. In particular, such multi-static observations allow the space-time ambiguity of gravity waves and vortical modes in the MLT to be resolved. Due to the technological feasibility of using passive remote stations with GPS disciplined rubidium clocks, such meteor radar networks can be deployed with fairly acceptable costs compared to other radars. The German MMARIA network is going to be the first regional meteor network, but other regions like Scandinavia are also suitable to build networks combining existing systems with new stations.

The recently developed wind analysis algorithms to resolve spatially variable winds in the MLT are also applicable to other multi-static measurements such as EISCAT_3D. It may even become possible to derive spatially resolved temperature maps from meteor radars. Such tomographic approaches are rather new to the study of atmospheric processes as multi-static observations were only rarely available in the past.

The EISCAT_3D radar facility is an international research infrastructure project under construction that by 2022 will consist of one 233 MHz, 5 MW transmitter/receiver and two additional receiver sites in Norway, Sweden and Finland (McCrea et al., 2015). EISCAT_3D will enable multistatic interferometric studies of the meteoroid-atmosphere interaction processes and determination of meteoroid orbits (Pellinen-Wannberg et al., 2016) while simultaneously studying variations in the background ionisation within the meteor zone using the incoherent scatter technique. During the 2002 Leonid storm, meteor-induced ionization was observed using the EISCAT UHF radar (Pellinen-Wannberg et al., 2014), but with the radar mode used at that time it only recorded time-integrated data, which did not allow head echo analysis of individual meteor events.

The EISCAT_3D multistatic capability will enable a search for high-altitude meteors without ambiguity issues present in monostatic and interferometric data due to their unresolved side lobes. It will also provide accurate velocity measurements and high enough sensitivity to unambiguously study the presence (or absence) of the small and slow meteor population predicted by models of the zodiacal cloud (Nesvorný et al., 2011a; Janches et al., 2017).

Thanks to their close proximity, it will be possible to combine the 233 MHz EISCAT_3D multi static observation volume with 53 MHz MAARSY interferometric measurements. For a short time frame, before the old EISCAT mainland systems are decommissioned, it will be possible to have monostatic 930 MHz EISCAT UHF measurements in the same volume as well. This three-frequency combination will enable unique reference data for better understanding of the head echo scattering process.

More simultaneous observations involving multiple radar frequencies are underway and needed in order to further develop scattering models. Marshall et al. (2017) investigated head echoes detected at multiple frequencies using CMOR. Head echo observations with SMR-type radar systems like CMOR, SAAMER (Janches et al., 2014a), and similar, demand meteors that are very bright (nearly fireball-class) in order to give rise to large enough RCS. This limits the detection rates to typically a few head echoes per day. However, future routine measurements of head echoes with such systems will enable significant advancement, as these radars are in continuous operation and, over time, will collect valuable statistics to investigate the characteristics of large RCS head echoes.

Future work also includes developing physics-based models taking into account both fragmentation and differential ablation and making a compelling connection between radar signatures and these processes. Finally, these new dynamical and physics-based models of meteoroids and meteors could be used to unify for the first time HPLA and SMR observations.

Acknowledgments

JK was supported by Swedish Research Council Project Grant 2012-4074. The work on MLT wind field dynamics with meteors has been partially funded by the WATILA Project (SAW-2015-IAP-1). The contributions of JDM to this chapter were supported under NSF Grants ATM 07-21613 and AGS 12-02019 to The Pennsylvania State University. EISCAT is an international association supported by research organisations in China (CRIRP), Finland (SA), Japan (NIPR and STEL), Norway (NFR), Sweden (VR), and the United Kingdom (NERC). The Jicamarca Radio Observatory is a facility of the Instituto Geofisico del Peru operated with support from the NSF AGS-1433968 through Cornell University.

References

Andrioli, V. F., Fritts, D. C., Batista, P. P., and Clemesha, B. R. 2013. Improved analysis of all-sky meteor radar measurements of gravity wave variances and momentum fluxes. *Ann. Geophys.*, **31**(5), 889–908.

Antonita, T. M., Ramkumar, G., Kumar, K. K., and Deepa, V. 2008. Meteor wind radar observations of gravity wave momentum fluxes and their forcing toward the Mesospheric Semiannual Oscillation. *J. Geophys. Res. (Atmos.)*, **113**(D12), D10115.

Baggaley, W. J. 1999. Changes in the background meteoroid mass distribution index with orbital characteristics. Pages 311–314 of: Baggaley, W. J., and Porubcan, V. (eds), *Proc. International Conf. held at Tatranska Lomnica, Slovakia, August 17–21, 1998. Meteoroids 1998.* Bratislava: Astronomical Institute, Slovak Academy of Sciences.

Baggaley, W. J. 2002. Radar observations. In: E. Murad and I. P. Williams (eds), *Meteors in the Earth's Atmosphere: Meteoroids and Cosmic Dust and Their Interactions with the Earth's Upper Atmosphere.* Cambridge, UK: Cambridge University Press, pp. 123–147.

Baggaley, W. J., Bennett, R. G. T., Steel, D. I., and Taylor, A. D. 1994a. The advanced meteor orbit radar facility - AMOR. *Quarterly J. R. Astron. Soc.*, **35**(Sept.), 293–320.

Baggaley, W. J., Bennett, R. G. T., Steel, D. I., and Taylor, A. D. 1994b. Radar studies of the orbits of small meteoroids. *Proc. Astron. Soc. Aust.*, **11**(Aug.), 151–156.

Becker, E. 2012. Dynamical control of the middle atmosphere. *Space Sci. Rev.*, **168**, 283–314.

Becker, E., and Vadas, S. L. 2018. Secondary gravity waves in the winter mesosphere: Results from a high-resolution global circulation model. *J. Geophys. Res. (Atmos.)*, **123**(Mar.), 2605–2627.

Belkovich, O. I., and Tohktasev, V. S. 1974a. Determination of the Quadrantid Incident Flux Density. Part I. *Bull. Astr. Inst. Czechosl.*, **25**, 112–115.

Belkovich, O. I., and Tohktasev, V. S. 1974b. Determination of the Quadrantid incident flux density. II. *Bull. Astr. Inst. Czechosl.*, **25**, 370–374.

Blaauw, R. C., Campbell-Brown, M. D., and Weryk, R. J. 2011a. A meteoroid stream survey using the Canadian Meteor Orbit Radar - III. Mass distribution indices of six major meteor showers. *Mon. Not. R. Astron. Soc.*, **414**(July), 3322–3329.

Blaauw, R. C., Campbell-Brown, M. D., and Weryk, R. J. 2011b. Mass distribution indices of sporadic meteors using radar data. *Mon. Not. R. Astron. Soc.*, **412**(Apr.), 2033–2039.

Borovička, J., Macke, R. J., Campbell-Brown, M. D. et al. 2019. Physical and Chemical Properties of Meteoroids. In: G. O. Ryabova, D. J. Asher, and M. D. Campbell-Brown (eds), *Meteoroids: Sources of Meteors on Earth and Beyond.* Cambridge, UK: Cambridge University Press, pp. 37–61.

Briczinski, S. J., Mathews, J. D., and Meisel, D. D. 2009. Statistical and fragmentation properties of the micrometeoroid flux observed at Arecibo. *J. Geophys. Res. (Space Phys.)*, **114**(Apr.), 4311.

Brosch, N., Häggström, I., and Pellinen-Wannberg, A. 2013. EISCAT observations of meteors from the sporadic complex. *Mon. Not. R. Astron. Soc.*, **434**(Oct.), 2907–2921.

Brown, P., Jones, J., Weryk, R. J., and Campbell-Brown, M. D. 2004. The velocity distribution of meteoroids at the Earth as measured by the Canadian Meteor Orbit Radar (CMOR). *Earth Moon Planets*, **95**(Dec.), 617–626.

Brown, P., Weryk, R. J., Wong, D. K., and Jones, J. 2008. A meteoroid stream survey using the Canadian Meteor Orbit Radar. I. Methodology and radiant catalogue. *Icarus*, **195**(May), 317–339.

Brown, P., Wong, D. K., Weryk, R. J., and Wiegert, P. 2010. A meteoroid stream survey using the Canadian Meteor Orbit Radar. II: Identification of minor showers using a 3D wavelet transform. *Icarus*, **207**(May), 66–81.

Brown, P. G., Stober, G., Schult, C. et al. 2017. Simultaneous optical and meteor head echo measurements using the Middle Atmosphere Alomar Radar System (MAARSY): Data collection and preliminary analysis. *Planet. Space Sci.*, **141**(July), 25–34.

Campbell-Brown, M., and Brown, P. G. 2015. A 13-year radar study of the η-Aquariid meteor shower. *Mon. Not. R. Astron. Soc.*, **446**(Feb.), 3669–3675.

Campbell-Brown, M., and Wiegert, P. 2009. Seasonal variations in the north toroidal sporadic meteor source. *Meteorit. Planet. Sci.*, **44**(Jan.), 1837–1848.

Campbell-Brown, M. D. 2004. Radar observations of the Arietids. *Mon. Not. R. Astron. Soc.*, **352**(Aug.), 1421–1425.

Campbell-Brown, M. D. 2008. High resolution radiant distribution and orbits of sporadic radar meteoroids. *Icarus*, **196**(July), 144–163.

Campbell-Brown, M. D., and Jones, J. 2006. Annual variation of sporadic radar meteor rates. *Mon. Not. R. Astron. Soc.*, **367**(Apr.), 709–716.

Campbell-Brown, M. D., Blaauw, R., and Kingery, A. 2016. Optical fluxes and meteor properties of the camelopardalid meteor shower. *Icarus*, **277**(Oct.), 141–153.

Ceplecha, Z., Borovička, J., Elford, W. G. et al. 1998. Meteor phenomena and bodies. *Space Sci. Rev.*, **84**(Sept.), 327–471.

Cervera, M. A., Elford, W. G., and Steel, D. I. 1997. A new method for the measurement of meteor speeds: The pre-t_0 phase technique. *Radio Sci.*, **32**, 805–816.

Chapin, E., and Kudeki, E. 1994. Radar interferometric imaging studies of long-duration meteor echoes observed at Jicamarca. *J. Geophys. Res.*, **99**, 8937–8949.

Chau, J. L., and Galindo, F. 2008. First definitive observations of meteor shower particles using a high-power large-aperture radar. *Icarus*, **194**(Mar.), 23–29.

Chau, J. L., and Woodman, R. F. 2004. Observations of meteor-head echoes using the Jicamarca 50 MHz radar in interferometer mode. *Atmos. Chem. Phys.*, **4**(Mar.), 511–521.

Chau, J. L, Woodman, R. F., and Galindo, F. 2007. Sporadic meteor sources as observed by the Jicamarca high-power large-aperture VHF radar. *Icarus*, **188**, 162–174.

Chau, J. L., Strelnikova, I., Schult, C. et al. 2014. Nonspecular meteor trails from non-field-aligned irregularities: Can they be explained by presence of charged meteor dust? *Geophys. Res. Lett.*, **41**(10), 3336–3343.

Chau, J. L., Stober, G., Hall, C. M. et al. 2017. Polar mesospheric horizontal divergence and relative vorticity measurements using multiple specular meteor radars. *Radio Sci.*, **52**(7), 811–828.

Close, S., Hunt, S. M., Minardi, M. J., and McKeen, F. M. 2000. Analysis of Perseid meteor head echo data collected using the Advanced Research Projects Agency Long-Range Tracking and Instrumentation Radar (ALTAIR). *Radio Sci.*, **35**, 1233–1240.

Close, S., Hunt, S. M., McKeen, F. M., and Minardi, M. J. 2002a. Characterization of Leonid meteor head echo data collected using the VHF-UHF Advanced Research Projects Agency Long-Range Tracking and Instrumentation Radar (ALTAIR). *Radio Sci.*, **37**(1), 010000–1.

Close, S., Oppenheim, M., Hunt, S., and Dyrud, L. 2002b. Scattering characteristics of high-resolution meteor head echoes detected at multiple frequencies. *J. Geophys. Res. (Space Phys.)*, **107**(Oct.), 9–1.

Close, S., Oppenheim, M., Hunt, S., and Coster, A. 2004. A technique for calculating meteor plasma density and meteoroid mass from radar head echo scattering. *Icarus*, **168**(Mar.), 43–52.

Close, S., Oppenheim, M., Durand, D., and Dyrud, L. 2005. A new method for determining meteoroid mass from head echo data. *J. Geophys. Res. (Space Phys.)*, **110**(Sept.), A09308.

Close, S., Brown, P., Campbell-Brown, M., Oppenheim, M., and Colestock, P. 2007. Meteor head echo radar data: Mass velocity selection effects. *Icarus*, **186**(Feb.), 547–556.

Close, S., Hamlin, T., Oppenheim, M., Cox, L., and Colestock, P. 2008. Dependence of radar signal strength on frequency and aspect angle of nonspecular meteor trails. *J. Geophys. Res.*, **113**(A06203).

Close, S., Kelley, M., Vertatschitsch, L. et al. 2011. Polarization and scattering of a long-duration meteor trail. *J. Geophys. Res. (Space Phys.)*, **116**(Jan.), A01309.

Close, S., Volz, R., Loveland, R. et al. 2012. Determining meteoroid bulk densities using a plasma scattering model with high-power large-aperture radar data. *Icarus*, **221**(Sept.), 300–309.

Conte, J. F., Chau, J. L., Stober, G. et al. 2017. Climatology of semidiurnal lunar and solar tides at middle and high latitudes: Interhemispheric comparison. *J. Geophys. Res. (Space Phys.)*, **122**(July), 7750–7760.

Das, S. S., Kumar, K. K., Veena, S. B., and Ramkumar, G. 2010. Simultaneous observation of quasi 16 day wave in the mesospheric winds and temperature over low latitudes with the SKiYMET radar. *Radio Sci.*, **45**(06), 1–11.

de Wit, R. J., Hibbins, R. E., Espy, P. J. et al. 2014. Observations of gravity wave forcing of the mesopause region during the January 2013 major Sudden Stratospheric Warming. *Geophys. Res. Lett.*, **41**(13), 4745–4752.

de Wit, R. J., Janches, D., Fritts, D. C., Stockwell, R. G., and Coy, L. 2017. Unexpected climatological behavior of MLT gravity wave momentum flux in the lee of the Southern Andes hot spot. *Geophys. Res. Lett.*, **44**(Jan.), 1182–1191.

Deepa, V., Ramkumar, G., Antonita, M., Kumar, K. K., and Sasi, M. N. 2006. Vertical propagation characteristics and seasonal variability of tidal wind oscillations in the MLT region over Trivandrum (8.5°N, 77°E): First results from SKiYMET Meteor Radar. *Ann. Geophys.*, **24**(11), 2877–2889.

Dimant, Y. S., and Oppenheim, M. M. 2017a. Formation of plasma around a small meteoroid: 1. Kinetic theory. *J. Geophys. Res. (Space Phys.)*, **122**(4), 4669–4696.

Dimant, Y. S., and Oppenheim, M. M. 2017b. Formation of plasma around a small meteoroid: 2. Implications for radar head echo. *J. Geophys. Res. (Space Phys.)*, **122**(4), 4697–4711.

Dyrland, M. E., Hall, C. M., Mulligan, F. J., Tsutsumi, M., and Sigernes, F. 2010. Improved estimates for neutral air temperatures at 90 km and 78°N using satellite and meteor radar data. *Radio Sci.*, **45**(4), RS4006.

Dyrud, L., and Janches, D. 2008. Modeling the meteor head echo using Arecibo radar observations. *J. Atmos. Solar-Terr. Phys.*, **70**(Sept.), 1621–1632.

Dyrud, L., Wilson, D., Boerve, S. et al. 2008a. Plasma and electromagnetic simulations of meteor head echo radar reflections. *Earth Moon Planets*, **102**(jun), 383–394.

Dyrud, L., Wilson, D., Boerve, S. et al. 2008b. Plasma and electromagnetic wave simulations of meteors. *Adv. Space Res.*, **42**(July), 136–142.

Dyrud, L. P., Kudeki, E., and Oppenheim, M. 2007. Modeling long duration meteor trails. *J. Geophys. Res.*, **112**(Dec.), 12307.

Egito, F., Andrioli, V. F., and Batista, P. P. 2016. Vertical winds and momentum fluxes due to equatorial planetary scale waves using all-sky meteor radar over Brazilian region. *J. Atmos. Solar-Terr. Phys.*, **149**, 108 – 119.

Elford, W. G. 2004. Radar observations of meteor trails, and their interpretation using Fresnel holography: A new tool in meteor science. *Atmos. Chem. Phys.*, **4**(June), 911–921.

Elford, W. G., and Campbell, L. 2001. Effects of meteoroid fragmentation on radar observations of meteor trails. Pages 419–423 of: Warmbein, B. (ed), *Proceedings of the Meteoroids 2001 Conference 6–10 August 2001, Kiruna, Sweden*. ESA Special Publication, vol. 495. Noordwijk: ESA.

Ellyett, C. D., and Davies, J. G. 1948. Velocity of meteors measured by diffraction of radio waves from trails during formation. *Nature*, **161**(Apr.), 596–597.

Ellyett, C. D., and Kennewell, J. A. 1980. Radar meteor rates and atmospheric density changes. *Nature*, **287**(Oct.), 521.

Evans, J. V. 1965. Radio-echo studies of meteors at 68-centimeter wavelength. *J. Geophys. Res.*, **70**(Nov.), 5395–5416.

Evans, J. V. 1966. Radar observations of meteor deceleration. *J. Geophys. Res.*, **71**(Jan.), 171–188.

Fritts, D. C., Janches, D., Iimura, H. et al. 2010. Southern Argentina Agile Meteor Radar: System design and initial measurements of large-scale winds and tides. *J. Geophys. Res. (Atmos.)*, **115**(Sept.), 18112.

Fritts, D. C., Janches, D., Hocking, W. K., Mitchell, N. J., and Taylor, M. J. 2012a. Assessment of gravity wave momentum flux measurement capabilities by meteor radars having different transmitter power and antenna configurations. *J. Geophys. Res. (Atmos.)*, **117**(D10). D10108.

Fritts, D. C., Janches, D., Iimura, H. et al. 2012b. Drake Antarctic Agile Meteor Radar first results: Configuration and comparison of mean and tidal wind and gravity wave momentum flux measurements with Southern Argentina Agile Meteor Radar. *J. Geophys. Res. (Atmos.)*, **117**(D2).

Fritts, D. C., and Alexander, M. J. 2003. Gravity wave dynamics and effects in the middle atmosphere. *Rev. Geophys.*, **41**(1), 1–64.

Fucetola, E. N. 2012. *Determining meteoroid properties using head echo observations from the Jicamarca radio observatory*. Ph.D. thesis, Boston University.

Fujiwara, Y., Ueda, M., Shiba, Y. et al. 1998. Meteor luminosity at 160 km altitude from TV observations for bright Leonid meteors. *Geophys. Res. Lett.*, **25**, 285–288.

Fujiwara, Y., Kero, J., Abo, M., Szasz, C., and Nakamura, T. 2016. MU radar head echo observations of the 2012 October Draconid outburst. *Mon. Not. R. Astron. Soc.*, **455**(Jan.), 3273–3280.

Galligan, D. P., and Baggaley, W. J. 2002a. Wavelet enhancement for detecting shower structure in radar meteoroid data I. methodology. Pages 42–47 of: Green, S. F., Williams, I. P., McDonnell, J. A. M., and McBride, N. (eds), *Proc. IAU Colloq. 181: Dust in the Solar System and Other Planetary Systems*. COSPAR Colloquia Ser., vol. 15. Oxford, UK: Pergamon.

Galligan, D. P., and Baggaley, W. J. 2002b. Wavelet enhancement for detecting shower structure in radar meteoroid data II. Application to the AMOR data. Pages 48–60 of: Green, S. F., Williams, I. P., McDonnell, J. A. M., and McBride, N. (eds), *Proc. IAU Colloq. 181: Dust in the Solar System and Other Planetary Systems*. COSPAR Colloquia Ser., vol. 15. Oxford, UK: Pergamon.

Galligan, D. P., and Baggaley, W. J. 2004. The orbital distribution of radar-detected meteoroids of the Solar system dust cloud. *Mon. Not. R. Astron. Soc.*, **353**(Sept.), 422–446.

Gao, B., and Mathews, J. D. 2015. High-altitude radar meteors observed at Jicamarca Radio Observatory using a multibaseline interferometric technique. *Mon. Not. R. Astron. Soc.*, **452**(Oct.), 4252–4262.

Greenhow, J. S., and Watkins, C. D. 1964. The characteristics of meteor trails observed at a frequency of 300 Mc/s. *J. Atmos. Solar-Terr. Phys.*, **26**(May), 539–542.

Hajdukova, M., Sterken, V., and Wiegert, P. 2019. Interstellar meteoroids. In: G. O. Ryabova, D. J. Asher, and M. D. Campbell-Brown (eds), *Meteoroids: Sources of Meteors on Earth and Beyond*. Cambridge, UK: Cambridge University Press, pp. 235–252.

Hall, C. M., Aso, T., and Tsutsumi, M. 2002. An examination of high latitude upper mesosphere dynamic stability using the Nippon/Norway Svalbard Meteor Radar. *Geophys. Res. Lett.*, **29**(Apr.), 1280–1282.

Hall, C. M., Aso, T., Tsutsumi, M., Höffner, J., and Sigernes, F. 2004. Multi-instrument derivation of 90 km temperatures over Svalbard (78°N 16°E). *Radio Sci.*, **39**(6), RS6001.

Hall, C. M., Dyrland, M. E., Tsutsumi, M., and Mulligan, F. J. 2012. Temperature trends at 90 km over Svalbard, Norway (78°N 16°E), seen in one decade of meteor radar observations. *J. Geophys. Res. (Atmos.)*, **117**(Apr.), D08104.

Hawkins, G. S. 1963. The Harvard radio meteor project. *Smithsonian Contrib. Astrophys.*, **7**, 53–62.

He, M., Chau, J. L., Stober, G. et al. 2018. Relations between semidiurnal tidal variants through diagnosing the zonal wavenumber using a phase dierencing technique based on two ground-based detectors. *J. Geophys. Res. (Atmos.)*, **123**(Apr.), 4015–4026.

Herlofson, N. 1951. Plasma resonance in ionospheric irregularities. *Arkiv för fysik*, **3**(15), 247–297.

Hey, J. S., Parsons, S. J., and Stewart, G. S. 1947. Radar observations of the Giacobinids meteor shower, 1946. *Mon. Not. R. Astron. Soc.*, **107**, 176–183.

Hocking, W. K. 1999. Temperatures using radar-meteor decay times. *Geophys. Res. Lett.*, **26**(21), 3297–3300.

Hocking, W. K. 2000. Real-time meteor entrance speed determinations made with interferometric meteor radars. *Radio Sci.*, **35**(Sept.), 1205–1220.

Hocking, W. K. 2005. A new approach to momentum flux determinations using SKiYMET meteor radars. *Ann. Geophys.*, **23**(7), 2433–2439.

Hocking, W. K., Thayaparan, T., and Jones, J. 1997. Meteor decay times and their use in determining a diagnostic mesospheric Temperature-pressure parameter: Methodology and one year of data. *Geophys. Res. Lett.*, **24**(23), 2977–2980.

Hocking, W. K., Fuller, B., and Vandepeer, B. 2001. Real-time determination of meteor-related parameters utilizing modern digital technology. *J. Atmos. Solar-Terr. Phys.*, **63**, 155–169.

Holdsworth, D. A., Reid, I. M., and Cervera, M. A. 2004. Buckland Park all-sky interferometric meteor radar. *Radio Sci.*, **39**(5), RS5009.

Holdsworth, D. A., Elford, W. G., Vincent, R. A. et al. 2007. All-sky interferometric meteor radar meteoroid speed estimation using the Fresnel transform. *Ann. Geophys.*, **25**(Mar.), 385–398.

Holdsworth, D. A., Murphy, D. J., Reid, I. M., and Morris, R. J. 2008. Antarctic meteor observations using the Davis MST and meteor radars. *Adv. Space Res.*, **42**(July), 143–154.

Jacobi, Ch., Fröhlich, K., Viehweg, C., Stober, G., and Krüschner, D. 2007. Midlatitude mesosphere/lower thermosphere meridional winds and temperatures measured with meteor radar. *Adv. Space Res.*, **39**(8), 1278–1283.

Jacobi, Ch., Hoffmann, P., Placke, M., and Stober, G. 2011. Some anomalies of mesosphere/lower thermosphere parameters during the recent solar minimum. *Adv. Radio Sci.*, **9**, 343–348.

Janches, D., Mathews, J. D., Meisel, D. D., Getman, V. S., and Zhou, Q.-H. 2000a. Doppler studies of near-antapex UHF radar micrometeors. *Icarus*, **143**(Feb.), 347–353.

Janches, D., Mathews, J. D., Meisel, D. D., and Zhou, Q.-H. 2000b. Micrometeor observations using the Arecibo 430 MHz Radar. I. Determination of the ballistic parameter from measured doppler velocity and deceleration results. *Icarus*, **145**(May), 53–63.

Janches, D., Meisel, D. D., and Mathews, J. D. 2001. Orbital properties of the Arecibo micrometeoroids at Earth interception. *Icarus*, **150**(Apr.), 206–218.

Janches, D., Nolan, M. C., and Sulzer, M. 2004. Radiant measurement accuracy of micrometeors detected by the Arecibo 430 MHz Dual-Beam Radar. *Atmos. Chem. Phys.*, **4**(Apr.), 621–626.

Janches, D., Heinselman, C. J., Chau, J. L., Chandran, A., and Woodman, R. 2006. Modeling the global micrometeor input function in the upper atmosphere observed by high power and large aperture radars. *J. Geophys. Res. (Space Phys.)*, **111**(A10), A07317.

Janches, D., Dyrud, L. P., Broadley, S. L., and Plane, J. M. C. 2009. First observation of micrometeoroid differential ablation in the atmosphere. *Geophys. Res. Lett.*, **36**(Mar.), L06101.

Janches, D., Hormaechea, J. L., Brunini, C., Hocking, W., and Fritts, D. C. 2013. An initial meteoroid stream survey in the southern hemisphere using the Southern Argentina Agile Meteor Radar (SAAMER). *Icarus*, **223**(Apr.), 677–683.

Janches, D., Hocking, W., Pifko, S. et al. 2014a. Interferometric meteor head echo observations using the Southern Argentina Agile Meteor Radar. *J. Geophys. Res. (Space Phys.)*, **119**(Mar.), 2269–2287.

Janches, D., Plane, J. M. C., Nesvorný, D. et al. 2014b. Radar detectability studies of slow and small zodiacal dust cloud particles. I. The case of Arecibo 430 MHz meteor head echo observations. *Astrophys. J.*, **796**(Nov.), 41.

Janches, D., Swarnalingam, N., Plane, J. M. C. et al. 2015. Radar Detectability studies of slow and small zodiacal dust cloud particles: II. A study of three radars with different sensitivity. *Astrophys. J.*, **807**(July), 13.

Janches, D., Swarnalingam, N., Carrillo-Sanchez, J. D. et al. 2017. Radar detectability studies of slow and small Zodiacal dust cloud particles. III. The role of sodium and the head echo size on the probability of detection. *Astrophys. J.*, **843**(July), 1.

Jenniskens, P., Baggaley, J., Crumpton, I. et al. 2018. A survey of southern hemisphere meteor showers. *Planet. Space Sci.*, **154**(May), 21–29.

Jones, J. 1977. Meteor radiant distribution using spherical harmonic analysis. *Bull. Astr. Inst. Czechosl.*, **28**, 272–277.

Jones, J., and Brown, P. 1993. Sporadic meteor radiant distributions - Orbital survey results. *Mon. Not. R. Astron. Soc.*, **265**(Dec.), 524.

Jones, J., and Jones, W. 2006. Meteor radiant activity mapping using single-station radar observations. *Mon. Not. R. Astron. Soc.*, **367**(Apr.), 1050–1056.

Jones, J., and Webster, A. R. 1991. Visual and radar studies of meteor head echoes. *Planet. Space Sci.*, **39**(June), 873–878.

Jones, J., Webster, A. R., and Hocking, W. K. 1998. An improved interferometer design for use with meteor radars. *Radio Sci.*, **33**(1), 55–65.

Jones, J., Brown, P., Ellis, K. J. et al. 2005. The Canadian Meteor Orbit Radar: System overview and preliminary results. *Planet. Space Sci.*, **53**(Apr.), 413–421.

Kaiser, T. R. 1960. The determination of the incident flux of radio-meteors. *Mon. Not. R. Astron. Soc.*, **121**, 284–298.

Kelley, M. C. 2004. A new explanation for long-duration meteor radar echoes: Persistent charged dust trains. *Radio Sci.*, **39**(Apr.), 2015.

Kelley, M. C., Alcala, C., and Cho, J. Y. N. 1998. Detection of a meteor contrail and meteoric dust in the Earth's upper mesosphere. *J. Atmos. Solar-Terr. Phys.*, **60**(Feb.), 359–369.

Kelley, M. C., Williamson, C.H.K., and Vlasov, M.N. 2013. Double laminar and turbulent meteor trails observed in space and simulated in the laboratory. *J. Geophys. Res.*, **118**(6), 3622–3625.

Kero, J. 2008. *High-resolution meteor exploration with tristatic radar methods*. Ph.D. thesis, Swedish Institute of Space Physics, Kiruna, Sweden.

Kero, J., Szasz, C., Pellinen-Wannberg, A. et al. 2008a. Determination of meteoroid physical properties from tristatic radar observations. *Ann. Geophys.*, **26**(Aug.), 2217–2228.

Kero, J., Szasz, C., Wannberg, G., Pellinen-Wannberg, A., and Westman, A. 2008b. On the meteoric head echo radar cross section angular dependence. *Geophys. Res. Lett.*, **35**(Apr.), L07101.

Kero, J., Szasz, C., Pellinen-Wannberg, A. et al. 2008c. Three-dimensional radar observation of a submillimeter meteoroid fragmentation. *Geophys. Res. Lett.*, **35**, L04101.

Kero, J., Szasz, C., Nakamura, T. et al. 2011. First results from the 2009–2010 MU radar head echo observation programme for sporadic and shower meteors: The Orionids 2009. *Mon. Not. R. Astron. Soc.*, **416**(Oct.), 2550–2559.

Kero, J., Szasz, C., Nakamura, T. et al. 2012a. A meteor head echo analysis algorithm for the lower VHF band. *Ann. Geophys.*, **30**, 639–659.

Kero, J., Fujiwara, Y., Abo, M., Szasz, C., and Nakamura, T. 2012b. MU radar head echo observations of the 2011 October Draconids. *Mon. Not. R. Astron. Soc.*, **424**(Aug.), 1799–1806.

Kero, J., Szasz, C., Nakamura, T. et al. 2012c. The 2009–2010 MU radar head echo observation programme for sporadic and shower meteors: Radiant densities and diurnal rates. *Mon. Not. R. Astron. Soc.*, **425**(Sept.), 135–146.

Kero, J., Szasz, C., and Nakamura, T. 2013. MU head echo observations of the 2010 Geminids: Radiant, orbit, and meteor flux observing biases. *Ann. Geophys.*, **31**(3), 439–449.

Koten, P., Spurný, P., Borovička, J. et al. 2006. The beginning heights and light curves of high-altitude meteors. *Meteorit. Planet. Sci.*, **41**(Sept.), 1305–1320.

Kumar, K. K., Ramkumar, G., and Shelbi, S. T. 2007. Initial results from SKiYMET meteor radar at Thumba (8.5°N, 77°E): 1. Comparison of wind measurements with MF spaced antenna radar system. *Radio Sci.*, **42**(Dec.), RS6008.

Li, G., Ning, B., Hu, L. et al. 2012. A comparison of lower thermospheric winds derived from range spread and specular meteor trail echoes. *J. Geophys. Res. (Space Phys.)*, **117**(A3), A03310.

Li, G., Ning, B., Chu, Y.-H. et al. 2014. Structural evolution of long-duration meteor trail irregularities driven by neutral wind. *J. Geophys. Res. (Space Phys.)*, **119**(12), 10,348–10,357.

Lima, L. M., Araujo, L. R., Alves, E. O., Batista, P. P., and Clemesha, B. R. 2015. Variations in meteor heights at 22.7°S during solar cycle 23. *J. Atmos. Solar-Terr. Phys.*, **133**, 139–144.

Lindblad, B. A. 1976. Meteor radar rates and the solar cycle. *Nature*, **259**(Jan.), 99–101.

Liu, L., Liu, H., Chen, Y. et al. 2017. Variations of the meteor echo heights at Beijing and Mohe, China. *J. Geophys. Res. (Space Phys.)*, **122**(1), 1117–1127.

Livesey, N. J., Snyder, W. Van, Read, W. G., and Wagner, P. A. 2006. Retrieval algorithms for the EOS Microwave limb sounder (MLS). *IEEE Trans. Geosci. Remote. Sens.*, **44**(5), 1144–1155.

Lovell, A. C. B., and Clegg, J. A. 1948. Characteristics of radio echoes from meteor trails: I. The intensity of the radio reflections and electron density in the trails. *Proc. Phys. Soc.*, **60**(May), 491–498.

Luebken, F.-J., Hillert, W., Lehmacher, G. et al. 1994. Intercomparison of density and temperature profiles obtained by lidar, ionization gauges, falling spheres, datasondes and radiosondes during the DYANA campaign. *J. Atmos. Solar-Terr. Phys.*, **56**(13), 1969–1984.

Malhotra, A., and Mathews, J. D. 2011. A statistical study of meteoroid fragmentation and differential ablation using the Resolute Bay Incoherent Scatter Radar. *J. Geophys. Res. (Space Phys.)*, **116**(Apr.), A04316.

Malhotra, A., Mathews, J. D., and Urbina, J. 2007. Multi-static, common volume radar observations of meteors at Jicamarca. *Geophys. Res. Lett.*, **34**(Dec.), L24103.

Marshall, R. A., and Close, S. 2015. An FDTD model of scattering from meteor head plasma. *J. Geophys. Res. (Space Phys.)*, **120**(7), 5931–5942.

Marshall, R. A., Brown, P., and Close, S. 2017. Plasma distributions in meteor head echoes and implications for radar cross section interpretation. *Planet. Space Sci.*, **143**, 203–208.

Mathews, J. D. 2004. Radio science issues surrounding HF/VHF/UHF radar meteor studies. *J. Atmos. Solar-Terr. Phys.*, **66**(Feb.), 285–299.

Mathews, J. D., Meisel, D. D., Hunter, K. P., Getman, V. S., and Zhou, Q. 1997. Very high resolution studies of micrometeors using the Arecibo 430 MHz radar. *Icarus*, **126**(Mar.), 157–169.

Mathews, J. D., Briczinski, S. J., Meisel, D. D., and Heinselman, C. J. 2008. Radio and meteor science outcomes from comparisons of meteor radar observations at AMISR Poker Flat, Sondrestrom and Arecibo. *Earth Moon Planets*, **102**, 365–372.

Mathews, J. D., Briczinski, S. J., Malhotra, A., and Cross, J. 2010. Extensive meteoroid fragmentation in V/UHF radar meteor observations at Arecibo Observatory. *Geophys. Res. Lett.*, **37**(Feb.), L04103.

Matthias, V., Hoffmann, P., Rapp, M., and Baumgarten, G. 2012. Composite analysis of the temporal development of waves in the polar MLT region during stratospheric warmings. *J. Atmos. Solar-Terr. Phys.*, **90**(Dec.), 86–96.

McCormack, J., Hoppel, K., Kuhl, D. et al. 2017. Comparison of mesospheric winds from a high-altitude meteorological analysis system and meteor radar observations during the boreal winters of 2009–2010 and 2012–2013. *J. Atmos. Solar-Terr. Phys.*, **154**, 132–166.

McCrea, I., Aikio, A., Alfonsi, L. et al. 2015. The science case for the EISCAT_3D radar. *Prog. Earth. Planet. Sci.*, **2**(Dec.), 21.

McKinley, D. W. R. 1961. *Meteor Science and Engineering*. McGraw-Hill Series in Engineering Sciences. New York: McGraw-Hill Book Company, Inc.

McKinley, D. W. R., and Millman, P. M. 1949. A phenomenological theory of radar echoes from meteors. *Proc. of the IRE*, **37**(4), 364–375.

Megner, L., and Gumbel, J. 2009. Charged meteoric particles as ice nuclei in the mesosphere: Part 2. A feasibility study. *J. Atmos. Solar-Terr. Phys.*, **71**(Aug.), 1236–1244.

Meisel, D. D., Janches, D., and Mathews, J. D. 2002. The size distribution of Arecibo interstellar particles and its implications. *Astrophys. J.*, **579**(Nov.), 895–904.

Michell, R.G., DeLucac, M., Janches, D., Chen, R., and Samara, M. 2019. Simultaneous optical and dual-frequency radar observations of small mass meteors at Arecibo. *Planet. Space Sci.*, **166**, 1–8 doi:10.1016/j.pss.2018.07.015.

Mitchell, N. J., Pancheva, D., Middleton, H. R., and Hagan, M. E. 2002. Mean winds and tides in the Arctic mesosphere and lower thermosphere. *J. Geophys. Res. (Space Phys.)*, **107**(Jan.), 1004.

Nesvorný, D., Jenniskens, P., Levison, H. F. et al. 2010. Cometary origin of the zodiacal cloud and carbonaceous micrometeorites. Implications for hot debris disks. *Astrophys. J.*, **713**(Apr.), 816–836.

Nesvorný, D., Janches, D., Vokrouhlický, D. et al. 2011a. Dynamical model for the zodiacal cloud and sporadic meteors. *Astrophys. J.*, **743**(Dec.), 129–144.

Nesvorný, D., Vokrouhlický, D., Pokorný, P., and Janches, D. 2011b. Dynamics of dust particles released from Oort cloud comets and their contribution to radar meteors. *Astrophys. J.*, **743**(Dec.), 37.

Nielsen, K. P., Röttger, J., and Sigernes, F. 2001. Simultaneous measurements of temperature in the upper mesosphere with an Ebert-Fastie Spectrometer and a VHF meteor radar on Svalbard (78°N, 16°E). *Geophys. Res. Lett.*, **28**(5), 943–946.

Oppenheim, M. M., and Dimant, Y. S. 2015. First 3-D simulations of meteor plasma dynamics and turbulence. *Geophys. Res. Lett.*, **42**(3), 681–687.

Oppenheim, M. M., Sugar, G., Slowey, N. O. et al. 2009. Remote sensing lower thermosphere wind profiles using non-specular meteor echoes. *Geophys. Res. Lett.*, **36**(9), L09817.

Oppenheim, M. M., Arredondo, S., and Sugar, G. 2014. Intense winds and shears in the equatorial lower thermosphere measured by high-resolution nonspecular meteor radar. *J. Geophys. Res. (Space Phys.)*, **119**(3), 2178–2186.

Pellinen-Wannberg, A., and Wannberg, G. 1994. Meteor observations with the European Incoherent Scatter UHF radar. *J. Geophys. Res.*, **99**(A6), 11379–11390.

Pellinen-Wannberg, A. K., Häggström, I., Carrillo Sánchez, J. D., Plane, J. M. C., and Westman, A. 2014. Strong E region ionization caused by the 1767 trail during the 2002 Leonids. *J. Geophys. Res. (Space Phys.)*, **119**(Sept.), 7880–7888.

Pellinen-Wannberg, A., Kero, J., Häggström, I., Mann, I., and Tjulin, A. 2016. The forthcoming EISCAT_3D as an extra-terrestrial matter monitor. *Planet. Space Sci.*, **123**(Apr.), 33–40.

Picone, J. M., Hedin, A. E., Drob, D. P., and Aikin, A. C. 2002. NRLMSISE-00 empirical model of the atmosphere: Statistical comparisons and scientific issues. *J. Geophys. Res. (Space Phys.)*, **107**(A12), 1468.

Placke, M., Hoffmann, P., Becker, E. et al. 2011. Gravity wave momentum fluxes in the MLT Part II: Meteor radar investigations at high and midlatitudes in comparison with modeling studies. *J. Atmos. Solar-Terr. Phys.*, **73**(9), 911–920.

Plane, J. M. C. 2012. Cosmic dust in the earth's atmosphere. *Chemical Society Reviews*, **41**(Apr.), 6507–6518.

Plane, J. M. C., Feng, W., and Dawkins, E. C. M. 2015. The mesosphere and metals: Chemistry and changes. *Chem. Rev.*, **115**(10), 4497–4541.

Pokorný, P., and Brown, P. G. 2016. A reproducible method to determine the meteoroid mass index. *Astron. Astrophys.*, **592**(Aug.), A150.

Pokorný, P., Vokrouhlický, D., Nesvorný, D., Campbell-Brown, M., and Brown, P. 2014. Dynamical model for the toroidal sporadic meteors. *Astrophys. J.*, **789**(July), 25–45.

Pokorný, P., Janches, D., Brown, P. G., and Hormaechea, J. L. 2017. An orbital meteoroid stream survey using the Southern Argentina Agile MEteor Radar (SAAMER) based on a wavelet approach. *Icarus*, **290**(July), 162–182.

Popova, O., Borovička, J., and Campbell-Brown, M. 2019. Modelling the Entry of Meteoroids. In: G. O. Ryabova, D. J. Asher, and M. D. Campbell-Brown (eds), *Meteoroids: Sources of Meteors on Earth and Beyond*. Cambridge, UK: Cambridge University Press, pp. 9–36.

Rapp, M. 2009. Charging of mesospheric aerosol particles: The role of photodetachment and photoionization from meteoric smoke and ice particles. *Ann. Geophys.*, **27**(6), 2417–2422.

Rapp, M., and Lübken, F.-J. 2004. Polar mesosphere summer echoes (PMSE): Review of observations and current understanding. *Atmos. Chem. Phys.*, **4**(Dec), 2601.

Rapp, M., Plane, J. M. C., Strelnikov, B. et al. 2012. In situ observations of meteor smoke particles (MSP) during the Geminids 2010: Constraints on MSP size, work function and composition. *Ann. Geophys.*, **30**(Dec.), 1661–1673.

Remsberg, E. E., Marshall, B. T., Garcia-Comas, M. et al. 2008. Assessment of the quality of the Version 1.07 temperature-versus-pressure profiles of the middle atmosphere from TIMED/SABER. *J. Geophys. Res. (Atmos.)*, **113**(D17), D17101.

Rezac, L., Jian, Y., Yue, J. et al. 2015. Validation of the global distribution of CO_2 volume mixing ratio in the mesosphere and lower thermosphere from SABER. *J. Geophys. Res. (Atmos.)*, **120**(23), 12,067–12,081.

Röttger, J. 2000. Radar investigations of the mesosphere, stratosphere and the troposphere in Svalbard. *Adv. Polar Upper Atmosph. Res.*, **14**, 202–220.

Ryabova, G. O. 2008. Calculation of the incident flux density of meteors by numerical integration I. *WGN*, **36**(Dec.), 120–123.

Ryabova, G. O. 2009. Calculation of the incident flux density of meteors by numerical integration II. *WGN*, **37**(Apr.), 63–67.

Sandford, D. J., Muller, H. G., and Mitchell, N. J. 2006. Observations of lunar tides in the mesosphere and lower thermosphere at Arctic and middle latitudes. *Atmos. Chem. Phys.*, **6**(12), 4117–4127.

Sansom, E. K., Bland, P., Paxman, J., and Towner, M. 2015. A novel approach to fireball modeling: The observable and the calculated. *Meteorit. Planet. Sci.*, **50**(Aug.), 1423–1435.

Sato, T., Nakamura, T., and Nishimura, K. 2000. Orbit Determination of Meteors Using the MU Radar. *IEICE Trans. Comm.*, **E83-B**(9), 1990–1995.

Schult, C., Stober, G., Chau, J. L., and Latteck, R. 2013. Determination of meteor-head echo trajectories using the interferometric capabilities of MAARSY. *Ann. Geophys.*, **31**(Oct.), 1843–1851.

Schult, C., Stober, G., Keuer, D., and Singer, W. 2015. Radar observations of the Maribo fireball over Juliusruh: Revised trajectory and meteoroid mass estimation. *Mon. Not. R. Astron. Soc.*, **450**(2), 1460–1464.

Schult, C., Stober, G., Janches, D., and Chau, J. L. 2017. Results of the first continuous meteor head echo survey at polar latitudes. *Icarus*, **297**(Nov.), 1–13.

Schult, C., Brown, P., Pokorný, P., Stober, G., and Chau, J. L. 2018. A meteoroid stream survey using meteor head echo observations from the Middle Atmosphere ALOMAR Radar System (MAARSY). *Icarus*, **309**(July), 177–186.

Schwartz, M. J., Lambert, A., Manney, G. L. et al. 2008. Validation of the Aura Microwave Limb Sounder temperature and geopotential height measurements. *J. Geophys. Res. (Atmos.)*, **113**(D15). D15S11.

Sekanina, Z. 1970a. Statistical model of meteor streams. I. Analysis of the model. *Icarus*, **13**(Nov.), 459–474.

Sekanina, Z. 1970b. Statistical model of meteor streams. II. Major showers. *Icarus*, **13**(Nov.), 475–493.

Sekanina, Z. 1973. Statistical model of meteor streams. III. Stream search among 19303 radio meteors. *Icarus*, **18**(Feb.), 253–284.

Sekanina, Z., and Southworth, R. B. 1975. Physical and dynamical studies of meteors. Meteor-fragmentation and stream-distribution studies, NASA Contractor Report CR-2615. Cambridge, MA: Smithsonian Astrophysical Observatory.

Skolnik, Merrill I. 1962. *Introduction to Radar Systems*, International student edn. Tokyo: McGraw-Hill Kogakusha, Ltd.

Sparks, J. J., Janches, D., Nicolls, M. J., and Heinselman, C. J. 2009. Seasonal and diurnal variability of the meteor flux at high latitudes observed using PFISR. *J. Atmos. Solar-Terr. Phys.*, **71**(May), 644–652.

Spurný, P., Betlem, H., Jobse, K., Koten, P., and van't Leven, J. 2000. New type of radiation of bright Leonid meteors above 130 km. *Meteorit. Planet. Sci.*, **35**(Sept.), 1109–1115.

Spurný, P., Shrbený, L., Borovička, J. et al. 2014. Bright Perseid fireball with exceptional beginning height of 170 km observed by different techniques. *Astron. Astrophys.*, **563**(Mar.), A64.

Steel, D. I., and Elford, W. G. 1991. The height distribution of radio meteors - Comparison of observations at different frequencies on the basis of standard echo theory. *J. Atmos. Solar-Terr. Phys.*, **53**(May), 409–417.

Stober, G., and Chau, J. L. 2015. A multistatic and multifrequency novel approach for specular meteor radars to improve wind measurements in the MLT region. *Radio Sci.*, **50**(5), 431–442.

Stober, G., Jacobi, Ch., and Fr K. 2008. Meteor radar temperatures over Collm (51.3°N, 13°E). *Adv. Space Res.*, **42**(7), 1253–1258.

Stober, G., Jacobi, C., Matthias, V., Hoffmann, P., and Gerding, M. 2012. Neutral air density variations during strong planetary wave activity in the mesopause region derived from meteor radar observations. *J. Atmos. Solar-Terr. Phys.*, **74**, 55–63.

Stober, G., Matthias, V., Brown, P., and Chau, J. L. 2014. Neutral density variation from specular meteor echo observations spanning one solar cycle. *Geophys. Res. Lett.*, **41**(19), 6919–6925.

Stober, G., Matthias, V., Jacobi, C. et al. 2017. Exceptionally strong summer-like zonal wind reversal in the upper mesosphere during winter 2015/16. *Ann. Geophys.*, **35**(3), 711–720.

Stober, G., Chau, J. L., Vierinen, J., Jacobi, C., and Wilhelm, S. 2018. Retrieving horizontally resolved wind fields using multistatic meteor radar observations. *Atmos. Meas. Tech.*, **11**(Aug.), 4891–4907.

Sugar, G., Oppenheim, M. M., Bass, E., and Chau, J. L. 2010. Nonspecular meteor trail altitude distributions and durations observed by a 50 MHz high-power radar. *J. Geophys. Res. (Space Phys.)*, **115**(A12), A12334.

Szasz, C., Kero, J., Meisel, D. D. et al. 2008. Orbit characteristics of the tristatic EISCAT UHF meteors. *Mon. Not. R. Astron. Soc.*, **388**(July), 15–25.

Takahashi, H., Nakamura, T., Tsuda, T., Buriti, R. A., and Gobbi, D. 2002. First measurement of atmospheric density and pressure by meteor diffusion coefficient and airglow OH temperature in the mesopause region. *Geophys. Res. Lett.*, **29**(8), 1165.

Taylor, A. D. 1995. The Harvard Radio Meteor Project meteor velocity distribution reappraised. *Icarus*, **116**(July), 154–158.

Taylor, A. D., and Elford, W. G. 1998. Meteoroid orbital element distributions at 1 AU deduced from the Harvard Radio Meteor Project observations. *Earth Planets Space*, **50**(June), 569–575.

Thomas, E., Horányi, M., Janches, D. et al. 2016. Measurements of the ionization coefficient of simulated iron micrometeoroids. *Geophys. Res. Lett.*, **43**(Apr.), 3645–3652.

Vierinen, J., Fentzke, J., and Miller, E. 2014. An explanation for observations of apparently high-altitude meteors. *Mon. Not. R. Astron. Soc.*, **438**(Mar.), 2406–2412.

Vierinen, J., Chau, J. L., Pfeffer, N., Clahsen, M., and Stober, G. 2016. Coded continuous wave meteor radar. *Atmos. Meas. Tech.*, **9**(2), 829–839.

Vincent, R. A., Kovalam, S., Reid, I. M., and Younger, J. P. 2010. Gravity wave flux retrievals using meteor radars. *Geophys. Res. Lett.*, **37**(July), L14802.

Vondrak, T., Plane, J. M. C., Broadley, S., and Janches, D. 2008. A chemical model of meteoric ablation. *Atmos. Chem. Phys.*, **8**(Dec.), 7015–7031.

Waldteufel, P., and Corbin, H. 1979. On the Analysis of Single-Doppler Radar Data. *J. Appl. Meteor.*, **18**(2), 532–542.

Wannberg, G., Pellinen-Wannberg, A., and Westman, A. 1996. An ambiguity-function-based method for analysis of Doppler decompressed radar signals applied to EISCAT measurements of oblique UHF-VHF meteor echoes. *Radio Sci.*, **31**, 497–518.

Wannberg, G., Westman, A., and Pellinen-Wannberg, A. 2011. Meteor head echo polarization at 930 MHz studied with the EISCAT UHF HPLA radar. *Ann. Geophys.*, **29**(6), 1197–1208.

Webster, A. R., Brown, P. G., Jones, J., Ellis, K. J., and Campbell-Brown, M. D. 2004. Canadian Meteor Orbit Radar (CMOR). *Atmos. Chem. Phys.*, **4**(May), 679–684.

Westman, A., Wannberg, G., and Pellinen-Wannberg, A. 2004. Meteor head echo altitude distributions and the height cutoff effect studied with the EISCAT HPLA UHF and VHF radars. *Ann. Geophys.*, **22**, 1575–1584.

Wiegert, P., Vaubaillon, J., and Campbell-Brown, M. 2009. A dynamical model of the sporadic meteoroid complex. *Icarus*, **201**(May), 295–310.

Williams, E. R., Wu, Y.-J., Chau, J., and Hsu, R.-R. 2017. Intercomparison of radar meteor velocity corrections using different ionization coefficients. *Geophys. Res. Lett.*, **44**(June), 5766–5773.

Williams, I. P., Jopek, T. J., Rudawska, R., Tóth, J., and Kornoš, L. 2019. Minor Meteor Showers and the Sporadic Background. In: G. O. Ryabova, D. J. Asher, and M. D. Campbell-Brown (eds), *Meteoroids: Sources of Meteors on Earth and Beyond*. Cambridge, UK: Cambridge University Press, pp. 210–234.

Ye, Q., Brown, P. G., Campbell-Brown, M. D., and Weryk, R. J. 2013. Radar observations of the 2011 October Draconid outburst. *Mon. Not. R. Astron. Soc.*, **436**(Nov.), 675–689.

Ye, Q., Wiegert, P. A., Brown, P. G., Campbell-Brown, M. D., and Weryk, R. J. 2014. The unexpected 2012 Draconid meteor storm. *Mon. Not. R. Astron. Soc.*, **437**(Feb.), 3812–3823.

Yee, J., and Close, S. 2013. Plasma turbulence of nonspecular trail plasmas as measured by a high-power large-aperture radar. *J. Geophys. Res. (Atmos.)*, **118**(24), 13,449–13,462.

Yi, W., Xue, X., Chen, J. et al. 2016. Estimation of mesopause temperatures at low latitudes using the Kunming meteor radar. *Radio Sci.*, **51**(3), 130–141.

Younger, J. P., Reid, I. M., Vincent, R. A., Holdsworth, D. A., and Murphy, D. J. 2009. A southern hemisphere survey of meteor shower radiants and associated stream orbits using single station radar observations. *Mon. Not. R. Astron. Soc.*, **398**(Sept.), 350–356.

Younger, J. P., Lee, C. S., Reid, I. M. et al. 2014. The effects of deionization processes on meteor radar diffusion coefficients below 90 km. *J. Geophys. Res. (Atmos.)*, **119**(Aug.), 10.

Younger, J. P., Reid, I. M., Vincent, R. A., and Murphy, D. J. 2015. A method for estimating the height of a mesospheric density level using meteor radar. *Geophys. Res. Lett.*, **42**(14), 6106–6111.

Yu, Y., Wan, W., Ren, Z. et al. 2015. Seasonal variations of MLT tides revealed by a meteor radar chain based on Hough mode decomposition. *J. Geophys. Res. (Space Phys.)*, **120**(Aug.), 7030–7048.

Zhu, Q., Dinsmore, R., Gao, B., and Mathews, J. D. 2016. High-resolution radar observations of meteoroid fragmentation and flaring at the Jicamarca Radio Observatory. *Mon. Not. R. Astron. Soc.*, **457**(Apr.), 1759–1769.

Zvolankova, J. 1983. Dependence of the observed rate of meteors on the Zenith distance of the radiant. *Bull. Astr. Inst. Czechosl.*, **34**(Mar.), 122–128.

4

Meteors and Meteor Showers as Observed by Optical Techniques

Pavel Koten, Jürgen Rendtel, Lukáš Shrbený, Peter Gural, Jiří Borovička, and Pavel Kozak

4.1 Introduction

Optical observations are one of the most important sources of data for meteor science. They include visual, photographic, video, and spectroscopic observations. Visual observation was the only method used for centuries. Today, visual data collection is composed of a network of amateur astronomers, and it remains an important branch of meteor astronomy in terms of monitoring daily shower activity. Photography became an important tool in the second half of the twentieth century, later followed by television and video. Direct and spectral observational techniques progressed significantly in the past two decades. Multitudes of cameras have now been deployed, and new networks established. Simultaneously, new image processing methods were developed with a focus on greater automation and acceleration of the data reduction pipeline.

Today, all these techniques provide valuable data on meteors and meteor showers. Multi-instrumental campaigns can significantly increase the scientific value of observations (for example, Jenniskens and Vaubaillon, 2008). Several groups around the world simultaneously use different kinds of instruments to acquire more data on meteor events. The combination of different data sets can provide more information about such events than photographic or video records alone (for example Spurný et al., 2014b).

The importance of optical observations has been demonstrated in many branches of meteor astronomy. Visual and video observations of meteor shower activity are often used for validation of models of meteoroid stream formation and dynamic evolution. Models of meteoroid structure and fragmentation are based on photographic and video data. Meteorite falls are often observed with optical techniques from multiple sites to determine strewn field locations. Video cameras also record so-called "anomalous meteors" showing interesting features like extremely high beginning heights or transverse jets.

This chapter is organised as follows: First, we summarise the current status and recent progress in optical observational techniques including both visual and instrumental observations as well as data processing. Then recent results obtained by optical observations in the main fields of meteor science are described. For this purpose, the three fields are defined as (1) meteoroid streams and meteor showers, (2) meteoroid structure and composition, and (3) fireballs and meteorite falls. Only the most important results are mentioned here since specific chapters of this book are dedicated to these topics. Finally, some possible future developments in the field are summarised.

4.2 Optical Observational Techniques and Methods

In this section we describe techniques for visual and instrumental meteor observations in the optical range and concepts of data analysis to derive physical parameters of individual meteors as well as meteor showers and meteoroid streams.

4.2.1 Visual Observations

In spite of the availability of instrumental methods, visual meteor observations remain an important source for monitoring meteor shower activity. The technique is well established and calibrated (Koschack and Rendtel, 1990a). The primary shower data, Zenithal Hourly Rate (ZHR) and population index r (indicating the magnitude distribution of the meteors), allow us to derive physical stream parameters such as the mass index s and the meteoroid incident flux density Q (hereinafter flux) or number density.

The general advantages of the method are:

- the global distribution of observers, covering most longitude ranges (good temporal coverage)
- the large sample (in most cases)
- the possibility of recovering data from archives and past publications to extend the database
- impact on the public, especially for shower peak periods

Of course, the accuracy of a single visual meteor detection cannot be compared with modern imaging methods. This is especially true for determining the radiant positions of weak showers. However, numerous analyses have shown that visual data can be used best to calculate the activity from known or expected sources, and to determine the flux and mass index across the stream passage. It has also been possible to detect very short-term fluctuations within dense showers (e.g. 013 LEO[1], Singer, 2000). The availability of visual meteor data covering many decades allows us to obtain information about the evolution of meteor showers over at least a century. Last but not least, visual and video data have been used as independent data sets to search for minor structures within streams or to calibrate visual and video data against one another to improve each of their reliability. A summary of recent visual meteor work is given by Rendtel (2017).

---

[1] IAU Meteor Data center code for Leonids

Extensive investigations of the way visual observations are made and analysed have been carried out and published by Koschack and Rendtel (1990a,b) and are described in the IMO's[2] Meteor Observers Handbook (Rendtel and Arlt, 2016a). Procedures to select and analyse huge data samples are described e.g. in Arlt and Rendtel (2006).

From visual meteor counts, we may calculate the population index r and ZHR. The population index r can be calculated from the distribution of the magnitudes of the observed meteors, taking into account the detection probabilities $p(\Delta m)$ for each magnitude range (Koschack and Rendtel, 1990a). The approach of Arlt (1998) uses a Monte Carlo simulation of artificial magnitude distributions $n(\Delta m)$ for a given r which most closely matches the observed magnitude distribution. Typical values of r are close to 3.0 for sporadic meteors (Rendtel, 2004b) and most weak showers, and $2.0 < r < 2.5$ for most dense shower regions. From theoretical assumptions, Richter (2018) finds that r should generally be between 2 and 3.

The ZHR is the number of shower meteors N seen by a single observer under reference conditions (vertical incidence of meteors, i.e. shower radiant in the zenith, $h_R = 90°$, limiting stellar magnitude LM of +6.5 and unlimited field of view ($c_F = 1.00$)):

$$\text{ZHR} = \frac{c_F \times N \times r^{(6.5-\text{LM})}}{T \times \sin^\gamma h_R}. \tag{4.1}$$

$c_F = 1/(1-k)$ where k is the percentage of obscured field of view, and T is the effective observing time. The zenith coefficient γ accounts for the effects of different entry angles on the transfer of energy into luminosity. Bellot Rubio (1995) has shown that single body ablation theory gives $\gamma = 1.83$; Zvolankova (1983) derived $\gamma = 1.47$ from data analysis. Bellot-Rubio also showed that for visual data, $\gamma = 1$ is appropriate. This has been confirmed by various analyses when data obtained from different locations have been combined. Rendtel (2007a) demonstrated this using data from observers east and west of the Atlantic Ocean during the Orionids.

ZHRs appear overcorrected for elevations $h_R = 20°$ to $45°$ if $\gamma > 1$ is applied (Rendtel, 2007a); hence $\gamma = 1.0$ is used. The error margins are more affected by the scattering of conditions among all count intervals. Choosing very strict selection parameters reduces the sample and may introduce gaps in the time series. Experience proves that it is best to exclude only significant outliers, deviating more than 50% from the average. This requires a successive approximation. The temporal resolution – and thus the definition of peaks or peak widths – depends on the sample. It is necessary to adopt an optimum bin length for each section of a given data set to ensure that it contains a sufficient number of shower meteors and is short enough to avoid smoothing the profiles. In principle, the accuracy is half the step length between successive data points. Data from shower returns occurring under poor conditions (e.g. near Full Moon) shows a tendency of underestimated ZHRs. Until now, no final clue why the perception changes if the sky background is bright, is given. The standard LM-correction handles losses due to reduced sky transparency well but is less reliable in the case of scattered background illumination.

Assuming a constant conversion of kinetic energy from the meteoroid into luminosity, the mass index is $s = 1 + 2.5b \log r$ where $b = 0.92$ according to Verniani (1973). In Koschack and Rendtel (1990a,b), the procedure to calculate the number density ϱ (meteoroids in a cube with 1000 km edge length) of a stream is described. It is different from the flux Q only by a factor of the atmospheric entry velocity.

The flux can also be obtained from video meteor observations, and both methods have been calibrated against one another on several occasions, both for major and minor shower activity. This way we can ensure that the fluxes calculated from visual and from video data are comparable and consistent and can be used to verify minor features detected in activity profiles (Molau and Barentsen, 2014; Blaauw et al., 2016).

4.2.2 Photographic and Video Observations

All instrumental observations of meteors began at the end of 19th century, when the first known photograph of a meteor was captured on 1885 November 27 in Prague by Ladislaus Weinek during observations of the Andromedid meteor shower (Weinek, 1886). Scientific recording of meteors on photographic plates was developed and improved during subsequent decades, followed by television and video techniques in the second half of the twentieth-century (Ceplecha et al., 1998). Hawkes (2002) provides an excellent review that covers all the methods and associated processing developments up to the early 2000s.

The successful double-station photographic observation of a meteorite fall on 1959 April 7, and recovery of four meteorites near Příbram in the former Czechoslovakia (Ceplecha, 1961), initiated systematic observational programs – photographic fireball networks. Photographic observations have provided the most precise and accurate data on meteors. Standard video cameras usually have lower resolution than photographic cameras. Trajectories and orbits calculated from video observations are not as precise as those for photographic experiments. On the other hand, video observations typically detect fainter meteors and cover the temporal evolution of each meteor event.

A basic optical instrument consists of the camera (analogue or digital) and the objective lens. An image intensifier can be added to enhance the sensitivity of a video camera. In this case, the camera is coupled with a second lens focused on the intensifier output.

Photographic and video cameras are used for many kinds of studies. A single camera can be used for monitoring meteor shower activity, studying meteor light curves or the mass distribution. A good example of this kind of monitoring activity is the single station survey network IMO Video Meteor Network, which consists of independent stations around the world that operate year-round. The network can detect brief enhancements of meteor activity, such as the short outburst of the July γ Draconids (184 GDR) on 2016 July 28 (Molau et al., 2016). Continuous monitoring of the sky with this network since 1999 allowed Molau and Rendtel (2009) to perform an automated radiant search on a large, consistent sample. This includes flux profiles for all established showers.

Adding another camera at a different geographical location and pointing both at the same volume of atmosphere

---
[2] International Meteor Organization

Table 4.1. *Examples of meteor camera video systems (in operation in 2018)*

System	Camera	Lens	FPS	FOV [°]	Resolution [arcmin/px]	LM (stellar)	LM (meteor)	Ref.
ASGARD	Ex-view HAD	1.6 mm f/1.4	30 i	180	17	+1.0	−2.0	1
FRIPON	asA1300-30gm	1.25 mm f/2	25 p	180	12	+0.0	−1.0	2
SonotaCo	Watec 100N	6 mm f/0.8	30 i	56	5.2	+5.0	+4.0	3
CMN	1004X	4 mm f/1.2	25 i	64	5	+3.2	+2.1	4
RMS	IMX 290	4 mm f/1.2	30 p	64	2.9	+5.3	+3.7	5
CAMS	Ex-view HAD II	12 mm f/1.2	30 i	28	2.6	+5.4	+3.8	6
Intensified	GEN III + COHU	25 mm f/0.85	30 i	25	2.3	+8.5	+6.7	7
Intensified	NUVU EMCCD	50 mm f/1.2	32 p	15	0.9	+12.0	+8.5	8
KOWA	Ex-view HAD II	55 mm f/1.0	25 p	5	0.5	+10.2	+6.7	9
MAIA	JAI CM-040	50 mm f/1.4	61.15 p	50	3.8	+6.0	+4.5	10
Intensified	DMK 23G445	135 mm f/2.0	30 p	20	1.0	+9.0	+7.0	11
AMOS	DMK41AU02	15 mm f/2.8	20 p	180	6.8	+5.0	+4.0	12

References: [1] Weryk et al. (2008), [2] Colas et al. (2015), [3] Sonotaco (2009), [4] Šegon et al. (2015), [5] Vida et al. (2018a), [6] Jenniskens et al. (2011), [7] and [8] P. Gural personal communication, [9] Šegon et al. (2015), [10] Vitek et al. (2011), Koten et al. (2014b), [11] this work, [12] Tóth et al. (2015)
FPS = full frames per second for interleaved (i) or progressive scan (p), FOV = field-of-view, LM = limiting magnitude.

significantly increases the scientific value of the experiment. The double station experiment allows the atmospheric trajectories and heliocentric orbit to be calculated, the absolute brightness to be measured and photometric mass of the meteoroid to be calculated (Koten et al., 2004). This data is a valuable source for modelling the meteoroid composition and structure (Campbell-Brown and Koschny, 2004).

Prior to the turn of the century, observations were rather strictly divided between photographic and video. Later, with the improvement of digital technology after 2000 the boundary between the two methods has become less distinct. Today, the term "photographic" also includes still digital cameras.

There are a number of camera configurations possible in a meteor camera system depending on the science goals. The system setup may bear on the type of data reduction techniques applied. Video systems can involve either intensified or non-intensified systems which have significantly different noise characterization based on camera gain settings and the "generation" of the intensifier. Video cameras nominally operate at a frame rate of 25 or 30 frames per second, but may also be gated with short integration times, operate at higher frame rates, or be configured for longer integration times of 1 to 90 seconds. The focal plane may be a standard 0.3-megapixel video camera, a large format 4Kx4K system, or high definition (HD) 720p or 1080p system, among other formats. The cameras may operate in an interleaved mode where there is alternating collection of odd and even rows, or, as in the case with modern digital systems, be progressive scan with all rows integrated at once (global shutter) or with time-delayed row integration (rolling shutter). Each of these scenarios can influence the choice of the data processing algorithm used for pre-processing, background clutter suppression, meteor detection, track measurement extraction, and false alarm mitigation.

The availability of sensitive and relatively cheap cameras has allowed more people and teams to join the video meteor community. Double and multi-station facilities involve not only professional, but also amateur astronomers. Several video networks are currently in operation – CAMS in California (Jenniskens et al., 2011) and Benelux (Roggemans et al., 2014), SOMN in Canada (Weryk et al., 2008), the NASA All-Sky Fireball Network (Cooke and Moser, 2012), the SonotaCo network in Japan (Sonotaco, 2009), the Spanish meteor network SPMN (Madiedo and Trigo-Rodríguez, 2007), the Croatian Meteor Network CMN (Andreič and Šegon, 2010), AMOS in Slovakia (Zigo et al., 2013), EDMOND in Central Europe[3], FRIPON in France (Colas et al., 2015) etc. These networks serve as an excellent tool for the identification of new meteor showers and shower candidates. For example, 60 new showers were identified from CAMS observations taken up to the end of March 2013 (Jenniskens et al., 2016). Examples of current analog and digital video meteor camera systems, from all-sky to telescopic, are given in Table 4.1.

On the other hand, there are dedicated cameras used for a very specific kind of observation. One example of such a camera is the Canadian Automated Meteor Observatory (Weryk et al., 2013), which is able to track meteors in real time using a very narrow field camera (Figure 4.1). Data are useful for detailed studies of the motion of fragments (Stokan and Campbell-Brown, 2014; Borovička et al., 2019) or for testing of current dustball ablation models (Campbell-Brown et al., 2013). As another example, a fast speed camera recorded a bright Leonid meteor (magnitude −3) at 1000 frames per second and revealed an ablation cloud around the meteor particle, which resembles a shock wave (Stenbaek-Nielsen and Jenniskens, 2004).

It is not only dedicated scientific video camera records which can be used for research. With an increasing number of security, surveillance and dashboard cameras, the number of causal meteor detections is also increasing. The well-known Chelyabinsk superbolide was recorded by hundreds of such cameras (Borovička et al., 2016). The video records from three

---
[3] http://cement.fireball.sk, www.daa.fmph.uniba.sk

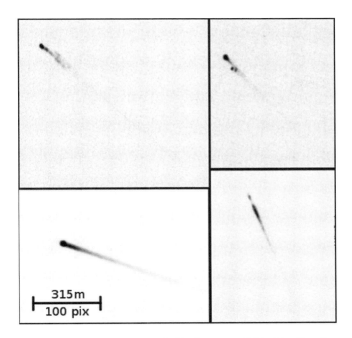

Figure 4.1. Four meteors recorded in the narrow field of the Canadian Automated Meteor Observatory (CAMO) tracking system. At a scale of a few metres per pixel, fragmentation is obvious in more than 90% of meteors observed, and there are a wide range of morphologies seen (Campbell-Brown et al., 2013).

Figure 4.2. Spectrum of terminal flare of an Orionid fireball taken at the Ondřejov Observatory on 2015 October 24. DSLR camera Canon EOS 5D Mark II equipped with Sigma 15 mm F2.8 fisheye lens and a plastic holographic diffraction grating with 1000 grooves per mm was used. Since the grating was not blazed, spectra of identical intensities were captured on both sides of the fireball. The brightest spectral lines belong to neutral magnesium and sodium.

security cameras were used to determine the trajectory and orbit of the Košice meteorites (Borovička et al., 2013b).

Spectroscopy is also among the traditional methods of meteor studies. Meteor spectra were observed as early as the nineteenth century with the use of visual prism spectrographs. New techniques were used as they became available: Photography at the beginning of the twentieth century, diffraction gratings in the 1930s, sensitive TV technology in the 1970s. Most of the time, nevertheless, meteor spectroscopy remained a minor field in astronomy and even within meteor astronomy. The history of meteor spectroscopy up to the end of the twentieth century is described e.g. in Ceplecha et al. (1998) and references therein.

The spectroscopy of meteors is different from the spectroscopy of other targets in astronomy since meteors appear unpredictably at random positions in the sky, move fast, and disappear quickly. Continuous monitoring of the sky with slitless objective spectrographs is therefore needed. The spectrograph is a normal imaging camera with a diffraction element (prism or grating) placed in front of the lens. Since a meteor is a point-like object or a narrow streak, the series of its monochromatic images form the meteor spectrum.

An ideal meteor spectrograph would have a large field of view, high resolution, high sensitivity, and wide spectral range. In practice, all these requirements cannot be fulfilled simultaneously. Classical analog photographic cameras had high resolution thanks to a large image sensor (photographic plate or film up to 30 cm in size) combined with lenses with relatively large focal length. The field of view was moderate and sensitivity was low. As a result, the number of useful spectra grew slowly over the years during the photographic era. The introduction of video cameras provided high sensitivity but much lower resolution.

A comparison of photographic and video spectra is given in Borovička (2016).

Table 4.2 lists the approximate parameters of selected currently running spectroscopic projects. The resolution was estimated from published spectra, and gives the separation of the closest spectral lines which can be resolved in the first spectral order under good geometric conditions (when meteor flight is nearly perpendicular to the direction of dispersion). The field of view is given for a single camera. Some projects use multiple cameras, in some cases with different parameters (e.g. SMART). Note that there are other spectroscopic programs not listed in Table 4.2, see e.g. Ward (2015), Koukal et al. (2015), Koschny et al. (2015), Dubs and Schlatter (2015) or Wiśniewski et al. (2017). An example of a spectrum from a DSLR camera with a fisheye lens is given in Figure 4.2. Identification of the spectral lines in this Orionid spectrum is shown in Figure 4.3. The European Fireball Network (EN) started to deploy such cameras to selected stations in order to routinely get spectra of bright fireballs.

4.2.3 Photographic Fireball Networks

Despite the expansion of video networks, photographic fireball networks still represent the top scientific instruments, especially in the terms of precision of the meteor data. We will therefore dedicate more attention to them in this chapter. In the past, several photographic networks were established to observe meteors. The European Fireball Network (EN) (Ceplecha and Rajchl, 1965; Ceplecha et al., 1973) is still in operation in a significantly modernised form (Spurný et al., 2007; Spurný, 2016). Two other historical networks, the Prairie Network (McCrosky and Boeschenstein, 1965) and MORP (Halliday, 1973), ceased operations a long time ago. On the other hand, two new photographic camera networks have been established in the last 15 years: the Desert Fireball Network (DN) in Australia (Bland, 2004)

Table 4.2. *Parameters of some currently running meteor spectroscopy projects*

Project Description	Field of View	Limiting Magnitude	Scale	Resolution	Exposure Time	Spectral Range	Ref.
Ondřejov, analog photo, f = 360 mm, 18 × 24 cm	36 × 24°	−5	4.6 nm/mm	0.25 nm	hours	360–660 nm	1
Ondřejov, intensified video, 720 × 576 px	50°	+2	3 nm/px	9 nm	0.02 s	380–900 nm	2
SMART, Watec video, 720 × 576 px	50 × 40°	−3	1.3 nm/px	3 nm	0.02 s	370–800 nm	3
CAMSS, Watec video, 640 × 480 px	30 × 22°	+1	1.1 nm/px	2 nm	0.033 s	370–880 nm	4
AMOS, intensified video, 1600 × 1200 px	100°	−2	1.3 nm/px	3 nm	0.08 s	370–900 nm	5
Miyazaki, Sony α7s 4K video, 3840 × 2160 px	70 × 50°	0	0.7 nm/px	2 nm	0.033 s	400–800 nm	6
EN, Canon EOS 6D fish-eye photo, 5472 × 3648 px	140 × 90°	−7	0.4 nm/px	1 nm	30 s	370–880 nm	7

References: [1] Borovička (2016) [2] Vojáček et al. (2015) [3] Madiedo (2017) [4] Jenniskens et al. (2014) [5] Rudawska et al. (2016) [6] Maeda and Fujiwara (2016) [7] Ryabova et al. (2019).

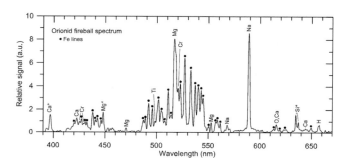

Figure 4.3. Identification of spectral lines seen in the Orionid fireball spectrum which was reproduced in Figure 4.2. Line intensities have not been calibrated. Most lines belong to neutral iron and are marked by a bullet. Lines of Ca^+, Mg^+, Si^+, and H belong to the high temperature meteoric component, which is not present (or is much fainter) in slow fireballs and in faint meteors. The oxygen line (here mixed with Ca) is of atmospheric origin. Brighter atmospheric lines of O and N lie in the infrared. The intensity of atmospheric lines increases with meteor velocity.

and the Tajikistan Fireball Network in Tajikistan (Babadzhanov et al., 2009). The parameters of photographic fireball networks operating in the last 15 years are summarised in Table 4.3.

4.2.3.1 Czech Fireball Network

The Czech Fireball Network, the founding part of the EN, started regular operation of all-sky cameras in autumn of 1963 and has remained in full operation to the present. Significant improvement of the network began in 2003 with the replacement of manual all-sky cameras with the new Autonomous Fireball Observatories (AFO) (Spurný et al., 2007). The AFO imaging system consists of a Zeiss Distagon fish-eye objective ($f/3.5$, $f = 30$ mm) and a large-format sheet film.

The next significant change in the network occurred in 2014 when high-resolution Digital Autonomous Fireball Observatories (DAFO) were gradually installed (Spurný, 2016). The DAFO imaging system consists of two full-frame Canon 6D digital cameras and Sigma fish-eye lenses ($f/3.5$, $f = 8$ mm) equipped with an electronic liquid crystal display (LCD) shutter for meteor speed determination. In the standard regime, 16 breaks of meteor image per second (Figure 4.5) and 35 s exposures are used. DAFO – as well as AFO – are also equipped with all-sky radiometers with a time resolution of 5000 samples per second (Figure 4.4). The DAFO system achieves the same absolute positional accuracy as the AFO system and is able to record three times more fireballs than the AFO system (Spurný et al., 2017a).

4.2.3.2 Other Countries in the EN

Three DAFO cameras are also in operation in Slovakia, and one in Austria (Spurný et al., 2017a). The German part of the EN is still equipped with all-sky mirror type cameras with Leica film cameras looking down and taking images of a convex mirror. One Leica camera was replaced by a digital Canon EOS 5D in 2006 (Flohrer et al, 2006; Oberst et al., 2010). Similar digital tests were also undertaken by the Dutch Meteor Society starting in 2005 when Canon EOS 300D and 350D digital cameras were placed above a convex mirror (Jobse, 2005). These cameras were later also tested with all-sky lenses (Jobse, 2009).

4.2.3.3 Other Networks in Europe

Various types of Canon digital single-lens reflex (DSLR) cameras with various all-sky lenses (mainly Sigma $f/2.8$, $f = 4.5$ mm) placed in weather-resistant cases were tested, with or without rotating shutters placed in front of the lenses, since 2005 as a part of the Polish Fireball Network (PFN) (Olech et al., 2006) and since 2008 in Benelux (Biets, 2008; Jobse, 2009;

Table 4.3. *Parameters of some photographic fireball networks that have been in operation in the last 15 years*

Project Description	Exposure Time	LM (meteor)	Scale	Coverage [km^2]	Trajectory Uncertainty [m]	Speed Uncertainty [m s^{-1}]	Ref.
Analog EN	hours	-4	2°/mm	10^6	10–20	10	1
Digital EN	35 s	-2	3′/px	5×10^5	10–20	10	2
Analog DN	hours	-4	2°/mm	2×10^5	10–20	10	3
Digital DN	29 s	0.5	2′/px	3×10^6	<50	<100	4
Analog TN	hours	-4	2°/mm	1.1×10^4	10–20	10	5

References: [1] Ceplecha and Rajchl (1965); Oberst et al. (1998), [2] Spurný et al. (2017a), [3] Bland (2004); Bland et al. (2012); Spurný et al. (2012), [4] Howie et al. (2017a,b), [5] Kokhirova et al. (2015)
EN = European Network, DN = Desert Network, TN = Tajikistan Network, LM = limiting magnitude

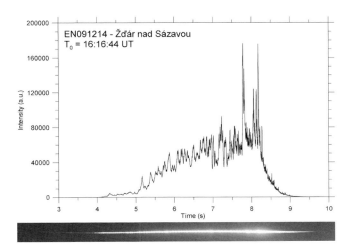

Figure 4.4. Radiometric light curve and detail of photograph of the Žďár nad Sázavou fireball recorded by DAFO at station Polom in the Czech part of EN on 2014 December 9.

van Leuteren, 2010; Nijland, 2012). These digital observatories are manually operated on clear nights. Unfortunately, published information about these digital observatories is scant.

A different approach has been taken in the Spain Meteor Network (SPMN). With the increasing size of CCD chips, they developed an all-sky CCD camera in 2002 in order to find alternatives to the all-sky film camera. The all-sky CCD camera consists of a Kodak KAF168001 chip (30 second exposure) and Nikkon fish-eye objective ($f/3.5, f = 16$ mm). Velocity data were deduced from video observation. Every station was equipped with two of these automated CCD cameras. The accuracy of the system is two to four times lower than that of the AFO system (Trigo-Rodríguez et al., 2003a, 2004). A rotating shutter close to the focal plane was implemented in 2006 and the uncertainty in the velocity is of the order of 0.2–0.3 km s^{-1} (Trigo-Rodríguez et al., 2007). Since the network in Spain moved to video observations, only one photographic CCD camera is in operation at present (Madiedo et al., 2014a; Blanch et al., 2017).

4.2.3.4 Desert Fireball Network

The only photographic fireball network in the Southern Hemisphere is the Desert Fireball Network (DN) in Australia. The DN was designed to recover meteorites with accurate orbits. A small network of three stations was established in December 2005. The fourth station was set up in November 2007 (Bland, 2004; Bland et al., 2008, 2012; Spurný et al., 2012). The stations of the DN were originally equipped with autonomous film cameras: Desert Fireball Observatories (DFO). The DFO imaging system consisted of a Zeiss Distagon fish-eye objective ($f/3.5, f = 30$ mm) and a large-format sheet film. All DFOs were equipped with a rotating shutter close to the focal plane to determine fireball velocity and an all-sky brightness sensor (radiometer) with a sampling rate of 500 measurements per second to determine the accurate time of fireball passage, its duration, and detailed light curve. The DFO is a modification of the AFO that was used in the Czech part of the EN (Spurný et al., 2007).

Starting in 2015 the DFOs were replaced by a few tens of digital cameras and the network expanded in the next few years (Towner et al., 2012; Sansom et al., 2015). The Automated Digital Fireball Observatory (ADFO) is equipped with a Nikon D800E (later replaced with the D810) still digital camera and Samyang fish-eye lens ($f/3.5, f = 8$ mm) as a primary imaging system and a Watec 902H video camera for additional fireball imagery - especially of fragmentation events. The ADFO system uses a liquid crystal shutter, modulated, during 29 s long exposures, according to a de Bruijn sequence (10 breaks of two different lengths of meteor image per second in a sequence allowing precise timing of 20 points per second (Figure 4.5) - the dash starts and ends) for both meteor velocity determination and absolute fireball timing (Howie et al., 2017a). 49 ADFOs have been deployed up to December 2016 (Bland et al., 2014; Towner et al., 2015; Sansom et al., 2017; Howie et al., 2017a,b).

4.2.3.5 Tajikistan Fireball Network

The only photographic fireball network in Asia was the Tajikistan Fireball Network. Systematic double-station photographic observations of fireballs have been carried out using manual all-sky cameras since 2006, and from 2008 until 2013 five stations situated in the southern part of Tajikistan were in regular operation (Kokhirova et al., 2015). The imaging system consisted of a Zeiss Distagon fisheye objective ($f/3.5, f = 30$ mm) and a large-format sheet film, which is the same as in the Czech part of the EN. Symmetrical two-blade shutters very close to the focal plane of the camera creating 10, 12, or 37/3 breaks of meteor image per second were used (Babadzhanov et al., 2009; Kokhirova and Borovička, 2011). The stations were also

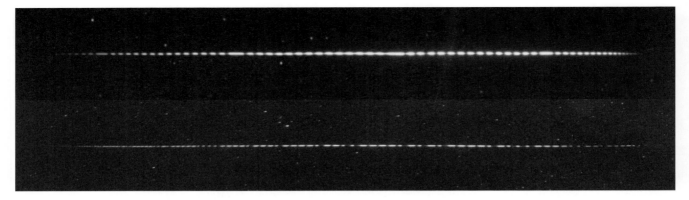

Figure 4.5. Comparison of digital fireball images with different liquid crystal shutter sequences. The upper image has been taken by the Digital Autonomous Fireball Observatory in the Czech part of the European Fireball Network and shows regular occultation of fireball image with double-length dash as the time stamp every whole second (absolute timing together with brightness detector). The bottom image has been taken by the Automated Digital Fireball Observatory in the Desert Fireball Network in Australia and shows irregular occultation of fireball image caused by de Bruijn sequence, which provides absolute fireball timing for fireballs lasting longer than the subsequence length, which corresponds to 0.9 s (Howie et al., 2017b). Both the fireballs flew from left to right.

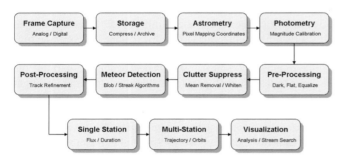

Figure 4.6. Data reduction processing flow comprising multiple steps to go from capturing photons to computing and analysing Solar System orbits.

equipped with manually operated Nikon D2X and Nikon D300 digital cameras with the Nikkor fisheye lenses ($f/2.8$, $f = 10.5$ mm) since 2008. The exposure time was 30 s. Moreover, one digital camera was equipped with a rotating shutter in front of the objective giving 15 breaks per second (Kokhirova and Borovička, 2011). Times of fireball passages were determined either using the method of combination of negatives of the fixed and guided cameras, or from the digital image of the object (Kokhirova et al., 2015).

4.2.4 Video Data Processing

The detection and processing of meteors recorded with video have been continuously developing and parallel advances in imaging hardware and data interfaces. Many of the techniques and steps have been utilised by one processing pipeline or another. The most important thing is that all the steps are properly done. For a truly complete video meteor system, the end-to-end processing chain starts with the capture of meteor line emission and produces an estimation of a meteoroid's trajectory through the Solar System (Figure 4.6).

Among the most important steps are the astrometric and photometric calibrations. The aim of the first is to map the focal plane pixels to stellar equatorial coordinates (or more nominally gnomic or standard coordinates). Often referred to as an astrometric solution, this can entail anything from a simple affine transform, cubic polynomial, or 13 parameter fitting functions representing the warping model for narrow, median FOV, or all-sky imagers respectively (Steyaert, 1990; Wray, 1967; Borovicka et al., 1995). Recently work by Bannister (2013) has reduced the all-sky parameter set to just 8 parameters. Automating the astrometric calibration of fixed pointing cameras using a single manually fitted solution which is updated each night is important for reducing labour in large networks.

Photometric calibration has advanced incrementally in the past 15 years, with attempts to account for the imaging system's spectral response (Jenniskens et al., 2011) and the full calibration of a camera in the lab before deployment (Swift et al., 2004). Aperture photometry is still typically used, where a relationship is fit between background-subtracted, spatially integrated star and meteor counts. Many video sensors are particularly sensitive in the red part of the spectrum, and care must be taken that the reddest spectral classes of stars do not bias the magnitude fit. With advances in the near infra-red sensitivity of some newer off-the-shelf surveillance cameras, there are also potential biases in the magnitude-mass relationship normally used, due to inconsistent calibration between the human visible range and the camera spectral range (Šegon et al., 2018).

When the image pre-processing is finished, another important step is meteor detection. Numerous detection algorithms have been used to identify meteors in video sequences and they can be categorised into two main groups: blob detectors and line detectors. Blob or cluster detection algorithms are used for all-sky imagery due to the very short nature of the meteor streaks visible in each frame. The small meteor extent arises from the large angular viewing size of the pixels given short focal length lenses coupled with commercial video cameras (Blaauw, 2014). This has recently been extended to moderate FOV meteor video systems (Gural, 2016) due to the dramatic speed advantages of pixel clustering over streak detection. A very fast clustering technique was first pioneered in a meteor tracking system based on fast steering mirrors. It required very low time latency for detection to identify, steer towards, and track meteors before they faded away (Gural et al., 2004). Small scale orientation kernels fall between a blob detector and streak detector, for example, the software package MetRec, which employed 5x5 pixel kernels (Molau and Gural, 2005). Line detection algorithms may use variants of the classic Hough transform such as localised pixel pairing used in MeteorScan (Gural, 1999, 2007), 4D Hough transform (Trayner et al., 1997), and the phase coded disk for binary (Clode et al., 2004) or Hueckel transform using greyscale pixel values (Hueckel, 1973) for single point position and orientation mapping to Hough space. Subspace algorithms such as ESPRIT (Aghajan and Kailath, 1994) may also be a future candidate for line detection.

Once meteor detections are made and the propagating track parameters have been estimated, the next step is to mitigate false alarms. In MeteorScan this involves a matched filter maximum likelihood test, whereas a system like CAMS uses human verification of tracks. Attempts have been made to both manually and automatically fit the leading edge of meteor tracks to avoid these issues (Kozak, 2008).

With the large scale deployment of multiple camera networks around the world within the past decade, multi-station processing has become a critical step in the data reduction pipeline. The classic method for this is the intersecting planes method, which dates back over a century (Davidson, 1937; Porter, 1942; Whipple and Jacchia, 1957; Wray, 1967; Ceplecha, 1987; Betonvil, 2006) and was enhanced by fitting a 3D straight line to the camera's measurement rays (Borovička, 1990). Either procedure was followed by velocity estimation using the computed radiant. This was further improved by adding a propagation model to the 3D line, given each camera's frame rate and fitting all the trajectory parameters at once: radiant direction, entry speed, starting position, and timing offsets between cameras (Gural, 2012). The latest work in progress is attempting to employ an adaptable propagation model to better fit the deceleration behaviour of actual meteors (Egal et al., 2017). The final computational step is to extrapolate the atmospheric trajectory to a Keplerian orbit in the Solar System (Ceplecha, 1987). This requires accounting for diurnal and light aberrations plus station motion on a spinning Earth (Wray, 1967), and propagation outside Earth's sphere of influence (Clark and Wiegert, 2011).

Amateurs and professionals use a number of meteor analysis programs, some of which are in the public domain. A selection is listed in Table 4.4. One desired result of the data reduction is a set of orbits that can be used in comet ejection studies, meteoroid stream evolution, and flux modelling at the Earth. These data are often compiled and published in catalogues associated with each major network (CAMS[4], SonotaCo[5], CMN[6], EDMOND[7]).

[4] http://cams.seti.org/
[5] http://sonotaco.jp/doc/SNM/index.html
[6] http://cmn.rgn.hr/
[7] http://cement.fireball.sk

Table 4.4. *Some of current programs for the processing of the meteor records.*

Software	Purpose	Source
MeteorScan	Detection	Gural (1999)
MetRec	Detection	www.metrec.org
UFOCapture	Detection	Sonotaco (2017a)
HandyAVI	Detection	www.handyavi.com
ASGARD	Detection	http://meteor.uwo.ca/~weryk/asgard/
dMAIA	Detection	Koten et al. (2014b)
UFOOrbit	Calculations	Sonotaco (2017a)
MORB	Calculations	Ceplecha et al. (2000)
CAMS software suite	End-to-end	Jenniskens et al. (2011)
UWO processing pipeline	End-to-end	http://meteor.uwo.ca/
Falling Star	End-to-end	Kozak (2008)

4.3 Results of Optical Observations of Meteors

There are many goals of regular meteor observations. First, there is continuous monitoring of meteor activity. This is also connected with studies of meteoroid stream structure and dynamical properties, as well as parent body linkages. Second, the structure and composition of meteoroids and the physics of their atmospheric entry are also of interest. Third, observations of fireballs, the dynamics of their atmospheric passage, the possibility of meteorite falls and their recovery, as well as their potential to probe asteroid families, is of interest.

4.3.1 Meteoroid Streams and Meteor Showers

Investigation of meteor showers is one of the main topics of meteor science. For most major showers, the IMO provides live graphs on their web page. The data are based on reports of visual observers; this enables them to provide analyses for most major annual shower returns. These data are provided and described in the Meteor Shower Workbook (Rendtel, 2014b) and the annual IMO Meteor Shower Calendar.

Since 2013, the list of showers for optical observations has been extended to daytime showers. The Daytime Arietids (171 ARI) and Daytime Sextantids (221 DSX) are assumed to be strong enough to be detectable at dawn, and their radiants are known with sufficient accuracy (Rendtel, 2014a). The data requires careful consideration of the rapidly changing conditions at local dawn. The prospects of the instrumental observation of daylight fireballs have also been studied (Egal et al., 2016).

While the IMO Meteor Observers Handbook (Rendtel and Arlt, 2016a) and the Meteor Shower Workbook (Rendtel, 2014b) are meant to be a compilation of instructions, general information, and data collection; the annual IMO Meteor Shower Calendar is widely used as a reference for almanacs and e.g. for the annual International Geophysical Calendar.

Table 4.5. *Extreme particle population – lowest population index r*

Shower	Year	r	Reference
Leonids (20 revolution)	1998	1.19 ± 0.02	Arlt (1998)
September ϵ–Perseids	2013	1.45 ± 0.15	Rendtel et al. (2014)
Orionids	2006	1.58 ± 0.08	Rendtel (2007a)
Leonids (background component)	2001	1.60 ± 0.05	Arlt et al. (2001)
Geminids (end of peak plateau)	2004	1.73 ± 0.04	Arlt and Rendtel (2006)
Aurigids	2007	1.74 ± 0.08	Rendtel (2007b)
Perseids	1993	1.75 ± 0.05	Rendtel (1993)

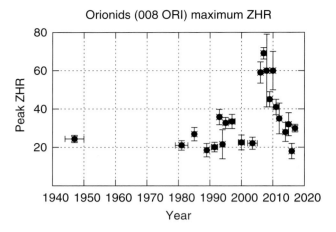

Figure 4.7. Maximum Orionid ZHR per return from visual data stored in the Visual Meteor DataBase of the IMO since 1988, from visual data re-analysed from systematic reports or publications. A horizontal error bar indicates that we averaged data from consecutive years for one data point.

When investigating the activity of meteor showers we can distinguish between long term studies of regular shower activity and peculiar events like outbursts and dust trail encounters.

4.3.1.1 Long-Term Studies

One of the advantages of visual data is the extent of the data set, covering decades. These data sets usually do not include information about observing conditions (in terms of limiting magnitude, field of view etc.) and also lack data on meteor magnitudes for each shower. Hence, we cannot derive r-values for each return. Instead, we need to apply more recent values, and suppose that the variations along the stream are not too strong. Given the results in Section 4.2.1, this may not be the case. However, many reports include information about non-shower meteors, which may be used to calibrate the shower ZHR. This also holds for series which start before or extend after the peak period, so that a background level can be established. The assumption of the "standard" r-values seems reasonable except for cases where observers explicitly mention anomalies like e.g. "many fireballs". During all analyses of shower returns and outbursts, we have found extremely low values of r, like those listed in Table 4.5. For regular returns, the variations around the established averages are small. This has been found e.g. for the Perseids in the period 1874–2001 (Belkovich and Ishmukhametova, 2006) and the Lyrids between 1900 and 2007 (Belkovich and Ishmukhametova, 2010).

Orionids (008 ORI): This shower has been investigated to look for the earlier occurrence of resonant meteoroids with respective rate enhancements, which should have been observable about 70 years before the 2006–2009 events. Figure 4.7 shows the maximum ZHRs per return obtained from visual data stored in the IMO's VMDB since 1988. Data from years prior to that have been described in detail in Rendtel (2007a).

Geminids (004 GEM): Here we can trace the evolution of the general activity level and the shape of the activity profile continuously since 1984 with consistent data. Since the shower is quite active and no other significant source is active at the same time, we may use old publications and reports to derive ZHR information over earlier decades. This data mining requires

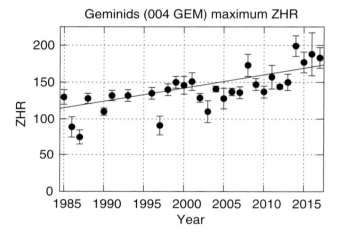

Figure 4.8. Maximum ZHR of the Geminids from 1984 to 2017 and a linear fit showing the rate increase.

careful calibration as described earlier in this section. We find (i) a remarkably constant general profile since 1984 and (ii) rather little ZHR variation from one return to the next (Rendtel, 2004a). The maximum ZHR level has increased significantly over the past ≈ 35 years and, according to modelling, should continue to do so in the coming years (Ryabova and Rendtel, 2018). Figure 4.8 shows the data including the 2017 return (preliminary analysis). An investigation of reports from the first half of the twentieth century is still ongoing.

κ-Cygnids (012 KCG): This shower has been studied in detail by Koseki (2014), showing that it is part of a complex. Higher activity was observed in 2007 and 2014. A roughly 7-year periodicity in rate enhancements has been suspected (Koseki, 2014; Moorhead, 2015). Applying a wavelet analysis to the re-analysed visual plotting data collected in the period 1975–2014 (Figure 4.9) did not reveal any such periodicity. Details of the wavelet analysis are discussed in (Rendtel & Arlt, 2016b).

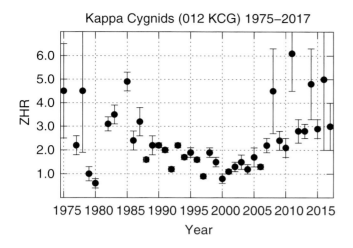

Figure 4.9. Maximum ZHR of the κ-Cygnids between 1975 and 2017. The data samples of the 2008, 2011, and 2016 returns are small due to moonlit conditions and lower attention to the Perseids in these years. The 1985 value, however, is well established.

April Lyrids (006 LYR): Belkovich (2011) discusses periodicities in the Lyrid peak activity between 1900 and 2007 which may be associated with perturbations by Jupiter.

The IAU Meteor Data Center (IAU MDC) was established in 1982. Since then, it has collected data on photographic meteor orbits. The latest version of the database contains the orbital and geophysical parameters of 4873 meteors (Neslušan et al., 2014). These long-term data on shower fireballs allow statistical work on the fireball orbits.

4.3.1.2 Peculiar Shower Activity

Numerous outbursts and dust trail encounters of many meteor showers have been observed and well documented. This includes both major and minor showers. There have been several events caused by lesser known meteor showers. In the following, we briefly summarise results concerning the activity of showers derived from visual data and instrumental data.

Orionids (008 ORI): The activity of this shower has been documented over decades. Usually, we find a broad maximum around October 21 with typical ZHRs of about 15–30. In 2006–2009, enhanced rates (ZHR 40–70; see Figure 4.10) due to meteoroids trapped in a 1:6-resonance with Jupiter (Sato and Watanabe, 2007) have been observed. The peculiarity of these meteoroids was the significant difference in their mass distribution, causing a large number of fireballs, which is also obvious from the low population index $r < 2.0$ compared to $r = 2.5$ during regular returns. The fireball activity originated from a very compact geocentric radiant defined by $\alpha = 95.10°\pm0.10°, \delta = 15.50°\pm0.06°$ (Spurný and Shrbený, 2008). Additionally, some early peaks with ZHR of the order of the main maximum ZHR have been recorded on October 17/18 in 1993 and 1998.

Leonids (013 LEO): One of the main topics at the turn of century was the series of Leonid meteor outbursts and storms occurring between 1998 and 2002, which was associated with the recent perihelion passage of the parent comet 55P/Tempel-Tuttle in 1998 (McNaught and Asher, 1999). Each event attracted a lot of attention and a number of ground-based and airborne obser-

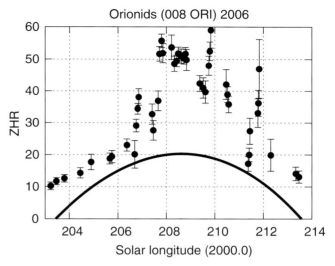

Figure 4.10. ZHR profile of the maximum period of the 2006 Orionids. The solid line shows the fitted average ZHR curve for the well-observed Orionid returns in 1993 and 1995, which can be regarded as a reference for an average return.

vational campaigns were carried out (Jenniskens and Buttow, 1999; Jenniskens et al., 2000b). These campaigns were often multi-instrumental.

After 2002 the activity of the Leonid meteor shower decreased again to its usual annual level, but two small outbursts were predicted (Vaubaillon et al., 2005) for 2006 and 2009. According to the visual observation network, the last two dust trail encounters reached a peak ZHR of about 80 (total ZHR, with a background rate of approximately 20 in 2006 and 15 in 2009). Details of the 2006 return have been discussed in detail by Arlt and Barentsen (2006). This encounter with material ejected in 1932 was observed almost exactly at the predicted time (Jenniskens et al., 2008a; Koten et al., 2008).

For the 2009 return, Jenniskens et al. (2009) give peaks at $\lambda_\odot = 235°.35, 235°.50$ and $235°.67$ which may be associated with encounters with the 1466 and 1533 dust trails of comet 55P/Tempel-Tuttle. A video experiment carried out in Tajikistan showed that the enhanced activity was observed within the predicted time, and the broader peak was consistent with the age of the filaments (Koten et al., 2011).

The present analysis (Figure 4.11) confirms both the well-defined central peak at $\lambda_\odot = 235°.497$ and the peak at $235°.67$. Two peaks at $235°.375$ and $235°.593$ are defined by just one data point and are less certain. The latter peak is based on a substantial sample and may be considered real. A similar very short peak has been discussed for the 2016 Quadrantids (Rendtel et al., 2016).

October Draconids (009 DRA): The Draconid meteor shower is one of the most interesting meteor showers, producing spectacular storms in 1933 and again in 1946, as well as several other outbursts in the past century.

The parent comet, 21P/Giacobini-Zinner, reached perihelion in 2005. No strong activity was expected, but the CMOR radar detected significant activity of mainly faint meteors (Campbell-Brown et al., 2006). The descending branch of the activity curve was observed by video in the Czech Republic (Koten et al.,

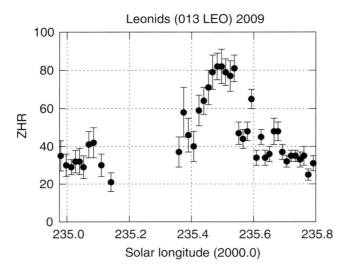

Figure 4.11. ZHR profile of the 2009 Leonid return with dust trail encounters discussed in the text. Besides the main and the late maxima there are two peaks marked by only one data point but calculated from a reasonable sample.

2007). The observed radiants of video meteors, as well as one fireball, were in agreement with a model given in Campbell-Brown et al. (2006). Atmospheric properties of the meteors were consistent with the very fragile nature of the Draconid meteoroids.

On the other hand, strong activity was expected for the next return of the parent comet. Vaubaillon et al. (2011) and Maslov (2011) predicted very high activity, but not a storm, for 2011 October 8. Two Falcon aircraft flew above northern Europe and the Atlantic to make double station observations (Vaubaillon et al., 2015). The measured main peak of the activity occurred around 20:15±0:05 UT, which was consistent with the model prediction, as well as with IMO network visual observations. Moreover, one of the instruments detected meteors associated with material ejected from the parent comet before 1900 (Figure 4.12), and thus confirmed a prediction of the model, even though it was based on uncertain pre-1900 cometary data (Koten et al., 2014).

According to the visual observers contributing to the IMO network, the peak had a ZHR of 280 and was found at the predicted time at $\lambda_\odot = 195°035$, i.e. October 8, 20^h13^m UT (Figure 4.13). The FWHM[8] of the peak was 80 min, and the peak was slightly asymmetric. To determine FWHM, a third order polynomial was fitted to the ascending and descending branches of the activity profile. Both the quantity and quality of the data are good, despite the fact that the shower occurred under moonlit conditions. A detailed comparison of the visual observations with video flux data (Molau et al., 2012) shows that the results obtained by both methods are well calibrated and consistent well within the error margins.

Fully automated flux density measurements operated for the first time by the IMO Video Meteor Network detected a primary maximum at 20:09±0:05 UT. Another small sub-maximum

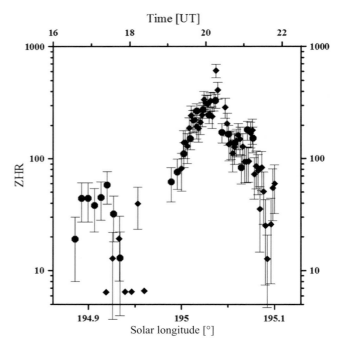

Figure 4.12. Draconid 2011 outburst profile observed by two video cameras deployed on the aircraft mission (Koten et al., 2014). The first peak around 17:20 UT was caused by pre-1900 material. The main peak after 20 UT is consistent with the visual data (Figure 4.13).

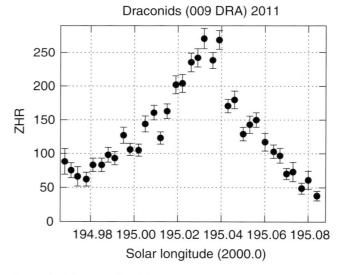

Figure 4.13. ZHR profile of the 2011 Draconid outburst. A population index of $r = 2.80$ was used for this analysis. The ZHR was above 140 between 1920–2035 UT (FWHM 75±5 min) with a steeper descent.

was revealed at 19:34±0:07 UT and may be connected with the 1907 trail (Molau and Barentsen, 2014). The geocentric parameters of the observed meteors from this and other sources are summarised in Table 4.6. Results of all the listed campaigns agreed well with the model predictions.

Another entirely unexpected outburst on 2012 October 8 was completely different, as it was comprised of mainly faint meteors which were recorded by the CMOR radar. The moderate outburst

---

[8] Full width at half maximum

Table 4.6. *Geocentric radiants and velocities from the Draconid 2011 outburst*

$\alpha_G[°]$	$\delta_G[°]$	v_G [km s^{-1}]	Source
262.6 ± 2.2	55.7 ± 0.9	20.7 ± 0.7	[1]
263.0 ± 0.4	55.3 ± 0.3	20.8	[2]
262.3 ± 0.4	55.4 ± 0.3	20.8	[3]
263.2 ± 0.05	55.7 ± 0.05	20.8 ± 0.05	[4]
263.3 ± 1.5	55.6 ± 1.0		[5]
263.2	55.8	20.9	[6]

References: [1] Šegon et al. (2014b) [2] Trigo-Rodríguez et al. (2013) [3] Koten et al. (2015) [4] Borovička et al. (2014) [5] Tóth et al. (2012) [6] Vaubaillon et al. (2011) (model).

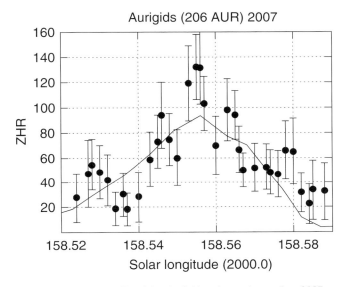

Figure 4.14. ZHR profile of the Aurigid outburst observed on 2007 August 31, based on all count intervals of ≤ 5 min duration. The solid line shows the smoother profile for all intervals ≤ 15 min.

was also observed by video cameras in central Europe and 36 multi-station Draconid meteors were recorded (Tóth et al., 2014).

Aurigids (206 AUR): The Aurigid annual meteor shower is usually very weak but experienced several strong outbursts in the past century. On 2007 September 1, an outburst was predicted (Jenniskens and Vaubaillon, 2008) and well observed under moonlit conditions. The analysis (Rendtel, 2007b) shows a sharp peak at $\lambda_\odot = 158°556 \pm 0°003$, i.e. $11^h 20^m$ UT ±3 min, and a peculiarly low population index of $r = 1.74 \pm 0.08$ which meant that a significant fraction of meteors were bright. The shower profile shown in Figure 4.14 (single points) has the maximum temporal resolution. The number per bin becomes small and many of the features are only observed at marginal statistical significance. These features would require an independent data set for confirmation. The solid line shows a smoother profile calculated with lower temporal resolution. We see a lower peak ZHR, but can better characterise the general profile, which is slightly asymmetric (ascent 27, descent 18 minutes), with a FWHM of 45±5 minutes.

The Aurigid Multi-Instrument Aircraft Campaign was carried out using two Gulfstream aircraft. The peak of activity was recorded 18 minutes earlier than predicted. The maximum rate (ZHR) was about 130, short of the predicted 200 (Jenniskens and Vaubaillon, 2008). The results were confirmed by ground-based double-station video observations reported by Atreya and Christou (2009). The orbits of the double station meteors were found to be in good agreement with that of the shower parent comet C/1911 N1 (Kiess).

September ϵ-Perseids (208 SPE): An unexpected outburst was recorded by visual, video and radar observers on 2013 September 9. Molau et al. (2013) analysed data obtained with the IMO Video Meteor Network and identified almost 300 members of the September ϵ-Perseid shower. The outburst started very quickly with a steep ascending branch and reached its peak at 22:00±0:05 UT. The descending branch was shallower, but the duration of the outburst was still short, with a FWHM of only 41±5 minutes. Data recorded with the EDMOND network also showed a peak around 22:05±0:10 UT, followed by a decrease of activity with another peak at 22:40±0:10 UT (Gajdoš et al., 2014). The European fireball network observations of this event were reported by Shrbený et al. (2016). Madiedo et al. (2018) analysed the properties of the meteoroids and compared them with other meteor showers. The meteoroids consist of regular cometary material and are associated with an unknown long-period comet.

Visual ZHRs have been used for calibration of radio forward signatures of the ϵ-Perseids and ϵ-Eridanids (209 EER) in 2016 (Rendtel et al., 2017).

Phoenicids (254 PHO): This shower has only been detected in 1956. A return of activity was predicted for 2014 December 2 using a dust trail simulation based on the 1956 event (Sato et al., 2017). Observations confirmed activity at the calculated time with a ZHR of about 16 at the predicted position.

Perseids (007 PER): Modelling results for the 2016 return of this shower included a generally enhanced background rate of 150–160 due to a shift of the stream closer to the Earth's orbit because of Jupiter (Rendtel, 2015). At $\lambda_\odot = 139°436$ an increase in small meteoroids (and therefore higher r) from the 1-revolution trail was expected to cause an increase of the ZHR by about 10. Brighter meteors from the 4-revolution trail were expected at $\lambda_\odot = 139°470$. Later, between $\lambda_\odot = 139°49$ and $139°66$, an encounter with a stream section dominated by meteoroids from the 2-revolution trail was predicted – all well before the broad nodal maximum around $\lambda_\odot = 140°0$. Most of these components were detected in visual data (Miskotte and Vandeputte, 2017).

Visual data from the 2018 return revealed a clear signature of the filament listed by Jenniskens (2006) on August 12 close to 20h UT. We find a minimum of $r = 1.68$ near 18:30 UT ($\lambda_\odot \approx 139.74°$). The neighboring r-values are 1.71–1.79 with a rather poor temporal resolution of about 0.12°. Within the error margins, a peak ZHR = 135±14 occurs at 18:50 UT ($\lambda_\odot \approx 139.76°$). This corresponds to a peak of roughly 1–2 hours in width, as the neighboring ZHR values are 103±11 (Rendtel et al., 2018).

Quadrantids (010 QUA): In 2016, minor features in the ascending branch defined by only single values at $282°884$ have been found at identical positions in visual and video data (see figure 3 in Rendtel et al., 2016). Since enhanced activity was also found in radio forward scatter data, we are confident that the very short increase in rate and flux is real. This is also of interest for the single-point peaks in the Leonid profile shown in Figure 4.11.

Ursids (015 URS): A strong outburst of the Ursid meteor shower occurred on 2000 December 22. The outburst was recorded by double station video and photographic observations in California. A total of 431 Ursids were detected by video cameras. The ZHR reached a level of 100. The time of maximum matched the expected encounter times with the dust trails from 1405 and 1392. For the first time, the link between the Ursid meteor shower and comet 8P/Tuttle was confirmed (Jenniskens et al., 2002a).

The Ursids returned on 2014 December 22–23 when a strong but short peak of activity was recorded around midnight by the IMO Video Meteor Network (Molau et al., 2015; Moreno-Ibáñez et al., 2017). According to the models, the activity was connected with dust released from the parent comet in 1392 and 1405.

October Camelopardalids (281 OCT): A relatively weak outburst was recorded by video cameras from a radiant around $\alpha_G = 164°, \delta_G = +79°$ on 2005 October 5. The newly detected October Camelopardalid shower was identified as the debris of an unknown long period or Halley-type comet (Jenniskens et al., 2005).

Comae Berenicids (020 COM): On 2006 December 24–25 the video cameras of the Spanish Meteor Network detected high activity from the Comae Berenicids. The maximum flux was equivalent to the visual ZHR = 60±25. This activity was ten times higher than expected for this meteor shower (Madiedo and Trigo-Rodríguez, 2008).

Southern Taurids (002 STA): The shower usually produces low activity in October and November. Enhanced activity occurs when a swarm of meteoroids locked in a 7:2 resonance with Jupiter is encountered. Such an event happened in 2015. The European fireball network recorded 144 Taurid fireballs. Spurný et al. (2017a) recognised that 113 of them belong to the newly discovered branch of this shower. The masses of observed meteoroids range from 0.1 g to 1000 kg. The asteroids 2015 TX$_{24}$ and 2005 UR, both of diameters 200–300 m, are hypothesised to be members of the new branch.

4.3.2 Structure and Composition of Meteoroids

White light and spectral optical observations using photographic and video cameras provide crucial data for the investigation of the internal structure and composition of meteoroids as well as for models of the meteoroid atmospheric entry process. A larger number of interesting events have been recorded due to the expansion of observation networks, and more precise data has become available for modelling. A number of so-called anomalous meteors have been recorded in the past decade or two. Such meteors exhibit unusual features and their observations provide additional data for models.

4.3.2.1 Structure and Fragmentation

ReVelle and Ceplecha (2001) derived the dependence of the ablation and shape-density coefficients, and of the luminous efficiency, on various time-dependent parameters using PN and EN fireballs. The fragmentation model, which allows arbitrary fragmentation points and two possible types of fragmentation – into large pieces and into a cluster of small fragments – was presented by Ceplecha and ReVelle (2005). It also allows the ablation coefficient, the shape density coefficient, and the luminous efficiency to vary with time while fitting both the light curve and the dynamics. Values of these parameters with (intrinsic) and without (apparent) taking fragmentation into account are defined. The fragmentation model explains the difference between dynamic and photometric masses, which has been an unexplained problem in meteor science for a long time.

Spurný and Ceplecha (2008) proposed triboelectricity, induced in meteoroids during their atmospheric entry, as the main process to adequately explain the meteor phenomenon, with hypersonic aerodynamic processes being only of secondary importance. This proposal was made on the basis of three years of observations from the EN and the DN with the AFO system, when precise atmospheric trajectories together with detailed light curves of fireballs from radiometers were available for the first time.

Babadzhanov and Kokhirova (2009) published mean mineralogical and bulk densities of meteoroids belonging to nine meteoroid streams and the sporadic background on the basis of the results of the Tajikistan fireball network, processed according to the theory of quasi-continuous fragmentation. Two meteorite-dropping Geminids have been reported (Madiedo et al., 2013a; Spurný and Borovička, 2014).

Analyses of thirteen meteorite falls with precise atmospheric trajectories showed that the bulk strength inferred from the observations is typically of the order of ten to a hundred times lower than the tensile strength of recovered samples (Popova et al., 2011). The authors conclude that metre-scale meteoroids are highly fractured before entering the atmosphere and can break up under stresses of a few megapascals.

The deceleration and light curves of six double-station Draconid meteors observed during the 2005 enhanced activity were modelled using a newly developed thermal erosion model (Borovička et al., 2007). It was found that the meteoroids were porous aggregates of grains with a porosity of about 90% and bulk densities of 300 kg m^{-3}. Individual millimetre-sized meteoroids consisted of tens of thousands to a million grains. The same model was applied to a sporadic meteor on a cometary orbit observed by video and spectral cameras. The meteoroid structure was significantly different from the Draconid meteoroids, with larger grains and larger erosion energy (Borovička et al., 2008).

Campbell-Brown et al. (2013) compared two different ablation models – the thermal disruption model (Campbell-Brown and Koschny, 2004) and thermal erosion model (Borovička et al., 2007). Both models were used to fit the brightness and deceleration curves and the predicted wakes were compared with the wakes observed by the CAMO telescopic system. Whereas both models provide satisfactory results for wide field cameras measurement, both also predict far more wake than was observed. The authors conclude that the models of meteoroid fragmentation in the atmosphere need significant improvements. The CAMO data provide important constraints for future models.

An interesting feature was found for the Camelopardalid meteoroids (451 CAM) observed in 2014. While the fainter Camelopardalid meteors were found to be stronger in comparison with other meteors with the same velocity, the brightest meteors were weaker than typical. Campbell-Brown et al. (2016) proposed that large Camelopardalids are weak conglomerates of more refractory grains, which are easily disrupted in space, and therefore the shower is rich in small particles and depleted in large ones.

4.3.2.2 Composition and Spectra

The last two decades saw progress mostly in the spectroscopy of fainter meteors observed by video spectrographs. As described in Borovička et al. (1999), meteor spectra contain emission lines of both meteoric and atmospheric origin, molecular bands, and continuum radiation. Borovička et al. (2005) compared the intensities of meteoric lines of Mg, Na, and Fe in 97 mostly sporadic meteors and found large differences in abundances of these three elements in small, millimetre-sized, meteoroids. This kind of analysis was later performed for low-resolution spectra of both faint and bright meteors by various authors, e.g. Vojáček et al. (2015), Rudawska et al. (2016), Maeda and Fujiwara (2016) and Madiedo (2017) and references therein. The results are discussed in more detail in Chapter 2 (Borovička et al., 2019).

A number of papers presented low-resolution spectra of meteors from various meteor showers. They include the Quadrantids (Koten et al., 2006b; Ward, 2015; Madiedo et al., 2016b), Geminids (Borovička, 2010), Leonids (Borovička et al., 1999; Abe et al., 2000; Kokhirova and Borovička, 2011), Taurids (Matlovič et al., 2017), α-Capricornids (Madiedo et al., 2014b), Draconids during the 2011 outburst (Madiedo et al., 2013b; Borovička et al., 2014; Rudawska et al., 2014) and Camelopardalids during the 2014 outburst (Madiedo et al., 2014c). The catalog of Vojáček et al. (2015) also contains spectra of major meteor showers. Most shower meteors have nearly normal spectra with a tendency for Fe underabundance in cometary meteoroids. The Quadrantids and especially Geminids exhibit large variations of Na content.

Another aspect which can be well studied from video spectra is differential ablation. In some cases, a meteor spectrum changes strongly with time. Typically, the Na line is bright at the beginning and faint or absent toward the end of the meteor. Sodium is a volatile element and, as shown theoretically by Janches et al. (2009), can be evaporated from small grains earlier than Mg, Fe and other main constituents. Refractory elements (Ca, Al) evaporate last. The effect of preferential ablation of Na (Figure 4.15) has been observed in Leonids (Borovička et al., 1999), Quadrantids (Koten et al., 2006b) and Draconids (Borovička et al., 2014) but the effect varies strongly from one meteor to another even within the same meteor shower. This is evidence of large differences in structure and fragmentation behaviour of individual meteoroids. Differential ablation has also been studied using multiple cameras with narrow wavelength filters (Bloxam and Campbell-Brown, 2017).

Video spectroscopy is convenient to observe the spectra of meteor trains. The evolution of a bright Leonid train was studied by Borovička and Jenniskens (2000). It was shown that the

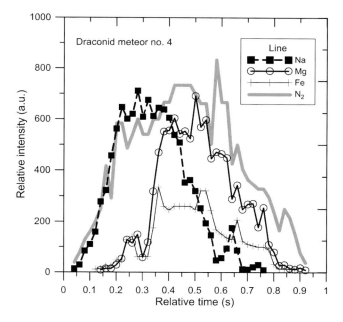

Figure 4.15. The effect of differential ablation seen in Draconid meteor no. 4 from Borovička et al. (2014). The meteor reached absolute magnitude 2.3. Sodium started and ended its ablation much sooner than magnesium and iron. The bands of molecular nitrogen of atmospheric origin have the widest light curve. Line intensities are not on a common scale. Note that not all Draconid meteors showed differential ablation.

atomic level population was not in thermal equilibrium during the afterglow phase just after the meteoroid passage due to a paucity of electron-atom collisions. The excitation temperature dropped from ~5000 K to 1200 K within two seconds. The energy for radiation was the remnant heat from the meteoroid disintegration. In some meteors, there is another phase, which can last longer, when the energy for line radiation is provided by recombination. Finally, the longest phase, which can last for tens of minutes, has a featureless spectrum, probably of molecular origin. The identification of responsible molecules is uncertain, possible candidates are NO_2 or FeO. More details and discussions can be found in Borovička (2006) and references therein. Newer examples of train spectra can be found e.g. in Spurný et al. (2014b) or Madiedo (2015).

Spurný et al. (2014b) presented the best-to-date meteor spectrum above an altitude of 130 km. The spectrum consists of only atmospheric emissions of O, N, and N_2, which is consistent with the sputtering theory for radiation of meteors at high altitudes (Popova et al., 2007).

Though some authors have tried to derive atomic abundances from low-resolution spectra, reliable results need high-resolution multi-line spectra available with brighter meteors. Such data allow the excitation temperature and density of the radiating gas to be computed at various points along the trajectory. The classical work of Borovička (1993) was followed by Trigo-Rodríguez et al. (2003b), Borovička (2004, 2005) and Jenniskens (2007), who used higher order spectra from a CCD spectrograph. Typical temperatures of the main spectral component are 4000–6000 K. The abundances of the following elements were derived in at least some spectra: Fe,

Mg, Na, Ca, Cr, Mn, Al, Si, Ti, Ni, Co, Li, H. The results are discussed in Chapter 2 (Borovička et al., 2019). Unfortunately, the derived abundances do not directly reflect the composition of the meteoroid. The first reason is incomplete evaporation of refractory elements. Part of Al, Ca, Ti remains in the solid or liquid phase and does not contribute to gas radiation. Borovička (2005) showed that evaporation was more complete after a short meteor flare. It is also more complete at low altitudes (Borovička and Spurný, 1996) and in the high temperature ($\sim$10,000 K) component of fast meteors, which is probably connected with the meteor shock wave (Borovička, 2004).

Another complication is that some atoms can be bound in molecules. Berezhnoy and Borovička (2010) computed the expected abundances of molecules in meteor gas under equilibrium conditions and found that a significant part of Si, Ti, and C can be in molecular form (in oxides) even at temperatures above 4000 K. The oxides cannot be directly observed since SiO and CO do not have strong bands in visible wavelengths and TiO is a minor constituent. FeO, on the other hand, is commonly seen in meteor spectra. Borovička and Berezhnoy (2016) analysed radiation of FeO, CaO, AlO, and MgO in the Benešov superbolide. Emission was strong in the wake and in the dust cloud left behind the bolide. Other molecules firmly detected in meteor spectra are atmospheric N_2 (e.g. Jenniskens et al., 2004a) and N_2^+ (e.g. Abe et al., 2005).

Sometimes, high resolution meteor spectra have been detected serendipitously by large telescopes during observations of other targets. Such spectra were presented e.g. by Jenniskens et al. (2004b), Kasuga et al. (2007) and Passas et al. (2016).

Of special interest are two Leonid spectra obtained from space covering the ultraviolet region, reported by Jenniskens et al. (2002b) and Carbary et al. (2003), respectively. The latter spectrum goes down to 110 nm and contains lines of atomic carbon. Since carbon does not have strong lines in the visible region, this is the best evidence for carbon in meteoroids. Infrared spectra of Leonids were reported by Taylor et al. (2007).

4.3.2.3 Anomalous Meteors

The increasing sensitivity and resolution of video systems have also led to the detection of peculiar meteors exhibiting some kind of unusual behaviour. Such meteors are uncommon due to their photometric features, appearance, trajectory or heliocentric orbit.

Grazing meteors: At present, there are just a few reliable detections of meteors that passed through the Earth's atmosphere and moved back into space on a different heliocentric orbit. In all cases, these were rather bright meteors. Historically, the first such case was a fireball observed over the west of the USA and Canada on 1972 August 10 (Ceplecha, 1994).

More recently, a fireball of magnitude -8 was detected by the Japanese video network SonotaCo in March 2006 travelling over a distance of about 1000 km (Abe et al., 2006). A meteor of -4 magnitude was detected over Spain on 2012 June 10 (Madiedo et al., 2016a). Over 17 s the meteor moved a distance of 510 km at a velocity of 29 km s^{-1}. Its height in perigee was estimated to be 98 km. A relatively faint meteor of magnitude $+3$ was observed over Ukraine in September 2003. Having an initial velocity 62.9 km s^{-1}, the meteor moved 426 km after perigee (at an altitude of 101.7 km) for 6.8 s before entering the fields of view of TV cameras. The recorded trajectory covered 35 km, during which the mass loss was estimated to be 5×10^{-3} g (Kozak and Watanabe, 2017).

Extreme beginning heights: The classical model of meteor flight predicts initial luminosity at altitudes of up to 130 km. It was mainly the Leonid storms which changed this view, though the first report of meteors with beginning heights up to 160 km appears earlier (Fujiwara et al., 1998). It quickly became clear that such behaviour appears mainly among members of the high-velocity showers – 1 Perseid was detected at 136.8 km (Hajduková et al., 1995), 3 Leonids at 137.9 km, 144.4 km and 130.9 km (Campbell et al., 2000), more Leonids up to 195 km (Spurný et al., 2000a), 2 η-Aquariids at 150.2 km, 133.8 km, and 1 Perseid at 149.0 km (Koten et al., 2001), 1 Leonid at 174$\pm$8 km (Gährken and Michelberger, 2003), a few Leonids at altitudes 135–145 km (Kozak et al., 2007), an Orionid meteor at 164.8 km (Olech et al., 2013) and another Perseid at 170 km (Spurný et al., 2014b). This last is the best-documented case since it was recorded by video, photographic and spectral cameras in addition to having persistent train. An exception to the high speed requirement is a Lyrid meteor (geocentric velocity $v_G = 45.3$ km s^{-1}) with a beginning height of 136.8 km (Koten et al., 2001). An overview of meteors with extreme beginning heights is given by Koten et al. (2006a), where suddenly terminating meteors are also reported.

Spurný et al. (2000a) propose a qualitative explanation of collisional excitation of atmosphere molecules (mainly O I and N_2) at altitudes of 200 km, in which the luminous efficiency of the process is much higher than during ablation. Spurný et al. (2000b) note the diffusive, comet-like structure in images of some bright Leonids in the altitude range of 200–135 km, with intermediate phase between 135 and 125 km, and formation consistent with the classical ablation model with meteor train below 125 km. In addition, they mention luminous transverse jets up to a few kilometres long at altitudes above 130 km. Both the diffuse structure of a meteor coma and transverse jets up to a few kilometres were confirmed independently by LeBlanc et al. (2000). Similar results were presented after observations of the Leonid meteor shower in Mg I (Taylor et al., 2000). A quantitative mechanism of sputtering meteoroid surface by the flow of air molecules at ultra-high altitudes, which explain the diffuse structure of meteor coma, was developed by Popova et al. (2007).

Features on light curves: Television and video observations show that some meteors exhibit anomalies on their light curves. While relatively small oscillations in a meteor light curve can be explained by fragmentation, which is observed from time to time with long focus lens observational instruments (Gorbanev, 2009; Stokan and Campbell-Brown, 2014), the practically complete temporal disappearance of a meteor is more difficult to explain (Figure 4.16). Such a meteor was described, for example, by Roberts et al. (2014), where the authors explained the observation using the dustball model, in which the solid components of the body with different masses are significantly separated before the beginning of intensive evaporation.

Meteor clusters: A rare phenomenon is the detection of ultra-short outbursts of meteor activity, pointing to a cluster structure of the meteoroid stream. They have been observed mainly during Leonid outbursts and storms. Between 100 and 150 meteors

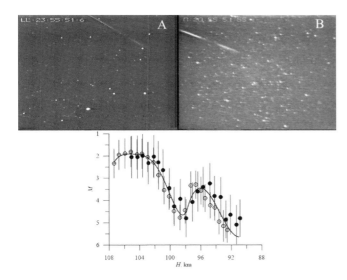

Figure 4.16. Bimodal light curve of a meteor from the Perseid meteor shower, detected in Kyiv (Ukraine) on 1991 August 14, UT = 20:55:51 using super-isocon TV systems. The white circles correspond to observation point A, black to B.

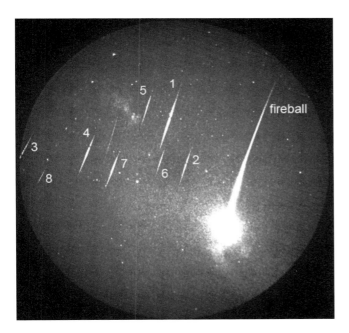

Figure 4.17. Composite image shows the September ϵ-Perseid fireball and eight faint meteor trails. The meteors were separated by 14 to 104 km (Koten et al., 2017).

were recorded within 2 s during the 1997 Leonids (Kinoshita et al., 1999). Watanabe et al. (2002) observed at least fifteen meteors within four seconds during the 2002 Leonids. Watanabe et al. (2003) reported thirty-eight meteors within two seconds during the 2002 Leonid campaign. They also demonstrated that such small clusters a few hundred kilometres in size had to be ejected from a comet nucleus during its last passage through perihelion before the encounter with Earth.

A bright fireball from the September ϵ-Perseid shower on 2016 September 9 was accompanied by eight fainter meteors on parallel trajectories within less than two seconds (Figure 4.17). The cluster of meteors was interpreted to be the result of the orbital fragmentation of a larger meteoroid which happened no more than two or three days before the encounter with the Earth at a distance smaller than ~ 0.08 AU from the Earth (Koten et al., 2017).

4.3.3 Bolides and Meteorite Falls

When a meteoroid entering the atmosphere is larger than about 20 cm (for 15 km s^{-1}), its mass is not usually fully ablated before it decelerates down to about 3 km s^{-1} and its remnant falls as a meteorite (Ceplecha et al., 1998). Such cases are of major scientific interest, not only because it is the cheapest way to sample interplanetary material from the Solar system. As mentioned in Section 4.2.3, some camera networks were established primarily with the goal of recovering meteorites (e.g. DN or FRIPON).

Bright fireballs, as very remarkable events, also usually attract broad public attention. There are several ways for the public to report a fireball observation to scientists. The IMO Fireball report form has been collecting such observations since 2015 (Hankey and Perlerin, 2015). The form has been translated into twenty-eight languages. The "Fireballs in the Sky" application uses the internal GPS and smartphone sensors for a more accurate witness report of observed fireball (Sansom et al., 2016). Its database is directly linked to the DN.

4.3.3.1 Meteorite Falls

An exceptional position among meteorite falls is occupied by the Chelyabinsk meteorite fall on 2013 February 15. It was the largest impact on the Earth since the 1908 Tunguska event (Popova et al., 2013). The atmospheric trajectory and heliocentric orbit were determined on the basis of video records. The orbit was found to be similar to the orbit of the two-kilometre-diameter asteroid 86039 (1999 NC$_{43}$) (Borovička et al., 2013a). The impact was caused by an asteroid about 19 metres in diameter entering the atmosphere at 19 km s^{-1}; the estimated energy of the airburst was equivalent to approximately 500 kt of TNT (Brown et al., 2013). The shockwave hit a densely populated area, and more than a thousand people sought medical assistance in hospitals. The type of injury did not depend on the distance from the asteroid trajectory (Kartashova et al., 2018).

The number of recoveries of meteorites with known orbits has significantly increased in the last decade. As of 2019 March 22 the total number of these has reached 29[9]. Only those cases published in a peer-reviewed journal or pre-published in a conference abstract or a thesis are included in this number. Although all these cases are scientifically interesting events, here only the most important are mentioned.

A bright fireball was observed above the Austrian-German border on 2002 April 6, followed by multiple meteorites falling in the vicinity of the Neuschwanstein castle. The heliocentric orbit of its parent was remarkably similar to the orbit of the Příbram meteoroid. The chemical classifications and cosmic-ray exposure ages of the two meteorites are quite different,

---

[9] www.meteoriteorbits.info/

which implies a heterogeneous stream (Spurný et al., 2003). Subsequent video observational campaigns showed that there are no faint meteors in the same orbit (Koten et al., 2014c).

The Desert Fireball Network recorded the first achondrite fall with a known orbit on 2007 July 20. Bunburra Rockhole was delivered from an unusual Aten-type orbit. The initial mass of the meteoroid was 22 kg, and three meteorites of a total mass of 339 g were recovered. They were classified as anomalous basaltic meteorites (Bland et al., 2009; Spurný et al., 2012).

Among other meteorite falls with known orbits are two carbonaceous chondrites – Maribo (Haack et al., 2012) and Sutter's Mill (Jenniskens et al., 2012). Both cases had very similar orbits, initial velocities above 28 km s^{-1}, both originated from large (multi-metre) meteoroids and showed early fragmentation and high end heights (Borovička et al., 2015b).

The Košice meteorite fall in eastern Slovakia on 2010 February 28 was imaged by three security video cameras from Hungary. The meteoroid, with an estimated mass of 3500 kg, entered the atmosphere with a velocity of 15 km s^{-1}. A maximum brightness of absolute stellar magnitude about -18 was reached at a height of 36 km. A number of meteorites were recovered in the predicted fall area during official searches (Borovička et al., 2013b).

Tables 1, 2 and 3 in Borovička et al. (2015b) contain a comprehensive list of instrumentally observed meteorite falls. The Chelyabinsk meteorite fall from 2013 is the last case in this list. Other falls with published orbits which occurred in the last five years are summarised in Table 4.7.

It is not only new falls that have provided new data. Spurný et al. (2014a) published a paper on a re-analysis of the Benešov superbolide. They showed that in some special cases it is still possible to predict and find meteorites a long time after the fall. They also demonstrated that larger meteoroids can be very complicated compositionally, since the Benešov meteoroid consisted of at least three different types of material – LL3.5, H5, and primitive achondrite.

The 2014 January 13 fireball observed above the Czech Republic probably resulted in the fall of very small meteorites – which were not searched for – but it was very interesting for another reason. The modelling of its atmospheric deceleration and fragmentation showed that a meteoroid with an initial mass of 5 kg entered the atmosphere with an Earth-relative velocity of 11 km s^{-1}. The low-resolution spectrum confirmed its natural origin. Back integrations yield 92%–98% probability that the object was a temporarily captured orbiter of the Earth (Clark et al., 2016).

4.3.3.2 Predicted Impacts

On 2008 October 6, a small asteroid was discovered by the Catalina Sky Survey. It received the designation 2008 TC$_3$. The Minor Planet Center announced that this object will impact the Earth only 20h after its discovery (McGaha et al., 2008). 12 hours before the impact, NASA JPL predicted that the impact would occur in the Nubian Desert in Sudan. Jenniskens et al. (2008c) found that the asteroid broke up at an altitude of 37 km. A dedicated search campaign recovered more than 6000 meteorites – named Almahata Sitta – with a total mass of 10.7 kg. Using almost 900 astrometric observations taken before the impact, Farnocchia et al. (2017) estimated the trajectory of 2008 TC$_3$ and its ground track. Borovička and Charvát (2009) analysed observations by the Meteosat 8 weather satellite, which recorded the fireball and the associated dust cloud, and estimated that the bulk porosity of 2008 TC$_3$ was of the order of 50%, i.e. higher than the porosity of the recovered Almahata Sitta meteorites. 2008 TC$_3$ was the first ever predicted impact of a near-Earth asteroid.

The second such case was the impact of asteroid 2014 AA, which entered the Earth's atmosphere on 2014 January 2 only twenty-one hours after it was discovered, again by the Catalina Sky Survey (Farnocchia et al., 2016). A combination of ground-based optical astrometry and infrasound data determined the impact point was at longitude of 44°.2 W and a latitude of 13°.1 N with a 1σ uncertainty of 140 km (central Atlantic ocean). The estimated minimum mass was 22.6 t. The third asteroid detected pre-impact was 2018 LA (Kowalski et al., 2018), which was discovered eight hours before it entered the atmosphere over South Africa (de la Fuente Marcos and de la Fuente Marcos, 2018).

A different type of predicted impact is the re-entry of interplanetary spacecraft into the Earth's atmosphere, usually in the form of small capsules carrying collected samples. The Sample Return Capsule of NASA's Stardust mission entered the atmosphere at a speed of 12.8 km s^{-1} on 2006 January 15. NASA's DC-8 airborne laboratory carried several imagers and spectrometers to an altitude of 11.9 km to record the re-entry of this artificial fireball. The spectra served for comparison with spectra of low-velocity natural meteors (e.g. Jenniskens et al., 2010).

The Hayabusa mission ended not only with a sample return but also with the re-entry of the spacecraft itself on 2010 June 13 (Fujita et al., 2011). The trajectories and velocities of the spacecraft, its fragments, and the capsule were derived from

Table 4.7. *Instrumentally observed meteorite falls after Chelyabinsk (with published orbits)*

Name	Date	Classification	Source
Annama	2014 April 18	H5	Kohout et al. (2015)
Žďár nad Sázavou	2014 December 9	L3	Spurný (2016)
Murilli	2015 October 27	H5	Howie et al. (2017b)
Ejby	2016 February 6	H5/6	Spurný et al. (2017b)
Stubenberg	2016 March 6	LL6	Spurný et al. (2016)
Dingle Dell	2016 October 31	LL5	Devillepoix et al. (2018)

photographic records (Borovička et al., 2011), and spectra were taken in near-ultraviolet and visible light (Abe et al., 2011).

4.4 Future Work

All observational techniques will continuously improve with time. Cameras with better resolution and higher dynamical range will become available. The volume of recorded data will also increase. New methods for data acquisition, storing and processing will be necessary.

For visual observations, the IMO website allows the submission of reports of visual data, which go directly into the database after being approved. Combining samples obtained by different techniques (e.g. visual, video, radio forward scatter, radar) allows us to confirm the reliability of detected features and to extend the findings to a larger mass range. Such multiple datasets are essential for calibration purposes as well.

Currently, a new and comprehensive program package is being prepared to allow shower analyses on a broader basis, ensuring the continuity of analyses of global visual data provided from 1988 onwards. The comparison of predictions and model calculations with observational results need to be continued. Interactions of observers and model authors are necessary to modify and adapt parameters, and finally to improve models.

In the field of photographic observations, the advantage of digital photographic systems is the availability of recorded fireball data in real time, which minimises the time between a meteorite fall and the first visit to the impact area. This fact increases the probability of a successful meteorite recovery. On the other hand, digital photographic systems are still not able to record fireballs in the same high dynamic range and resolution as film cameras. Future improvements of digital photographic systems should cover the implementation of digital chips with higher dynamic range and resolution.

For photographic and video observations, the volume of data is growing with more and larger networks of cameras. It has become evident in the meteor community that greater automation is needed in the processing pipeline, as well as for final product aggregation and analysis. Many of the processing steps outlined in Section 4.2.4 are effectively automated. However, steps like user confirmation of tracks, event coincidence aggregation and trajectory acceptance often still employ human labour. Recent work in applying machine learning techniques (Decicco et al., 2018) shows promise in taking the human out of the loop. This has included having a computer application learn from thousands of images to separate meteor tracks from clouds and aircraft, and applying realisable geometric constraints to calculated trajectories (Gural, 2018).

On the other hand, automation should not be the only goal for observers. The recent work of Hajduková et al. (2017) shows that the calculated velocities of video meteors in several databases are underestimated in comparison with photographic and radio meteors, probably due to the methods used for positional and velocity measurements. Such discrepancies may result in a misidentification of new meteor showers as well as biasing models. The quantity of the data should not exceed its quality. Precise determination of the meteor velocity is crucial and we can expect new methods will be developed (Egal et al., 2017; Vida et al., 2018b).

We can expect that new instruments with higher resolution and sensitivity will contribute to the development of better models of meteoroid ablation and fragmentation. The role of the instruments like CAMO will be crucial for these efforts.

The re-entry of several interplanetary spacecraft occurred during the past fifteen years, and was observed by dedicated airborne missions. The impacts of natural objects, on the other hand, are being predicted just hours before they happen. We can expect that earlier detection of such objects will allow airborne observation campaign to be carried out, which will observe the atmospheric entry of small asteroids. Such an event will enable ground-based observations of the asteroids in the interplanetary space to be compared with the results of fireball observations, and improve the calibration of both kinds of instruments.

Acknowledgements

This work was partly supported by grant no. 16-00761S from GA ČR. The authors also thank to all observers submitting their reports to the IMO's VMDB.

References

Abe, S., Yano, H., Ebizuka, N., and Watanabe, J.-I. (2000). First results of high-definition TV spectroscopic observations of the 1999 Leonid meteor shower. *Earth Moon and Planets*, **82**, 369–377.

Abe, S., Yano, H., Ebizuka, N. et al. (2003). Twin peaks of the 2002 Leonid Meteor Storm observed in the Leonid MAC Airborne Mission. *Publications of the Astronomical Society of Japan*, **55**, 559–565.

Abe, S., Ebizuka, N., Murayama, H. et al. (2004). Video and photographic spectroscopy of 1998 and 2001 Leonid persistent trains from 300 tO 930 nm. *Earth Moon and Planets*, **95**, 265–277.

Abe, S., Ebizuka, N., Yano, H., Watanabe, J., and Borovička, J. (2005). Detection of the N_2^+ first negative system in a bright Leonid fireball. *The Astrophysical Journal*, **618**, L141–L144.

Abe, S., Borovička, J., Spurný, P. et al. (2006). Earth-grazing fireball on March 29, 2006. *European Planetary Science Congress*, **1**, EPSC2006–486.

Abe, S., Fujita, K., Kakinami, Y. et al. (2011). Near-ultraviolet and visible spectroscopy of HAYABUSA spacecraft re-entry. *Publications of the Astronomical Society of Japan*, **63**, 1011–1021.

Aghajan, H., and Kailath, T. (1994). SLIDE: Subspace Based Line Detection. *IEEE Transactions of Pattern Analysis and Machine Intelligence*, **16**, 1057–1073.

Andreić, Z., and Šegon, D. (2010). The first year of Croatian Meteor Network. Pages 16–23 of: Kaniansky, S., and Zimnikoval, P. (eds), *Proceedings of the International Meteor Conference, Šachtička, Slovakia, 18–21 September, 2008*. Hove, BE: International Meteor Organization.

Arlt, R. (1998). Bulletin 13 of the International Leonid Watch: The 1998 Leonid meteor shower. *WGN, Journal of the International Meteor Organization*, **26**, 239–248.

Arlt, R. (2003). Bulletin 19 of the International Leonid Watch: Population index study of the 2002 Leonid meteors. *WGN, Journal of the International Meteor Organization*, **31**, 77–87.

Arlt, R., and Barentsen, G. (2006). Bulletin 21 of the International Leonid Watch: Global analysis of visual observations of the 2006 Leonid meteor shower. *WGN, Journal of the International Meteor Organization*, **34**, 163–168.

Arlt, R., and Rendtel, J. (2006). The activity of the 2004 Geminid meteor shower from global visual observations. *Monthly Notices of the Royal Astronomical Society*, **367**, 1721–1726.

Arlt, R., Rendtel, J., Brown, P. et al. (1999). The 1998 outburst and history of the June Boötid meteor shower. *Monthly Notices of the Royal Astronomical Society*, **308**, 887–896.

Arlt, R., Kac, J., Krumov, V., Buchmann, A., and Verbert, J. (2001). Bulletin 17 of the International Leonid Watch; first global analysis of the 2001 Leonid storms. *WGN, Journal of the International Meteor Organization*, **29**, 187–194.

Atreya, P., and Christou, A. A. (2009). The 2007 Aurigid meteor outburst. *Monthly Notices of the Royal Astronomical Society*, **393**, 1493–1497.

Babadzhanov, P. B., and Kokhirova, G. I. (2009). Densities and porosities of meteoroids. *Astronomy and Astrophysics*, **495**, 353–358.

Babadzhanov, P. B., Kokhirova, G. I., Borovička, J., and Spurný, P. (2009). Photographic observations of fireballs in Tajikistan. *Solar System Research*, **43**, 353–363.

Bannister, S., Boucheron, L., and Voelz, D. (2013). A numerical analysis of a frame calibration method for Video-Based All-Sky Camera Systems. *Publications of the Astronomical Society of the Pacific*, **125**, 1108–1118.

Belkovich, O., and Ishmukhametova, M. (2006). Mass distribution of Perseid meteoroids. *Solar System Research*, **40**, 208–213.

Belkovich, O., and Ishmukhametova, M., (2010). Mass distribution of lyrid meteoroids. *Solar System Research*, **44**, 320–326.

Belkovich, O., Ishmukhametova, M. and Kondrat'eva, E. (2011). On activity periods of the April Lyrids. *Solar System Research*, **45**, 529–538.

Bellot Rubio, L. R. (1995). Effects of a dependence of meteor brightness on the entry angle. *Astronomy and Astrophysics*, **301**, 602–608.

Berezhnoy, A. A., and Borovička, J. (2010). Formation of molecules in bright meteors. *Icarus*, **210**, 150–157.

Bettonvil, E. J. A. (2006). Least squares estimation of a meteor trajectory and radiant with a Gauss-Markov model. Pages 63–77 of: Bastiaens, L., Verbert, J., Wislez, J.-M., and Verbeeck, C. (eds), *Proceedings of the International Meteor Conference, Oostmalle, Belgium, 15–18 September, 2005*. Hove, BE: International Meteor Organization.

Bettonvil, F. (2010). Digital All-sky cameras V: Liquid crystal optical shutters. Pages 14–18 of: Andreič, Z., and Kac, J. (eds), *Proceedings of the International Meteor Conference, Porec, Croatia, 24–27 September, 2009*. Hove, BE: International Meteor Organization.

Biets, J. M. (2008). De all-sky camera te Wilderen. *eRadiant, Journal of the Dutch Meteor Society*, **4**, 147–151. available at: http://adsabs.harvard.edu/abs/2008eRad....4..147B

Blaauw, R., Campbell-Brown, M., Cooke, W. et al. (2014). Automated optical meteor fluxes and preliminary results of major showers. *Asteroids, Comets, Meteors Conference*, ACM2013.

Blaauw, R., Campbell-Brown, M., and Kingery, A. (2016). Optical meteor fluxes and application to the 2015 Perseids. *Monthly Notices of the Royal Astronomical Society*, **463**, 441–448.

Blanch, E., Trigo-Rodríguez, J. M. , Madiedo, J. M., et al. (2017). A 2016 North Taurid Bolide Brighter as the Full Moon Imaged in the Framework of the Spanish Fireball Network. *48th Lunar and Planetary Science Conference*. LPI Contribution 1964, 1418.

Bland, P. A. (2004). Fireball cameras: The Desert Fireball Network. *Astronomy & Geophysics*, **45**, 5.20–5.23.

Bland, P. A., Spurný, P., Shrbený, L. et al. (2008). The Desert Fireball Network: First results of two years systematic monitoring of fireballs over the Nullarbor Desert of South Western Australia. *Asteroids, Comets, Meteors Conference*. LPI Contribution 1405, 8246.

Bland, P. A., Spurný, P., Towner, M. C. et al. (2009). An Anomalous Basaltic Meteorite from the Innermost Main Belt. *Science*, **325**, 1525–1527.

Bland, P. A., Spurný, P, Bevan, A. W. R., et al. (2012). The Australian Desert Fireball Network: a new era for planetary science. *Australian Journal of Earth Sciences*, **59**, 177–187.

Bland, P. A., Towner, M. C., Paxman, J. P. et al. (2014). Digital Expansion of the Desert Fireball Network. *77th Annual Meeting of the Meteoritical Society*. LPI Contribution 1800, 5287.

Bloxam, K., and Campbell-Brown, M. (2017). A spectral analysis of ablating meteors. *Planetary and Space Science*, **143**, 28–33.

Borovička, J. (1990). The comparison of two methods of determining meteor trajectories from photographs. *Bulletin of Astronomical Institute of Czechoslovakia*, **41**, 391–396.

Borovička, J. (1993). A fireball spectrum analysis. *Astronomy and Astrophysics*, **279**, 627–645.

Borovička, J. (2004). Elemental abundances in Leonid and Perseid meteoroids. *Earth Moon and Planets*, **95**, 245–253.

Borovička, J. (2005). Spectral investigation of two asteroidal fireballs. *Earth Moon and Planets*, **97**, 279–293.

Borovička, J. (2006). Meteor trains – Terminology and physical interpretation. *Journal of the Royal Astronomical Society of Canada*, **100**, 194–198.

Borovička, J. (2010). Spectroscopic Analysis of Geminid Meteors. Pages 42–51 of: Rendtel, J., and Vaubaillon, J. (eds), *Proceedings of the International Meteor Conference, Bareges, France, 7–10 June, 2007*. Hove, BE: International Meteor Organization.

Borovička, J. (2016). Photographic spectra of fireballs. Pages 34–38 of: Roggemans, A., and Roggemans, P. (eds), *Proceedings of the International Meteor Conference, Egmond, the Netherlands, 2–5 June 2016*. Hove, BE: International Meteor Organization.

Borovička, J., and Berezhnoy, A. A. (2016). Radiation of molecules in Benešov bolide spectra. *Icarus*, **278**, 248–265.

Borovička, J., and Charvát, Z. (2009). Meteosat observation of the atmospheric entry of 2008 TC$_3$ over Sudan and the associated dust cloud. *Astronomy and Astrophysics*, **507**, 1015–1022.

Borovička, J., and Jenniskens, P. (2000). Time resolved spectroscopy of a Leonid fireball afterglow. *Earth Moon and Planets*, **82**, 399–428.

Borovička, J., and Spurný, P. (1996). Radiation study of two very bright terrestrial bolides and an application to the comet S-L 9 collision with Jupiter. *Icarus*, **121**, 484–510.

Borovička, J., Spurný, P., and Brown, P. (2015b). Small Near-Earth Asteroids as a Source of Meteorites. In: P. Michel, F. E. DeMeo, and W. F. Bottke, eds., Asteroids IV. Tucson: University of Arizona Press, pp. 257–280.

Borovička, J., Spurný, P., and Keclíková, J. (1995). A new positional astrometric method for All-sky cameras. *Astronomy and Astrophysics Supplement Series*, **112**, 173–178.

Borovička, J., Spurný, P., and Koten, P. (2007). Atmospheric deceleration and light curves of Draconid meteors and implications for the structure of cometary dust. *Astronomy and Astrophysics*, **473**, 661–672.

Borovička, J., Štork, R., and Boček, J. (1999). First results from video spectroscopy of 1998 Leonid meteors. *Meteoritics and Planetary Science*, **34**, 987–994.

Borovička, J., Koten, P., Spurný, P., Boček, J., and Štork, R. (2005). A survey of meteor spectra and orbits: Evidence for three populations of Na-free meteoroids. *Icarus*, **174**, 15–30.

Borovička, J., Koten, P., Spurný, P., and Štork, R. (2008). Analysis of a low density meteoroid with enhanced sodium. *Earth Moon and Planets*, **102**, 485–493.

Borovička, J., Abe, S., Shrbený, L., Spurný, P., and Bland, P. (2011). Photographic and radiometric observations of the HAYABUSA re-entry. *Publications of the Astronomical Society of Japan*, **63**, 1003–1009.

Borovička, J., Spurný, P., Brown, P. et al. (2013a). The trajectory, structure and origin of the Chelyabinsk asteroidal impactor. *Nature*, **503**, 235–237.

Borovička, J., Tóth, J., Igaz, A. et al. (2013b). The Košice meteorite fall: Atmospheric trajectory, fragmentation, and orbit. *Meteoritics and Planetary Science*, **48**, 1757–1779.

Borovička, J., Koten, P., Shrbený, L., Štork, R., and Hornoch, K. (2014). Spectral, photometric, and dynamic analysis of eight Draconid meteors. *Earth Moon and Planets*, **113**, 15–31.

Borovička, J., Spurný, P., Šegon, D. et al. (2015). The instrumentally recorded fall of the Krizevci meteorite, Croatia, February 4, 2011. *Meteoritics and Planetary Science*, **50**, 1244–1259.

Borovička, J., Shrbený, L., Kalenda, P. et al. (2016). A catalog of video records of the 2013 Chelyabinsk superbolide. *Astronomy and Astrophysics*, **585**, A90.

Borovička, J., Macke, R. J., Campbell-Brown, M. D. et al. (2019) Physical and chemical properties of meteoroids. In: G. O. Ryabova, D. J. Asher, and M. D. Campbell-Brown, eds. *Meteoroids: Sources of Meteors on Earth and Beyond*. Cambridge, UK: Cambridge University Press, pp. 37–62.

Brosch, N., Helled, R., Polishook, D., Almoznino, E., and David, N. (2004). Meteor light curves: The relevant parameters. *Monthly Notices of the Royal Astronomical Society*, **355**, 1, 111–119.

Brown, P., Campbell, M., Ellis, K. J. et al. (2000). Global ground-based electro-optical and radar observations of the 1999 Leonid shower: First results. *Earth Moon and Planets*, **82**, 167–190.

Brown, P., Campbell, M., Suggs, R. et al. (2002). Video and radar observations of the 2000 Leonids: Evidence for a strong flux peak associated with 1932 ejecta?. *Monthly Notices of the Royal Astronomical Society*, **335**, 473–479.

Brown, P. G., Assink, J. D., Astiz, L. et al. (2013). A 500-kiloton airburst over Chelyabinsk and an enhanced hazard from small impactors. *Nature*, **503**, 238–241.

Campbell, M. D., Brown, P. G., LeBlanc, A. G. et al. (2000). Image-intensified video results from the 1998 Leonid shower: I. Atmospheric trajectories and physical structure. *Meteoritics and Planetary Science*, **35**, 1259–1267.

Campbell-Brown, M. D., and Koschny, D. (2004). Model of the ablation of faint meteors. *Astronomy and Astrophysics*, **418**, 751–758.

Campbell-Brown, M. D., Blaauw, R., and Kingery, A. (2016). Optical fluxes and meteor properties of the Camelopardalid meteor shower. *Icarus*, **277**, 141–153.

Campbell-Brown, M., Vaubaillon, J., Brown, P., Weryk, R., and Arlt, R. (2006). The 2005 Draconid outburst. *Astronomy and Astrophysics*, **451**, 339–344.

Campbell-Brown, M. D., Borovička, J., Brown, P. G., and Stokan, E. (2013). High-resolution modelling of meteoroid ablation. *Astronomy and Astrophysics*, **557**, A41.

Carbary, J. F., Morrison, D., Romick, G. J., and Yee, J.-H. (2003). Leonid meteor spectrum from 110 to 860 nm. *Icarus*, **161**, 223–234.

Ceplecha, Z. (1953). Meteor photographs. *Bulletin of the Astronomical Institutes of Czechoslovakia*, **4**, 55–59.

Ceplecha, Z. (1957). Photographic Geminids 1955. *Bulletin of the Astronomical Institutes of Czechoslovakia*, **8**, 51–61.

Ceplecha, Z. (1961). Multiple fall of Příbram meteorites photographed. 1. Double-station photographs of the fireball and their relations to the found meteorites. *Bulletin of the Astronomical Institutes of Czechoslovakia*, **12**, 21–47.

Ceplecha, Z., and Rajchl, J. (1965). Programme of fireball photography in Czechoslovakia. *Bulletin of the Astronomical Institute of Czechoslovakia*, **16**, 15–22.

Ceplecha, Z., and ReVelle, D. O. (2005). Fragmentation model of meteoroid motion, mass loss, and radiation in the atmosphere. *Meteoritics and Planetary Science*, **40**, 35–54.

Ceplecha, Z., Ježková, M., Boček, J., Kirsten, T, and Kiko, J. (1973). Data on Three Significant Fireballs Photographed within the European Network in 1971. *Bulletin of the Astronomical Institutes of Czechoslovakia*, **24**, 13–22.

Ceplecha Z. (1987). Geometric, dynamic, orbital and photometric data on meteoroids from photographic fireball networks. *Bulletin of Astronomical Institute of Czechoslovakia*, **38**, 222–234.

Ceplecha, Z. (1994). Earth-grazing daylight fireball of August 10, 1972. *Astronomy and Astrophysics*, **283**, 287–288.

Ceplecha, Z., Borovička, J., Elford, W. G. et al. (1998). Meteor phenomena and bodies. *Space Science Reviews*, **84**, 327–471.

Ceplecha Z., Spurny P., and Borovička J. (2000). MORB Software to Determine Meteoroid Orbits, Ondrejov Observatory, Czech Republic.

Clark, D., and Wiegert, P. (2011). A numerical comparison with the Ceplecha analytical meteoroid orbit determination method. *Meteoritics and Planetary Science*, **46**, 1217–1225.

Clark, D. L., Spurný, P., Wiegert, P. et al. (2016). Impact detections of temporarily captured natural satellites. *The Astronomical Journal*, **151**, 135.

Clode, S., Zelniker, E., Kootsookos, P., Vaughan, I., and Clarkson, L. (2004). A phase coded disk approach to thick curvilinear line detection. Pages 1147–1150 of: *2004 XII. European Signal Processing Conference, Vienna, Austria, 6–10 September 2004*. IEEE.

Colas, F., Zanda, B., Vaubaillon, J. et al. (2015). French fireball network FRIPON. Pages 37–40 of: Rault, J.-L., and Roggemans, P. (eds), *Proceedings of the International Meteor Conference, Mistelbach, Austria, 27–30 August 2015*. Hove, BE: International Meteor Organization.

Cooke, W. J. and Moser, D. E. (2012). The status of the NASA All Sky Fireball Network. Pages 9–12 of: Gyssens, M., and Roggemans, P. (eds), *Proceedings of the International Meteor Conference, Sibiu, Romania, 15–18 September, 2011*. Hove, BE: International Meteor Organization.

Davidson M. (1937). The computation of the real paths of meteors. *Journal of the British Astronomical Association*, **46**, 3, 292–300.

De Cicco, M., Zoghbi, S., Stapper, A., et al. (2018). Artificial intelligence techniques for automating the CAMS processing pipeline to direct the search for long-period comets. Pages 64–69 of: Gyssens, M., and Rault, J.-L. (eds), *Proceedings of the International Meteor Conference, Petnica, Serbia, 21–24 September, 2017*. Hove, BE: International Meteor Organization.

de la Fuente Marcos, C. and de la Fuente Marcos, R. (2018). On the Pre-impact Orbital Evolution of 2018 LA, Parent Body of the Bright Fireball Observed Over Botswana on 2018 June 2. *Research Notes of the American Astronomical Society*, **2**, 57.

Devillepoix, H. A. R., Sansom, E. K., Bland, P. A. et al. (2018). The Dingle Dell meteorite: a Halloween treat from the Main Belt. *Meteoritics and Planetary Science*, **53**, 2212–2227.

Dubs, M., and Schlatter, P. (2015). A practical method for the analysis of meteor spectra. *WGN, Journal of the International Meteor Organization*, **43**, 94–101.

Egal, A., Kwon, M.-K., Colas, F., Vaubaillon, J., and Marmo, C. (2016). The challenge of meteor daylight observations. Pages 73–75 of: Roggemans, A., and Roggemans, P. (eds), *Proceedings of the International Meteor Conference, Egmond, the Netherlands, 2–5 June 2016*. Hove, BE: International Meteor Organization.

Egal, A., Gural, P. S., Vaubaillon, J., Colas, F., and Thuillot, W. (2017). The challenge associated with the robust computation of meteor velocities from video and photographic records. *Icarus*, **294**, 43–57.

Faloon, A. J., Thaler, J. D., and Hawkes, R. L. (2004). Searching for light curve evidence of meteoroid structure and fragmentation. *Earth Moon and Planets*, **95**, 289–295.

Farnocchia, D., Chesley, S. R., Brown, P. G., and Chodas, P. W. (2016). The trajectory and atmospheric impact of asteroid 2014 AA. *Icarus*, **274**, 327–333.

Farnocchia, D., Jenniskens, P., Robertson, D. K. et al. (2017). The impact trajectory of asteroid 2008 TC_3. *Icarus*, **294**, 218–226.

Flohrer, J., Oberst, J., Heinlein, D., Grau, T., and Spurný, P. (2006). The European Fireball Network - Current status of the all-sky cameras in Germany. *European Planetary Science Congress 2006*, **1**, EPSC2006-518.

Fujita, K., Yamamoto, M.-Y., Abe, S. et al. (2011). An Overview of JAXA's ground-observation activities for HAYABUSA reentry. *Publications of the Astronomical Society of Japan*, **63**, 961–969.

Fujiwara, Y., Ueda, M., Shiba, Y. et al. (1998). Meteor luminosity at 160 km altitude from TV observations for bright Leonid meteors. *Geophysical Research Letters*, **25**, 285–288.

Gährken, B. and Michelberger, J. (2003). A bright, high altitude 2002 Leonid. *WGN, Journal of the International Meteor Organization*, **31**, 137–138.

Gajdoš, Š., Tóth, J., Kornoš, L. et al. (2014). The September epsilon Perseids in 2013. *WGN, Journal of the International Meteor Organization*, **42**, 48–56.

Gorbanev, Y. M. (2009). Odessa television meteor patrol. *Odessa Astronomical Publications*, **22**, 60–67.

Gural, P. (1999). MeteorScan: Documentation and User's Guide Version 2.2, Sterling, VA, USA.

Gural, P. (2007). Algorithms and software for meteor detection. *Earth, Moon, and Planets*, **102**, 269–275.

Gural, P. (2012). A new method of meteor trajectory determination applied to multiple unsynchronized video cameras. *Meteoritics and Planetary Science*, **47**, 1405–1418.

Gural, P. (2016). A fast meteor detection algorithm. Pages 96–104 of: Roggemans, A., and Roggemans, P. (eds), *Proceedings of the International Meteor Conference, Egmond, the Netherlands, 2–5 June 2016*. Hove, BE: International Meteor Organization.

Gural, P. (2018). New developments in meteor processing algorithms. Pages 55–63 of: Gyssens, M., and Rault, J.-L. (eds), *Proceedings of the International Meteor Conference, Petnica, Serbia, 21–24 September, 2017*. Hove, BE: International Meteor Organization.

Gural, P., and Jenniskens, P. (2000). Leonid Storm Flux Analysis from One Leonid Mac Video AL50R. *Earth Moon and Planets*, **82**, 221–247.

Gural, P., Jenniskens, P., and Varros, G. (2004). Results from the AIM-IT Meteor Tracking System. *Earth Moon and Planets*, **95**, 541–552.

Haack, H., Grau, T., Bischoff, A., et al. (2012). Maribo-A new CM fall from Denmark. *Meteoritics and Planetary Science*, **47**, 30–50.

Hajduková, M., Kruchinenko, V. G., Kazantsev, A. M. et al. (1995). Perseid meteor stream 1991–1993 from TV observations in Kiev. *Earth Moon and Planets*, **68**, 297–301.

Hajduková, Jr. M., Koten, P., Kornoš, L., and Tóth, J. (2017). Meteoroid orbits from video meteors. The case of the Geminid stream. *Planetary and Space Science*, **143**, 89–98.

Halliday, I. (1973). Photographic fireball networks. Pages 1–8 of: Hemenway, C. L., Millman, P. M., and Cook, A. F. (eds), *Proceedings of IAU Colloq. 13, Albany, NY, 14–17 June 1971*. NASA Special Publication 319.

Hankey, M., and Perlerin, V. (2015). IMO's new online fireball form. *WGN, Journal of the International Meteor Organization*, **43**, 2–7.

Hawkes, R. L. (1993). Television meteors. Pages 227–234 of: Štohl, J. and Williams, I. P. (eds), *Meteoroids and their Parent Bodies, Smolenice, Slovakia, July 6–12, 1992*. Bratislava: Astronomical Institute, Slovak Academy of Sciences.

Hawkes, R. L. (2002). Detection and analysis procedures for visual, photographic and image intensified CCD meteor observations. In: E. Murad and I. Williams, eds, *Meteors in the Earth's Atmosphere: Meteoroids and Cosmic Dust and their Interactions with the Earth's Upper Atmosphere*. Cambridge, UK: Cambridge University Press, pp 97–122.

Hawkes, R. L. and Jones, J. (1986). Electro-optical meteor observation techniques and results. *Quarterly Journal of the Royal Astronomical Society*, **27**, 569–589.

Howie, R., Samsom, E., Bland, P. et al. (2015). Precise fireball trajectories using liquid crystal shutters and de Bruijn sequences. In *46th Lunar and Planetary Science Conference*, Woodlands, Texas, **1832**, 1743.

Howie, R. M., Paxman, J., Bland, P. A. et al. (2017a). Submillisecond fireball timing using de Bruijn timecodes. *Meteoritics and Planetary Science*, **52**, 1669–1682.

Howie, R. M., Paxman, J., Bland, P. A. et al. (2017b). How to build a continental scale fireball camera network. *Experimental Astronomy*, **43**, 237–266.

Hueckel M. (1973). A local visual operator which recognizes edges and lines. *Journal of the Association for Computing Machinery*, **20**, 634–647

Janches, D., Dyrud, L. P., Broadley, S. L., and Plane, J. M. C. (2009). First observation of micrometeoroid differential ablation in the atmosphere. *Geophysical Research Letters*, **36**, L06101.

Jenniskens, P. (1998). Activity of the 1998 Leonid shower from the video records. *Meteoritics and Planetary Science*, **34**, 959–968.

Jenniskens, P. (2006). *Meteor Showers and their Parent Comets*. Cambridge, UK: Cambridge University Press.

Jenniskens, P. (2007). Quantitative meteor spectroscopy: Elemental abundances. *Advances in Space Research*, **39**, 491–512.

Jenniskens, P. and Butow, S. J. (1999). The 1998 Leonid multi-instrument aircraft campaingn – an early review. *Meteoritics and Planetary Science*, **34**, 933–943.

Jenniskens, P., and Vaubaillon, J. (2008). Predictions for the Aurigid Outburst of 2007 September 1. *Earth Moon and Planets*, **102**, 157–167.

Jenniskens, P., Crawford, C., Butow, S. J., et al. (2000a). Lorentz shaped comet dust trail cross section from new hybrid visual and video meteor counting technique – Implications for future Leonid storm encounters. *Earth Moon and Planets*, **82**, 191–208.

Jenniskens, P., Butow, S. J., and Fonda, M. (2000b). The 1999 Leonid Multi-Instrument Aircraft Campaign - An Early Review. *Earth Moon and Planets*, **82**, 1–26.

Jenniskens, P., Lyytinen, E., de Lignie, M. C. et al. (2002a). Dust Trails of 8P/Tuttle and the Unusual Outbursts of the Ursid Shower. *Icarus*, **159**, 197–209.

Jenniskens, P., Tedesco, E., Murthy, J. et al. (2002b). Spaceborne ultraviolet 251-384 nm spectroscopy of a meteor during the 1997 Leonid shower. *Meteoritics and Planetary Science*, **37**, 1071–1078.

Jenniskens, P., Laux, C. O., Wilson, M. A., and Schaller, E. L. (2004). The mass and speed dependence of meteor air plasma temperatures. *Astrobiology*, **4**, 81–94.

Jenniskens, P., Jehin, E., Cabanac, R. A. et al. (2004). Spectroscopic anatomy of a meteor trail cross section with the European Southern Observatory Very Large Telescope. *Meteoritics and Planetary Science*, **39**, 609–616.

Jenniskens, P., Moilanen, J., Lyytinen, E. et al. (2005). The 2005 October 5 outburst of October Camelopardalids. *WGN, Journal of the International Meteor Organization*, **33**, 125–128.

Jenniskens, P., de Kleer, K., Vaubaillon, J. et al. (2008a). Leonids 2006 observations of the tail of trails: Where is the comet fluff?. *Icarus*, **196**, 171–183.

Jenniskens, P., Brower, J., Martsching, P. et al. (2008b). September Perseid meteors 2008. *Central Bureau Electronic Telegram* 1501.

Jenniskens, P., Shaddad, M. H., Numan, D. et al. (2008c). The impact and recovery of asteroid 2008 TC_3. *Nature*, **458**, 485–488.

Jenniskens, P., Vaubaillon, J., Atreya, P., and Barentsen, G. (2009). Leonid meteors 2009. *Central Bureau Electronic Telegram* 2046.

Jenniskens, P., Wilson, M. A., Winter, M. et al. (2010). Resolved CN band profile of Stardust capsule radiation at peak heating. *Journal of Spacecraft and Rockets*, **47**, 873–877.

Jenniskens, P., Gural, P., Dynneson, L. et al. (2011). CAMS: Cameras for Allsky Meteor Surveillance to establish minor meteor showers. *Icarus*, **216**, 40–61.

Jenniskens, P., Fries, M. D., Yin, Q.-Z. et al. (2012). Radar-enabled recovery of the Sutter's Mill Meteorite, a carbonaceous chondrite regolith breccia. *Science*, **338**, 1583–1587.

Jenniskens, P., Gural, P., and Berdeu, A. (2014) CAMSS: A spectroscopic survey of meteoroid elemental abundances. Pages 117–124 of: Jopek, T. J., Rietmeijer, F. J. M., Watanabe, J., and Williams, I. P., (eds), *Proc. International Conf. held at A. M. University in Poznań, Poland, August 26–30, 2013, Meteoroids 2013*. Poznań: Wydanictwo Naukowe UAM.

Jenniskens, P., Nénon, Q., Gural, P. et al. (2016). CAMS newly detected meteor showers and the sporadic background. *Icarus*, **266**, 384–409.

Jopek, T., Rudawska, R. and Bartczak, P. (2008). Meteoroid stream searching: The use of vectorial elements. *Earth, Moon and Planets*, **102**, 73–78.

Jobse, K. (2005) Digi-All-Sky. *eRadiant, Journal of the Dutch Meteor Society*, **1**, 101–106. available at: http://adsabs.harvard.edu/abs/2005eRad....1..101J

Jobse, K. (2009). http://cyclops.klaasjobse.nl/images/allsky/

Kartashova, A. P., Popova, O. P., Glazachev, D. O. et al. (2018). Study of injuries from the Chelyabinsk airburst event. *Planetary and Space Science*, **160**, 107–114.

Kasuga, T., Iijima, T., and Watanabe, J. (2007). Is a 2004 Leonid meteor spectrum captured in a 182 cm telescope? *Astronomy and Astrophysics*, **474**, 639–645.

Kinoshita, M., Maruyama, T., and Sagayama, T. (1999). Preliminary activity of Leonid meteor storm observed with a video camera in 1997. *Geophysical Research Letters*, **26**, 41–44.

Kohout, T., Gritsevich, M., Lyytinen, E. et al. (2015). Annama H5 Meteorite Fall: Orbit, trajectory, recovery, petrology, noble gases, and cosmogenic radionuclides. *78th Annual Meeting of the Meteoritical Society*, **1856**, 5209.

Kokhirova, G. I., and Borovička, J. (2011). Observations of the 2009 Leonid activity by the Tajikistan fireball network. *Astronomy and Astrophysics*, **533**, A115.

Kokhirova, G. I., Babadzhanov, P. B., and Khamroev, U. Kh. (2015). Tajikistan fireball network and results of photographic observations. *Solar System Research*, **49**, 275–283.

Koschack, R., and Rendtel, J. (1990a). Determination of spatial number density and mass index from visual meteor observations (I). *WGN, Journal of the International Meteor Organization*, **18**, 44–58.

Koschack, R., and Rendtel, J. (1990b). Determination of spatial number density and mass index from visual meteor observations (II). *WGN, Journal of the International Meteor Organization*, **18**, 119–140.

Koschny, D., and Zender, J. (2000). Comparing meteor number fluxes from ground-based and airplane-based video observations. *Earth Moon and Planets*, **82**, 209–220.

Koschny, D., Albin, T., Drolshagen, E. et al. (2015). Current activities at the ESA/ESTEC Meteor Research Group. Pages 204–208 of: Rault, J.-L., and Roggemans, P. (eds), *Proceedings of the International Meteor Conference, Mistelbach, Austria, 27–30 August 2015*. Hove, BE: International Meteor Organization.

Koseki, M. (2014). Various meteor scenes II: Cygnid-Draconid Complex (κ-Cygnids)/ *WGN, Journal of the International Meteor Organization*, **42**, 181–197.

Koten, P., Spurný, P., Borovička, J., and Štork, R. (2001). Extreme beginning heights for non-Leonid meteors. Pages 119–122 of: Warmbein, B. (ed), *Proceedings of the Meteoroids 2001 Conference, 6–10 August 2001, Kiruna, Sweden*. ESA Special Publication, **495**, Noordwijk: ESA.

Koten, P., Borovička, J., Spurný, P. et al. (2004). Atmospheric trajectories and light curves of shower meteors. *Astronomy and Astrophysics*, **428**, 683–690.

Koten, P., Spurný, P., Borovička, J. et al. (2006a). The beginning heights and light curves of high-altitude meteors. *Meteoritics and Planetary Science*, **41**, 1305–1320.

Koten, P., Borovička, J., Spurný, P. et al. (2006b). Double station and spectroscopic observations of the Quadrantid meteor shower and the implications for its parent body. *Monthly Notices of the Royal Astronomical Society*, **366**, 1367–1372.

Koten, P., Borovička, J., Spurný, P. and Štork, R. (2007). Optical observations of enhanced activity of the 2005 Draconid meteor shower. *Astronomy and Astrophysics*, **466**, 729–735.

Koten, P., Borovička, J., Spurný, P. et al. (2008). Video Observations of the 2006 Leonid Outburst. *Earth Moon and Planets*, **102**, 151–156.

Koten, P., Borovička, J., and Kokhirova, G. I. (2011) . Activity of the Leonid meteor shower on 2009 November 17. *Astronomy and Astrophysics*, **528**, A94.

Koten, P., Vaubaillon, J., Tóth, J. et al. (2014a). Three Peaks of 2011 Draconid Activity Including that Connected with Pre-1900 Material. *Earth Moon and Planets*, **112**, 15–31.

Koten, P., Páta, P., Fliegel, K., and Vítek, S. (2014b). Detection of meteors by the MAIA system. Pages 53–56 of: Gyssens, M., Roggemans, P., and Zoladek, P. (eds), *Proceedings of the International Meteor Conference, Poznań, Poland, 22–25 August 2013*. Hive, BE: International Meteor Organization.

Koten, P., Vaubaillon, J., Čapek, D. et al. (2014c). Search for faint meteors on the orbits of Příbram and Neuschwanstein meteorites. *Icarus*, **239**, 244–252.

Koten, P., Vaubaillon, J., Margonis, A. et al. (2015). Double station observation of Draconid meteor outburst from two moving aircraft. *Planetary and Space Science*, **118**, 112–119.

Koten, P., Čapek, D., Spurný, P. et al. (2017). September epsilon Perseid cluster as a result of orbital fragmentation. *Astronomy and Astrophysics*, **600**, A74.

Koukal, J., Gorková, S., Srba, J. et al. (2015). Meteor spectra in the EDMOND database. Pages 149–154 of: Rault, J.-L., and Roggemans, P. (eds), *Proceedings of the International Meteor Conference, Mistelbach, Austria, 27–30 August 2015*. Hove, BE: International Meteor Organization.

Kowalski, R. A., Africano, B. M., Christensen, E. J. et al. (2018). 2018 LA. *Minor Planet Electronic Circulars*, 2018-L04. available at: www.minorplanetcenter.net/mpec/K18/K18L04.html

Kozak, P. (2008). "Falling Star": Software for Processing of Double Station TV Meteor Observations. *Earth, Moon and Planets*, **102**, 277–283.

Kozak, P. M., and Watanabe, J. (2017). Upward-moving low-light meteor – I. Observation results. *Monthly Notices of the Royal Astronomical Society*, **467**, 793–801.

Kozak, P., Rozhilo, O., Kruchynenko, V. et al. (2007). Results of processing of Leonids-2002 meteor storm TV observations in Kyiv. *Advances in Space Research*, **39**, 619–623.

LeBlanc, A. G., Murray, I. S., Hawkes, R. L. et al. (2000). Evidence for transverse spread in Leonid meteors. *Monthly Notices of the Royal Astronomical Society*, **313**, L9–L13.

Lindemann, F. A., and Dobson, G. M. B. (1923). Note on the photography of meteors. *Monthly Notices of the Royal Astronomical Society*, **83**, 163–166.

Lyytinen E. J., and Van Flandern T. (2000). Predicting the strength of Leonid outbursts. *Earth Moon and Planets*, **82**, 149–166.

Madiedo, J. M. (2015). Spectroscopy of a κ-Cygnid fireball afterglow. *Planetary and Space Science*, **118**, 90–94.

Madiedo, J. M. (2017). Automated systems for the analysis of meteor spectra: The SMART Project. *Planetary and Space Science*, **143**, 238–244.

Madiedo, J. M., and Trigo-Rodríguez, J. M. (2008). Multi-station video orbits of minor meteor showers. *Earth Moon and Planets*, **102**, 133–139.

Madiedo, J. M., Trigo-Rodríguez, J. M., Castro-Tirado, A. J. et al. (2013a). The Geminid meteoroid stream as a potential meteorite dropper: A case study. *Monthly Notices of the Royal Astronomical Society*, **436**, 2818–2823.

Madiedo, J. M., Trigo-Rodríguez, J. M., Konovalova, N. et al. (2013b). The 2011 October Draconids outburst – II. Meteoroid chemical abundances from fireball spectroscopy. *Monthly Notices of the Royal Astronomical Society*, **433**, 571–580.

Madiedo, J. M., Ortiz, J. L., Trigo-Rodríguez, J. M. et al. (2014a). Analysis of bright Taurid fireballs and their ability to produce meteorites. *Icarus*, **231**, 356–364.

Madiedo, J. M., Trigo-Rodríguez, J. M., Ortiz, J. L. et al. (2014b). Orbit and emission spectroscopy of α-Capricornid fireballs. *Icarus*, **239**, 273–280.

Madiedo, J. M., Trigo-Rodríguez, J. M., Zamorano, J. et al. (2014c). Orbits and emission spectra from the 2014 Camelopardalids. *Monthly Notices of the Royal Astronomical Society*, **445**, 3309–3314.

Madiedo, J. M., Espartero, F., Castro-Tirado, A. J. et al. (2016a). Earth-grazing fireball from the Daytime ζ-Perseid shower observed over Spain on 2012 June 10. *Monthly Notices of the Royal Astronomical Society*, **460**, 917–922.

Madiedo, J. M., Espartero, F., Trigo-Rodríguez, J. M. et al. (2016b). Observations of the Quadrantid meteor shower from 2008 to 2012: Orbits and emission spectra. *Icarus*, **275**, 193–202.

Madiedo, J. M., Zamorano, J., Trigo-Rodríguez, J. M. et al. (2018). Analysis of the September ϵ-Perseid outburst in 2013. *Monthly Notices of the Royal Astronomical Society*, **480**, 2501–2507.

Maeda, K., and Fujiwara, Y. (2016). Meteor spectra using high definition video camera. Pages 167–170 of: Roggemans, A., and Roggemans, P. (eds), *Proceedings of the International Meteor Conference, Egmond, the Netherlands, 2–5 June 2016*. Hove, BE: International Meteor Organization.

Maslov, M. (2011). Future Draconid outbursts (2011–2100). *WGN, Journal of the International Meteor Organization*, **39**, 64–47.

Matlovič, P., Tóth, J., Rudawska, R., and Kornoš, L. (2017). Spectra and physical properties of Taurid meteoroids. *Planetary and Space Science*, **143**, 104–115.

McCrosky, R. E., and Boeschenstein, H. Jr., (1965). The Prairie Meteorite Network. *SAO Special Report*, **173**, 1–24.

McGaha, J. E., Jacques, C., Pimentel, E. et al. (2008). 2008 TC3. *Minor Planet Electronic Circulars*, 2008-T50. available at: www.minorplanetcenter.net/iau/mpec/K08/K08T50.html

McNaught, R. H., and Asher, D. J. (1999). Leonid dust trails and meteor storms. *WGN, Journal of the International Meteor Organization*, **27**, 85–102.

Miskotte, K., and Vandeputte, M. (2017). The magnificent outburst of the 2016 Perseids, the analysis. *eMeteorNews*, **3**, 61–69. available at: http://adsabs.harvard.edu/abs/2017eMetN...2...61M

Molau, S., and Barentsen, G. (2014). Real-time flux density measurements of the 2011 Draconid Meteor Outburst. *Earth, Moon, and Planets*, **112**, 1–5.

Molau, S., and Gural, P. (2005). A review of meteor detection and analysis software. *WGN, Journal of the International Meteor Organization*, **33**, 15–20.

Molau, S., and Rendtel, J. (2009). A comprehensive list of meteor showers obtained from 10 years of observations with the IMO Video Meteor Network. *WGN, Journal of the International Meteor Organization*, **37**, 98–121.

Molau, S., Rendtel, J., and Bellot-Rubio, L. R., (2000). Video observations of Leonids 1999. *Earth Moon and Planets*, **87**, 1–10.

Molau, S., Gural, P., and Okamura, O. (2002). Comparison of the "American" and the "Asian" 2001 Leonid Meteor Storm. *WGN, Journal of the International Meteor Organization*, **30**, 3–21.

Molau, S., Kac, J., Berko, E. et al. (2012). Results of the IMO Video Meteor Network – October 2011. *WGN, Journal of the International Meteor Organization*, **40**, 41–47.

Molau, S., Kac, J., Crivello, S. et al. (2013). Results of the IMO Video Meteor Network - September 2013. *WGN, Journal of the International Meteor Organization*, **41**, 207–211.

Molau, S., Crivello, S., Goncalves, R. et al. (2016). Results of the IMO Video Meteor Network – July 2016. *WGN, Journal of the International Meteor Organization*, **44**, 205–210.

Molau, S., Kac, J., Crivello, S. et al. (2015). Results of the IMO Video Meteor Network – December 2014. *WGN, Journal of the International Meteor Organization*, **43**, 62–68.

Moorhead, A. V., Brown, P. G., Spurný, P. et al. (2015). The 2014 KCG meteor outburst: Clues to a parent body. *The Astronomical Journal*, **150**, 122.

Moreno-Ibáñez, M., Trigo-Rodríguez, J. M., Madiedo, J. M., et al. (2017). Multi-instrumental observations of the 2014 Ursid meteor outburst. *Monthly Notices of the Royal Astronomical Society*, **468**, 2206–2213.

Murray, I. S., Hawkes, R. L., and Jenniskens, P. (2000). Comparison of 1998 and 1999 Leonid light curve morphology and meteoroid structure. *Earth Moon and Planets*, **82–83**, 351–367.

Neslušan, L., Porubčan, V., and Svoreň, J. (2014). IAU MDC Photographic Meteor Orbits Database: Version 2013. *Earth Moon and Planets*, **111**, 105–114.

Nijland, J. (2012). Waarnemingen Puimichel, Zuid-Oost Frankrijk. *eRadiant, Journal of the Dutch Meteor Society*, **4**, 94–97. available at: www.dmsweb.org

Oberst, J., Molau, S., Heinlein, D. et al. (1998). The "European Fireball Network": Current status and future prospects. *Meteoritics and Planetary Science*, **33**, 49–56.

Oberst, J., Heinlein, D., Grau, T., and Flohrer, J. (2010). The European Fireball Network 2009 – Status and Results of Cameras in Germany. *European Planetary Science Congress 2010*, **5**, EPSC2010-302. available at: https://meetingorganizer.copernicus.org/EPSC2010/EPSC2010-302-1.pdf

Olech, A., Zoladek, P., Wisniewski, M., et al. (2006). Polish Fireball Network. Pages 53–56 of: Bastiaens, L., Verbert, J., Wislez, J.-M., and Verbeeck, C. (eds), *Proceedings of the International Meteor Conference, Oostmalle, Belgium, 15–18 September 2005*. Hove, BE: International Meteor Organization.

Olech, A., Żołądek, P., Wiśniewski, M. et al. (2013). PF191012 Myszyniec – highest Orionid meteor ever recorded. *Astronomy and Astrophysics*, **557**, A89.

Olivier, C. P. (1937). Results of the Yale photographic meteor work, 1893–1909. *Astronomical Journal*, **46**, 41–57.

Passas, M., Madiedo, J. M., and Gordillo-Vázquez, F. J. (2016). High resolution spectroscopy of an Orionid meteor from 700 to 800 nm. *Icarus*, **266**, 134–141.

Popova, O. P., Strelkov, A. S., and Sidneva, S. N. (2007). Sputtering of fast meteoroids' surface. *Advances in Space Research*, **39**, 567–573.

Popova, O., Borovička, J., Hartmann, W. K. et al. (2011). Very low strengths of interplanetary meteoroids and small asteroids. *Meteoritics and Planetary Science*, **46**, 1525–1550.

Popova, O. P., Jenniskens, P., Emel'yanenko, V. et al. (2013). Chelyabinsk Airburst, Damage Assessment, Meteorite Recovery, and Characterization. *Science*, **342**, 1069–1073.

Porter J. G. (1942). The reduction of meteor observations. *Memoirs of the British Astronomical Association*, **34**, 37–64.

Rendtel, J. (1993). Perseids 1993: a first analysis of global data. *WGN, Journal of the International Meteor Organization*, **21**, 235–239.

Rendtel, J. (2004a). Evolution of the geminids observed over 60 years. *Earth, Moon and Planets*, **95**, 27–32.

Rendtel, J. (2004b). The population index of sporadic meteors. Pages 114–122 of: Triglav-Čekada, M., and Trayner C. (eds), *Proceedings of the International Meteor Conference, Bollmannsruh, Germany, 19–21 September 2003*. Hove BE: International Meteor Organization.

Rendtel, J. (2007a). Three days of enhanced Orionid activity in 2006 – meteoroids from a resonance region? *WGN, Journal of the International Meteor Organization*, **35**, 41–45.

Rendtel, J. (2007b). Visual observations of the Aurigid peak on 2007 September 1. *WGN, Journal of the International Meteor Organization*, **35**, 108–112.

Rendtel, J. (2014a). Daytime meteor showers. Pages 93–97 of: Rault, J.-L., and Roggemans, P. (eds), *Proceedings of of the International Meteor Conference, Giron, France, 18–21 September 2014*. Hove BE: International Meteor Organization.

Rendtel, J. (ed.) (2014b). *Meteor Shower Workbook 2014*. Hove, BE: International Meteor Organization.

Rendtel, J. (ed.) (2015). IMO Meteor Shower Calendar 2016. *IMO_INFO* **2**, 1–24.

Rendtel, J. (2017). Review of amateur meteor research. *Planetary and Space Science*, **143**, 7–11.

Rendtel, J., and Arlt, R. (eds) (2016a). *Handbook for Meteor Observers*. Hove BE: International Meteor Organization.

Rendtel, J., and Arlt, R. (2016b). Kappa Cygnid rate variations over 41 years. *WGN, Journal of the International Meteor Organization*, **44**, 62–66.

Rendtel, J., Brown, P., and Molau, S. (1996). The 1995 outburst and possible origin of the alpha-Monocerotid meteoroid stream. *Monthly Notices of the Royal Astronomical Society*, **279**, L31–L36.

Rendtel, J., Lyytinen, E., Molau, S., and Barentsen, G. (2014). Peculiar activity of the September epsilon-Perseids on 2013 September 9. *WGN, Journal of the International Meteor Organization*, **42**, 40–47.

Rendtel, J., Ogawa, H., and Sugimoto, H. (2016). Quadrantids 2016: Observations of a short pre-maximum peak. *WGN, Journal of the International Meteor Organization*, **44**, 101–107.

Rendtel, J., Ogawa, H., and Sugimoto, H. (2017). Meteor showers 2016: Review of predictions and observations. *WGN, Journal of the International Meteor Organization*, **45**, 49–55.

Rendtel, J. et al. (2018). Analysis of visual data of the 2018 Perseids. *WGN, Journal of the International Meteor Organization*, **47**, 18–25

ReVelle, D. O., and Ceplecha, Z. (2001). Bolide physical theory with application to PN and EN fireballs. Pages 507–512 of: Warmbein, B. (ed), *Proceedings of the Meteoroids 2001 Conference, 6–10 August 2001, Kiruna, Sweden*. ESA Special Publication, **495**, Noordwijk: ESA.

Richter, J. (2018). About the mass and magnitude distributions of meteor showers. *WGN, Journal of the International Meteor Organization*, **46**, 34–38.

Roberts, I. D., Hawkes, R. L., Weryk, R. J. et al. (2014). Meteoroid structure and ablation implications from multiple maxima meteor light curves. Pages 155–162 of: Jopek, T. J., Rietmeijer, F. J. M., Watanabe, J. and Williams, I. P., (eds), *Proc. International Conf. held at A. M. University in Poznań, Poland, August 26–30, 2013, Meteoroids 2013*. Poznań: Wydanictwo Naukowe UAM.

Roggemans, P., Betlem, H., Bettonvil, F. et al. (2014). The Benelux CAMS Network – status July 2013. Pages 173–175 of: Gyssens, M., Roggemans, P., and Zoladek, P. (eds), *Proceedings of the International Meteor Conference, Poznań, Poland, 22–25 August 2013*. Hove, BE: International Meteor Organization.

Rudawska, R., Zender, J., Jenniskens, P. et al. (2014). Spectroscopic observations of the 2011 Draconids meteor shower. *Earth Moon and Planets*, **112**, 45–57.

Rudawska, R., Tóth, J., Kalmančok, D., Zigo, P., and Matlovič, P. (2016). Meteor spectra from AMOS video system. *Planetary and Space Science*, **123**, 25–32.

Ryabova, G. O. and Rendtel, J. (2018). Increasing Geminid meteor shower activity. *Monthly Notices of the Royal Astronomical Society*, **475**, L77–L80.

Sansom, E. K., Bland, P. A., Paxman, J. A and Towner, M. (2015). A novel approach to fireball modeling: The observable and the calculated. *Meteoritics and Planetary Science*, **50**, 1423–1435.

Sansom, E., Ridgewell, J., Bland, P. et al. (2016). Meteor reporting made easy – The Fireballs in the Sky smartphone app. Pages 267–269 of: Roggemans, A., and Roggemans, P. (eds), *Proceedings of the International Meteor Conference, Egmond, the Netherlands, 2–5 June 2016*. Hove, BE: International Meteor Organization.

Sansom, E. K., Rutten, M. G., and Bland, P. A. (2017). Analyzing meteoroid flights using particle filters. *The Astronomical Journal*, **153**, 87.

Sato, M., and Watanabe, J. I. (2007). Origin of the 2006 Orionid outburst. *Publications of the Astronomical Society of Japan*, **59**, L21–L24.

Sato, M., Watanabe, J. I., Tsuchiya, C. et al. (2017). Detection of the Phoenicids meteor shower in 2014. *Planetary and Space Science*, **143**, 132–137.

Šegon, D., Gural, P., Andreič, Ž. et al. (2014a). A parent body search across several video meteor data bases. Pages 251–262 of: Gyssens, M., Roggemans, P., and Zoladek, P. (eds), *Proceedings of the International Meteor Conference, Poznań, Poland, 22–25 August 2013*. Hove BE: International Meteor Organization.

Šegon, D., Andreič, Ž., Gural, P. et al. (2014b). Draconids 2011: Outburst observations by the Croatian Meteor Network. *Earth Moon and Planets*, **112**, 33–44.

Šegon, D., Andreić, Ž., Korlević, K. et al. (2015). Croatian Meteor Network: ongoing work 2014 – 2015. Pages 51–57 of: Rault, J.-L., and Roggemans, P. (eds), *Proceedings of the International Meteor Conference, Mistelbach, Austria, 27–30 August 2015*. Hove, BE: International Meteor Organization.

Šegon, D., Vaubaillon, J., Gural, P. et al. (2017). Dynamical modeling validation of parent bodies associated with newly discovered CMN meteor showers. *Astronomy and Astrophysics*, **598**, A15.

Šegon, D., Vukič, M., Šegon, M. et al. (2018). Meteors in the near-infrared as seen in the Ondrejov Catalogue of Representative Meteor Spectra. Pages 106–107 of: Gyssens, M., and Rault, J.-L. (eds), *Proceedings of the International Meteor Conference, Petnica, Serbia, 21–24 September, 2017*. Hove, BE: International Meteor Organization.

Shiba, Y. (1995). Radiant point, trajectory and velocity from single photographic or TV Observation. *Earth, Moon, and Planets*, **68**, 503–508.

Shrbený, L., and Spurný, P. (2013). Determination of atmospheric velocity of bright meteors on the basis of high-resolution light curves. *Astronomy and Astrophysics*, **550**, A31.

Shrbený, L., Spurný, P., and Borovička, J. (2016). September epsilon Perseids observed by the Czech Fireball Network in 2013 and 2015. *Meteoroids 2016 conference*, 9. available at: www.cosmos.esa.int/documents/653713/1000954/03_ORAL_Shrbeny.pdf

Singer, W., Molau, S., Rendtel, J. et al. (2000). The 1999 Leonid meteor storm: Verification of rapid activity variations by observations at three sites. *Monthly Notices of the Royal Astronomical Society*, **318**, L25–L29.

Sonotaco (2009). A meteor shower catalog based on video meteor observations in 2007–2008. *WGN, Journal of the International Meteor Organization*, **37**, 55–62.

Sonotaco (2017a). http://sonotaco.com/e_index.html

Spurný, P. (2016). Instrumentally documented meteorite falls: two recent cases and statistics from all falls. Pages 69–79 of: Chesley, S. R., Morbidelli, A., Jedicke, R., Farnocchia, D. (eds), *Asteroids: New Observations, New Models, Proceedings of the IAU Symposium 318*. Cambridge UK: Cambridge University Press.

Spurný, P. and Borovička, J. (2002). The autonomous all-sky photographic camera for meteor observation. Pages 257–259 of: Warmbein, B. (ed), *Proceedings of Asteroids, Comets, Meteors Conference, 29 July–2 August 2002, Berlin, Germany*. ESA Special Publication, **495**, Noordwijk: ESA.

Spurný, P., and Borovička, J. (2014). Precise multi-instrument data on exceptional fireballs recorded over Central Europe in the period 2012–2014. *Asteroids, Comets, Meteors 2014 COnference, Helsinki, Finland*, 526. available at: www.helsinki.fi/acm2014/pdf-material/Day-2/Session-2/Room-2/SPURNY-26B0.pdf

Spurný, P., and Ceplecha, Z. (2008). Is electric charge separation the main process for kinetic energy transformation into the meteor phenomenon? *Astronomy and Astrophysics*, **489**, 1, 449–454.

Spurný, P., and Shrbený, L. (2008). Exceptional Fireball Activity of Orionids in 2006. *Earth, Moon, and Planets*, **102**, 141–150.

Spurný, P., Betlem, H., van Leven, J., and Jenniskens, P. (2000a). Atmospheric behavior and extreme beginning heights of the thirteen brightest photographic Leonid meteors from the ground based expedition to China. *Meteoritics and Planetary Science*, **35**, 243–249.

Spurný, P., Betlem, H., Jobse, K. et al. (2000b). New type of radiation of bright Leonid meteors above 130 km. *Meteoritics and Planetary Science*, **35**, 1109–1115.

Spurný, P., Oberst, J., and Heinlein, D. (2003). Photographic observations of Neuschwanstein, a second meteorite from the orbit of the Príbram chondrite. *Nature*, **423**, 151–153.

Spurný, P., Borovička, J., and Shrbený, L. (2007). Automation of the Czech part of the European fireball network: equipment, methods and first results. Pages 121–130 of: Valsecchi, G. B., Vokrouhlický, D., and Milani, A. (eds), *Near Earth Objects, our Celestial Neighbors: Opportunity and Risk, Proceedings of IAU Symposium 236*. Cambridge UK: Cambridge University Press.

Spurný, P., Borovička, J., Kac, J. et al. (2010). Analysis of instrumental observations of the Jesenice meteorite fall on April 9, 2009. *Meteoritics and Planetary Science*, **45**, 1392–1407.

Spurný, P., Bland, P. A., Shrbený, L. et al. (2011). The Mason Gully Meteorite Fall in SW Australia: Fireball Trajectory and Orbit from Photographic Records. *74th Annual Meeting of the Meteoritical Society. Published in Meteoritics and Planetary Science Supplement*, id. 5101. available at: www.lpi.usra.edu/meetings/metsoc2011/pdf/\5101.pdf

Spurný, P., Bland, P. A., Shrbený, L. et al. (2012). The Bunburra Rockhole meteorite fall in SW Australia: fireball trajectory, luminosity, dynamics, orbit, and impact position from photographic and photoelectric records. *Meteoritics and Planetary Science*, **47**, 163–185.

Spurný, P., Haloda, J., Borovička, J., Shrbený, L., and Halodová, P. (2014a). Reanalysis of the Benesov bolide and recovery of polymict breccia meteorites - old mystery solved after 20 years. *Astronomy and Astrophysics*, **570**, A39.

Spurný, P., Shrbený, L., Borovička, J. et al. (2014b) . Bright Perseid fireball with exceptional beginning height of 170 km observed by different techniques. *Astronomy and Astrophysics*, **563**, A64.

Spurný, P., Borovička, J., Haloda, J. et al. (2016). Two Very Precisely Instrumentally Documented Meteorite Falls: Žďár nad Sázavou and Stubenberg - Prediction and Reality. *79th Annual Meeting of the Meteoritical Society*, id. 6221. available at: www.hou.usra.edu/meetings/metsoc2016/pdf/6221.pdf

Spurný, P., Borovička, J., Mucke, H., and Svoreň, J. (2017a). Discovery of a new branch of the Taurid meteoroid stream as a real source of potentially hazardous bodies. *Astronomy and Astrophysics*, **605**, A68.

Spurný, P., Borovička, J., Baumgarten, G. et al. (2017b). Atmospheric trajectory and heliocentric orbit of the Ejby meteorite fall in Denmark on February 6, 2016. *Planetary and Space Science*, **143**, 192–198.

Stenbaek-Nielsen, H. C., and Jenniskens, P. (2004). A "shocking" Leonid meteor at 1000 fps. *Advances in Space Research*, **33**, 1459–1465.

Steyaert, C. (1990). *Photographic Astrometry: Theory and Practice*. Hove, BE: International Meteor Organization.

Stokan, E., and Campbell-Brown, M. D. (2014). Transverse motion of fragmenting faint meteors observed with the Canadian Automated Meteor Observatory. *Icarus*, **232**, 1–12.

Swift, W., Suggs, R., and Cooke, W. (2004). Meteor44 Video Meteor Photometry. *Earth, Moon, and Planets*, **95**, 533–540.

Taylor, M. J., Gardner, R. C., Murray, I. S., and Jenniskens, P. (2000). Jet-like structures and wake in Mg I (518 nm) images of 1999 Leonid storm meteors. *Earth Moon and Planets*, **82–83**, 379–389.

Taylor, M. J., Jenniskens, P., Nielsen, K., and Pautet, D. (2007). First 0.96–1.46 micron near-IR spectra of meteors. *Advances in Space Research*, **39**, 544–549.

Tóth, J., Piffl, R., Koukal, J. et al. (2012) Video observation of Draconids 2011 from Italy. *WGN, Journal of the International Meteor Organization*, 40, 117–121.

Tóth, J., Koukal, J., Piffl, R. et al. (2014). Draconids 2012 – Unexpected outburst. Pages 179–180 of: Gyssens, M., Roggemans, P., and Zoladek, P. (eds), *Proceedings of the International Meteor Conference, Poznań, Poland, 22–25 August 2013*. Hove, BE: International Meteor Organization.

Tóth, J., Kornoš, L., Zigo, P. et al. (2015). All-sky Meteor Orbit System AMOS and preliminary analysis of three unusual meteor showers. *Planetary and Space Science*, **118**, 102–106.

Towner, M. C., Bland, P. A., Spurný, P. et al. (2012). Towards a Digital Desert Fireball Network for Meteorite Recovery. *75th Annual Meeting of the Meteoritical Society. Published in Meteoritics and Planetary Science Supplement*, id. 5123. available at: www.lpi.usra.edu/meetings/metsoc2012/pdf/5123.pdf

Towner, M. C., Bland, P. A., Cupák, M. et al. (2015). Initial Results from the DFN. *LPI Contribution No. 1832*, 1693. available at: www.hou.usra.edu/meetings/lpsc2015/pdf/1693.pdf

Trayner, C., Haynes, B., and Bailey, N. (1997). Automatic Detection of Meteors from CCD Images. Pages 103–113 of: Knöfel, A., and McBeath, A. (eds), *Proceedings of the International Meteor Conference, Apeldoorn, the Netherlands, 19–22 September 1996*. Hove, BE: International Meteor Organization.

Triglav-Čekada, M., and Arlt, R. (2005). Radiant ephemeris of the Taurid meteor complex. *WGN, Journal of the International Meteor Organization*, **33**, 41–58.

Trigo-Rodríguez, J. M., Castro-Tirado, A., Llorca, J. et al. (2003a). A superbolide recorded by the Spanish Fireball Network. *WGN, Journal of the International Meteor Organization*, **31**, 49–52.

Trigo-Rodríguez, J. M., Llorca, J., Borovička, J., and Fabregat, J. (2003b). Chemical abundances determined from meteor spectra: I Ratios of the main chemical elements. *Meteoritics and Planetary Science*, **38**, 1283–1294.

Trigo-Rodríguez, J. M., Castro-Tirado, A. J., Llorca, J. et al. (2004). The development of the Spanish Fireball Network using a new All-Sky CCD system. *Earth, Moon, and Planets*, **95**, 553–567.

Trigo-Rodríguez, J. M., Madiedo, J. M., Castro-Tirado, A, J. et al. (2007). Spanish Meteor Network: 2006 continuous monitoring results. *WGN, Journal of the International Meteor Organization*, **35**, 13–22.

Trigo-Rodríguez, J. M., Madiedo, J. M., Williams, I. P. et al. (2013). The 2011 October Draconids outburst – I. Orbital elements, meteoroid fluxes and 21P/Giacobini-Zinner delivered mass to Earth. *Monthly Notices of the Royal Astronomical Society*, **433**, 560–570.

van Leuteren, P. (2010). All-Sky Twente. *eRadiant, Journal of the Dutch Meteor Society*, **6**, 68–72. available at: www.dmsweb.org

Vaubaillon, J., Colas, F., and Jorda, L. (2005). A new method to predict meteor showers: II. Application to the Leonids. *Astronomy and Astrophysics*, **439**, 761–770.

Vaubaillon, J., Watanabe, J., Sato, M. et al. (2011). The coming 2011 Draconid meteor shower. *WGN, Journal of the International Meteor Organization*, **39**, 59–63.

Vaubaillon, J., Koten, P., Margonis, A. et al. (2015). The 2011 Draconids: The first European airborne meteor observation campaign. *Earth Moon and Planets*, **114**, 137–157.

Verniani, F. (1973). An Analysis of the physical parameters of 5759 Faint Radio Meteors. *Journal of Geophysical Research*, **78**, 8429–8462.

Vida D., Mazur M. J., Šegon, D. et al. (2018). First results of a Raspberry Pi-based meteor camera system. *WGN, Journal of the International Meteor Organization*, **46**, 71–78.

Vida, D., Brown, P. G., and Campbell-Brown, M. (2018). Modelling the measurement accuracy of pre-atmosphere velocities of meteoroids. *Monthly Notices of the Royal Astronomical Society*, **479**, 4307–4319.

Vítek, S., Fliegel, K., Páta, P. et al. (2011). MAIA: Technical Development of a Novel System for Video Observations of Meteors. *Acta Polytechnica*, **51**, 109–111.

Vojáček, V., Borovička, J., Koten, P. et al. (2015). Catalogue of representative meteor spectra. *Astronomy and Astrophysics*, **580**, A67.

Ward, B. (2015). Meteor spectroscopy during the 2015 Quadrantids. *WGN, Journal of the International Meteor Organization*, **43**, 102–105.

Watanabe, J.-I., Sekiguchi, T., Shikura, M. et al. (2002). Wide-field TV observation of the Leonid meteor storm in 2001: Main peak over Japan. *Publications of the Astronomical Society of Japan*, **54**, L23–L26.

Watanabe, J.-I., Tabe, I., Hasegawa, H. et al. (2003). Meteoroide clusters in Leonids: Evidence of fragmentation in Space. *Publications of the Astronomical Society of Japan*, **55**, L23–L26.

Weinek, L. (1886). Sternschnuppenfall 1885 Nov. 27, Mittheilungen vom Cap der guten Hoffnung, aus Helsingfors, Pulkowa, Bonn, Moncalieri, Prag und Tuschkau bei Pilsen. *Astronomische Nachrichten*, **113**, 374.

Weryk, R. J., Brown, P. G., Domokos, A. et al. (2008). The Southern Ontario All-sky Meteor Camera Network. *Earth Moon and Planets*, **102**, 241–246.

Weryk, R. J., Campbell-Brown, M. D., Wiegert, P. A. et al. (2013). The Canadian Automated Meteor Observatory (CAMO): System overview. *Icarus*, **225**, 614–622.

Whipple, F. L. (1938). Photographic Meteor Studies. *Proceedings of the American Philosophical Society*, **79**, 499–548.

Whipple, F. L. (1954). Photographic meteor orbits and their distribution in space. *Astronomical Journal*, **59**, 201–217.

Whipple, F. L., and Jacchia, L. G. (1957). Reduction methods for photographic meteor trails. *Smithsonian Contributions to Astrophysics*, **1**, 183–206.

Wiśniewski, M., Żołądek, P., Olech, A. et al. (2017). Current status of Polish Fireball Network. *Planetary and Space Science*, **143**, 12–20.

Wray, J. (1967), *The Computation of Orbits of Doubly Photographed Meteors*, Albuquerque: The University of New Mexico Press.

Zigo, P., Tóth, J., and Kalmančok, D. (2013). All-Sky Meteor Orbit System (AMOS). Pages 18–20 of: Gyssens, M., and Roggemans, P. (eds), *Proceedings of the International Meteor Conference, La Palma, Canary Islands, Spain, 20–23 September 2012*. Hove, BE: International Meteor Organization.

Zubović, D., Vida, D., Gural, P., and Šegon, D. (2015). Advances in the Development of a Low Cost Video Meteor Station. Pages 94–97 of: Rault, J.-L., and Roggemans, P. (eds), *Proceedings of the International Meteor Conference, Mistelbach, Austria, 27–30 August 2015*. Hove, BE: International Meteor Organization.

Zvolánková, J. (1983). Dependence of the observed rate of meteors on the Zenith distance of the radiant. *Bulletin of the Astronomical Institutes of Czechoslovakia*, **34**, 122–128.

Part III

Meteors on the Moon and Planets

5

Extra-Terrestrial Meteors

Apostolos Christou, Jérémie Vaubaillon, Paul Withers, Ricardo Hueso, and Rosemary Killen

5.1 Introduction

The beginning of the space age sixty years ago brought about a new era of discovery for the science of astronomy. Instruments could now be placed above the atmosphere, allowing access to new regions of the electromagnetic spectrum and unprecedented angular and spatial resolution. But the impact of spaceflight was nowhere as important as in planetary and space science, where it now became possible – and this is still the case, uniquely among astronomical disciplines – to physically touch, sniff and directly sample the bodies and particles of the solar system. A cursory reading of the chapters in this volume will show that meteor astronomy has progressed in leaps and bounds since the era of visual observations. Indeed, it has evolved into a true analytical science, on a par with sister disciplines such as cometary science and meteoritics. Unfortunately, instrumentation specifically designed to detect meteors on planets other than the Earth has yet to fly on a planetary mission.

Two factors have now brought the era of exo-meteor astronomy closer than ever before. One is the serendipitous detection of meteors and their effects in the relatively thick atmospheres of Mars and Jupiter as well as the tenuous exospheres of Mercury and the Moon, developments of sufficient import to warrant dedicated sections in this chapter. The other is technological advances in the detection of short-lived luminous phenomena in planetary atmospheres and a new understanding of the unique operational aspects of off-Earth meteor surveys. Throughout this chapter we use the term "extraterrestrial meteor" or "exo-meteor". We base our use of this terminology on the distinction between meteoroids and meteors. While "meteoroid" refers to particulate matter in interplanetary space, a "meteor" is a *phenomenon* – the light emitted during the entry of a meteoroid into an atmosphere – that is intrinsic to that atmosphere. Therefore, the term "extraterrestrial meteor" as previously used in the literature (Selsis et al., 2005) and the principal subject matter of this chapter, uniquely defines meteors in planetary atmospheres other than the Earth's.

Why study these meteors? Despite the current rapid progress of Earth-based meteor astronomy, our ability to do useful science will always be limited by the simple fact that we can only sample meteoroid populations with orbits that cross the Earth's. Yet, there is every reason to expect, for instance, that the solar system is criss-crossed with a web of dust streams spread along cometary orbits, all-but-invisible except where the particles are packed densely enough to be detectable, typically near the comet itself (Sykes and Walker, 1992; Reach and Sykes, 2000; Gehrz et al., 2006). Extending meteor observations to other planetary bodies allows us to map out these streams and investigate the nature of comets whose meteoroid streams do not intersect the Earth. Observations of showers corresponding to the same stream at two or more planets will allow to study a stream's cross-section. In addition, the models used to extract meteoroid parameters from the meteor data are fine-tuned, to a certain degree, for Earth's atmosphere. Processing exo-meteor data with these same models will be particularly useful in demonstrating their robustness or uncovering as-yet-unknown limitations that will impact our Earth-centred understanding of the meteor phenomenon. Last but not least, observations of exo-meteors will promote the safety of deep space missions, both crewed and robotic, by identifying regions where the meteoroid flux is high enough to pose significant risk to people or machines. It is fortuitous that the atmospheres of both our neighbouring planets lend themselves to meteor observations. Mars combines a predominantly clear atmosphere and a solid surface to serve as a stable observation platform. It is also the target of a vigorous international exploration program. Venus may, at first glance, not appear as appealing; apart from the extreme environmental conditions, the Venusian sky is perpetually hidden from the surface by planet-encircling cloud layers (Esposito et al., 1983). Yet this still leaves the possibility of observing meteors from above the atmosphere, a technique already employed in observing meteor showers at the Earth (Jenniskens et al., 2000).

5.2 Theoretical Expectations

Single body meteoroid ablation in the Martian atmosphere has been modelled in several works (Apshtein et al., 1982; Adolfsson et al., 1996; McAuliffe, 2006). Adolfsson et al. (1996) compared the height and brightness of meteors at Mars and the Earth, assuming that these properties do not sensitively depend on atmospheric composition. They found that, in general, meteors reach their maximum brightness within the same range of atmospheric density (Figure 5.1). Martian meteors ablate between altitudes of 90 and 50 km; fast (30 km s^{-1}), low-density (0.3 g cm^{-3}) meteoroids in the two atmospheres generate meteors of similar brightness while slower, denser (3 g cm^{-3}) particles produce significantly fainter meteors at Mars. They were able to reproduce their numerical results with the intensity law

$$I_m \propto m \mathrm{v}^{3+n}/H \tag{5.1}$$

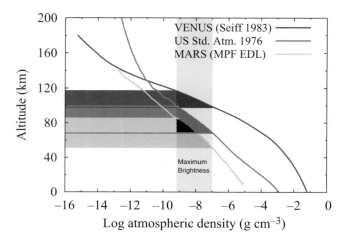

Figure 5.1. Atmospheric density-height profiles for different planets. From top to bottom: Venus ("nightside" atmospheric model by Seiff (1983) – dark grey line), Earth (US Standard Atmosphere 1976 – moderate grey line) and Mars (*Mars Pathfinder* entry profile – light grey line). The horizontal bands indicate the respective height ranges – corresponding to the same density range – at which meteors reach maximum luminosity. Adapted from Christou (2004b).

where m is the meteoroid's mass, v its atmospheric impact speed, H the density scale height and n depends on the meteoroid type as described in Adolfsson et al. (1996). Christou (2004b) applied the same principles to the case of Venus, finding that, everything else being equal, Venusian meteors would be as bright or brighter than at the Earth but also shorter-lived. Using the same argument as Adolfsson et al., whereby the meteoroid ablation rate peaks between $10^{-9.2}$ and 10^{-7} g cm^{-3}, meteors would appear between 100 and 120 km in the venusian atmosphere and above the haze layer (Esposito et al., 1983), boding well for meteor searches from Venus orbit. McAuliffe and Christou (2006) confirmed these earlier conclusions with a numerical ablation model that takes into account radiative surface cooling and heat conduction within the meteoroid, effects not considered in Christou (2004b). Additional numerical work has been carried out in the context of the presence of layers of meteoric metals – in both neutral and ionic form – in the upper atmospheres of these planets with emphasis on Mars (e.g. Pesnell and Grebowsky, 2000, see also Section 5.5). A first step for such studies has been to determine the mass deposition rate of individual atomic metal species (Fe, Mg) from the ablating meteoroid as a function of altitude. The deposition rates of these species represent different components of the gross ablation rate of the meteoroid, equivalent to meteor brightness in the single body ablation model. The results indicate – as in the earlier works – that the maximum brightness of Martian meteors is reached ~10 km lower in the atmosphere and over a similar or somewhat broader range of altitudes compared to terrestrial meteors. Similar calculations were done as part of a recent study (Frankland et al., 2017) of O_2 removal efficiency at Venus by CO_2 oxidation on meteoric remnant particles. These show, as for Mars, that the maximum deposition rate of meteoric metals agrees with the earlier predictions from single body ablation modelling, but also indicate that significant ablation may be taking place at altitudes as high as 130–135 km.

5.3 Observational Record

As mentioned in the Introduction, no direct optical detection of exo-meteors has been achieved to-date. A transient event observed by the Optical and Ultraviolet Visible Spectrometer (*OUVS*) instrument onboard the *Pioneer Venus Orbiter* on 1979 February 17 (Huestis and Slanger, 1993) was interpreted as a serendipitous detection of a meteor trail but remains unconfirmed. A detection of a meteor associated with a known Jupiter-family comet by the dual-eye Pancam imager onboard the *Spirit* rover on Mars was reported by Selsis et al. (2005). Later work by Domokos et al. (2007) showed that the claimed 2005 detection was most likely a cosmic ray hit on the Pancam detector. In addition, the non-availability of a broadband or clear filter on both Pancam eyes severely limited the scope of the search, which did not yield any definite detections. This highlights the need for dedicated instruments to carry out exo-meteor observations.

Although not a search for exo-meteors per se, it is worth mentioning here a successful recording of a meteor shower from space. The $13°\times10°$ Wide Field Imager (WFI) channel of the Ultraviolet and Visible Imagers and Spectrographic Imagers (UVISI) suite on board the *Midcourse Space eXperiment* (MSX) satellite logged 29 meteor detections during an effective observing time of ~20 min on 1997 November 17. The Leonid flux derived from these observations was $(5.5 \pm 0.9) \times 10^{-2}$ km^{-2} hr^{-1} down to a limiting absolute magnitude of -1.5^m and a population index of r = 1.7 (Jenniskens et al., 2000). It demonstrates that detection of even moderately strong meteor showers with instruments not specifically suited to the task is possible from orbit.

5.4 Predictions

5.4.1 Observability

For Mars, Adolfsson et al. (1996) estimated the flux of meteors of absolute visual magnitude[1] $-1^m - +4^m$ to be 50% of that at the Earth. Domokos et al. (2007) placed an upper flux limit of 4.4×10^{-6} km^{-2} hr^{-1} for > 4g meteoroids, consistent with the Adolfsson et al. prediction (4×10^{-7} km^{-2} hr^{-1}) derived by their scaling of the Earth flux according to the Grün et al. (1985) model. Such flux estimates, though approximate, may be used to estimate sporadic meteor detection rates for a given camera system from specific vantage points. For instance, the expected detection rate by a meteor camera system specifically designed for operation in space (Smart Panoramic Optical Sensor Head or SPOSH; Oberst et al., 2011) is between 1 and 7 meteors per orbit, or between 14 and 74 per Earth day (~11 nightside passes) for a 400 km altitude circular orbit at a limiting apparent magnitude of $+2^m$ (Christou et al., 2012). Bouquet et al. (2014) find a similar detection rate for a SPOSH-like system operating in Earth orbit.

For Venus, Beech and Brown (1995) investigated the feasibility of observing bright fireballs (entry mass of 10^4–10^7 kg)

---

[1] The magnitude a visual observer would perceive of a meteor at 100 km distance and at the zenith.

in the venusian atmosphere from the Earth. They found, for instance, that events with apparent magnitude $+10^m$ – equivalent to an absolute magnitude of -24^m – or brighter will occur every three weeks, whereas fainter, $+15^m$ events of -19^m meteor absolute magnitude will occur every two days. The authors pointed out that the flux could be higher at times when streams rich in fireball-producing meteoroids cross the orbit of Venus and advocated long-term monitoring programmes to determine whether such fireball-rich streams exist. Interestingly, Hansell et al. (1995) reported the detection of seven flashes of $(1-20) \times 10^9$ J in luminous energy from narrowband observations at 777 nm with an effective observing time of 3 hr, a rate of 2.7×10^{-12} km^{-2} sec^{-1} which they attributed to lightning. This is $6 \times (2 \times 24h)/3h = 100$ times the Beech and Brown rate for -20^m fireballs assuming a 10^{-3} luminous efficiency.

5.4.2 Showers and Outbursts

It has been known since ancient times that meteor activity is not uniformly random but tends to concentrate around specific times of the year. The first association between a comet and a meteor shower was made 150 years ago (Schiaparelli, 1867) with the first systematic survey of meteor shower activity carried out in the middle of the twentieth century (Lovell, 1949; Aspinal et al., 1949). Repeating the same exercise on other planets, we are faced with the added difficulty that there are essentially no observations, but a multitude of potential parent bodies that may be producing strong meteor activity or nothing at all. To illustrate the problem, consider the case of (3200) Phaethon associated with one of the most prolific annual showers, the December Geminids (Whipple, 1983). The nature of this object has been debated for decades (Ryabova, 2015, 2018). Recent observations from Sun-orbiting spacecraft indicate activity of an exceptional type, not related to release of dust grains from within a sublimating matrix (Li and Jewitt, 2013). Arguably, the motivation to study and explain Phaethon arose because of the "happy coincidence" whereby the Geminid stream intersects the orbit of the Earth. The distance of the comet orbit to the planet orbit (Δ) offers a necessary, but not sufficient, criterion for a comet to produce an observable meteor shower. Because it is computationally expedient to calculate it for a large number of orbits, it has been employed time and again in the literature (Terentjeva, 1993; Christou and Beurle, 1999; Treiman and Treiman, 2000; Larson, 2001; Christou, 2004b; Selsis et al., 2004; Neslusan, 2005). A byproduct of these works is that we now have a fairly strong grasp on the population characteristics of planet-approaching comets for Mars, Venus and the outer planets as compared to the Earth's. They allowed the best candidate parent bodies to be identified as input to more sophisticated techniques.

The observational record at the Earth teaches us that the minimum approach – or its projection on the orbit plane – cannot be the sole discriminator for meteor activity. While it works for e.g. the Perseids in August (109P/Swift-Tuttle), Leonids in November (55P/Tempel-Tuttle) and Lyrids in April (C/1861 G1 (Thatcher)) in the sense that Δ takes particularly small values for these objects, it fails in several prominent cases such as the Taurids (2P/Encke), η Aquariids and Orionids (1P/Halley)

where Δ is of order 0.1 au or greater. An important additional factor is the dynamical type of the orbit. Because of the efficient action of Jupiter in rapidly scattering the orbits of dust grains away from the comet's, annually recurring meteor activity from Jupiter Family Comets is typically weak or non-existent and large numbers of meteors appear only during outbursts e.g. the Draconids in October (21P/Giacobini-Zinner) or the Bootids in June (7P/Pons-Winnecke).

To identify the best shower candidates, Christou (2010) refined the Δ-search technique by exploiting the tendency for strong meteor showers to be associated with Halley-type and Encke-type comets, rather than Jupiter-family and Long-period comets. The final distillation of the best candidates among potential parent bodies and their distribution along the planetary orbit is illustrated in Figure 5.2. Note the relative lack of objects on the left half of each orbit, suggesting a semi-annual modulation of the shower frequency. This is likely a consequence of the method since the observed level of terrestrial meteor activity is a contributing factor in the predictions. Future surveys at Mars and Venus should confirm this.

Arguably, modern meteor forecasting was brought on by (a) the advent of cheap computing power, and (b) adopting the so-called trail model of cometary dust evolution (Kondrateva and Reznikov, 1985; McNaught and Asher, 1999) where populations of meteoroids ejected from the comet at a given perihelion passage maintain their cohesiveness in space as distinct "trails" of particles over many orbital revolutions. The key result is that the dynamical evolution of trails is *deterministic*, so meteor outbursts can be reliably forecasted with numerical simulations of large number of test particles to serve as tracers of the dust evolution. The potential of this approach was spectacularly demonstrated during the Leonid meteor storms in 1999 and 2001. Guided by trail model predictions, observational campaigns are now routinely organised in advance to study debris reaching the Earth from the same comet at different years, from different comets as well as from different comet types (Wiegert et al., 2011; Koten et al., 2015; Ye et al., 2015).

It was only a matter of time until the trail method found application in exo-meteor shower prediction. It was used by Vaubaillon and Christou (2006) and Christou et al. (2007) to identify dust trail encounters between Jupiter-Family comets 45P/Honda-Mrkos-Pajdusakova with Venus and 76P/West-Kohoutek-Ikemura with Mars respectively. Christou et al. (2008) simulated dust ejection from comet 1P/Halley $\sim$5,000 yr ago to form a model of the stream and study how the characteristics of the corresponding shower differ from Venus to Earth and to Mars.

The same approach was used by Christou and Vaubaillon (2011) on a study of larger scope, namely to simulate the meteoroid streams of the parent body candidates in Christou (2010). They found that particles from many of these comets physically approach Mars and Venus and do so year after year, suggesting the presence of annually-recurring activity. The typical efficiency of particle delivery – in other words the fraction of particles physically encountering a planet out of those ejected – is $\sim 10^{-4}$ per comet per planetary year. For six of these, however, the model dust distribution cross-section appears highly inhomogeneous, capable of producing outbursts of activity well above the annual level. We show one such case, the stream of

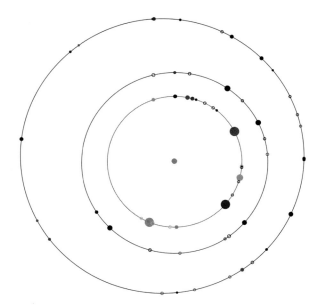

Figure 5.2. Locations of meteoroid streams encountering the orbits of Earth (middle circle), Venus (inner circle) and Mars (outer circle) based on data from Christou (2010). The First Point of Aries is towards the right. Open symbols correspond to Encke-type Comets, filled symbols to Intermediate Long Period and Halley-Type Comets. The size of each symbol is proportional to the encounter speed while the brightness indicates the solar elongation of the radiant. (A black and white version of this figure will appear in some formats. For the colour version, please refer to the plate section.)

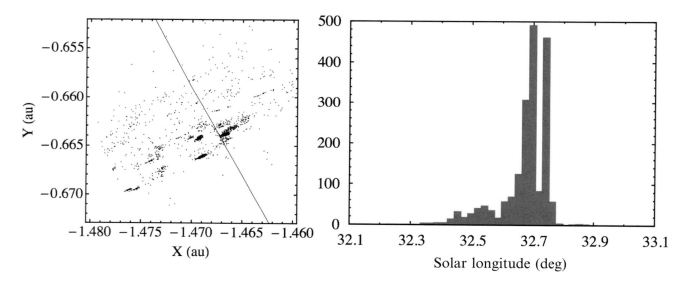

Figure 5.3. Left: Nodes of test particles from comet C/2007 H2 (Skiff) that encounter Mars between the years 2000 and 2050. The points represent Cartesian heliocentric J2000 coordinates and units of au, as the particles cross the martian orbit plane. The black curve is the orbit of Mars, with the planet's direction of motion being from top to bottom. Note the numerous concentrations of particles within the stream's cross-section. Right: Histogram of those particles on the left panel that approach the planet's orbit to within 0.005 au as a function of solar longitude in degrees. Each bin corresponds to a time interval of one hour. After Christou and Vaubaillon (2011).

comet C/2007 H2 (Skiff), in Figure 5.3. The dust distribution is reminiscent of that of the Leonids at the Earth and suggests that our planet is not alone in possessing annual showers that produce outbursts on certain years. Tables 5.1 and 5.2 show forecasts for Mars and Venus for the coming years, in addition to those in the literature (e.g., Christou, 2010; Christou and Vaubaillon, 2011). Interestingly, the width of a meteoroid trail decreases when its heliocentric distance decreases, leading to higher Zenithal Hourly Rate[2] (ZHR) and shorter duration showers at Venus than at Mars. However, as pointed out by Christou (2004a), since there are more Mars-crossing than Venus-crossing comets, we expect more meteor showers at Mars than at Earth and Venus. These two factors combine and we observe that the number

[2] The number of shower meteors that a visual observer would see on a clear night during one hour with the shower radiant at the zenith.

Table 5.1. *Meteor showers at Mars in the next few years, from the method described in Vaubaillon et al. (2005) and Vaubaillon and Christou (2006). "parent body" is the parent body of the stream causing the shower, "d" the minimum distance between the center of the stream cross-section and the planet (negative if the stream crosses inside the planet's orbit and positive otherwise), "Date" the epoch of the shower, α and δ the sky location of the radiant in J2000 Earth equatorial coordinates, V_p the planetocentric velocity and "conf_id" the confidence index as defined by Vaubaillon (2017).*

Parent Body	d (au)	Date (UT)	α (deg)	δ (deg)	V_p (km sec^{-1})	ZHR	conf_id
252P/	−0.0260	2019-11-17T07:32	70.2	−45.0	14.8	28	GYO1/60CU22.8
3D/	0.0004	2019-12-11T22:27	20.1	−2.0	22.6	22	SYO0/1CE0.0
49P/	−0.0006	2019-06-11T13:02	22.3	−67.0	16.4	1164	SYO1/1CE0.0
P/2012 S2	0.0616	2020-12-23T07:44	254.8	−40.8	16.4	917	SYO0/1CE0.0
156P/	−0.0856	2020-11-06T13:56	311.5	−79.5	12.6	67	GYO0/43CU0.0
156P/	−0.0807	2020-11-06T20:36	312.4	−79.7	12.6	25	SYO0/1CE0.0
156P/	−0.0855	2020-11-06T20:31	314.8	−79.7	12.6	115	SYO0/1CE0.0
156P/	−0.0970	2020-10-31T14:47	292.3	−74.1	11.6	138	SYO0/1CE0.0
156P/	−0.0698	2020-11-06T09:52	320.1	−81.8	13.1	57	SYO0/1CE0.0
10P/	−0.0358	2020-06-16T05:31	244.8	25.6	10.4	231	GYO20/56CU0.0
C/2007 H2	0.0009	2021-10-28T16:58	39.9	−55.2	34.4	122	GYO0/4CU0.0
304P/	0.0035	2021-01-25T01:47	280.8	−19.1	13.5	610	SYO0/1CE0.0
304P/	−0.0057	2021-01-31T01:44	280.8	−19.3	13.9	1640	SYO0/1CE0.0

Table 5.2. *Future meteor showers at Venus. See Table 5.1 for explanations.*

Parent Body	d (au)	Date (UT)	α (deg)	δ (deg)	V_p (km sec^{-1})	ZHR	conf_id
2008 BO$_{16}$	−0.0739	2019-08-18T08:36	295.6	−11.7	24.9	1	GYO1/9CU0.0
2008 BO$_{16}$	−0.0620	2019-06-16T13:58	287.8	−11.1	24.8	1	SYO0/1CE0.0
2008 ED$_{69}$	−0.0006	2019-07-09T04:58	247.5	63.0	22.9	1	GYO0/4CU12.7
2013 CT$_{36}$	−0.0377	2019-10-12T01:55	261.1	21.9	12.5	8	GYO0/19CU0.0
2013 CT$_{36}$	−0.0620	2019-06-16T13:58	287.8	−11.1	24.8	4	SYO0/1CE0.0
2013 CT$_{36}$	0.0190	2019-12-08T00:21	241.5	27.2	12.6	1	SYO0/1CE0.0

of showers at Venus is quite small, and the ZHR low, simply because the odds of encountering the planet are lower for Venus than for Mars.

The most recent work to model the meteoroid influx from individual comets on a planetary body has focused on the very close approach of comet C/2013 A1 (Siding Spring) to Mars (Vaubaillon et al., 2014a,b; Moorehead et al., 2014; Tricarico et al., 2014; Farnocchia et al., 2014) and an enhancement of the release rate of metallic species in Mercury's tenuous exosphere probably caused by particles from 2P/Encke (Christou et al., 2015; Killen and Hahn, 2015). These are described in more detail in Sections 5.5.4 and 5.7 respectively.

A question linked to the orbital simulations is that of the ensemble atmospheric properties of the hypothetical showers and their detectability. This problem admits to Monte Carlo simulations and has been investigated either parametrically, by varying pertinent detector properties such as field-of-view (FOV), limiting magnitude and orbital versus surface-based vantage point (McAuliffe, 2006; McAuliffe and Christou, 2006) or by using specific instruments as the baseline (Christou et al., 2012). To indicate what is possible, we quote here some results for the annual Leonid shower from Appendix C of McAuliffe. A zenith-pointed instrument with 60° FOV and limiting magnitude of $+2^m$ will detect 70 meteors per hour and 51 meteors per orbital revolution respectively from the Earth's surface and from Earth orbit. The respective figures for Mars are 25 and 15 while from Venus orbit, 215 detections per orbital revolution are expected. For orbital detection, a lesson that applies to all these planets is the crucial role of the shower population index, i.e. the slope of the meteoroid magnitude distribution and the speed. For a given Earth ZHR, fast showers with shallow magnitude distribution yield the highest number of detections. In this case, a large field of view is preferable to high sensitivity on account of the numerous bright meteors occurring far from nadir, a finding also true for sporadics (Bouquet et al., 2014). Surface-based meteor cameras at Mars will have to contend with the varying amount of atmospheric dust. While they may achieve several tens to a few hundreds of detections per hour at times when the atmosphere is clear (dust optical depth $\tau = 0.5$), during periods of high dust loading ($\tau = 3.0$) the detection rate drops to a few per hour at best.

In conclusion, work done so far suggests that (a) with the exception of a surface camera at Venus, camera systems specifically designed for meteor work will be no less useful at Venus and Mars than at the Earth, and that (b) one should expect comparable detection rates for orbital and surface-based meteor surveys.

5.5 Ionospheric Layers from Meteor Ablation

5.5.1 Overview and State-Of-The-Art

Meteoroid ablation deposits all species common in meteoroids, such as O, Na, Mg, Al, Si, K, Ca, Ti, and Fe, into the upper atmosphere (Plane et al., 2018). Since oxygen is already present in a typical atmosphere, trace amounts of meteoroid-derived oxygen have little effect. The metal species, however, are exotic constituents of atmospheres. They behave differently from the major constituents and trace amounts of metal species can have noticeable effects. These metallic species are deposited at altitudes where the neutral atmosphere is already weakly ionised by sunlight's ultraviolet photons. They can themselves be ionised by sunlight, by chemical reactions with existing atmospheric ions, or directly during ablation. Metallic species tend to form atomic ions like Mg^+ and Fe^+, whereas the ambient ionosphere at ablation altitude is dominated by molecular ions (Schunk and Nagy, 2009). They are destroyed by transport downwards into denser regions of the atmosphere where atomic metal ions undergo three-body reactions to form molecular ions that are quickly neutralised by dissociative recombination with an electron (Molina-Cuberos et al., 2008).

Since atomic ions cannot dissociatively recombine like molecular ions, atomic metal ions tend to be long-lived and a slow production rate of atomic ions can maintain a significant plasma population. Consequently, meteoroids affect the structure, chemistry, dynamics, and energetics of Earth's ionosphere. Compositional profiles have been obtained by fifty (as of 2002) sub-orbital rocket flights with in situ mass spectrometers (Grebowsky and Aikin, 2002). They consistently show a metal ion layer a few km wide located between 90 km and 100 km. Observed properties are highly variable. The many ground-based ionosondes in continuous operation regularly detect post-sunset narrow plasma layers that contain long-lived atomic metal ions at the same altitudes. These are called Sporadic E layers. Models of terrestrial metal ion layers rely on wind shear in a strong and inclined magnetic field to organise ions into narrow layers (e.g., Carter and Forbes, 1999).

Do meteoroids also produce similar layers of metal ions in the ionospheres of other solar system objects? The basic input exists throughout the solar system as meteoroids enter all solar system atmospheres at orbital speeds. However, ionospheric chemistry and magnetically-controlled plasma dynamics will differ between Earth and other objects (Molina-Cuberos et al., 2008; Schunk and Nagy, 2009). Theoretical models of the effects of meteoroids have been developed for many solar system ionospheres (e.g., Lyons, 1995; Pesnell and Grebowsky, 2000; Kim et al., 2001; Molina-Cuberos et al., 2003; Moses and Bass, 2000; Whalley and Plane, 2010; Molina-Cuberos et al., 2008, and references therein). In general, these models predict that the ablation of meteoroids should produce a narrow plasma layer a few neutral scale heights below the main ionospheric peak. For simplicity, these models generally focus on Mg and Fe, which are cosmochemically abundant, well-known from Earth observations, and readily ionised, but other metal species will also be present.

Section 5.5.2 presents observations of candidate metal ion layers throughout the solar system. Section 5.5.3 describes *MAVEN* observations of metal ions at Mars. Section 5.5.4 discusses the ionospheric effects of the encounter of Mars with comet C/2013 A1 (Siding Spring) in 2014.

5.5.2 Pre-MAVEN Observations of Possible Metal Ion Layers

Radio occultation experiments have been responsible for most measurements of extra-terrestrial ionospheres (e.g., Withers, 2010). These measure vertical profiles of electron density, but cannot provide any compositional information. They have observed narrow layers of plasma near predicted ablation altitudes on many objects (e.g., Waite and Cravens, 1987; Withers et al., 2013). However, since many processes can produce, transport, and destroy ionospheric plasma, definitive confirmation of the presence of plasma of meteoric origin in an extraterrestrial ionosphere requires the detection of metal ions. Since most ionospheric observing methods detect electrons, rather than ions, direct compositional information is rare for planetary ionospheres. Furthermore, in situ compositional measurements by retarding potential analysers and ion mass spectrometers are generally made at altitudes above the putative altitudes of metal ion layers.

Here we summarise observations of potential metal ion layers in solar system ionospheres, with the exclusion of observations made at Mars by the *MAVEN* mission. Since *MAVEN* is the only mission to provide relevant ion compositional information, *MAVEN*'s observations and their implications will be presented in Section 5.5.3.

At Venus, Pesnell and Grebowsky (2000) suggested that some nightside *Pioneer Venus Orbiter* (PVO) electron density profiles contain meteoric layers, and Butler and Chamberlain (1976) had earlier suggested that meteoroid influx could produce the surprisingly dense nightside ionosphere. However, recent work has not favoured these earlier conclusions (Fox and Kliore, 1997; Withers et al., 2008). On the dayside, Witasse and Nagy (2006) suggested that two PVO electron density profiles near the terminator (Solar Zenith Angle (SZA) of 85.6° and 91.6°) contain meteoric layers (Withers et al., 2008). Withers et al. (2013) identified 13 candidate meteoric layers at Venus in *Mariner 10*, *Venera 9/10*, and *Pioneer Venus Orbiter* data. However, the most convincing detection of a plasma layer whose altitude is consistent with ions derived from meteoroids is that of Pätzold et al. (2009), who used *Venus Express* radio occultation electron density profiles. They identified 18 instances of low-altitude plasma layers, but only at solar zenith angles between 55° and 90°. Typical peak plasma densities of 10^{10} m^{-3} are reached between 110 and 120 km altitude, peak electron densities increase with decreasing solar zenith angle, and layer shapes are symmetric with respect to peak altitude. This work was based on early *Venus*

Express observations in 2006–2007; a comprehensive survey of the full *Venus Express* dataset has not yet been published. The in situ *Pioneer Venus Orbiter* Retarding Potential Analyzer (ORPA) and Ion Mass Spectrometer (OIMS) did not report any detections of metal ions at their altitudes above 150 km (e.g., Brace et al., 1983, and references therein).

At Mars, Fox (2004) suggested that a Mars Global Surveyor electron density profile displayed a plasma layer near 90 km with a density of 5×10^9 m^{-3} that "could be attributed to meteoric ions". Pätzold et al. (2005) surveyed 120 *Mars Express* electron density profiles and found that 10 of them contained a low-altitude plasma layer between 65 km and 110 km with an average peak electron density of 8×10^9 m^{-3}. Subsequently, Withers et al. (2008) conducted a comprehensive survey of the entire set of 5,600 electron density profiles observed by *Mars Global Surveyor*. They found low-altitude plasma layers in 71 of the 5,600 profiles, a significantly lower detection rate than for *Mars Express*. An example is shown in Figure 5.4. The mean altitude of the meteoric layer was 91.7 ± 4.8 km. The mean peak electron density in the plasma layer was $(1.33 \pm 0.25) \times 10^{10}$ m^{-3} and the mean width of the layer was 10.3 ± 5.2 km. Pandya and Haider (2012) investigated these low-altitude plasma layers using the integrated total electron content between 80 km and 105 km, found a higher occurrence rate of candidate low-altitude layers than did Withers et al. (2008), and concluded that the total electron content in this altitude range increased by a factor of 1.5–3.0 when these layers were present. They suggested that some observed increases in total electron content were associated with the predicted occurrence of meteor showers, but did not show that these increases happened in multiple Mars Years. Searches for potential metal ion layers in earlier Mars datasets have been inconclusive. Using published images of electron density profiles, Withers et al. (2013) identified one candidate from *Mariner 7* and seven candidates from *Mariner 9*. Yet when the actual data from the *Mariner 9* profiles were acquired and analysed, Withers et al. (2015) concluded that "no meteoric layers have been firmly identified in the *Mariner 9* dataset". The two pre-MAVEN observations of ionospheric composition, made by retarding potential analysers on the two Viking Landers, did not detect any metal ions in their profiles that extended down to 110 km (Hanson et al., 1977). In a series of conference abstracts, Aikin and Maguire (2005) and Maguire and Aikin (2006) reported the detection of infrared emissions from the Mg$^+$CO$_2$ ion, but this work has not yet passed through the peer-review process. We note that considerable effort has been expended to search for correlations between temporal variations in the properties of these low-altitude plasma layers and the predicted occurrence of meteor showers, but no significant relationships have been identified (Espley et al., 2007; Withers et al., 2008; Pandya and Haider, 2012; Withers et al., 2013, and Section 5.4).

At Jupiter, *Pioneer 10*, *Pioneer 11*, *Voyager*, and *Galileo* radio occultation profiles commonly show plasma layers that have widths of 10 km and densities of 10^{10}–10^{11} m^{-3} at altitudes of a few hundred kilometers (Fjeldbo et al., 1975; Hinson et al., 1997, 1998b; Yelle and Miller, 2004). At Saturn, *Pioneer 11*, *Voyager*, and *Cassini* radio occultation profiles commonly show narrow plasma layers with densities of 10^{10} m^{-3} around 1000

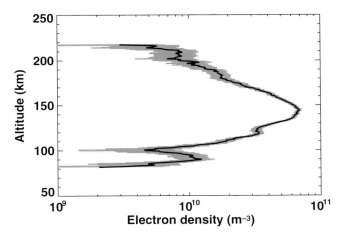

Figure 5.4. Mars Global Surveyor Radio Science profile 5127R00A.EDS showing a low-altitude plasma layer between 80 and 100 km. It was measured at latitude 66.0°N, longitude 2.4°E, 14.4 h LST, Solar Longitude = 206.8°, and SZA = 81.6° on 2005 May 7. The nominal profile is the solid line, and 1σ uncertainties in the electron densities are indicated by the grey region. After Withers et al. (2008).

km altitude (Lindal et al., 1985; Nagy et al., 2006; Kliore et al., 2009). Similar narrow plasma layers were also seen by *Voyager 2* at Uranus and Neptune (Lindal et al., 1987; Tyler et al., 1989; Lindal, 1992). *Cassini* was able to make in situ measurements of the composition of Saturn's ionosphere during its proximal orbits, but was not able to search for metal ions. The high speeds of these atmospheric passes restricted the *Cassini* ion mass spectrometer (INMS) to observations at masses below 7 daltons (pers. comm., Moore, 2018).

At Titan, several *Cassini* radio occultation profiles have been classified as "disturbed" (T31N, T31X, T57X) (Kliore et al., 2008, 2011). The two T31 profiles display a broad plasma layer at 500–600 km altitude with peak density of 2–3 $\times 10^9$ m^{-3}. The published T57 profile is not shown below 800 km (Kliore et al., 2011). Several hypotheses have been proposed for the plasma layer at 500–600 km, including meteoroid ablation (Molina-Cuberos et al., 2001, 2008) and ion precipitation (Cravens et al., 2008). Relative to the neutral scale height, this plasma layer on Titan is somewhat broader than the putative metal ion layers identified on other solar system objects. *Cassini* in situ observations by its mass spectrometer (INMS) that extend down to altitudes around 900 km have not detected metal ions (Vuitton et al., 2009; Cravens et al., 2010).

No features in the two *Voyager 2* radio occultation electron density profiles from Triton have been suggested to be metal ion layers (Tyler et al., 1989). With a surface pressure of 1–2 Pa, it is not clear that Triton's atmosphere is sufficiently dense to ablate meteoroids before they impact the icy surface (e.g., Moses and Bass, 2000). No features in radio occultation electron density profiles of the Galilean satellites have been suggested to be metal ion layers. There are ten *Galileo* profiles for Ganymede (Kliore et al., 2001a,b), six *Galileo* profiles for Europa (Kliore et al., 1997), eight *Galileo* profiles for Callisto (Kliore et al., 2002), and two *Pioneer 10* and ten *Galileo* profiles from Io (Kliore et al., 1974, 1975; Hinson et al., 1998a). Due to the rarefied neutral atmospheres of these satellites, little ablation of meteoroids will occur prior to meteoroid impact. Furthermore, since these atmospheres are ballistic, not collisional, ablated metal species will not be suspended in the atmosphere (but see Section 5.7).

Narrow plasma layers have been observed below the main ionospheric peak in many solar system ionospheres. These have often been interpreted as being metal ion layers caused by meteoroid ablation, but direct evidence for the presence of metal ions is lacking. It is worth considering why the meteoroid hypothesis for the origin of these plasma layers is quite widely accepted, despite the absence of direct composition measurements. Two factors seem significant: (1) the existence of meteoroid-derived metal ion layers on Earth that appear analogous to these extraterrestrial plasma layers; and (2) the paucity of verifiable non-meteoroid explanations for these low-altitude plasma layers. Yet recent observations of ionospheric composition at Mars have shed new light on these issues.

5.5.3 *MAVEN* Observations of Metal Ions at Mars

The *MAVEN* spacecraft is a Mars orbiter that makes extensive measurements of the ionosphere (Jakosky et al., 2015). Two *MAVEN* instruments are able to detect metal ions. In situ observations by the mass spectrometer (NGIMS; Mahaffy et al., 2014; Benna et al., 2015) are sensitive to metal ions above spacecraft periapsis (nominally 150 km, occasionally as low as 120 km) and remote sensing observations by the ultraviolet spectrometer (IUVS; McClintock et al., 2015) are sensitive to Mg^+ ions down to approximately 80 km altitude. Both instruments have detected metal ions.

The NGIMS instrument found that metal ions Na^+, Mg^+, and Fe^+ are continuously present down to the lowest altitudes sampled (120–130 km) (Grebowsky et al., 2017). Densities of Mg^+ and Fe^+ are, on average, approximately equal. This was unexpected – "one might expect Fe^+ to be less dominant with increasing altitude because of the gravitational mass separation anticipated for diffusion processes" (Grebowsky et al., 2017). Ionospheric models also predicted that Mg^+ would be appreciably more abundant than Fe^+ (Whalley and Plane, 2010). However, the Mg^+/Fe^+ ratio on an individual orbit may vary significantly from its long-term average. Densities of Mg^+ and Fe^+ are, on average, proportional to the density of the dominant neutral constituent, CO_2. This was also unexpected – on Earth, "observations made above the main ionospheric metal ion layer are often characterised by complex layers associated with electrodynamic sources, with no clear trend of ordered metal ion concentration decreases with increasing altitude" (Grebowsky et al., 2017). Furthermore, "isolated metal ion layers mimicking Earth's sporadic E layers occur despite the lack of a strong magnetic field as required at Earth" (Grebowsky et al., 2017). These metal ion layers are seen at altitudes around 140–170 km and have widths of 10 km.

The in situ NGIMS observations were unable to address how the trend of exponential increase in metal ion density with decreasing altitude continued below the 120 km limit of periapsis. Remote sensing observations are required to do so. The IUVS instrument found that Mg^+ ions are continuously present in the ~75–125 km altitude range (Figure 5.5, top panel; Crismani et al., 2017). Crismani et al. (2017) reported that

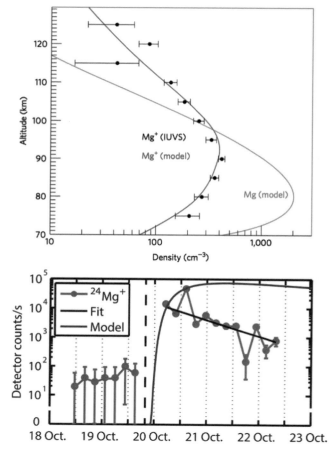

Figure 5.5. Top: Mg^+ altitude profiles derived from *MAVEN* IUVS at orbit 3040 compared with the model prediction. Note that Mg is not detected despite large predicted concentrations. From Crismani et al. (2017). Bottom: Temporal evolution of the abundances of Mg^+ measured by *MAVEN*/NGIMS at periapsis from 2014 October 18 to 2014 October 23. The exponential decay can be fitted by a time constant of 1.8 days. The dashed line marks the predicted time of maximum flux of C/2013 A1 dust. Predicted signal levels were derived by a 1-D model. Error bars reflect 3 × standard deviation of the sampled data due to counting statistics. Instrument background is 10–100 counts/s. From Benna et al. (2015). (A black and white version of this figure will appear in some formats. For the colour version, please refer to the plate section.)

Mg^+ ions formed a layer with peak density 2.5×10^8 m^{-3}, peak altitude of 90 km, and full-width at half-maximum of approximately 30 km. The layer shape appears symmetric with distance from the peak, distinct from the strikingly asymmetric shape of a Chapman layer.

Based on Grebowsky et al. (2017) and Crismani et al. (2017), the NGIMS and IUVS observations of Mg^+ appear to be consistent, but further work is needed to synthesise them into a coherent picture of the vertical profile of Mg^+ density from 75 to 175 km.

Both radio occultation experiments and *MAVEN* IUVS have observed a plasma layer at 90 km altitude. However, radio occultation experiments observe a sporadic layer of electrons with peak density in excess of 10^{10} m^{-3} and width of 10 km, whereas IUVS observes a continuous layer of

Mg$^+$ ions with peak density that is fifty times smaller and width that is three times larger. The persistent presence of Mg$^+$ ions in IUVS observations is consistent with terrestrial experience, where metal ions are a ubiquitous feature of the ionosphere. They derive predominantly from steady influxes of sporadic meteoroids, rather than shower meteoroids whose influx varies greatly with time. Shower meteoroids, although visually striking, contribute only a small fraction of the steady-state mass flux (e.g. Brown et al., 2008). Readers should note the potential for confusion between the "sporadic" occurrence of low-altitude electron density layers in radio occultation profiles and the "sporadic" meteoroid population, which supplies a relatively stable influx of interplanetary dust.

If the sporadic low-altitude plasma layers seen by radio occultation experiments are generated by meteoroid influx, then their chemistry must be such that metal ions like Mg$^+$ rapidly transform into other ion species that are much longer-lived. That possibility is extremely inconsistent with current understanding of the effects of meteoroids on planetary atmospheres (e.g. Whalley and Plane, 2010). Furthermore, their irregular occurrence must be reconciled with the constant influx of interplanetary dust. The other possibility is that the sporadic low-altitude plasma layers seen by radio occultation experiments are caused by other mechanisms, such as enhancements in the precipitation rate of charged particles of suitable energy to deposit their energy at these altitudes.

In light of the *MAVEN* IUVS observations, we judge that the cause of the sporadic low-altitude plasma layers seen at Mars by radio occultation experiments is currently unknown. Given the similarities between the ionospheres of Venus and Mars, caution should also be applied to the interpretation of the analogous layers in the ionosphere of Venus. However, given the many major differences between giant planets and terrestrial planets, these Mars findings should not necessarily be extended directly to the giant planets. Nevertheless, it is important to note that direct evidence for metal ions in giant planet ionospheres is absent.

5.5.4 Encounter of Mars with Comet C/2013 A1 (Siding Spring)

On 2014 October 19, Mars experienced a remarkably close encounter with comet C/2013 A1 (Siding Spring) at a distance of approximately 135,000 km (e.g. Withers, 2014, and references therein). As Mars passed through the comet's coma, the dust influx on Mars increased substantially. This was expected to affect the distribution of metal ions in the planet's ionosphere. This encounter occurred when several spacecraft were operational at Mars and able to perform relevant observations. *MAVEN*, which arrived at Mars only weeks earlier, searched for metal ions using the NGIMS and IUVS instruments. The MARSIS topside radar sounder on *Mars Express* measured the peak density of the ionosphere. The SHARAD radar sounder on Mars Reconnaissance Orbiter measured the total electron content of the ionosphere.

MAVEN IUVS observed enhanced densities of Mg$^+$ (Figure 5.5, bottom panel) and Fe$^+$ ions after the encounter (Schneider et al., 2015). (Note that Fe$^+$ was not discussed by Crismani et al. (2017) in their survey of the long-term behaviour of Mg$^+$ ions.) A typical vertical profile of Mg$^+$ density several hours after closest approach revealed a layer with a peak density on the order of 10^{10} m^{-3}, peak altitude of 120 km, symmetric shape, and full-width at half-maximum of approximately 10 km. Relative to observations under normal conditions, the peak density is larger, the peak altitude is higher, and the width is narrower. The increase in peak density is expected from the increase in dust flux. The increase in peak altitude is expected from the increase in dust speed from sporadic meteoroids at a few km s^{-1} to cometary dust at 56 km s^{-1}. The decrease in peak width has not been explained. Schneider et al. (2015) found that Mg$^+$ densities were enhanced globally for 1–2 days.

The *MAVEN* NGIMS instrument observed enhanced densities of a tremendous range of metal ions (singly ionised Na, Mg, Al, K, Ti, Cr, Mn, Fe, Co, Ni, Cu, and Zn, and possibly Si and Ca) at 185 km after the encounter (Benna et al., 2015). Observations did not extend below 185 km, which was the periapsis altitude at this early stage of the mission. In these observations, the most abundant metal ions were Na$^+$, Mg$^+$, and Fe$^+$, but in different proportions to those seen in normal circumstances. Here Na$^+$ was the most abundant ion (roughly three times as abundant as Fe$^+$) and Mg$^+$ was the next most abundant (roughly two times as abundant as Fe$^+$). That may reflect compositional differences between fresh dust from comet C/2013 A1 (Siding Spring) and background interplanetary dust. The Mg$^+$ density at 185 km shortly after the encounter ($153.6 \pm 7.5 \times 10^6$ m^{-3}) was increased from its normal value by approximately three orders of magnitude, which illustrates the dramatic impact of this comet on the distribution of metal species in the ionosphere and atmosphere. The Mg$^+$ density at 185 km decreased exponentially after the encounter with a time constant of 1.8 days.

Electron densities in the ionosphere of Mars were observed by other orbital instruments during the encounter. The MARSIS topside radar sounder on *Mars Express* observed peak electron densities of 1.5–2.5×10^{11} m^{-3} at 80–100 km altitude and solar zenith angles of 75–110° (Gurnett et al., 2015). Under normal conditions, the peak electron density and peak altitude at these solar zenith angles are 0.1–1.0×10^{11} m^{-3} and greater than 140 km, respectively. Although the ionospheric composition was not simultaneously observed for these electron density observations, it seems likely that metal ions constituted a significant fraction of the total ion density.

The SHARAD radar on Mars Reconnaissance Orbiter measured the vertical total electron content of the ionosphere at solar zenith angles of 94° and 113° shortly after the encounter (Restano et al., 2015). It found total electron content values on the order of 4×10^{15} m^{-2} (94°) and 2×10^{15} m^{-2} (113°). Under normal conditions, total electron contents at these solar zenith angles are approximately 2×10^{15} m^{-2} (94°) and 0.5×10^{15} m^{-2} (113°). The corresponding peak electron density can be estimated from the fact that the total electron content is usually on the order of $4NH$ where N is the peak electron density and H is the neutral scale height (10 km) (Withers, 2011). The inferred peak electron densities are 10^{11} m^{-3} (94°) and 5×10^{10} m^{-3} (113°), which are reasonably consistent with the MARSIS peak density measurements when allowance is made for plausible spatial and temporal variations in ionospheric conditions during this unusual event.

The overall picture of the ionosphere of Mars suggested by the *MAVEN* IUVS, *MAVEN* NGIMS, *Mars Express* MARSIS, and Mars Reconnaissance Orbiter SHARAD observations is of enhanced plasma densities at relatively low altitudes, altered ion composition, and substantial spatial and temporal variations. Realistic numerical simulations of the global-scale, time-varying picture of ionospheric properties during the cometary encounter have not yet been conducted. Such simulations, though technically challenging, are necessary to synthesise the disparate measurements available for this unique event.

In conclusion, metal ion layers are likely to exist in all planetary ionospheres, but the narrow electron density layers that have been observed throughout the solar system and suggested as being meteoric have not been proven to contain metal ions. In view of the difficulty of in situ sampling of composition at relevant pressure levels, remote sensing by ultraviolet spectroscopy offers the most promising path for detecting metal ions in planetary ionospheres.

5.6 Impact Flashes at Jupiter

The giant planet Jupiter is observed every year by thousands of amateur astronomers who acquire video observations of its atmosphere. Their images provide a nearly continuous observational record that is widely used to study the atmospheric dynamics of the planet (Hueso et al., 2010b, 2018b; Mousis et al., 2014). Since the year 2010 some of these video observations have resulted in the serendipitous discovery of energetic flashes of light that last from one to a few seconds and with visual brightness comparable to stars of magnitude $+6^m$ or brighter. Although Jupiter impacts visible from Earth are caused by asteroids or comets, not meteoroids, they do nevertheless produce meteor phenomena. The first of these optical flashes was discovered on 2010 June 3 and was simultaneously recorded by two observers (Anthony Wesley from Australia discovered the flash and Christopher Go from Philippines confirmed this with his own observation acquired at the same time). This bolide appeared as a bright flash that lasted about two seconds and did not leave any observable trace on the atmosphere afterwards. A large follow-up campaign of observations using telescopes such as the Very Large Telescope (VLT) or Hubble Space Telescope (HST) did not find any debris in the atmosphere (Hueso et al., 2010a). The analysis of the light-curve of the observations yielded an estimated energy of the impact of $(1.9-14) \times 10^{14}$ J which in turn corresponds to an object of $1.05-7.80 \times 10^5$ kg with diameter in the range 4.7–18 m depending on the density (Hueso et al., 2010a, 2013). Since then, other instances of energetic flashes on Jupiter have been found, totalling five different events observed by a total of twelve amateur observers. The light-curves of these events have been analysed by Hueso et al. (2013) and Hueso et al. (2018a) and interpreted as being caused by the impact of large meteoroids with diameters of 5–20 m. We note that these actually qualify as small asteroids or cometary fragments rather than meteoroids according to the IAU definition[3]. The estimated masses and energies released in Jupiter's

---

[3] www.iau.org/public/themes/meteors_and_meteorites/

Table 5.3. *Summary of impacts on Jupiter*

Date	Mass (kg)	Reference
81-03-05	11	Cook and Duxbury (1981)
94-07-16 – 24	1.0×10^{12}	Hammel et al. (1995), Harrington et al. (2004)
09-07-19	6.0×10^{10}	Sánchez-Lavega et al. (2010)
10-06-03	$1.05-7.80 \times 10^3$	Hueso et al. (2010a, 2013)
10-08-20	$2.05-6.10 \times 10^3$	Hueso et al. (2013)
12-09-10	$5.00-9.50 \times 10^3$	Hueso et al. (2013)
16-03-17	$4.03-8.05 \times 10^3$	Hueso et al. (2018b)
17-05-26	$0.75-1.30 \times 10^3$	Hueso et al. (2018b)

atmosphere by these impacts are summarised in Table 5.3 where we also compare with previous impacts in Jupiter: the meteor observed by *Voyager 1* in 1981 (Cook and Duxbury, 1981), the impacts of the D/1993 F2 (Shoemaker-Levy 9) fragments in 1994 (Hammel et al., 1995; Harrington et al., 2004) and the 500-m diameter object that impacted Jupiter in 2009 (Sánchez-Lavega et al., 2010), the latter also recorded by Anthony Wesley one year before his discovery of the bright flash.

Remarkably, each of these bolides was first noted by a single observer who detected the flash and issued an alert. The alert quickly resulted in reports from other observers who found the flashes in their video observations acquired on the same night and at the precise time of the reported flash. This occurs because most video recordings are obtained over several minutes and it is difficult to manually scrutinise video observations frame by frame in search of an unremarkable short-lived feature such as a flash. Software tools that process video observations to detect flashes are now available at, e.g., www.astrosurf.com/planetessaf/doc/project_detect.php and http://pvol2.ehu.eus/psws/jovian_impacts/ (André et al., 2018). However, the availability of these tools has not produced further detections, probably due to their limited usage up to now.

The brightest flash occurred in 2012 September and is discussed in Hueso et al. (2013). Figure 5.6 shows the frame with the peak of the flash, a synthetic image of the planet made by stacking the different frames where the flash is visible and a light-curve of this event. Analysis of the light-curve following photometric calibration of the video indicates a mass of about $5.00-9.50 \times 10^5$ kg for the impactor.

Hueso et al. (2013) present a numerical simulation of the airbursts produced by objects equivalent to the 2012 impact and colliding with Jupiter at velocities of 60 km s^{-1}. Objects of this size begin to break up at about 3 mbar pressure (120 km above the 1-bar level) and deposit most of their energy at the 5–6 mbar level (100 km height) where they break up in successive events. These result in light-curves that may show significant structure if observed with sufficient time resolution.

The latest analysis of the frequency of such impacts on Jupiter results in an estimate of 4–25 detectable events per year (Hueso et al., 2018a). Given that these flashes have to be found right at the moment they occur, and we may only observe one side of Jupiter and for a limited period of time every year, the total number of events on Jupiter's atmosphere should be 10–70 impacts per year. This impact rate is in basic agreement

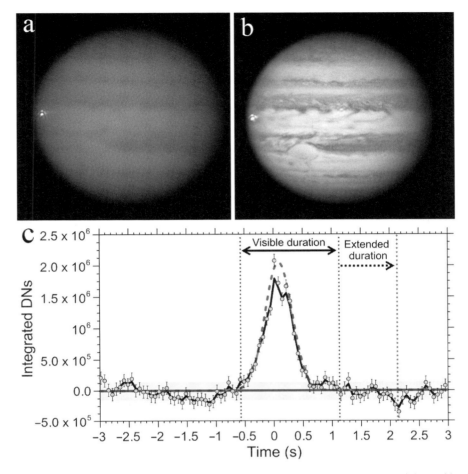

Figure 5.6. The 2012 September impact on Jupiter. This was the most energetic flash of light at Jupiter. (a) Brightest video frame in the recording; (b) Image composite made by stacking all frames where the flash is visible and 1000 frames for the planet. (c) Light-curve of the impact. The impact partially saturated some pixels at the peak of its brightness. Points represent the flash intensity measured over the different frames, the dashed line is a gaussian fit to the data and the continuous line is a weighted average of the photometric values. Further evidence of the flash is visible in the data at the level of the photometric noise. Based on data from Hueso et al. (2013).

with expectations from dynamical studies of comets (Levison et al., 2000). The contribution of impacts of this size range to the abundance of well-characterised exogenic species (mainly O atoms in CO and H_2O molecules; Lellouch et al. 2002; Moses and Poppe 2017) in Jupiter's upper atmosphere is probably limited (Hueso et al., 2018b).

We expect that the number of detections in the next few years will increase. On the one hand, the new generation of sensitive cameras used by amateur astronomers will allow detection of impacts by slightly smaller objects. On the other hand, Jupiter oppositions in the next few years will occur in spring and summer in the North hemisphere, where most of the regular observers of Jupiter are located. Additionally, improvements in the software tools used by the amateur community to systematically search for more flashes are underway, and will improve the detectability of these events, providing better constraints on the flux of impacts on Jupiter. Finally, similar impacts are also expected to occur on Saturn (Tiscareno et al., 2013). Their discovery from ground-based observations is not beyond the capability of equipment used by amateurs to discover bolide impacts on Jupiter.

5.7 Impact-Derived Species within Airless Body Exospheres

Objects such as the Moon, Mercury and asteroids that lack atmospheres in the traditional sense, host instead tenuous Surface-Bounded Exospheres (SBEs), the result of a delicate balance between poorly-understood sources and sinks.

Models of SBEs have extremely varied estimates of the importance of surface Impact Vaporisation (IV). Although this is an established field of study (Melosh, 1989; Pierazzo and et al., 2008; Hermalyn and Schultz, 2010), uncertainties regarding the importance of impact vaporisation on extraterrestrial bodies include the uncertainty in impact rates for both interplanetary dust and larger meteoroids and comets (Cremonese et al., 2016), the relative amounts of melt and vapour produced in an impact (Pierazzo et al., 1995), the temperature of the vapour – which affects escape rates (e.g. Cintala, 1992; Rivkin and Pierazzo, 2005), the relative amount of neutral versus ionised ejecta (Hornung et al., 2000), and the gas-surface interaction of the downwelling ejecta (Yakshinskiy and Madey, 2005). The importance of impact vaporisation as a source of exospheric neutrals has been constrained in part by observation of the escaping component of the exospheres – the Mercurian tail (Schmidt et al., 2012), the Ca exosphere of Mercury (Burger et al., 2014) and the lunar extended exosphere and tail (Wilson and Smith, 1999). These results all depend critically on the assumed temperature or velocity distribution of the initial vapour plume, which has been variously assumed to be between 1,500 and 5,000 K or non-thermal. They also depend critically on the assumed photoionisation rate. Values of the Na photoionisation rate have varied by a factor of three, and values of the Ca photoionisation rate have varied by a factor of 4.4 (e.g. see Killen et al., 2018). This obviously introduces a huge uncertainty in the escape rate, and hence the source process.

Colaprete et al. (2016) conclude, based on observations of the UV spectrometer onboard the *LADEE* mission, that there is a pronounced role for meteoroid impact vaporisation and surface exchange in determining the composition of surface-bounded exospheres. Most interestingly, the simulations show that a release of ejecta from a single injection will persist in the SBE-surface system for much longer than the ionisation lifetime. Their nominal model shows that about 30% of Na released is still adsorbed on the surface after 100 days (3 lunations). Residence times in the lunar environment of 45–90 days (mainly on the lunar surface) can be expected before escape to the solar wind, which would explain the long-term smooth increase and decrease in the Na column density observed as the result of meteoroid streams. This is the result of each particle residing in the soil for approximately an ionisation lifetime (i.e., several days) between bounces, combined with the many bounces that it has to take before being lost from the SBE. Although Leblanc and Johnson (2010) conclude that thermal desorption and Photon-Stimulated Desorption (PSD) are the dominant source processes for the Na exosphere of Mercury, they introduce a source term equivalent to the amount of fresh material required to maintain the SBE. This source is not identified but is required to maintain a surficial reservoir of adsorbed atoms containing at least 2000 times as many atoms as the exospheric content.

A pronounced increase in K was observed by *LADEE* during the Geminid meteor shower while no increase was seen in Na. Given that Na is more volatile than K, it is not obvious that the peaks are the result of the meteoroid streams as opposed to whatever is causing the monthly variation. Given the long residence time of Na on the surface deduced by both Leblanc and Johnson (2003) and Colaprete et al. (2016), it has been suggested that impacts are the primary source of atoms from the regolith to the extreme surface, and these atoms feed the subsequent release by photons or thermal processes. If this is the case, then micrometeoritic impacts play the dominant role in maintaining the SBEs while the less energetic processes such as PSD and thermal desorption serve to keep the atoms in play until they are destroyed by photoionisation.

Schmidt et al. (2012) studied the extended sodium tail of Mercury and concluded that both photon-stimulated desorption and micrometeoroid impacts are required to simulate the ~20% loss of Mercury's sodium atmosphere, depending on orbital phase, and that the two mechanisms are jointly responsible for the observed comet-like tail as driven by solar radiation pressure. Roughly three times as many atoms would have to be ejected by PSD than by IV at 3,000 K to accomplish a similar loss rate. The velocity distribution for PSD was required to have a low energy maximum (900 K) and high energy tail similar to that measured by Yakshinskiy and Madey (2004) for electron-stimulated desorption from a lunar sample at 100 K.

The distant sodium tail of the Moon has been observed by looking in the direction opposite the sun (Wilson and Smith, 1999). They concluded that the Na escape rate increased by a factor of 2–3 during the most intense period of the 1998 Leonid meteor shower. The changes in the lunar exosphere itself were not quantified, as the escape rates only constrain that fraction of the velocity distributions above $2.1\,km\,s^{-1}$. Nevertheless, the observation is evidence for a strong influence of meteor impacts on the lunar sodium exosphere and its escape rate. The calcium exosphere of Mercury exhibits several attributes that point to meteoritic impact as the source of the calcium exosphere (Figure 5.7). First, the calcium source peaks strongly on the dawn side of the planet, exhibiting the same morphology seen in the lunar dust observed by the *LADEE* spacecraft (Szalay et al., 2018; Szalay and Horányi, 2015; Janches et al., 2018). The variation of the calcium exosphere with True Anomaly Angle (TAA) was modeled independently by Killen and Hahn (2015) and Pokorný et al. (2018), indicating that the calcium source rate could be the result of impacts from the interplanetary dust disk except for a dramatic increase near TAA of $20°$–$30°$. This increase was attributed to the intersection of Mercury's orbit and that of comet Encke (Killen and Hahn, 2015) and further shown to be consistent with the evolution of the dust stream by the influence of planetary perturbations and Poynting-Robertson drag (Christou et al., 2015).

Although much of the literature concerning impact vaporisation on airless bodies relates to the ejection of sodium, potassium or calcium, the origin of schreibersite, a phosphide common to impact breccias at all Apollo sites, has been proposed to be a meteoritic contaminant, or alternatively produced in situ by reduction on the lunar surface. Pasek (2015) proposed that schreibersite and other siderophilic P phases have an origin from impact vaporisation of phosphates at the lunar oxygen fugacity. Phosphorus has not been observed in any SBE, but that may be due to the difficulty in observing it.

5.8 Exo-Meteor Observations: Future Prospects

In the near future, we will continue to rely on serendipity for in situ detection of exo-meteors, either directly or indirectly. The *MAVEN* orbiter continues to monitor the metal content of Mars' upper atmosphere for increases of similar magnitude and character to the C/2013 A1 (Siding Spring) event that would indicate passage through a comet's dust stream. At the surface, the *Opportunity* and *Curiosity* rovers regularly image the Martian sky, but confirmation of a martian meteor will require the fortuitous interruption of the trail by a foreground object, such as a rock, a mountain or part of the spacecraft structure (Christou et al., 2007). The *InSight* lander that touched down on the Martian surface in late 2018 is equipped with seismometers that will detect atmospheric entry of large meteoroids (Stevanović et al., 2017). Optical cameras are also present on the spacecraft, but are not designed for night-time observations and therefore

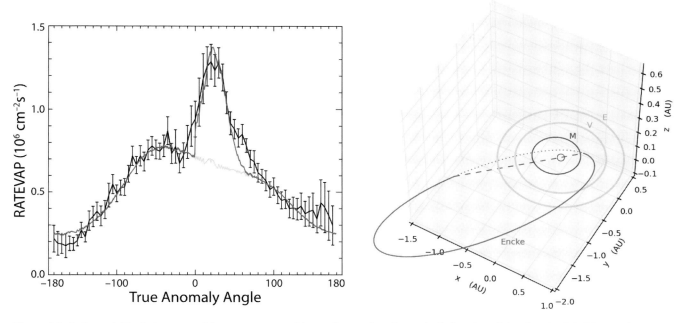

Figure 5.7. Left panel: Total planetary calcium source rate to Mercury's exosphere is a periodic function of the planet's TAA. Mercury is at perihelion when TAA = 0° and at aphelion when TAA = ±180°. The black curve is this total rate summed over the planet at each TAA, derived from observations obtained by the *MESSENGER* MASCS spectrometer 2011 March – 2013 March (from Burger et al., 2014). The dark grey line is the modeled contribution from a cometary dust stream with peak density at TAA = 25° plus that due to an interplanetary dust-disk. The light grey line is the contribution from the disk. Adapted from Killen and Hahn (2015). Right panel: Encke's orbit (dark grey) along with those of Mercury (grey circle M), Venus and Earth (grey circles V and E). After Killen and Hahn (2015). (A black and white version of this figure will appear in some formats. For the colour version, please refer to the plate section.)

will most likely not deliver many meteor images. Closer to the Sun, the JAXA *Akatsuki* spacecraft finally entered orbit around Venus in 2015 November, at the end of an extended cruise period. The Lightning and Airglow Camera (LAC) is a 12° FOV high-speed imaging system onboard the orbiter that searches for rapidly-varying luminous phenomena in the nighttime Venusian atmosphere (Takahashi et al., 2008). The instrument has been active since 2016 December, observing Venus for ~30 min every 10-day orbit (Takahashi et al., 2017) and may detect bright meteors if, or when, their frequency is sufficiently high. In the meantime, valuable operational experience in this type of observation is being gained by ongoing – as well as planned – spaceborne monitoring of the atmosphere of our own planet for meteor events (Arai et al., 2014; Abdellaoui et al., 2017). These efforts are complemented by ground-based monitoring of the atmosphere of Jupiter by amateur and professional astronomers for impact flashes, possibly to be extended to the other giant planets as well as the Venusian nightside atmosphere. Numerical modelling and supporting laboratory work is also likely to continue. Many aspects of recent *MAVEN* observations of metal ions at Mars have yet to be reproduced or explained by numerical models. In particular, rate coefficients for many relevant chemical reactions involving metal species and ambient atmospheric molecules are not well-known. Previous laboratory work has naturally focused on reactions involving the nitrogen and oxygen species prevalent in Earth's atmosphere, not the carbon and hydrogen species common in other atmospheres.

References

Abdellaoui, G., Abe, S., Acheli, A. et al. 2017. Meteor studies in the framework of the JEM-EUSO program. *Planet. Space Sci.*, **143**, 245–255.

Adolfsson, L., Gustafson, B. A. S., and Murray, C. D. 1996. The Martian atmosphere as a meteoroid detector. *Icarus*, **119**, 144–152.

Aikin, A. C., and Maguire, W. C. 2005. Detection in the infrared of $Mg^+ \cdot CO_2$ ion produced via meteoritic material in the martian atmosphere. *American Astronomical Society DPS Meeting*, **37**, 33.37.

André, N., Grande, M., Achilleos, N. et al. 2018. Virtual Planetary Space Weather Services offered by the Europlanet H2020 Research Infrastructure. *Planet. Space Sci.*, **150**, 50–59.

Apshtein, E. Z., Pylyugin, N. N., and Varmanian, N. 1982. Investigation of meteor ablation in the atmospheres of the Earth, Mars, and Venus. *Kossm. Issl.*, **20**, 730–735 (in russian).

Arai, T., Kobayashi, M., Yamada, M., and Matsui, T., COMETSS project team 2014. Meteor observation HDTV camera onboard the International Space Station. *45th Lunar. Planet. Sci. Conf.*, Abstract 1610.

Aspinal, A., Clegg, J. A., and Lovell, B. 1949. The daytime meteor streams of 1948 - I. Measurement of the activity and radiant positions. *Mon. Not. R. Astron. Soc.*, **109**, 352–358.

Beech, M., and Brown, P. 1995. On the visibility of bright Venusian fireballs from Earth. *Earth Moon Planets*, **68**, 171–179.

Benna, M., Mahaffy, P. R., Grebowsky, J. M. et al. 2015. Metallic ions in the upper atmosphere of Mars from the passage of comet C/2013 A1 (Siding Spring). *Geophys. Res. Lett.*, **42**, 4670–4675.

Bouquet, A., Baratoux, D., Vaubaillon, J. et al. 2014. Simulation of the capabilities of an orbiter for monitoring the entry of interplanetary matter into the terrestrial atmosphere. *Planet. Space Sci.*, **103**, 238–249.

Brace, L. H., Taylor, H. A., Gombosi, T. I. et al. 1983. The ionosphere of Venus: Observations and their interpretations. In: D. M. Hunten, L. Colin, T. M. Donahue, and V. I. Moroz (eds), *Venus*. Tucson: University of Arizona Press, pp. 779–840.

Brown, P., Weryk, R. J., Wong, D. K. et al. 2008. A meteoroid stream survey using the Canadian Meteor Orbit Radar I. Methodology and radiant catalogue. *Icarus*, **195**, 317–339.

Burger, M., Killen, R. M., McClintock, W. E. et al. 2014. Seasonal variations in Mercury's dayside calcium exosphere. *Icarus*, **133**, 51–58.

Butler, D. M., and Chamberlain, J. W. 1976. Venus' night side ionosphere – Its origin and maintenance. *J. Geophys. Res.*, **81**, 4757–4760.

Carter, L. N., and Forbes, J. M. 1999. Global transport and localized layering of metallic ions in the upper atmosphere. *Ann. Geophys.*, **17**, 190–209.

Christou, A. A. 2004a. Predicting martian and venusian meteor shower activity. *Earth Moon Planets*, **95**, 425–431.

Christou, A. A. 2004b. Prospects for meteor shower activity in the venusian atmosphere. *Icarus*, **168**, 23–33.

Christou, A. A. 2010. Annual meteor showers at Venus and Mars: lessons from the Earth. *Mon. Not. R. Astron. Soc.*, **402**, 2759–2770.

Christou, A. A., Oberst, J., Koschny, D. et al. 2007. Comparative studies of meteoroid-planet interaction in the inner solar system. *Planet. Space Sci.*, **55**, 2049–2062.

Christou, A. A., and Beurle, K. 1999. Meteoroid streams at Mars: possibilities and implications. *Planet. Space Sci.*, **47**, 1475–1485.

Christou, A. A., and Vaubaillon, J. 2011. Numerical modeling of cometary meteoroid streams encountering Mars and Venus. Pages 26–30 of: Cooke, W. J., Moser, D. E., Hardin, B. F., and Janches, J. (eds), *Meteoroids: The Smallest Solar System Bodies, Proceedings of the Meteoroids Conference held in Breckenridge, CO, USA, May 24–28, 2010*. NASA/CP-2011-216469.

Christou, A. A., Vaubaillon, J., and Withers, P. 2007. The dust trail complex of comet 79P/du Toit-Hartley and meteor outbursts at Mars. *Astron. Astrophys*, **471**, 321–329.

Christou, A. A., Vaubaillon, J., and Withers, P. 2008. The P/Halley Stream: Meteor Showers on Earth, Venus and Mars. *Earth Moon Planets*, **102**, 125–131.

Christou, A. A., Oberst, J., Elgner, S. et al. 2012. Orbital observations of meteors in the Martian atmosphere using the SPOSH camera. *Planet. Space Sci.*, **60**, 229–235.

Christou, A. A., Killen, R. M., and Burger, M. H. 2015. The meteoroid stream of comet Encke at Mercury: Implications for MErcury Surface, Space ENvironment, GEochemistry, and Ranging observations of the exosphere. *Geophys. Res. Lett.*, **42**, 7311–7318.

Cintala, M. J. 1992. Impact-induced thermal effects in the Lunar and Mercurian regoliths. *J. Geophys. Res.*, **97**, 947–973.

Colaprete, A., Sarantos, M., Wooden, D. H. et al. 2016. How surface composition and meteoroid impacts mediate sodium and potassium in the lunar exosphere. *Science*, **351**, 249–252.

Cook, A. F., and Duxbury, T. C. 1981. A fireball in Jupiter's atmosphere. *J. Geophys. Res.*, **86**, 8815–8817.

Cravens, T. E., Robertson, I. P., Ledvina, S. A. et al. 2008. Energetic ion precipitation at Titan. *Geophys. Res. Lett.*, **35**, L03103.

Cravens, T. E., Yelle, R. V., Wahlund, J.-E., Shemansky, D. E., and Nagy, A. F. 2010. Composition and structure of the ionosphere and thermosphere. In: R. H. Brown, J.-P. Lebreton, and J. H. Waite (eds), *Titan from Cassini-Huygens*. Cambridge University Press/Springer: Dordrecht, Heidelberg, London, New York.

Cremonese, G., Borin, P., Lucchetti, A., Marzari, E., and Bruno, M. 2016. Micrometeoroids flux on the Moon. *Astron. Astrophys.*, **551**, id. A27.

Crismani, M. M. J., Schneider, N. M., Plane, J. M. C. et al. 2017. Detection of a persistent meteoric metal layer in the Martian atmosphere. *Nat. Geosci.*, **10**, 401–404.

Domokos, A., Bell, J. F. III, Lemmon, M. T. et al. 2007. Measurement of the meteoroid flux at Mars. *Icarus*, **191**, 141–150.

Espley, J. R., Farrell, W. M., Brain, D. A. et al. 2007. Absorption of MARSIS radar signals: Solar energetic particles and the daytime ionosphere. *Geophys. Res. Lett.*, **34**, L9101.

Esposito, L. W., Knollenberg, R. G., Marov, M. Ya., Toon, O. B., and Turco, R. P. 1983. The clouds and hazes of Venus. In: D. M. Hunten, L. Colin, T. M. Donahue, and V. I. Moroz (eds), *Venus*. Tucson: University of Arizona Press, pp. 484–564.

Farnocchia, D., Chesley, S. R., Chodas, P. W. et al. 2014. Trajectory analysis for the nucleus and dust of comet C/2013 A1 (Siding Spring). *Astrophys. J.*, **790**, id. 114.

Fjeldbo, G., Kliore, A., Seidel, B., Sweetnam, D., and Cain, D. 1975. The Pioneer 10 radio occultation measurements of the ionosphere of Jupiter. *Astron. Astrophys.*, **39**, 91–96.

Fox, J. L. 2004. Advances in the aeronomy of Venus and Mars. *Adv. Space Res.*, **33**, 132–139.

Fox, J. L., and Kliore, A. J. 1997. Ionosphere: Solar cycle variations. In: S. W. Bougher, D. M. Hunten, and R. J. Phillips (eds), *Venus II*. Tucson: University of Arizona Press, pp. 161–188.

Frankland, V., James, A. D., Carrillo-Sánchez, J. D. et al. 2017. CO oxidation and O_2 removal on meteoric material in Venus' atmosphere. *Icarus*, **296**, 150–162.

Gehrz, R. D., Reach, W. T., Woodward, C. E., and Kelley, M. S. 2006. Infrared observations of comets with the Spitzer Space Telescope. *Adv. Space Res.*, **38**, 2031–2038.

Grebowsky, J. M., and Aikin, A. C. 2002. In situ measurements of meteoric ions. In: E. Murad and I. P. Williams (eds), *Meteors in the Earth's Atmosphere*. New York: Cambridge University Press, pp. 189–214.

Grebowsky, J. M., Benna, M., Plane, J. M. C. et al. 2017. Unique, non-Earthlike, meteoritic ion behavior in upper atmosphere of Mars. *Geophys. Res. Lett.*, **44**, 3066–3072.

Grün, E., Zook, H. A., Fechtig, H., and Giese, R. H. 1985. Collisional balance of the meteoritic complex. *Icarus*, **62**, 244–272.

Gurnett, D. A., Morgan, D. D., Persoon, A. M. et al. 2015. An ionized layer in the upper atmosphere of Mars caused by dust impacts from comet Siding Spring. *Geophys. Res. Lett.*, **42**, 4745–4751.

Hammel, H. B., Beebe, R. F., Ingersoll, A. P. et al. 1995. HST imaging of atmospheric phenomena created by the impact of Comet Shoemaker-Levy 9. *Science*, **267**, 1288–1296.

Hansell, S. A., Wells, W. K., and Hunten, D. M. 1995. Optical detection of lightning on Venus. *Icarus*, **117**, 345–351.

Hanson, W. B., Sanatani, S., and Zuccaro, D. R. 1977. The Martian ionosphere as observed by the Viking retarding potential analyzers. *J. Geophys. Res.*, **82**, 4351–4363.

Harrington, J., de Pater, I., Brecht, S. H. et al. 2004. *Lessons from Shoemaker-Levy 9 about Jupiter and planetary impacts*. In: F. Bagenal, T. E. Dowling, and W. B. McKinnon (eds), *Jupiter: The Planet, Satellites, and Magnetosphere*. New York: Cambridge University Press, pp. 159–184.

Hermalyn, B., and Schultz, P. 2010. Early-stage ejecta velocity distribution for vertical hypervelocity impacts into sand. *Icarus*, **209**, 866–870.

Hinson, D. P., Flasar, F. M., Kliore, A. J. et al. 1997. Jupiter's ionosphere: Results from the first Galileo radio occultation experiment. *Geophys. Res. Lett.*, **24**, 2107–2110.

Hinson, D. P., Kliore, A. J., Flasar, F. M. et al. 1998a. Galileo radio occultation measurements of Io's ionosphere and plasma wake. *J. Geophys. Res.*, **103**, 29343–29358.

Hinson, D. P., Twicken, J. D., and Karayel, E. T. 1998b. Jupiter's ionosphere: New results from Voyager 2 radio occultation measurements. *J. Geophys. Res.*, **103**, 9505–9520.

Hornung, K., Malama, Y. G., and Kestenboim, K. S. 2000. Impact vaporization and ionization of cosmic dust particles. *Astrophys. Space Sci.*, **274**, 355–363.

Hueso, R., Wesley, A., Go, C. et al. 2010a. First Earth-based Detection of a Superbolide on Jupiter. *Astrophys. J.*, **721**, L129–L133.

Hueso, R., Legarreta, J., Pérez-Hoyos, S. et al. 2010b. The international outer planets watch atmospheres node database of giant-planet images. *Planet. Space Sci.*, **58**, 1152–1159.

Hueso, R., Pérez-Hoyos, S., Sánchez-Lavega, A. et al. 2013. Impact flux on Jupiter: From superbolides to large-scale collisions. *Astron. Astrophys.*, **560**, A55.

Hueso, R., Delcroix, M., Sánchez-Lavega, A. et al. 2018a. Small impacts on the giant planet Jupiter. *Astron. Astrophys.*, **617**, A68.

Hueso, R., Juaristi, J., Legarreta, J. et al. 2018b. The Planetary Virtual Observatory and Laboratory (PVOL) and its integration into the Virtual European Solar and Planetary Access (VESPA). *Planet. Space Sci.*, **150**, 22–35.

Huestis, D. L., and Slanger, T. G. 1993. New perspectives on the Venus nightglow. *J. Geophys. Res.*, **98**, 10839–10847.

Jakosky, B. M., Lin, R. P., Grebowsky, J. M. et al. 2015. The Mars Atmosphere and Volatile Evolution (MAVEN) Mission. *Space Sci. Rev.*, **195**, 3–48.

Janches, D., Pokorny, P., Sarantos, M. et al. 2018. Constraining the ratio of micrometeoroids from short- and long-period comets at 1 AU from LADEE observations of the lunar dust cloud. *Geophys. Res. Lett.*, **45**, 1713–1722.

Jenniskens, P., Tedesco, E., and Murthy, J. 2000. 1997 Leonid shower from space. *Earth Moon Planets*, **82–83**, 305–312.

Killen, R. M., and Hahn, J. M. 2015. Impact vaporization as a possible source of Mercury's calcium exosphere. *Icarus*, **250**, 230–237.

Killen, R. M., Burger, M. H., Vervack, Jr., R. J., and Cassidy, T. A. 2019. Understanding Mercury's exosphere: Models derived from MESSENGER observations. In: S. C. Solomon, L. R. Nittler, and B. J. Anderson (eds), *Mercury: The View after MESSENGER*. Cambridge University Press, pp. 407–429.

Kim, Y. H., Pesnell, W. D., Grebowsky, J. M., and Fox, J. L. 2001. Meteoric ions in the ionosphere of Jupiter. *Icarus*, **150**, 261–278.

Kliore, A., Cain, D. L., Fjeldbo, G., Seidel, B. L., and Rasool, S. I. 1974. Preliminary results on the atmospheres of Io and Jupiter from the Pioneer 10 S-Band occultation experiment. *Science*, **183**, 323–324.

Kliore, A. J., Fjeldbo, G., Seidel, B. L. et al. 1975. The atmosphere of Io from Pioneer 10 radio occultation measurements. *Icarus*, **24**, 407–410.

Kliore, A. J., Hinson, D. P., Flasar, F. M., Nagy, A. F., and Cravens, T. E. 1997. The ionosphere of Europa from Galileo radio occultations. *Science*, **277**, 355–358.

Kliore, A. J., Anabtawi, A., and Nagy, A. F. 2001a. The ionospheres of Europa, Ganymede, and Callisto. *AGU Fall Meeting Abstracts*, P12B–0506.

Kliore, A. J., Anabtawi, A., Nagy, A. F., and Galileo Radio Propagation Science Team. 2001b. The ionospheres of Ganymede and Callisto from Galileo radio occultations. *Bull. Am. Astron. Soc.*, **33**, 1084.

Kliore, A. J., Anabtawi, A., Herrera, R. G. et al. 2002. Ionosphere of Callisto from Galileo radio occultation observations. *J. Geophys. Res.*, **107**(A11), 1407.

Kliore, A. J., Nagy, A. F., Marouf, E. A. et al. 2008. First results from the Cassini radio occultations of the Titan ionosphere. *J. Geophys. Res.*, **113**, A09317.

Kliore, A. J., Nagy, A. F., Marouf, E. A. et al. 2009. Midlatitude and high-latitude electron density profiles in the ionosphere of Saturn obtained by Cassini radio occultation observations. *J. Geophys. Res.*, **114**, A04315.

Kliore, A. J., Nagy, A. F., Cravens, T. E., Richard, M. S., and Rymer, A. M. 2011. Unusual electron density profiles observed by Cassini radio occultations in Titan's ionosphere: Effects of enhanced magnetospheric electron precipitation? *J. Geophys. Res.*, **116**, A11318.

Kondrateva, E. D., and Reznikov, E. A. 1985. . *Solar Sys. Res.*, **19**, 96–101.

Koten, P., Vaubaillon, J., Margonis, A. et al. 2015. Double station observation of Draconid meteor outburst from two moving aircraft. *Planet. Space Sci.*, **118**, 112–119.

Larson, S. L. 2001. Determination of meteor showers on other planets using comet ephemerides. *Astron. J.*, **121**, 1722–1729.

Leblanc, F., and Johnson, R. E. 2003. Mercury's sodium exosphere. *Icarus*, **164**, 261–281.

Leblanc, F., and Johnson, R. E. 2010. Mercury exosphere I. Global circulation model of its sodium component. *Icarus*, **209**, 280–300.

Lellouch, E., Bézard, B., Moses, J. I. et al. 2002. The Origin of Water Vapor and Carbon Dioxide in Jupiter's Stratosphere. *Icarus*, **159**, 112–131.

Levison, H. F., Duncan, M. J., Zahnle, K., Holman, M., and Dones, L. 2000. NOTE: Planetary Impact Rates from Ecliptic Comets. *Icarus*, **143**, 415–420.

Li, J., and Jewitt, D. 2013. Recurrent perihelion activity in 3200 Phaethon. *Astron. J.*, **145**, 154–162.

Lindal, G. F. 1992. The atmosphere of Neptune — An analysis of radio occultation data acquired with Voyager 2. *Astron. J.*, **103**, 967–982.

Lindal, G. F., Sweetnam, D. N., and Eshleman, V. R. 1985. The atmosphere of Saturn — An analysis of the Voyager radio occultation measurements. *Astron. J.*, **90**, 1136–1146.

Lindal, G. F., Lyons, J. R., Sweetnam, D. N., Eshleman, V. R., and Hinson, D. P. 1987. The atmosphere of Uranus — Results of radio occultation measurements with Voyager 2. *J. Geophys. Res.*, **92**, 14987–15001.

Lovell, B. 1949. Radio astronomy. *Popular Astronomy*, **57**, 273–276.

Lyons, J. R. 1995. Metal ions in the atmosphere of Neptune. *Science*, **267**, 648–651.

Maguire, W. C., and Aikin, A. C. 2006. Infrared signature of meteoritic material in the Martian atmosphere from MGS/TES limb observations. *American Astronomical Society DPS Meeting*, **38**, 60.23.

Mahaffy, P. R., Benna, M., King, T. et al. 2015. The Neutral Gas and Ion Mass Spectrometer on the Mars Atmosphere and Volatile Evolution mission. *Space Sci. Rev.*, **195**, 49–73.

McAuliffe, J. P., and Christou, A. A. 2006. Simulating meteor showers in the Martian atmosphere. Pages 155–160 of: Bastiaens, L., Verbert, J., Wislez, J.-M., Verbeeck, C. (eds), *Proceedings of the International Meteor Conference, Oostmalle, Belgium, 15–18 September, 2005*. Hove, BE: International Meteor Organisation.

McAuliffe, J.P. 2006. *Modelling meteor phenomena in the atmospheres of the terrestrial planets*. Ph.D. thesis, Queen's University of Belfast.

McClintock, W. E., Schneider, N. M., Holsclaw, G. M. et al. 2015. The Imaging Ultraviolet Spectrograph (IUVS) for the MAVEN mission. *Space Sci. Rev.*, **195**, 75–124.

McNaught, R. H., and Asher, D. J. 1999. Leonid dust trails and meteor storms. *WGN, J. Int. Meteor Org.*, **27**, 85–102.

Melosh, H. J. 1989. *Impact Cratering: A Geologic Process*. Oxford: Oxford University Press.

Molina-Cuberos, G. J., Lammer, H., Stumptner, W. et al. 2001. Ionospheric layer induced by meteoric ionization in Titan's atmosphere. *Planet. Space Sci.*, **49**, 143–153.

Molina-Cuberos, G. J., Witasse, O., Lebreton, J.-P., Rodrigo, R., and López-Moreno, J. J. 2003. Meteoric ions in the atmosphere of Mars. *Planet. Space Sci.*, **51**, 239–249.

Molina-Cuberos, J. G., López-Moreno, J. J., and Arnold, F. 2008. Meteoric layers in planetary atmospheres. *Space Sci. Rev.*, **137**, 175–191.

Moorehead, A. V., Wiegert, P. A., Cooke, W. J. 2014. The meteoroid fluence at Mars due to Comet C/2013 A1 (Siding Spring). *Icarus*, **231**, 13–21.

Moses, J. I., and Bass, S. F. 2000. The effects of external material on the chemistry and structure of Saturn's ionosphere. *J. Geophys. Res.*, **105**, 7013–7052.

Moses, J. I., and Poppe, A. R. 2017. Dust ablation on the giant planets: Consequences for stratospheric photochemistry. Icarus, **297**, 33–58.

Mousis, O., Hueso, R., Beaulieu, J.-P. et al. 2014. Instrumental methods for professional and amateur collaborations in planetary astronomy. *Exp. Astron.*, **38**, 91–191.

Nagy, A. F., Kliore, A. J., Marouf, E. et al. 2006. First results from the ionospheric radio occultations of Saturn by the Cassini spacecraft. *J. Geophys. Res.*, **111**, A06310.

Neslusan, L. 2005. The potential meteoroid streams crossing the orbits of terrestrial planets. *Contr. Astron. Obs. Skalnate Pleso*, **35**, 163–179.

Oberst, J., Flohrer, J., Elgner, S. et al. 2011. The Smart Panoramic Optical Sensor Head (SPOSH) - A camera for observations of transient luminous events on planetary night sides. *Planetary and Space Sci.*, **59**, 1–9.

Pandya, B. M., and Haider, S. A. 2012. Meteor impact perturbation in the lower ionosphere of Mars: MGS observations. *Planet. Space Sci.*, **63**, 105–109.

Pasek, M. A. 2015. Phosphorus as a lunar volatile. *Icarus*, **255**, 18–23.

Pätzold, M., Tellmann, S., Häusler, B. et al. 2005. A sporadic third layer in the ionosphere of Mars. *Science*, **310**, 837–839.

Pätzold, M., Tellmann, S., Häusler, B. et al. 2009. A sporadic layer in the Venus lower ionosphere of meteoric origin. *Geophys. Res. Lett.*, **36**, L05203.

Pesnell, W. D., and Grebowsky, J. 2000. Meteoric magnesium ions in the Martian atmosphere. *J. Geophys. Res*, **105**, 1695–1707.

Pierazzo, E., and et al. 2008. Validation of numerical codes for impact and explosion cratering: Impacts on strengthless and metal targets. *Met. Planet. Sci.*, **43**, 1917–1938.

Pierazzo, E., Vickery, A. M., and Melosh, H. J. 1995. A re-evaluation of impact melt/vapor production. *26th Lunar Planet. Sci. Conf.*, Abstract 1119.

Plane, J. M. C., Flynn, G. J., Määttänen, A. et al. 2018. Impacts of cosmic dust on planetary atmospheres and surfaces. *Space Sci. Rev.*, **214**, id. 23.

Pokorný, P., Sarantos, M., and Diego, J. 2018. A comprehensive model of the meteoroid environment around Mercury. *Astrophys. J.*, **863**, id. 31.

Reach, W. T., and Sykes, M. V. 2000. The formation of Encke meteoroids and dust trail. *Icarus*, **148**, 80–94.

Restano, Marco, Plaut, Jeffrey J., Campbell, Bruce A. et al. 2015. Effects of the passage of Comet C/2013 A1 (Siding Spring) observed by the Shallow Radar (SHARAD) on Mars Reconnaissance Orbiter. *Geophys. Res. Lett.*, **42**, 4663–4669.

Rivkin, A. S., and Pierazzo, E. 2005. Investigating the impact evolution of hydrated asteroids. *36th Lunar. Planet. Sci. Conf.*, Abstract 2014.

Ryabova, G. O. 2015. Could the Geminid meteoroid stream be the result of long-term thermal fracture? *European Planetary Science Congress.* **10**, EPSC2015-754.

Ryabova, G. O. 2018. Could the Geminid meteoroid stream be the result of long-term thermal fracture? *Mon. Not. R. Astron. Soc.*, **479**, 1017–1020.

Sánchez-Lavega, A., Wesley, A., Orton, G. et al. 2010. The Impact of a Large Object on Jupiter in 2009 July. *Astrophys. J.*, **715**, L155–L159.

Schiaparelli, G. V. 1867. *Note e Riflessioni intorno Alla Teoria Astronomica delle Stelle Cadenti*. Firenze: Stamperia Reale.

Schmidt, C. A., Baumgardner, J., Mendillo, M., and Wilson, J. K. 2012. Escape rates and variability constraints for high-energy sodium sources at Mercury. *J. Geophys. Res.*, **117**, A03301.

Schneider, N. M., Deighan, J. I., Stewart, A. I. F. et al. 2015. MAVEN IUVS observations of the aftermath of the Comet Siding Spring meteor shower on Mars. *Geophys. Res. Lett.*, **42**, 4755–4761.

Schunk, R. W., and Nagy, A. F. 2009. *Ionospheres, second edition*. New York: Cambridge University Press.

Seiff, A. 1983. Appendix A. Models of Venus atmospheric structure. In: D. M. Hunten, L. Colin, T. M. Donahue, and V. I. Moroz (eds), *Venus*. Tucson: University of Arizona Press, pp. 1045–1048.

Selsis, F., Brillet, J., and Rapaport, M. 2004. Meteor showers of cometary origin in the Solar System: Revised predictions. *Astron. Astrophys.*, **416**, 783–789.

Selsis, F., Lemmon, M. T., Vaubaillon, J., and Bell, J. F. 2005. Extraterrestrial meteors: A Martian meteor and its parent comet. *Nature*, **435**, 581.

Stevanović, J., Teanby, N. A., Wookey, J. et al. 2017. Bolide airbursts as a seismic source for the 2018 Mars InSight mission. *Space Sci. Rev.*, **211**, 525–545.

Sykes, M. V., and Walker, R. G. 1992. Cometary dust trails. I. Survey. *Icarus*, **95**, 180–210.

Szalay, J. R., and Horányi, M. 2015. Annual variation and synodic modulation of the sporadic meteoroid flux to the Moon. *Geophys. Res. Lett.*, **42**, 10,580–10,584.

Szalay, J. R., Poppe, A. R., Agarwal, J. et al. 2018. Dust phenomena relating to airless bodies. *Space Sci. Rev.*, **214**, id. 98.

Takahashi, Y., Yoshida, J., Yair, Y., Imamura, T., and Nakamura, M. 2008. Lightning detection by LAC onboard the Japanese Venus Climate Orbiter, Planet-C. *Space Sci. Rev.*, **137**, 317–334.

Takahashi, Y., Sato, M., and Imai, M. 2017. Hunt for optical lightning flash in Venus using LAC onboard Akatsuki spacecraft. *Geophysical Research Abstracts*, **19**, EGU2017-11381.

Terentjeva, A. 1993. Meteor bodies near the orbit of Mars. In. Roggemans, P. (ed), *Proceedings of the International Meteor Conference*, Puimichel, France, 23–26 September 1993. Mechelen, BE: International Meteor Organisation, pp. 97–105

Tiscareno, M. S., Mitchell, C. J., Murray, C. D. et al. 2013. Observations of ejecta clouds produced by impacts onto Saturn's rings. *Science*, **340**, 460–464.

Treiman, A. H., and Treiman, J. S. 2000. Cometary dust streams at Mars: Preliminary predictions from meteor streams at Earth and from periodic comets. *J. Geophys. Res.*, **105**, 24571–24582.

Tricarico, P., Samarasinha, N. H., Sykes, M. V. et al. 2014. Delivery of dust grains from Comet C/2013 A1 (Siding Spring) to Mars. *Astrophys. J.*, **787**, id. L35.

Tyler, G. L., Sweetnam, D. N., Anderson, J. D. et al. 1989. Voyager radio science observations of Neptune and Triton. *Science*, **246**, 1466–1473.

Vaubaillon, J. 2017. A confidence index for forecasting of meteor showers. *Planet. Space Sci.*, **143**, 78–82.

Vaubaillon, J., and Christou, A. A. 2006. Encounters of the dust trails of comet 45P/Honda-Mrkos-Padjusakova with Venus in 2006. *Astron. Astrophys.*, **451**, L5–L8.

Vaubaillon, J., Colas, F., and Jorda, L. 2005. A new method to predict meteor showers. I. Description of the model. *Astron. Astrophys.*, **439**, 751–760.

Vaubaillon, J., Maquet, L., and Soja, R. J. 2014a. Meteor hurricane at Mars on 2014 October 19 from comet C/2013 A1, 2014. *Mon. Not. R. Astron. Soc.*, **439**, 3294–3299.

Vaubaillon, J., and Soja, J., Maquet, L. et al 2014b. Update on recent-past and near-future meteor shower outbursts on Earth and on Mars. Pages 134–135 of: Rault, J.-L., Roggemans, P. (eds), *Proceedings of the International Meteor Conference, Giron, France, 18–21 September, 2014*. Hove, BE: International Meteor Organisation.

Vuitton, V., Yelle, R. V., and Lavvas, P. 2009. Composition and chemistry of Titan's thermosphere and ionosphere. *Philos. T. R. Soc. A*, **367**, 729–741.

Waite, Jr., J. H., and Cravens, T. E. 1987. Current review of the Jupiter, Saturn, and Uranus ionospheres. *Adv. Space Res.*, **7**, 119–134.

Whalley, C. L., and Plane, J. M. C. 2010. Meteoric ion layers in the Martian atmosphere. *Faraday Discussions*, **147**, 349–368.

Whipple, F. L. 1983. 1983 TB and the Geminid Meteors. *IAU Circ.*, **3881**, 1.

Wiegert, P. A., Brown, P. G., Weryk, R. J., and Wong, D. K. 2011. The Daytime Craterids, a radar-detected meteor shower outburst from hyperbolic comet C/2007 W1 (Boattini). *Mon. Not. R. Astron. Soc.*, **414**, 668–676.

Wilson, J. K. S., and Smith, S. M. 1999. Modeling an enhancement of the lunar sodium tail during the Leonid meteor shower of 1998. *Geophys. Res. Lett.*, **26**, 1645–1648.

Witasse, O., and Nagy, A. F. 2006. Outstanding aeronomy problems at Venus. *Planet. Space Sci.*, **54**, 1381–1388.

Withers, P. 2010. Prediction of uncertainties in atmospheric properties measured by radio occultation experiments. *Adv. Space Res.*, **46**, 58–73.

Withers, P. 2011. Attenuation of radio signals by the ionosphere of Mars: Theoretical development and application to MARSIS observations. *Radio Sci.*, **46**, 2004.

Withers, P. 2014. Predictions of the effects of Mars's encounter with comet C/2013 A1 (Siding Spring) upon metal species in its ionosphere. *Geophys. Res. Lett.*, **41**, 6635–6643.

Withers, P., Mendillo, M., Hinson, D. P., and Cahoy, K. 2008. Physical characteristics and occurrence rates of meteoric plasma layers detected in the Martian ionosphere by the Mars Global Surveyor Radio Science Experiment. *J. Geophys. Res.*, **113**, A12314.

Withers, P., Christou, A. A., and Vaubaillon, J. 2013. Meteoric ion layers in the ionospheres of Venus and Mars: Early observations and consideration of the role of meteor showers. *Adv. Space Res.*, **52**, 1207–1216.

Withers, P., Weiner, S., and Ferreri, N. R. 2015. Recovery and validation of Mars ionospheric electron density profiles from Mariner 9. *Earth, Planets, and Space*, **67**, 194.

Yakshinskiy, B. V., and Madey, T. E. 2004. Photon-stimulated desorption of Na from a lunar sample: Temperature-dependent effects. *Icarus*, **168**, 53–59.

Yakshinskiy, B. V., and Madey, T. E. 2005. Temperature-dependent DIET of alkalis from SiO2 films: Comparison with a lunar sample. *Surface Sci.*, **593**, 202–209.

Ye, Q. -Z., Brown, P. G., Bell, C. et al. 2015. Bangs and Neteors from the quiet comet 15P/Finlay. *Astrophys. J.*, **814**, 79.

Yelle, R. V., and Miller, S. 2004. Jupiter's thermosphere and ionosphere. In: F. Bagenal, T. E. Dowling, and W. B. McKinnon (eds), *Jupiter: The Planet, Satellites, and Magnetosphere*. New York: Cambridge University Press, pp. 185–218.

6

Impact Flashes of Meteoroids on the Moon

José M. Madiedo, José L. Ortiz, Masahisa Yanagisawa, Jesús Aceituno and Francisco Aceituno

6.1 Introduction

Most meteoroids, once released from their parent bodies, move in the Solar System in orbits very similar to those of the objects from which they come. In time, these orbits are gradually modified due to the gravitational influence of the planets (mainly Jupiter) and forces derived from different nongravitational effects (Burns et al., 1979; Kapisinsky, 1984). Particles within meteoroid streams move in similar orbits, and were typically released from their progenitors less than several thousand years ago (see Chapter 7, Vaubaillon et al., 2019). In contrast, sporadic meteoroids were released from their parent objects tens of thousands of years ago, or even longer, and so each orbit differs very significantly from that of the body of origin (Jenniskens, 2006).

Both the Earth and the Moon continuously intercept meteoroids. However, the physical processes that take place when these particles collide with these two celestial bodies are very different. In the case of our planet, the atmosphere does not allow most meteoroids to reach the Earth's surface. Meteoroids that come from objects belonging to our Solar System enter the atmosphere at velocities ranging between 11.2 and 72.8 km s^{-1} (Jenniskens, 2006). Under these conditions, particles collide with molecules in the upper atmosphere and the ablation phenomenon takes place (see Chapter 1, Popova et al., 2019). In a few tenths of a second the surface of the meteoroid is heated above 2000°C and the particle loses mass in the form of solid fragments and hot fluid matter. A part of the kinetic energy of the meteoroid is then transformed into light and heat, producing the luminous phenomenon known as a *meteor*. It is precisely the analysis of this interaction with air molecules that has allowed us to gather most of the information that is known about these particles. In that sense, the atmosphere can be considered as a large sensor that provides information about the flux and properties of meteoroids that impact it (see Chapter 4, Koten et al., 2019). In this way, the flux of meteoroids with diameters ranging between 0.1 and 1 m has been estimated by several researchers using meteor-observing stations based on photographic or CCD devices (see, for example, Ceplecha, 1988; Halliday et al., 1996; Trigo-Rodriguez et al., 2008; Madiedo et al., 2014a). For diameters ranging from 1 to 10 m the flux has been obtained by combining observations made by military satellites and infrasound detectors (Brown et al., 2002).

More recently, a technique has been developed to obtain information about the meteoroid environment of the Earth-Moon system based on the analysis of the impact flashes produced when meteoroids hit the lunar surface (see, e.g., Ortiz et al., 1999, 2006; Madiedo and Ortiz, 2018a). These particles are completely destroyed during these collisions, giving rise to brief flashes that can be recorded from Earth by means of telescopes, as was demonstrated by Artem'eva et al. (2000, 2001). In this way the Moon can be employed as a large area detector to characterize the meteoroid environment at 1 AU (Szalay and Horányi, 2015). In particular, the results derived from the observations of lunar impact flashes have been employed to estimate the impact flux of incoming bodies to the Earth and the Moon (see e.g. Ortiz et al., 2006; Suggs et al., 2014; Madiedo et al., 2014b), and the mass distribution of meteoroids impacting the lunar surface (Betzler and Borges, 2015).

In addition to energy radiated as light, part of the electromagnetic emission resulting from meteoroid impacts on the Moon is in the radio range (Volvach et al., 2009; Kesaraju et al., 2016). Seismic waves are another effect of the bombardment of meteoroids on the lunar surface. From the analysis of data recorded by the Apollo lunar seismic network, Oberst and Nakamura (1991) proposed that the source of small (mass < 1 kg) meteoroids hitting the Moon might be primarily cometary, while the parent bodies of larger particles might be near-earth asteroids and short-period comets.

Meteoroid collisions on the Moon's surface have been found to influence the dust distribution in the lunar environment (see e.g. Popel et al., 2016a,b, 2017, 2018a,b; Szalay et al., 2018a). The variability of the dust in the lunar exosphere was studied by means of the Lunar Dust Experiment (LDEX), which employed an impact ionization dust detector onboard the Lunar Atmosphere and Dust Environment Explorer (LADEE) mission (Horányi et al., 2014). This experiment analyzed the contributions of the two mechanisms that give rise to dust around the Moon: the continuous impact of micrometeoroids on the lunar soil and the levitation of small grains from the lunar ground due to near-surface electric fields. Horányi et al. (2015) observed a permanent and asymmetric dust cloud around the Moon. This is produced by ejecta generated by impacts of high-speed cometary meteoroids moving on eccentric heliocentric orbits. The density of this cloud was also found to increase during the activity period of major meteor showers, especially the Geminids. In the framework of LDEX, Szalay et al. (2018b) analyzed the activity on the Moon of the Geminid meteoroid stream in December 2013. The observations, which were performed using the surface density of impact ejecta as a proxy for meteoroid activity, suggested that the dust detector aboard LADEE recorded the ejecta produced by Geminid meteoroids with radii ranging between 2 and 20 mm.

The first systematic attempts to identify impact flashes due to the collision of large meteoroids on the lunar surface using telescopes equipped with CCD cameras date back to 1997 (Ortiz et al., 1999). Since then, this technique has unequivocally detected impact flashes during the period of maximum activity of several major meteor showers (see, for example, Ortiz et al., 2000; Cudnik et al., 2002; Ortiz et al., 2002; Yanagisawa and Kisaichi, 2002; Cooke et al., 2006; Yanagisawa et al., 2006; Suggs et al., 2014; Madiedo et al., 2015a,b), and has also identified flashes of sporadic origin (Ortiz et al., 2006; Suggs et al., 2008; Madiedo et al., 2015b, 2014b). The results obtained from the analysis of these events depend critically on a parameter called luminous efficiency, which currently is not known with enough accuracy. This luminous efficiency is the fraction of the kinetic energy of the meteoroid that is converted into light during the impact. In this chapter we review some of the knowledge derived thus far from the analysis of meteoroid impact flashes on the Moon.

6.2 Description of the Observational Technique

Since the first systematic observations of lunar impact flashes with CCD devices (Ortiz et al., 1999), the detection of these events has been carried out by means of telescopes equipped with high-sensitivity cameras that monitor a part of the night side of the Moon. This method provides a very important advantage over the technique of recording meteoroids interacting with the Earth's atmosphere by means meteor-observing stations: the lunar surface area monitored by a single instrument is much larger than that covered in the atmosphere from a meteor-observing station.

In general, this method can be employed when the illuminated fraction of the lunar disk ranges between approximately 5 and 60%; i.e., during the waxing and waning phases. For fractions lower than 5% the Moon appears so close to the Sun that the effective observing time is too short. When the fraction is larger than 60%, the light coming from the dayside is excessive, and the scattered light masks dimmer flashes and most features on the Moon's surface. The larger the illuminated fraction, the more critical the atmospheric transparency for optimal observing conditions since, for instance, high clouds, dust, and ice particles will increase stray light from the sunlit area.

To record the lightcurve of these lunar flashes, it is preferable to use video cameras. Impact flashes are very brief; most of them are contained in just one or two video frames when cameras operating at 25 frames per second are employed. Identification by eye is not practical, and computer software is required to automatically detect impact flash candidates. An appropriate time-inserting device, which stamps accurate date and time information on every recorded image, is also important in order to know exactly when these flashes occur.

Observations performed with just one camera are not reliable, so any results obtained in this way must be treated with extreme care. It is necessary to employ at least two cameras monitoring the same area on the Moon in order to discard false positives produced by other phenomena, such as electrical noise or cosmic rays impacting the CCD sensor. In general, this is done by employing more than one telescope aimed simultaneously at the same region of the lunar disk. An alternative is to employ

Figure 6.1. The white rectangle shows a typical field of view covered during the monitoring of Moon impact flashes, and the orientation of the CCD sensor in relation to the lunar terminator.

just one telescope equipped with a beam splitter, so the light is split into two different cameras. In this way, if a given event is recorded by just one of the cameras, it can be automatically discarded. However, there are other sources of false positives, such as the reflection of sunlight from artificial satellites and space debris located between the Earth and the Moon. These can produce glints that can be confused with lunar impact flashes. To overcome this problem, it is necessary to employ at least two telescopes at different locations, since, because of parallax, these flares will appear at different positions on the lunar disk. In some cases, when these glints last long enough to move clearly against the lunar disk, they can be easily discarded without employing multi-location observations.

During a typical observing session the telescopes are tracked at lunar rate. To maximize the monitored lunar surface area, each camera is oriented so that the lunar equator is perpendicular to the longest side of the CCD sensor. The terminator should be avoided in order to prevent saturation of the CCD and also to avoid excessive scattered light from the illuminated side of the Moon (Figure 6.1). Under these conditions, lunar features are easily identified in the earthshine, and these can be used to determine the selenographic coordinates (i.e., latitude and longitude on the lunar surface) of impact flashes.

When no major meteor showers are active, the telescopes are pointed at an arbitrary region on the Moon surface in order to cover a common maximum area. However, during the activity period of major showers, which in principle increase the probability of events detection, the telescopes should be aimed to the area where meteoroids producing these showers can impact. The determination of this area and the detectability of events produced by a given meteoroid stream are described in Sections 6.2 and 6.5.1, respectively.

6.2.1 Example 1: The MIDAS Project

As a continuation of the lunar impacts survey started in 1997 by Ortiz et al., 1999, 2000, a renewed project is being conducted to monitor the night side of the Moon by means of high-sensitivity video cameras (Madiedo et al., 2010, 2015a,b; Madiedo and Ortiz, 2018b). This project is called MIDAS (Moon Impacts Detection and Analysis System), and it is being operated from three observatories in Spain: Sevilla, La Hita and La Sagra. Between 2002 and 2015 several monitoring campaigns were also conducted in both the near-infrared and visible bands with the 1.25 m and the 3.5 m telescopes at the Calar Alto Observatory, and also with the 2 m Liverpool telescope located at La Palma (Canary Islands) (Madiedo and Ortiz, 2018c).

At Sevilla this survey operates two 0.36 m Schmidt-Cassegrain telescopes, but also four smaller ones. Three of these instruments have an aperture of 0.28 m, and the fourth one has an aperture of 0.24 m. They employ monochrome 8-bit high-sensitivity CCD video cameras. GPS time inserters are used to stamp date and time information on every video frame. In addition, f/3.3 focal reducers are used in order to increase the lunar surface area monitored. Since December 2014, some of these instruments are equipped with filters to perform observations in different spectral bands. With this configuration, MIDAS became the first system that can determine the temperature of lunar impact flashes (Madiedo et al., 2018). In 2013, an additional Newtonian telescope with an aperture of 0.4 m was installed at La Hita Observatory. This also employs an 8-bit CCD video camera. Two additional 0.36 m Schmidt-Cassegrain telescopes equipped with 12-bit CMOS video cameras started operation at La Sagra in 2015.

To detect impact candidates and perform data reduction of confirmed flashes, the MIDAS software (Moon Impacts Detection and Analysis Software) was developed (Madiedo et al., 2010, 2015b).

6.2.2 Example 2: Observations at UEC

Observations at the University of Electro-Communications (UEC) in Tokyo (Japan), are conducted with two telescopes: a Newtonian telescope with an aperture of 0.45 m and a focal length of 2.025 m, and a smaller Schmidt-Cassegrain telescope with an aperture of 0.28 m and a focal length of 2.8 m. A f/3.3 focal reducer is attached to the smaller telescope, which makes its effective focal length of about 0.92 m. Initially, analog video cameras with output digitized at 8 bits were employed, and an impact flash detection software similar to those in the MIDAS Project (Madiedo et al., 2015b) and the LunarScan package developed by NASA (Suggs et al., 2008) was used until 2016. Since that year, digital CMOS and CCD video cameras have been employed. Thus, the smaller telescope is equipped with a 14-bit gray scale CCD camera, and a 12-bit gray scale CMOS camera is attached to the larger telescope. A diffraction grating with 70 grooves per millimeter is attached to the cover glass of this CMOS sensor to obtain low resolution spectra of impact flashes. The flashes detected by this camera show both the 0th and 1st order spectrum.

6.2.3 Example 3: The NASA Lunar Impact Monitoring System

NASA's Marshall Space Flight Center (MSFC) has been conducting systematic monitoring of lunar impact flashes employing small telescopes since 2006 (Suggs et al., 2008, 2014). For this purpose, two Schmidt-Cassegrain telescopes with an aperture of 0.35 m are operated at MSFC's Automated Lunar and Meteor Observatory. From 2008 to 2010, some observations were also performed with a 0.5 m Ritchey Chretien telescope. To allow parallax discrimination between lunar impact flashes and sunglints from satellites and space debris, in September 2007 this survey set up a third 0.35 m telescope, located in an observatory near Chickamauga, Georgia, at 125.5 km from MSFC.

These instruments employ high-sensitivity analog CCD video cameras (with output digitized at 8 bits) and f/3.3 focal reducers to maximize the field of view. Accurate date and time information is stamped on every video frame by means of GPS time inserters. For some observations in 2009 and 2010 an InGaAs near-infrared video camera was also used on one of the telescopes.

The detection of lunar impact flashes from previously digitized and stored video images are performed by means of the LunarScan software package (Suggs et al., 2008). After detection and confirmation, another computer program, LunaCon, is used to perform photometric analysis and the calculation of the lunar area within the field of view of the telescopes (Swift et al., 2008).

6.2.4 Example 4: The NELIOTA Project

The NEO Lunar Impacts and Optical TrAnsients (NELIOTA) project, which is funded by the European Space Agency (ESA), started operation in February 2017. The NELIOTA campaign was designed to last for twenty-two months. One of the primary aims of this project is the determination of the temperature of lunar impact flashes (Bonanos et al., 2018).

This survey employs a single telescope to conduct the monitoring of the night side of the Moon: the 1.2 m Kryoneri telescope located at the National Observatory of Athens, Greece. This device was recently modified to convert it into a prime-focus instrument with a focal ratio of f/2.8.

The NELIOTA setup consists of two identical fast-frame CMOS cameras observing simultaneously in the photometric R- and I-bands by means of Cousin filters. These devices are installed at the primary focus of the telescope, where a dichroic beam-splitter directs the light onto the two cameras.

NELIOTA provides a web-based database where detected impact flashes are listed, so that they are made available to the scientific community. This database can be found under the following URL: https://neliota.astro.noa.gr/DataAccess.

6.3 Impact Geometry

The determination of the geometry associated with lunar impact flashes is fundamental to the analysis of these events. For instance, this allows us to know which area on the Moon can be impacted by meteoroids belonging to a given meteoroid stream and also provides the impact angle. It must be taken into account that for sporadic meteoroids, which may come from any direction, the impact angle will be unknown. When this angle is necessary for data analysis, the most likely value for this parameter, 45°, is usually employed (see, e.g., Ortiz et al., 2006). Meteoroids belonging to the sporadic background can impact anywhere on the lunar disk. On the contrary, impactors associated with meteoroid streams can only impact at positions where their meteoric tube intercepts the Moon, and their impact angle can be determined from the position of the so-called subradiant: the projection of the stream radiant on the lunar surface. The impact angle ϕ is the angular distance between the position of the subradiant and the position measured for the impact flash (Bellot-Rubio et al., 2000a,b). The impact angle θ with respect to the local horizontal, which is denoted by θ, is given by:

$$\theta = 90° - \phi \tag{6.1}$$

The position of the subradiant (Φ, Λ) can be calculated from the radiant (α, δ) of the meteor shower, the Earth's position in its orbit around the Sun (given by the solar longitude λ) and the Moon's position with respect to our planet. Bellot-Rubio et al. (2000b) developed a method to determine the subradiant by considering, for simplicity, that the average inclination of the lunar equator with respect to the ecliptic is zero and that the rotation axis of the Moon is aligned with the north ecliptic pole. The actual value for this angle is about 1.5°. According to this method, in the first step the radiant equatorial coordinates are transformed into ecliptical coordinates (l, b), which define the direction of the trajectory of meteoroids in the Earth's environment. The ecliptic coordinates of Earth are given by $l_E = \lambda + 180°$ and $b_E = 0°$. The position of the Moon with respect to our planet is given by the phase angle χ, which is zero for the New Moon, 90° for the first quarter, and so on. As can be seen in Figure 6.2, the selenographic longitude Φ of the subradiant is given by $\Phi = l - l_E - \chi$, and the selenographic latitude Λ coincides with the ecliptic latitude of the radiant, b. It must be taken into account that selenographic longitudes are measured counterclockwise from the central meridian as seen from the north lunar pole.

The total area on the night side of the Moon visible from Earth and perpendicular to the direction of impacting meteoroids, $A_\perp$, is a measure of the efficiency of impact flash detection. This parameter depends on the lunar phase and the position of the subradiant. The method proposed by Bellot-Rubio et al. (2000b) obtains this area by employing Monte Carlo techniques.

6.4 Impact Energy and Luminous Efficiency

The observed brightness of an impact flash allows us to estimate the radiated power P, in W m^{-2}, by means of the following equation:

$$P = \kappa \times 10^{-M/2.5} \Delta \lambda \tag{6.2}$$

where κ is the flux density in W m^{-2} μm^{-1} for a magnitude 0 source (Bessel et al., 1998), M is the (time dependent) apparent magnitude of the flash and $\Delta \lambda$, in μm, is the width of the spectral interval within which the emitted energy is to be computed. Equation (6.2) assumes that the spectrum of the flash is constant within the spectral interval considered (i.e., that the flux density does not depend on wavelength). A more appropriate expression, which takes into consideration a non-flat flash spectrum, is:

$$P = \int_{\lambda_0 - \Delta\lambda/2}^{\lambda_0 + \Delta\lambda/2} I(\lambda) d\lambda \tag{6.3}$$

where $I(\lambda)$ is the flux density and λ_0 the effective midpoint of the spectral window. But since no full spectrum has been obtained to date for a lunar impact flash and the functional form of $I(\lambda)$ is unknown, in most cases it is assumed that flux density is constant in the spectral range employed.

With the lightcurve of the flash, which defines the evolution with time of its apparent magnitude M, and by integrating the value of P along the flash duration, the integrated flux of energy observed from Earth, E_d, is obtained. This flux allows to estimate the energy E_r emitted within the spectral interval defined by $\Delta\lambda$ on the lunar surface by means of the relationship

$$E_r = E_d \pi f R^2 \tag{6.4}$$

where R is the Earth-Moon distance at the instant when the flash occurs, and f is a factor which encapsulates the anisotropy degree of the light emission process. Whereas $f = 2$ if this emission takes place isotropically on the lunar surface, $f = 4$ if this process occurs at high altitude on the lunar soil (Bellot-Rubio et al., 2000a,b). The impactor kinetic energy E can then be obtained as:

$$E = E_r/\eta, \tag{6.5}$$

In this equation, the parameter η is the luminous efficiency, that is, the fraction of the kinetic energy of the impactor that is converted into radiated energy in the collision in the spectral range

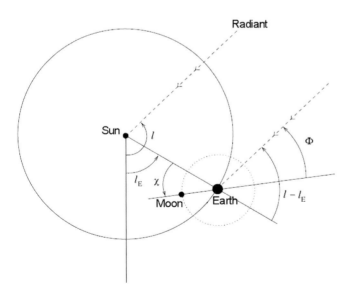

Figure 6.2. Geometry for the calculation of the selenographic coordinates of the radiant. For clarity, the Earth-Moon distance has been exaggerated. The radiant direction is indicated by means of dashed lines.

corresponding to $\Delta\lambda$. The efficiency defined by this relationship should not be confused with the luminous efficiency employed in meteor studies (see, e.g., Ceplecha et al., 1998), since the concepts refer to different physical phenomena.

The term 'luminous' refers to the so-called 'luminous range', which was arbitrarily defined to lie between 400 and 900 nm in early works on the luminous efficiency (see, e.g., Bellot-Rubio et al., 2000a,b; Ortiz et al., 2000; Yanagisawa et al., 2006). This corresponds to the range of sensitivity of typical unfiltered CCD detectors employed in these works. Note, however, that this standard is valid independent of the particular detector employed. The luminous efficiencies derived from the above-mentioned works corresponded to this 500 nm wide luminous range. Suggs et al. (2014) follow a different approach to determine the energy of impact flashes from the peak R magnitude of these events, and by considering a width for the filter passband of 160.07 nm. These authors estimate this R magnitude from unfiltered observations, performing color corrections.

For filtered observations performed in other spectral ranges, such as the R-band, I-band, V-band or other filters, the fraction of the kinetic energy converted into radiation in those wavelength ranges is called radiation efficiency, and is usually denoted as η_R, η_I, η_V, etc., to clearly distinguish it from η (Madiedo et al., 2018; Madiedo and Ortiz, 2018d). For instance, for the Johnson-Cousins U, V, R, and I ranges the radiation efficiencies (denoted by η_U, η_V, η_R and η_I, respectively) would correspond to the following spectral windows by taking into account the respective effective wavelength midpoint and full width at half maximum (FWHM): η_U to the $(365 - 33, 365 + 33)$ nm range; η_V to $(551 - 44, 551 + 44)$ nm; η_R to $(658 - 69, 658 + 69)$ nm; and η_I would correspond to $(806 - 74.5, 806 + 74.5)$ nm. It is very important to emphasize that the radiation efficiency depends on the spectral range, i.e, the energy radiated by the flash at a given frequency is a function of λ. So, to determine the kinetic energy of the impactor from the integrated observed energy flux E_d (Equations (6.4) and (6.5)) it is of paramount importance to employ a value for η corresponding to the same spectral range as that defined by $\Delta\lambda$. Otherwise one would come to the nonsensical conclusion that the kinetic energy of the meteoroid does not only depend on the impactor, but also on the properties of the camera and filter employed to record and measure the brightness of the impact flash.

Currently the luminous efficiency is not known very accurately, and several estimates have been obtained and employed by different works (Bellot-Rubio et al., 2000a; Ortiz et al., 2002; Swift et al., 2011). The best estimate obtained by Ortiz et al. and Bellot-Rubio et al. from the analysis of Leonid (see e.g. Bellot-Rubio et al., 2000a; Ortiz et al., 2002) and Geminid lunar impact flashes (Ortiz et al., 2015) is $\eta = 2 \times 10^{-3}$, close to the $\eta = 1.5 \times 10^{-3}$ value used by other investigators (Swift et al., 2011; Bouley et al., 2012). Madiedo et al. (2018) estimated the value of the radiation efficiency in the I-band, and found a value of 4.7×10^{-3} for η_I, which is higher, by around 56%, than the luminous efficiency η.

According to Swift et al. (2011), the luminous efficiency η depends exponentially on the impactor velocity V:

$$\eta = 1.5 \times 10^{-3} \exp(-(9.3/V)^2) \qquad (6.6)$$

In this equation, the impact velocity must be provided in km s^{-1}. This expression was derived from laboratory hypervelocity impacts, and from the value of η estimated by Moser et al. (2011) from observed lunar impact flashes that were associated with the Geminid, the April Lyrid, and the Taurid meteoroid streams. Equation (6.6) predicts a weak dependence of η with V for impact velocities larger than about 27 km s^{-1}. This relationship was derived by combining results obtained from laboratory experiments and lunar impact flash monitoring techniques, where the velocities and impactor masses regimes were very different. In this sense, some studies show that η would not depend on impactor velocity for lunar impact flashes associated to different meteoroid streams (see, e.g., Ortiz et al., 2015).

The light emitted by a Moon impact flash contains light generated by several processes. Part of the light is produced during the collision itself, while the rest is generated during the subsequent cooling process. In early works about the luminous efficiency it was considered that it is difficult to establish when light emission due to the collision between the impactor and the target concludes and when the cooling phase and its associated emitted light begins. For instance, during the camera exposure time the radiation corresponding to the first process could be spread over more than one video frame and contaminated by the secondary emission from cooling. In addition, a part of the light emitted during the collision could be absorbed by the impact debris and re-emitted later on during the cooling phase. For this reason, the first definition given for η considered that all the light is emitted at the expense of the kinetic energy of the impactor and therefore that the whole light curve of the flash should be taken into account to derive the energy E_r radiated by the event (Bellot-Rubio et al., 2000a,b; Ortiz et al., 2000). Other authors do not follow this convention and only take into account the peak magnitude of the flash to derive E_r (see e.g. Suggs et al., 2014). These authors consider that the peak energy of the flash would be released during the impact process itself, and that the rest of the lightcurve contains radiation that is emitted during cooling and therefore should not be taken into account. The two criteria are usually employed, but lead to different values for E_r. Since the kinetic energy of the impactor should be the same in both cases, according to Equation (6.5) these criteria lead to different values for the luminous efficiency even if the same spectral interval is employed. Because of this, and to avoid any confusion, it has been proposed to distinguish between 'integral' and 'instantaneous' luminous efficiencies. The first corresponds to the definition of η given in the above-mentioned early works, while the instantaneous efficiency corresponds to the second criterion.

In the first works published by Ortiz et al. and Bellot-Rubio et al. (see e.g. Ortiz et al., 1999; Bellot-Rubio et al., 2000a,b; Ortiz et al., 2000), impact flashes and comparison stars were observed without any filters. Observations were done this way in order to maximize the sensitivity of the experimental setup. Since there is no suitable standard catalog for the entire luminous wavelength range that goes from 400 to 900 nm, these authors were forced to use star catalogs in narrower wavelength ranges. For instance, the Johnson-Cousins V band or the Johnson-Cousins R band could be selected, since these are the standard photometric bands closer to the peak sensitivity of the CCD devices employed and for which there are good stellar catalogs. These authors chose the V band and used the derived flash magnitudes to obtain the flux density at the central wavelength of the photometric system. That flux density was employed to estimate the radiated energy in their luminous wavelength range. The brightness B measured for impact flashes and stars is given by:

$$B = k \int_0^\infty F(\lambda) f_{sensor}(\lambda) d\lambda \quad (6.7)$$

where $F(\lambda)$ is the spectral energy flux from the impact flashes or stars, $f_{sensor}(\lambda)$ is the spectral sensitivity of the camera sensor, and k is a constant. The spectral sensitivity can be expressed as

$$f_{sensor}(\lambda) = R(\lambda) f_{const} \quad (6.8)$$

where f_{const} is the sensitivity of the sensor that is constant over all wavelengths and has a peak value of $f_{sensor}(\lambda)$. $R(\lambda)$ is the response function of the observation system (the V band in this case). Ortiz et al. approximated $R(\lambda)$ as follows:

$$R(\lambda) = 1 \quad \text{for} \quad 400 \text{ nm} < \lambda < 900 \text{ nm}$$
$$R(\lambda) = 0 \quad \text{for other wavelengths} \quad (6.9)$$

because the CCD video cameras employed were most sensitive in this luminous wavelength range. It should be noted however that

$$k \int_0^\infty F(\lambda) R(\lambda) f_{const} d\lambda = k \int_{400\,nm}^{900\,nm} F(\lambda) f_{const} d\lambda \quad (6.10)$$

is an approximation. Nevertheless, $f_{sensor}(\lambda)$ for the cameras used by different researchers has a similar wavelength dependence. So it could be claimed that almost all the observations and the experiments have been conducted so far with approximately a "typical CCD video camera spectral band." For instance, the luminous efficiency obtained by Bellot-Rubio et al. (2000b) is $\eta = 2.0 \times 10^{-3}$, and the value for η estimated by Moser et al. (2011) and Swift et al. (2011) by means of observations performed without any filter yielded 1.5×10^{-3} at high impact velocities. The difference between them is smaller than the uncertainties in these values and is not significant.

Photometric calibrations, which are employed to derive the flux density at a particular wavelength, can be improved if color corrections are employed. This, however, is not done by some authors, who claim that the shape of the spectra of impact flashes is currently unknown, and so the appropriate way such a correction should be performed is not clear (see, e.g., Bellot-Rubio et al., 2000a,b; Ortiz et al., 2006; Yanagisawa et al., 2006; Madiedo et al., 2014b). Other authors (see, e.g., Suggs et al., 2014) employ color terms by assuming that impact flashes behave as a blackbody with a temperature of 2,800 K on the basis of a simulation performed by Nemtchinov et al. (1998).

6.4.1 Calculation of the Luminous Efficiency

Laboratory experiments have been employed to estimate the luminous efficiency for projectiles impacting on different target materials. These experiments showed that this efficiency depends on the impactor mass and speed (see, e.g., Oberst et al., 2012). However, laboratory hypervelocity tests are performed at velocities below 10 km s^{-1}, with typical impactor masses of the order of 10^{-16} kg. These correspond to a different regime from projectiles producing lunar impact flashes detectable from Earth and are performed in small chambers; the extrapolation to the analysis of lunar impact flashes is questionable.

The standard procedure to estimate the luminous efficiency corresponding to lunar impact flashes was described in Bellot-Rubio et al. (2000a) and Bellot-Rubio et al. (2000b). According to this approach, the number N of expected impact flashes above an energy E_d is then given by

$$N(E_d) = \int_{t_o}^{t_o+\Delta t} F(m_o, t) \left(\frac{2f\pi R^2}{\eta m_o V^2} E_d\right)^{1-s} A dt, \quad (6.11)$$

where t_o and $t_o + \Delta t$ define the observing time interval, V is the impact velocity, E_d is the time-integrated optical energy flux of the flash observed from Earth, m_0 is the mass of a shower meteoroid producing on Earth a meteor of magnitude +6.5 (Jenniskens, 1994), $F(m_o,t)$ is the flux of meteoroids on the Moon with mass higher than m$_o$ at time t, A is the projected area of the observed lunar surface (i.e., visible from Earth) perpendicular to the meteoroid stream considered, and s is the mass index of the shower. The mass index is related to the population index r (the ratio of the number of meteors with magnitude $M + 1$ or less to the number of meteors with magnitude M or less) by means of the equation

$$s = 1 + 2.5 \log(r) \quad (6.12)$$

If the flux of meteoroids can be regarded as constant during the observation, Equation (6.11) can be simplified, yielding

$$N(E_d) = F(m_o) \Delta t \left(\frac{2f\pi R^2}{\eta m_o V^2} E_d\right)^{1-s} A \quad (6.13)$$

In this relationship, $F(m_o)$ is the above-mentioned constant value of the meteoroid flux on the Moon. To derive η, the value of $N(E_d)$ given by Equations (6.11) or (6.13) must be compared with the observations.

This technique assumes that the mass index s (and so the population index r) for meteoroids producing meteors on Earth and meteoroids giving rise to detectable flashes on the Moon is the same. It also relies on the assumption that the meteoroid flux on the Moon at the time of observation is known. This flux differs from the flux on Earth because of the different gravitational focusing effect for both bodies. The gravitational focusing factor Φ for particles moving at velocity V in the vicinity of a given central body with escape velocity V_{esc} is defined by

$$\Phi = 1 + \frac{V_{esc}^2}{V^2} \quad (6.14)$$

The different gravitational focusing effect for the Moon and the Earth is given by the quotient γ between the gravitational focusing factors for the two bodies Φ_{Earth} and Φ_{Moon}, respectively:

$$\gamma = \Phi_{Moon}/\Phi_{Earth} \quad (6.15)$$

The meteoroid flux on Earth F' is then γ^{-1} times larger than the flux on the Moon F. The mass index and the shower flux F' on

Earth (and, so, the flux F on the Moon) can be obtained from the analysis of meteors ablating in the atmosphere.

6.5 Meteoroid Impact Velocity and Mass

Once its kinetic energy is known, the impactor mass m can be obtained by:

$$m = 2EV^{-2}, \qquad (6.16)$$

However, one of the drawbacks of the observational technique described in this chapter is that it does not allow the impact velocity to be obtained, since this method only provides the position vector of the impact flash, but not the velocity vector of the impactor. This problem can be solved if the meteoroid orbit is known. Since particles belonging to a given meteoroid stream can be regarded as mono-kinetic with a geocentric velocity V_g (see e.g. Jenniskens, 2006), the impact velocity of these meteoroids can be obtained by means of the following procedure. First, from the known position of the stream radiant and the modulus of the geocentric velocity V_g, the geocentric velocity vector $\vec{V}_g$ is obtained. Next, the heliocentric velocity vector $\vec{V}_h$ of the meteoroid is calculated from the following relationship:

$$\vec{V}_h = \vec{V}_E + \vec{V}_g. \qquad (6.17)$$

where the heliocentric velocity vector of Earth $\vec{V}_E$ can be obtained, for instance, by querying the JPL Horizons on-line ephemeris generator (http://ssd.jpl.nasa.gov/?horizons). The heliocentric velocity of the Moon $\vec{V}_M$ can be also determined from this ephemeris. Then, the selenocentric velocity vector $\vec{V}_S$ of the meteoroid is given by

$$\vec{V}_S = \vec{V}_h - \vec{V}_M, \qquad (6.18)$$

and from the modulus of $\vec{V}_S$ the modulus V of the impact velocity can be inferred by taking into account the escape velocity of the Moon V_{EM} (2.4 km s^{-1}):

$$V^2 = V_S^2 + V_{EM}^2. \qquad (6.19)$$

For flashes produced by sporadic meteoroids, since V_g will be unknown, an average value can be employed. However, the value for the average impact velocity on the Moon of sporadics is currently an open question. Some authors, on the basis of the statistics of a large meteoroid orbit database performed by Steel (1996), employ $V = 17$ km s^{-1} (see, e.g., Ortiz et al., 1999; Brown et al., 2002; Ortiz et al., 2006). The justification for this selection is that the mass and size range of meteoroids producing lunar impact flashes detectable from our planet corresponds to that of meteoroids producing fireballs in the Earth's atmosphere. But other authors, on the basis of the analysis performed by McNamara et al. (2004), consider a lunar impact velocity $V = 24$ km s^{-1} (see, e.g., Suggs et al., 2014). However, it is important to note that the study of McNamara et al. (2004) employed radar techniques, which are biased towards much smaller meteoroids in the statistics for the calculation of this mean velocity. In this sense, the analysis performed by Stuart and Binzel (2004), which is based on the LINEAR survey of the Near-Earth Object population, yields an average velocity of objects in the Earth's environment of 20.4 km s^{-1}. This is equivalent to an impact velocity on the Moon $V = 17.8$ km s^{-1}.

6.6 Associating Lunar Impact Flashes with Meteoroid Streams: The Probability Parameter

Meteor-observing stations monitoring meteoroids ablating in the Earth's atmosphere provide information that allows us to establish if these particles belong to a given meteoroid stream or to the sporadic background. It is well-known that the simultaneous observation of a given meteor from at least two different locations allows us to obtain the velocity vector of the meteoroid, and also the radiant and the heliocentric orbit of the particle (Ceplecha, 1987). Once these orbital parameters are known, it is possible to employ the so-called dissimilarity criteria (see, e.g., Williams, 2011; Jenniskens, 2008; Madiedo et al., 2013) to link the progenitor meteoroid to a specific meteoroid stream (see Chapter 9, Williams et al., 2019, for a review of these criteria).

However, associating a lunar impact flash with a given meteoroid stream is not trivial. The monitoring of these flashes provides the position vector of the impact location, but not the velocity vector of the projectile. Since this velocity vector is unknown, the orbital parameters of the meteoroid cannot be calculated. This leads to an important conclusion: for lunar impact flashes recorded with this technique, the association of these events with specific meteoroid streams cannot be done unambiguously. We can only talk about the probability that the progenitor meteoroid is associated with a given meteoroid stream.

To analyze lunar impact flashes it is very important to establish the source of the meteoroids that hit the Moon. With this information the impact velocity can be estimated and the meteoroid mass can be also calculated. But this is also important to obtain the luminous efficiency associated with meteoroids belonging to a given stream and, by comparing the results obtained for different swarms, to study the possible dependence of the luminous efficiency with the impact velocity.

Since we only can talk about the probability that an impact flash is related to a given meteoroid stream, we need to define a parameter to quantify this probability and the quality of these associations. However, in most studies carried out by different researchers, the association of recorded impact flashes with a given meteoroid swarm has been performed by assuming that those events produced outside the activity period of major meteor showers were produced by sporadic meteoroids. On the contrary, flashes identified around the activity peak of these showers were automatically assigned to the corresponding meteoroid stream, provided that the impact geometry was compatible with the stream producing the shower (Ortiz et al., 2000; Cudnik et al., 2002; Ortiz et al., 2002; Yanagisawa et al., 2006; Yanagisawa and Kisaichi, 2002; Cooke et al., 2006; Ortiz et al., 2006; Suggs et al., 2008; Moser et al., 2011). The problem with this approach is that it does not provide any measure of the reliability of these associations.

In the framework of the lunar impact flash monitoring project developed by NASA, Suggs et al. (2014) defined the so-called figure of merit (FOM) in order to quantify the possibility of associating impact flashes with a specific stream. The assumption is made that all meteor showers have a similar mass index. This factor is given by the relationship

$$FOM = ZHR \times FOM_{time} \cos(\phi) \qquad (6.20)$$

where ZHR is the zenithal hourly rate of the meteor shower. The parameter ϕ is the angular distance between the subradiant and the position where the flash occurs. This angular distance coincides with the impact angle as measured with respect to the local vertical. The parameter FOM_{time} is defined by:

$$FOM_{time} = 1 - (\lambda_{max} - \lambda_{flash})/(\lambda_{max} - \lambda_o) \qquad (6.21)$$

if the flash is produced before the activity peak on Earth of the meteor shower produced by this stream, and by

$$FOM_{time} = 1 - (\lambda_{flash} - \lambda_{max})/(\lambda_f - \lambda_{max}) \qquad (6.22)$$

if the flash is produced after that peak. In these relationships the parameters λ_{flash}, λ_{max}, λ_o and λ_f are, respectively, the solar longitude corresponding to the instant when the flash is produced, to the peak of the meteor shower, to the beginning of the activity of the shower and to the end of that activity period. According to this approach, an impact flash can be associated with the meteoroid stream that gives the highest value for the figure of merit. The FOM factor, however, does not provide the probability of association since the figure of merit is not bounded from above (as a probability would always be). A key parameter in Equation (6.20) is the zenithal hourly rate, which by definition is obtained by taking into account all meteoroids producing meteors of magnitude +6.5 or brighter (Jenniskens, 2006). However, not all meteoroids that can produce a meteor in the Earth's atmosphere can also give rise to an impact flash on the lunar surface detectable from Earth. The experimental results show that masses derived for meteoroids producing lunar impact flashes are several order of magnitudes larger than those producing magnitude +6.5 meteors on Earth. For a given meteoroid stream with geocentric velocity V_g, the mass m_o of meteoroids producing magnitude +6.5 meteors on Earth can be estimated, for instance, from the following equations (Hughes, 1987):

$$M = 40 - 2.5 \log(\Gamma) \qquad (6.23)$$

$$\Gamma = 7.7 \times 10^{-10} m_0^{0.92} V_g^{3.91} \qquad (6.24)$$

where Γ is the electron density produced in the atmosphere as a consequence of the ablation process and M, the meteor magnitude, must be set to +6.5. For instance, this mass yields 5.0×10^{-8} kg for Perseid meteoroids moving at $V_g = 59$ km s^{-1}, 2.4×10^{-8} kg for Leonids moving at $V_g = 70$ km s^{-1}, and 5.0×10^{-6} kg for sporadic meteoroids with an average velocity in the vicinity of Earth of 20 km s^{-1} (Brown et al., 2002). However, the masses corresponding to impact flashes recorded on the Moon are several orders of magnitudes larger than these figures (see, e.g., Ortiz et al., 2006; Yanagisawa et al., 2006; Swift et al., 2011). This means that the minimum kinetic energy E_{min} to produce a detectable impact flash on the Moon is much higher than the kinetic energy of a fraction of the meteoroids included in the computation of hourly rates on Earth. To address this issue, Suggs et al. (2014) employ in their calculations ZHR values determined for a list of showers appearing in Cook (1973), where only meteors produced by larger meteoroids were taken into account. ZHR values not included in Cook's catalog were taken from Jenniskens (1994).

Madiedo et al. (2015a) developed a technique that makes it possible, for the first time, to obtain a probability parameter to determine the most likely source of impact flashes by correlating the results from the lunar impact monitoring with the data obtained from the monitoring of meteor activity on Earth. According to this approach, the association probability parameter p is defined as

$$p^{ST} = \frac{N^{ST}}{N^{ST} + N^{OTHER}}, \qquad (6.25)$$

where p^{ST} is the probability (between 0 and 1) that the flash is associated with a given meteoroid stream, N^{ST} the number of impacts per unit time that can be produced by that stream on the Moon, and N^{OTHER} the number of impacts per unit time that can be produced by the rest of the available sources (other streams and the sporadic background) at the same area on the lunar surface and at the same time. The flash can then be associated with the source that provides the highest value for the probability parameter p. Of course $N^{ST} = 0$ for those flashes located out of the area on the Moon that can be impacted by meteoroids from the stream. The calculation of the probability parameter relies on the same assumption as the method of Bellot-Rubio et al., 2000a, described in Section 6.4.1, to estimate the luminous efficiency: that the mass index s (and so the population index r) for meteoroids producing meteors on Earth and meteoroids giving rise to detectable flashes on the Moon is the same. If only one meteoroid stream can significantly contribute to the impact flux, N^{OTHER} can be substituted by N^{SPO}, the number of impacts per unit time produced by the sporadic background:

$$p^{ST} = \frac{N^{ST}}{N^{ST} + N^{SPO}}. \qquad (6.26)$$

If for simplicity it is assumed that only one meteoroid stream is producing activity, the probability that the flash is associated with a sporadic meteoroid is

$$p^{SPO} = 1 - p^{ST} = \frac{N^{SPO}}{N^{ST} + N^{SPO}}. \qquad (6.27)$$

The probability parameter p^{ST} is then given by:

$$p^{ST} = \frac{\nu^{ST} \gamma^{ST} \cos(\phi) \Xi ZHR^{ST}_{Earth}(max) 10^{-b|\lambda - \lambda_{max}|}}{\nu^{SPO} \gamma^{SPO} HR^{SPO}_{Earth} + \nu^{ST} \gamma^{ST} \cos(\phi) \Xi ZHR^{ST}_{Earth}(max) 10^{-b|\lambda - \lambda_{max}|}} \qquad (6.28)$$

where ϕ is the impact angle with respect to the local vertical, HR^{SPO}_{Earth} is the average hourly rate of sporadic events on Earth, λ is the solar longitude corresponding to the instant of detection of the impact flash, and $ZHR^{ST}_{Earth}(max)$ is the shower peak ZHR on Earth, which corresponds to the time defined by the solar longitude λ_{max}. For meteor showers exhibiting non-symmetrical ascending and descending activity profiles or several maxima, Equation (6.28) should be modified according to the expressions given by Jenniskens (1994). For the average hourly rate of sporadic events the value $HR^{SPO}_{Earth} = 10$ meteors h^{-1} can be assumed (Dubietis and Arlt, 2010). The typical values of both parameters $ZHR_{Earth}^{ST}(max)$ and λ_{max} for different meteor showers can be found, for instance, in Jenniskens, (2006). Nevertheless, it should be taken into account that meteor showers tend to exhibit variations from one year to the next, and so it is convenient to obtain these quantities for the period corresponding to the detection of the impact flashes being analyzed.

The factor Ξ in Equation (6.28) takes into account that the distances from the Earth and the Moon to the meteoric filament will in general be different, and this would give rise to a different density of stream meteoroids for both bodies. Under the assumption that this filament can be approximated by a tube where the meteoroid density decreases linearly from its central axis, the following definition can be adopted for Ξ,

$$\Xi = \frac{d_{Earth}}{d_{Moon}}, \qquad (6.29)$$

where d_{Earth} and d_{Moon} are the distance from the center of the meteoric tube to the Earth and the Moon, respectively. The parameter γ^{SPO} is the quotient between the gravitational focusing factors of the Earth and the Moon for particles belonging to the sporadic background (Equations (6.14) and (6.15)). For sporadic meteoroids, with an average velocity of 20 km s^{-1} in the Earth-Moon environment (Brown et al., 2002), this velocity-dependent focusing effect given by Equation (6.14) is higher for the Earth by a factor of 1.3 (Ortiz et al., 2006), and so Equation (6.15) yields $\gamma^{SPO} = 0.77$. The value of γ for a given meteoroid stream depends on the velocity of these particles and is denoted by γ^{ST}.

The two factors ν^{ST} and ν^{SPO}, which correspond to the meteoroid stream and the sporadic background respectively, take into account that only those particles with a kinetic energy above a threshold value denoted by E_{min} are capable of producing lunar impact flashes detectable from Earth. These parameters are calculated in the following way:

$$\nu = \left(\frac{m_o V^2}{2}\right)^{s-1} E_{min}^{1-s}, \qquad (6.30)$$

where V is the impact velocity, m_o is the mass of a shower meteoroid producing on Earth a meteor of magnitude +6.5, and s is the mass index of the shower.

If at the time of detection of the impact flash n additional meteoroid streams with significant contribution to the impact rate (and, of course, also with compatible impact geometry) must be considered, the denominator in Equation (6.28) must be modified as follows:

$$p^{ST} = \frac{\nu^{ST} \gamma^{ST} \cos(\phi) \Xi ZHR_{Earth}^{ST}(max) 10^{-b|\lambda-\lambda_{max}|}}{\kappa + \nu^{SPO} \gamma^{SPO} HR^{SPO} Earth} \\ + \nu^{ST} \gamma^{ST} \cos(\phi) \Xi ZHR_{Earth}^{ST}(max) 10^{-b|\lambda-\lambda_{max}|}$$

(6.31)

where the factor κ given by the expression

$$\kappa = \sum_{i=1}^{n} \nu_i^{ST} \gamma_i^{ST} \cos(\phi_i) \Xi ZHR_{i,Earth}^{ST}(max) 10^{-b_i|\lambda_i-\lambda_{i,max}|}$$

(6.32)

accounts for the contribution of these additional n streams.

6.6.1 Calculation of the Kinetic Energy Threshold

By definition, the threshold kinetic energy E_{min} appearing in Equations (6.28) and (6.31) corresponds to the minimum radiated energy E_{r_min} on the Moon detectable from observations on Earth. Thus, according to Equation (6.5), we have:

$$E_{r\_min} = \eta E_{min}. \qquad (6.33)$$

This minimum radiated energy on the Moon, in turn, corresponds to the maximum apparent magnitude for detectable impact flashes m_{max}. These parameters depend, among other factors, on the experimental setup employed. Thus, the larger the telescope aperture and the higher the camera sensitivity, the higher will be the maximum magnitude and the lower the threshold kinetic energy.

To obtain E_{min}, the value for m_{max} must be estimated. This can be obtained from experimentation. For instance, for the system that is currently operated from Sevilla in the framework of the MIDAS Project, m_{max} is about +10 in the visual range. The radiated power P corresponding to m_{max} can then be calculated from Equation (6.2). If a video camera is employed to record the impact flashes, when this power is multiplied by the minimum integrating time of the camera, the minimum integrated flux of energy observed from Earth, E_{dmin}, is obtained. Then, the value of E_{r_min} can be calculated from Equation (6.4), which in turn, according to Equation (6.33) provides the value of E_{min}. Note that, according to Equation (6.4), the minimum radiated energy for flash detectability (and so the minimum kinetic energy of the impactor) depends on the Earth-Moon distance R. Thus, as expected, these energies are higher for larger R. So, the kinetic energy threshold is time-dependent.

6.6.2 Behavior of the Probability Parameter

For simplicity, to analyze how the shower association probability depends on the different variables included in its definition, it will be considered that just one meteoroid stream and the sporadic background can significantly contribute to the impact flux on the Moon. In this case, the probability parameter is given by Equation (6.28) and its behavior can be shown in Figures 6.3 and 6.4. These plots have been obtained assuming $\phi = 45°$ (since this is the most likely value for the impact angle), $\Xi = 1$ and a population index for sporadics $r = 3.0$. A maximum magnitude of +10 for flashes detectable from Earth has been assumed. This is consistent with the typical minimum brightness of impact flashes recorded by surveys employing small telescopes, as for instance those operated in the framework of the MIDAS Project.

As can be seen in Figure 6.3, when the values of the population index r of the meteoroid stream and the luminous efficiency η are fixed, the probability p that an impact flash is produced by a particle for that stream decreases as the geocentric velocity of the meteoroids increases. This effect is more important for streams with low zenithal hourly rates. The decrease of p with increasing V_g is due to the fact that, as implied by Equations (6.23) and (6.24), the higher the geocentric velocity of the meteoroid, the lower m_o will be (the mass for meteoroids giving rise to mag. +6.5 meteors on Earth). In spite of the fact that the increase in V_g tends to increase the kinetic energy of the projectile, the effect

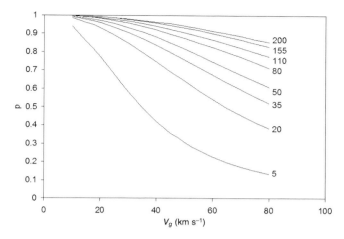

Figure 6.3. Variation of the probability parameter p, defined by Equation (6.28), with the geocentric velocity V_g (in km s^{-1}) for different values of the zenithal hourly rate ZHR (in meteors h^{-1}). The plots have been obtained by considering $r = 2.5$, $\phi = 45°$, $\Xi = 1$ and $\eta = 2 \times 10^{-3}$.

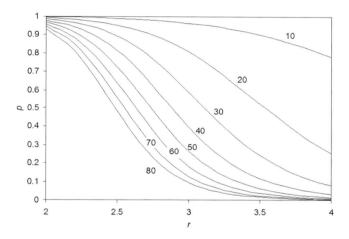

Figure 6.4. Variation with the population index r, for different values of the geocentric velocity V_g (in km s^{-1}), of the probability parameter p defined by Equation (6.28). The plots have been obtained by considering $ZHR = 30$ meteors h^{-1}, $\phi = 45°$, $\Xi = 1$ and $\eta = 2 \times 10^{-3}$.

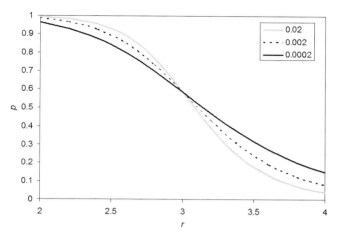

Figure 6.5. Variation of the probability parameter p given by Equation (6.28) with the population index r for different values of the luminous efficiency η. The plots have been obtained by considering $ZHR = 30$ meteors h^{-1}, $V_g = 30$ km s^{-1}, $\phi = 45°$ and $\Xi = 1$.

particles in the stream. Consequently, there is a decrease in the number of meteoroids with a kinetic energy large enough (i.e., exceeding the threshold value E_{min}) to produce lunar impact flashes detectable from Earth. Figure 6.5 shows the effect of the luminous efficiency η on the probability parameter p for fixed values of the geocentric velocity and the zenithal hourly rate of a shower. By definition, larger values of η correspond to a higher fraction of the impactor kinetic energy that is converted into light. So when the luminous efficiency increases, the minimum kinetic energy required to produce an impact flash detectable from Earth decreases. As Figure 6.5 shows, this implies that for streams with a population index lower than the population index considered for sporadics ($r = 3.0$ in this case), p is higher. This is because in this case the fraction of meteoroids with larger mass in the stream is larger and, so, the number of meteoroids with kinetic energy larger than the minimum threshold value E_{min} is also larger. On the contrary, for streams with $r > 3.0$ we have the opposite situation, since in this case the fraction of meteoroids with lower mass is higher.

6.6.3 Constraints for the Association between Impact Flashes and Meteoroid Streams

From the previous discussion, it is apparent that the value of the association probability depends critically on the value of several parameters of the meteor shower produced on Earth by the corresponding meteoroid stream. For instance, it is necessary to know the peak date (given by λ_{max}), the maximum zenithal hourly rate and the population index. The usual values for these parameters can be obtained from bibliographic sources (see, for instance, Jenniskens, 2006), or they can be found in on-line databases, such as the International Astronomical Union (IAU) meteor shower database (www.astro.amu.edu.pl/~jopek/MDC2007/). However, these are subject to change from one year to the next, and it is preferable to estimate them by performing meteor observations in the Earth's atmosphere in parallel to the monitoring of lunar impact flashes. This is fundamental,

of the decrease in m_o is more important, so that the net result is a decrease in the kinetic energy of meteoroids that produce meteors of magnitude 6.5 in the atmosphere. Consequently, the minimum mass for particles exceeding the threshold kinetic energy E_{min} for flash detectability increases. So, for the same values of r, η and the zenithal hourly rate, an increase in V_g gives rise to a decrease of the frequency for detectable lunar impact flashes. So at higher V_g, the zenithal hourly rate must also be higher to compensate for this effect. The effect of the population index r on the probability parameter, for fixed values of the zenithal hourly rate and the luminous efficiency, is shown in Figure 6.4. As can be seen, an increase in r gives rise to a decrease in p, and this decrease is more important for meteoroid streams with higher geocentric speeds. This is expected since, according to the definition of the population index, larger values of r correspond to a higher fraction of less massive for instance, in those cases where activity outbursts occur. The observation of meteors in the Earth's atmosphere during the peak activity of the shower produced by a meteoroid stream provides a value of $ZHR_{Earth}^{ST}(\max)$ and λ_{max}. The observations performed on a given date corresponding to a solar longitude λ provide the value of the zenithal hourly rate with shower activity profiles in Jenniskens (1994). The population index r, and so the mass index s, can also vary along the activity period of a meteor shower. This effect has been observed, for instance, for the Geminids (see, e.g., Plavcová, 1962; Šimek, 1973; Bel'kovich et al., 1982; Rendtel, 2004; Arlt and Rendtel, 2006; Ortiz et al., 2015). So, because of the strong effect of this parameter on the value of the probability factor, it is also fundamental to monitor possible changes in r. The same holds for the hourly rate corresponding to sporadic events. From these considerations one concludes that there is an important synergy between systems for lunar impact flash monitoring and those employed to monitor meteor activity in the atmosphere.

According to the IAU meteor shower database, there are about 680 meteoroid streams producing meteor showers on Earth (www.astro.amu.edu.pl/~jopek/MDC2007/). However, the activity of most of these is very low, with very small values of the peak zenithal hourly rate (in many cases, even below 1 meteor h^{-1}). In addition, for most of these showers the information obtained to-date is not enough to characterize their activity period, the peak ZHR and the population index (Jenniskens, 2006). Again, this shows the important role of systems that monitor meteor activity in the atmosphere (see, e.g., Ortiz et al., 2015; Madiedo et al., 2015b). Nevertheless, it is expected that most of the recorded lunar impact flashes will be associated with meteoroid streams that give rise to major meteor showers on Earth, with an activity higher than that of the sporadic meteor component.

For simplicity, as in the previous section, we will also consider here the situation where just one stream and the sporadic meteoroid background can contribute significantly to the impact flux on the Moon. In this case, in order to associate an impact flash to the meteoroid stream the value of p^{ST} given by Equation (6.28) should be at least above 0.5, and the larger this value the higher the confidence level of the association.

Of course, the association probability is expected to be highest around the peak activity of the meteor shower produced by the meteoroid stream, and to decrease as we go far away from this peak, since the ZHR decreases. To determine for which meteoroid streams it is possible to associate detectable lunar impact flashes, the value for the minimum zenithal hourly rate that gives rise to an association probability of at least 50% can be estimated from Equation (6.28). The dependence of this minimum ZHR on the geocentric velocity is shown in Figure 6.6 for different values of the luminous efficiency (from 2×10^{-4} to 2×10^{-2}) and for a population index $r = 2.5$ for the meteoroid stream. In the calculations, it has been assumed that $\Xi = 1$, $\phi = 45°$ (the most likely value for the impact angle), an Earth-Moon distance $R = 384,000$ km, a maximum magnitude of +10 for flashes detectable from Earth, and a population index of 3.0 for sporadics. The values for the minimum ZHR necessary for an association probability p of 68, 95 and 99.7% are shown, respectively, in Figures 6.7, 6.8 and 6.9. These correspond to a confidence level in this association of 1-σ, 2-σ and 3-σ,

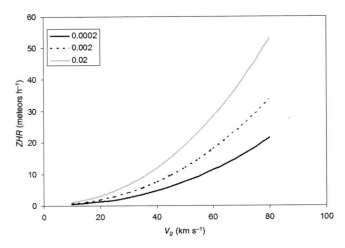

Figure 6.6. Variation with the geocentric velocity V_g of the minimum zenithal hourly rate ZHR that gives a probability $p = 50\%$ in Equation (6.28), using a population index $r = 2.5$. The calculations have been performed for luminous efficiencies equal to 2×10^{-2}, 2×10^{-3} and 2×10^{-4}.

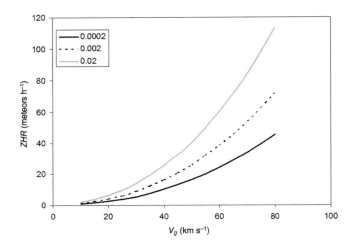

Figure 6.7. Variation with the geocentric velocity V_g of the minimum zenithal hourly rate ZHR that gives a probability $p = 68\%$ in Equation (6.28), using a population index $r = 2.5$. The calculations have been performed for luminous efficiencies equal to 2×10^{-2}, 2×10^{-3} and 2×10^{-4}.

respectively. As Figure 6.6 shows, the minimum zenithal hourly rate necessary to link a lunar impact flash with a meteoroid stream (i.e., to obtain $p = 0.5$) increases as the geocentric velocity of the meteoroids increases. For the same value of the geocentric velocity this minimum ZHR increases as the luminous efficiency increases. Figures 6.7, 6.9 and 6.9 show that this tendency holds for association probabilities of 68% (confidence level of 1-σ), 95% (2-σ) and 99.7% (3-σ), although the minimum zenithal hourly rate necessary to claim this association is higher for higher imposed confidence levels. Since, as mentioned in Section 6.6, the probability parameter depends strongly on the population index, it is also convenient to find out how this minimum ZHR depends on r. Figure 6.10 shows, for a luminous efficiency $\eta = 2 \times 10^{-3}$ and different values for the population index, the variation with the geocentric velocity

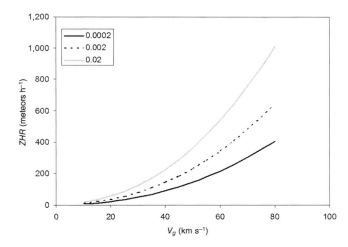

Figure 6.8. Variation with the geocentric velocity V_g of the minimum zenithal hourly rate ZHR that gives a probability $p = 95\%$ in Equation (6.28), using a population index $r = 2.5$. The calculations have been performed for luminous efficiencies equal to 2×10^{-2}, 2×10^{-3} and 2×10^{-4}.

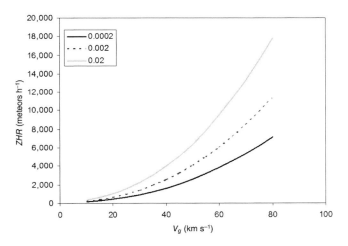

Figure 6.9. Variation with the geocentric velocity V_g of the minimum zenithal hourly rate ZHR that gives a probability $p = 99.7\%$ in Equation (6.28), using a population index $r = 2.5$. The calculations have been performed for luminous efficiencies equal to 2×10^{-2}, 2×10^{-3} and 2×10^{-4}.

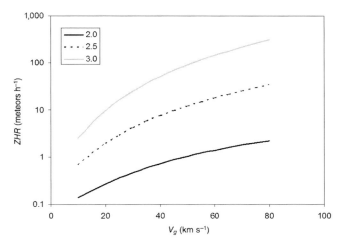

Figure 6.10. Variation with the geocentric velocity V_g of the minimum zenithal hourly rate ZHR that gives a probability $p = 50\%$ in Equation (6.28). The calculations have been performed for values of r of 2.0, 2.5 and 3.0, considering $\eta = 2 \times 10^{-3}$.

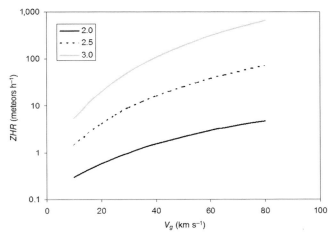

Figure 6.11. Variation with the geocentric velocity V_g of the minimum zenithal hourly rate ZHR that gives a probability $p = 68\%$ in Equation (6.28). The calculations have been performed for values of r of 2.0, 2.5 and 3.0, considering $\eta = 2 \times 10^{-3}$.

of the minimum zenithal hourly rate that allows an association probability of 50% to be obtained. As can be seen, for the same value of V_g the minimum ZHR increases strongly as r increases. Thus, the differences in ZHR are almost one order of magnitude when r changes from 2.0 to 2.5 and from 2.5 to 3.0. If the required confidence level for the association is incremented to 68, 95 and 99.7% (Figures 6.11, 6.12 and 6.13, respectively), the minimum values of ZHR again increase strongly.

As an application of the information contained in Figures 6.6–6.9, Table 6.1 shows, for major meteor showers, the value of the zenithal hourly rate that allows a lunar impact flash to be associated to a given meteoroid stream with probabilities that range from 50 to 99.7%. To calculate these values a luminous efficiency $\eta = 2 \times 10^{-3}$ has been assumed. It has also been assumed that the population index for meteoroids producing detectable impact flashes coincides with the population index derived from the observation of meteor activity on Earth. The conclusion derived from Table 6.1 is that the peak activity produced by some streams is too low to allow lunar impact flashes to be associated with them with a probability equal or higher than 50%. This is the case, for instance, for the Ursids and the South δ-Aquariids. It is also the case for the Orionids, assuming a population index $r = 2.9$. However, during those periods where the Orionids have exhibited an outburst, different researchers have obtained for this shower values for r significantly lower, of about 2.0 (Jenniskens, 1995) or even 1.4 (Trigo-Rodriguez et al., 2007). These values of r decrease the minimum zenithal hourly rate to reach an association probability of 50% at about 2 meteors h^{-1} (for $r = 2.0$) and around 1 meteor h^{-1} (for $r = 1.4$). In the same way, with a value of $r = 2.5$, the Leonids would need a minimum ZHR of about 26 meteors h^{-1} to reach an association

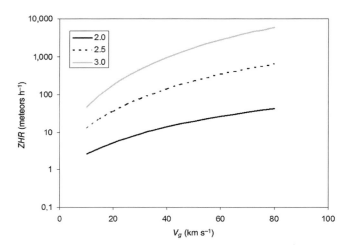

Figure 6.12. Variation with the geocentric velocity V_g of the minimum zenithal hourly rate ZHR that gives a probability $p = 95\%$ in Equation (6.28). The calculations have been performed for values of r of 2.0, 2.5 and 3.0, considering $\eta = 2 \times 10^{-3}$.

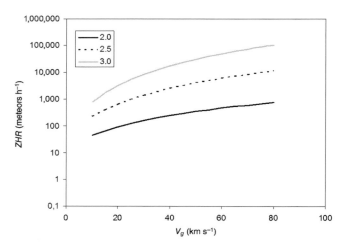

Figure 6.13. Variation with the geocentric velocity V_g of the minimum zenithal hourly rate ZHR that gives a probability $p = 99.7\%$ in Equation (6.28). The calculations have been performed for values of r of 2.0, 2.5 and 3.0, considering $\eta = 2 \times 10^{-3}$.

probability $p = 50\%$, a value much higher the nominal peak ZHR of about 15 meteors h^{-1} for this shower. However the Leonids are characterized by frequent activity outbursts (see, e.g., Table 4a in Jenniskens, 2006), where population indices of about 1.3 have been measured, with peak ZHR values as high as several hundreds of meteors h^{-1}. In some cases, the peak ZHR of the Leonids has even ranged between 13,000 and 35,000 meteors h^{-1}.

Table 6.2 focuses on some minor meteor showers. As can be seen, the situation is more unfavorable for meteoroid streams that give rise to these showers. Thus, the minimum ZHR necessary to associate with a lunar impact flash with one of these streams with a probability of 50% exceeds in most cases the value for the peak zenithal hourly rate. The exceptions are the α Centaurids and the June Bootids, with a population index equal to 2.0 and 2.2, respectively. For the rest of the showers in this Table, only in those cases where an activity outburst takes place would it be possible to have a peak ZHR large enough to be able to establish an association with a lunar impact flash with enough confidence level. Nevertheless, it is important to note that most of the population indices published for minor showers are quite uncertain.

This analysis shows, again, the need for a systematic monitoring of meteor activity on Earth in order to obtain the activity levels and the population index. One of the consequences is that previous associations performed by surveys that did not take into account these considerations should be reviewed. Special care should be taken with parameters derived from the analysis of events assigned to a given meteoroid stream or to the sporadic background just on the basis of the coincidence of the detection time with the activity peak of a given shower. These include impactor masses m, impact velocities V, impact angles ϕ and crater sizes derived from equations involving m, V or ϕ (see Section 6.7). And, of course, the luminous efficiency η and, consequently, the dependence of this parameter on meteor velocity. The same considerations hold for those flashes associated to meteoroid sources on the basis of the factor of merit FOM defined by Equation (6.20) (Suggs et al., 2014).

6.7 Crater Size

Craters produced by meteoroids impacting the Moon belong to the category of simple craters. These are characterized by their bowl shape, and are common for crater diameters below 15 km on the Moon (Melosh, 1989). Simple craters have a raised edge on the pre-impact level of the terrain, and their lower part is covered with materials that come from the walls of the transient cavity produced at the instant of the collision. When measuring their diameter it is necessary to distinguish between the so-called apparent diameter (D_A), which uses as a reference the terrain level before the impact, and the rim-to-rim diameter D, which is the diameter measured at the top of the crater rim (Figure 6.14). Different equations, called scaling laws, can be employed to estimate the size of craters produced by the impact of meteoroids on the lunar surface. The result obtained for the crater diameter depend strongly on the scaling law used. This size is also affected by the value considered for the luminous efficiency, since this parameter plays a key role and the uncertainty in this efficiency is very high.

For collisions produced at high velocities (of the order of several kilometers per second or more), the final size of the crater is in general much larger than the size of the impactor (see, e.g., Melosh, 1989; Holsapple and Housen, 2007). The theoretical models employed to describe the formation of impact craters show that the crater size can be calculated by means of powers of different parameters that describe both the impactor and the target. Besides, these magnitudes depend on a factor that is defined by the expression $AU^\mu \rho_p^\nu$, where A is the impactor radius, ρ_p its density and U the normal component of the impact velocity (Holsapple and Schmidt, 1987; Holsapple, 1993; Housen and Holsapple, 2011). For impacts produced on the lunar surface the value of the exponents in this expression are $\mu = 0.4$ and

Table 6.1. *Zenithal hourly rate (ZHR) necessary to link, with a probability of 50, 68, 95 and 99.7%, a lunar impact flash with particles belonging to streams producing major meteor showers on Earth. For the streams listed here, the values for the activity period, the solar longitude corresponding to the activity peak λ_{max} (J2000.0), maximum zenithal hourly rate ZHR_{max} and geocentric velocity V_g were obtained from Jenniskens (2006). The stream abbreviation is the three-letter code in the IAU meteor shower database.*

Stream (abbreviation)	Activity period	λ_{max} (°)	ZHR_{max} (meteors h^{-1})	V_g (km s^{-1})	r	ZHR (meteors h^{-1})			
						$p = 50\%$	$p = 68\%$	$p = 95\%$	$p = 99.7\%$
Quadrantids (QUA)	December 28–January 12	283.3 (January 4)	120	41	2.1	1.3	2.7	24	425
April Lyrids (LYR)	April 16–April 25	32.4 (April 22)	18	49	2.1	1.7	3.6	33	576
η-Aquariids (ETA)	April 19–May 28	46.9 (May 6)	40	66	2.4	14	29	263	4597
South δ-Aquariids (SDA)	July 12–August 23	127.0 (July 30)	18	41	3.2	107	227	2034	35570
α-Capricornids (CAP)	July 3–August 15	127.0 (July 30)	5	25	2.5	2.9	6.2	56	976
Perseids (PER)	July 17–August 24	140.2 (August 13)	100	59	2.0	1.3	2.9	26	456
Orionids (ORI)	October 2–November 7	208.6 (October 21)	15	67	2.9	120	278	2486	43486
South Taurids (STA)	September 16–December 29	224.0 (November 5)	8	27	2.3	1.5	3.2	29	513
North Taurids (NTA)	September 25–December 19	224.0 (November 5)	4	29	2.3	1.7	3.7	33	583
Leonids (LEO)	20 November 6–November 30	235.1 (November 18)	15	71	2.5	26	55	496	8685
Geminids (GEM)	December 4–December 17	262.1 (December 14)	120	33	2.5	5.7	12	109	1911
Ursids (URS)	December 17–December 26	270.7 (December 23)	10	35	3.0	31	67	597	10445

Table 6.2. *Zenithal hourly rate (ZHR) necessary to link, with a probability of 50, 68, 95 and 99.7%, a lunar impact flash with particles belonging to the main streams producing minor meteor showers on Earth. For the streams listed here, the values for the activity period, the population index r, the solar longitude corresponding to the activity peak λ_{max} (J2000.0), maximum zenithal hourly rate ZHR_{max} and geocentric velocity V_g were obtained from Jenniskens (2006). The stream abbreviation is the three-letter code in the IAU meteor shower database.*

Stream (abbreviation)	Activity period	λ_{max} (°)	ZHR_{max} (meteors h^{-1})	V_g (km s^{-1})	r	ZHR (meteors h^{-1})			
						$p = 50\%$	$p = 68\%$	$p = 95\%$	$p = 99.7\%$
δ-Leonids (DLE)	January 21–March 12	334.7 (February 23)	1	20	3.0	10	20	182	3180
α-Centaurids (ACE)	January 28–February 21	319.2 (February 2)	6	56	2.0	1.2	3	24	126
Virginids (VIR)	February 18–March 25	354.0 (March 14)	5	30	3.0	25	56	472	2456
γ-Normids (GNO)	February 25–March 28	354.5 (March 15)	6	56	2.4	10	21	188	978
η-Lyrids (ELY)	May 3–May 15	48.0 (May 9)	3	43	3.0	61	130	1166	6077
June Bootids (JBO)	June 22–July 2	95.7 (June 27)	1	18	2.2	0.5	1.1	10	54
κ-Cygnids (KCG)	August 3–August 24	145.0 (August 18)	3	25	3.0	16	34	304	1585
Aurigids (AUR)	August 25–September 8	158.6 (September 1)	6	66	2.5	22	67	425	2213
September ε-Perseids (SPE)	September 5–September 21	166.7 (September 9)	5	64	3.0	173	366	3278	17080
October Draconids (DRA)	20 October 6–October 10	195.4 (October 9)	1	20	2.6	3	6	52	271
δ-Aurigids (DAU)	October 10–October 18	198.0 (October 11)	2	54	3.0	173	366	3278	17080
ε-Geminids (EGE)	October 14–October 27	205.0 (October 18)	3	70	3.0	218	464	4150	21622
Andromedids (AND)	October 8–November 22	232.0 (November 14)	2	16	3.0	6	12	112	585
Monocerotids (MON)	November 27–December 17	257.0 (December 9)	2	42	3.0	58	133	1098	5720
σ-Hydrids (HYD)	December 3–December 15	260.0 (December 12)	3	58	3.0	133	283	2532	13195
Comae Berenicids (COM)	December 12–December 23	264.0 (December 16)	3	65	3.0	180	382	3414	17790

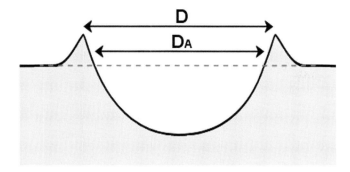

Figure 6.14. Section of a simple crater. The dashed line corresponds to the pre-impact terrain level. D_A: apparent diameter; D: rim-to-rim diameter.

$\nu = 0.333$, and the apparent diameter of the crater is given by the relationship (Holsapple, 1993):

$$D_A = 2K_r \left(\frac{\pi_v m}{\rho_t} \right)^{1/3} \quad (6.34)$$

The adimensional parameter π_v in this equation is calculated in the following way:

$$\pi_v = K_1 \left[\frac{Ag}{U^2} \left(\frac{\rho_t}{\rho_p} \right)^{\frac{6\nu-2-\mu}{3\mu}} + \left[\frac{K_2 Y}{\rho_t U^2} \left(\frac{\rho_t}{\rho_p} \right)^{\frac{6\nu-2}{3\mu}} \right]^{\frac{2+\mu}{2}} \right]^{\frac{-3\mu}{2+\mu}} \quad (6.35)$$

The normal component U of the impact velocity V is given by

$$U = V \sin(\theta) \quad (6.36)$$

In these expressions magnitudes are expressed in the mks system, with $K_1 = 0.2$, $K_2 = 0.75$, $K_r = 1.1$ and $Y = 1000$ Pa; m is the impactor mass, g the acceleration of gravity on the lunar surface, θ the impact angle with respect to the local horizontal, and ρ_p and ρ_t the projectile and target density, respectively. For the target bulk density, that of lunar regolith is usually employed. This is expected to be between 1.3 and 1.5 g cm^{-3} (Melosh, 1989). The value of the impactor density for different meteoroid streams can be found, for instance, in Babadzhanov and Kokhirova (2009). The rim-to-rim crater diameter D is obtained by multiplying the apparent diameter D_A by 1.3 (Housen et al., 1983).

An alternative relationship that has been employed by some researchers (see, e.g., Bouley et al., 2012; Suggs et al., 2014) to estimate the size of impact craters is the equation proposed by Gault (Gault, 1974; Melosh, 1989), which is valid for craters with diameters of up to about 100 meters in loose soil or regolith:

$$D = 0.25 \rho_p^{1/6} \rho_t^{-0.5} E^{0.29} \sin(\theta)^{1/3} \quad (6.37)$$

In this equation, where parameter values must be also entered in mks units, E is the kinetic energy of the meteoroid, ρ_p and ρ_t are the impactor and target bulk densities, respectively, and θ is the impact angle with respect to the horizontal.

The derived sizes found so far from different surveys are up to several meters; too small for ground-based observatories to identify. Lunar orbiting spacecraft, however, can be employed to take images of the impact regions to recognize fresh craters and study them. The experimental measurement of their diameter would allow us to better constrain the luminous efficiency, which is a poorly characterized parameter and plays a key role in the analysis of these impact events.

6.8 Impact Plume Temperature

Nemtchinov et al. (1998b) performed a numerical simulation to obtain the distribution of temperature in impact plumes produced by the collision of meteoroids on the lunar surface. In that work, impactor velocities between 15 and 30 km/s were considered, with luminous efficiencies ranging between 3×10^{-4} and 1×10^{-5}. One of the conclusions derived from these simulations is that the emission spectrum of lunar impact flashes would differ from that of a blackbody for a constant temperature, since it would contain emission lines derived from the chemical composition of the impactor and the lunar soil. However, this has not been experimentally tested since, as mentioned in Section 6.3, to date no full emission spectrum for a lunar impact flash has been obtained. But under the assumption that the intensity distribution of these events follows Planck's law, the blackbody temperature of impact plumes can be estimated from impact flash photometry (see, e.g., Bonanos et al., 2018; Madiedo et al., 2018; Madiedo and Ortiz, 2018c). For this purpose, the flash must be simultaneously observed in two different spectral bands. If this is done and the magnitudes of the flash in the two bands, for example, the visual V and infrared I bands, are determined with sufficient accuracy, the evolution with time of the impact plume blackbody temperature T is obtained as follows. The spectral irradiance of the flash is expressed as a function of wavelength λ as

$$F_f(\lambda) = g_f B(T, \lambda) \quad (6.38)$$

where g_f is a constant, and

$$B(T, \lambda) = \frac{2\pi h c^2}{\lambda^5} \frac{1}{\exp(\frac{hc}{\lambda kT}) - 1} \quad (6.39)$$

is the spectral irradiance from the blackbody and has units e.g. Wm^{-2}m^{-1}. The V and I magnitudes M_V and M_I are then defined as

$$M_V - 0 = -2.5 \log_{10} \frac{g_f \int_0^\infty B(T, \lambda) R_V(\lambda) d\lambda}{\int_0^\infty F_{A0} R_V(\lambda) d\lambda} \quad (6.40)$$

$$M_I - 0 = -2.5 \log_{10} \frac{g_f \int_0^\infty B(T, \lambda) R_I(\lambda) d\lambda}{\int_0^\infty F_{A0} R_I(\lambda) d\lambda} \quad (6.41)$$

where $F_{A0}(\lambda)$ is the spectral irradiance of a 0-magnitude star of spectral type A0. The V and I response functions are denoted by $R_V(\lambda)$ and $R_I(\lambda)$, respectively. Equations (6.40) and (6.41) are combined as

$$M_V - M_I = -2.5 \log_{10} \left(\frac{\int_0^\infty B(T, \lambda) R_V(\lambda) d\lambda}{\int_0^\infty F_{A0} R_V(\lambda) d\lambda} \times \frac{\int_0^\infty F_{A0} R_I(\lambda) d\lambda}{\int_0^\infty B(T, \lambda) R_I(\lambda) d\lambda} \right) \quad (6.42)$$

$F_{A0}(\lambda)$ can be approximated by a blackbody spectrum in the wavelength range including the V and I bands as

$$F_{A0}(\lambda) = g_{A0} B(T_{A0}, \lambda) \quad (6.43)$$

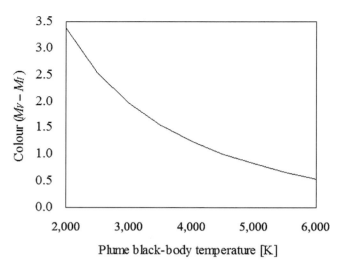

Figure 6.15. Relationship between the impact plume blackbody temperature and the color index (difference in magnitude between V-band and I-band).

where g_{A0} is a constant and T_{A0} is the effective temperature of A0 type stars. Then, Equation (6.42) leads to

$$M_V - M_I = -2.5 \log_{10} \left(\frac{\int_0^\infty B(T,\lambda) R_V(\lambda) d\lambda}{\int_0^\infty B(T_{A0},\lambda) R_V(\lambda) d\lambda} \times \frac{\int_0^\infty B(T_{A0},\lambda) R_I(\lambda) d\lambda}{\int_0^\infty B(T,\lambda) R_I(\lambda) d\lambda} \right) \quad (6.44)$$

A color index $M_V - M_I$ is thus related to the impact plume blackbody temperature T, and this temperature can be obtained from the index. The relationship is shown in Figure 6.15 where $T_{A0} = 10000$ K, and $R_V(\lambda)$ and $R_I(\lambda)$ are from Figure 1 in Bessel (2005).

6.9 Individual Events

The first unambiguous lunar impact flashes observed by means of ground-based telescopes were recorded on 1999 November 18, coinciding with the 1999 Leonid storm. A total of 10 flashes were recorded during the peak activity of the Leonid meteor shower, with apparent magnitudes ranging from 3 to 7 (Dunham et al., 2000; Ortiz et al., 2000; Yanagisawa and Kisaichi, 2002). The duration of most of these events was < 0.1 s and it was concluded that the impactors belonged to the Leonid meteoroid stream. This association was established on the basis of the coincidence of the observing date with the maximum activity period of the Leonids and a compatible impact geometry. During the 2001 Leonids, six additional impact flashes were recorded by Ortiz et al. (2002) and Cudnik et al. (2002) on November 18–19.

The Perseids were the second meteoroid stream for which a link was established with a lunar impact flash. The first flash associated with a Perseid meteoroid was a magnitude 9.5 event recorded on 2004 August 11 at 18h28m27s UTC, in the framework of a survey organized in Japan (Yanagisawa et al., 2006). A link with the Perseids was established because the closest approach of the Earth to the dust trail originating from the 1862 activity of Comet 109P/Swift-Tuttle (the parent body of the Perseid meteoroid stream) was predicted to occur that day at about 21:00 UTC 2004 (Lyytinen and Van Flandern, 2004). Yanagisawa et al. (2006) found that the Moon encountered that trail at around 18:00 UTC. In addition, the impact geometry for Perseid meteoroids was compatible with the selenographic coordinates determined for this event. Additional impact flashes associated with this stream were observed during the 2005 Perseids (Ortiz et al., 2006), and also during the 2012 and 2013 activity peaks of this shower with association probabilities > 95% (Madiedo et al., 2015a).

The association of these impact events with specific meteoroid streams provided their impact velocity on the Moon. For this reason the luminous efficiency for lunar impact flashes associated with these streams could be derived by employing the method described in Bellot-Rubio et al. (2000b). This yielded $\eta = 2.0 \times 10^{-3}$ for the Leonids (Bellot-Rubio et al., 2000a; Ortiz et al., 2002) and $\eta = 1.8 \times 10^{-3}$ for the Perseids (Madiedo et al., 2015a).

Moon impact flashes recorded during the peak activity of the 2007 Geminids were reported by Yanagisawa et al. (2008), but luminous efficiencies were not derived from them. This parameter for the Geminids was estimated by Ortiz et al. (2015) from a series of impact flashes observed during the activity period of the 2007 and the 2013 Geminids. The link between these events and the Geminid meteoroid stream was established from Equations (6.28) and (6.31), with association probabilities ranging from 95 to 96% for the 2007 Geminids and from 65 to 91% for the 2013 Geminids. The average luminous efficiency was $\eta = 2.1 \times 10^{-3}$, a value consistent with the luminous efficiencies obtained for the Leonids and the Perseids. The impact velocity on the Moon for meteoroids belonging to the Geminids, the Perseids and the Leonids is about 35 km s^{-1}, 59 km s^{-1}, and 70 km s^{-1}, respectively, so this result suggests that the luminous efficiency does not strongly depend on impact velocity at least for speeds > 35 km s^{-1}.

In order to obtain a high sensitivity in the recording system, the majority of the observations of lunar impact flashes have been performed with unfiltered cameras. Some observations, however, have been performed in the near infrared (NIR) by Ortiz et al. (in preparation), Suggs et al. (2014), Madiedo et al. (2018), Madiedo and Ortiz (2018c) and Bonanos et al. (2018). The first confirmed impact flash that allowed the value of the luminous efficiency to be estimated in the NIR was the event observed by Madiedo et al. (2018) on 2015 March 25 at 21h 00m 16.80 ±0.01 s UTC. This event also provided the first determination of an impact plume temperature. According to this work, the radiation efficiency η_I in this region of the electromagnetic spectrum is higher, by around 56%, than the luminous efficiency in the visible band. This experimental result agrees with the analysis performed by Oberst et al. (2012), who suggested that much of the emitted energy would be in the infrared.

To date, several hundred lunar impact flashes have been recorded from ground-based small telescopes with high-sensitivity video cameras (see, e.g., Bouley et al., 2012; Suggs et al., 2014; Madiedo et al., 2015b). The vast majority of these last only a fraction of a second. However, two events stand out for their extraordinary brightness and duration. These took place on 2013 March 17 (Suggs et al., 2014) and 2013 September 11 (Madiedo et al., 2014b), respectively. 'Before' and 'after' images of the lunar region where these impacts took place made it possible to identify the fresh craters associated with these events. These lunar impact flashes are described in Sections 6.9.1 and 6.9.2.

6.9.1 The 2013 March 17 Impact Event

This event was observed at 3h 50m 54.312s UTC by two telescopes with a diameter of 0.35 m operating at NASA's Marshall Space Flight Center (MSFC). The flash had a peak magnitude in R band of 3.0 ± 0.4, with a duration of around 1 s. The event was observed by employing unfiltered cameras, and this R magnitude was estimated by employing a magnitude conversion procedure described in Suggs et al. (2014). In this magnitude conversion process, it is assumed that the radiation emitted by impact flashes corresponds to that of a blackbody with a temperature of 2,800 K, on the basis of the modelling performed by Nemtchinov et al. (1998).

The impactor was assumed to be a meteoroid associated to the Virginid Meteor Complex (Suggs et al., 2014; Moser et al., 2014, 2015) on the basis that this shower was observed to be active that night by the NASA All Sky Fireball Network (Cooke and Moser, 2012) and the Southern Ontario Meteor Network (Weryk et al., 2008). A likely outburst was reported by Jenniskens et al. (2016), who observed twelve meteors from the η-Virginids (EVI) on the night of 2013 March 17. This result is consistent with the peak ZHR of 4 meteors h^{-1} measured that night by Madiedo (Madiedo et al., in preparation). The normal activity of this minor shower is lower, with a typical peak ZHR of 1.5 meteors h^{-1}, a population index $r = 3.0$, and meteoroids moving with a geocentric velocity of about 25 km s^{-1} (Jenniskens, 2006). Since no other meteor shower contributed significantly to meteor activity on that night, the probability parameter was estimated for the EVI by considering that this impact flash could be also produced by the sporadic background. In this way Equation (6.28) yields $p^{ST} = 22\%$. According to this, the most likely source for this lunar impact flash was the sporadic background, with a probability of 78%. The link with the η-Virginids is questionable when the quantitative approach based on the association probability parameter is employed (Ortiz et al., 2015). Nevertheless, according to the assumed link with the Virginids, the impactor velocity was $V = 25.6$ km s^{-1}, its mass $m = 16$ kg, and the impact angle with respect to the local horizontal $\theta = 56°$. By using the Holsapple (1993) and Gault (1974) scaling equations, the apparent and rim-to-rim diameter for the crater formed by this impact event were estimated to range between 9 and 15 m and 12 and 20 m, respectively (Suggs et al., 2014). The estimated position for this crater was 20.60 ± 0.17 N, 23.92 ± 0.30 W (Moser et al., 2014).

A new crater with a 18.8 m rim-to-rim diameter was found in Mare Imbrium by the Lunar Reconnaissance Orbiter (LRO) probe, which has been in a polar orbit around the Moon since June 2009 (http://lunar.gsfc.nasa.gov/). The Lunar Reconnaissance Orbiter camera (LROC) team located this feature at the selenographic coordinates 20.7135° N, 24.3302° W (Robinson et al., 2015), about 13 km from the location predicted. This was the first time that a fresh crater was found on the Moon associated with a lunar impact flash recorded and confirmed by means of ground-based telescopes. The subsequent re-calculation of the flash coordinates on the lunar disk with a new technique reduced this positional error to approximately 3 km (Moser et al., 2015).

6.9.2 The 2013 September 11 Impact Event

On 2013 September 11, with a six-day-old Moon, two of the Schmidt-Cassegrain telescopes of the MIDAS project in Spain recorded an extraordinary flash on the lunar surface (Madiedo et al., 2014b). The aperture of these telescopes, located at Sevilla, was 0.36 m and 0.28 m, respectively. The event, which is shown in Figure 6.16, was spotted at 20h 07m 28.68 ± 0.01 s UTC at the selenographic coordinates 17.2 ± 0.2°S, 20.5 ± 0.2°W, in the western part of Mare Nubium. The flash reached a peak apparent brightness equivalent to a stellar magnitude of 2.9 ± 0.2, with a total duration of 8.3 s. The recordings from both instruments confirmed that the flash was produced by the impact of a meteoroid on the Moon, since it was simultaneously imaged at the same selenographic coordinates by both telescopes, and the centroid of the flash did not experience any relative motion with respect to the above-mentioned position during this time. This eliminated other potential sources such as a satellite or sunglints from space debris. This is the brightest and longest confirmed lunar impact flash recorded to date on the lunar surface. The time evolution of the magnitude of the flash is plotted in Figure 6.17. As this graph shows, the event exhibited a very rapid decrease of luminosity, with flash magnitude increasing from 2.9 to 8 (i.e. about 5 magnitudes) in about 0.25 s. This brightness decay rate is similar to that found by Yanagisawa and Kisaichi (2002) for their brightest lunar impact flash, and also similar to the decay rate seen in the lightcurve of the brightest 2001 Leonid flash recorded by Ortiz et al. (2002), although in the latter case the decay was not smooth and seems somewhat longer.

To establish a quantitative link between this impact flash and a meteoroid stream, it was noted that this event took place

Figure 6.16. Lunar impact flash recorded on 2013 September 11 at 20h 07m 28.68 ± 0.01 s UTC by one of the 0.36 m telescopes operating from Sevilla in the framework of the MIDAS Project.

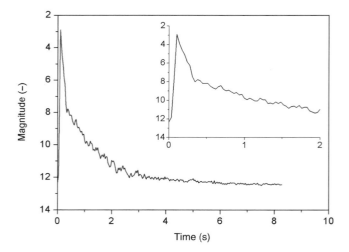

Figure 6.17. Time evolution of the apparent magnitude of the lunar impact flash recorded on 2013 September 11. The insert shows the evolution of the magnitude during the first two seconds.

close to the activity peak of the September ϵ-Perseids. This is a minor shower with a population index $r = 3.0$ and a maximum zenithal hourly rate of about 5 meteors h^{-1}. This peak is reached between September 9 and 12. Meteoroids giving rise to this shower move with a geocentric velocity of about 64 km s^{-1} (Jenniskens, 2006). The selenographic coordinates estimated for this impact event were found to be compatible with the impact geometry for this meteoroid stream. As can be seen in Table 6.2, the minimum ZHR necessary to link a lunar impact flash with the September ϵ-Perseids with a probability of 50% is of around 173 meteors h^{-1}, a value well above the above-mentioned peak ZHR. However, the September ϵ-Perseids exhibited an activity outburst around 2013 September 9, with a peak ZHR of the order of 100 meteors h^{-1} with $r = 1.5$ (Jenniskens, 2013; Rendtel et al., 2014; Gajdos et al., 2014). This value of the population index is significantly lower than the typical $r = 3.0$ value for this shower (Jenniskens, 2006). Nevertheless, the activity dropped very rapidly, and between 10 and 12 September a maximum ZHR of about 5 meteors h^{-1} with $r = 2.5$ was measured by Madiedo. These values coincide with those found by Rendtel et al. (2014). Since no other meteor shower contributed significantly to meteor activity on that date, the calculation of the probability parameter for the September ϵ-Perseids considering that this flash could be also produced by the sporadic component yields (Equation (6.28)) $p^{ST} = 13\%$. Therefore, the most likely source for this lunar impact flash is the sporadic background, with a probability of 87% (Madiedo and Ortiz, 2018e).

The kinetic energy estimated for the projectile, assuming a luminous efficiency of 2×10^{-3} and an impact velocity on the Moon for sporadics $V = 17$ km s^{-1} (Ortiz et al., 1999), is $(6.5 \pm 1.0) \times 10^{10}$ J, equivalent to 15.6 ± 2.5 tons of TNT. The estimated impactor mass was 450 ± 75 kg.

Images of the crater produced by this event (Figure 6.18) were also obtained by LRO. The LROC team targeted the reported selenographic coordinates of the flash and acquired several images over a few months until the crater was found in images acquired on 2014 March 16 and 2014 April 13. The measured

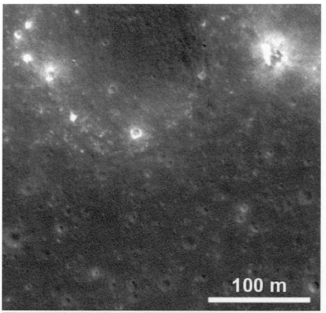

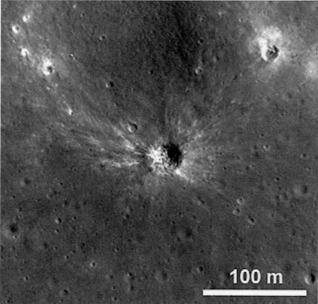

Figure 6.18. LRO pre- and post- impact photographs (upper and lower image, respectively) of the lunar region where the the 11 September 2013 impact event took place. Credits: NASA/GSFC/Arizona State University.

rim-to-rim diameter of this crater was 34 m. These images were published by LROC on the following website: http://lroc.sese.asu.edu/posts/810. Despite the poor atmospheric transparency at the time of the observation, which made it very difficult to identify features on the lunar surface (see Figure 6.16), the location originally predicted by the MIDAS software for the new crater produced by this event differed from the position measured by LRO (17.167 °S, 20.401 °W) by only 2 km (Mark Robinson, personal communication). This positional error is lower than the error determined for the 2013 March 17 event described in Section 6.9.1.

An analysis of the crater size of this impact compared to the brightness measurement was done in Ortiz et al. (2015) and a good agreement was found between measurement and theoretical estimates for luminous efficiencies of 2×10^{-3} or smaller.

6.10 Future Work

The observation of lunar impact flashes by means of Earth-based telescopes has several important drawbacks. The effective monitoring time is limited by weather conditions and by the amount of light coming from the diurnal side that is scattered in the atmosphere. As mentioned in Section 6.2, this stray light limits observations to those nights where the illuminated fraction of the lunar disk is below 60%. These issues could be addressed by employing imaging devices aboard probes orbiting the Moon or located in the vicinity of the Moon, e.g., at the L1 or L2 Earth-Moon Lagrangian points (Tost et al., 2006; Fuse et al., 2019; Koschny and McAuliffe, 2009). These probes would also allow us to detect impact flashes on the far side of the Moon during the lunar night. In this way, the effective observing time and the total area monitored could be increased significantly.

In addition, as mentioned in Section 6.4, the value of the luminous efficiency, as well as of the radiation efficiencies in other spectral bands, is not yet known with enough accuracy. It is desirable to continue performing observations that allow the parameter to be better constrained. This could be done, for instance, by performing new observations of fresh craters produced by events with recorded impact flashes. In this way, the measured and predicted diameters could be compared. Since the uncertainty in this parameter is very high, its value can be modified to match almost any possible crater diameter measurement, including the 2013 March 17 and 2013 September 11 events discussed here (Ortiz et al., 2015).

Another interesting point is the possibility of obtaining high-resolution emission spectra of lunar impact flashes. This could help to solve several open questions: for instance, if the radiation emitted by these flashes corresponds to that of a blackbody or not. Chemical species present in the impactor could be also identified by means of these spectra. In this way, as is done for meteoroids ablating in the Earth's atmosphere, the nature of the impactor (metallic, chondritic, etc.) could be established.

Acknowledgements

JMM and JLO acknowledge support from projects AYA2014-61357-EXP (MINECO), AYA2015-68646-P (MINECO/FEDER), Proyecto de Excelencia de la Junta de AndalucÃa, J.A. 2012-FQM1776, and from FEDER.

References

Arlt, R., and Rendtel, J. 2006. The activity of the 2004 Geminid meteor shower from global visual observations. *Monthly Notices of the Royal Astronomical Society*, **367**, 1721–1726.

Artem'eva, N. A., Kosarev, I. B., Nemchinov, I. V, Trubetskaya, I. A., and Shuvalov, V. V. 2000. Evaluation of light flashes caused by impacts of small comets on the surface of the moon. *Solar System Research*, **34**, 453–459.

Artem'eva, N. A., Kosarev, I. B., Nemchinov, I. V, Trubetskaya, I. A., and Shuvalov, V. V. 2001. Light flashes xaused by Leonid meteoroid impacts on the Lunar surface. *Solar System Research*, **35**, 177–180.

Babadzhanov, P. B., and Kokhirova, G. I. 2009. Densities and porosities of meteoroids. *Astronomy and Astrophysics*, **495**, 353–358.

Bel'kovich, O. I., Sulejmanov, N. I., and Tokhtas'ev, V. S. 1982. The Geminid shower from radar, photographic and visual observations. In: O. I. Bel'kovich, P. B. Babadzhanov, V. A. Bronshten, and N. I. Sulejmanov (eds), *Meteor matter in the interplanetary space, Proceedings of the All-USSR Symposium Held in Kazan*, USSR, Sept. 9–11, 1980. Moskow-Kazan, pp. 88–101

Bellot-Rubio, L. R., Ortiz, J. L., and Sada, P. V. 2000a. Luminous efficiency in hypervelocity impacts from the 1999 lunar leonids. *The Astrophysical Journal*, **542**, L65–L68.

Bellot-Rubio, L. R., Ortiz, J. L., and Sada, P. V. 2000b. Observation and interpretation of meteoroid impact flashes on the moon. *Earth, Moon, and Planets*, **82/83**, 575–598.

Bessel, M. S. 2005. Standard photometric systems. *Annual Review of Astronomy and Astrophysics*, **43**, 293–336.

Bessel, M. S., Castelli, F., and Plez, B. 1998. Model atmospheres broad-band colors, bolometric corrections and temperature calibrations for O-M stars. *Astronomy and Astrophysics*, **333**, 231–250.

Betzler, A. S., and Borges, E. P. 2015. Non-extensive statistical analysis of meteor showers and lunar flashes. *Monthly Notices of the Royal Astronomical Society*, **447**, 765–771.

Bonanos, A., Avdellidou, C., Liakos, A. et al. 2018. NELIOTA: First temperature measurement of lunar impact flashes. *Astronomy and Astrophysics*, **612**, id.A76.

Bouley, S., Baratoux, D., Vaubaillon, J. et al. 2012. Power and duration of impact flashes on the Moon: Implication for the cause of radiation. *Icarus*, **218**, 115–124.

Brown, P., Spalding, R. E., Revelle, D. O., Tagliaferri, E., and Worden, S. P. 2002. The flux of small near-Earth objects colliding with the Earth. *Nature*, **420**, 294–296.

Burns, J. A., Lamy, P. L., and Soter, S. 1979. Radiation forces on small particles in the solar system. *Icarus*, **40**, 1–48.

Ceplecha, Z. 1987. Geometric, dynamic, orbital and photometric data on meteoroids from photographic fireball networks. *Bulletin of the Astronomical Institutes of Czechoslovakia*, **38**, 222–234.

Ceplecha, Z. 1988. Earth's influx of different populations of sporadic meteoroids from photographic and television data. *Bulletin of the Astronomical Institutes of Czechoslovakia*, **39**, 221–236.

Ceplecha, Z., Borovivka, J., Elford, W. G. et al. 1998. Meteor phenomena and bodies. *Space Science Reviews*, **84**, 327–471.

Cook, A. F. 1973. A working list of meteor streams. In: C. L. Hemenway, P. M. Millman, and A. F. Cook (eds), *Evolutionary and Physical Properties of Meteoroids, Proceedings of IAU Colloq. 13, held in Albany, NY, 14–17 June 1971*, NASA SP-319. Washington, DC: National Aeronautics and Space Administration, pp. 183–191.

Cooke, W. J., and Moser, D. E. 2012. The status of the NASA All Sky Fireball Network. In: M. Gyssens, and P. Roggemans (eds), *Proceedings of the International Meteor Conference, 30th IMC, Sibiu, Romania, 2011*, Hove, BE: International Meteor Organization, pp. 9–12.

Cooke, W. J, Suggs, R. M, and Swift, W. R. 2006. A probable taurid impact on the moon. *37th Annual Lunar and Planetary Science Conference, March 13–17, 2006, League City, Texas*, id. 1731.

Cudnik, B. M., Dunham, D. W., Palmer, D. M. et al. 2002. Ground-based observations of high velocity impacts on the moon's surface – the lunar leonid phenomena of 1999 and 2001. *33rd Annual Lunar and Planetary Science Conference, March 11–15, 2002, Houston, Texas,* id.1329.

Dubietis, A., and Arlt, R. 2010. Periodic variability of visual sporadic meteor rates. *Earth, Moon, and Planets*, **106**, 2–4.

Dunham, D. W., Cudnik, Palmer, D. M., Sada, P. V. et al. 2000. The first confirmed videorecordings of lunar meteor impacts. *31st Annual Lunar and Planetary Science Conference, March 13–17, 2000, Houston, Texas,* id.1547.

Fuse, R., Abe, S., Yanagisawa, M., Funase, R., and Yano, H. 2019. Space-based observation of lunar impact flashes. *Transactions of the Japan Society for Aeronautical and Space Sciences,* in press.

Gajdos, S., Toth, J., Kornos, L., Koukal, J., and Piffl, R. 2014. The September epsilon Perseids in 2013. *WGN, Journal of the International Meteor Organization*, **42**, 48–56.

Gault, D. E. 1974. Impact cratering. In: R. Greeley and P. H. Schultz (eds), *A Primer in Lunar Geology*. Washington, DC: National Aeronautics and Space Administration, pp. 137–175.

Halliday, I., Griffin, A. A., and Blackwell, A. T. 1996. Detailed data for 259 fireballs from the Canadian camera network and inferences concerning the influx of large meteoroids. *Meteoritics and Planetary Science*, **31**, 185–217.

Holsapple, K. A. 1993. The scaling of impact processes in planetary sciences. *Annual Review of Earth and Planetary Sciences*, **21**, 333–373.

Holsapple, K. A., and Housen, K. R. 2007. A crater and its ejecta: An interpretation of Deep Impact. *Icarus*, **187**, 345–356.

Holsapple, K. A., and Schmidt, R. M. 1987. Point source solutions and coupling parameters in cratering mechanics. *Journal of Geophysical Research*, **92**, 6350–6376.

Horányi, M., Sternovsky, Z., Lankton, M. et al. 2014. The lunar dust experiment (LDEX) onboard the lunar atmosphere and dust environment explorer (LADEE) mission. *Space Science Reviews*, **185**, 93–113.

Horányi, M., Szalay, J. R., Kempf, S. et al. 2015. A permanent, asymmetric dust cloud around the Moon. *Nature*, **522**, 324–326.

Housen, K. R., and Holsapple, K. A. 2011. Ejecta from impact craters. *Icarus*, **211**, 856–875.

Housen, K. R., Schmidt, R. M., and Holsapple, K. A. 1983. Crater ejecta scaling laws–Fundamental forms based on dimensional analysis. *Journal of Geophysical Research*, **88**, 2485–2499.

Hughes, D. W. 1987. P/Halley dust characteristics–A comparison between Orionid and Eta Aquarid meteor observations and those from the flyby spacecraft. *Astronomy and Astrophysics*, **187**, 879–888.

Jenniskens, P. 1994. Meteor stream activity I. The annual streams. *Astronomy and Astrophysics*, **287**, 990–1013.

Jenniskens, P. 1995. Meteor stream activity. 2: Meteor outbursts. *Astronomy and Astrophysics*, **295**, 206–235.

Jenniskens, P. 2006. *Meteor Showers and their Parent Comets*. Cambridge, UK: Cambridge University Press.

Jenniskens, P. 2008. Meteoroid streams that trace to candidate dormant comets. *Icarus*, **194**, 13–22.

Jenniskens, P. 2013. September epsilon perseids 2013. *Central Bureau Electronic Telegrams*, **3652**(Sept.).

Jenniskens, P., Nenon, Q., Albers, J. et al. 2016. The established meteor showers as observed by CAMS. *Icarus*, **266**, 331–354.

Kapisinsky, I. 1984. Nongravitational effects affecting small meteoroids in interplanetary space. *Contributions of the Astronomical Observatory Skalnate Pleso*, **12**, 99–111.

Kesaraju, S., Mathews, J. D., Vierinen, J., Perillat, P., and Meisel, D. D. 2016. A search for meteoroid lunar impact generated electromagnetic pulses. *Earth, Moon, and Planets*, **119**, 1–21.

Koschny, D., and McAuliffe, J. 2009. Estimating the number of impact flashes visible on the Moon from an orbiting camera. *Meteoritics and Planetary Science*, **44**, 1871–1875.

Koten, P., Rendtel, J., Shrbený, L. et al. 2019. Meteors and Meteor Showers as Observed by Optical Techniques. In: G. O. Ryabova, D. J. Asher, and M. D. Campbell-Brown (eds), *Meteoroids: Sources of Meteors on Earth and Beyond*. Cambridge, UK: Cambridge University Press, pp. 90–115.

Lyytinen, E., and Van Flandern, T. 2004. Perseid one-revolution outburst in 2004. *WGN, Journal of the International Meteor Organization*, **32**, 51–53.

Madiedo, J. M., and Ortiz, J. L. 2018a. Lunar Impact Flashes, Causes and Detection. In: Brian Cudnik (ed), *Encyclopedia of Lunar Science*. Cham: Springer. Available at: https://doi.org/10.1007/978-3-319-05546-6_113-1.

Madiedo, J. M., and Ortiz, J. L. 2018b. MIDAS System. In: Brian Cudnik (ed), *Encyclopedia of Lunar Science*. Cham: Springer. Available at: https://doi.org/10.1007/978-3-319-05546-6_128-1.

Madiedo, J. M., and Ortiz, J. L. 2019. Lunar impact flashes, temperature. In: Brian Cudnik (ed), *Encyclopedia of Lunar Science*. Cham: Springer (Switzerland: Springer Nature, 2019). Available at: https://doi.org/10.1007/978-3-319-05546-6_222-1.

Madiedo, J. M., and Ortiz, J. L. 2018d. Lunar Impact Events, Luminous Efficiency, and Energy of. In: Brian Cudnik (ed), *Encyclopedia of Lunar Science*. Cham: Springer. Available at: https://doi.org/10.1007/978-3-319-05546-6_116-1.

Madiedo, J. M., and Ortiz, J. L. 2018e. Lunar Impact Event: The 11 September 2013. In: Brian Cudnik (ed), *Encyclopedia of Lunar Science*. Cham: Springer. Available at: https://doi.org/10.1007/978-3-319-05546-6_127-1.

Madiedo, J. M., Trigo-Rodríguez, J. M., Ortiz, J. L., and Morales, N. 2010. Robotic systems for meteor observing and Moon impact flashes detection in Spain. *Advances in Astronomy*, id. 167494.

Madiedo, J. M., Trigo-Rodriguez, J. M., Williams, I. P., Ortiz, J. L., and Cabrera-Cano, J. 2013. The Northern Chi Orionid meteoroid stream and possible association with the potentially hazardous asteroid 2008XM1. *Monthly Notices of the Royal Astronomical Society*, **431**, 2464–2470.

Madiedo, J. M., Ortiz, J. L., Trigo-Rodríguez, J. M. et al. 2014a. Analysis of two superbolides with a cometary origin observed over the Iberian Peninsula. *Icarus*, **233**, 27–35.

Madiedo, J. M., Ortiz, J. L., Morales, N., and Cabrera-Cano, J. 2014b. A large lunar impact blast on 2013 September 11. *Monthly Notices of the Royal Astronomical Society*, **439**, 2364–2369.

Madiedo, J. M., Ortiz, J. L., Organero, F. et al. 2015a. Analysis of Moon impact flashes detected during the 2012 and 2013 Perseids. *Astronomy and Astrophysics*, **577**, id.A118.

Madiedo, J. M., Ortiz, J. L., Morales, N., and Cabrera-Cano, J. 2015b. MIDAS: Software for the detection and analysis of lunar impact flashes. *Planetary and Space Science*, **111**, 105–115.

Madiedo, J. M., Ortiz, J. L., and Morales, N. 2018. The first observations to determine the temperature of a lunar impact flash and its evolution. *Monthly Notices of the Royal Astronomical Society*, **480**, 5010–5016.

McNamara, H., Jones, J., Kauffman, B. et al. 2004. Meteoroid Engineering Model (MEM): A meteoroid model for the inner Solar System. *Earth, Moon, and Planets*, **95**, 123–139.

Melosh, H. J. 1989. *Impact Cratering : A Geologic Process*. New York: Oxford University Press.

Moser, D. E., Suggs, R. M., Swift, W. R. et al. 2011. Luminous efficiency of hypervelocity meteoroid impacts on the Moon derived from the 2006 Geminids, 2007 Lyrids, and 2008 Taurids. In: W. J. Cooke, D. E. Moser, B. F. Hardin, and D. Janches (eds), *Meteoroids: The Smallest Solar System Bodies (Meteoroids 2010)*.

Washington, D. C.: National Aeronautics and Space Administration, pp. 142–154.

Moser, D. E., Suggs, R., and Suggs, R. J. 2014. Large meteoroid impact on the Moon on 17 March 2013. In: In: K. Muinonen, A. Penttil, M. Granvik et al. (eds), *Proceedings of the Asteroids, Comets, Meteors 2014 Conference*. Helsinki: University of Helsinki, p. 386

Moser, D. E., Suggs, R., and Suggs, R. J. 2015. A bright lunar impact flash linked to the Virginid Meteor Complex. In: *Abstract for the Stanford Meteor Environments and Effects (SMEE) Workshop, held in Stanford, California, 14-16 July 2015*.

Nemtchinov, I. V., Shuvalov, V. V., Artem'eva, N. A. et al. 1998b. Light flashes caused by meteoroid impacts on the Lunar surface. *Solar System Research*, **32**, 99–114.

Nemtchinov, I. V., Shuvalov, V. V., Artemieva, N. A. et al. 1998. Light impulse created by meteoroids impacting the Moon. *29th Annual Lunar and Planetary Science Conference, March 16-20, 1998, Houston, TX*, id.1032.

Oberst, J., and Nakamura, Y. 1991. A search for clustering among the meteoroid impacts detected by the Apollo lunar seismic network. *Icarus*, **91**, 315–325.

Oberst, J., Christou, A., Suggs, R. et al. 2012. The present-day flux of large meteoroids on the lunar surface–A synthesis of models and observational techniques. *Planetary and Space Science*, **74**, 179–193.

Ortiz, J. L., Aceituno, F. J., and Aceituno, J. 1999. A search for meteoritic flashes on the Moon. *Astronomy and Astrophysics*, **343**, L57–L60.

Ortiz, J. L., Aceituno, F. J., and Aceituno, J. 2000. Optical detection of meteoroidal impacts on the Moon. *Nature*, **405**, 921–923.

Ortiz, J. L., Quesada, J. A., Aceituno, J., Aceituno, F. J., and Bellot-Rubio, L. R. 2002. Observation and Interpretation of Leonid impact flashes on the Moon in 2001. *The Astrophysical Journal*, **576**, 567–573.

Ortiz, J. L., Aceituno, F. J., Quesada, J. A. et al. 2006. Detection of sporadic impact flashes on the Moon: Implications for the luminous efficiency of hypervelocity impacts and derived terrestrial impact rates. *Icarus*, **184**, 319–326.

Ortiz, J. L., Madiedo, J. M., Morales, N., Santos-Sanz, P., and Aceituno, F. J. 2015. Lunar impact flashes from Geminids: Analysis of luminous efficiencies and the flux of large meteoroids on Earth. *Monthly Notices of the Royal Astronomical Society*, **454**, 344–352.

Plavcová, Z. 1962. Radio-echo observations of the Geminid meteor stream in 1959. *Bulletin of the Astronomical Institutes of Czechoslovakia*, **13**, 176–176.

Popel, S. I., Golub', A. P., Lisin, E. A. et al. 2016a. Impacts of fast meteoroids and the separation of dust particles from the surface of the Moon. *Journal of Experimental and Theoretical Physics Letters*, **103**, 563–567.

Popel, S. I., Golub', A. P., Lisin, E. A. et al. 2016b. Meteoroid impacts and dust particles in near-surface lunar exosphere. *Journal of Physics Conference Series*, **774**, id. 012175.

Popel, S. I., Golub', A. P., Zelenyi, L. M., and Horányi, M. 2017. Impacts of fast meteoroids and a plasma-dust cloud over the lunar surface. *Journal of Experimental and Theoretical Physics Letters*, **105**, 635–640.

Popel, S. I., Golub', A. P., Zelenyi, L. M., and Horányi, M. 2018a. Dusty plasmas in the lunar exosphere: Effects of meteoroids. *Journal of Physics Conference Series*, **946**, id. 012142.

Popel, S. I., Zelenyi, L. M., Golub', A. P., and Dubinskii, A. Yu. 2018b. Lunar dust and dusty plasmas: Recent developments, advances, and unsolved problems. *Planetary and Space Science*, **156**, 71–84.

Popova, O., Borovička, J., and Campbell-Brown, M. D. 2019. Modelling the entry of meteoroids. In: G. O. Ryabova, D. J. Asher, and M. D. Campbell-Brown (eds), *Meteoroids: Sources of Meteors on Earth and Beyond*. Cambridge, UK: Cambridge University Press, pp. 9–36.

Rendtel, J. 2004. Evolution of the Geminids observed over 60 years. *Earth, Moon, and Planets*, **95**, 27–32.

Rendtel, J., Lyytinen, E., Molau, S., and Barentsen, G. 2014. Peculiar activity of the September epsilon-Perseids on 2013 September 9. *WGN, Journal of the International Meteor Organization*, **42**, 40–47.

Robinson, M. S., Boyd, A. K., Denevi, B. W. et al. 2015. New crater on the Moon and a swarm of secondaries. *Icarus*, **252**, 229–235.

Šimek, M. 1973. A radio observation of the Geminids 1959–1969. Overdense echoes. *Bulletin of the Astronomical Institutes of Czechoslovakia*, **24**, 213–213.

Steel, D. 1996. Meteoroid orbits. *Space Science Reviews*, **78**, 507–553.

Stuart, J. S., and Binzel, R. P. 2004. Bias-corrected population, size distribution, and impact hazard for the near-Earth objects. *Icarus*, **170**, 295–311.

Suggs, R. M., Cooke, W. J., Suggs, R. J., Swift, W. R, and Hollon, N. 2008. The NASA lunar impact monitoring program. *Earth, Moon, and Planets*, **102**, 293–298.

Suggs, R. M., Moser, D. E., Cooke, W. J., and Suggs, R. J. 2014. The flux of kilogram-sized meteoroids from lunar impact monitoring. *Icarus*, **238**, 23–26.

Swift, W., Suggs, R., and Cooke, B. 2008. Algorithms for lunar flash video search, measurement, and archiving. *Earth Moon Planets*, **102**, 299–303.

Swift, W. R., Moser, D. E., Suggs, R. M., and Cooke, W. J. 2011. An exponential luminous efficiency model for hypervelocity impact into regolith. In: W. J. Cooke, D. E. Moser, B. F. Hardin, and D. Janches (eds), *Meteoroids: The Smallest Solar System Bodies*. Washington, D.C.: National Aeronautics and Space Administration, pp. 125–141.

Szalay, Jamey R., and Horányi, M. 2015. Annual variation and synodic modulation of the sporadic meteoroid flux to the Moon. *Geophysical Research Letters*, **42**, 10,580–10,584.

Szalay, J. R., Poppe, A. R., Agarwal, J. et al. 2018a. Dust phenomena relating to airless bodies. *Space Science Reviews*, **214**, 98.

Szalay, Jamey R., Pokorný, P., Jenniskens, P., and Horányi, M. 2018b. Activity of the 2013 Geminid meteoroid stream at the Moon. *Monthly Notices of the Royal Astronomical Society*, **474**, 4225–4231.

Tost, W., Oberst, J., Flohrer, J., and Laufer, R. 2006. Lunar impact flashes: History of observations and recommendations for future campaigns. *European Planetary Science Congress 2006*, **1**, EPSC2006-546.

Trigo-Rodriguez, J. M., Madiedo, J. M., Llorca, J. et al. 2007. The 2006 Orionid outburst imaged by all-sky CCD cameras from Spain: Meteoroid spatial fluxes and orbital elements. *Monthly Notices of the Royal Astronomical Society*, **380**, 126–132.

Trigo-Rodriguez, Josep M., Madiedo, José M., Gural, Peter S. et al. 2008. Determination of meteoroid orbits and spatial fluxes by using high-resolution All-Sky CCD cameras. *Earth, Moon, and Planets*, **102**, 231–240.

Vaubaillon, J., Neslušan, L., Sekhar, A., Rudawska, R., and Ryabova, G. 2019. From Parent Body to Meteor Shower: the Dynamics of Meteoroid Streams. In: G. O. Ryabova, D. J. Asher, and M. D. Campbell-Brown (eds), *Meteoroids: Sources of Meteors on Earth and Beyond*. Cambridge, UK: Cambridge University Press, pp. 161–186.

Volvach, A. E., Berezhnoy, A. A., Foing, B. et al. 2009. Search for radio flashes caused by collisions of meteoroids with the moon. *Kinematics and Physics of Celestial Bodies*, **25**, 194–197.

Weryk, R. J., Brown, P. G., Domokos, A. et al. 2008. The Southern Ontario All-sky Meteor Camera Network. *Earth, Moon, and Planets*, **102**, 241–246.

Williams, I. P., Jopek, T. J., Rudawska, R., Tóth, J., and Kornoš, L. 2019. Minor Meteor Showers and the Sporadic Background. In: G. O. Ryabova, D. J. Asher, and M. D. Campbell-Brown (eds), *Meteoroids: Sources of Meteors on Earth and Beyond*. Cambridge, UK: Cambridge University Press, pp. 210–234.

Williams, I. P. 2011. The origin and evolution of meteor showers and meteoroid streams. *Astronomy and Geophysics*, **52**, 2.20–2.26.

Yanagisawa, M., and Kisaichi, N. 2002. Lightcurves of 1999 Leonid impact flashes on the Moon. *Icarus*, **159**, 31–38.

Yanagisawa, M., Ohnishi, K., Takamura, Y. et al. 2006. The first confirmed Perseid lunar impact flash. *Icarus*, **182**, 489–495.

Yanagisawa, M., Ikegami, H., Ishida, M. et al. 2008. Lunar Impact Flashes by Geminid Meteoroids in 2007. *71st Annual Meeting of the Meteoritical Society*, **43**, id.5169.

Part IV

Interrelations

7

From Parent Body to Meteor Shower: The Dynamics of Meteoroid Streams

Jérémie Vaubaillon, Luboš Neslušan, Aswin Sekhar, Regina Rudawska, and Galina O. Ryabova

7.1 Introduction

The link between meteors and comets was first established in the nineteenth century by measuring the orbit from the location of the radiant and testing several hypotheses on the semi-major axis of known comets. A simple comparison of a meteor typical (or mean) orbit and that of a known comet revealed a striking similarity for a few of them. Hence, the link between the Perseids and comet 109P/Swift-Tuttle, as well as the Leonids and comet 55P/Tempel-Tuttle was demonstrated by Schiaparelli (1867). Adams (1867) concluded that the Leonids had a > 33 years periodicity, which was remarkably similar to the orbital period of comet 55P. However, such a simple approach also led to a huge disappointment in 1899 when a Leonid meteor storm was predicted, in spite of the work by Stoney and Downing (1899) showing the influence of the giant planets on the stream.

Today, knowing that meteoroids are blown away from a comet nucleus, physical considerations tell us that the ejection speed driven by the outgassing process cannot exceed the gas thermal speed and represents only a small fraction of the orbital speed. We also know that once ejected, nothing ties up meteoroids to their parent body. The natural consequence is that meteoroid orbits are, at first, very similar to the parent one. The generation and later evolution of meteoroids in the Solar System is the topic of this chapter. Namely, the questions explored here include the following list. How are meteoroids generated in the Solar System? What happens to them once they are released from their parent bodies? How long do they typically survive? How can we find the parent body of a meteoroid stream? Which parents are today clearly identified and which are still under consideration? Why are some parent bodies still under debate after many years of discussion? How do resonances and relativistic effects affect stream structures? Such are the questions this chapter is addressing.

The goal of this chapter is to provide the reader with an overview of the latest work in the study of the dynamics of meteoroid streams in the Solar System, as well as some of the current debate related to e.g. the determination of the very existence of meteor showers and their parenthood with comets or asteroids (also discussed in Williams et al., 2019; see Chapter 9). A previous review with a similar goal was presented by Williams (2002).

Dynamical studies allow us to determine the parent body, age, origin, evolution and fate of meteoroid streams. Some meteor showers that at first look unrelated are now understood as several parts of a complicated and ever-evolving process. Similarly, some parent bodies (comets or asteroids) are suspected to be related, or at least share some dynamical history with one or several meteoroid streams in our times (see also Kasuga and Jewitt, 2019; see Chapter 8). The dynamics of meteoroid streams provides us with a broad understanding of the connection between meteoroids, comets and asteroids, i.e. making the links between other Solar System bodies neater and closer.

We focus on 1) the basics of the dynamics of meteoroids, 2) several scenarios meteoroid streams may encounter, 3) the fate and lifetime expectancy of those streams, 4) the way meteor showers are defined (dynamically), and finally 5) describe some of the famous and recent advances in demonstrating the parenthood of meteor showers with comets or asteroids.

7.2 The Making of a Meteoroid Stream

A meteoroid is typically the product of a comet outgassing or a collision between at least two asteroids. When we model a meteoroid ejection from a cometary nucleus, three key parameters are considered: the position of the ejection point on the reference orbit, the ejection velocity value and direction. Since the initial conditions influence the whole history of a given meteoroid, we examine here how they are ejected, what are the forces acting on them, and how they translate into orbital elements and evolution.

7.2.1 Ejection Process

When meteoroid stream modeling made its first steps, attempts to estimate the ejection velocity V_{ej} from observations of meteors were numerous. The formulae giving the changes of the orbital elements caused by the ejection were derived by Plavec (1955, 1957) and used later in many studies. However the errors in determining the meteoroid orbit's parameters were (and still are) too large, so the ejection velocity determinations gave no reliable results. For details and references, see reviews by Williams (2001) and Ryabova (2006). Rudawska et al. (2005) came to the same negative conclusion and also showed that it is almost impossible to retrieve the epoch of ejection from the measurement of orbits.

The estimate of the V_{ej} value from cometary trail modeling is another possible resource. Such a method was proposed by Müller et al. (2001) and is based on the "dust shell" approach introduced by Finson and Probstein (1968) and the dust trail theory in the form introduced by McNaught and Asher (1999). The authors elaborated a technique allowing to calculate the set of all possible trajectories that can reach the Earth at the same time. The method did not find an application because 1) it is rather complicated and 2) it needs the meteor shower mass distribution and the dust production rate of its parent comet as a function of heliocentric distance. Alas, at present there is no meteoroid stream for which both these parameters are reliably known. The mass distribution and dust production rate are most often measured for micron-size meteoroids, from today's cometary coma observations (de Almeida et al., 2007). This might be used as a proxy for larger meteoroids, but under the condition that the mass distribution index s_m[1] is constant over several orders of magnitude, which is not certain at all. Direct observation of mm- to dm-size meteoroids was performed in infrared wavelengths by the Spitzer telescope, or *in situ* measurements (e.g. Schulz et al., 2004; Reach et al., 2007; Vaubaillon and Reach, 2010; Kelley et al., 2013; Boissier et al., 2014; Fulle et al., 2016; Ott et al., 2017). If meteoroid stream detection is feasible with such methods, their number remains quite small compared to optical detections.

Today, the forward approach[2] is dominating the modeling of meteoroid streams. Theoretically, the best agreement between a model and a meteoroid stream's parameters obtained from observations should be achieved by varying the free parameters of the model. However, the numerical modeling is rather an expensive procedure, requiring a large amount of computer time. Therefore it is reasonable and understandable that modelers use just one physical model, and their choice is now confined to two models: the classical (Whipple, 1951, with its modification by Jones 1995; Hughes 2000; Ma et al. 2002) and models by (Crifo, 1995; Crifo and Rodionov, 1997, see details of models and modifications in Ryabova 2013). An attempt to compare various models showed that it is not possible to recommend one model only (see review by Ryabova, 2013).

A distinction is to be made between comet tail, comet trail and comet meteoroid stream. Whereas the tail refers to the gas escaping the nucleus, the trails of the stream are composed of meteoroids. A (meteoroid) trail is ejected from the comet at one single perihelion passage. The ensemble of trails ejected at several returns makes the (meteoroid) stream. A subsection of a stream (e.g., trapped in a resonance) might be referred to as a swarm. Although the existence of streams was known from meteor showers, their first direct observation as a structure in the Solar System was performed by Sykes et al. (1986) using the IRAS telescope. Kresák (1993) considered a meteoroid stream as a cometary dust trail, and that is quite correct. The first direct optical observation of comet 22P/Kopff's stream was performed by Ishiguro et al. (2002). In comet astronomy, however, the term "cometary dust trail" means a narrow structure seen in the images along the projection of the cometary orbit. A comparative analysis of cometary dust trail modeling for five comets (2P/Encke, 4P/Faye, 17P/Holmes, 22P/Kopff, 67P/Churyumov-Gerasimenko) has shown that the results obtained have large variations and differ from the results obtained by, say, the classical Whipple model (Ryabova, 2013).

The summary of the present section is the following: no single model of ejection velocity can be recommended, mainly by lack of strong constraints and generalization to all comets. Physical characteristics of the comets and their activity are essentially different, and it is quite possible that every comet needs its own model. We may tell only, that the limit for the ejection velocity of meteoroids is the gas velocity. How large may this velocity be? From observations of comet 1P/Halley, the maximal value of outflow velocity of H_2O is 2 km s^{-1} near perihelion (Bockelee-Morvan et al., 1990). The modeling also showed that the outflow velocity of gases can reach 3 km s^{-1} at a heliocentric distance of 0.14 AU (Combi et al., 2004). However the usual (i.e., not catastrophic) expected ejection speed of large particles producing detectable meteors is of several hundred or several tens of meters per second.

7.2.2 Forces Acting on Meteoroids

Like most bodies in the Solar System, meteoroids revolve around the Sun on Keplerian orbits, perturbed by the gravitation of planets and non-gravitational forces.

7.2.2.1 Newtonian Gravitation and Non-Gravitational Forces

For accurate dynamical considerations, the gravitational influence of the Sun, the eight planets, Pluto and the Moon are considered today. Since the mathematical formulation of such a multi-body problem does not exist, the usual way to tackle the matter is to perform numerical simulations. This requires us to compute and sum the gravitational forces acting on an individual meteoroid from each perturbing body, for each and every integration time step. The ephemeris of the planets might be computed or directly taken from an already existing ephemeris (Giorgini et al., 1996; Viswanathan et al., 2018).

A schematic overview of the non-gravitational forces acting on a meteoroid is given in Figure 7.1. Several authors (Burns et al., 1979; Kapišinský, 1984; Olsson-Steel, 1987; Klačka, 2014; Christou et al., 2015) produce a clear and detailed overview of the radiation pressure and Poynting-Robertson (PR) forces acting on a meteoroid, describing each of their radial and transverse components.

Very often, the factor describing the net result of absorption, scattering of light and solar wind (e.g. denoted Q in Burns et al., 1979) is considered close or equal to unity. In reality, this factor is highly unknown and might even change with time for a given meteoroid. However, given the wide range of initial conditions of meteoroids, changing this factor does not dramatically change the overall behavior of a meteoroid stream.

The radiation pressure is as strong as the gravitation of the Sun ($\beta = F_{non-grav}/F_{grav} = 1$) for a ~ 1 μm size meteoroid having a bulk density of ~ 600 kg m^{-3}. For mm-size, β is down to $\sim 10^{-3}$. PR-drag is $\sim 10^{-4}$ the radiation pressure in modulus.

The out of the plane non-gravitational force is usually neglected, except for the seasonal Yarkovsky–Radzievskii force (hereafter referred to simply as "Yarkovsky": Öpik, 1951;

---

[1] Usually $2 < s_m < 4$.
[2] When we are calculating parameters of a model from the observational data – this is an inverse problem. When we define [construct] a model using our knowledge about an object or a phenomenon (e.g. a comet and a meteoroid stream) and calculate "the observational data" – this is a forward problem. Considering that an inverse problem is almost always ill-posed, the forward approach here is justified and generally more simple.

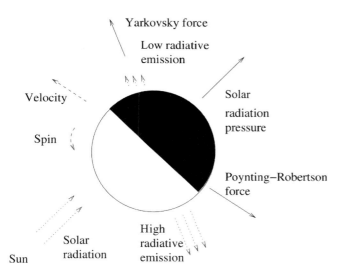

Figure 7.1. Recapitulation of the non-gravitational forces acting on a meteoroid (from Vaubaillon et al., 2005a, reproduced with permission © ESO).

Radzievskii, 1952). The so called YORP (Yarkovsky–O'Keefe–Radzievskii–Paddack) force results from the torque induced by radiative forces on irregularly shaped meteoroids. It changes the rotational state of the particle rather than directly changing its semi-major axis. Ryabova (2014) elaborated some formulae to compute the average changes of a meteoroid's orbital elements from such radiative forces. In theory, for fast-rotating meteoroids the dispersion in semi-major axis may increase by a factor of 3. Because the YORP force is mass-dependent, its effect increases the mass separation in a stream. However, this force is not strong enough to shift the whole meteoroid stream in a measurable way so far. It would be quite a challenge to measure such an effect on such tiny particles.

7.2.2.2 Relativistic Forces: General Relativistic Effects

Long term, general relativistic (GR) effects play a role in the evolution of meteoroid streams. The GR perihelion precession rate primarily depends on the perihelion distance and the orbital period. As shown in Equation (7.1), a combination of both moderately small perihelion distance and moderately small orbital period increases the GR precession rate.

GR precession in ω (change in argument of pericenter) can be computed using (Weinberg, 1972, p. 197):

$$\Delta\omega = \frac{6\pi GM}{a(1-e^2)} = \frac{6\pi GM}{q(1+e)} \quad (7.1)$$

in radians/revolution. Standard celestial mechanics notations used are G: universal gravitation constant, M: mass of Sun, a: semi-major axis, e: eccentricity, q: perihelion distance.

Long term, GR precession is measurable for some meteoroid streams with $q \leq 0.4$ AU (i.e. q of Mercury) and $a \leq 1$ AU (i.e., a of Earth) (Sekhar, 2013). For the record, GR precession was measured for Mercury (Einstein, 1915) and the Earth (Weinberg, 1972). The q and a phase space of multiple meteoroid streams falls between these values.

Present results show that $\Delta\omega$ due to GR precession is critical for these low-q meteoroid streams ($q \leq 0.15$ AU Sekhar, 2013).

Ignoring this value can make substantial changes in the forecast of meteor showers.

Past and recent works (Fox et al., 1982; Galushina et al., 2015) showed that GR precession has a medium level of perturbations in the long-term evolution of asteroid 3200 Phaethon, and this in turn applies to the Geminid meteoroid stream as well.

A subtler and smaller effect is the Lense-Thirring effect due to dragging of space-time due to a rotating body. Because the system has a rotating central mass (i.e., the Sun is rotating on its own axis), it affects the local space-time continuum and leads to small changes in ω and nodal longitude Ω of the low-q meteoroid particle. The change in $\Delta\omega$ due to the Lense-Thirring effect is typically four orders of magnitude smaller than the change in $\Delta\omega$ due to the GR precession (Iorio, 2005). Hence this effect is usually ignored in calculations of meteoroid dynamics.

7.2.3 Formation of a Meteoroid Stream

If we discount the gravitational influence of the planets, what is the net effect of the ejection velocity and non-gravitational forces? The radiation pressure acts as if the Sun's gravity were lower, and therefore increases the semi-major axis of a meteoroid, compared to its parent orbit. Different size leads to different β (Section 7.2.2.1), which translates into a different semi-major axis. The ejection velocity changes the semi-major axis as well. The natural consequence of the emission of many different meteoroids, of differing sizes and ejection velocities, is a quasi-continuous spread of the meteoroids along the parent orbit (Klačka, 2014). Meteoroids spend most of their time at aphelion, just like any other body in the Solar System. As a consequence, at a given time, most meteoroids are located closer to aphelion than perihelion.

The Poynting–Robertson drag acts on long time scales (typically a few tens of thousands of years) and causes the meteoroid to lose energy and spiral to the Sun (Williams, 2002). The smaller the meteoroid, the faster it sinks into the Sun. The Yarkovsky force either causes an increase or a decrease of the semi-major axis, depending on which side it acts. Note that because of the change of spin state (due to, e.g., YORP effect), the Yarkovsky influence changes with time. In addition, its effect depends on the size of the meteoroid as well. However, as discussed in Section 7.2.2.1, both diurnal and seasonal effects are hardly detectable in meteoroids. The left hand side of Figure 7.3 (see Section 7.3.2) shows the natural and most common configuration of a meteoroid stream with respect to its parent comet (i.e., in a trailing position).

7.3 Short-Term Evolution of Meteoroid Streams

Once ejected, the most powerful disturbing effect on a meteoroid orbit is close encounters with planets. In addition, mean motion resonances (MMRs) act on a time scale of several revolutions, affecting a stream in a relatively short time scale.

7.3.1 Close Encounters

Close encounters between a stream and a planet last from a few hours (with e.g. the Earth) to a few days (with giant planets). Figure 7.2 (top) illustrates the consequence of repeated close

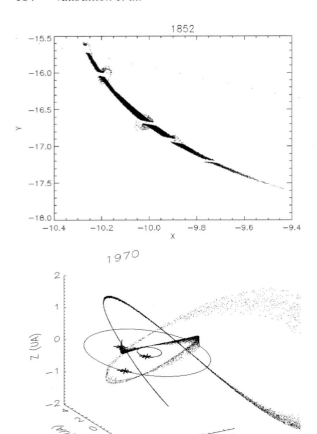

Figure 7.2. Top: Gaps in a Leonid trail (ejected from 55P in 1767), created by close encounter with the Earth. Bottom: The meteoroid stream ejected from comet 67P in 1947, as seen in 1970. Close encounters with Jupiter have highly changed the parent comet orbit, but also the stream (Figures 3 and 4 from Vaubaillon et al., 2006).

encounters of a trail with the Earth. Gaps are created, from the gravitational pull caused by the mass of the planet. Figure 7.2 (bottom) illustrates the consequence of a close encounter on one single trail (resulting from a single passage of the comet). Several different parts of the stream were differently perturbed during the close encounter. As a natural result, each part of the trail has seen its orbits changed in a unique way. The trail can hardly be recognized as a single entity now, although it came from a single return of its parent body. After a few dozen revolutions, accompanied sometimes by other close encounters, such a trail is spread throughout the entire inner Solar System, as well as beyond Jupiter. Such a scenario is common among Jupiter family streams.

7.3.2 The Reversal Process

As explained in Section 7.2.3, the natural tendency of a meteoroid stream is to be trailing from the parent body since most meteoroids have a semi-major axis larger than that of the comet. However, Vaubaillon et al. (2004, see their Figure 1) showed that close encounters with giant planets can reverse such a picture, as illustrated in Figure 7.3. A close encounter with Jupiter lowers down the semi-major axis of the particles. As a consequence, their period is lower than that of the parent body. After a few revolutions, instead of trailing the comet, the stream is now leading it.

7.3.3 Two-Body Resonances with Giant Planets

7.3.3.1 Resonances and Meteoroid Streams

The MMRs can be broadly classified into interior and exterior resonances (Peale, 1976) depending on the orbit sizes and geometry. The orbits of meteoroids, especially those in the inner Solar System, are often influenced by resonant action of a major planet through interior resonances. For meteoroids originating from Halley-type comets, exterior resonances come into play.

Resonances induce periodic gravitational nudges making multiple bodies evolve in similar ways over long time scales (~kyr). This leads to a pattern and symmetry in the long term orbits of the bodies trapped in such resonances (Murray and Dermott, 1999). Moreover resonances are extremely effective in avoiding close encounters with the body (mostly giant planets in this case) driving the resonance mechanism in itself. This is a very useful property in terms of the stability of the meteoroid stream sub-structures surviving in the Solar System for a long time without dispersion and preserving dense trail structures for a long time.

When such super-dense clusters intersect the Earth, it leads to meteor outbursts and/or storms. Hence orbital resonances have a direct role in meteor science in creating meteor outbursts and spectacular storms (Asher and Izumi, 1998). A few well-observed examples of annual showers that have shown meteor outbursts due to Jovian resonances are Leonids, Orionids and Perseids. Not necessarily all the meteor outbursts or storms in the past or present are connected with resonances though; sometimes it could just be due to young dust trails (like in the case of the famed Leonid meteor storm in 1833) which have not had enough time for dispersion and scattering.

On the other hand, Williams (1997) showed that Uranian resonances can lead to depletion and removal of meteoroids from some parts of the Leonid stream. This in turn hints at the possibility that, sometimes, diminished activity in some years might be connected with some peculiar resonances. This work introduced a totally new line of thought connected with resonances and meteor showers. We hereafter discuss Jovian and Saturnian resonances.

7.3.3.2 Two-Body Jovian Resonances

The influence of two-body Jovian resonances in meteoroid streams was tackled by analytical, semi-analytical and numerical approaches.

Froeschlé and Scholl (1986) pointed out that the orbits of the Quadrantids are close to the interior 2:1 MMR with Jupiter. Asher et al. (1999) found that the trail ejected from 55P/Tempel-Tuttle was trapped in the 5:14 Jovian resonance and caused the 1998 Leonid meteor storm. Following these authors, Vaubaillon et al. (2006) explored mechanisms leading to orphaned streams from resonances.

Examining the Taurid complex, Asher (1991) and Asher (1994) searched for evidence of meteoroids in the 7:2 MMR

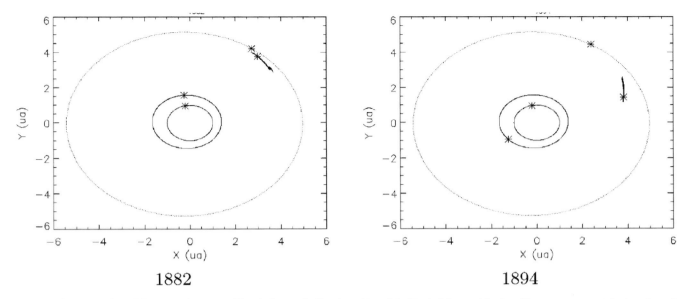

Figure 7.3. Illustration of the reversal process. The circles symbolize the orbits of the Earth, Mars and Jupiter. The stars represent the location of the planets and the parent comet (9P/Tempel 1). Initially (left) the stream is trailing its parent comet, but after a close encounter it is leading the comet (right) (from Vaubaillon et al., 2004).

and noted that sub-structures can also be, in principle, explained with the help of the 3:1, 4:1, 10:3, and 11:3 MMRs with Jupiter. These resonances correspond to orbital periods that are typical of Taurids, Jovian MMRs having a crucial role in the long-term evolution of various sub-stream structures of the stream.

Wu and Williams (1992a) use a small set of Leonid test particles to present the configuration of several meteoroids with respect to their parent body, as well as the influence of periodic perturbations on the trail. This shows the importance of taking such periodic and secular effects into account.

Gaps were found in the distribution of the reciprocal semi-major axes of Perseid meteor orbits extracted from the IAU MDC (Meteor Data Center) database (Wu and Williams, 1995). They were perceived as due to the gravitational perturbations of major planets and their positions correspond to the MMRs with the planets. Svoreň et al. (2006) also presented the existence of gaps in semi-major axis and showed that seventeen resonant filaments in the Perseid stream correlated with Jovian and Saturnian resonances.

Jenniskens (2006) gives a compilation of multiple Jovian resonances correlated with past, present and future meteor outbursts and storms. Similarly, Rendtel (2007) and Sato and Watanabe (2007) present simulations correlating the 1:6 Jovian resonant Orionids with enhanced meteor activity in 2006 and anticipated enhanced activity for the years 2006–2010 coming from the same resonant stream. Sekhar and Asher (2014a) and Sekhar (2014) looked into the long-term dynamics of 5:14 and 4:11 Jovian resonances in the Leonids and 1:6 and 2:13 Jovian resonances in the Orionids. The size and scale of resonant zones, the survival times of resonant particles, and the duration of past, present (and future) enhanced activity were compared with observations of Orionids and Leonids.

Christou et al. (2008) discuss in detail about the occurrence of showers due to 1P/Halley on multiple planets (i.e., Venus, Earth and Mars). The role of Jovian resonance in creating enhanced activity is discussed here.

Soja et al. (2011) and Emel'yanenko (2001) employ analytical and semi-analytical techniques to understand Jovian resonances in multiple streams. A detailed table in Emel'yanenko (2001) gives the nominal resonant locations, resonant libration widths and possible resonant configurations in multiple streams showing both interior and exterior resonances.

7.3.3.3 Two-Body Saturnian Resonances

Usually Jovian resonances are stronger than Saturnian resonances. However, the latter become dominant (see Figure 7.4) when the nearby Jovian resonances are weak and unstable, or when the order of a Saturnian resonance is smaller than that of the adjacent Jovian resonances (Sekhar, 2014).

One of the biggest challenges in distinguishing between Saturnian and near-Jovian resonances is to correctly identify and eliminate Jovian resonances masquerading as fake Saturnian resonances due to the Great Inequality problem (Brouwer and van Woerkom, 1950). This is a well-known near-2:5 commensurability between Saturn and Jupiter (Milani and Knežević, 1990) which has been studied in detail for decades.

The safe method is to find all the nearby Jovian resonances and eliminate them one by one by checking the evolution of resonant arguments using D'Alembert rules (Murray and Dermott, 1999). Order and strength of resonance have a direct relationship when it comes to long-term stream dynamics. For this, one has to calculate the nearby Jovian and Saturnian resonances around the semi-major axis of the comet at perihelion and compare between various orders of resonances. Different combinations of resonant arguments have to be considered and their corresponding evolution is verified for libration or circulation (Murray and Dermott, 1999). This is a time-consuming task if the comet has been populating the stream over many perihelion passages in the past (like, e.g., comet Halley).

Brown (1999) discusses briefly the Saturnian and Uranian resonances in Leonids and Perseids with respect to multiple

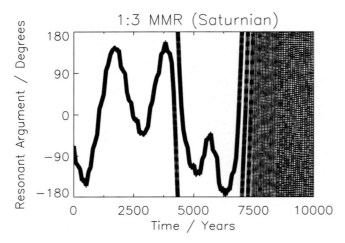
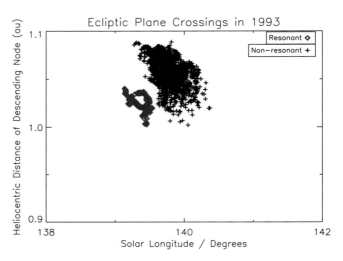

Figure 7.4. Libration of 1:3 (Saturnian) resonant argument for an Orionid meteoroid, confirming the presence of 1:3 MMR with Saturn in the Orionid stream. The order of this Saturnian resonance is smaller than adjacent Jovian resonances and hence the long term influence of Saturn is stronger than that of Jupiter for these particles. Saturnian resonances lead to clustering of meteoroids thereby preserving dense dust trails for a few kyr (figure 1a from Sekhar and Asher, 2013).

Figure 7.5. Ecliptic plane crossings of Perseids: heliocentric distance of descending node versus solar longitude (J2000) in AD 1993, for three-body resonant meteoroids (2000 gray diamonds, showing dense structures) and non-resonant (2000 black crosses, showing significant dispersion) ones. Integration started at the 69 BC perihelion time of parent comet 109P/Swift-Tuttle (figure 6b from Sekhar et al., 2016). (A black and white version of this figure will appear in some formats. For the colour version, please refer to the plate section.)

possible resonant locations in these streams. However, Saturnian resonances in meteoroid streams were not looked into in detail until recently (Sekhar and Asher, 2013). Strong evidence of Saturnian 1:3 MMR in the Orionids and 8:9 MMR in the Leonids was found.

Analyzing the difference in long-term evolution between resonant (Saturnian) and non-resonant meteoroid particles clearly shows that Saturnian resonances can retain dense stream structures without dispersion for long time scales ($\sim$ a few kyr). More studies are required to identify Saturnian resonances in other streams and thereby correlate with observed enhanced activity during present times.

7.3.4 Three-Body Resonances with Giant Planets

Unlike the two-body resonances involving two simple integer ratios (where comparing orders and strengths is easier, Murray and Dermott, 1999), nomenclature and comparisons of multiple three-body mean motion resonances (3BMMR) in a stream are tricky (Yoder and Peale, 1981). For three-body resonances to occur, counterintuitively, the bodies need not be resonant with each other in pairs (Greenberg, 1975). Their identification needs long and detailed simulation involving each and every combination of resonant configurations in the order of strengths using, e.g., the disturbing function algorithm developed by Gallardo (2014). Various combinations of resonant arguments have to be checked for libration (to confirm three-body resonance) or circulation (to rule out three-body resonance).

The first historical example of three-body resonance in the Solar System is the famous Laplacian resonance found in Galilean moons Io, Europa and Ganymede (Laplace, 1799). Sekhar et al. (2016) for the first time showed a unique three-body resonance in a meteoroid stream. Three-body resonances are linked with small stability islands surrounded by seas of chaos (Beaugé et al., 2008) and hence the ejected meteoroid particles getting constrained in a small window of orbital parameters is essential for this particular resonance to work efficiently. Therefore, the perihelion distance of the parent comet and the range of ejection velocities involved are crucial for efficiently populating these three-body resonant zones. Theoretically, there can be a series of overlapping three-body resonances in different meteoroid streams, but in reality only very few three-body resonances have the strength, stability and survival times to affect the long-term orbital evolution of stream structures. Out of these relatively high-strength three-body resonances, only a few of them can be populated (depending crucially on the semi-major axis range) by the parent body at particular epochs. This highlights the point that multiple conditions have to match in the case of a particular stream to exhibit strong and stable three-body resonant structures which could cause observable meteor outbursts or storms in future. As of now, only the Perseids were found to exhibit this phenomenon in a distinct and distinguishable way (Sekhar et al., 2016). 1:4S:10J resonance was found in the Perseid stream – i.e., for every one orbital period of a resonant Perseid particle, Saturn makes four revolutions and Jupiter makes ten revolutions respectively. The orbital periods of the Perseids trapped in this unique resonance are around 115 years. Numerical simulations show that this resonance is stable and active for the order of a few kyr. It is efficient in making super-dense dust trails survive for a few kyr (see Figure 7.5). In addition, some of the bright and spectacular 1993 Perseids might have originated from this three-body resonance. Indeed, the timing of observed enhanced activity and dust trail-Earth intersection timing from modeling match neatly.

A detailed analysis using the entire list of streams and parent bodies is necessary to find more possible examples (not restricted to the 1:4S:10J case alone) of this rather interesting combination of Jovian and Saturnian resonances, and unveil the cause of past or future meteor shower outbursts. There is much scope for future work in this area.

7.4 The Fate of Meteoroid Streams

We are now interested in long-term dynamical effects influencing meteoroids.

7.4.1 Lost Parent

Among the 112 established meteor showers listed at the IAU Meteor Data Center (Jopek and Jenniskens, 2011; Jopek and Kaňuchová, 2014), only a few dozen have a clearly identified parent body. The difficulty of linking a given meteoroid stream with a parent body is explained in detail in Section 7.5. Vaubaillon et al. (2006) have reviewed the different mechanisms leading to an "orphan meteoroid stream" (OMS). The most efficient way, by far, is close encounters with giant planets, especially Jupiter, because of the drastic and rapid change of orbital elements. Complete disruption of a cometary nucleus is another rapid way to create OMS. Slow variations of cometary orbital elements because of the outgassing forces have quite a weak influence on a time scale of a few hundred years. However, this still makes it hard to link a given stream with its potential parents if the required time scale is of the order of typically several thousand years, because of the multiplicity of possible dynamical scenarios such non-gravitational forces induce.

7.4.2 Circulation of Orbital Angular Elements

Secular precession in argument of pericenter brings meteoroids close to nodal points intersecting planet's orbit(s). Babadzhanov et al. (2008a,b) have pointed out a mechanism responsible for several potential meteor showers at a given planet (Earth in particular) from a single stream. The circulation of the argument of pericenter brings meteoroid-orbit nodal points to intersect a planet orbit at two different arcs of the planet orbit, with two possible velocity vectors at each. Figure 7.6 illustrates the process. Adding the Kozai mechanism (see Section 7.4.3), a given stream might produce up to eight different showers on Earth.

7.4.3 GR Precession and Kozai Mechanism in Meteoroid Orbits

The Kozai mechanism (Kozai, 1962) (see also Kasuga and Jewitt, 2019, Section 8.5.4) is correlated changes of i, e and ω, accompanied with change of Ω as a natural consequence. Such a mechanism in streams induced by Jupiter was presented by Sekhar et al. (2017), although theoretically it can be induced (to a smaller degree) by other massive planets as well.

The Kozai mechanism induced by Jupiter happens in the evolution of the Quadrantid stream (Wiegert and Brown, 2005) and other streams (Sekhar et al., 2017). The evolution of streams described in Sekhar et al. (2017) also shows that the GR precession rate may be significantly enhanced due to the secular Kozai mechanism.

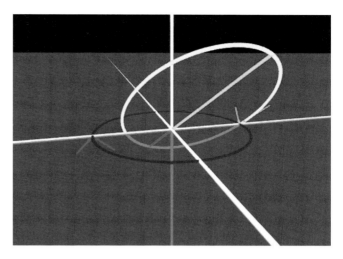

Figure 7.6. Circulation of the argument of pericenter causing four meteor showers from a single stream (personal communication from Paul Wiegert, University of Western Ontario). (A black and white version of this figure will appear in some formats. For the colour version, please refer to the plate section.)

A good example of this phenomenon (due to such Kozai-induced Sun-grazing and Sun-colliding phases) can be seen in the orbital evolution of comet 96P/Machholz 1 (also discussed in Kasuga and Jewitt, 2019, see Section 8.3.2) which was directly linked with the Marsden group of sungrazing comets, the Daytime Arietids (IAU #171 ARI) and the Southern δ-Aquariids (#005 SDA) (Sekanina and Chodas, 2005).

7.4.4 Meteoroid Collision Lifetime

The collision lifetime of a meteoroid is still a topic of debate today. It is usually constrained by either global models of the Earth's meteoroid environment, or observational results that do not fit a simple view of a meteoroid trail (Grün et al., 1985; Jenniskens et al., 2008). Estimates are performed using lab experiments on collisions, global dynamical models fitting the sporadic background, or from a given trail observed at different times (Brisset et al., 2016; Nesvorný et al., 2011; Jenniskens et al., 2008).

According to Grün et al. (1985), the particles larger than $\sim 10^{-8}$ kg and moving in the Earth's vicinity and in low-inclination orbits are depleted on a time scale of 10^4 years. Jenniskens (2006, p. 539) estimates an upper lifetime of millimeter-sized particles to be about 62 kyr. Wiegert et al. (2009) argued that the particles in more inclined and larger orbits have a much longer collisional lifetime than claimed by Grün et al. In their simulations, they considered the evolutionary periods from 0.1 to 2.5 Myr. Soja et al. (2016) and Drolshagen et al. (2017), by considering the meteoroid flux at the Earth, questioned the Grün et al. (1985) distribution and, as a consequence, the collisional lifetime. Jenniskens (2017) argues that there might be a size effect on this quantity. The very nature of meteoroids might have an influence as well (Jenniskens et al., 2008). The first ever definitely observed spontaneous fragmentation of meteoroids in space (i.e., outside the Earth's atmosphere) was achieved by Koten et al. (2017). This is a good and necessary step toward a better understanding of the collisional lifetime, but more investigations are needed to create a consensus on this question.

7.4.5 The Age of a Stream and Its Lifetime Expectancy

In general, the difference between the orbits of meteoroids and that of their parent body keeps increasing with time. After a long period of time, this difference is so large that it is impossible to dynamically distinguish the relationship between a meteoroid and its parent. Once the meteoroids of a given stream are spread out such that no particular feature can be recognized, they are regarded as part of the sporadic background. However, such a scenario might not be true for an aging stream controlled by a resonance preventing the meteoroid orbits from dispersing, and keeping them in a confined orbital corridor (see Section 7.3.3).

The age of a meteoroid stream obviously varies from stream to stream and was estimated for a few streams only. Arter and Williams (1997), using precise IAU MDC photographic database data, derived an age of the April Lyrids of $\sim 1.5 \cdot 10^6$ years. Jenniskens et al. (1997) found that the main component of the Quadrantid meteoroid stream is only about 500 years old. Later, the core of the Quadrantid meteoroid stream, consisting of particles originating from asteroid 196 256 (preliminary designation 2003 EH_1), was found to be even younger: $\sim$ 200–300 years (Wiegert and Brown, 2004, 2005; Abedin et al., 2015). On the other hand, the bulk of observational characteristics of all showers associated with the 96P – 196 256 complex can be explained by an age ranging from 10 000 BCE and 20 000 BCE (Abedin et al., 2018). Earlier, Abedin et al. (2017) demonstrated that the age and the formation mechanism of the Arietids are consistent with a continuous cometary activity of 96P/Machholz 1 over a time span of $\sim$ 12 000 years.

How long can a trail ejected in a stream survive to be observable today? Certainly this depends on the forces controlling the stream orbital evolution. For example, Williams and Ryabova (2011) found that a trail generated in the Geminid stream is lasting for thousands of years, while one in the Quadrantid stream is lost after only 500 years (mainly due to the proximity with the orbit of Jupiter). Orbital resonances may also prevent the meteoroid orbits from dispersing (see Section 7.3.3).

7.4.6 JFC-Stream versus HTC-Stream

From Section 7.3.1 it is clear that repeated close encounters with Jupiter significantly increase the dispersion of meteoroids in the Solar System. This is what most often happens to meteoroid streams of Jupiter family cometary (JFC) origin. However, some of these are less influenced by the giant planet simply because their orbit does not extend as far as 5 AU (e.g.: stream ejected from comet 2P/Encke, although one might argue that this is not really a JFC). The fate of JFC meteoroids is rather simple: they are either ejected in the outer Solar System (or even outside the Solar System by Jupiter), or, after times ranging between ten years and a hundred thousand years, they sink into the Sun, because of the Poynting–Robertson drag. In reality, the role of resonances as well as close encounters might make this simple scenario more complex, by e.g. sending a part of a trail into a resonant orbit. Whatever the mechanism, the net effect is that the stream is spread with time, no matter what other processes may act to confine it (resonance or reversal process).

Streams created by Halley-type comets (HTC) or even long-period comets have, at first, a completely different scenario. Most often, they do not intersect with giant planets, making their evolution much smoother, as well as much longer. Perseid meteoroids stay on almost constant orbits for at least 100 000 years (Brown and Jones, 1998). Typically, unless one part of the stream encounters a planet, the main evolutionary component is the PR-drag. The first effect is to lower the eccentricity (orbit circularization: Liou and Zook, 1996). Then the particle takes a few hundred thousand years to spiral down to the Sun. On its way, the chance of experiencing a close encounter with any planet increases. By doing so, the net effect of such encounters is to increase the eccentricity. Then the following evolution is similar to Jupiter family streams.

7.4.7 From Stream to Sporadic Meteoroids

What is the difference between a stream meteoroid and a sporadic meteoroid? Basically, the former is associated with other meteoroids, via either its orbital elements or its radiant and geocentric velocity (*stricto censu* this is equivalent, but as a and e might be loosely constrained, one might prefer to use other parameters). Because of all the phenomena described in Section 7.3, the very existence of a stream is limited in time. The time scale strongly depends on the parent orbit (crossing other planets or not), as well as the size of the particles. Sometimes the scale can be very short because some parent bodies release meteoroids with a large dispersion of semi-major axes. These particles immediately become members of the sporadic background. An example is the theoretical stream ejected from comet 126P/IRAS (Tomko and Neslušan, 2012) (preliminary designation used by Tomko and Neslušan was 126P/1996 P1). The dispersion of the radiants is shown in Figure 7.7.

However, to classify a meteor as sporadic also depends on the observer's ability to compare its orbit (or radiant + velocity) to

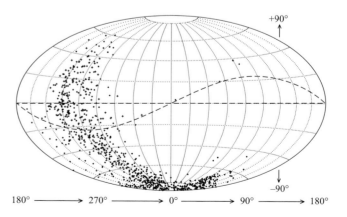

Figure 7.7. The radiant distribution of the theoretical shower of comet 126P/IRAS. The radiants of the simulated particles after 400 (1000) orbital periods of the parent are shown with crosses (small bullets). Orbital period of 126P is 13.3 years. The sinusoid-like curve illustrates the ecliptic. (Taken from Tomko and Neslušan (2012).)

other meteors. This ability has dramatically changed over the past few years because of several factors. First, the number of dedicated systematic surveys has grown up these past few years. Second, the number of meteor orbit catalogs taking into account both professional and amateur contributions has considerably increased the number of data publicly available (Williams et al., 2019, see Section 9.4).

The immediate outcome of such an amount of data is that new showers are recognized, which *de facto* decreases the number of sporadic meteoroids. What will the future look like? Will there be a point where all meteors will be regarded as part of a stream? Dynamics tell us (Section 7.3) that the orbital memory of a given meteoroid is lost after a more or less long period of time (but sometimes very short in case of a close encounter). So in all cases, sporadic meteors will always exist just because it is impossible to trace back the origin of all meteoroids beyond a certain time (which depends on the very orbit of the meteoroid). Now the next challenge is to be able to distinguish between a shower meteor and an accidental coincidence of orbit similarity (see also Section 7.5.1).

7.4.8 On the Use of Estimating the Meteoroid Environment in the Solar System

Today the meteoroid sporadic background is estimated for several reasons. Scientifically, this allows one to estimate the collisional lifetime of meteoroids (Soja et al., 2016) and the age and replenishing rate of the zodiacal light (Nesvorný et al., 2011). More practically, space agencies are interested in the protection of spacecraft from collisions with meteoroids. For this purpose, several models of the sporadic background were built by McNamara et al. (2004); Dikarev et al. (2005); Soja et al. (2015). It is worth mentioning that Soja et al. (2016) have included all the freshly ejected meteoroid trails in their model, producing then the most complete view of the meteoroid environment of the Solar System as we know it today.

7.5 Linking Meteoroid Streams with Parent Bodies

The search for the parent body of a meteoroid stream usually implies one of two approaches. Either the current orbits of known comets or asteroids are compared with those of many meteoroids, or one or several meteor showers are predicted and compared to observed showers. In any case, to know the parent body allows us to study the origin and age of a given stream, and often provides us with the big picture explaining the current meteoroid environment.

7.5.1 Link Based on Orbital Elements Comparison

Shortly after their ejection (Section 7.2) meteoroids move in orbits slightly different from that of the parent. This initial orbital similarity does not persist forever. As soon as a meteoroid leaves its parent body, its orbit is constantly perturbed (see Section 7.3). Each orbit is independent but evolves in a similar way to the other meteoroids ejected by the parent body. Hence, the existence of a structure composed of meteoroids having similar orbits – a meteoroid stream – is maintained over a long period of time, i.e., as long as the perturbations still allow one to define the stream (Ryabova, 1999; Ryabova et al., 2008).

The first identification of the parent bodies of meteoroid streams was, thus, naturally based on a similarity between the mean orbit of the corresponding meteor shower and orbit of the parent body, which necessarily passes close to our planet (within ~ 0.15 AU) (see also Williams et al., 2019, Section 9.2). Doing so, the generic relationship between the Perseids (IAU #007) – 109P/Swift-Tuttle, Orionids (#008) and η-Aquariids (#031) – 1P/Halley, Leonids (#013) – 55P/Tempel-Tuttle, April Lyrids (#006) – C/1861 G1 (Thatcher), October Draconids (#009) – 21P/Giacobini-Zinner, Andromedids (#018) – 3D/Biela, Geminids (#004) – 3200 Phaethon, and several others was found. The simple similarity criterion was also recently used by Pokorný et al. (2017) to identify the parent bodies of twenty-six showers in the IAU MDC list and thirty-four new showers they found.

Of particular interest are meteor showers originating from asteroids having a comet-like orbit (Jewitt, 2012). Indeed, the presence of such a meteoroid stream tells us that any such celestial body is most certainly an extinct comet, from both dynamical and physical considerations. Such studies allow us to link different bodies of the Solar System and unveil their origins. Comets are thought to be remnants of the material formed in the very early stage of the Solar System. Thus, meteoroids supplied by comets give us important information about the physical conditions at the dawn of our planetary system. Thus, determining the parent body of a meteoroid enables us to study the origin of the Solar System as well as its evolution by linking the small bodies among each other (see Figures 2 and 3 in Valsecchi et al., 1999).

The evaluation of orbital similarity was also used to study the meteor showers impacting other terrestrial planets (Christou et al., 2019, see Section 5.4).

7.5.1.1 Methods of Meteoroid-Stream Identification

These methods are based on three components: (1) a dynamical similarity function (e.g. D-criterion), (2) a similarity threshold, D_c, and (3) a cluster analysis algorithm. Williams et al. (2019, see Section 9.2) provide all details regarding the commonly used similarity functions.

Two orbits O_i and O_j are considered associated if $D(O_i, O_j) < D_c$, with D_c a given threshold. Once the distance function and the similarity threshold are defined, a meteoroid stream can be detected thanks to a cluster analysis algorithm. The one used by Southworth and Hawkins (1963) is a single neighbor linking technique. Starting with a given orbit, any other orbit obeying $D < D_c$ is merged in a stream. Sekanina (1976), Neslušan et al. (1995), and Welch (2001) proposed an iterative method. A meteoroid stream is a set of orbits contained in a sphere of orbital parameters. The radius of the sphere is defined by D_c, while the centre lies at the mean orbit of the stream.

A completely different method, called "method of indices", was proposed by Svoreň et al. (2000). Different ranges of N dynamical parameters are divided into a number of equidistant intervals and an index consisting of the serial numbers of the intervals relevant to a given meteor is assigned to the meteor.

The meteoroids with the same indices are regarded as the members of the same meteoroid stream.

Yet another method defines a meteoroid stream by using the wavelet transform technique from a huge dataset of radar radiants and velocities (Galligan and Baggaley, 2002; Brown et al., 2008) or a method detecting meteor showers using a density-based spatial clustering method (Sugar et al., 2017).

7.5.1.2 Break-Point Method of Shower Identification

Individual showers differ from each other by the number of their members and their dispersion of orbits. In an optimal selection of shower members, one cannot consider the same value of D_c when separating the shower in the way described in Section 7.5.1.1. In the remainder of Section 7.5.1.2, a method to distinguish the most compact core of a shower by determining the appropriate threshold value of the similarity parameter (Neslušan et al., 1995, 2013a) is briefly described. A similar approach was recently developed by Moorhead (2016).

The method, known as the "break-point method", is based on considerations by Sekanina (1970a,b, 1973, 1976). Let us suppose a data set of meteor orbits containing a well-defined shower. Let us further consider a fixed value of the (Southworth and Hawkins, 1963) D-discriminant, D_{SH}, called D_{lim}, and compute the value D_{SH} between the orbit of each meteor and the mean orbit of the shower (which is known at least approximately). If $D_{SH} \leq D_{lim}$, the meteor is selected as a member of the shower.

In reality, the mean orbit of the shower is unknown. Nevertheless, we often know a predicted mean orbit or the orbit of the parent body. After the shower meteors for a given D_{lim} are selected, a new mean orbit is calculated and a new selection of shower meteors is performed. This iterative process is repeated until the difference between new and previous mean orbits is negligible (in practice, until $D_{SH}^{new} - D_{SH}^{old} < 10^{-5}$).

The iteration is performed for a consecutive set of D_{lim} values and the number of shower meteors $N = f(D_{lim})$ is examined. If the considered data set contains exclusively the meteors of a single shower, N would steadily and linearly increase with D_{lim} (the dark gray dashed curve in Figure 7.8). Once D_{lim} is higher than the phase space distribution of the shower meteors, N would show a plateau behavior, starting at $D_{lim} = D_{bp}$ and a further increase of D_{lim} would have no effect on N. The curve is said to be "broken" and becomes constant at the point corresponding to D_{bp}.

If the data set contains sporadic meteors only, the dependence $N = N(D_{lim})$ would be almost constant, with a small number of shower meteors for low values of D_{lim}. For larger D_{lim}, N would increase steadily (the lighter gray dotted curve in Figure 7.8). In a real database, shower meteors are mixed with sporadic meteors, therefore the actual $N = N(D_{lim})$ dependence is the superposition of the dependence on shower and sporadic meteors. This is shown as the black thin solid line in Figure 7.8. We see that the "break point" (marked with an arrow in the figure) is located at the beginning of a "plateau" in the $N = N(D_{lim})$-dependence. The selection of the shower using D_{lim} corresponding to the break point provides us with the most compact part of the given shower.

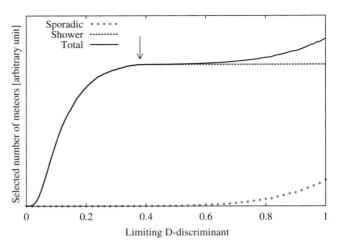

Figure 7.8. The dependence of the number of meteors selected from a database on the threshold value of the D-discriminant. The black solid curve shows the dependence for the selected number from a real database. This dependence is the superposition of the selected shower meteors (dark gray dashed curve) and sporadic background meteors (lighter gray dotted curve). Taken from Neslušan et al. (2013a). (A black and white version of this figure will appear in some formats. For the colour version, please refer to the plate section.)

7.5.2 Link Based on Modeling of Long-Term Dynamics

7.5.2.1 The Case for a Robust Modeling

Historically, the worst defeat of modern astronomy occurred because the gravitational pull from Jupiter and Saturn of the Leonid stream was not correctly taken into account when predicting the 1899 shower. This happened again in 1998 when the main outburst occurred sooner than expected (Arlt, 1998), although the orbit of the parent body comet 55P/Tempel-Tuttle was known with high accuracy. This outlines the need for a model that takes into account the meteoroids as individual entities, distinct from the stream parent. Given the range of initial conditions a relatively high number of numerical simulations are usually necessary to perform the prediction of meteor showers.

Right after their ejection, all meteoroids of a given trail are physically close to each other, and their orbits are all very similar. The part of interplanetary space (a tube) in which all these orbits are situated is also called a corridor (Neslušan et al., 2013b). Planetary perturbations, especially those of Jupiter, might change the orbits of meteoroids in a systematic way, in addition to close encounters. This happens when the orbits of the stream are in a vicinity of a resonance with a planet (see Section 7.3.3).

The systematic change of meteoroid orbits may result in a division of an original corridor into two or more well separated and still compact corridors. The original stream thus acquires a filamentary structure. If more than a single filament crosses the orbit of the Earth, several meteor showers originating from the same parent occur. A good example of such a splitting is the stream associated with asteroid 196 256. The evolution of the longitude of ascending node, Ω, is shown in Figure 7.9. At first, the particles of the stream have an initial $\Omega \sim 112°$. Soon after their ejection, two filaments appear: the first with an

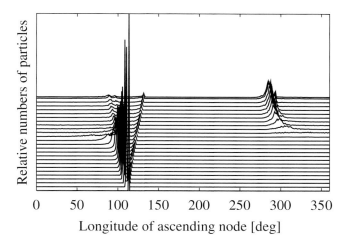

Figure 7.9. The distribution of longitude of ascending node as a function of time for a theoretical stream released from asteroid 196 256 at time 4952 years before present. The bottom curve illustrates the distribution at the time of the release, while successive curves above illustrate the distribution separated with an interval of 200 years. The upper-most curve represents the distribution 4952 years after the start of the numerical integration (taken from Kaňuchová and Neslušan, 2007, reproduced with permission © ESO).

increasing and the second with a decreasing mean Ω-value. After 3200 years, a third significant peak appears, at $\Omega \sim 300°$ and decreasing to $\sim 280°$.

A stream meteoroid might be so perturbed that further dynamics is no longer determined by its initial dynamical state. Its new orbit might be stable enough that it stays in such an orbital configuration for a long time, compared to meteoroids in a transition regime. Such resulting streams can only be studied via their full modeling and following up their dynamical evolution. In modern approaches, several methods are used.

7.5.2.2 Quick Overview of Different Models of Meteoroid Streams

Here is a quick review of several recent models of meteoroid streams. Due to time and space limitations, the following list is not an exhaustive one.

Ryabova (1989, 2016) was interested in the dynamics of the Geminids and created her own model of dynamical evolution, firstly based on polynomial approximation of the planetary dynamics, then considering thorough numerical simulation.

Kondrateva and Reznikov (1985) correctly predicted a Leonid meteor storm by modeling the meteoroid stream ejected from comet 55P/Tempel-Tuttle. They were independently followed later by Wu and Williams (1996) and McNaught and Asher (1999), who looked for the circumstances putting a meteoroid at the Earth at the time corresponding to the outburst. Quickly, many other models were developed. Lyytinen and Van Flandern (2000) considered a loosely bound model of a comet, and Vaubaillon et al. (2005b) used extensive numerical simulations. Sato and Watanabe (2010), following a similar method (post)-predicted the Phoenicids. Jenniskens (2006) compiled the prediction of most meteor showers for the following 50 years by using the results from all these authors. Wiegert et al. (2013) and Ye et al. (2014) use MOID (Minimum Orbital Intersection Distance) to derive Earth intersecting meteoroids. The sources of the toroidal shower were investigated by Pokorný et al. (2014).

Generalizing the approach, meteoroid models of the whole Solar System were constructed by Moser and Cooke (2008), followed by Wiegert et al. (2009) to reconcile radar and optical flux measurements of the sporadic background. The latter found that a parent body on an orbit similar to comet 2P/Encke has a major contribution. Nesvorný et al. (2011) found that JFC comets were the main source of the smallest sporadic meteoroids. Soja et al. (2016) considered most known cometary streams over short and medium-term time scales, and revised the collision frequency of meteoroids. The influence of Sun-grazing comets on the generation of meteoroid streams was studied by Sekhar and Asher (2014b).

7.5.2.3 Method Based on the Circulation of Orbital Elements

The main motivation to model a meteoroid stream is either to predict the corresponding meteor shower(s) or to study the stream structure. A relatively simple model was developed by e.g. Babadzhanov et al. (2008a,b, and references therein, as well as Section 7.4.2) in the case of a Jupiter-family stream. As the positions of meteoroids differ from that of the parent at any given time, the meteoroids experience different planetary perturbations, implying that the length of the cycles of variations of the angular orbital elements is also different. As a result, and as a first approximation, the sum of all meteoroid orbits corresponds to every stage of the evolutionary cycle. In other words, meteoroid orbits are similar to any orbit the parent body occupies over a full cycle. They produce a meteor shower when the heliocentric distance of the node is about 1 AU, i.e., when $q(1+e) \approx 1 - e\cos\omega$ or $q(1+e) \approx 1 + e\cos\omega$. For any given pair of (q, e), there are four possible values of argument of perihelion, ω. As a result, a stream might produce a nighttime shower with northern and southern branches at the pre-perihelion passage and a daytime shower with northern and southern branches at post-perihelion. If e and q of the parent body undergo strong changes during a whole cycle (through, e.g., Kozai resonance), then two sets of ω values are possible, leading to a total of eight possible meteor showers (see also Figure 7.6).

In order to identify such crossings, the orbital changes over one cycle of variation of the argument of perihelion need to be investigated. Hence, the orbit of the body is integrated over a long enough time to find out the period of its libration or circulation cycle. Together with the nominal orbit of the parent, a set of clones is studied as well.

7.5.2.4 Models Based on the Prediction of Meteor Showers

Another method to follow the evolution of a theoretical stream was worked out by Neslušan (1999), improved by Tomko and Neslušan (2012), and later used many times (Kaňuchová and Neslušan, 2007; Neslušan et al., 2013b,c, 2014a; Neslušan and Hajduková, 2014; Jakubík and Neslušan, 2015; Tomko and Neslušan, 2016; Hajduková and Neslušan, 2017). A set of models of a meteoroid stream, characterized with two free parameters (evolutionary time of the stream, t_{ev} and PR-effect parameter β) are considered. The well-known orbit of the parent body is integrated backward until an arbitrarily chosen perihelion passage which is nearest to time t_{ev}. A cloud of, typically, 10 000 test particles is released and integrated forward in time, until the present day. Then the particles whose orbits put them within 0.05 AU of the Earth's orbit are selected to predict the mean characteristics of the expected shower.

The distributions of t_{ev} and β are found by comparing the predicted and observed showers. The agreement implies that the particles with a given set of t_{ev} and β are also present in the real stream. A better description of the latter is achieved by superposing several successful models. For each predicted shower, a search for its corresponding real counterpart is performed in the list of meteor showers or searched for in a meteor database (see Section 7.6.3 and Williams et al., 2019).

7.5.3 Additional Notes

7.5.3.1 Note on the Prediction of Meteor Showers

The missed predictions of the Leonid outburst in 1899 and 1998 were a source of great disappointment for many people. It did not help either that, indeed, some later predictions were too optimistic (e.g., Vaubaillon et al., 2014). So, do such predictions work or not? Well, as often, it depends. When the parent body orbit is well known and stable, and has proved to produce heavily populated trails (such as e.g. 55P/Tempel-Tuttle), the forecasting works very well. However, most such cases are the easy cases and are more or less covered now (Jenniskens, 2006), although updates are always welcome. Nowadays, the research effort deals with poorly known parents and harder cases. A review of the consequences of the specific modeling circumstances on meteor shower forecasting and an attempt to qualify them was provided by Vaubaillon (2017).

7.5.3.2 Note on the Entropy of a Meteoroid Stream

Unless a meteoroid stream mean orbit is controlled by resonances, the natural dispersion of individual members increases with time (see Section 7.4). Therefore, when the stream is backward integrated, it may seem that the dispersion should decrease and point to the initial orbits, i.e., to the stream source. If this was the case, it would be straightforward to derive the origin and age of almost any meteoroid stream.

Unfortunately, the second law of thermodynamics, stating that the entropy of a dynamical system is always increasing, does not allow such a search. Specifically, the law is symmetrical with respect to time. Therefore the entropy increases from an arbitrary initial time to any final time, regardless of the time arrow. In other words, the dispersion of the orbits in a stream increases, paradoxically, regardless of whether we predict its dynamical evolution forward or backward in time.

In a theoretical model, the exact orbits of particles are integrated. If the integration of these orbits is performed forward in time, their dispersion increases. If it is then turned back, the dispersion decreases back to its starting point. This is not, however, valid for real meteoroids, whose orbits are determined with a relatively high uncertainty. Any backward integration results in an increase of orbit dispersion. Thus, it cannot reveal an indication of the relationship between well-known stream members and their parent body. Levison and Duncan (1994) also described this problem in the context of statistical incoming and outcoming comets into the orbital phase space of short-period comets. Rudawska et al. (2012) showed that even in numerical integrations, a slight uncertainty added to a set of meteoroid orbits all originating from a single trail can hardly point to their time of ejection.

It is worth mentioning that this result is especially true for meteoroids, for which the orbital elements are poorly constrained by measurements (Egal et al., 2017; Vida et al., 2018). However, for accurately measured asteroid orbits, backward integrations have allowed one to determine the age of a family from the time of the disruption of the main fragment a few million years ago (Nesvorný et al., 2002a,b).

Despite such limitations, many parent bodies were identified. In the next section, we outline some successes and current debates on both the parenthood and the ages of several streams.

7.6 Known Relationships between Meteoroid Streams and Parent Bodies: Review of Recent Progress

We deal now with the specific identifications of meteor showers associated to particular parent bodies via numerical simulations. We first provide an updated list of known parent bodies, and then focus on four major cases that illustrate the scientific knowledge and methods in this matter.

7.6.1 Recapitulation of All Known Parents

A summary of all identifications, performed via different modeling and including references, is given in Table 7.1. It is worth mentioning that most of these associations were identified after 2000, illustrating the (r)evolution dynamical studies coupled with large datasets have brought to meteor science. As a reminder, the identification of the parent body is often the first step in the forecasting of future meteor shower outbursts.

7.6.2 The Quadrantid (#0010) Complex, Comet 96P/Machholz 1 and Asteroid 196 256 (2003 EH$_1$)

A short time after their discoveries, both comet 96P and asteroid 196 256 were suggested to be the parent body of the Quadrantids (McIntosh, 1990; Jenniskens, 2004). Later, it was revealed that they are, most probably, the parents of a complicated meteoroid complex that causes several meteor showers at the Earth.

Comet 96P, as well as asteroid 196 256, occupy an orbital phase space shared with the Quadrantid stream. Even before their discoveries, the dynamics of the Quadrantids was investigated. Since the late seventies, it is well known that the inclination of Quadrantid orbits might drastically change, from $\sim 71^o$ down to $\sim 12^o$ (Williams et al., 1979; Hughes et al., 1979). Similarly, the perihelion distance is reduced from ~ 0.98 AU to 0.1 AU.

The vicinity to the 2:1 MMR with Jupiter causes the stream to split, within a time scale of 10^3 years, into filaments with distinct dynamical evolution (Froeschlé and Scholl, 1986). The complicated structure of the Quadrantid complex occurs due to the fact that the stream intersects the Earth's orbit four times and each of these parts can have both northern and southern branches. Specifically, the whole stream consists of Quadrantids, Northern and Southern δ-Aquariids, as well as Ursids and Carinids (Babadzhanov and Obrubov, 1987; Babadzhanov et al., 1991; Wu and Williams, 1992b).

According to McIntosh (1990), the perturbed behavior of 96P is similar to that of the Quadrantid stream except that their respective 4000-year orbit-libration cycles are shifted from each other by 2000 years. In addition, McIntosh noticed that the orbital evolution of the δ-Aquariid and Arietid meteoroid streams were also consistent with this behavior, except for a different phase shift.

The parent comet could be captured at its last close approach with Jupiter about 2200 years ago. If so, the resulting stream evolved enough to produce most of the features of the presently observed Quadrantid/Arietid/Southern δ-Aquariid complex (Jones and Jones, 1993).

196 256 was found in the corridor of the Quadrantid filament of the complex (Jenniskens, 2004) and, hence, suggested to be the parent body of the Quadrantids. Williams et al. (2004) subsequently summarized the situation with the Quadrantid parent with special attention to 196 256. Around AD 1500, the orbit of the latter was very similar to the mean orbit of the Quadrantids.

Subsequently, the relationship between asteroid 196 256 and the Quadrantids was studied by many authors (recently e.g. Porubčan and Kornoš, 2005; Ryabova and Nogami, 2005; Wiegert and Brown, 2005; Kaňuchová and Neslušan, 2007; Neslušan et al., 2013b,c; Kasuga and Jewitt, 2015). Specifically, Wiegert and Brown (2005) found as a part of the whole complex the following comets: 226P/Pigott-LINEAR-Kowalski (1783 W1), 5D/Brorsen, 206P/Barnard-Boattini (1892 T1), 96P/Machholz 1, and 141P/Machholz 2; asteroids 1994 JX, 1999 LT$_1$, 2000 PG$_3$, 2002 AR$_{129}$, 2002 KF$_4$, 2002 UO$_3$, 2003 YS$_1$, and 2004 BZ$_{74}$; as well as meteor showers: Arietids and δ-Aquariids. Wiegert and Brown (2005) considered that the Quadrantids had not been observed prior to 1835, and several other objects are in orbits close to the orbital phase space of asteroid 196 256. However, Hasegawa (1993) found, in the Chinese astronomical records, the activity of a meteor shower on 1798 December 31, which was identified with the Quadrantids. Lee et al. (2009) noticed a remarkable similarity of the orbits of all Quadrantids, the orbit of comet C/1490 Y1 (as determined by Hasegawa (1979)), and asteroid 196 256. The optical colors of 196 256's surface are consistent with the mean colors of dead or dormant cometary nuclei (Kasuga and Jewitt, 2015).

The Quadrantid stream could be reproduced with almost the same structure, considering either comet 96P or asteroid 196 256 (Kaňuchová and Neslušan, 2007). If non-gravitational effects are ignored, the prediction considering 96P only is consistent with the observed mean orbit of the Quadrantids, while that considering 196 256 is slightly different. The nominal orbits of the comet and the asteroid were extraordinarily similar during some short intervals of their circulation cycles (also the bodies themselves closely approached each other several times). This fact, however, does not necessarily imply a common physical origin of both objects, from a common progenitor via splitting.

Instead, there is a possibility that each object was independently captured by Jupiter into the same orbital phase space. The remarkable orbital similarity would then occur as a consequence of Jupiter's perturbing action.

Babadzhanov et al. (2008a) found eight orbits of asteroid 196 256 crossing the Earth's orbit within one circulation cycle. These orbits correspond to eight observed showers: Quadrantids, Carinids, Northern δ-Aquariids, Southern δ-Aquariids, η-Piscids and β-Arietids, α-Piscids, α-Draconids, and Puppid-Velids. For the latter, they identified the real showers in meteor catalogs using a threshold $D_{SH} = 0.25$ (Section 7.5.1).

In contrast, Neslušan et al. (2013b) stated that two of eight crossing points of the 96P orbit during its libration cycle are so close to two other ones that the corresponding showers cannot be distinguished. Hence, they predicted six showers from 96P's stream. The radiants of the predicted showers are shown in Figure 7.10 (upper plot). Four of these showers were reliably identified with the well-known major showers: daytime Arietids, Quadrantids, Southern δ-Aquariids, and Northern δ-Aquariids. In addition, they identified one of the predicted filaments of the northern branch of the daytime Arietids, and another one corresponding to the southern branch of this shower was also therefore predicted. Its characteristics vaguely resemble those of the α-Cetids, which were found to be a part of the 96P complex by Babadzhanov and Obrubov (1993).

Considering the mean orbits of the groups of Sun-grazing comets derived by Sekanina and Chodas (2005), the models by Neslušan et al. (2013b) predicted a similarity between the Arietids and the Marsden group of Sun-grazing comets. A relationship between the 96P complex and the Kracht group is questionable. Sekhar and Asher (2014b) confirmed the relationship between the Arietids (#171) and the Marsden group, but have not discussed the relationship of the stream with other groups (Kracht, Kreutz, Meyer, or any other one).

The Arietids, Southern and Northern δ-Aquariids, and the predicted southern branch of the Arietids have radiant areas located near the ecliptic and can therefore be classified as ecliptic showers (Jenniskens, 2006). The radiant area of the Quadrantids is located close to the north ecliptic pole and this shower can thus be classified as a toroidal shower. The sixth shower predicted by Neslušan et al. (2013b) is the toroidal counterpart of the Quadrantids and is located close to the south ecliptic pole. However, the lack of meteors in this part of the sky does not confirm the existence of this shower. Its identification with the κ-Velids, mentioned by Babadzhanov and Obrubov (1992) and Babadzhanov (1994) as a part of the 96P complex, is uncertain.

It is interesting to represent the radiant in a geocentric, Earth apex-centered longitude frame, where the geocentric ecliptic longitude λ_g is replaced with longitude $\lambda_A = \lambda_g - (\lambda + 270°)$, where λ is the meteor's solar longitude. In such a frame, the following symmetry of radiant areas with respect to the Earth apex is observed (see Figure 7.11): the Southern δ-Aquariids are symmetrical with the northern branch of the Arietids, the Northern δ-Aquariids with the southern branch of Arietids, and the Quadrantids with their southern toroidal counterpart. This symmetry reflects the circulation of similar mean orbital planes of symmetric pairs. Neslušan et al. (2013c) found the period of the circulation cycle of the argument of perihelion and the longitude of the node to be ~8200 years for comet 96P. During

Table 7.1. *A brief review of known generic relationships between meteor showers and their parent bodies found via a robust modeling of the stream and following its dynamical evolution. The identification of the showers given in parentheses is uncertain, or was not confirmed in the subsequent study(-ies) by other authors. Shower numbers # are IAU MDC identifiers.*

Parent Body	Associated Showers	Source of Information
2P/Encke, 2201 Oljato, 4184 Cuno, 4197 (1982 TA), 4341 Poseidon, 5025 P–L, 5143 Heracles, 5731 Zeus, 6063 Jason, 8201 (1994 AH$_2$), 16960 (1998 QS$_{52}$), 1990 HA, 1991 BA, 1991 GO, 1991 TB$_2$, 1993 KA$_2$, 1995 FF, 1996 RG$_3$, 1996 SK, 1998 VD$_{31}$, 1999 VK$_{12}$, 1999 VR$_6$, 2002 XM$_{35}$, 2003 QC$_{10}$, 2003 UL$_3$, 2003 WP$_{21}$, 2004 TG$_{10}$, (2101 Adonis), (2212 Hephaistos), (4496 Mithra)	Taurid Complex (*considered showers*: Northern Taurids Southern Taurids β-Taurids ζ-Perseids Southern Piscids o-Orionids)	Steel et al. (1991) Babadzhanov (1999, 2001) Babadzhanov et al. (2008b) Porubčan et al. (2006)
12P/Pons-Brooks	December κ-Draconids, # 336 (Northern June Aquilids, # 164)	Tomko & Neslušan (2016)
96P/Machholz 1 and 196256 (2003 EH$_1$)	daytime Arietids Southern δ-Aquariids, # 5 Northern δ-Aquariids, # 26 Quadrantids, # 10 (Ursids) =? (November ι-Draconids or December α-Draconids) (Carinids) =? (θ-Carinids) (κ-Velids, # 784) (α-Cetids) =? (daytime λ-Taurids) (η-Piscids) (β-Arietids) (α-Piscids) (α-Draconids) (Puppid-Velids)	Babadzhanov and Obrubov (1987) Babadzhanov and Obrubov (1992, 1993) Babadzhanov et al. (1991) Gonczi et al. (1992) Froeschlé et al. (1993) Williams et al. (2004) Porubčan and Kornoš (2005) Ryabova and Nogami (2005) Wiegert and Brown (2005) Kaňuchová and Neslušan (2007) Babadzhanov et al. (2008b) Lee et al. (2009) Neslušan et al. (2013a,b)
122P/de Vico	*unknown*	Tomko (2014)
126P/IRAS	*directly to sporadic*	Tomko and Neslušan (2012)
161P/Hartley-IRAS	*unknown*	Tomko and Neslušan (2012)
169P/NEAT = = 2002 EX$_{12}$	α-Capricornids, #1 (χ-Capricornids, # 114)	Jenniskens and Vaubaillon (2010) Kasuga et al. (2010)
255P/Levy	α-Cepheids	Šegon et al. (2017)
C/1917 F1 (Mellish)	December Monocerotids, # 19 (November Orionids, # 250) (April ρ-Cygnids, # 348)	Porter (1952) McCrosky and Posen (1961) Lindblad (1971); Kresáková (1974) Drummond (1981); Ohtsuka (1989) Lindblad and Olsson-Steel (1990) Vereš et al. (2011) Neslušan and Hajduková (2014)
C/1964 N1 (Ikeya)	July ξ-Arietids	Šegon et al. (2017)

Table 7.1. *(Cont.)*

C/1979 Y1 (Bradfield)	July Pegasids, # 175 = = *(former)* July γ-Pegasids, # 462 *(new)* α-Microscopiids *(transitory)* (γ-Bootids, # 104) (Southern α-Pegasids, # 522)	Rendtel et al. (1995, p. 169) Jenniskens (2006) Hajduková and Neslušan (2017)
2101 Adonis, 2008 BO$_{16}$, 2011 EC$_{41}$, 2013 CT$_{36}$	σ-Capricornids χ-Sagittariids Capricornids-Sagittariids χ-Capricornids	Babadzhanov et al. (2014a) Babadzhanov et al. (2015a)
2329 Orthos	ω-Herculids γ-Draconids =? J-Draconids =? =? ζ-Draconids ι-Draconids =? η-Draconids ζ-Bootids	Babadzhanov (1994, 1995)
3200 Phaethon (1983 TB)	Geminids, # 4 Sextantids, # 221 (Canis Minorids) (δ-Leonids)	Hughes (1983); Whipple (1983) Fox et al. (1984, 1985) Hunt et al. (1985) Kramer and Shestaka (1986) Bel'kovich and Riabova (1989) Williams and Wu (1993a) Gustafson (1989) Adolfsson and Gustafson (1991) Babadzhanov (1994); Beech (2002) Beech et al. (2003); Ryabova (2008b) de León et al. (2010) Jakubík and Neslušan (2015)
3552 Don Quixote (1983 SA)	κ-Lyrids, # 464 μ-Draconids, # 470	Rudawska and Vaubaillon (2015)
1997 GL$_3$, 2000 PG$_3$, 2002 GM$_2$, 2002 JC$_9$	Piscids	Babadzhanov et al. (2008a)
2001 MG$_1$ (and 2004 LA$_{12}$)	κ-Cygnids	Jones et al. (2006)
2001 XQ	66 Draconids, # 541	Šegon et al. (2017)
2002 JS$_2$, 2002 PD$_{11}$, 2003 MT$_9$	ι-Aquariids	Babadzhanov et al. (2009)
2003 HP$_2$, 2006 WX$_{29}$, 2007 VH$_{189}$, 2007 WT$_3$, 2007 WY$_3$, 2008 UM$_1$	χ-Scorpiids δ-Scorpiids β-Librids σ-Librids	Babadzhanov et al. (2013)
2004 CK$_{39}$	Northern ν-Virginids Southern ν-Virginids Northern Virgiinids Southern Virginids	Babadzhanov et al. (2012)
2007 CA$_{19}$	Northern η-Virginids Southern η-Virginids	Babadzhanov et al. (2014b) Babadzhanov et al. (2015b)
2008 ED$_{69}$	κ-Cygnids, # 12	Jenniskens and Vaubaillon (2008)
2008 GV	ψ-Draconids, # 754	Šegon et al. (2017)
2009 SG$_{18}$	κ-Cepheids, # 751	Šegon et al. (2017)
2009 WN$_{25}$	November Draconids, # 753	Šegon et al. (2017)
2015 TB$_{145}$	April Cygnids = ζ-Cygnids, # 40	Kokhirova et al. (2017)

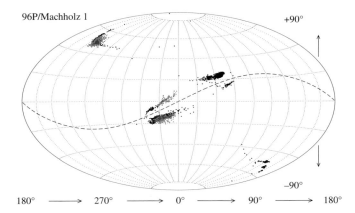

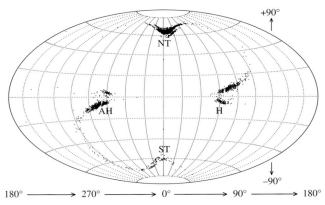

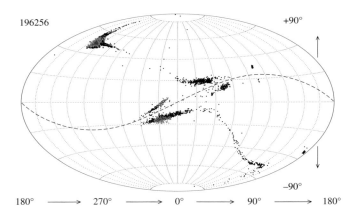

Figure 7.10. The distribution of the predicted and observed radiants of the meteoroid complex generated by 96P/Machholz 1 (upper plot) and asteroid 196 256 (lower plot). The black dots illustrate the radiants of the test particles approaching the Earth's orbit within 0.05 AU at the end of their followed orbital evolution, and considered as causing meteor showers. The dark gray (medium gray) dots show the radiants of real meteors of three of the 96P-complex showers, Quadrantids, Northern δ-Aquariids, and Southern δ-Aquariids, identified in the IAU MDC photographic (SonotaCo video) databases (taken from Neslušan et al., 2014b). (A black and white version of this figure will appear in some formats. For the colour version, please refer to the plate section.)

Figure 7.11. The ecliptic-toroidal structure of the meteoroid complex associated with asteroid 196 256. The predicted radiants are shown in an ecliptic coordinate frame with an origin at the Earth apex. Abbreviations NT, ST, H, and AH stand for the northern toroidal, southern toroidal, helion, and antihelion directions, respectively. NT can be identified with the Quadrantids, H with the Arietids (both well-known northern and predicted southern strands), and AH with the Northern and Southern δ-Aquariids (taken from Neslušan et al., 2013c, reproduced with permission © ESO).

this period, the perihelion distance, eccentricity, and inclination exhibit two similar libration cycles that last ∼4100 years.

Similar conclusions can also be drawn for the stream ejected from asteroid 196 256 (Neslušan et al., 2013c). The latter has practically the same ecliptic-toroidal structure and six showers are predicted. The theoretical radiants are shown in Figure 7.10 (bottom). Comparing the models for various evolutionary times, t_{ev}, and for both parent bodies, a significant difference is observed at the beginning of their dynamical evolution: while the daytime Arietids are caused by the stream of 96P, the Quadrantids come from the stream of 196 256 for $t_{ev} = 500$ years. Later, other showers of the complex, gradually appear in both cases. After about 2500 years, the streams originating from both parent bodies are practically indistinguishable. The Kozai cycle for 196 256 is similar to that of 96P and was estimated to last about 7500 years by Wiegert and Brown (2005).[3]

The core of the Quadrantid stream originates from 196 256 and is relatively young. Its estimated lifetime is 200–300 years (Wiegert and Brown, 2004, 2005). However, no recent activity of 196 256 was observed (Kasuga and Jewitt, 2015); therefore, a release of meteoroid particles from this parent body was episodic if it is actually the parent body of Quadrantids. The rest of the Quadrantids, as well as the other showers of the complex, can be explained by meteoroids mainly ejected from 96P/Machholz 1 between 10 000 BCE and 20 000 BCE (Abedin et al., 2018).

Abedin et al. (2018) suggested that Jupiter may have captured a proto-comet 96P circa 20 000 BCE and a subsequent major breakup around 100–950 CE resulted in the formation of the Marsden group of sungrazing comets. They also concluded that the meteor showers identified by Babadzhanov and Obrubov (1992) as the α-Cetids, Ursids, and Carinids correspond to the daytime λ-Taurids, November ι-Draconids or December α-Draconids, and the θ-Carinids.

It is worth mentioning that other objects were suggested as the parent body of the Quadrantids before the discovery of 96P and 196 256. Hasegawa (1979) derived the orbital elements of several historical comets and concluded that comet C/1490 Y1 might be the parent comet of the Quadrantids. Williams and Wu (1993b) concluded that Hasegawa's hypothesis is consistent with the data if the eccentricity of the comet is assumed to be 0.77 (instead of considering a parabolic orbit). Searching

---

[3] According to the modeling performed by Neslušan et al. (2013b), the first Quadrantid meteoroids could hit the Earth about 2000 years after their release from 96P's surface. If we combined this result with the suggestion claimed by Jones and Jones (1993) that 96P might be captured to the orbital phase space – where it is currently situated – about 2200 years ago, then it would be possible to explain why no Quadrantids were observed prior to approximately AD 1800. However, no meteoroids of this shower could then originate from 196 256.

for other Quadrantid parents, Williams and Collander-Brown (1998) suggested that, besides 96P, asteroid 5496 (1973 NA) could contribute to the stream. Further information about the Quadrantids and their parent bodies is also presented by Kasuga and Jewitt (2019, Section 8.3.2).

7.6.3 July Pegasids (#0175) and Comet C/1979 Y1 (Bradfield)

The July Pegasids are a weak meteor shower with a short activity period, observed by many surveys (Rendtel et al., 1995, p. 169; Molau and Rendtel, 2009; Ueda, 2012; Andreić et al., 2013; Gural, 2011; Jenniskens et al., 2011, and references therein). Jenniskens (2006, table 7, p. 715) suggested comet C/1979 Y1 (Bradfield) as the possible parent body. Later, Holman and Jenniskens (2012) also proposed comet C/1771 A1 as a possible parent body, suggesting that both C/1979 Y1 and C/1771 A1 may either be the same comet or originate from a common progenitor at the time of the formation of the July Pegasid shower.

Hajduková and Neslušan (2017) modeled the stream ejected by C/1979 Y1 and found it crosses the Earth's orbit twice, causing two potential showers: one before (July Pegasids) and one after (α-Microscopiids) perihelion. This is analogous to the stream of comet 1P/Halley which creates the η-Aquariid and Orionid showers. The α-Microscopiids are a daytime shower since the angular distance of their radiant from the Sun is only $\sim 62°$. It should be active in the middle of January. The radiant location of the particles from C/1979 Y1 not significantly influenced by the non-gravitational PR-effect, are symmetrical with respect to the Earth's apex. The mean orbits of both predicted showers are almost identical implying that the showers are caused by the meteoroids orbiting the Sun in a single filament.

Because of orbital-elements similarity with the July Pegasids, the meteor shower #462, July γ-Pegasids was identified as similar to the July Pegasids (Kornoš et al., 2014; Rudawska and Jenniskens, 2014) and was subsequently removed from the IAU MDC list. Similarly, Andreić et al. (2013) deemed the Southern α-Pegasids (#522) identical to the July Pegasids (#175) and recommended it to be removed from the list.

Theoretical radiants for $\beta = 10^{-5}$ and a range of values of evolutionary times, t_{ev}, are shown in Figure 7.12a. The size of the radiant location increases with increasing evolutionary time, showing that the spatial corridor of the July Pegasids is not delimited by planetary resonances. The dispersion of the shower members increases with time, therefore a normal aging of the shower is observed. Comparing the theoretical and observed radiant locations, the July Pegasids seem younger than 10 kyr (Figure 7.12). For $t_{ev} > 10$ kyr the size of the radiant area does not match its observed counterpart. If the stream is modeled for larger β-values (i.e. a significant PR-effect), the modeled radiant area is shifted and does not match the observed one.

Modeling the meteoroid stream ejected from C/1979 Y1 and considering relatively large values of P-R-effect $\beta \geq 0.0045$, a shower is rarely predicted (Hajduková and Neslušan, 2017). As expected, a stronger P-R effect deflects the particles away from the Earth-intersecting trajectories. However, a complete deflection does not always occur. Sometimes, the stream is

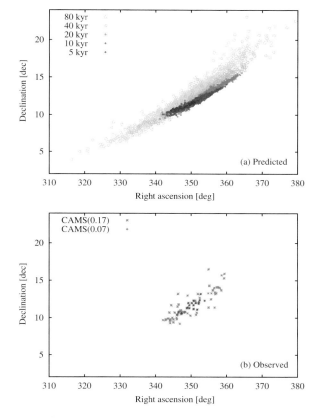

Figure 7.12. The positions of geocentric radiants of theoretical particles (upper plot) corresponding to the July Pegasids, IAU #175, and real July Pegasids (lower plot) separated from the CAMS video database (medium gray asterisks for $D_{lim} = 0.17$ and dark gray crosses for $D_{lim} = 0.07$). The theoretical radiants were obtained for $\beta = 10^{-5}$ and $t_{ev} = 5, 10, 20, 40,$ and 80 kyr. The positions for the different t_{ev} are distinguished using the different marks and shades of gray (taken from Hajduková and Neslušan, 2017, reproduced with permission © ESO). (A black and white version of this figure will appear in some formats. For the colour version, please refer to the plate section.)

deflected away from the Earth's orbit in its arc corresponding to the July Pegasid-α-Microscopiid filament, but it starts to cross this orbit at the other arc. In contrast to the July Pegasid-α-Microscopiid filament, a new filament occurs only for specific combinations of t_{ev} and β. Therefore, it is expected to survive only during a limited period and, likely, to consist of particles that have a narrow interval of sizes. Such a filament is referred to as a "transitory filament". It causes a "transitory shower", in contrast to the July Pegasid-α-Microscopiid filament, predicted in many models and regarded as the filament causing the annual showers.

Specifically, two largely different transitory filaments occur for $0.005 \leq \beta \leq 0.008$ and $0.005 \leq \beta \leq 0.010$, but only after a relatively long time $t_{ev} \geq 40$ kyr. These filaments differ from each other, especially by their geocentric velocity: 50.0 versus 37.7 km s^{-1}. One transitory filament can vaguely be identified with γ-Bootids (#104). It is worth mentioning that the radiant areas of transitory filaments are never symmetrical with respect to the Earth's apex.

7.6.4 Geminids (#0004) and Asteroid 3200 Phaethon (1983 TB)

Soon after the discovery of asteroid 3200 Phaethon (initially designated 1983 TB), Hughes (1983) and Whipple (1983) almost simultaneously reported its possible relationship with the Geminid stream. Subsequently, several studies proved that Phaethon is the parent body of the Geminids (Fox et al., 1984, 1985; Hunt et al., 1985; Kramer and Shestaka, 1986; Williams and Wu, 1993a). Everyone noticed the similarity between its orbit and that of the stream.

In the period 1989–2003, several authors found that the age of the Geminids must be relatively low, 500–4000 years (Gustafson, 1989; Adolfsson and Gustafson, 1991; Beech, 2002; Ryabova, 1999; Beech et al., 2003). Beech (2002) also determined the tensile strengths of the meteoroids of $\sim 3 \cdot 10^5$ Pa and interpreted it as being supportive of the argument that the parent body to the Geminid stream, 3200 Phaethon, is an aged cometary nucleus.

The debate about Phaethon's nature is still going on (see also Kasuga and Jewitt, 2019, Section 8.3.1). Is it an activated asteroid, a dormant comet, or a dead comet? What is the dynamical pathway that led it to its present specific orbit: small size (semi-major axis $a < 2$ AU), and very small perihelion distance ($q < 0.15$ AU)? However the very existence of the Geminid meteoroid stream, *observed at the Earth*, and its very peculiar structure place a constraint on the origin of Phaethon. Bimodality of the Geminids' activity profile (see details in Ryabova, 2001) and its high stability (Rendtel, 2004), as well as the absence of the nodal shift with time (McIntosh and Šimek, 1980; Porubčan et al., 1980), must be explained.

Ryabova (2007, 2008a) presented a mathematical model explaining these and other features of the Geminids. The hypothesized life scenario of the comet-candidate hypothesis is as follows: 1) capture of a Sun-skirting orbit, 2) catastrophic gas and dust release with strong transformation of the comet's orbit, 3) conversion to the asteroid. The Geminids' structure is in good agreement with this theory.

A dynamical pathway from the Pallas family to Phaethon requires > 0.3 Myr for delivery to the present orbit (de León et al., 2010; Todorović, 2018). However it does not solve the origin of the Geminids.

Since its discovery, Phaethon had never shown any trace of activity until observed in 2009 by Jewitt and Li (2010), showing evidence of weak activity. It was again observed in 2012 (Li and Jewitt, 2013) and 2016 (Hui and Li, 2017). Thermal fracture/decomposition of the surface was considered the most probable mechanism for the activity (Jewitt and Li, 2010), but it cannot be the main source of the Geminids (Ryabova, 2018).

Two more asteroids, namely 155 140 (2005 UD) and 1999 YC were considered as a candidates for the Phaethon-Geminid complex (Ryabova, 2008b, see the review and the references herein). A preliminary study (Ryabova et al., 2014) showed that 2005 UD may be a remnant of the progenitor from which Phaethon and/or 1999 YC split, i.e. 1999 YC is not an immediate relative of Phaethon, and this is supported by the results of Ohtsuka et al. (2008).

A recent modeling of the Phaethon meteoroid stream on a timespan of 100 kyr has shown that the PR-effect is important to explain the dynamics of the Geminid stream (Jakubík and Neslušan, 2015). Namely, if the PR-effect is neglected, the corridor of the stream is predicted to be clearly aside of its observed counterpart. Only after considering a significant influence of the PR-effect, the modeled and observed streams become identical, whereby the border of the Ω-distribution is $\approx 254.5°$, in agreement with the photographically observed border. The PR-effect parameter $\beta \approx 0.017$ for the particles at the border.

7.6.5 Parent Asteroids of the Taurid Meteor Complex (#0002, #0017)

The primary parent body of the Taurids is most probably comet 2P/Encke. However, the Taurid stream structure consists of several filaments hitting the Earth during at least four months of the year. Because of this complex structure and the absence of more comets in the Taurid orbital elements phase space, additional parent bodies were searched among near-Earth asteroids (NEAs). Steel et al. (1991) investigated if the Taurid complex might originate from a giant comet, entering the inner Solar System some time in the past 10 000–20 000 years. However, in the absence of other explanations, the origin of the Taurid complex from a giant comet is possible only if unrealistically large ejection velocities are assumed.

Babadzhanov et al. (2008c, and references therein) proposed several NEAs as other parent bodies of the Taurids (Table 7.1). In addition, several showers were regarded as part of the Taurid complex: along with Northern and Southern Taurids were also the β-Taurids and the ζ-Perseids. Porubčan et al. (2006) pointed out a possible association between seven Taurid filaments and nine NEAs. The most probable of these associations are the Southern Piscids (b) with 2003 QC_{10}, Northern Taurids (a) with 2004 TG_{10}, o-Orionids with 2003 UL_3, and Northern Taurids (b) with 2002 XM_{35}.[4] The list of the Taurid-complex NEAs was revisited several times and its last version may neither be exact nor definitive (see also Kasuga and Jewitt, 2019, Section 8.3.4). Our knowledge of the Taurid complex and its parent bodies remains open.

7.7 Future Work

Probably the biggest challenge today is the measurement of meteoroid orbits. A meteor lasts a few seconds at most, enabling us to sample its trajectory over a *few dozen km*, while its orbit might span several *AU*, i.e. *several hundred million km*. It is therefore no surprise that the derivation of a meteoroid orbit is a highly challenging task. Any dynamicist's dream would be, e.g., to distinguish two trails from the 2001 Leonid outburst (Arlt et al., 2001) or to detect resonant meteoroids, only from the measurement of orbital elements. This would require a meteor-speed precision of a few *dm/s* at least, and a semi-major axis accuracy better than 0.01 AU. Both seem unreachable today, not only because of technical limitations, but also because of the very method to derive

---

[4] Small letters "a" and "b" were used by the authors to distinguish between the filaments of the Southern Piscids and the Taurids.

the orbital elements and the influence of atmospheric drag (Egal et al., 2017; Vida et al., 2018). However, with the recent progress of data reduction, it would be worth revisiting all published meteor orbits, in order to help revise our vision of the existence of meteor showers, as well as to reduce the searched orbital elements space phase of potential parents. Better orbits will also help the identification of the showers from large datasets, and of parent bodies.

Complete cometary models able to fully describe the ejection of meteoroids from comet nuclei can be constructed by fitting the physical parameters to existing and thorough observational data (Maquet et al., 2012). The comparison of models of the cometary dust trail and the associated meteoroid stream will be worth investigating.

Possible dynamical perturbations usually neglected in the dynamics of meteoroid streams include: minor planets Ceres and Pluto, massive asteroids (e.g. Vesta, Pallas, etc.), the asteroid belt (modeled as a ring, if any), galactic tide (if one wants to study Oort cloud meteoroids) and passing stars. It would be an interesting challenge as well to define an experiment to measure the impact of the Yarkovsky effect on individual meteoroids as well as on a stream as a whole.

As of now, various authors have only looked into two-body resonances in sub-stream structures due to Jupiter, Saturn and Uranus. It is possible to find such mechanisms (on a smaller scale and strength) in sub-stream parts due to other planets as well. Such comparisons will help us understand the finer nuances of resonances in the Solar System in terms of resonant libration widths, overlapping resonances, survival time scales, chaotic behavior between resonant boundaries etc. In practice it would help to understand various clustering phenomena of small particles (due to resonances), which matters for the safety of artificial satellites.

Chances are that additional three-body resonances will be identified in meteoroid streams and meteor outbursts will be found due to unique 3BMMRs. Such identifications will help in understanding the similarities and differences between two-body resonances and three-body resonances from a dynamical, as well as observational, point of view.

Because three-body resonances are strongly connected with small stability islands surrounded by chaotic spaces, it would be useful to find whether there are cases where three-body resonances deplete and depopulate parts of streams, thus leading to diminished meteor activity for certain showers in some years. A detailed study using chaos indicators and ranking chaotic levels would be a good future step in this direction.

Similarly, the influence of general relativity (GR) is still difficult to measure directly on meteoroid streams with present technology. However, the real key in future is to find the difference in long-term evolution of low perihelion distance streams with and without GR corrections. This will enable us to indirectly measure the effects of GR for sub-structures inside the stream and correlate with real observations by comparing when they intersect the Earth using the time of intersection and duration of enhanced activity. Subtler effects like the Lense-Thirring effect are even harder to measure, but with the accuracy of observations, measurements and clocks increasing rapidly every year, the time is not far off when we will be able to verify seconds of delay or advance between various sub-stream structures hitting the Earth after they pass close to the Sun repeatedly for a long time.

Another task for the future is finding the long-term effect of the Kozai mechanism in meteoroid streams. Because the Kozai mechanism (also see Kasuga and Jewitt, 2019, Section 8.5.4) is effective in the production of Sun-grazers and making comets go into Sun-skirting, Sun-grazing and Sun-colliding trajectories, it would be worthwhile to check whether there could be sub-stream structures (ejected from such comets) falling into the Sun on a regular basis due to this dynamical effect. Rather than just focusing on Jupiter alone, it would be insightful to find Kozai effects induced by other giant planets as well. This could help one compare between Kozai effects in multiple small bodies induced by multiple massive bodies in the Solar System.

Even the definition of meteor shower might change in the future, given the new tools developed to identify such features in ever-growing meteoroid orbit databases. On one hand, the amount of data reveals more showers, and on the other hand, dynamical works as described here tell us that several showers regarded as different and unrelated are just parts of a single meteoroid stream evolutionary path.

Currently, there are 27714 meteor orbits with the aphelia larger than 50 AU in public meteor databases,[5] and 51906 hyperbolic orbits.[6] If there is another planet beyond Neptune, these orbits have to be influenced by it and the data carry information about its mass and position. The long-term modeling of meteoroid streams can help us to reveal a massive object in the trans-neptunian region, if any exists. The very existence of interstellar meteoroids is the topic of Hajduková et al. (2019; see also Chapter 10).

Acknowledgements

Sekhar acknowledges the Crater Clock (235058/F20) project based at CEED (through the Centres of Excellence scheme project number 223272 funded by the Research Council of Norway) and thanks USIT UNINETT Sigma2 computational resource allocation through the Stallo cluster for numerical simulations. Neslušan acknowledges a partial support by VEGA - the Slovak Grant Agency for Science (grant No. 2/0037/18). Ryabova acknowledges support by Tomsk State University International Competitiveness Improvement Programme for 2013–2020. Vaubaillon is thankful to CINES (Montpelier, France) supercomputer facility.

References

Abedin, A., Spurný, P., Wiegert, P. et al. 2015. On the age and formation mechanism of the core of the Quadrantid meteoroid stream. *Icarus*, **261**, 100–117.

---

[5] 342, 5272, 14904, 6811, and 385, respectively in IAU MDC photographic and video – CAMS, SonotaCo, EDMOND, and radio-meteor databases, consulted in January 2018.
[6] 575, 13914, 27243, 8166, and 2008 respectively.

Abedin, A., Wiegert, P., Pokorný, P., and Brown, P. 2017. The age and the probable parent body of the daytime Arietid meteor shower. *Icarus*, **281**, 417–443.

Abedin, A., Wiegert, P., Janches, D. et al. 2018. Formation and past evolution of the showers of 96P/Machholz complex. *Icarus*, **300**, 360–385.

Adams, J. C. 1867. On the orbit of the November meteors. *Monthly Notices of the Royal Astronomical Society*, **27**, 247–252.

Adolfsson, L., and Gustafson, B. Å. S. 1991. Geminid meteoroids and the probability for cometary activity on Phaethon. *Bulletin of the American Astronomical Society*, **23**, 1141.

Andreić, Ž., Šegon, D., Korlević, K. et al. 2013. Ten possible new showers from the Croatian Meteor Network and SonotaCo datasets. *WGN, the Journal of the International Meteor Organization*, **41**, 103–108.

Arlt, R. 1998. Bulletin 13 of the International Leonid Watch: The 1998 Leonid Meteor Shower. *WGN, the Journal of the International Meteor Organization*, **26**, 239–248.

Arlt, R., Kac, J., Krumov, V., Buchmann, A., and Verbert, J. 2001. Bulletin 17 of the International Leonid Watch: First global analysis of the 2001 Leonid storms. *WGN, the Journal of the International Meteor Organization*, **29**, 187–194.

Arter, T. R., and Williams, I. P. 1997. The mean orbit of the April Lyrids. *Monthly Notices of the Royal Astronomical Society*, **289**, 721–728.

Asher, D. J. 1991. *The Taurid Meteoroid Complex*. D.Phil. thesis, Oxford University.

Asher, D. 1994. Meteoroid swarms and the Taurid Complex. Pages 88–91 of: Roggemans, P. (ed), *Proceedings of the International Meteor Conference, Puimichel, France, 1993*. Mechelen, BE: International Meteor Organization.

Asher, D., and Izumi, K. 1998. Taurid Swarm Appearing in 1998? *WGN, the Journal of the International Meteor Organization*, **26**, 217.

Asher, D. J., Bailey, M. E., and Emel'yanenko, V. V. 1999. Resonant meteoroids from Comet Tempel-Tuttle in 1333: The cause of the unexpected Leonid outburst in 1998. *Monthly Notices of the Royal Astronomical Society*, **304**, L53–L56.

Babadzhanov, P. B. 1994. Asteroids and their meteor showers. In: Kozai, Y., Binzel, R. P., and Hirayama, T. (eds), *75 Years of Hirayama Asteroid Families: The Role of Collisions in the Solar System History*. Astronomical Society of the Pacific Conference Series, vol. 63, pp. 168–185.

Babadzhanov, P. B. 1995. Orthos' meteor showers. *Earth, Moon, and Planets*, **68**, 165–170.

Babadzhanov, P. B. 1999. Near-Earth asteroids associated with meteor showers. Pages 185–190 of: Baggaley, W. J., and Porubčan, V. (eds), *Meteoroids 1998, Proceedings of the International Conference Held at Tatranska Lomnica, Slovakia, August 17–21, 1998*. Bratislava: Astronomical Institute of the Slovak Academy of Sciences.

Babadzhanov, P. B. 2001. Search for meteor showers associated with Near-Earth asteroids. I. Taurid Complex. *Astronomy and Astrophysics*, **373**, 329–335.

Babadzhanov, P. B., and Obrubov, I. V. 1987. Evolution of meteoroid streams. *Publications of the Astronomical Institute of the Czechoslovak Academy of Sciences*, **67**, 141–150.

Babadzhanov, P. B., and Obrubov, Y. V. 1992. Comet Machholz 1986 VIII and the Quadrantid meteoroid swarm–orbital evolution and relationship. *Solar System Research*, **26**, 288.

Babadzhanov, P. B., and Obrubov, Y. V. 1993. Unknown meteor showers of comet Machholz and asteroid Phaethon. *Solar System Research*, **27**, 194–200.

Babadzhanov, P. B., Obrubov, Y. V., and Pushkarev, A. N. 1991. Evolution of the Quadrantid meteoroid swarm. *Solar System Research*, **25**, 63.

Babadzhanov, P. B., Williams, I. P., and Kokhirova, G. I. 2008a. Meteor showers associated with 2003EH1. *Monthly Notices of the Royal Astronomical Society*, **386**, 2271–2277.

Babadzhanov, P. B., Williams, I. P., and Kokhirova, G. I. 2008b. Near-Earth asteroids among the Piscids meteoroid stream. *Astronomy and Astrophysics*, **479**, 249–255.

Babadzhanov, P. B., Williams, I. P., and Kokhirova, G. I. 2008c. Near-Earth objects in the Taurid complex. *Monthly Notices of the Royal Astronomical Society*, **386**, 1436–1442.

Babadzhanov, P. B., Williams, I. P., and Kokhirova, G. I. 2009. Near-Earth asteroids among the Iota Aquariids meteoroid stream. *Astronomy and Astrophysics*, **507**, 1067–1072.

Babadzhanov, P. B., Williams, I. P., and Kokhirova, G. I. 2012. Near-Earth object 2004CK39 and its associated meteor showers. *Monthly Notices of the Royal Astronomical Society*, **420**, 2546–2550.

Babadzhanov, P. B., Williams, I. P., and Kokhirova, G. I. 2013. Near-Earth asteroids among the Scorpiids meteoroid complex. *Astronomy and Astrophysics*, **556**, A25.

Babadzhanov, P. B., Kokhirova, G. I., and Khamroev, U. K. 2014a. The Capricornids asteroid-meteoroid complex. Pages 199–204 of: Jopek, T. J., Rietmeijer, F. J. M., Watanabe, J., and Williams, I. P. (eds), *Proceedings of the Meteoroids 2013 Conference, Aug. 26–30, 2013, A. M. University, Poznań, Poland*. Adam Mickiewicz University Press.

Babadzhanov, P., Kokhirova, G., and Obrubov, Y. 2014b. The near-Earth asteroid 2007 CA$_{19}$ as a parent of the Eta-Virginids meteoroid stream. Page 42 of: Muinonen, K., Penttilä, A., Granvik, M. et al. (eds), *Asteroids, Comets, Meteors – Book of Abstracts*. Helsinki: University of Helsinki.

Babadzhanov, P. B., Kokhirova, G. I., and Khamroev, U. K. 2015a. The σ-Capricornids complex of near-Earth objects. *Advances in Space Research*, **55**, 1784–1791.

Babadzhanov, P. B., Kokhirova, G. I., and Obrubov, Y. V. 2015b. The potentially hazardous asteroid 2007CA19 as the parent of the η-Virginids meteoroid stream. *Astronomy and Astrophysics*, **579**, A119.

Beaugé, C., Giuppone, C. A., Ferraz-Mello, S., and Michtchenko, T. A. 2008. Reliability of orbital fits for resonant extrasolar planetary systems: the case of HD82943. *Monthly Notices of the Royal Astronomical Society*, **385**, 2151–2160.

Beech, M. 2002. The age of the Geminids: a constraint from the spin-up time-scale. *Monthly Notices of the Royal Astronomical Society*, **336**, 559–563.

Beech, M., Illingworth, A., and Murray, I. S. 2003. Analysis of a "flickering" Geminid fireball. *Meteoritics and Planetary Science*, **38**, 1045–1051.

Bel'kovich, O. I., and Riabova, G. O. 1989. The formation of the Geminid meteor stream upon the disintegration of the cometary nucleus. *Astronomicheskii Vestnik*, **23**, 157–163.

Bockelee-Morvan, D., Crovisier, J., and Gerard, E. 1990. Retrieving the coma gas expansion velocity in P/Halley, Wilson (1987 VII) and several other comets from the 18-cm OH line shapes. *Astronomy and Astrophysics*, **238**, 382–400.

Boissier, J., Bockelée-Morvan, D., Biver, N. et al. 2014. Gas and dust productions of Comet 103P/Hartley 2 from millimetre observations: Interpreting rotation-induced time variations. *Icarus*, **228**, 197–216.

Brisset, J., Heißelmann, D., Kothe, S., Weidling, R., and Blum, J. 2016. Submillimetre-sized dust aggregate collision and growth properties. Experimental study of a multi-particle system on a suborbital rocket. *Astronomy and Astrophysics*, **593**, A3.

Brouwer, D., and van Woerkom, A. J. J. 1950. The secular variations of the orbital elements of the principal planets. *Astron. Papers Amer. Eph. Naut. Almanac*, **13**, Pt. 2, pp. 81–107.

Brown, P. G. 1999. Evolution of Two Periodic Meteoroid Streams: The Perseids and Leonids. Ph.D. thesis, University of Western Ontario.

Brown, P., and Jones, J. 1998. Simulation of the formation and evolution of the Perseid meteoroid stream. *Icarus*, **133**, 36–68.

Brown, P., Weryk, R. J., Wong, D. K., and Jones, J. 2008. A meteoroid stream survey using the Canadian Meteor Orbit Radar. I. Methodology and radiant catalogue. *Icarus*, **195**, 317–339.

Burns, J. A., Lamy, P. L., and Soter, S. 1979. Radiation forces on small particles in the solar system. *Icarus*, **40**, 1–48.

Christou, A. A., Vaubaillon, J., and Withers, P. 2008. The P/Halley stream: meteor showers on Earth, Venus and Mars. *Earth, Moon, and Planets*, **102**, 125–131.

Christou, A. A., Killen, R. M., and Burger, M. H. 2015. The meteoroid stream of comet Encke at Mercury: Implications for MErcury Surface, Space ENvironment, GEochemistry, and Ranging observations of the exosphere. *Geophysical Research Letters*, **42**, 7311–7318.

Christou, A., Vaubaillon, J., Withers, P., Hueso, R., and Killen, R. 2019. Chapter 5: Extra-terrestrial meteors. Ryabova, G. O., Asher, D. J., and Campbell-Brown, M. D. (eds), *Meteoroids: Sources of Meteors on Earth and Beyond*. Cambridge, UK: Cambridge University Press, pp.119–135.

Combi, M. R., Harris, W. M., and Smyth, W. H. 2004. Gas dynamics and kinetics in the cometary coma: theory and observations. Pages 523–552 of: Festou, M. C., Keller, H. U., and Weaver, H. A. (eds), *Comets II*. Tucson: University of Arizona Press.

Crifo, J. F. 1995. A general physicochemical model of the inner coma of active comets. 1: Implications of spatially distributed gas and dust production. *The Astrophysical Journal*, **445**, 470–488.

Crifo, J. F., and Rodionov, A. V. 1997. The dependence of the circumnuclear coma structure on the properties of the nucleus. *Icarus*, **127**, 319–353.

de Almeida, A. A., Sanzovo, G. C., Singh, P. D. et al. 2007. On the relationship between visual magnitudes and gas and dust production rates in target comets to space missions. *Advances in Space Research*, **39**, 432–445.

de León, J., Campins, H., Tsiganis, K., Morbidelli, A., and Licandro, J. 2010. Origin of the near-Earth asteroid Phaethon and the Geminids meteor shower. *Astronomy and Astrophysics*, **513**, A26.

Dikarev, V., Grün, E., Baggaley, J. et al. 2005. The new ESA meteoroid model. *Advances in Space Research*, **35**, 1282–1289.

Drolshagen, G., Koschny, D., Drolshagen, S., Kretschmer, J., and Poppe, B. 2017. Mass accumulation of Earth from interplanetary dust, meteoroids, asteroids and comets. *Planetary and Space Science*, **143**, 21–27.

Drummond, J. D. 1981. A test of comet and meteor shower associations. *Icarus*, **45**, 545–553.

Egal, A., Gural, P. S., Vaubaillon, J., Colas, F., and Thuillot, W. 2017. The challenge associated with the robust computation of meteor velocities from video and photographic records. *Icarus*, **294**, 43–57.

Einstein, A. 1915. Erklärung der Perihelionbewegung der Merkur aus der allgemeinen Relativitätstheorie. *Sitzungsberichte der Königlich Preussischen Akademie der Wissenschaften*, **47**(2), 831–839.

Emel'yanenko, V. V. 2001. Resonance structure of meteoroid streams. Pages 43–45 of: Warmbein, B. (ed) *Proceedings of the Meteoroids 2001 Conference in Kiruna, Sweden, 6–10 August 2001*. ESA SP-495. Noordwijk: ESA Publications Division.

Finson, M. J., and Probstein, R. F. 1968. A theory of dust comets. I. Model and equations. *The Astrophysical Journal*, **154**, 327–352.

Fox, K., Williams, I. P., and Hughes, D. W. 1982. The evolution of the orbit of the Geminid meteor stream. *Monthly Notices of the Royal Astronomical Society*, **200**, 313–324.

Fox, K., Williams, I. P., and Hughes, D. W. 1984. The 'Geminid' asteroid (1983 TB) and its orbital evolution. *Monthly Notices of the Royal Astronomical Society*, **208**, 11P–15P.

Fox, K., Williams, I. P., and Hunt, J. 1985. The past and future of 1983 TB and its relationship to the Geminid meteor stream. Pages 143–148 of: Carusi, A., and Valsecchi, G. B. (eds), *IAU Colloq. 83: Dynamics of Comets: Their Origin and Evolution*. Astrophysics and Space Science Library, vol. 115. Dordrecht: D. Reidel Publishing Co.

Froeschlé, C., and Scholl, H. 1986. Gravitational splitting of Quadrantid-like meteor streams in resonance with Jupiter. *Astronomy and Astrophysics*, **158**, 259–265.

Froeschlé, C., Gonczi, R., and Rickmam, H. 1993. New results on the connection between comet P/Machholz and the Quadrantid meteor streams: Poynting-Robertson drag and chaotic motion. Pages 169–172 of: Štohl, J., and Williams, I. P. (eds), *Meteoroids and Their Parent Bodies*. Bratislava: Astronomical Institute of the Slovak Academy of Sciences.

Fulle, M., Marzari, F., Della Corte, V. et al. 2016. Evolution of the dust size distribution of Comet 67P/Churyumov-Gerasimenko from 2.2 au to perihelion. *The Astrophysical Journal*, **821**, 19.

Gallardo, T. 2014. Atlas of three body mean motion resonances in the Solar System. *Icarus*, **231**, 273–286.

Galligan, D. P., and Baggaley, W. J. 2002. Wavelet enhancement for detecting shower structure in radar meteoroid data I. methodology. Page 42 of: Green, S. F., Williams, I. P., McDonnell, J. A. M., and McBride, N. (eds), *IAU Colloq. 181: Dust in the Solar System and Other Planetary Systems*. Oxford: Pergamon.

Galushina, T. Y., Ryabova, G. O., and Skripnichenko, P. V. 2015. The force model for asteroid (3200) Phaethon. *Planetary and Space Science*, **118**, 296–301.

Giorgini, J. D., Yeomans, D. K., Chamberlin, A. B. et al. 1996. JPL's on-line solar system data service. Page 1158 of: *AAS/Division for Planetary Sciences Meeting Abstracts #28. Bulletin of the American Astronomical Society*, **28**, 1158.

Gonczi, R., Rickman, H., and Froeschlé, C. 1992. The connection between Comet P/Machholz and the Quadrantid meteor. *Monthly Notices of the Royal Astronomical Society*, **254**, 627–634.

Greenberg, R. 1975. On the Laplace relation among the satellites of Uranus. *Monthly Notices of the Royal Astronomical Society*, **173**, 121–129.

Grün, E., Zook, H. A., Fechtig, H., and Giese, R. H. 1985. Collisional balance of the meteoritic complex. *Icarus*, **62**, 244–272.

Gural, P. S. 2011. The California All-sky Meteor Surveillance (CAMS) System. Pages 28–31 of: Asher, D. J., Christou, A. A., Atreya, P., and Barentsen, G. (eds), *Proceedings of the International Meteor Conference, Armagh, Northern Ireland, 2010*. Hove, BE: International Meteor Organization.

Gustafson, B. A. S. 1989. Geminid meteoroids traced to cometary activity on Phaethon. *Astronomy and Astrophysics*, **225**, 533–540.

Hajduková, M., and Neslušan, L. 2017. Regular and transitory showers of comet C/1979 Y1 (Bradfield). *Astronomy and Astrophysics*, **605**, A36.

Hajduková, M., Sterken, V., and Wiegert, P. 2019. Chapter 10: Interstellar Meteoroids. Ryabova, G. O., Asher, D. J., and Campbell-Brown, M. D. (eds), *Meteoroids: Sources of Meteors on Earth and Beyond*. Cambridge, UK: Cambridge University Press, pp. 235–252.

Hasegawa, I. 1979. Orbits of ancient and medieval comets. *Publications of the Astronomical Society of Japan*, **31**, 257–270.

Hasegawa, I. 1993. Historical records of meteor showers. Pages 209–223 of: Štohl, J., and Williams, I. P. (eds), *Meteoroids and Their Parent Bodies*. Bratislava: Astronomical Institute of the Slovak Academy of Science.

Holman, D., and Jenniskens, P. 2012. Confirmation of the Northern Delta Aquariids (NDA, IAU #26) and the Northern June Aquilids (NZC, IAU #164). *WGN, the Journal of the International Meteor Organization*, **40**, 166–170.

Hughes, D. W. 1983. Astronomy: 1983 TB and the Geminids. *Nature*, **306**, 116.

Hughes, D. W. 2000. On the velocity of large cometary dust particles,. *Planetary and Space Science*, **48**, 1–7.

Hughes, D. W., Williams, I. P., and Murray, C. D. 1979. The orbital evolution of the Quadrantid meteor stream between AD 1830 and 2030. *Monthly Notices of the Royal Astronomical Society*, **189**, 493–500.

Hui, M.-T., and Li, J. 2017. Resurrection of (3200) Phaethon in 2016. *The Astronomical Journal*, **153**, 23.

Hunt, J., Williams, I. P., and Fox, K. 1985. Planetary perturbations on the Geminid meteor stream. *Monthly Notices of the Royal Astronomical Society*, **217**, 533–538.

Iorio, L. 2005. Is it possible to measure the Lense-Thirring effect on the orbits of the planets in the gravitational field of the Sun? *Astronomy and Astrophysics*, **431**, 385–389.

Ishiguro, M., Watanabe, J., Usui, F. et al. 2002. First detection of an optical dust trail along the orbit of 22P/Kopff. *The Astrophysical Journal*, **572**, L117–L120.

Jakubík, M., and Neslušan, L. 2015. Meteor complex of asteroid 3200 Phaethon: Its features derived from theory and updated meteor data bases. *Monthly Notices of the Royal Astronomical Society*, **453**, 1186–1200.

Jenniskens, P. 2004. 2003 EH_1 is the Quadrantid shower parent comet. *The Astronomical Journal*, **127**, 3018–3022.

Jenniskens, P. 2006. *Meteor Showers and their Parent Comets*. Cambridge, UK: Cambridge University Press.

Jenniskens, P. 2017. Meteor showers in review. *Planetary and Space Science*, **143**, 116–124.

Jenniskens, P., and Vaubaillon, J. 2008. Minor planet 2008 ED69 and the Kappa Cygnid meteor shower. *The Astronomical Journal*, **136**, 725–730.

Jenniskens, P., and Vaubaillon, J. 2010. Minor planet 2002 EX_{12} (=169P/NEAT) and the Alpha Capricornid shower. *The Astronomical Journal*, **139**, 1822–1830.

Jenniskens, P., Betlem, H., de Lignie, M., Langbroek, M., and van Vliet, M. 1997. Meteor stream activity. V. The Quadrantids, a very young stream. *Astronomy and Astrophysics*, **327**, 1242–1252.

Jenniskens, P., de Kleer, K., Vaubaillon, J. et al. 2008. Leonids 2006 observations of the tail of trails: Where is the comet fluff? *Icarus*, **196**, 171–183.

Jenniskens, P., Gural, P. S., Dynneson, L. et al. 2011. CAMS: Cameras for Allsky Meteor Surveillance to establish minor meteor showers. *Icarus*, **216**, 40–61.

Jewitt, D. 2012. The active asteroids. *The Astronomical Journal*, **143**, 66.

Jewitt, D., and Li, J. 2010. Activity in Geminid parent (3200) Phaethon. *The Astronomical Journal*, **140**, 1519–1527.

Jones, D. C., Williams, I. P., and Porubčan, V. 2006. The Kappa Cygnid meteoroid complex. *Monthly Notices of the Royal Astronomical Society*, **371**, 684–694.

Jones, J. 1995. The ejection of meteoroids from comets. *Monthly Notices of the Royal Astronomical Society*, **275**, 773–780.

Jones, J., and Jones, W. 1993. Comet Machholz and the Quadrantid meteor stream. *Monthly Notices of the Royal Astronomical Society*, **261**, 605–611.

Jopek, T. J., and Jenniskens, P. M. 2011. The Working Group on meteor showers nomenclature: A history, current status and a call for contributions. Pages 7–13 of: Cooke, W. J., Moser, D. E., Hardin, B. F., and Janches, D. (eds), *Meteoroids: The Smallest Solar System Bodies*. NASA/CP-2011-216469.

Jopek, T. J., and Kaňuchová, Z. 2014. Current status of the IAU MDC Meteor Showers Database. Pages 353–364 of: Jopek, T. J., Rietmeijer, F. J. M., Watanabe, J., and Williams, I. P. (eds), *Proceedings of the Meteoroids 2013 Conference, Aug. 26–30, 2013, A. M. University, Poznań, Poland*. Poznań: Adam Mickiewicz University Press.

Kaňuchová, Z., and Neslušan, L. 2007. The parent bodies of the Quadrantid meteoroid stream. *Astronomy and Astrophysics*, **470**, 1123–1136.

Kapišinský, I. 1984. Nongravitational effects affecting small meteoroids in interplanetary space. *Contributions of the Astronomical Observatory Skalnate Pleso*, **12**, 99–111.

Kasuga, T., and Jewitt, D. 2015. Physical observations of (196256) 2003 EH1, presumed parent of the Quadrantid meteoroid stream. *The Astronomical Journal*, **150**, 152.

Kasuga, T., and Jewitt, D. 2019. Chapter 8: Asteroid–Meteoroid complexes. Ryabova, G. O., Asher, D. J., and Campbell-Brown, M. D. (eds), *Meteoroids: Sources of Meteors on Earth and Beyond*. Cambridge, UK: Cambridge University Press, pp. 187–209.

Kasuga, T., Balam, D. D., and Wiegert, P. A. 2010. Comet 169P/NEAT(=2002 EX_{12}): The parent body of the α-Capricornid meteoroid stream. *The Astronomical Journal*, **140**, 1806–1813.

Kelley, M. S., Fernández, Y. R., Licandro, J. et al. 2013. The persistent activity of Jupiter-family comets at 3–7 AU. *Icarus*, **225**, 475–494.

Klačka, J. 2014. Solar wind dominance over the Poynting-Robertson effect in secular orbital evolution of dust particles. *Monthly Notices of the Royal Astronomical Society*, **443**, 213–229.

Kokhirova, G. I., Babadzhanov, P. B., and Khamroev, U. H. 2017. On a possible cometary origin of the object 2015TB145. *Planetary and Space Science*, **143**, 164–168.

Kondrateva, E. D., and Reznikov, E. A. 1985. Comet Tempel-Tuttle and the Leonid meteor swarm. *Astronomicheskii Vestnik*, **19**, 144–151.

Kornoš, L., Matlovič, P., Rudawska, R. et al. 2014. Confirmation and characterization of IAU temporary meteor showers in EDMOND database. Pages 225–233 of: Jopek, T. J., Rietmeijer, F. J. M., Watanabe, J., and Williams, I. P. (eds), *Proceedings of the Meteoroids 2013 Conference, Aug. 26–30, 2013, A. M. University, Poznań, Poland*. Poznań: Adam Mickiewicz University Press.

Koten, P., Čapek, D., Spurný, P. et al. 2017. September epsilon Perseid cluster as a result of orbital fragmentation. *Astronomy and Astrophysics*, **600**, A74.

Kozai, Y. 1962. Secular perturbations of asteroids with high inclination and eccentricity. *The Astronomical Journal*, **67**, 591–598.

Kramer, E. N., and Shestaka, I. S. 1986. The age, origin and evolution of Geminids' meteor swarm. *Kinematika i Fizika Nebesnykh Tel*, **2**, 81–86.

Kresák, L. 1993. Cometary dust trails and meteor storms. *Astronomy and Astrophysics*, **279**, 646–660.

Kresáková, M. 1974. Meteors of periodic Comet Mellish and the Geminids. *Bulletin of the Astronomical Institutes of Czechoslovakia*, **25**, 20–33.

Laplace, P. S. 1799. Beweis des Satzes, dass die anziehende Kraft bey einem Weltkörper so groß seyn könne, dass das Licht davon nicht ausströmen kann. *Allgemeine Geographische Ephemeriden*, **4**(1), 1–6.

Lee, K.-W., Yang, H.-J., and Park, M.-G. 2009. Orbital elements of comet C/1490 Y1 and the Quadrantid shower. *Monthly Notices of the Royal Astronomical Society*, **400**, 1389–1393.

Levison, H. F., and Duncan, M. J. 1994. The long-term dynamical behavior of short-period comets. *Icarus*, **108**, 18–36.

Li, J., and Jewitt, D. 2013. Recurrent perihelion activity in (3200) Phaethon. *The Astronomical Journal*, **145**, 154.

Lindblad, B. A. 1971. Meteor streams. Pages 287–297 of: Kondratyev, K. Y., Rycroft, M. J., and Sagan, C. (eds), *Space Research XI*.

Lindblad, B. A., and Olsson-Steel, D. 1990. The Monocerotid meteor stream and Comet Mellish. *Bulletin of the Astronomical Institutes of Czechoslovakia*, **41**, 193–200.

Liou, Jer-Chyi, and Zook, Herbert A. 1996. Comets as a source of low eccentricity and low inclination interplanetary dust particles. *Icarus*, **123**, 491–502.

Lyytinen, E. J., and Van Flandern, T. 2000. Predicting the strength of Leonid outbursts. *Earth, Moon, and Planets*, **82**, 149–166.

Ma, Y., Williams, I. P., and Chen, W. 2002. On the ejection velocity of meteoroids from comets. *Monthly Notices of the Royal Astronomical Society*, **337**, 1081–1086.

Maquet, L., Colas, F., Jorda, L., and Crovisier, J. 2012. CONGO, model of cometary non-gravitational forces combining astrometric and production rate data. Application to Comet 19P/Borrelly. *Astronomy and Astrophysics*, **548**, A81.

McCrosky, R. E., and Posen, A. 1961. Orbital elements of photographic meteors. *Smithsonian Contributions to Astrophysics*, **4**, 15–84.

McIntosh, B. A. 1990. Comet P/Machholz and the Quadrantid meteor stream. *Icarus*, **86**, 299–304.

McIntosh, B. A., and Šimek, M. 1980. Geminid meteor stream: Structure from 20 years of radar observations. *Bulletin of the Astronomical Institutes of Czechoslovakia*, **31**, 39–50.

McNamara, H., Jones, J., Kauffman, B. et al. 2004. Meteoroid Engineering Model (MEM): A meteoroid Model for the inner Solar System. *Earth, Moon, and Planets*, **95**, 123–139.

McNaught, R. H., and Asher, D. J. 1999. Leonid dust trails and meteor storms. *WGN, the Journal of the International Meteor Organization*, **27**, 85–102.

Milani, A., and Knežević, Z. 1990. Secular perturbation theory and computation of asteroid proper elements. *Celestial Mechanics and Dynamical Astronomy*, **49**, 347–411.

Molau, S., and Rendtel, J. 2009. A comprehensive list of meteor showers obtained from 10 years of observations with the IMO Video Meteor Network. *WGN, the Journal of the International Meteor Organization*, **37**, 98–121.

Moorhead, A. V. 2016. Performance of D-criteria in isolating meteor showers from the sporadic background in an optical data set. *Monthly Notices of the Royal Astronomical Society*, **455**, 4329–4338.

Moser, D. E., and Cooke, W. J. 2008. Updates to the MSFC meteoroid stream model. *Earth, Moon, and Planets*, **102**, 285–291.

Müller, M., Green, S. F., and McBride, N. 2001. Constraining cometary ejection models from meteor storm observations. Pages 47–54 of: Warmbein, B. (ed), *Meteoroids 2001 Conference*. ESA Special Publication, vol. 495.

Murray, C. D., and Dermott, S. F. 1999. *Solar System Dynamics*. Cambridge, UK: Cambridge University Press.

Neslušan, L. 1999. Comets 14P/Wolf and D/1892 T1 as parent bodies of a common, α-Capricornids related, meteor stream. *Astronomy and Astrophysics*, **351**, 752–758.

Neslušan, L., and Hajduková, M. 2014. The meteor-shower complex of comet C/1917 F1 (Mellish). *Astronomy and Astrophysics*, **566**, A33.

Neslušan, L., Svoreň, J., and Porubčan, V. 1995. A procedure of selection of meteors from major streams for determination of mean orbits. *Earth, Moon, and Planets*, **68**, 427–433.

Neslušan, L., Svoreň, J., and Porubčan, V. 2013a. The method of selection of major-shower meteors revisited. *Earth, Moon, and Planets*, **110**, 41–66.

Neslušan, L., Kaňuchová, Z., and Tomko, D. 2013b. The meteor-shower complex of 96P/Machholz revisited. *Astronomy and Astrophysics*, **551**, A87.

Neslušan, L., Hajduková, M., and Jakubík, M. 2013c. Meteor-shower complex of asteroid 2003 EH1 compared with that of comet 96P/Machholz. *Astronomy and Astrophysics*, **560**, A47.

Neslušan, L., Porubčan, V., and Svoreň, J. 2014a. IAU MDC photographic meteor orbits database: Version 2013. *Earth, Moon, and Planets*, **111**, 105–114.

Neslušan, L., Hajduková, M., Tomko, D., Kaňuchová, Z., and Jakubík, M. 2014b. The prediction of meteor showers from all potential parent comets. Pages 139–145 of: Rault, J.-L., and Roggemans, P. (eds), *Proceedings of the International Meteor Conference, Giron, France, 18–21 September 2014*. Hove, BE: International Meteor Organization.

Nesvorný, D., Morbidelli, A., Vokrouhlický, D., Bottke, W. F., and Brož, M. 2002a. The Flora family: A case of the dynamically dispersed collisional swarm? *Icarus*, **157**, 155–172.

Nesvorný, D., Bottke, Jr., W. F., Dones, L., and Levison, H. F. 2002b. The recent breakup of an asteroid in the main-belt region. *Nature*, **417**, 720–771.

Nesvorný, D., Vokrouhlický, D., Pokorný, P., and Janches, D. 2011. Dynamics of dust particles released from Oort Cloud Comets and their contribution to radar meteors. *The Astrophysical Journal*, **743**, 37.

Ohtsuka, K. 1989. The December Monocerotids and P/Mellish. *WGN, the Journal of the International Meteor Organization*, **17**, 93–96.

Ohtsuka, K., Arakida, H., Ito, T., Yoshikawa, M., and Asher, D. J. 2008. Apollo asteroid 1999 YC: Another large member of the PGC? *Meteoritics and Planetary Science Supplement*, **43**, 5055.

Olsson-Steel, D. 1987. Theoretical meteor radiants of Earth-approaching asteroids and comets. *Australian Journal of Astronomy*, **2**, 21–35.

Öpik, E. J. 1951. Collision probability with the planets and the distribution of interplanetary matter. *Proceedings of the Royal Irish Academy, Section A*, **54**, 165–199.

Ott, T., Drolshagen, E., Koschny, D. et al. 2017. Dust mass distribution around comet 67P/Churyumov-Gerasimenko determined via parallax measurements using Rosetta's OSIRIS cameras. *Monthly Notices of the Royal Astronomical Society*, **469**, S276–S284.

Peale, S. J. 1976. Orbital resonances in the solar system. *Annual Review of Astronomy and Astrophysics*, **14**, 215–246.

Plavec, M. 1955. Ejection theory of the meteor shower formation I. orbit of an ejected meteor. *Bulletin of the Astronomical Institutes of Czechoslovakia*, **6**, 20–23.

Plavec, M. 1957. Vznik a rana vyvojova stadia meteorickyck roju. On the origin and early stages of the meteor streams. *Publications of the Astronomical Institute of the Czechoslovak Academy of Sciences*, **30**, 1–93.

Pokorný, P., Vokrouhlický, D., Nesvorný, D., Campbell-Brown, M., and Brown, P. 2014. Dynamical model for the toroidal sporadic meteors. *The Astrophysical Journal*, **789**, 25.

Pokorný, P., Janches, D., Brown, P. G., and Hormaechea, J. L. 2017. An orbital meteoroid stream survey using the Southern Argentina Agile MEteor Radar (SAAMER) based on a wavelet approach. *Icarus*, **290**, 162–182.

Porter, J. K. 1952. *Comets and Meteor Streams*. London: Chapman and Hall.

Porubčan, V., and Kornoš, L. 2005. The Quadrantid meteor stream and 2003 EH1. *Contributions of the Astronomical Observatory Skalnate Pleso*, **35**, 5–16.

Porubčan, V., Kornoš, L., and Williams, I. P. 2006. The Taurid complex meteor showers and asteroids. *Contributions of the Astronomical Observatory Skalnate Pleso*, **36**, 103–117.

Porubčan, V., Kresáková, M., and Štohl, J. 1980. Geminid meteor shower: Activity and magnitude distribution. *Contributions of the Astronomical Observatory Skalnate Pleso*, **9**, 125–144.

Radzievskii, V. V. 1952. On the influence of an anisotropic re-emission of solar radiation on the orbital motion of asteroids and meteorites. *Astronomicheskii Zhurnal*, **29**, 162–170.

Reach, W. T., Kelley, M. S., and Sykes, M. V. 2007. A survey of debris trails from short-period comets. *Icarus*, **191**, 298–322.

Rendtel, J., Arlt, R. and McBeath, A. (eds). 1995. *Handbook for Visual Meteor Observers*. Potsdam: International Meteor Organization, p. 169.

Rendtel, J. 2004. Evolution of the Geminids observed over 60 years. *Earth, Moon, and Planets*, **95**, 27–32.

Rendtel, J. 2007. Three days of enhanced Orionid activity in 2006 – Meteoroids from a resonance region? *WGN, the Journal of the International Meteor Organization*, **35**, 41–45.

Rudawska, R., and Jenniskens, P. 2014. New meteor showers identifed in the CAMS and SonotaCo meteoroid orbit surveys. Pages 217–224 of: Jopek, T. J., Rietmeijer, F. J. M., Watanabe, J., and Williams, I. P. (eds), *Proceedings of the Meteoroids 2013 Conference, Aug. 26–30, 2013, A. M. University, Poznań, Poland*. Poznań: Adam Mickiewicz University Press.

Rudawska, R., and Vaubaillon, J. 2015. Don Quixote-A possible parent body of a meteor shower. *Planetary and Space Science*, **118**, 25–27.

Rudawska, R., Jopek, T. J., and Dybczyński, P. A. 2005. The changes of the orbital elements and estimation of the initial velocities of stream meteoroids ejected from comets and asteroids. *Earth, Moon, and Planets*, **97**, 295–310.

Rudawska, R., Vaubaillon, J., and Atreya, P. 2012. Association of individual meteors with their parent bodies. *Astronomy and Astrophysics*, **541**, A2.

Ryabova, G. O. 1989. Effect of secular perturbations and the Poynting-Robertson effect on structure of the Geminid meteor stream. *Solar System Research*, **23**, 158–165.

Ryabova, G. O. 1999. Age of the Geminid meteor stream (review). *Solar System Research*, **33**, 224–238.

Ryabova, G. O. 2001. Mathematical model of the Geminid meteor stream formation. Pages 77–81 of: Warmbein, B. (ed), *Meteoroids 2001 Conference*. ESA Special Publication, vol. 495.

Ryabova, G. O. 2006. Meteoroid streams: mathematical modelling and observations. Pages 229–247 of: Lazzaro, D., Ferraz Mello, S., and Fernández, J. A. (eds), *Asteroids, Comets, Meteors*. IAU Symposium, vol. 229.

Ryabova, G. O. 2007. Mathematical modelling of the Geminid meteoroid stream. *Monthly Notices of the Royal Astronomical Society*, **375**, 1371–1380.

Ryabova, G. O. 2008a. Model radiants of the Geminid meteor shower. *Earth, Moon, and Planets*, **102**, 95–102.

Ryabova, G. O. 2008b. Origin of the (3200) Phaethon - Geminid meteoroid stream complex. Page 226 of: *European Planetary Science Congress 2008*.

Ryabova, G. O. 2013. Modeling of meteoroid streams: The velocity of ejection of meteoroids from comets (a review). *Solar System Research*, **47**, 219–238.

Ryabova, G. O. 2014. Averaged changes in the orbital elements of meteoroids due to Yarkovsky–Radzievskij force. Pages 160–161 of: Knežević, Z., and Lemaître, A. (eds), *Complex Planetary Systems, Proceedings of the International Astronomical Union*. IAU Symposium, vol. 310.

Ryabova, G. O. 2016. A preliminary numerical model of the Geminid meteoroid stream. *Monthly Notices of the Royal Astronomical Society*, **456**, 78–84.

Ryabova, G. O. 2018. Could the Geminid meteoroid stream be the result of long-term thermal fracture? *Monthly Notices of the Royal Astronomical Society*, **479**, 1017–1020.

Ryabova, G., and Nogami, N. 2005. Asteroid 2003 EH1 and the Quadrantid meteoroid stream. Pages 63–65 of: Triglav-Čekada, M., Kac, J., and McBeath, A. (eds), *Proceedings of the International Meteor Conference, Varna, Bulgaria, 2004*. Hove, BE: Intenational Meteor Organization

Ryabova, G., Avdyushev, V., Koschny, D., and Williams, I. 2014. Asteroid (3200) Phaethon and the Geminid meteoroid stream complex. In: Muinonen, K., Penttilä, A., Granvik, M. et al. (eds), *Asteroids, Comets, Meteors – Book of Abstracts*. Helsinki: University of Helsinki, p. 481. www.helsinki.fi/acm2014/pdf-material/ACM2014.pdf.

Ryabova, G. O., Pleshanova, A. V., and Konstantinov, V. S. 2008. Determining the age of meteor streams with the retrospective evolution method. *Solar System Research*, **42**, 335–340.

Sato, M., and Watanabe, J.-i. 2007. Origin of the 2006 Orionid outburst. *Publications of the Astronomical Society of Japan*, **59**, L21–L24.

Sato, M., and Watanabe, J.-i. 2010. Forecast for Phoenicids in 2008, 2014, and 2019. *Publications of the Astronomical Society of Japan*, **62**, 509–513.

Schiaparelli, J. V. 1867. Sur la relation qui existe entre les comètes et les étoiles filantes. *Astronomische Nachrichten*, **68**, 331–332.

Schulz, R., Stüwe, J. A., and Boehnhardt, H. 2004. Rosetta target comet 67P/Churyumov-Gerasimenko. Postperihelion gas and dust production rates. *Astronomy and Astrophysics*, **422**, L19–L21.

Šegon, D., Vaubaillon, J., Gural, P. S. et al. 2017. Dynamical modeling validation of parent bodies associated with newly discovered CMN meteor showers. *Astronomy and Astrophysics*, **598**, A15.

Sekanina, Z. 1970a. Statistical model of meteor streams. I. Analysis of the model. *Icarus*, **13**, 459–474.

Sekanina, Z. 1970b. Statistical model of meteor streams. II. Major showers. *Icarus*, **13**, 475–493.

Sekanina, Z. 1973. Statistical model of meteor streams. III. Stream search among 19303 radio meteors. *Icarus*, **18**, 253–284.

Sekanina, Z. 1976. Statistical model of meteor streams. IV. A study of radio streams from the synoptic year. *Icarus*, **27**, 265–321.

Sekanina, Z., and Chodas, P. W. 2005. Origin of the Marsden and Kracht Groups of sunskirting comets. I. Association with Comet 96P/Machholz and its interplanetary complex. *Astrophysical Journal Supplement Series*, **161**, 551–586.

Sekhar, A. 2013. General relativistic precession in meteoroid orbits. *WGN, the Journal of the International Meteor Organization*, **41**, 179–183.

Sekhar, A. 2014. Evolution of Halley-type Comets and Meteoroid Streams. Ph.D. thesis, Queen's University Belfast.

Sekhar, A., and Asher, D. J. 2013. Saturnian mean motion resonances in meteoroid streams. *Monthly Notices of the Royal Astronomical Society*, **433**, L84–L88.

Sekhar, A., and Asher, D. J. 2014a. Resonant behavior of comet Halley and the Orionid stream. *Meteoritics and Planetary Science*, **49**, 52–62.

Sekhar, A., and Asher, D. J. 2014b. Meteor showers on Earth from sungrazing comets. *Monthly Notices of the Royal Astronomical Society*, **437**, L71–L75.

Sekhar, A., Asher, D. J., and Vaubaillon, J. 2016. Three-body resonance in meteoroid streams. *Monthly Notices of the Royal Astronomical Society*, **460**, 1417–1427.

Sekhar, A., Asher, D. J., Werner, S. C., Vaubaillon, J., and Li, G. 2017. Change in general relativistic precession rates due to Lidov–Kozai oscillations in Solar system. *Monthly Notices of the Royal Astronomical Society*, **468**, 1405–1414.

Soja, R. H., Baggaley, W. J., Brown, P., and Hamilton, D. P. 2011. Dynamical resonant structures in meteoroid stream orbits. *Monthly Notices of the Royal Astronomical Society*, **414**, 1059–1076.

Soja, R. H., Sommer, M., Herzog, J. et al. 2015. Characteristics of the dust trail of 67P/Churyumov-Gerasimenko: an application of the IMEX model. *Astronomy and Astrophysics*, **583**, A18.

Soja, R. H., Schwarzkopf, G. J., Sommer, M. et al. 2016. Collisional lifetimes of meteoroids. Pages 284–286 of: Roggemans, A., and Roggemans, P. (eds), *Proceedings of the International Meteor Conference, Egmond, the Netherlands, 2–5 June 2016*. Hove, BE: International Meteor Organization.

Southworth, R. B., and Hawkins, G. S. 1963. Statistics of meteor streams. *Smithsonian Contributions to Astrophysics*, **7**, 261–285.

Steel, D. I., Asher, D. J., and Clube, S. V. M. 1991. The Taurid Complex: Giant comet origin? Pages 327–330 of: Levasseur-Regourd, A. C., and Hasegawa, H. (eds), *IAU Colloq. 126: Origin and Evolution of Interplanetary Dust*. Astrophysics and Space Science Library, vol. 173. Dordrecht: Kluwer.

Stoney, G. J., and Downing, A. M. W. 1899. Perturbations of the Leonids. *Popular Astronomy*, **7**, 227–233.

Sugar, G., Moorhead, A., Brown, P., and Cooke, W. 2017. Meteor shower detection with density-based clustering. *Meteoritics & Planetary Science*, **52**, 1048–1059.

Svoreň, J., Neslušan, L., and Porubčan, V. 2000. A search for streams and associations in meteor databases. Method of indices. *Planetary and Space Science*, **48**, 933–937.

Svoreň, J., Kaňuchová, Z., and Jakubík, M. 2006. Filaments within the Perseid meteoroid stream and their coincidence with the location of mean-motion resonances. *Icarus*, **183**, 115–121.

Sykes, M. V., Lebofsky, L. A., Hunten, D. M., and Low, F. 1986. The discovery of dust trails in the orbits of periodic comets. *Science*, **232**, 1115–1117.

Todorović, N. 2018. The dynamical connection between Phaethon and Pallas. *Monthly Notices of the Royal Astronomical Society*, **475**, 601–604.

Tomko, D. 2014. Prediction of evolution of meteor shower associated with comet 122P/de Vico. *Contributions of the Astronomical Observatory Skalnate Pleso*, **44**, 33–42.

Tomko, D., and Neslušan, L. 2012. Search for new parent bodies of meteoroid streams among comets. I. Showers of Comets 126P/1996 P1 and 161P/2004 V2 with radiants on southern Sky. *Earth, Moon, and Planets*, **108**, 123–138.

Tomko, D., and Neslušan, L. 2016. Meteoroid stream of 12P/Pons-Brooks, December κ-Draconids, and Northern June Aquilids. *Astronomy and Astrophysics*, **592**, A107.

Ueda, M. 2012. Orbits of the July Pegasid meteors observed during 2008 to 2011. *WGN, the Journal of the International Meteor Organization*, **40**, 59–64.

Valsecchi, G. B., Jopek, T. J., and Froeschlé, C. 1999. Meteoroid stream identification: a new approach-I. Theory. *Monthly Notices of the Royal Astronomical Society*, **304**, 743–750.

Vaubaillon, J. 2017. A confidence index for forecasting of meteor showers. *Planetary and Space Science*, **143**, 78–82.

Vaubaillon, J. J., and Reach, W. T. 2010. Spitzer Space Telescope observations and the particle size distribution of Comet 73P/Schwassmann-Wachmann 3. *The Astronomical Journal*, **139**, 1491–1498.

Vaubaillon, J., Lamy, P., and Jorda, L. 2004. Meteoroid streams associated to Comets 9P/Tempel 1 and 67P/Churyumov-Gerasimenko. *Earth, Moon, and Planets*, **95**, 75–80.

Vaubaillon, J., Colas, F., and Jorda, L. 2005a. A new method to predict meteor showers. I. Description of the model. *Astronomy and Astrophysics*, **439**, 751–760.

Vaubaillon, J., Colas, F., and Jorda, L. 2005b. A new method to predict meteor showers. II. Application to the Leonids. *Astronomy and Astrophysics*, **439**, 761–770.

Vaubaillon, J., Lamy, P., and Jorda, L. 2006. On the mechanisms leading to orphan meteoroid streams. *Monthly Notices of the Royal Astronomical Society*, **370**, 1841–1848.

Vaubaillon, J., Maquet, L., and Soja, R. 2014. Meteor hurricane at Mars on 2014 October 19 from comet C/2013 A1. *Monthly Notices of the Royal Astronomical Society*, **439**, 3294–3299.

Vereš, P., Kornoš, L., and Tóth, J. 2011. Meteor showers of comet C/1917 F1 Mellish. *Monthly Notices of the Royal Astronomical Society*, **412**, 511–521.

Vida, D., Brown, P. G., and Campbell-Brown, M. 2018. Modelling the measurement accuracy of pre-atmosphere velocities of meteoroids. *Monthly Notices of the Royal Astronomical Society*, **479**, 4307–4319.

Viswanathan, V., Fienga, A., Minazzoli, O. et al. 2018. The new lunar ephemeris INPOP17a and its application to fundamental physics. *Monthly Notices of the Royal Astronomical Society*, **476**, 1877–1888.

Weinberg, S. 1972. *Gravitation and Cosmology: Principles and Applications of the General Theory of Relativity*. New York: Wiley.

Welch, P. G. 2001. A new search method for streams in meteor data bases and its application. *Monthly Notices of the Royal Astronomical Society*, **328**, 101–111.

Whipple, F. L. 1951. A comet model. II. Physical relations for comets and meteors. *The Astrophysical Journal*, **113**, 464–474.

Whipple, F. L. 1983. 1983 TB and the Geminid meteors. *International Astronomical Union Circulars*, **3881**.

Wiegert, P., and Brown, P. 2004. The core of the Quadrantid meteoroid stream is two hundred years old. *Earth, Moon, and Planets*, **95**, 81–88.

Wiegert, P., and Brown, P. 2005. The Quadrantid meteoroid complex. *Icarus*, **179**, 139–157.

Wiegert, P., Vaubaillon, J., and Campbell-Brown, M. 2009. A dynamical model of the sporadic meteoroid complex. *Icarus*, **201**, 295–310.

Wiegert, P. A., Brown, P. G., Weryk, R. J., and Wong, D. K. 2013. The return of the Andromedids meteor shower. *The Astronomical Journal*, **145**, 70.

Williams, I. P. 1997. The Leonid meteor shower: why are there storms but no regular annual activity? *Monthly Notices of the Royal Astronomical Society*, **292**, L37–L40.

Williams, I. P. 2001. The determination of the ejection velocity of meteoroids from cometary nuclei. Pages 33–42 of: Warmbein, B. (ed), *Proceedings of the Meteoroids 2001 Conference, 6–10 August 2001, Kiruna, Sweden*, ESA SP-495. Noordwijk: ESA Publications Division.

Williams, I. P. 2002. The evolution of meteoroid streams. Pages 13–32 of: Murad, E., and Williams, I. P. (eds), *Meteors in the Earth's Atmosphere*. Cambridge, UK: Cambridge University Press.

Williams, I. P., and Collander-Brown, S. J. 1998. The parent of the Quadrantid meteoroid stream. *Monthly Notices of the Royal Astronomical Society*, **294**, 127–138.

Williams, I. P., and Ryabova, G. O. 2011. Meteor shower features: are they governed by the initial formation process or by subsequent gravitational perturbations? *Monthly Notices of the Royal Astronomical Society*, **415**, 3914–3920.

Williams, I. P., and Wu, Z. 1993a. The Geminid meteor stream and asteroid 3200 Phaethon. *Monthly Notices of the Royal Astronomical Society*, **262**, 231–248.

Williams, I. P., and Wu, Z. 1993b. The Quadrantid meteoroid stream and Comet 1491I. *Monthly Notices of the Royal Astronomical Society*, **264**, 659–664.

Williams, I. P., Murray, C. D., and Hughes, D. W. 1979. The long-term orbital evolution of the Quadrantid meteor stream. *Monthly Notices of the Royal Astronomical Society*, **189**, 483–492.

Williams, I. P., Ryabova, G. O., Baturin, A. P., and Chernitsov, A. M. 2004. The parent of the Quadrantid meteoroid stream and asteroid 2003 EH1. *Monthly Notices of the Royal Astronomical Society*, **355**, 1171–1181.

Williams, I. P., Jopek, T. J., Rudawska, R., Tóth, J., and Kornoš, L. 2019. Chapter 9: Minor Meteor Showers and the Sporadic Background. Ryabova, G. O., Asher, D. J., and Campbell-Brown, M. D. (eds), *Meteoroids: Sources of Meteors on Earth and Beyond*. Cambridge, UK: Cambridge University Press, pp. 210–234.

Wu, Z., and Williams, I. P. 1992a. Formation of the Leonid meteor stream and storm. Pages 661–665 of: Harris, A. W., and Bowell, E. (eds), *Asteroids, Comets, Meteors 1991*. Houston: Lunar and Planetary Institute.

Wu, Z., and Williams, I. P. 1992b. On the Quadrantid meteoroid stream complex. *Monthly Notices of the Royal Astronomical Society*, **259**, 617–628.

Wu, Z., and Williams, I. P. 1995. Gaps in the distribution of semimajor axes of the Perseid meteors. *Monthly Notices of the Royal Astronomical Society*, **276**, 1017–1023.

Wu, Z., and Williams, I. P. 1996. Leonid meteor storms. *Monthly Notices of the Royal Astronomical Society*, **280**, 1210–1218.

Ye, Q., Wiegert, P. A., Brown, P. G., Campbell-Brown, M. D., and Weryk, R. J. 2014. The unexpected 2012 Draconid meteor storm. *Monthly Notices of the Royal Astronomical Society*, **437**, 3812–3823.

Yoder, C. F., and Peale, S. J. 1981. The tides of Io. *Icarus*, **47**, 1–35.

8

Asteroid–Meteoroid Complexes

Toshihiro Kasuga and David Jewitt

8.1 Introduction

Physical disintegration of asteroids and comets leads to the production of orbit-hugging debris streams. In many cases, the mechanisms underlying disintegration are uncharacterized, or even unknown. Therefore, considerable scientific interest lies in tracing the physical and dynamical properties of the asteroid-meteoroid complexes backwards in time, in order to learn how they form.

Small solar system bodies offer the opportunity to understand the origin and evolution of the planetary system. They include the comets and asteroids, as well as the mostly unseen objects in the much more distant Kuiper belt and Oort cloud reservoirs. Observationally, asteroids and comets are distinguished principally by their optical morphologies, with asteroids appearing as point sources, and comets as diffuse objects with unbound atmospheres (comae), at least when near the Sun. The principal difference between the two is thought to be the volatile content, especially the abundance of water ice. Sublimation of ice in comets drives a gas flux into the adjacent vacuum while drag forces from the expanding gas are exerted on embedded dust and debris particles, expelling them into interplanetary space. Meteoroid streams, consisting of large particles ejected at low speeds and confined to move approximately in the orbit of the parent body, are one result.

When the orbit of the parent body intersects that of the Earth, meteoroids strike the atmosphere and all but the largest are burned up by frictional heating, creating the familiar meteors (Olmsted, 1834). Later, the phenomenon has been realized as ablation by shock wave radiation heating (Bronshten, 1983). The first established comet-meteoroid stream relationships were identified by G. Schiaparelli and E. Weiss in 1866 (see Ceplecha et al., 1998). In the last twenty years a cometary meteoroid stream theory has been established, enabling accurate shower activity prediction of both major (e.g. Leonids, Kondrat'eva et al., 1997; McNaught and Asher, 1999) and minor showers (2004 June Boötids, Vaubaillon et al., 2005). This theory deals with the perturbed motion of streams encountering the Earth after the ejection from relevant parent bodies. This improved theory has provided striking opportunities for meteor shower studies of orbital trajectories, velocities and compositions, resulting in a revolution in meteor science.

Some meteoroid streams seem to be made of debris released from asteroids. The notion that not all stream parents are comets is comparatively old, having been suggested by Whipple (1939a,b, 1940) (see also Olivier, 1925; Hoffmeister, 1937). The Geminid meteoroid stream (GEM/4, from Jopek and Jenniskens, 2011)[1] and asteroid 3200 Phaethon are probably the best-known examples (Whipple, 1983). In such cases, it appears unlikely that ice sublimation drives the expulsion of solid matter, raising following general question: what produces the meteoroid streams? Suggested alternative triggers include thermal stress, rotational instability and collisions (impacts) by secondary bodies (Jewitt, 2012; Jewitt et al., 2015). Any of these, if sufficiently violent or prolonged, could lead to the production of a debris trail that would, if it crossed Earth's orbit, be classified as a meteoroid stream or an "Asteroid-Meteoroid Complex", comprising streams and several macroscopic, split fragments (Jones, 1986; Voloshchuk and Kashcheev, 1986; Ceplecha et al., 1998).

The dynamics of stream members and their parent objects may differ, and dynamical associations are not always obvious. Direct searches for dynamical similarities employ a distance parameter, D_{SH}, which measures the separation in orbital element space by comparing q (perihelion distance), e (eccentricity), i (inclination), Ω (longitude of the ascending node), and ω (argument of perihelion) (Southworth and Hawkins, 1963). A smaller D_{SH} indicates a closer degree of orbital similarity between two bodies, with an empirical cut-off for significance often set at $D_{SH} \lesssim 0.10-0.20$ (Williams et al., 2019, Section 9.2.2). The statistical significance of proposed parent-shower associations has been coupled with D_{SH} (Wiegert and Brown, 2004; Ye et al., 2016). Recent models assess the long-term dynamical stability for high-i and -e asteroids. Ohtsuka et al. (2006, 2008a) find the Phaethon-Geminid Complex (PGC) and the Icarus complex together using as criteria the C_1 (Moiseev, 1945) and C_2 (Lidov, 1962) integrals. These are secular orbital variations expressed by

$$C_1 = (1 - e^2)\cos^2(i) \tag{8.1}$$

$$C_2 = e^2 (0.4 - \sin^2(i)\sin^2(\omega)). \tag{8.2}$$

So-called time-lag theory is utilized to demonstrate long-term orbital evolution of complex members. When a stream-complex is formed, the orbital energies ($\propto a^{-1}$, where a is the semimajor axis) of ejected fragments are expected to be slightly different from the energy of the precursor. The motions of the released objects are either accelerated or decelerated relative to the precursor under gravitational perturbations (possibly including non-gravitational perturbations), effectively causing a time lag, Δt, in the orbital evolution to arise. Both C_1 and C_2 are approximately invariant during dynamical evolution,

[1] IAU Meteor Data Center, Nomenclature.

Table 8.1. *Orbital properties*

Complex	Object	a^a	e^b	i^c	q^d	ω^e	Ω^f	Q^g	P_{orb}^h	C_1^i	C_2^i	T_J^j
Geminids	Phaethon	1.271	0.890	22.253	0.140	322.174	265.231	2.402	1.43	0.178	0.274	4.509
	2005 UD	1.275	0.872	28.682	0.163	207.573	19.746	2.387	1.44	0.184	0.267	4.504
	1999 YC	1.422	0.831	38.226	0.241	156.395	64.791	2.603	1.70	0.191	0.234	4.114
Quadrantids	2003 EH$_1$	3.123	0.619	70.838	1.191	171.361	282.979	5.056	5.52	0.066	0.146	2.065
	96P/Machholz 1	3.018	0.959	59.975	0.125	14.622	94.548	5.911	5.24	0.020	0.324	1.939
Capricornids	169P/NEAT	2.604	0.767	11.304	0.607	217.977	176.219	4.602	4.20	0.396	0.227	2.887
	P/2003 T12	2.568	0.776	11.475	0.575	217.669	176.465	4.561	4.12	0.382	0.232	2.894
	2017 MB$_1$	2.372	0.753	8.508	0.586	264.628	126.974	4.158	3.65	0.424	0.215	3.071
Taurids	2P/Encke	2.215	0.848	11.781	0.336	186.542	334.569	4.094	3.30	0.269	0.287	3.025
Taurids-Perseids	1566 Icarus	1.078	0.827	22.852	0.187	31.297	88.082	1.969	1.12	0.268	0.246	5.296
	2007 MK$_6$	1.081	0.819	25.138	0.196	25.466	92.887	1.966	1.12	0.270	0.246	5.284
Phoenicids	2003 WY$_{25}$	3.046	0.685	5.9000	0.961	9.839	68.931	5.132	5.32	0.525	0.188	2.816

Notes: Orbital data are obtained from NASA JPL HORIZONS (https://ssd.jpl.nasa.gov/horizons.cgi)
[a] Semimajor axis (AU)
[b] Eccentricity
[c] Inclination (degrees)
[d] Perihelion distance (AU)
[e] Argument of perihelion (degrees)
[f] Longitude of ascending node (degrees)
[g] Aphelion distance (AU)
[h] Orbital period (yr)
[i] Dynamical invariants: $C_1 = (1 - e^2) \cos^2(i)$, $C_2 = e^2 (0.4 - \sin^2(i) \sin^2(\omega))$
[j] Tisserand parameter with respect to Jupiter (asteroids with $T_J > 3.08$, comets with $T_J < 3.08$; $a < a_J(5.2$ AU$)$)

distinguishing the complex members. The PGC members (Phaethon, 2005 UD, 1999 YC), for example, dynamically follow the Lidov-Kozai mechanism based on secularly perturbed motion of the asteroids (Kozai, 1962).

Most known parent bodies are near-Earth Objects (NEOs), including both asteroids and comets. The comets include various sub-types: Jupiter-family comets (JFCs), Halley-type comets (HTCs) and Encke-type comets (Ye, 2018) (Table 8.1). A classification of parents and their associated streams has been proposed based on their inferred evolutionary stages (Babadzhanov and Obrubov, 1987, 1992a,b). Some streams originating from asteroids (e.g. Phaethon), or from comets (e.g. 96P, 2P) are the most evolved. For example, the Geminids, Quadrantids (QUA/10) and Taurid Complex (TAU/247) show stable secular variation of the orbital elements under mean motion resonance with Jupiter and the Kozai resonance, producing annual meteor showers. On the other hand, young streams are usually from JFCs and HTCs. JFC orbits, in particular, are chaotically scattered by frequent close encounters with Jupiter, tending to produce irregular streams, e.g. Phoenicids (PHO/254). HTCs orbit with widespread inclinations, including retrograde orbits that are absent in the JFCs, and may also generate regular showers, e.g. Leonids (LEO/13) and Perseids (PER/7).

Non-gravitational force effects can be important in the evolution of stream complexes but are difficult to model, since they depend on many unknowns such as the size, rotation, and thermal properties of the small bodies involved. In the last decade, video- and radar- based surveys of meteors have also been used to trace the trajectories back to potential parent NEOs, and so to find new stream complexes (e.g. Jenniskens, 2008a;

Brown et al., 2008a,b, 2010; Musci et al., 2012; Weryk and Brown, 2012; Rudawska et al., 2015; Jenniskens et al., 2016a,b; Jenniskens and Nénon, 2016; Ye et al., 2016) (Reviewed in Jenniskens, 2017).

Meteor spectroscopy provides some additional constraints on the composition of meteoroids (Millman and McKinley, 1963; Millman, 1980). Spectroscopy is typically capable of obtaining useful data for meteors having optical absolute magnitudes +3 to +4 or brighter, corresponding to meteoroid sizes $\gtrsim$ 1 mm (Lindblad, 1987; Ceplecha et al., 1998; Borovička et al., 2010). Meteor spectra consist primarily of atomic emission lines and some molecular bands in the visible to near-infrared wavelengths. The commonly identified neutral atoms are Mg I, Fe I, Ca I, and Na I, while the singly ionized atomic emissions of Ca II and Mg II also appear in some fast-moving meteors (e.g. Leonids with geocentric velocity $V_g \sim 72$ km s^{-1} impact much more quickly than the Geminids, with $V_g \sim 35$ km s^{-1}). The abundances, the excitation temperatures and the electron densities can be deduced for each spectrum, assuming the Boltzmann distribution for electron energies, but the measurements are difficult, and the resulting elemental abundances and/or intensity ratios (e.g. Nagasawa, 1978; Borovička, 1993; Borovička et al., 2005; Kasuga et al., 2005b) are scattered (summarized in Ceplecha et al., 1998; Kasuga et al., 2006b). Generally, most meteors are found to have solar abundance within factors of $\sim$3–4. Some elements are noticeably underabundant, probably affected by incomplete evaporation (Borovička et al., 1999; Trigo-Rodríguez et al., 2003; Kasuga et al., 2005b). In particular, sodium (Na) is a relatively abundant and moderately volatile element that is easily volatilized. As a result, the abundance

Table 8.2. *Physical properties*

Complex	Object	D_e^a	p_v^b	P_{rot}^c	$B-V^d$	$V-R^d$	$R-I^d$	a/b^e
Geminids	Phaethon[1]	5–6	0.09–0.13	3.604	0.59 ± 0.01	0.35 ± 0.01	0.32 ± 0.01	~1.45
	2005 UD[2]	1.3 ± 0.1	0.11^f	5.249	0.66 ± 0.03	0.35 ± 0.02	0.33 ± 0.02	1.45 ± 0.06
	1999 YC[3]	1.7 ± 0.2^g	0.09 ± 0.03^g	4.495	0.71 ± 0.04	0.36 ± 0.03	–	1.89 ± 0.09
Quadrantids	2003 EH$_1$[4]	4.0 ± 0.3	0.04^f	12.650	0.69 ± 0.01	0.39 ± 0.01	0.38 ± 0.01	1.50 ± 0.01
	96P/Machholz 1[5]	6.4	0.04^f	6.38	-	0.40 ± 0.03	-	≥ 1.4
Capricornids	169P/NEAT[6]	4.6 ± 0.6	0.03 ± 0.01	8.410	0.73 ± 0.02	0.43 ± 0.02	0.44 ± 0.04	1.31 ± 0.03
	P/2003 T12	–	–	–	–	–	–	–
	2017 MB$_1$[7]	0.52	–	6.69	–	–	–	~1.2
Taurids	2P/Encke[8]	4.8 ± 0.4	0.05 ± 0.02	11	0.73 ± 0.06	0.39 ± 0.06	–	1.44 ± 0.06
Taurids-Perseids	1566 Icarus[9]	1.0–1.3	0.30–0.50^h	2.273	0.76 ± 0.02	0.41 ± 0.02	0.28 ± 0.02	1.2–1.4
	2007 MK$_6$	0.18^i	0.40^f	–	–	–	–	–
Phoenicids	2003 WY$_{25}$[10]	≤ 0.32	0.04^f	-	–	–	–	–

Notes:
^{a} Effective diameter (km)
^{b} Geometric albedo
^{c} Rotational period (hr)
^{d} Color index. Solar colors are $B - V = 0.64 \pm 0.02$, $V - R = 0.35 \pm 0.01$ and $R - I = 0.33 \pm 0.01$ (Holmberg et al., 2006).
^{e} Axis ratio
^{f} Assumed value
^{g} $D_e = 1.4 \pm 0.1$ with the assumed $p_v = 0.11$ (Kasuga and Jewitt, 2008).
^{h} Extremely high $p_v > 0.7$ is given $D_e < 0.8$ km (Mahapatra et al., 1999; Harris and Lagerros, 2002).
^{i} Estimated from the absolute magnitude $H = 20.3$ (MPO386777) with the assumed $p_v = 0.40$.
[1] Green et al. (1985); Tedesco et al. (2004); Dundon (2005); Ansdell et al. (2014); Hanuš et al. (2016); Taylor et al. (2019)
[2] Jewitt and Hsieh (2006); Kinoshita et al. (2007)
[3] Kasuga and Jewitt (2008); Mainzer et al. (2011); Warner (2017)
[4] Kasuga and Jewitt (2015)
[5] Licandro et al. (2000); Meech et al. (2004); Lamy et al. (2004)
[6] Kasuga et al. (2010); Weissman et al. (2004); A'Hearn et al. (2005); DeMeo and Binzel (2008)
[7] Warner (2018)
[8] Kokotanekova et al. (2017); Lamy et al. (2004); Campins and Fernández (2002); Lowry and Weissman (2007); Fernández et al. (2000)
[9] Gehrels et al. (1970); Veeder et al. (1989); Chapman et al. (1994); Nugent et al. (2015); Miner and Young (1969); Harris (1998); De Angelis (1995); Dundon (2005)
[10] Jewitt (2006)

of Na in meteors is a good indicator of thermal evolution of meteoroids. Heating either during their residence in interplanetary space or within the parent bodies themselves can lead to a sodium depletion.

In this chapter, we give a brief summary of observational results on parents and their associated showers. We discuss the properties of specific complexes, and tabulate their dynamical and physical properties (Tables 8.1 and 8.2). We focus on the main meteoroid streams for which the properties and associations seem the most secure. Numerous additional streams and their less certain associations are discussed in the literature reviewed by Jenniskens (2006, 2008b); Borovička (2007); Ye (2018); Vaubaillon et al. (2019) (see Section 7.6).

8.2 General Properties

To date, twelve objects in the six asteroid-meteoroid complexes have been studied in detail. Figure 8.1 represents their distributions in the semimajor axis versus eccentricity plane, while Figure 8.2 shows semimajor axis versus inclination (Table 8.1 summarizes the orbital properties).

Traditionally, the Tisserand parameter with respect to Jupiter, T_J, is used to characterize the dynamics of small bodies (Kresak, 1982; Kosai, 1992). It is defined by

$$T_J = \frac{a_J}{a} + 2\left[(1 - e^2)\frac{a}{a_J}\right]^{1/2} \cos(i) \qquad (8.3)$$

where a, e and i are the semimajor axis, eccentricity and inclination of the orbit and $a_J = 5.2$ AU is the semimajor axis of the orbit of Jupiter. This parameter, which is conserved in the circular, restricted three-body problem, provides a measure of the close-approach speed to Jupiter. Jupiter itself has $T_J = 3$. Main belt asteroids have $a < a_J$ and $T_J > 3$, while dynamical comets from the Kuiper belt have $2 \leq T_J < 3$ and comets from the Oort cloud have $T_J < 2$. In principle, the asteroids and comets can also be distinguished compositionally. The main belt asteroids are generally rocky, non-icy objects that probably

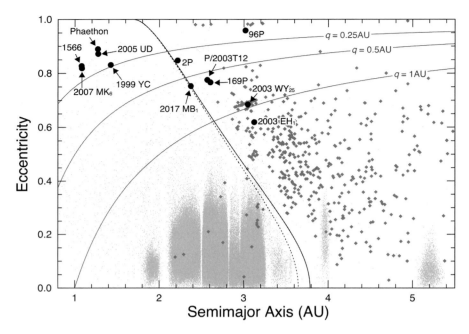

Figure 8.1. Distribution of the parent bodies (large numbered dots) in the semimajor axis versus orbital eccentricity plane (cf. Table 8.1). The distributions of asteroids (fine gray dots) and comets (diamonds) are shown for reference. The lines for $T_J = 3.08$ with $i = 0°$ (solid black curve) and with $i = 9°$ (dotted black curve) broadly separate asteroids and comets. Perihelion distances $q = 0.25, 0.5$ and 1 AU are shown as solid gray curves. (A black and white version of this figure will appear in some formats. For the colour version, please refer to the plate section.)

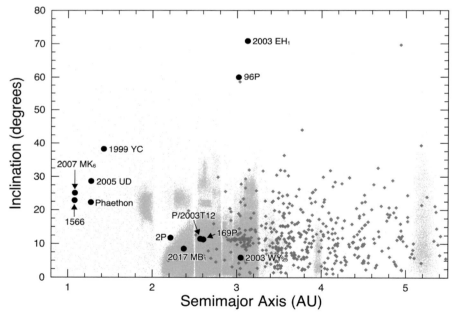

Figure 8.2. Same as Figure 8.1 but semimajor axis versus inclination. (A black and white version of this figure will appear in some formats. For the colour version, please refer to the plate section.)

formed inside the snow-line in the protoplanetary disk, while comets contain a larger ice fraction and formed beyond it. In practice, it is difficult or impossible to measure the compositions of most small bodies in the solar system, so that composition is not often a useful diagnostic.

The use of T_J as a discriminant breaks down near $T_J \simeq 3$, since the definition assumes that Jupiter's orbit is a circle, the gravity of other planets is neglected, and so on. Accordingly, a functional definition for the boundary employed here is $T_J \simeq 3.08$ (Jewitt et al., 2015), shown in Figure 8.1, where the solid curve represents T_J computed assuming $i = 0°$, while the dotted curve is the same but with $i = 9°$ (equal to the average inclination of 516,633 numbered objects in the JPL Horizons database). They are very similar, indicating the definition broadly works independently of i. This avoids chaotic cases caused by deviation of the real solar system from the circular, restricted three-body approximation. Also, comet 2P/Encke with $T_J \sim 3.03$ and the quasi-Hilda comets with $T_J \sim 2.9 - 3.04$ are appropriately classified with this criterion.

As seen in Figure 8.1, all of the parent bodies have $e \gtrsim 0.6$. Seven objects fall on the right side of the boundary with $T_J < 3.08$, corresponding to the region of comets. Mass-loss activity has been directly detected in most of them, although 2003 EH$_1$ and 2017 MB$_1$ have yet to show evidence for current activity. Objects on the left side of the diagram are classified in the region of the asteroids, with $T_J > 3.08$. In these objects, a range of physical processes appear to drive the mass loss. They have small $q \lesssim 0.25$ AU and are categorized as near-Sun objects (see Figure 8.1). Recurrent activity of Phaethon at perihelion, including the formation of a tail, has been reported.

In Figure 8.2, 2003 EH$_1$ and 96P show remarkably high-$i \gtrsim 60°$. Five objects with $a = 1.0 \sim 1.4$ AU have moderate i of $20° \sim 40°$. Another five objects at $a = 2.2 \sim 3.1$ AU have low-i, compatible with those of most main belt asteroids.

8.3 Known Asteroid-Meteoroid Complexes

The physical properties of the main asteroid-meteoroid complexes are listed in Table 8.2, and we review them focusing on the observational evidence.

8.3.1 Geminids – (3200) Phaethon

The Geminid meteor shower is one of the most active annual showers (Whipple, 1939b; Spurný, 1993). The shower currently has zenithal hourly rate (ZHR) – the number of meteors visible per hour under a clear-dark sky ($\lesssim$ limiting magnitude +6.5) – of ~ 120 and is expected to continue to increase to a peak ZHR ~ 190 in 2050 as the Earth moves deeper into the stream core (Jones and Hawkes, 1986; Jenniskens, 2006). The Geminids are dynamically associated with the near-Earth asteroid (3200) Phaethon (1983 TB) (Whipple, 1983) (Figure 8.3). A notable orbital feature of both is the small perihelion distance, $q \sim 0.14$ AU, raising the possibility of strong thermal processing of the surface. Indeed, the peak temperature at perihelion is ~ 1000 K (Ohtsuka et al., 2009).

Spectroscopy of Geminid meteors shows an extreme diversity in Na content, from strong depletion of Na abundance in some to sun-like values in others. Kasuga et al. (2005a) found a huge depletion in the Na/Mg abundance ratio of a Geminid meteor, with a value an order of magnitude smaller than the solar abundance ratio (Anders and Grevesse, 1989; Lodders, 2003). Line intensity ratios of Na I, Mg I, and Fe I emissions of the Geminids also show a wide range of Na line strengths, from undetectable to intense (Borovička et al., 2005). Studies of Geminid meteor spectra, reported since the 1950s (e.g. Millman, 1955; Russell et al., 1956; Harvey, 1973; Borovička, 2001; Trigo-Rodríguez et al., 2003, 2004), mostly fit this pattern.

As a summary for Na in the Geminids in the last decade, Kasuga et al. (2006b) investigated perihelion dependent thermal effects on meteoroid streams. The effect is supposed to alter the metal abundances from their intrinsic values in their parents, especially for temperature-sensitive elements: a good example

Figure 8.3. Composite R-band image of Phaethon taken at the University of Hawai'i 2.2-m telescope (UH2.2) on UT 2003 December 19. The full image width is $1'$ and this is a 3600 s equivalent integration, with Phaethon at heliocentric distance $R = 1.60$ AU, geocentric distance $\Delta = 1.39$ AU, and phase angle $\alpha = 0.3°$. The image shows no evidence of mass-loss activity. From Hsieh and Jewitt (2005).

is Na in alkaline silicate. As a result, meteoroid streams with $q \lesssim 0.1$ AU should be depleted in Na by thermal desorption, because the corresponding meteoroid temperature (characterized as blackbody) exceeds the sublimation temperature of alkaline silicates (~ 900 K for sodalite: $Na_4(AlSiO_4)_3Cl$). For this reason, the Na loss in Geminids ($q \sim 0.14$ AU) is most likely to be caused by thermal processes on Phaethon itself (see Section 8.6.1).

The parent body Phaethon (diameter 5–6 km, from Tedesco et al., 2004; Taylor et al., 2019) has an optically blue (so-called "B-type") reflection spectrum that distinguishes it from most other asteroids and from the nuclei of comets. Specifically, only ~ 1 in 23 asteroids is of B-spectral type and most cometary nuclei are slightly reddish, like the C-type asteroids or the (redder) D-type Jovian Trojans.

A dynamical pathway to another B-type, the large main-belt asteroid (2) Pallas, has been reported with $\sim 2\%$ probability (de León et al., 2010, see also Todorović, 2018). However, the colors of Phaethon and Pallas are not strictly identical, as would be expected if one were an unprocessed chip from the other, although both are blue. Color differences might result from preferential heating and modification of Phaethon, with its much smaller perihelion distance (0.14 AU versus ~ 2.1 AU for Pallas). Recently, Ohtsuka et al. (2006, 2008a) suggested the existence of a "Phaethon-Geminid Complex (PGC)" – consisting of a group of dynamically associated split fragments – and identified the 1 km-sized asteroids 2005 UD and 1999 YC as having a common origin with Phaethon (cf. Figure 8.4). Photometry of 2005 UD and 1999 YC revealed the optical colors (Figures 8.5, 8.6). The former is another rare blue object (Jewitt and

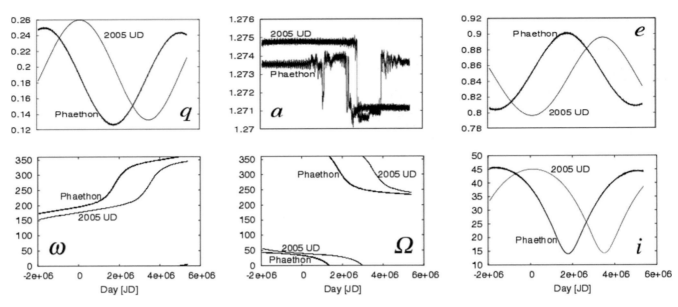

Figure 8.4. Dynamical evolution of (3200) Phaethon and 2005 UD for the six orbital elements, representing similar behavior with a time-shift of ∼4,600 yr. The abscissa shows time in Julian Terrestrial Date (JTD). From Ohtsuka et al. (2006), reproduced with permission © ESO.

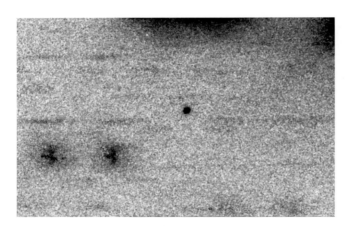

Figure 8.5. R-band image of 2005 UD in 1800 s exposure recorded using the UH2.2 on UT 2005 November 21. The region shown is ∼ 150″ in width and the distances and phase angle of the object were $R = 1.59$ AU, $\Delta = 0.96$ AU and $\alpha = 35.8°$, respectively. From Jewitt and Hsieh (2006).

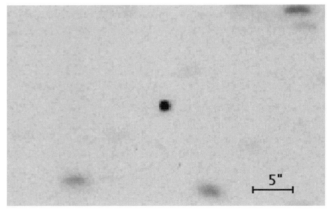

Figure 8.6. Keck telescope R-band image of 1999 YC on UT 2007 October 12 showing a point source with FWHM ∼ 0.65″ in 400 s integration, centered within a frame 40″ wide. Heliocentric, geocentric distances and phase angle were $R = 2.60$ AU, $\Delta = 1.82$ AU and $\alpha = 16.4°$, respectively. From Kasuga and Jewitt (2008).

Hsieh, 2006; Kinoshita et al., 2007) while the latter is spectrally neutral (Kasuga and Jewitt, 2008) (see Section 8.5.1).

The key question is how the Geminid meteoroid stream was produced from Phaethon. The extreme possibilities are that the Geminids are the products of a catastrophic event (for example an energetic collision, or a rotational disruption) or that they are produced in steady-state by continuing mass-loss from Phaethon (Figures 8.5 and 8.6).

In the steady-state case, the entire stream mass, $M_s \sim 10^{12} - 10^{13}$ kg (Hughes and McBride, 1989; Jenniskens, 1994; Blaauw, 2017)(cf. 1–2 orders larger in Ryabova, 2017), must be released over the last $\tau \sim 10^3$ years (the dynamical lifetime of the stream, cf. (Jones, 1978; Jones and Hawkes, 1986; Gustafson, 1989; Williams and Wu, 1993; Ryabova, 1999; Beech, 2002; Jakubík and Neslušan, 2015; Vaubaillon et al., 2019, Section 7.6.4)).

This gives $dM_s/dt \sim M_s/\tau = 30 - 300$ kg s^{-1}, comparable to the mass loss rates exhibited by active Jupiter family comets. However, while the Jupiter family comets are notable for their distinctive comae of ejected dust, Phaethon generally appears as a point source, devoid of coma or other evidence for on-going mass loss (Chamberlin et al., 1996; Hsieh and Jewitt, 2005; Wiegert et al., 2008; Jewitt et al., 2018) (cf. Figure 8.3).

Recently, this general picture has changed with the detection of mass loss using near-perihelion observations taken with the Solar Terrestrial Relations Observatory (STEREO) spacecraft in 2009, 2012 and 2016 (Jewitt and Li, 2010; Jewitt et al., 2013; Li and Jewitt, 2013; Hui and Li, 2017). In addition to factor-of-two brightening at perihelion relative to the expected phase-darkened inverse-square law brightness, a diffuse, linear tail has

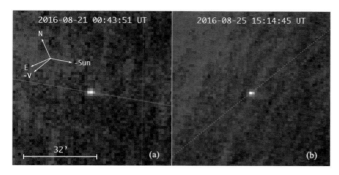

Figure 8.7. STEREO-A image taken on 2016 August 20 of (a) (3200) Phaethon and tail aligned on the position angles of the anti-solar direction and (b) the whole image sequence on the negative heliocentric velocity vector projected on the sky plane. Geometric parameters are $R \sim 0.14$ AU, Phaethon-STEREO distance ~ 0.9 AU, and the phase angle $\sim 120°$. The projected Sun-Phaethon line (the narrow white line across the left panel) is drawn to better illustrate that the direction of the tail was anti-solar, while the white dotted line across the right panel is the orientation. From Hui and Li (2017). (A black and white version of this figure will appear in some formats. For the colour version, please refer to the plate section.)

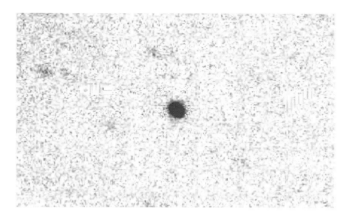

Figure 8.8. 2003 EH$_1$ in a 360 s R-band image taken at Keck on UT 2013 October 2. No coma or tail is visible on the object having an FWHM of $0.86''$ in the frame of $40'' \times 25''$. $R = 2.1$ AU, $\Delta = 2.0$ AU and $\alpha = 27.6°$. From Kasuga and Jewitt (2015).

been resolved, as shown in Figure 8.7. Because of its close association with the high temperatures experienced at perihelion, the activity is likely to result from thermal fracture or the desiccation of hydrated minerals. The observed particles from Phaethon are micron-sized, and are highly susceptible to solar radiation pressure sweeping. They are rapidly accelerated by radiation pressure and so cannot be retained in the Geminid stream (Jewitt and Li, 2010). In addition, their combined mass, $\sim 3 \times 10^5$ kg per perihelion, is at least $10^7 \times$ smaller than the stream mass, M_s (Jewitt et al., 2013). It is possible that much more mass is contained in larger particles which, however, present a small fraction of the scattering cross-section and which, therefore, are unsampled in the optical data from STEREO.

In this regard, larger particles (sizes > 10 μm) were recently reported in thermal emission at 25 μm (Arendt, 2014), but evidently contribute little to the optical scattering cross-section. The particles in the stream are estimated to be near mm-scale or larger (e.g. Blaauw, 2017), up to ~ 1 to 10 cm, as measured in lunar impacts (Yanagisawa et al., 2008; Suggs et al., 2014; Ortiz et al., 2015; Szalay et al., 2018). The limit to optical depth of Phaethon's trail is $\leq 3 \times 10^{-9}$ (Jewitt et al., 2018), consistent with those of cometary dust trails (10^{-9}–10^{-8}, from Sykes and Walker, 1992; Ishiguro et al., 2009) (see also Section 8.6.3). While continued long wavelength observations of Phaethon to detect large particles will be helpful (Jewitt et al., 2015), the true nature of Phaethon and the PGC complex objects may await spacecraft missions resembling NASA's "Deep Impact" (A'Hearn et al., 2005; Kasuga et al., 2006a, 2007a) and JAXA's "DESTINY+" (Sarli et al., 2018).

8.3.2 Quadrantids – 2003 EH$_1$

The Quadrantid meteor shower was first reported in 1835 (Quetelet, 1839) and appears annually in early January. The shower consists of two different components, the so-called young and old meteoroid streams, which represent a very short duration of core activity (lasting ~ 0.5 day) and a broader, longer-lived (~ 4 days) background activity (Wiegert and Brown, 2005; Brown et al., 2010, and references therein). The width of a meteoroid stream depends on its age, as a result of broadening by accumulated planetary perturbations. The small width of the Quadrantid core stream indicates ejection ages of only ~ 200–500 years (Jenniskens, 2004; Williams et al., 2004; Abedin et al., 2015), and there is even some suggestion that the first reports of meteoroid stream activity coincide with the formation of the stream. On the other hand, the broader background stream implies larger ages of perhaps $\sim 3,500$ years or more (Ohtsuka et al., 1995, 2008b; Kańuchová and Neslušan, 2007).

Two parent bodies of the Quadrantid complex have been proposed. The 4 km diameter Near-Earth Object (196256) 2003 EH$_1$ (hereafter 2003 EH$_1$), discovered on UT 2003 March 6 by the Lowell Observatory Near-Earth-Object Search (LONEOS) (Skiff, 2003), may be responsible for the young core stream (Jenniskens, 2004; Williams et al., 2004; Wiegert and Brown, 2005; Babadzhanov et al., 2008a; Jopek, 2011; Abedin et al., 2015). The orbit of 2003 EH$_1$ has $a = 3.123$ AU, $e = 0.619$, $i = 70°.838$ and $q = 1.191$ AU (Table 8.1). The $T_J (= 2.07)$ identifies it as a likely Jupiter-family comet, albeit one in which on-going activity has yet to be detected (Koten et al., 2006; Babadzhanov et al., 2008a; Borovička et al., 2010; Tancredi, 2014). The steady-state production rates $\lesssim 10^{-2}$ kg s^{-1} estimated from 2003 EH$_1$ at $R = 2.1$ AU are at least five orders of magnitude too small to supply the core Quadrantid stream mass $M_s \sim 10^{13}$ kg (Kasuga and Jewitt, 2015) (see Figure 8.8). Even at $q = 0.7 - 0.9$ AU a few hundred years ago, sublimation-driven activity from the entire body takes ~ 10s of years in the whole orbit, being hard to reconcile. In order to form the core Quadrantid stream, we consider episodic replenishment by an unknown process to be more likely.

Comet 96P/Machholz 1 has been suggested as the source of the older, broader part of the Quadrantid complex (Vaubaillon et al., 2019, Section 7.6.2), with meteoroids released 2,000–5,000 years ago (McIntosh, 1990; Babadzhanov and Obrubov, 1992b; Gonczi et al., 1992; Jones and Jones, 1993; Wiegert and

Brown, 2005; Abedin et al., 2017, 2018). Comet 96P currently has a small perihelion orbit ($a = 3.018$ AU, $e = 0.959$, $i = 59°.975$ and $q = 0.125$ AU from Table 8.1) substantially different from that of 2003 EH_1. Despite this, calculations show rapid dynamical evolution that allows the possibility that 2003 EH_1 is a fragment of 96P, or that both were released from a precursor body (together defining the Machholz complex: Sekanina and Chodas, 2005). One or both of these bodies can be the parents of the Quadrantid meteoroids (Kaňuchová and Neslušan, 2007; Babadzhanov et al., 2008a; Neslušan et al., 2013a,b, 2014; Vaubaillon et al., 2019, Section 7.6.2).

A notable dynamical feature of 2003 EH_1 is the strong evolution of the perihelion distance (Wiegert and Brown, 2005; Neslušan et al., 2013b; Fernández et al., 2014). Numerical integrations indicate that the minimum perihelion distance $q \sim 0.12$ AU ($e \sim 0.96$) occurred ~ 1500 yr ago (Neslušan et al., 2013b; Fernández et al., 2014), and the perihelion has increased approximately linearly with time from 0.2 AU 1000 years ago to the present-day value of 1.2 AU. At its recently very small (Phaethon-like) perihelion distance, it is reasonable to expect that the surface layers should have been heated to the point of fracture and desiccation (see Section 8.5.4).

As described in Section 8.3.1, the Phaethon-produced Geminid meteoroids ($q \sim 0.14$ AU) show extreme diversity in their Na abundance, from strong depletion to near sun-like Na content (Kasuga et al., 2005a; Borovička et al., 2005). Curiously, the Quadrantid meteoroids from the core stream are less depleted in Na than the majority of Geminid meteoroids (Koten et al., 2006; Borovička et al., 2010). The interpretation of this observation is unclear (see Section 8.6.2).

The optical colors of 2003 EH_1 are similar to, but slightly redder than, those of the Sun. They are most taxonomically compatible with the colors of C-type asteroids (Kasuga and Jewitt, 2015) (see Section 8.5.1).

8.3.3 Capricornids – 169P/NEAT

The α-Capricornids (CAP/1) are active from late July to early August, usually showing slow (~ 22 km s^{-1}) and bright meteors. The shower, with an ascending nodal intersection of $\omega = 270°$ with the Earth, is expected to be a twin stream also producing a daytime shower (Jenniskens, 2006). Because of the low entry velocities, the meteor plasma excitation temperature is $T_{ex} \lesssim 3600$K and no trace of high temperature gas (i.e. hot component of $T_{ex} \sim 10^4$ K) is found (Borovička and Weber, 1996). The metal contents of the α-Capricornids are unremarkable, being within a factor of a few of the Solar abundance (Borovička and Weber, 1996; Madiedo et al., 2014).

Recently Brown et al. (2010) suggested that the Daytime Capricornids-Sagittariids (DCS/115) are closely dynamically related to the α-Capricornids. One of the parent body candidates, comet 169P/NEAT, has been identified as the parent body of the α-Capricornid meteoroid stream by numerical simulations (Jenniskens and Vaubaillon, 2010). The object was discovered as asteroid 2002 EX_{12} by the NEAT survey in 2002 (cf. Warner and Fitzsimmons, 2005; Green, 2005) and was re-designated as 169P/NEAT in 2005 after revealing a cometary appearance (Green, 2005). The orbital properties (Levison, 1996, $T_J = 2.89$) and optical observations reveal that 169P/NEAT

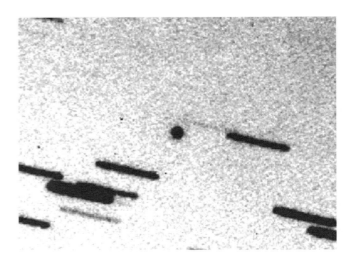

Figure 8.9. Comet 169P/NEAT in a 600 sec, r'-band image taken at the Dominion Astrophysical Observatory 1.8 m telescope on UT 2010 February 17. The frame size is $200'' \times 150''$. No coma or tail is visible on the object having an FWHM of $2.8''$. $R = 1.43$ AU, $\Delta = 0.47$ AU and $\alpha = 16.1°$. From Kasuga et al. (2010).

is a ~ 4 km diameter, nearly dormant Jupiter family comet with tiny mass-loss rate $\sim 10^{-2}$ kg s^{-1} (Kasuga et al., 2010) (Figure 8.9).

In the steady state, the stream mass $M_s \sim 10^{13}$–10^{15} kg and the age $\tau \sim 5{,}000$ yr (Jenniskens and Vaubaillon, 2010) together require a mass-loss rate four orders of magnitude larger than measured in 2010. In the case of 169P, cometary activity (a dust tail) was confirmed in 2005 (almost one orbital period before) and episodic mass-loss should be expected. This is a different case from other asteroidal parents of complexes. Kasuga et al. (2010) used the fractional change in the spin angular velocity to estimate the mass loss from 169P as $\sim 10^9$–10^{10} kg per orbit. With the M_s and τ, the conclusion is that the origin of the α-Capricornids meteoroid stream could be formed by the steady disintegration of 169P.

Other parent body candidates continue to be proposed for the Capricornids. P/2003 T12 (SOHO) was suggested to share a common parent with 169P, following a breakup ~ 2900 yr ago (Sosa and Fernández, 2015). Comet 169P is a large, almost inactive body (Kasuga et al., 2010), while P/2003 T12 seems to be a very small comet, with a sub-km radius nucleus (Sosa and Fernández, 2015) accompanied by dust-tails in near-Sun STEREO-B observations (Hui, 2013). The orbit of 2017 MB_1 was suggested to resemble that of the α-Capricornids meteor shower (Wiegert et al., 2017). 2017 MB_1 has not been reported to show any sign of mass-loss activity.

8.3.4 Taurid Complex – 2P/Encke

The Taurid meteor shower includes the Northern, the Southern and other small branches (Vaubaillon et al., 2019, Section 7.6.5), possibly originating from more than one parent body. The Taurids show protracted, low-level activity with many fireballs from September to December, peaking in early November every year. The Taurid meteoroid complex has been suggested to be formed by a disrupted giant comet (40 km-sized) 10^4 years ago (Clube and Napier, 1984, 1987; Asher et al., 1993), although the very recent break-up of such a large (i.e., rare) body is statistically unlikely. Comet 2P/Encke has, for a long time, been considered as the most probable parent of the shower (Whipple, 1940).

The Taurid complex has a dispersed structure with low inclination and perihelia between 0.2 and 0.5 AU. The low inclination of the stream orbit enhances the effect of planetary perturbations from the terrestrial planets (Levison et al., 2006), resulting in the observed, diffuse structure of the complex (Matlovič et al., 2017). Furthermore, 2P has a relatively small heliocentric distance (aphelion is $Q = 4.1$ AU) allowing it to stay mildly active around its orbit, and producing a larger spread in $q \sim 0.34$ AU, than could be explained from ejection at perihelion alone (Gehrz et al., 2006; Kokotanekova et al., 2017).

An unfortunate artifact of the low inclination of the Taurid complex ($i \sim 12°$, Table 8.1) is that many near-Earth asteroids are plausible parent bodies based on orbital dynamical calculations (e.g. Asher et al., 1993; Steel and Asher, 1996; Babadzhanov, 2001; Porubčan et al., 2004, 2006; Babadzhanov et al., 2008b). As a result, many of the proposed associations are likely coincidental (Spurný et al., 2017; Matlovič et al., 2017). Actually no spectroscopic linkage between 2P/Encke and the 10 potential Taurid-complex NEOs has been confirmed (the latter are classified variously as X, S, Q, C, V, O, and K-types) (Popescu et al., 2014; Tubiana et al., 2015), this being totally different from the case of the Phaethon-Geminid Complex. Here, we focus on the physical properties of the Taurid meteor shower and the most strongly associated parent, 2P/Encke.

Spectroscopic studies of some Taurid meteors find a carbonaceous feature (Borovička, 2007; Matlovič et al., 2017). The heterogeneity (large dispersion of Fe content) and low strength (0.02–0.10 MPa) of the Taurids (Borovička et al., 2019, Section 2.3.4) suggest a cometary origin, consistent with but not proving 2P as a parent (Borovička, 2007; Matlovič et al., 2017). Note that the current perihelion distance ($q \sim 0.34$ AU, where $T \sim 480$ K) is too large for strong thermal metamorphism to be expected (Borovička, 2007).

Comet 2P/Encke is one of the best characterized short-period comets, with published determinations of its rotation period, color, albedo and phase function (see reviews; Lamy et al., 2004; Kokotanekova et al., 2017). For example, the effective radius is 2.4 km, the average color indices, $B - V = 0.73 \pm 0.06$ and $V - R = 0.39 \pm 0.06$ (e.g., Lowry and Weissman, 2007), the rotational period is about 11 hr or 22 hr (but changing with time in response to outgassing torques (Belton et al., 2005; Kokotanekova et al., 2017, and references therein); with minimum axis ratio 1.4 (Fernández et al., 2000; Lowry and Weissman, 2007; see also Lamy et al., 2004).

Optically, 2P appears dust-poor because optically bright micron and sub-micron -sized dust particles are underabundant in its coma (Jewitt, 2004a). However, the dust / gas ratio determined from thermal emission is an extraordinary $\mu \sim 30$, suggesting a dust-rich body (Reach et al., 2000; Lisse et al., 2004) compared to, for example, Jupiter family comet 67P/Churyumov-Gerasimenko, where $\mu = 4 \pm 2$ (Rotundi et al., 2015). The total mass loss is 2–6 $\times 10^{10}$ kg per orbit, mostly in the form of large particles that spread around the orbit and give rise to 2P's thermal dust trail as the source of Taurid meteor showers (Asher and Clube, 1997; Reach et al., 2000)

Figure 8.10. The 24 μm infrared image of comet 2P/Encke obtained in 2004 June by the Spitzer Space Telescope. The image field of view is about 6' and centered on the nucleus. The near horizontal emission is produced by recent cometary activity, and the diagonal emission across the image is the meteoroid stream along the orbit (Kelley et al.,2006; see also Reach et al., 2007). Courtesy NASA/JPL-Caltech/M. Kelley (Univ. of Minnesota). (A black and white version of this figure will appear in some formats. For the colour version, please refer to the plate section.)

(see Figure 8.10). The entire structured stream mass, $M_s \sim 10^{14}$ kg (Asher et al., 1994), must have been released over the last $\tau \sim 5{,}000 - 20{,}000$ years (the dynamical lifetime of the stream, cf. Whipple, 1940; Babadzhanov and Obrubov, 1992a; Jenniskens, 2006). This gives $dM_s/dt \sim M_s/\tau = 200$–600 kg s^{-1}, a few times larger than the mass loss rates typically reported for active Jupiter family comets. Various small near-Earth objects and some meteorite falls have been linked with the orbit of the stream as potentially hazardous (Brown et al., 2013; Olech et al., 2017; Spurný et al., 2017).

8.3.5 Sekanina's (1973) Taurids-Perseids – Icarus

The Icarus asteroid family was reported as the first family found in the near-Earth region, which dynamically relates asteroids 1566 Icarus, 2007 MK$_6$ and Sekanina's (1973) Taurid-Perseid meteor shower (Ohtsuka et al., 2007) (see Figure 8.11).

Near-Earth Apollo asteroid 1566 Icarus (= 1949 MA) was discovered in 1949, having distinctive small $q = 0.19$ AU and high $i = 23°$ (Baade et al., 1950). The object has diameter $D_e \sim 1$ km (e.g. table 1, from Chapman et al., 1994), a moderately high albedo of 0.30–0.50 (cf. Gehrels et al., 1970; Veeder et al., 1989; Chapman et al., 1994; Nugent et al., 2015) and a short rotational period, ~ 2.273 hr (e.g. Miner and Young, 1969) (see also Harris and Lagerros, 2002). A reflection spectrum close to Q- or V-type asteroids is found (Gehrels et al., 1970; Hicks et al., 1998; Dundon, 2005) (see Section 8.5.1).

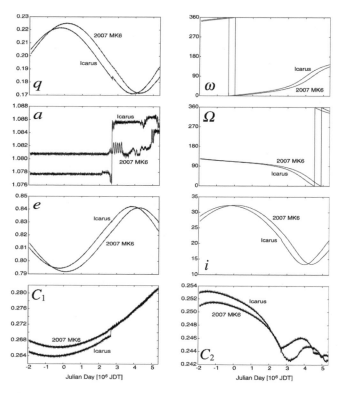

Figure 8.11. Dynamical evolution process of 1566 Icarus and 2007 MK$_6$ in JTD. The orbital elements (q, a, e, ω, Ω and i) and the C_1 and C_2 integrals are plotted (cf. Table 8.1). Time-shifting is only $\sim 1,000$ yr. From Ohtsuka et al. (2007).

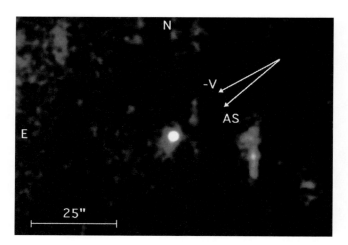

Figure 8.12. 2003 WY$_{25}$ (=289P/Blanpain) imaged in R-band, 500 sec integration at $R = 1.6$ AU, $\Delta = 0.7$ AU and $\alpha = 20.7°$ using UH2.2 on UT 2004 March 20. A faint coma is apparent, extending to the southeast. Arrows show the directions of the negative heliocentric velocity vector (marked "-V") and the anti-solar direction ("AS"). The estimated radius ~ 160 m is the smallest active cometary parent ever observed. From Jewitt (2006). (A black and white version of this figure will appear in some formats. For the colour version, please refer to the plate section.)

On the other hand, near-Earth asteroid 2007 MK$_6$ (= 2006 KT$_{67}$) was discovered in 2007 (Hill et al., 2007). Assuming an albedo like that of Icarus, then 2007 MK$_6$ is ~ 180 m in diameter as computed from the absolute magnitude $H = 20.3$ (MPO386777) (see Table 8.2). The breakup hypothesis from Icarus, if true, could be due to near a critical rotation period and thermal stress induced at small $q \sim 0.19$ AU (subsolar temperature ~ 900 K), which might be related to the production of the meteoroid stream (see Section 8.5.2). The Taurid-Perseid meteoroids can be dynamically related with the Icarus asteroid family ($D_{SH} \sim 0.08$), speculated to cross the Earth's orbit (Sekanina, 1973). The rare detection of the Taurid-Perseid meteor shower may result from the intermittent stream (swarm) due to very limited dust supply phase from the parent body.

8.4 Possible Complexes

Here, we describe two examples of less well-characterized complexes suspected to include stream branches and one or more parent bodies.

8.4.1 Phoenicids – Comet D/1819 W1 (Blanpain)

The Phoenicid meteor shower was first reported more than 50 years ago, on December 5 in 1956 (Huruhata and Nakamura, 1957). The lost Jupiter-family comet D/1819 W1 (289P/Blanpain) was promptly proposed as the potential source (Ridley, 1957, reviewed in Ridley, 1963). In 2003, the planet-crossing asteroid 2003 WY$_{25}$ was discovered (Ticha et al., 2003), with orbital elements resembling those of D/Blanpain (Foglia et al., 2005; Micheli, 2005). The related Phoenicids' activity in 1956 and 2014 (Watanabe et al., 2005; Jenniskens and Lyytinen, 2005; Sato et al., 2017; Tsuchiya et al., 2017), raised the possibility that 2003 WY$_{25}$ might be either the dead nucleus of D/Blanpain itself or a remnant of the nucleus surviving from an earlier, unseen disintegration. Jewitt (2006) optically observed asteroid 2003 WY$_{25}$, finding the radius of 160 m (an order of magnitude smaller than typical cometary nuclei), and revealing a weak coma consistent with mass-loss rates of 10^{-2} kg s^{-1} (Figure 8.12). The latter is too small to supply the estimated 10^{11} kg stream mass on reasonable timescales ($\leq 10,000$ yrs, Jenniskens and Lyytinen, 2005). Indeed, the mass of 2003 WY$_{25}$ (assuming density 1000 kg m^{-3} and a spherical shape) is $\sim 2 \times 10^{10}$ kg, smaller than the stream mass. Either the stream was produced impulsively by the final stages of the break-up of a once much larger precursor to 2003 WY$_{25}$, or another parent body may await discovery (Jewitt, 2006).

8.4.2 Andromedids – Comet 3D/Biela

The Andromedid meteor shower (AND/18) was firstly reported in 1798 (Hawkins et al., 1959). The dynamics were linked with Jupiter-family comet 3D/Biela (Kronk, 1988, 1999). The shower is proposed to result from continuous disintegration of 3D/Biela from 1842 until its sudden disappearance in 1852, resulting in irregular meteor shower appearances (e.g. Olivier, 1925; Cook, 1973; Jenniskens and Vaubaillon, 2007). The estimated stream mass is 10^{10} kg (Jenniskens and Lyytinen, 2005), however, the absence of parent candidates means that little can be determined about the production of the meteoroids. Nonetheless,

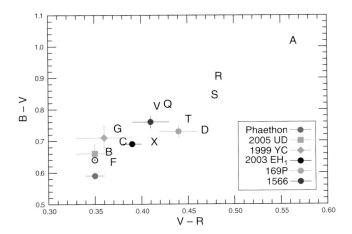
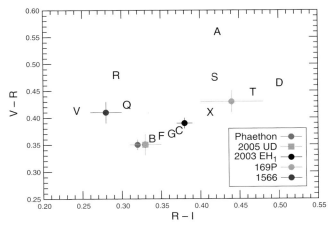

Figure 8.13. Color plots of $V - R$ vs. $B - V$ showing parent bodies (filled symbols) from Table 8.2, and Tholen taxonomic classifications (Tholen, 1984), as tabulated by Dandy et al. (2003). The color of the Sun (black circle) is also plotted. (A black and white version of this figure will appear in some formats. For the colour version, please refer to the plate section.)

Figure 8.14. The same as Figure 8.13 but in the $R - I$ vs. $V - R$ color plane. The color of the Sun is exactly coincident with that of 2005 UD. (A black and white version of this figure will appear in some formats. For the colour version, please refer to the plate section.)

the Andromedid meteor shower was actually detected by radar in 2011 and is numerically predicted to appear in the coming decades (Wiegert et al., 2013).

8.5 Parent Bodies

In this section we discuss group physical properties of the parent bodies (cf. Table 8.2). Most objects (e.g. 3200 Phaethon, 2005 UD, 1999 YC, 2003 EH$_1$ and 169P) show point-like images (Figures 8.3, 8.5, 8.6, 8.8, 8.9) from which we can be confident that the measured properties refer to the bare objects (or nuclei) alone. However, 2P/Encke, 2003 WY$_{25}$ and some other comets may sometimes be active (e.g. Figures 8.10, 8.12) leading to potential confusion between the properties of the nucleus and the near-nucleus comae.

8.5.1 Colors

Figures 8.13 and 8.14 show distributions of the colors of the parent bodies from Table 8.2. In addition, Tholen taxonomy classes are plotted from photometry of NEOs from Dandy et al. (2003). Here, 2P is not included because of the coma contamination suggesting mild activity during the whole orbit.

The asteroids of the PGC (3200 and 2005 UD, 1999 YC) show colors from nearly neutral to blue. Asteroids 3200 Phaethon and 2005 UD are classified as B-type asteroids (cf. Dundon, 2005; Jewitt and Hsieh, 2006; Kinoshita et al., 2007; Licandro et al., 2007; Kasuga and Jewitt, 2008; Jewitt, 2013; Ansdell et al., 2014), while 1999 YC is a C-type asteroid (Kasuga and Jewitt, 2008). Heterogeneity on the surfaces of Phaethon and 2005 UD may be due to intrinsically inhomogeneous composition, perhaps affected by hydration processes (Licandro et al., 2007), and by thermal alteration (Kinoshita et al., 2007). The rotational color variation of 2005 UD shows B-type for 75% of the rotational phase but C-type for the remainder (Kinoshita et al., 2007). The colors of the PGC objects are broadly consistent with being neutral-blue.

Optical colors of 2003 EH$_1$ are taxonomically compatible with those of C-type asteroids (Kasuga and Jewitt, 2015) (Figures 8.13 and 8.14). The $V - R$ color (0.39 ± 0.01) is similar to that of 96P ($V - R = 0.40 \pm 0.03$, from Licandro et al., 2000; Meech et al., 2004). We note that the optical colors of 2003 EH$_1$ are significantly less red than the average colors of cometary nuclei (Jewitt, 2002; Lamy et al., 2004). This could be a result of past thermal processing when the object had a perihelion far inside Earth's orbit. Indeed, the weighted mean color of 8 near-Sun asteroids having perihelion distances $\lesssim 0.25$ AU (subsolar temperatures $\gtrsim 800$ K) is $V - R = 0.36 \pm 0.01$ (Jewitt, 2013), consistent with the color of EH$_1$ (cf. Section 8.5.4).

The optical colors measured for 169P/NEAT are less red than D-type objects, as found in normal cometary nuclei and Trojans, but similar to those of T- and X- type asteroids (Figures 8.13 and 8.14). The near-infrared spectrum measurement (0.8–2.5 μm) classified 169P as a T-type asteroid based on the Bus taxonomy with $p_v = 0.03 \pm 0.01$ (DeMeo and Binzel, 2008). The T-type asteroids represent slightly redder-sloped visible wavelength spectra than those of C-type. Perhaps a refractory rubble mantle has formed on the 169P surface, driven by volatile sublimation, and red matter has been lost (Jewitt, 2002).

Asteroid 1566 Icarus is taxonomically classified as a Q- or V-type (Figures 8.13 and 8.14). These types suggest thermal evolution (perhaps at the level of the ordinary chondrites) relative to the more primitive carbonaceous chondrites. The formation process of the associated complex is unknown, but we speculate that processes other than comet-like sublimation of ice are responsible.

8.5.2 Dust Production Mechanism: Example of 1566 Icarus & 2007 MK$_6$

Here we consider possible dust production mechanisms from asteroidal parents. Figure 8.15 shows a diameter – rotation plot compiled from the data in Table 8.2. The rotational period of 1566 Icarus ($P_{\rm rot} = 2.273$ hr) is near the spin barrier period of ~ 2.2 hr (Warner et al., 2009; Chang et al., 2015). Asteroids

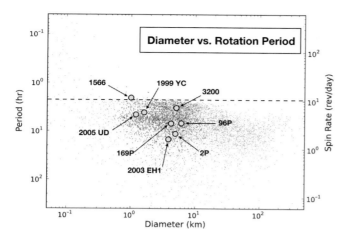

Figure 8.15. Diameter versus rotational period for asteroids (dots) and the parent bodies of meteoroid streams (large circles). The asteroid data are taken from Chang et al. (2015) while the parent body parameters are listed in Table 8.2. The horizontal dotted line shows the spin barrier period of ~2.2 hr (e.g. Warner et al., 2009) and asteroid 1566 Icarus rotates nearby.

rotating near or faster than this barrier are presumed to have been destroyed when centrifugal forces have overcome the gravitational and cohesive forces binding them together (Pravec et al., 2008).

The aftermaths of recent and on-going asteroid break-up have been identified observationally (e.g. P/2013 R3, Jewitt et al., 2014, 2017) and studied theoretically (Hirabayashi et al., 2014). Additionally, different mechanisms can operate together. Rotational instability in P/2013 R3, for instance, might have been induced by YORP torques, or by outgassing torques from sublimated ice, or by a combination of the two. Thermal disintegration, electrostatic ejection and radiation pressure sweeping may all occur together on near-Sun object 3200 Phaethon (Jewitt and Li, 2010). Figure 8.15 shows that 1566 rotates near the ~2.2 hr spin barrier period, implying a rotational breakup in the past. Both 1566 and its possible fragment 2007 MK$_6$ have small perihelia ($q \sim 0.19$ AU). We consider rotational instability as a possible cause of their past separation.

In principle, rotation rates of asteroids can be accelerated to critical limits by torques exerted from solar radiation through the YORP effect (Vokrouhlický et al., 2015). The YORP e-folding timescale of the spin, τ_Y, is estimated from the ratio of the rotational angular momentum, L, to the torque, T. The relation may be simply expressed as $\tau_Y \sim K D_e^2 R_h^2$ (Jewitt et al., 2015), where K is a constant, D_e is the asteroid diameter (km) and R_h is the heliocentric distance (AU). The value of constant K is sensitive to many unknown parameters (the body shape, surface texture, thermal properties and spin vector of the asteroid and so on), but can be experimentally estimated from published measurements of YORP acceleration in seven well-characterized asteroids (table 2 from Rozitis and Green, 2013). Scaling K to the bulk density of 1566 Icarus $\rho = 3400$ kg m^{-3} (V-type or ordinary chondrite from Wilkison and Robinson, 2000; Britt et al., 2002) and its rotation period $P_{rot} = 2.273$ hr, we find $K \sim 7 \times 10^{13}$ s km^{-2} AU^{-2}. The approximation is represented as (cf. equation (3) of Jewitt et al., 2015),

$$\tau_Y \text{ (Myr)} \approx 2 \left[\frac{D_e}{1 \text{ km}}\right]^2 \left[\frac{R_h}{1 \text{ AU}}\right]^2. \quad (8.4)$$

For 1566 with $D_e = 1$ km orbiting at $R_h \sim 1.08$ AU, Equation (8.4) gives $\tau_Y \approx 2$ Myr. This is two orders of magnitude smaller than the collisional lifetime of 1-km near-Earth asteroids (Bottke et al., 1994), suggesting that YORP torque spin-up is plausible.

Asteroids rotating faster than the spin-barrier cannot be held together by self-gravitation only, but require cohesive strength (e.g. Scheeres et al., 2010). The cohesive strength at rotational breakup of a body can be estimated by the dispersed fragmental sizes, initial separation speed, and the bulk density using equation (5) of Jewitt et al. (2015),

$$S \sim \rho \left(\frac{D'_e}{D_e}\right) (\Delta v)^2. \quad (8.5)$$

Both fragmental asteroids (1566 and 2007 MK$_6$) are assumed to have the same bulk density ρ, D_e and D'_e are diameters of 1566 and 2007 MK$_6$ respectively (see Table 8.2), and Δv is the excess velocity of escaping fragments, assumed comparable to the escape velocity from 1566. Adopting the same value for ρ (see previous paragraph) and substituting $(D'_e/D_e) = 0.18$ (the diameter ratio between MK$_6$ and 1566), and $\Delta v = 0.69$ m s^{-1}, we find $S \sim 290$ N m^{-2}.

This small value is comparable to strengths ~10–100 N m^{-2} modeled by a rubble-pile asteroid bounded by weak van der Waals forces (reviewed in Scheeres and Sánchez, 2018), but more than five orders of magnitude smaller than the values of typical competent rocks ($10^7 - 10^8$ N m^{-2}). A rotational breakup origin of 1566 and MK$_6$ is possible provided they have a weak, rubble-pile structure, as is thought likely for a majority of kilometer-sized asteroids as a result of past, non-destructive impacts.

Several processes could eject dust from the surface of 1566. Firstly, thermal disintegration can be induced by thermal expansion forces that make cracks on the surfaces of asteroids and produce dust particles. The characteristic speeds of dust particles produced by disintegration can be derived by conversion from thermal strain energy into kinetic energy of ejected dust particles. The necessary conversion efficiency, η, is given by (cf. equation (3) of Jewitt and Li, 2010)

$$\eta \sim \left(\frac{v_e}{\alpha \delta T}\right)^2 \left(\frac{\rho}{Y}\right), \quad (8.6)$$

where, $v_e = 0.69$ m s^{-1} is the escape velocity from 1566, $\alpha \sim 10^{-5}$ K^{-1} is the characteristic thermal expansivity of rock (Lauriello, 1974; Richter and Simmons, 1974), $\delta T \sim 450$ K is the temperature variation between the q and Q, and $Y = (1-10) \times 10^{10}$ N m^{-2} are typical Young's moduli for rock (Pariseau, 2006, p. 474). Again taking $\rho = 3400$ kg m^{-3}, we find $\eta \gtrsim 0.1-1$ % is needed for the velocities of ejected dust particles to surpass the escape velocity. This very small value of conversion efficiency is sufficient for most dust particles produced by thermal disintegration to be launched into interplanetary space.

Secondly, electrostatic forces caused by photoionization by solar UV can eject small particles. The critical size for a 1 km-diameter asteroid is $a_e \lesssim 4$ μm (equation (12) of Jewitt et al., 2015). Millimeter-sized particles cannot be electrostatically launched and this process may contribute little or nothing to meteoroid stream formation.

Finally, radiation pressure sweeping can remove small particles from an asteroid once they are detached from the surface by another process (i.e. once the surface contact forces are temporarily broken). Radiation pressure sweeping is most effective at small heliocentric distances. The critical size to be swept away, $a_{rad}(\mu m)$, is estimated by equating the net surface acceleration (gravitational and centripetal) with the acceleration due to radiation pressure, given by equation (6) of Jewitt and Li (2010)

$$a_{rad} \sim \frac{3 g_\odot}{2\pi R_{AU}^2 f^{1/2} D_e} \left[\frac{G\rho}{f^2} - \frac{3\pi}{P_{rot}^2} \right]^{-1}, \quad (8.7)$$

where $g_\odot$ is the gravitational acceleration to the Sun at 1 AU, R_{AU} is the heliocentric distance expressed in AU, f is the limit to the axis ratio (=a/b), G is the gravitational constant. We substitute $g_\odot = 0.006$ m s^{-2}, $R_{AU} = 0.187$ (Table 8.1), $f = 1.2$ (Table 8.2), $G = 6.67 \times 10^{-11}$ m^3 kg^{-1} s^{-2} and adopt the same values of D_e, ρ and P_{rot} (as when evaluating Equation (8.4)) into Equation (8.7), then obtain $a_{rad} \sim 4,500~\mu m \approx 5$ mm. This size is large enough to contribute meteoroid-sized particles to a stream (see Section 8.6.3).

In summary, asteroid 1566 Icarus is a possible product of rotational breakup and, given its small perihelion distance, potentially experiences a mass loss process similar to those inferred on Phaethon. Near-perihelion observations of 1566 and/or 2007 MK$_6$ may indeed show Phaethon-like mass-loss.

8.5.3 End State

Active objects on comet-like orbits with $T_J < 3.08$ (Table 8.1 and Figure 8.1) are presumed to be potential ice sublimators. The timescales for the loss of ice from a mantled body τ_{dv}, for the heat propagation into the interior of a body τ_c, and dynamical lifetime of short-period comets $\tau_{sp} \sim 10^5$–10^6 yr (Duncan et al., 2004) can be compared to predict an object's end state (Jewitt, 2004b).

The τ_{dv} is calculated using $\rho_n D_e / 2 f(dm/dt)$, where $\rho_n = 600$ kg m^{-3} is the cometary bulk density (Weissman et al., 2004) and $f = 0.01$ is the mantle fraction (A'Hearn et al., 1995). The orbit with averaged $\bar{a} \sim 2.7$ AU and $\bar{e} \sim 0.8$ from the seven objects has a specific mass loss rate of water ice $dm/dt \lesssim 10^{-4}$ kg m^{-2} s^{-1} (figure 6 in Jewitt, 2004b). Then we find $\tau_{dv} \gtrsim 10^4 D_e$ in yr, where D_e is an effective diameter in km (Table 8.2). The $\tau_c \sim 8.0 \times 10^4 D_e^2$ in yr is derived from the equation of heat conduction given by $D_e^2/4\kappa$, where $\kappa = 10^{-7}$ m^2 s^{-1} is the assumed thermal diffusivity of a porous object. The critical size of object to form an inactive, devolatilized surface is constrained by the relation $\tau_{dv} \lesssim \tau_c$, which gives $D_e \gtrsim 0.13$ km. Likewise, the size capable of containing ice in the interior of a body for the dynamical lifetime is given by the relation $\tau_c \gtrsim \tau_{sp}$, which gives $D_e \gtrsim 1.1$ km.

Most comet-like objects are expected to be dormant, with ice depleted from the surface region, but potentially still packed deep inside. We note that 2003 WY$_{25}$ of the Phoenicids is on its way to the dead state due to its small size ($D_e = 0.32$ km). The $\tau_c \sim 8,000$ yr suggests that solar heat can reach into the body

Table 8.3. *Closest approach to the Sun by the Lidov-Kozai Mechanism*

Complex	Object	e_{max}[a]	q_{min}[b]	T_{peak}[c]	Ref.[d]
Geminids	Phaethon	0.90	0.13	1100	~0.13[1]
	2005 UD	0.90	0.13	1100	0.13–0.14[1]
	1999 YC	0.89	0.16	1000	-
Quadrantids	2003 EH$_1$	0.96	0.12	1100	~0.12[2]
	96P/Machholz 1	0.99	0.03	2300	0.03–0.05[3]
Taurids-	1566 Icarus	0.85	0.16	1000	~0.17[4]
Perseids	2007 MK$_6$	0.85	0.16	1000	~0.17[4]

Notes: Near-Sun objects have perihelia $\lesssim 0.25$ AU.
[a] Maximum eccentricity (Equation (8.8))
[b] Minimum perihelion distance (AU) estimated by $\simeq a(1 - e_{max})$, where a is from Table 8.1.
[c] Peak temperature at q_{min} (K)
[d] Referred minimum perihelion distance (AU) from numerical integrations and analytical methods
[1] Ohtsuka et al. (1997, 2006) (cf. Figure 8.4)
[2] Neslušan et al. (2013b); Fernández et al. (2014) (see Section 8.3.2)
[3] Bailey et al. (1992); Sekanina and Chodas (2005); Abedin et al. (2018)
[4] Ohtsuka et al. (2007) (cf. Figure 8.11)

core approximately one or two orders of magnitude sooner than the end of the dynamical lifetime. The core temperature around the orbit (Jewitt and Hsieh, 2006), $T_{core} \sim 180$ K, exceeds the sublimation temperature of water ice 150 K (Yamamoto, 1985). The extremely weak activity of 2003 WY$_{25}$ may portend its imminent demise (cf. Section 8.4.1).

8.5.4 Lidov-Kozai Mechanism

The Lidov-Kozai mechanism[2] works on the secular dynamics of small solar system objects. Large-amplitude periodic oscillations of the e and i (in antiphase) are produced, whereas the a is approximately conserved, while the ω librates around $\pi/2$ or $3\pi/2$ if $C_2 < 0$ or circulates if $C_2 > 0$ (Kozai, 1962; Lidov, 1962).

Perihelia can be deflected into the vicinity of the Sun by this mechanism (on timescale ~1000s of years), perhaps causing physical alteration (or even breakup) due to enormous solar heating (Emel'yanenko, 2017). We find all parent bodies stay in the circulation region ($C_2 > 0$, see Table 8.1). The minimum perihelion distance q_{min} can be computed using maximum eccentricity, e_{max}, given by equations (5) and (28) of Antognini (2015)

$$e_{max} = \sqrt{1 - \frac{1}{6}\left(\zeta - \sqrt{\zeta^2 - 60 C_1}\right)}, \quad (8.8)$$

where $\zeta = 3 + 5(C_1 + C_2)$ (equation (31) of Antognini, 2015), C_1 and C_2 are from Table 8.1. With the obtained q_{min}, we find the Geminids (PGC), Quadrantids and Sekanina's (1973)

---
[2] In Chapter 7 (Vaubaillon et al., 2019) the Lidov-Kozai mechanism focuses on secular changes in e, i and ω to find how e and ω relate to whether an orbit intersects Earth's orbit to produce a meteor shower.

Taurids-Perseids complexes are near-Sun objects (Table 8.3). Among them, 2003 EH$_1$ turns itself into a near-Sun object with $q_{min} \sim 0.12$ AU (cf. Section 8.3.2), albeit $q \sim 1.2$ AU at present (see Figure 8.1). The peak temperature $\gtrsim 1000$ K is similar to that experienced by Phaethon (cf. 1566 Icarus), and could likewise cause strong thermal and desiccation stresses, cracking and alteration in EH$_1$, with the release of dust (Jewitt and Li, 2010; Molaro et al., 2015; Springmann et al., 2019). This example reminds us that, even in objects with q presently far from the Sun, we cannot exclude the action of extreme thermal processes in past orbits.

8.6 Meteors and Streams

8.6.1 Na Loss: Thermal

Sodium loss in Geminid meteors results from the action of a thermal process in or on the parent body, Phaethon (Section 8.3.1). Čapek and Borovička (2009) calculated the timescale for thermal depletion of Na from an assumed initial solar value down to 10% of solar abundance in Geminid meteoroids during their orbital motion in interplanetary space (using assumed albite (NaAlSi$_3$O$_8$) and orthoclase (KAlSi$_3$O$_8$) compositions). With particle diameters $\geq$ mm-scale, they found depletion timescales $\gtrsim 10^4$–10^5 yr (Figure 8.16), some one to two orders of magnitude longer than the stream age of $\lesssim 10^3$ yr. On the other hand, the dynamical lifetime of Phaethon, while very uncertain, is estimated to be $\sim 3 \times 10^7$ yr (de León et al., 2010). This is surely long compared to the age of the Geminid stream and long enough for Na to be thermally depleted from Phaethon.

In principle, other processes might affect the sodium abundance. Sputtering by the solar wind, photon-stimulated Na desorption (on Mercury and the Moon) (McGrath et al., 1986; Potter et al., 2000; Killen et al., 2004; Yakshinskiy and Madey, 2004) and cosmic ray bombardment (Sasaki et al., 2001) have been well studied. These processes act only on the surface, and are inefficient in removing Na from deeper layers (Čapek and Borovička, 2009).

8.6.2 Abundances versus Intensity Ratios

In the interpretation of meteor spectra, two types of evaluation methods are employed in the literature. Some investigators calculate elemental abundances while others use simple line intensity ratios. Here we describe advantages and disadvantages hidden in both methods.

Elemental abundances are quantitatively utilized for comparing with those of meteor showers and the solar abundances (Anders and Grevesse, 1989; Lodders, 2003). The derivation of abundances is influenced by the complicated physics of the ionized gas at the head of the meteor (Borovička, 1993; Kasuga et al., 2005b). For example, the Saha equation is needed to calculate the neutral versus ion balance, but this depends on the assumption of a plausible excitation temperature, T_{ex}, for the emitting region. This is particularly important for Na, which has a smaller first ionization energy (≈ 5.1 eV) compared to other species (e.g. Mg: ≈ 7.6 eV, Fe: ≈ 7.9 eV). The selection of the appropriate T_{ex} is problematic. Fireball-like spectra have been suggested to be the combination of some thermal components with typical $T_{ex} \sim 5{,}000$, 8,000 and 10,000 K, respectively (Borovička, 1993; Kasuga et al., 2005b, 2007b). Borovička (1993) considered Ca II lines in the hot component ($T_{ex} \sim 10{,}000$ K) and estimated electron density from the radiating volume of the meteor in the direction of its flight. This implies that Ca II may not reflect the actual electron density from their spectral emission profile. In Kasuga et al. (2005b), on the other hand, Ca II is taken from the main component ($T_{ex} \sim 5{,}000$ K) instead of the hot one, and they derive electron density reflected upon the measured spectral profile.

The definition of hot component theory has to satisfy the equality of total metal abundances (Ca/Mg) and pressure of the radiant gas between the main and the hot components (Borovička, 1993). However, the relation does not fit in most spectral data (e.g. Leonids). The Saha function is used to verify the definition but mostly finds negative values of electron density, which is clearly unrealistic (Kasuga et al., 2005b). This proves that the original hot component theory may go against its own definition (Borovička, 1993). Kasuga et al. (2005b) suggest that the Ca II lines do not always belong to the hot component, but instead to the main component. Plausibility is also found in their low excitation energies (Ca II ≈ 3.1 eV), which are compatible with those of other neutral metals (e.g. Na I ≈ 2.1 eV) identified in the main component (Kasuga et al., 2007c). On the other hand, the hot component primarily consists of species with high excitation energies $\gtrsim 10$ eV. Accordingly the Ca II lines are most likely to belong to the main component.

Given the difficulties in calculating absolute abundances, many researchers have employed simple line intensity ratios (e.g. Borovička et al., 2005). The method analyzes neutral atomic emission lines of Na I, Mg I, and (weak) Fe I only. Note that the intensity ratios do not directly reflect the elemental abundances due to no consideration for excitation properties of the elements and ions, including electron densities

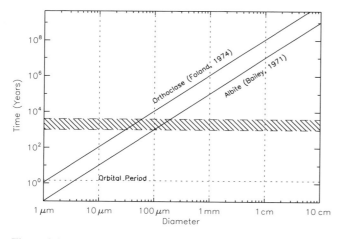

Figure 8.16. Timescale for escape of 90% of initial Na content from Geminid meteoroids in the stream, as a function of the meteoroid size. The low diffusion data for Na in orthoclase was used as a limit for the slowest loss, while the faster diffusion data for Na in albite (10× higher than for orthoclase) is considered as a more realistic value for the Geminids. The time interval in the shaded region corresponds to the estimated age of the Geminid meteoroid stream (1,000–4,000 yr). From Čapek and Borovička (2009).

(e.g. Borovička, 2001; Borovička et al., 2005; Koten et al., 2006). Laboratory spectroscopic experiment proves that intensity ratios are not informative of the abundance and cannot be used to determine the meteorite analogue (Drouard et al., 2018). Line ratios can be used, at best, to study the trend of elemental content in meteors which happen to possess similar physical conditions (including similar entry velocities, similar strengths and similar excitation temperatures). Line intensity ratios can suggest the trend of elemental content in meteor showers but cannot provide the abundances.

8.6.3 Meteoroid Streams

Physical properties of meteoroids in meteoroid streams or debris in dust trails whose orbits do not intersect that of the Earth can be revealed by thermal and optical observations. The streams (or trails) and meteoroids mostly consist of mm- to cm-scale compact aggregates, as estimated from the ratio of solar radiation pressure to solar gravity of $10^{-5} \lesssim \beta \lesssim 10^{-3}$ (e.g. Sykes and Walker, 1992; Reach et al., 2000, 2007; Ishiguro et al., 2002; Sarugaku et al., 2015).

Meteor ablation models in the Earth atmosphere are also available to estimate the size of meteoroids (Bronshten, 1983; Ceplecha et al., 1998). The classical models find typical meteors are 10 μm – 10 cm in size, however, with uncertainties caused by various parameters (e.g. luminous efficiency, ablation coefficient, and fragmentation) which are sensitive to the meteoroid entry speed, tensile strength and brightness (Ceplecha et al., 1998; Babadzhanov, 2002). Faint meteors are estimated to be 10μm $\sim$1 mm, but model improvements are needed to better represent fragmentation (Campbell-Brown and Koschny, 2004; Borovička et al., 2007).

8.6.4 Zodiacal Cloud

The zodiacal cloud is a circumsolar disk consisting of small dust particles supplied by comets and asteroids. The total mass is $\sim 4 \times 10^{16}$ kg, most ($\sim$90%?) of which is supplied by JFC disruptions and the rest by Oort cloud comets ($\lesssim$ 10%) and asteroids ($\lesssim$ 5%) (Nesvorný et al., 2010, 2011; Jenniskens, 2015). The supply rate needed to maintain the zodiacal cloud in steady-state is 10^3 to 10^4 kg s^{-1} (Nesvorný et al., 2011).

The fate of dust particles released from comets into the zodiacal cloud is traceable (cf. Grun et al., 1985). Sub-micron particles, with $\beta > 0.5$, are immediately blown out of the solar system on hyperbolic orbits by radiation pressure and are referred-to as β-meteoroids (Zook and Berg, 1975; Grun et al., 1985). The JFCs frequently disintegrate when near perihelion and form dust trails or meteoroid streams, mainly consisting of mm- to cm-sized dust particles (see Section 8.6.3). The collisional lifetime of mm-scale particles at 1 AU is estimated to be $\tau_{\text{col}} \gtrsim 10^5$ yr, modeled with the orbital distribution of sporadic meteors measured by radar (Nesvorný et al., 2011) (cf. $\tau_{\text{col}} \sim 10^4$ or 10^5 yr, Grun et al., 1985; Soja et al., 2016; Yang and Ishiguro, 2018). On the other hand, Poynting-Robertson (P-R) and solar-wind drag cause dust particles to spiral down to the Sun. The P-R drag timescale, τ_{PR}, to drift down from $\bar{a} \sim 2.7$ AU with $\bar{e} \sim 0.8$ (objects with $T_J < 3.08$, see Table 8.1 and Figure 8.1) to 1 AU around the Earth orbit ($e \sim 0.017$) is calculated using the equation in Wyatt and Whipple (1950) (cf. Dermott et al., 2002),

$$\tau_{\text{PR}} \simeq \frac{730}{\beta(1+sw)} \text{ yr}, \qquad (8.9)$$

where $\beta \lesssim 10^{-3}$ (see Section 8.6.3) for dust particles of radius $\gtrsim 1$ mm, with bulk density of 600 kg m^{-3} (Weissman et al., 2004), and $sw = 0.3$ is efficiency of solar-wind drag on a particle normalized to the P-R drag effect (Gustafson, 1994). The estimated P-R drag lifetime is $\tau_{\text{PR}} \gtrsim 6 \times 10^5$ yr. This being somewhat longer than τ_{col}, the mm-scale dust particles are subject to collisional disruption while spiraling down to the Sun by P-R drag. As a result of competition between these two effects (loss to Poynting-Robertson at small sizes, loss to collisional shattering at large sizes), the 100–200 μm dust particles are the most abundant in the zodiacal cloud (Love and Brownlee, 1993; Grun et al., 1985; Ceplecha et al., 1998; Nesvorný et al., 2010).

Nesvorný et al. (2011) noted that the collisional lifetime for mm-scale particles is long compared to the plausible lifetimes of most meteoroid streams ($\lesssim 10^4$ yr). They speculate that cm-scale particles are sources of smaller dust grains. Centimeter-scale particles are also released from JFCs. The sequence may result in a population of mm-sized or smaller particles, which could be more resistant to collisions. Recent meteor observations suggest a relative lack of large particles ($\sim$7 mm) (Jenniskens et al., 2016b), and also suggest that some of these larger particles disappear on timescales $\sim 10^4$ yr, not from collisions, but from other processes. Moorhead et al. (2017) finds a two-population sporadic meteoroid bulk density distribution suggesting that the physical character of freshly ejected dust particles could be altered over time. As another example, Rosetta dust collectors sampled both very pristine fluffy aggregates and compact particles ($\gtrsim 4$ cm in diameter) with a possible range of the dust bulk density from 400 to 3000 kg m^{-3} (Rotundi et al., 2015). This variety could have resulted from aggregate fragmentation into the denser collected grains as the spacecraft approached, while the packing effect is proposed as a plausible mechanism theory for fluffy dust particles released from comets (Mukai and Fechtig, 1983). The particle size could be reduced on a timescale of 10^4–10^5 yr, comparable with the meteoroid stream lifetime. The effect makes the bulk density increase from 600 kg m^{-3} to 3000 kg m^{-3}, corresponding to shrinking the particle size approximately down to half. This could be a potential explanation for disappearing larger-scale dust particles in the meteoroid streams.

8.7 Summary and Future Work

In the last decade, a growing understanding of parent bodies and meteoroid streams has been achieved by combining new physical observations and dynamical investigations. Still, even where the associations are relatively clear, most complexes have multiple potential parent bodies, and it remains unclear how the streams were formed.

Observationally, a major challenge is posed by the difficulty of measuring the physical characteristics of parent bodies, most of which are faint by virtue of their small size (typically $\lesssim$ a few km). They are also frequently observationally inaccessible because of their eccentric orbits, which cause them to spend most of the time far away near aphelion. Long-term surveillance of NEOs around their entire orbits might better reveal how and when parent bodies disintegrate and produce debris.

Dynamically, there are at least two challenging problems. One concerns the identification of parent bodies through the comparison of the orbital elements of meteoroids and potential parents by a D-criterion. Such methods work best for parents of young streams, where the effects of differential dynamical evolution are limited. However, in older systems, the dynamical elements have evolved enough to seriously undercut the use of the D-criteria. For this reason, for example, numerous Taurid parent bodies continue to be proposed. A key objective is to find a way to more reliably associate older meteoroid streams with their parent bodies. A second problem is the use of long-term dynamical simulations in which the initial conditions and/or potentially important non-gravitational effects are partly or wholly neglected.

We list key questions to be answered in the next decade.

1. Geminids: What process can act on ∼1000 yr timescales to produce the Geminid meteoroid stream? Phaethon appears dynamically associated with at least two kilometer-sized asteroids (2005 UD and 1999 YC) suggesting a past breakup or other catastrophe. But the likely timescale for such an event is ≫1000 yr. What caused the breakup and is it related to the Geminids? How many other PGC-related objects await discovery? Are Geminids represented in the meteorite collections and, if so, how can we identify them?
2. Quadrantids: Presumed parent 2003 EH$_1$ is currently inactive, but was recently as close to the Sun as is Phaethon at perihelion. Can residual mass loss in EH$_1$ be detected? Is the Quadrantid sodium abundance depleted as a result of the previously smaller perihelion? What physical difference is to be found in 96P, which has a near-Sun orbit even now?
3. Capricornids: Several parent bodies have been proposed including both active comets and inactive asteroids. Did they originate from a common precursor? Is asteroid 2017 MB$_1$ related?
4. Taurids: The prime parent body is 2P/Encke, but numerous additional parents with diverse properties continue to be proposed (mostly based on the D-criterion). How can we establish the relevance of these other objects to the Taurid stream? Can activity be detected? Is the D-criterion appropriate to judge?
5. Sekanina's (1973) Taurid-Perseids: Do 1566 Icarus and 2007 MK$_6$ share common physical properties? Is the sodium abundance in Taurid-Perseids depleted by solar heating due to the small perihelion?
6. Phoenicids, Andromedids and other minor complexes: Fragmentation is expected to produce a wide range of object sizes, with many bodies being too small to have been detected so far. What role can be played in the search for stream-related bodies by upcoming deep sky surveys, like the Large Synoptic Survey Telescope?
7. Which is the better index, the D-criterion (e.g. D_{SH}) or the dynamical invariants (C_1, C_2)?
8. How many near-Sun objects, driven by the Lidov-Kozai mechanism, exist?
9. What more can we learn from meteor spectroscopy, particularly of faint meteors?
10. Sporadic meteoroid populations tend to lose large dust particles (sizes $\gtrsim$ 7 mm) on timescales of 10^4 yr. Why?

Acknowledgements

We are grateful to David Asher for his enthusiastic guidance to improve this chapter. TK thanks Takaya Okamoto for wholehearted assistance with this study and also Junichi Watanabe, Hideyo Kawakita, Mikiya Sato and Chie Tsuchiya for support. We appreciate David Čapek, Jiří Borovička, Man-To Hui, Jing Li and Michael S. P. Kelley for figure contributions. We acknowledge reviews by Tadeusz Jopek and an anonymous reviewer. Lastly, we thank Galina Ryabova, again David Asher and Margaret Campbell-Brown for organizing this Meteoroidsbook project.

References

Abedin, A., Spurný, P., Wiegert, P. et al. 2015. On the age and formation mechanism of the core of the Quadrantid meteoroid stream. Icarus, **261**, 100–117.

Abedin, A., Wiegert, P., Pokorný, P., and Brown, P. 2017. The age and the probable parent body of the daytime arietid meteor shower. *Icarus*, **281**, 417–443.

Abedin, A., Wiegert, P., Janches, D. et al. 2018. Formation and past evolution of the showers of 96P/Machholz complex. *Icarus*, **300**, 360–385.

A'Hearn, M. F., Belton, M. J. S., Delamere, W. A. et al. 2005. Deep Impact: Excavating Comet Tempel 1. *Science*, **310**, 258–264.

A'Hearn, M. F., Millis, R. L., Schleicher, D. G., Osip, D. J., and Birch, P. V. 1995. The ensemble properties of comets: Results from narrowband photometry of 85 comets, 1976–1992. *Icarus*, **118**(Dec.), 223–270.

Anders, E., and Grevesse, N. 1989. Abundances of the elements – Meteoritic and solar. *Geochimica et Cosmochimica Acta*, **53**(Jan.), 197–214.

Ansdell, M., Meech, K. J., Hainaut, O. et al. 2014. Refined Rotational Period, Pole Solution, and Shape Model for (3200) Phaethon. *The Astrophysical Journal*, **793**(1), 50.

Antognini, J. M. O. 2015. Timescales of Kozai-Lidov oscillations at quadrupole and octupole order in the test particle limit. *Monthly Notices of the Royal Astronomical Society*, **452**(Oct.), 3610–3619.

Arendt, R. G. 2014. DIRBE Comet Trails. *The Astronomical Journal*, **148**(Dec.), 135.

Asher, D. J., and Clube, S. V. M. 1997. Towards a Dynamical History of 'Proto-Encke'. *Celestial Mechanics and Dynamical Astronomy*, **69**(Sept.), 149–170.

Asher, D. J., Clube, S. V. M., and Steel, D. I. 1993. Asteroids in the Taurid Complex. *Monthly Notices of the Royal Astronomical Society*, **264**(Sept.), 93–105.

Asher, D. J., Clube, S. V. M., Napier, W. M., and Steel, D. I. 1994. Coherent catastrophism. *Vistas in Astronomy*, **38**(Jan.), 1–27.

Baade, W., Cameron, R. C., and Folkman. 1950. 1949 MA = (1566) Icarus. *Minor Planet Circulars*, **347**(Jan.), 1.

Babadzhanov, P. B. 2001. Search for meteor showers associated with Near-Earth Asteroids. I. Taurid Complex. *Astronomy and Astrophysics*, **373**(July), 329–335.

Babadzhanov, P. B. 2002. Fragmentation and densities of meteoroids. *Astronomy and Astrophysics*, **384**(Mar.), 317–321.

Babadzhanov, P. B., and Obrubov, Y. V. 1987. Evolution of meteoroid streams. *Publications of the Astronomical Institute of the Czechoslovak Academy of Sciences*, **67**, 141–150.

Babadzhanov, P. B., and Obrubov, Y. V. 1992a. Evolution of short-period meteoroid streams. *Celestial Mechanics and Dynamical Astronomy*, **54**(Mar.), 111–127.

Babadzhanov, P. B., and Obrubov, Y. V. 1992b. P/Machholz 1986 VIII and Quadrantid meteoroid stream. Orbital evolution and relationship. In: Harris, A. W., and Bowell, E. (eds), *Asteroids, Comets, Meteors 1991. Proceedings of the international conference held at Northern Arizona University, Flagstaff, June 24–28, 1991*. Houston: Lunar and Planetary Institute, pp. 27–32.

Babadzhanov, P. B., Williams, I. P., and Kokhirova, G. I. 2008a. Meteor showers associated with 2003EH1. *Monthly Notices of the Royal Astronomical Society*, **386**, 2271–2277.

Babadzhanov, P. B., Williams, I. P., and Kokhirova, G. I. 2008b. Near-Earth objects in the Taurid complex. *Monthly Notices of the Royal Astronomical Society*, **386**, 1436–1442.

Bailey, M. E., Chambers, J. E., and Hahn, G. 1992. Origin of sungrazers: A frequent cometary end-state. *Astronomy and Astrophysics*, **257**(Apr.), 315–322.

Beech, M. 2002. The age of the Geminids: A constraint from the spin-up time-scale. *Monthly Notices of the Royal Astronomical Society*, **336**(Oct.), 559–563.

Belton, M. J. S., Samarasinha, N. H., Fernández, Y. R., and Meech, K. J. 2005. The excited spin state of Comet 2P/Encke. *Icarus*, **175**(May), 181–193.

Blaauw, R. C. 2017. The mass index and mass of the Geminid meteoroid stream as determined with radar, optical and lunar impact data. *Planetary and Space Science*, **143**(Sept.), 83–88.

Borovička, J. 1993. A fireball spectrum analysis. *Astronomy and Astrophysics*, **279**, 627–645.

Borovička, J. 2001. Video spectra of Leonids and other meteors. In: Warmbein, B. (ed), *Proceedings of the Meteoroids 2001 Conference, 6–10 August 2001, Kiruna, Sweden*. ESA SP-495, Noordwijk: ESA Publications Division, pp. 203–208.

Borovička, J. 2007 (May). Properties of meteoroids from different classes of parent bodies. In: Milani, A., Valsecchi, G. B. and Vokrouhlický, D. (eds), *Near Earth Objects, our Celestial Neighbors: Opportunity and Risk, Proceedings of IAU Symposium 236*. Cambridge, UK: Cambridge University Press, pp.107–120.

Borovička, J., and Weber, M. 1996. An α-Capricornid Meteor Spectrum. *WGN, Journal of the International Meteor Organization*, **24**(Apr.), 30–32.

Borovička, J., Stork, R., and Bocek, J. 1999. First results from video spectroscopy of 1998 Leonid meteors. *Meteoritics and Planetary Science*, **34**(Nov.), 987–994.

Borovička, J., Koten, P., Spurný, P., Boček, J., and Štork, R. 2005. A survey of meteor spectra and orbits: evidence for three populations of Na-free meteoroids. *Icarus*, **174**, 15–30.

Borovička, J. Spurný, P., and Koten, P. 2007. Atmospheric deceleration and light curves of Draconid meteors and implications for the structure of cometary dust. *Astronomy & Astrophysics*, **473**, 661–672.

Borovička, J., Koten, P., Spurný, P. et al. 2010. Material properties of transition objects 3200 Phaethon and 2003 EH$_1$. In: Fernandez, J. A., Lazzaro, D., Prialnik, D., and Schulz, R. (eds), *Icy Bodies of the Solar System Proceedings of IAU Symposium No. 263, 2009*. Cambridge, UK: Cambridge University Press, pp. 218–222.

Borovička, J., Macke, R. J., Campbell-Brown, M. D. et al. 2019. Physical and Chemical Properties of Meteoroids. In: G. O. Ryabova, D. J. Asher, and M. D. Campbell-Brown (eds), *Meteoroids: Sources of Meteors on the Earth and Beyond*. Cambridge, UK: Cambridge University Press.

Bottke, W. F., Jr., Nolan, M. C., Greenberg, R., and Kolvoord, R. A. 1994. Collisional lifetimes and impact statistics of near-Earth asteroids. In: T. Gehrels, M. S. Matthews, and A. M. Schumann (eds), *Hazards Due to Comets and Asteroids*. Tucson: University of Arizona Press, pp. 337–357.

Britt, D. T., Yeomans, D., Housen, K., and Consolmagno, G. 2002. Asteroid density, porosity, and structure. In: W. F. Bottke Jr., A. Cellino, P. Paolicchi, and R. P. Binzel (eds), *Asteroids III*. Tucson: University of Arizona Press, pp. 485–500.

Bronshten, V. A. 1983. *Physics of Meteoric Phenomena*. Dordrecht: D. Reidel Publishing Co.

Brown, P., Weryk, R. J., Wong, D. K., and Jones, J. 2008a. A meteoroid stream survey using the Canadian Meteor Orbit Radar. I. Methodology and radiant catalogue. *Icarus*, **195**, 317–339.

Brown, P., Weryk, R. J., Wong, D. K., and Jones, J. 2008b. The Canadian Meteor Orbit Radar meteor stream catalogue. *Earth, Moon, and Planets*, **102**(June), 209–219.

Brown, P., Wong, D. K., Weryk, R. J., and Wiegert, P. 2010. A meteoroid stream survey using the Canadian Meteor Orbit Radar. II: Identification of minor showers using a 3D wavelet transform. *Icarus*, **207**(May), 66–81.

Brown, P., Marchenko, V., Moser, D. E., Weryk, R., and Cooke, W. 2013. Meteorites from meteor showers: A case study of the Taurids. *Meteoritics and Planetary Science*, **48**(Feb.), 270–288.

Campbell-Brown, M. D., and Koschny, D. 2004. Model of the ablation of faint meteors. *Astronomy and Astrophysics*, **418**, 751–758.

Campins, H., and Fernández, Y. 2002. Observational constraints on surface characteristics of comet nuclei. *Earth, Moon, and Planets*, **89**(Oct.), 117–134.

Čapek, D., and Borovička, J. 2009. Quantitative model of the release of sodium from meteoroids in the vicinity of the Sun: Application to Geminids. *Icarus*, **202**(2), 361–370.

Ceplecha, Z., Borovička, J., Elford, W. G. et al. 1998. Meteor phenomena and bodies. *Space Science Reviews*, **84**(Sept.), 327–471.

Chamberlin, A. B., McFadden, L.-A., Schulz, R., Schleicher, D. G., and Bus, S. J. 1996. 4015 Wilson-Harrington, 2201 Oljato, and 3200 Phaethon: Search for CN emission. *Icarus*, **119**(Jan.), 173–181.

Chang, C.-K., Ip, W.-H., Lin, H.-W. et al. 2015. Asteroid spin-rate study using the intermediate Palomar Transient Factory. *The Astrophysical Journal Supplement Series*, **219**(Aug.), 27.

Chapman, C. R., Harris, A. W., and Binzel, R. 1994. Physical properties of near-Earth asteroids: Implications for the hazard issue. In: T. Gehrels, M. S. Matthews, and A. M. Schumann (eds), *Hazards Due to Comets and Asteroids*. Tucson: University of Arizona Press, pp. 537–549.

Clube, S. V. M., and Napier, W. M. 1984. The microstructure of terrestrial catastrophism. *Monthly Notices of the Royal Astronomical Society*, **211**(Dec.), 953–968.

Clube, S. V. M., and Napier, W. M. 1987. The cometary breakup hypothesis re-examined - A reply. *Monthly Notices of the Royal Astronomical Society*, **225**(Apr.), 55P–58P.

Cook, A. F. 1973. A working list of meteor streams. In: Hemenway, C. L., Millman, P. M., and Cook, A. F. (eds), *Evolutionary and Physical Properties of Meteoroids, Proceedings of IAU Colloq. 13, held in Albany, NY, 14–17 June 1971*. National Aeronautics and Space Administration SP–319. Washington, D. C.: National Aeronautics and Space Administration, pp. 183–191.

Dandy, C. L., Fitzsimmons, A., and Collander-Brown, S. J. 2003. Optical colors of 56 near-Earth objects: trends with size and orbit. *Icarus*, **163**(June), 363–373.

De Angelis, G. 1995. Asteroid spin, pole and shape determinations. *Planetary and Space Science*, **43**, 649–682.

de León, J., Campins, H., Tsiganis, K., Morbidelli, A., and Licandro, J. 2010. Origin of the near-Earth asteroid Phaethon and the Geminids meteor shower. *Astronomy and Astrophysics*, **513**(Apr.), A26.

DeMeo, F., and Binzel, R. P. 2008. Comets in the near-Earth object population. *Icarus*, **194**(Apr.), 436–449.

Dermott, S. F., Durda, D. D., Grogan, K., and Kehoe, T. J. J. 2002. Asteroidal dust. In: W. F. Bottke Jr., A. Cellino, P. Paolicchi, and R. P. Binzel (eds), *Asteroids III*. Tucson: University of Arizona Press, pp. 423–442.

Drouard, A., Vernazza, P., Loehle, S. et al. 2018. Probing the use of spectroscopy to determine the meteoritic analogues of meteors. *Astronomy and Astrophysics*, **613**(May), A54.

Duncan, M., Levison, H., and Dones, L. 2004. Dynamical evolution of ecliptic comets. In: M. C. Festou, H. U. Keller, and H. A. Weaver (eds), *Comets II*. Tucson: University of Arizona Press, pp. 193–204.

Dundon, L. R. 2005. *The enigmatic surface of (3200) Phaethon: Comparison with cometary candidates*. M.Phil. thesis, University of Hawaii at Manoa, Honolulu.

Emel'yanenko, V. V. 2017. Near-Sun asteroids. *Solar System Research*, **51**(Jan.), 59–63.

Fernández, J. A., Sosa, A., Gallardo, T., and Gutiérrez, J. N. 2014. Assessing the physical nature of near-Earth asteroids through their dynamical histories. *Icarus*, **238**(Aug.), 1–12.

Fernández, Y. R., Lisse, C. M., Ulrich Käufl, H. et al. 2000. Physical Properties of the Nucleus of Comet 2P/Encke. *Icarus*, **147**(Sept.), 145–160.

Foglia, S., Micheli, M., Ridley, H. B., Jenniskens, P., and Marsden, B. G. 2005. Comet D/1819 W1 (Blanpain) and 2003 WY25. *International Astronomical Union Circular*, **8485**(Feb.).

Gehrels, T., Roemer, E., Taylor, R. C., and Zellner, B. H. 1970. Minor planets and related objects. IV. Asteroid (1566) Icarus. *The Astronomical Journal*, **75**(Mar.), 186–195.

Gehrz, R. D., Reach, W. T., Woodward, C. E., and Kelley, M. S. 2006. Infrared observations of comets with the Spitzer Space Telescope. *Advances in Space Research*, **38**(Jan.), 2031–2038.

Gonczi, R., Rickman, H., and Froeschle, C. 1992. The connection between Comet P/Machholz and the Quadrantid meteor. *Monthly Notices of the Royal Astronomical Society*, **254**(Feb.), 627–634.

Green, D. W. E. 2005. Comets 169P/2002 EX$_{12}$ (NEAT), 170P/2005 M1 (Christensen). *International Astronomical Union Circular*, **8591**(Aug.).

Green, S. F., Meadows, A. J., and Davies, J. K. 1985. Infrared observations of the extinct cometary candidate minor planet (3200) 1983TB. *Monthly Notices of the Royal Astronomical Society*, **214**(June), 29P–36P.

Grun, E., Zook, H. A., Fechtig, H., and Giese, R. H. 1985. Collisional balance of the meteoritic complex. *Icarus*, **62**(May), 244–272.

Gustafson, B. A. S. 1989. Geminid meteoroids traced to cometary activity on Phaethon. *Astronomy and Astrophysics*, **225**(Nov.), 533–540.

Gustafson, B. A. S. 1994. Physics of zodiacal dust. *Annual Review of Earth and Planetary Sciences*, **22**, 553–595.

Hanuš, J., Delbo', M., Vokrouhlický, D. et al. 2016. Near-Earth asteroid (3200) Phaethon: Characterization of its orbit, spin state, and thermophysical parameters. *Astronomy and Astrophysics*, **592**(July), A34.

Harris, A. W. 1998. A thermal model for mear-Earth asteroids. *Icarus*, **131**(Feb.), 291–301.

Harris, A. W., and Lagerros, J. S. V. 2002. Asteroids in the thermal infrared. In: W. F. Bottke Jr., A. Cellino, P. Paolicchi, and R. P. Binzel (eds), *Asteroids III*. Tucson: University of Arizona Press, pp. 205–218.

Harvey, G. A. 1973. Spectral analysis of four meteors. *NASA Special Publication*, **319**, 103–129.

Hawkins, G. S., Southworth, R. B., and Steinon, F. 1959. Recovery of the Andromedids. *The Astronomical Journal*, **64**(June), 183–188.

Hicks, M. D., Fink, U., and Grundy, W. M. 1998. The unusual spectra of 15 near-Earth asteroids and extinct comet candidates. *Icarus*, **133**(May), 69–78.

Hill, R. E., Gibbs, A. R., and Boattini, A. 2007. 2007 MK_6. *Minor Planet Electronic Circulars*, **2007-M32**.

Hirabayashi, M., Scheeres, D. J., Sánchez, D. P., and Gabriel, T. 2014. Constraints on the physical properties of main belt Comet P/2013 R3 from its breakup event. *The Astrophysical Journal*, **789**(July), L12.

Hoffmeister, C. 1937. *Die Meteore, ihre kosmischen und irdischen Beziehungen*. Leipzig: Akademische Verlagsgesellschaft m.b.h.

Holmberg, J., Flynn, C., and Portinari, L. 2006. The colours of the Sun. *Monthly Notices of the Royal Astronomical Society*, **367**(Apr.), 449–453.

Hsieh, H. H., and Jewitt, D. 2005. Search for activity in 3200 Phaethon. *The Astronomical Journal*, **624**(May), 1093–1096.

Hughes, D. W., and McBride, N. 1989. The mass of meteoroid streams. *Monthly Notices of the Royal Astronomical Society*, **240**(Sept.), 73–79.

Hui, M.-T. 2013. Observations of Comet P/2003 T12 = 2012 A3 (SOHO) at large phase angle in STEREO-B. *Monthly Notices of the Royal Astronomical Society*, **436**(Dec.), 1564–1575.

Hui, M.-T., and Li, J. 2017. Resurrection of (3200) Phaethon in 2016. *The Astronomical Journal*, **153**(Jan.), 23.

Huruhata, M., and Nakamura, J. 1957. Meteoric shower observed on December 5, 1956 in the Indian Ocean. *Tokyo Astronomical Bulletin, 2nd Series*, **99**, 1053–1054.

Ishiguro, M., Watanabe, J., Usui, F. et al. 2002. First detection of an optical dust trail along the orbit of 22P/Kopff. *The Astrophysical Journal*, **572**(June), L117–L120.

Ishiguro, M., Sarugaku, Y., Nishihara, S. et al. 2009. Report on the Kiso cometary dust trail survey. *Advances in Space Research*, **43**(Mar.), 875–879.

Jakubík, M., and Neslušan, L. 2015. Meteor complex of asteroid 3200 Phaethon: its features derived from theory and updated meteor data bases. *Monthly Notices of the Royal Astronomical Society*, **453**(Oct.), 1186–1200.

Jenniskens, P. 1994. Meteor stream activity I. The annual streams. *Astronomy and Astrophysics*, **287**(July), 990–1013.

Jenniskens, P. 2004. 2003 EH_1 is the Quadrantid shower parent comet. *The Astronomical Journal*, **127**(May), 3018–3022.

Jenniskens, P. 2006. *Meteor Showers and their Parent Comets*. Cambridge, UK: Cambridge University Press.

Jenniskens, P. 2008a. Meteoroid streams that trace to candidate dormant comets. *Icarus*, **194**(Mar.), 13–22.

Jenniskens, P. 2008b. Mostly dormant comets and their disintegration into meteoroid streams: A review. *Earth, Moon, and Planets*, **102**(June), 505–520.

Jenniskens, P. 2015. Meteoroid streams and the zodiacal cloud. In: P. Michel, F. E. DeMeo, and W. F. Bottke (eds), *Asteroids IV*. Tucson: University of Arizona Press, pp. 281–295.

Jenniskens, P. 2017. Meteor showers in review. *Planetary and Space Science*, **143**(Sept.), 116–124.

Jenniskens, P., and Lyytinen, E. 2005. Meteor showers from the debris of broken comets: D/1819 W_1 (Blanpain), 2003 WY_{25}, and the Phoenicids. *The Astronomical Journal*, **130**(Sept.), 1286–1290.

Jenniskens, P., and Nénon, Q. 2016. CAMS verification of single-linked high-threshold D-criterion detected meteor showers. *Icarus*, **266**(Mar.), 371–383.

Jenniskens, P., and Vaubaillon, J. 2007. 3D/Biela and the Andromedids: Fragmenting versus sublimating comets. *The Astronomical Journal*, **134**(Sept.), 1037–1045.

Jenniskens, P., and Vaubaillon, J. 2010. Minor Planet 2002 EX_{12} (=169P/NEAT) and the Alpha Capricornid shower. *The Astronomical Journal*, **139**(May), 1822–1830.

Jenniskens, P., Nénon, Q., Gural, P. S. et al. 2016a. CAMS confirmation of previously reported meteor showers. *Icarus*, **266**(Mar.), 355–370.

Jenniskens, P., Nénon, Q., Gural, P. S. et al. 2016b. CAMS newly detected meteor showers and the sporadic background. *Icarus*, **266**(Mar.), 384–409.

Jewitt, D. C. 2002. From Kuiper Belt Object to cometary nucleus: The missing ultrared matter. *The Astronomical Journal*, **123**(Feb.), 1039–1049.

Jewitt, D. 2004a. Looking through the HIPPO: Nucleus and dust in Comet 2P/Encke. *The Astrophysical Journal*, **128**(Dec.), 3061–3069.

Jewitt, D. C. 2004b. From cradle to grave: the rise and demise of the comets. In: M. C. Festou, H. U. Keller, and H. A. Weaver (eds), *Comets II*. Tucson: University of Arizona Press, pp. 659–676.

Jewitt, D. 2006. Comet D/1819 W1 (Blanpain): Not dead yet. *The Astrophysical Journal*, **131**(Apr.), 2327–2331.

Jewitt, D. 2012. The active asteroids. *The Astrophysical Journal*, **143**, 66.

Jewitt, D. 2013. Properties of near-sun asteroids. *The Astrophysical Journal*, **145**(May), 133.

Jewitt, D., and Hsieh, H. 2006. Physical observations of 2005 UD: A mini-Phaethon. *The Astrophysical Journal*, **132**(Oct.), 1624–1629.

Jewitt, D., and Li, J. 2010. Activity in Geminid parent (3200) Phaethon. *The Astrophysical Journal*, **140**(Nov.), 1519–1527.

Jewitt, D., Li, J., and Agarwal, J. 2013. The dust tail of Asteroid (3200) Phaethon. *The Astrophysical Journal*, **771**(July), L36.

Jewitt, D., Agarwal, J., Li, J. et al. 2014. Disintegrating Asteroid P/2013 R3. *The Astrophysical Journal*, **784**(Mar.), L8.

Jewitt, D., Hsieh, H., and Agarwal, J. 2015. The active asteroids. In: P. Michel, F. E. DeMeo, and W. F. Bottke (eds), *Asteroids IV*. Tucson: University of Arizona Press, pp. 221–241.

Jewitt, D., Agarwal, J., Li, J. et al. 2017. Anatomy of an asteroid breakup: The Case of P/2013 R3. *The Astronomical Journal*, **153**, 223.

Jewitt, D., Mutchler, M., Agarwal, J., and Li, J. 2018. Hubble Space Telescope observations of 3200 Phaethon at closest approach. *The Astronomical Journal*, **156**(Nov.), 238.

Jones, J. 1978. On the period of the Geminid meteor stream. *Monthly Notices of the Royal Astronomical Society*, **183**(May), 539–546.

Jones, J. 1986. The effect of gravitational perturbations on the evolution of the Taurid meteor stream complex. *Monthly Notices of the Royal Astronomical Society*, **221**(July), 257–267.

Jones, J., and Hawkes, R. L. 1986. The structure of the Geminid meteor stream. II – The combined action of the cometary ejection process and gravitational perturbations. *Monthly Notices of the Royal Astronomical Society*, **223**(Dec.), 479–486.

Jones, J., and Jones, W. 1993. Comet Machholz and the Quadrantid meteor stream. *Monthly Notices of the Royal Astronomical Society*, **261**(Apr.), 605–611.

Jopek, T. J. 2011. Meteoroid streams and their parent bodies. *Memorie della Societa Astronomica Italiana*, **82**, 310–320.

Jopek, T. J., and Jenniskens, P. M. 2011. The Working Group on meteor showers nomenclature: A history, current status and a call for contributions. In: Cooke, W. J., Moser D. E. Hardin B. F., and Janches, D. (eds), *Meteoroids: The Smallest Solar System Bodies, Proceedings of the Meteoroids 2010 Conference held in Breckenridge, Colorado, USA, May 24–28, 2010*. NASA/CP-2011-216469. Huntsville, AL: National Aeronautics and Space Administration, pp. 7–13.

Kaňuchová, Z., and Neslušan, L. 2007. The parent bodies of the Quadrantid meteoroid stream. *Astronomy and Astrophysics*, **470**(Aug.), 1123–1136.

Kasuga, T., and Jewitt, D. 2008. Observations of 1999 YC and the breakup of the Geminid stream parent. *The Astronomical Journal*, **136**(Aug.), 881–889.

Kasuga, T., and Jewitt, D. 2015. Physical observations of (196256) 2003 EH1, presumed parent of the Quadrantid meteoroid stream. *The Astronomical Journal*, **150**(Nov.), 152.

Kasuga, T., Watanabe, J., and Ebizuka, N. 2005a. A 2004 Geminid meteor spectrum in the visible-ultraviolet region. Extreme Na depletion? *Astronomy and Astrophysics*, **438**(Aug.), L17–L20.

Kasuga, T., Yamamoto, T., Watanabe, J. et al. 2005b. Metallic abundances of the 2002 Leonid meteor deduced from high-definition TV spectra. *Astronomy and Astrophysics*, **435**(May), 341–351.

Kasuga, T., Watanabe, J.-I., and Sato, M. 2006a. Benefits of an impact mission to 3200 Phaethon: Nature of the extinct comet and artificial meteor shower. *Monthly Notices of the Royal Astronomical Society*, **373**(Dec.), 1107–1111.

Kasuga, T., Yamamoto, T., Kimura, H., and Watanabe, J. 2006b. Thermal desorption of Na in meteoroids. *Astronomy and Astrophysics*, **453**(2), L17–L20.

Kasuga, T., Sato, M., and Watanabe, J. 2007a. Creating an artificial Geminid meteor shower: Correlation between ejecta velocity and observability. *Advances in Space Research*, **40**, 215–219.

Kasuga, T., Iijima, T., and Watanabe, J. 2007b. Is a 2004 Leonid meteor spectrum captured in a 182 cm telescope? *Astronomy and Astrophysics*, **474**(Nov.), 639–645.

Kasuga, T., Watanabe, J., Kawakita, H., and Yamamoto, T. 2007c. The origin of the Ca(II) emission, in one of two plasma components, and the metallic abundances in a 2002 Leonid meteor spectrum. *Advances in Space Research*, **39**, 513–516.

Kasuga, T., Balam, D. D., and Wiegert, P. A. 2010. Comet 169P/NEAT(=2002 EX_{12}): The parent body of the α-Capricornid meteoroid stream. *The Astronomical Journal*, **140**(Dec.), 1806–1813.

Kelley, M. S., Woodward, C. E., Harker, D. E. et al. 2006. A Spitzer study of Comets 2P/Encke, 67P/Churyumov-Gerasimenko, and C/2001 HT50 (LINEAR-NEAT). *The Astrophysical Journal*, **651**(Nov.), 1256–1271.

Killen, R. M., Sarantos, M., Potter, A. E., and Reiff, P. 2004. Source rates and ion recycling rates for Na and K in Mercury's atmosphere. *Icarus*, **171**(Sept.), 1–19.

Kinoshita, D., Ohtsuka, K., Sekiguchi, T. et al. 2007. Surface heterogeneity of 2005 UD from photometric observations. *Astronomy and Astrophysics*, **466**(May), 1153–1158.

Kokotanekova, R., Snodgrass, C., Lacerda, P. et al. 2017. Rotation of cometary nuclei: New light curves and an update of the ensemble properties of Jupiter-family comets. *Monthly Notices of the Royal Astronomical Society*, **471**(Nov.), 2974–3007.

Kondrat'eva, E. D., Murav'eva, I. N., and Reznikov, E. D. 1997. On the forthcoming return of the Leonid Meteoric Swarm. *Solar System Research*, **31**, 489–492.

Kosai, H. 1992. Short-period comets and Apollo-Amor-Aten type asteroids in view of Tisserand invariant. *Celestial Mechanics and Dynamical Astronomy*, **54**(Mar.), 237–240.

Koten, P., Borovička, J., Spurný, P. et al. 2006. Double station and spectroscopic observations of the Quadrantid meteor shower and the implications for its parent body. *Monthly Notices of the Royal Astronomical Society*, **366**(Mar.), 1367–1372.

Kozai, Y. 1962. Secular perturbations of asteroids with high inclination and eccentricity. *The Astronomical Journal*, **67**(Nov.), 591–598.

Kresak, L. 1982. On the similarity of orbits of associated comets, asteroids and meteoroids. *Bulletin of the Astronomical Institutes of Czechoslovakia*, **33**(May), 104–110.

Kronk, G. W. 1988. *Meteor Showers. A Descriptive Catalog*. Hillside, NJ: Enslow Publishers.

Kronk, G. W. 1999. *Cometography: A Catalog of Comets, Volume 1: Ancient-1799*. Cambridge, UK: Cambridge University Press.

Lamy, P. L., Toth, I., Fernandez, Y. R., and Weaver, H. A. 2004. The sizes, shapes, albedos, and colors of cometary nuclei. In: M. C. Festou, H. U. Keller, and H. A. Weaver (eds), *Comets II*. Tucson: University of Arizona Press, pp. 223–264.

Lauriello, P.J. 1974. Application of a convective heat source to the thermal fracturing of rock. *International Journal of Rock Mechanics and Mining Sciences & Geomechanics Abstracts*, **11**(2), 75–81.

Levison, H. F. 1996. Comet Taxonomy. In: Rettig, T., and Hahn, J. M. (eds), *Completing the Inventory of the Solar System. Proceedings of a Symposium held in conjunction with the 106th Annual Meeting of the ASP held at Lowell Observatory, Flagstaff, Arizona, 25–30 June 1994*. Astronomical Society of the Pacific Conference Proceedings, volume 107. San Francisco: Astronomical Society of the Pacific, pp. 173–191.

Levison, H. F., Terrell, D., Wiegert, P. A., Dones, L., and Duncan, M. J. 2006. On the origin of the unusual orbit of Comet 2P/Encke. *Icarus*, **182**(May), 161–168.

Li, J., and Jewitt, D. 2013. Recurrent perihelion activity in (3200) Phaethon. *The Astronomical Journal*, **145**(June), 154.

Licandro, J., Tancredi, G., Lindgren, M., Rickman, H., and Hutton, R. G. 2000. CCD photometry of cometary nuclei, I: Observations from 1990–1995. *Icarus*, **147**(Sept.), 161–179.

Licandro, J., Campins, H., Mothé-Diniz, T., Pinilla-Alonso, N., and de León, J. 2007. The nature of comet-asteroid transition object (3200) Phaethon. *Astronomy and Astrophysics*, **461**(Jan.), 751–757.

Lidov, M. L. 1962. The evolution of orbits of artificial satellites of planets under the action of gravitational perturbations of external bodies. *Planetary and Space Science*, **9**(Oct.), 719–759.

Lindblad, B. A. 1987. Physics and Orbits of Meteoroids. Page 229–251 of: Fulchignoni, M., and Kresak, L. (eds), *The Evolution of the Small Bodies of the Solar System. Proceedings of the International School of Physics "Enrico Fermi", held at Villa Monastero, Varenna on Lake Como, Italy, August 5–10*, 1985. Amsterdam: North-Holland Physics Publications.

Lisse, C. M., Fernández, Y. R., A'Hearn, M. F. et al. 2004. A tale of two very different comets: ISO and MSX measurements of dust emission from 126P/IRAS (1996) and 2P/Encke (1997). *Icarus*, **171**(Oct.), 444–462.

Lodders, K. 2003. Solar System abundances and condensation temperatures of the Elements. *The Astrophysical Journal*, **591**(July), 1220–1247.

Love, S. G., and Brownlee, D. E. 1993. A direct measurement of the terrestrial mass accretion rate of cosmic dust. *Science*, **262**(5133), 550–553.

Lowry, S. C., and Weissman, P. R. 2007. Rotation and color properties of the nucleus of Comet 2P/Encke. *Icarus*, **188**, 212–223.

Madiedo, J. M., Trigo-Rodríguez, J. M., Ortiz, J. L., Castro-Tirado, A. J., and Cabrera-Caño, J. 2014. Orbit and emission spectroscopy of α-Capricornid fireballs. *Icarus*, **239**(Sept.), 273–280.

Mahapatra, P. R., Ostro, S. J., Benner, L. A. M. et al. 1999. Recent radar observations of asteroid 1566 Icarus. *Planetary and Space Science*, **47**(Aug.), 987–995.

Mainzer, A., Grav, T., Bauer, J. et al. 2011. NEOWISE observations of near-Earth objects: Preliminary results. *The Astrophysical Journal*, **743**(Dec.), 156.

Matlovič, P., Tóth, J., Rudawska, R., and Kornoš, L. 2017. Spectra and physical properties of Taurid meteoroids. *Planetary and Space Science*, **143**(Sept.), 104–115.

McGrath, M. A., Johnson, R. E., and Lanzerotti, L. J. 1986. Sputtering of sodium on the planet Mercury. *Nature*, **323**, 694–696.

McIntosh, B. A. 1990. Comet P/Machholz and the Quadrantid meteor stream. *Icarus*, **86**(July), 299–304.

McNaught, R. H., and Asher, D. J. 1999. Leonid dust trails and meteor storms. *WGN, Journal of the International Meteor Organization*, **27**(Apr.), 85–102.

Meech, K. J., Hainaut, O. R., and Marsden, B. G. 2004. Comet nucleus size distributions from HST and Keck telescopes. *Icarus*, **170**(Aug.), 463–491.

Micheli, M. 2005. Possibile correlazione tra l'asteroide 2003 WY$_{25}$, la cometa D/1819 W1 (Blanpain) e due sciami meteorici occasionali. *Astronomia. La rivista dell' Unione Astrofili Italiani*, **1**(Feb.), 47–53.

Millman, P. M. 1955. Photographic meteor spectra (Appendix 3); Orionid and Geminid observations at Montreal; A provisional supplementary list of meteor showers. *Journal of the Royal Astronomical Society of Canada*, **49**(Aug.), 169–173.

Millman, P. M. 1980. One hundred and fifteen years of meteor spectroscopy. Pages 121–127 of: Halliday, I., and McIntosh, B. A. (eds), *Solid Particles in the Solar System*. IAU Symposium, no. 90. Dordrecht D. Reidel Publishing Co.

Millman, P. M., and McKinley, D. W. R. 1963. Meteors. In: Kuiper, G. P., and Middlehurst, B. M. (eds), *The Moon, Meteorites and Comets*. Vol. IV of *The Solar System*. University of Chicago Press, pp. 674–773.

Miner, E., and Young, J. 1969. Photometric determination of the rotation period of 1566 Icarus. *Icarus*, **10**, 436–440.

Moiseev, N. D. 1945. On certain basic simplified schemes of celestial mechanics obtained with the aid of averaging different variants of the problem of three bodies (in Russian). *Trudy Gosudarstvennogo astronomicheskogo instituta im. Sternberga*, **15**, 75–99.

Molaro, J. L., Byrne, S., and Langer, S. A. 2015. Grain-scale thermoelastic stresses and spatiotemporal temperature gradients on airless bodies, implications for rock breakdown. *Journal of Geophysical Research (Planets)*, **120**(Feb.), 255–277.

Moorhead, A. V., Blaauw, R. C., Moser, D. E. et al. 2017. A two-population sporadic meteoroid bulk density distribution and its implications for environment models. *Monthly Notices of the Royal Astronomical Society*, **472**(Dec.), 3833–3841.

Mukai, T., and Fechtig, H. 1983. Packing effect of fluffy particles. *Planetary and Space Science*, **31**(June), 655–658.

Musci, R., Weryk, R. J., Brown, P., Campbell-Brown, M. D., and Wiegert, P. A. 2012. An optical survey for millimeter-sized interstellar meteoroids. *The Astrophysical Journal*, **745**, 161.

Nagasawa, K. 1978. Analysis of the spectra of Leonid meteors. *Annals of the Tokyo Astronomical Observatory*, **16**, 157–187.

Neslušan, L., Hajduková, M., and Jakubík, M. 2013a. Meteor-shower complex of asteroid 2003 EH1 compared with that of comet 96P/Machholz. *Astronomy and Astrophysics*, **560**(Dec.), A47.

Neslušan, L., Kaňuchová, Z., and Tomko, D. 2013b. The meteor-shower complex of 96P/Machholz revisited. *Astronomy and Astrophysics*, **551**(Mar.), A87.

Neslušan, L., Kaňuchová, Z., and Tomko, D. 2014. The ecliptic-toroidal structure of the meteor complex of comet 96P/Machholz. Pages 235–242 of: Jopek, T. J., Rietmeijer, F. J. M., Watanabe, J., and Williams, I. P. (eds), *Proc. International Conf. held at A.M. University in Poznań, Poland, August 26–30, 2013. Meteoroids 2013.* Poznań: Wydanictwo Naukowe UAM.

Nesvorný, D., Jenniskens, P., Levison, H. F. et al. 2010. Cometary origin of the zodiacal cloud and carbonaceous Micrometeorites. Implications for hot debris disks. *The Astrophysical Journal*, **713**(Apr.), 816–836.

Nesvorný, D., Janches, D., Vokrouhlický, D. et al. 2011. Dynamical model for the zodiacal cloud and sporadic meteors. *The Astrophysical Journal*, **743**(Dec.), 129.

Nugent, C. R., Mainzer, A., Masiero, J. et al. 2015. NEOWISE reactivation mission year one: Preliminary asteroid diameters and albedos. *The Astrophysical Journal*, **814**(Dec.), 117.

Ohtsuka, K., Yoshikawa, M., and Watanabe, J.-I. 1995. Impulse effects on the orbit of 1987 Quadrantid Swarm. *Publications of the Astronomical Society of Japan*, **47**(Aug.), 477–486.

Ohtsuka, K., Shimoda, C., Yoshikawa, M., and Watanabe, J.-I. 1997. Activity profile of the Sextantid meteor shower. *Earth, Moon, and Planets*, **77**(Apr.), 83–91.

Ohtsuka, K., Sekiguchi, T., Kinoshita, D. et al. 2006. Apollo asteroid 2005 UD: Split nucleus of (3200) Phaethon? *Astronomy and Astrophysics*, **450**(May), L25–L28.

Ohtsuka, K., Arakida, H., Ito, T. et al. 2007. Apollo asteroids 1566 Icarus and 2007 MK_6: Icarus family members? *The Astrophysical Journal*, **668**(Oct.), L71–L74.

Ohtsuka, K., Arakida, H., Ito, T., Yoshikawa, M., and Asher, D. J. 2008a. Apollo asteroid 1999 YC: Another large member of the PGC? *Meteoritics and Planetary Science Supplement*, **43**, 5055.

Ohtsuka, K., Yoshikawa, M., Watanabe, J. et al. 2008b. On the substantial spatial spread of the Quadrantid meteoroid stream. *Earth, Moon, and Planets*, **102**(June), 179–182.

Ohtsuka, K., Nakato, A., Nakamura, T. et al. 2009. Solar-radiation heating effects on 3200 Phaethon. *Publications of the Astronomical Society of Japan*, **61**, 1375–1387.

Olech, A., Żołądek, P., Wiśniewski, M. et al. 2017. Enhanced activity of the Southern Taurids in 2005 and 2015. *Monthly Notices of the Royal Astronomical Society*, **469**(Aug.), 2077–2088.

Olivier, C. P. 1925. *Meteors*. Baltimore: Williams & Wilkins.

Olmsted, D. 1834. Observations of the meteors of November 13, 1833. *American Journal of Science*, **25**, 354–411.

Ortiz, J. L., Madiedo, J. M., Morales, N., Santos-Sanz, P., and Aceituno, F. J. 2015. Lunar impact flashes from Geminids: Analysis of luminous efficiencies and the flux of large meteoroids on Earth. *Monthly Notices of the Royal Astronomical Society*, **454**(Nov.), 344–352.

Pariseau, G. W. 2006. *Design Analysis in Rock Mechanics*. London: Taylor and Francis.

Popescu, M., Birlan, M., Nedelcu, D. A., Vaubaillon, J., and Cristescu, C. P. 2014. Spectral properties of the largest asteroids associated with Taurid Complex. *Astronomy and Astrophysics*, **572**(Dec.), A106.

Porubčan, V., Williams, I. P., and Kornoš, L. 2004. Associations between asteroids and meteoroid streams. *Earth, Moon, and Planets*, **95**(Dec.), 697–712.

Porubčan, V., Kornoš, L., and Williams, I. P. 2006. The Taurid complex meteor showers and asteroids. *Contributions of the Astronomical Observatory Skalnate Pleso*, **36**(June), 103–117.

Potter, A. E., Killen, R. M., and Morgan, T. H. 2000. Variation of lunar sodium during passage of the Moon through the Earth's magnetotail. *Journal of Geophysical Research*, **105**(June), 15073–15084.

Pravec, P., Harris, A. W., Vokrouhlický, D. et al. 2008. Spin rate distribution of small asteroids. *Icarus*, **197**(Oct.), 497–504.

Quetelet, A. 1839. *Catalogue des principales apparitions d'étoiles filantes*. Bruxelles: M. Hayez, imprimeur.

Reach, W. T., Sykes, M. V., Lien, D., and Davies, J. K. 2000. The formation of Encke meteoroids and dust trail. *Icarus*, **148**(Nov.), 80–94.

Reach, W. T., Kelley, M. S., and Sykes, M. V. 2007. A survey of debris trails from short-period comets. *Icarus*, **191**(Nov.), 298–322.

Richter, D., and Simmons, G. 1974. Thermal expansion behavior of igneous rocks. *International Journal of Rock Mechanics and Mining Sciences & Geomechanics Abstracts*, **11**(10), 403–411.

Ridley, H. B. 1957. *Circulars of the British Astronomical Association*, **382**. Cited by Ridley (1963).

Ridley, H. B. 1963. The Phoenicid meteor shower of 1956 December 5. *Monthly Notes of the Astronomical Society of South Africa*, **22**(Jan.), 42–49.

Rotundi, A., Sierks, H., Della Corte, V. et al. 2015. Dust measurements in the coma of comet 67P/Churyumov-Gerasimenko inbound to the Sun. *Science*, **347**(1), aaa3905.

Rozitis, B., and Green, S. F. 2013. The strength and detectability of the YORP effect in near-Earth asteroids: A statistical approach. *Monthly Notices of the Royal Astronomical Society*, **430**(Apr.), 1376–1389.

Rudawska, R., Matlovič, P., Tóth, J., and Kornoš, L. 2015. Independent identification of meteor showers in EDMOND database. *Planetary and Space Science*, **118**(Dec.), 38–47.

Russell, J. A., Sadoski, M. J. Jr., Sadoski, D. C., and Wetzel, G. F. 1956. The spectrum of a Geminid meteor. *Publications of the Astronomical Society of the Pacific*, **68**(Feb.), 64–69.

Ryabova, G. O. 1999. Age of the Geminid meteor stream (Review). *Solar System Research*, **33**(Jan.), 224–238.

Ryabova, G. O. 2017. The mass of the Geminid meteoroid stream. *Planetary and Space Science*, **143**(Sept.), 125–131.

Sarli, B. V., Horikawa, M., Yam, C. H., Kawakatsu, Y., and Yamamoto, T. 2018. DESTINY+ trajectory design to (3200) Phaethon. *Journal of the Astronautical Sciences*, **65**(Mar.), 82–110.

Sarugaku, Y., Ishiguro, M., Ueno, M., Usui, F., and Reach, W. T. 2015. Infrared and optical imagings of the Comet 2P/Encke dust cloud in the 2003 return. *The Astrophysical Journal*, **804**(May), 127.

Sasaki, S., Nakamura, K., Hamabe, Y., Kurahashi, E., and Hiroi, T. 2001. Production of iron nanoparticles by laser irradiation in a simulation of lunar-like space weathering. *Nature*, **410**(Mar.), 555–557.

Sato, M., Watanabe, J.-i., Tsuchiya, C. et al. 2017. Detection of the Phoenicids meteor shower in 2014. *Planetary and Space Science*, **143**(Sept.), 132–137.

Scheeres, D. J., Hartzell, C. M., Sánchez, P., and Swift, M. 2010. Scaling forces to asteroid surfaces: The role of cohesion. *Icarus*, **210**(Dec.), 968–984.

Scheeres, D. J., and Sánchez, P. 2018. Implications of cohesive strength in asteroid interiors and surfaces and its measurement. *Progress in Earth and Planetary Science*, **5**(Dec.), 25.

Sekanina, Z. 1973. Statistical model of meteor streams. III. Stream search among 19303 radio meteors. *Icarus*, **18**(Feb.), 253–284.

Sekanina, Z., and Chodas, P. W. 2005. Origin of the Marsden and Kracht Groups of sunskirting comets. I. Association with Comet 96P/Machholz and its interplanetary complex. *The Astrophysical Journal Supplement Series*, **161**(2), 551–586.

Skiff, B. A. 2003. 2003 EH_1. *Minor Planet Electronic Circulars*, **2003-E27**.

Soja, R. H., Schwarzkopf, G. J., Sommer, M. et al. 2016 (Jan.). Collisional lifetimes of meteoroids. In: Roggemans, A. and Roggemans, P. (eds), *Proceedings of the International Meteor Conference, Egmond, the Netherlands, 2–5 June 2016*. Hove, BE: International Meteor Organization, pp. 284–286.

Sosa, A., and Fernández, J. A. 2015. Comets 169P/NEAT and P/2003 T12 (SOHO): Two possible fragments of a common ancestor? *IAU General Assembly*, **22**(Aug.), 2255583.

Southworth, R. B., and Hawkins, G. S. 1963. Statistics of meteor streams. *Smithsonian Contributions to Astrophysics*, **7**, 261–285.

Springmann, A., Lauretta, D. S., Klaue, B. et al. 2019. Thermal alteration of labile elements in carbonaceous chondrites. *Icarus*, **324**, 104–119.

Spurný, P. 1993. Geminids from photographic records. In: Stohl, J., and Williams, I. P. (eds), *Meteoroids and Their Parent Bodies, Proceedings of the International Astronomical Symposium held at Smolenice, Slovakia, July 6–12, 1992*. Bratislava: Astronomical Institute, Slovak Academy of Sciences, pp. 193–196.

Spurný, P., Borovička, J., Mucke, H., and Svoreň, J. 2017. Discovery of a new branch of the Taurid meteoroid stream as a real source of potentially hazardous bodies. *Astronomy and Astrophysics*, **605**(Sept.), A68.

Steel, D. I., and Asher, D. J. 1996. The orbital dispersion of the macroscopic Taurid objects. *Monthly Notices of the Royal Astronomical Society*, **280**(June), 806–822.

Suggs, R. M., Moser, D. E., Cooke, W. J., and Suggs, R. J. 2014. The flux of kilogram-sized meteoroids from lunar impact monitoring. *Icarus*, **238**(Aug.), 23–36.

Sykes, M. V., and Walker, R. G. 1992. Cometary dust trails. I Galina: – Survey. *Icarus*, **95**(Feb.), 180–210.

Szalay, J. R., Pokorný, P., Jenniskens, P., and Horányi, M. 2018. Activity of the 2013 Geminid meteoroid stream at the Moon. *Monthly Notices of the Royal Astronomical Society*, **474**(Mar.), 4225–4231.

Tancredi, G. 2014. A criterion to classify asteroids and comets based on the orbital parameters. *Icarus*, **234**(May), 66–80.

Taylor, P. A., Rivera-Valentín, E. G., Benner, L. A. M. et al. 2019. Arecibo radar observations of near-Earth asteroid (3200) Phaethon during the 2017 apparition. *Planetary and Space Science*, **167**, 1–8. DOI: 10.1016/j.pss.2019.01.009

Tedesco, E. F., Noah, P. V., Noah, M., and Price, S. D. 2004. IRAS Minor Planet Survey V6.0. *NASA Planetary Data System*, IRAS-A-FPA-3-RDR-IMPS-V6.0.

Tholen, D. J. 1984. *Asteroid taxonomy from cluster analysis of photometry*. Ph.D. thesis, University of Arizona, Tucson.

Ticha, J., Tichy, M., Kocer, M. et al. 2003. 2003 WY_{25}. *Minor Planet Electronic Circulars*, **2003-W41**(Nov.).

Todorović, N. 2018. The dynamical connection between Phaethon and Pallas. *Monthly Notices of the Royal Astronomical Society*, **475**(Mar.), 601–604.

Trigo-Rodríguez, J. M., Llorca, J., Borovicka, J., and Fabregat, J. 2003. Chemical abundances determined from meteor spectra: I Ratios of the main chemical elements. *Meteoritics and Planetary Science*, **38**(Aug.), 1283–1294.

Trigo-Rodríguez, J. M., Llorca, J., and Fabregat, J. 2004. Chemical abundances determined from meteor spectra – II. Evidence for enlarged sodium abundances in meteoroids. *Monthly Notices of the Royal Astronomical Society*, **348**(Mar.), 802–810.

Tsuchiya, C., Sato, M., Watanabe, J.-i. et al. 2017. Correction effect to the dispersion of radiant point in case of low velocity meteor showers. *Planetary and Space Science*, **143**(Sept.), 142–146.

Tubiana, C., Snodgrass, C., Michelsen, R. et al. 2015. 2P/Encke, the Taurid complex NEOs and the Maribo and Sutter's Mill meteorites. *Astronomy and Astrophysics*, **584**(Dec.), A97.

Vaubaillon, J., Arlt, R., Shanov, S., Dubrovski, S., and Sato, M. 2005. The 2004 June Bootid meteor shower. *Monthly Notices of the Royal Astronomical Society*, **362**(Oct.), 1463–1471.

Vaubaillon, J., Neslušan, L., Sekhar, A., Rudawska, R., and Ryabova, G. 2019. From Parent Body to Meteor Shower: the Dynamics of Meteoroid Streams. In: G. O. Ryabova, D. J. Asher, and M. D. Campbell-Brown (eds), *Meteoroids: Sources of Meteors on Earth and Beyond*. Cambridge, UK: Cambridge University Press, pp. 161–186.

Veeder, G. J., Hanner, M. S., Matson, D. L. et al. 1989. Radiometry of near-Earth asteroids. *The Astronomical Journal*, **97**(Apr.), 1211–1219.

Vokrouhlický, D., Bottke, W. F., Chesley, S. R., Scheeres, D. J., and Statler, T. S. 2015. The Yarkovsky and YORP effects. In: P. Michel, F. E. DeMeo, and W. F. Bottke (eds), *Asteroids IV*. Tucson: University of Arizona Press, pp. 509–531.

Voloshchuk, I. I., and Kashcheev, B. L. 1986. Study of the structure of a meteor complex near the Earth's orbit. I - Description of the model. *Solar System Research*, **19**(Apr.), 213–216.

Warner, B. D. 2017. Near-Earth asteroid lightcurve analysis at CS3-Palmer Divide Station: 2016 October–December. *Minor Planet Bulletin*, **44**(Apr.), 98–107.

Warner, B. D. 2018. Near-Earth asteroid lightcurve analysis at CS3-Palmer Divide Station: 2017 July–October. *Minor Planet Bulletin*, **45**(Jan.), 19–34.

Warner, B. D., and Fitzsimmons, A. 2005. Comet P/2002 EX_12 (NEAT). *International Astronomical Union Circular*, **8578**(Aug.).

Warner, B. D., Harris, A. W., and Pravec, P. 2009. The asteroid lightcurve database. *Icarus*, **202**(1), 134–146.

Watanabe, J.-I., Sato, M., and Kasuga, T. 2005. Phoenicids in 1956 revisited. *Publications of the Astronomical Society of Japan*, **57**(Oct.), L45–L49.

Weissman, P. R., Asphaug, E., and Lowry, S. C. 2004. Structure and density of cometary nuclei. In: M. C. Festou, H. U. Keller, and H. A. Weaver (eds), *Comets II*. Tucson: University of Arizona Press, pp. 337–357.

Weryk, R. J., and Brown, P. G. 2012. Simultaneous radar and video meteors—I: Metric comparisons. *Planetary and Space Science*, **62**(Mar.), 132–152.

Whipple, F. L. 1939a. Photographic meteor studies I. *Publications of the American Astronomical Society*, **9**, pp. 136–137.

Whipple, F. L. 1939b. Photographic study of the Geminid meteor shower. *Publications of the American Astronomical Society*, **9**, p. 274.

Whipple, F. L. 1940. Photographic Meteor Studies. III. The Taurid Shower. *Proceedings of the American Philosophical Society*, **83**(Oct.), 711–745.

Whipple, F. L. 1983. 1983 TB and the Geminid meteors. *International Astronomical Union Circular*, **3881**(Oct.).

Wiegert, P., and Brown, P. 2004. The problem of linking minor meteor showers to their parent bodies: Initial considerations. *Earth, Moon, and Planets*, **95**(Dec.), 19–26.

Wiegert, P., and Brown, P. 2005. The Quadrantid meteoroid complex. *Icarus*, **179**(Dec.), 139–157.

Wiegert, P. A., Houde, M., and Peng, R. 2008. An upper limit on gas production from 3200 Phaethon. *Icarus*, **194**(Apr.), 843–846.

Wiegert, P. A., Brown, P. G., Weryk, R. J., and Wong, D. K. 2013. The return of the Andromedids meteor shower. *The Astronomical Journal*, **145**(Mar.), 70.

Wiegert, P., Clark, D., Campbell-Brown, M., and Brown, P. 2017. Minor planet 2017 MB_1 and the Alpha Capricornids meteor shower. *Central Bureau Electronic Telegrams*, **4415**(Aug.).

Wilkison, S. L., and Robinson, M. S. 2000. Bulk density of ordinary chondrite meteorites and implications for asteroidal internal structure. *Meteoritics and Planetary Science*, **35**(Nov.), 1203–1213.

Williams, I. P., and Wu, Z. 1993. The Geminid meteor stream and asteroid 3200 Phaethon. *Monthly Notices of the Royal Astronomical Society*, **262**(May), 231–248.

Williams, I. P., Ryabova, G. O., Baturin, A. P., and Chernitsov, A. M. 2004. The parent of the Quadrantid meteoroid stream and asteroid 2003 EH1. *Monthly Notices of the Royal Astronomical Society*, **355**(Dec.), 1171–1181.

Williams, I. P., Jopek, T. J., Rudawska, R., Tóth, J., and Kornoš, L. 2019. Minor Meteor Showers and the Sporadic Background. In: G. O. Ryabova, D. J. Asher, and M. D. Campbell-Brown (eds), *Meteoroids: Sources of Meteors on Earth and Beyond*. Cambridge, UK: Cambridge University Press, pp. 210–234.

Wyatt, S. P., and Whipple, F. L. 1950. The Poynting-Robertson effect on meteor orbits. *The Astrophysical Journal*, **111**(Jan.), 134–141.

Yakshinskiy, B. V., and Madey, T. E. 2004. Photon-stimulated desorption of Na from a lunar sample: Temperature-dependent effects. *Icarus*, **168**(Mar.), 53–59.

Yamamoto, T. 1985. Formation environment of cometary nuclei in the primordial solar nebula. *Astronomy and Astrophysics*, **142**(Jan.), 31–36.

Yanagisawa, M., Ikegami, H., Ishida, M. et al. 2008. Lunar impact flashes by Geminid meteoroids in 2007. *Meteoritics and Planetary Science Supplement*, **43**(Aug.), 5169.

Yang, H., and Ishiguro, M. 2018. Evolution of cometary dust particles to the orbit of the Earth: Particle size, shape, and mutual collisions. *The Astrophysical Journal*, **854**(Feb.), 173.

Ye, Q.-Z. 2018. Meteor showers from active asteroids and dormant comets in near-Earth space: A review. *Planetary and Space Science*, **164**, 7–12.

Ye, Q.-Z., Brown, P. G., and Pokorný, P. 2016. Dormant comets among the near-Earth object population: A meteor-based survey. *Monthly Notices of the Royal Astronomical Society*, **462**(Nov.), 3511–3527.

Zook, H. A., and Berg, O. E. 1975. A source for hyperbolic cosmic dust particles. *Planetary and Space Science*, **23**(Jan.), 183–203.

9

Minor Meteor Showers and the Sporadic Background

Iwan P. Williams, Tadeusz J. Jopek, Regina Rudawska, Juraj Tóth, and Leonard Kornoš

9.1 Introduction

Meteors have been observed since antiquity and can be seen every night at an average rate of around five every hour. This rate is however far from constant, with values in excess of 10 000 being recorded on rare occasions. Astronomers in China, Korea and Japan started recording times and locations on the sky more than two millennia ago and these records have been chronicled by Hasegawa (1993). Until the nineteenth century, the views of Aristotle that meteors were caused by combustion in the atmosphere generally prevailed in Western scientific circles. However several fortuitous events, including spectacular displays of the Leonid (#013[1]) showers in 1799 and 1833 together with the break-up of comet 3D/Biela, producing the Andromedids (#018), moved meteor studies firmly into the astronomical field. It became generally accepted that meteors were the visible trail caused by the ablation of an extraterrestrial particle (meteoroid) in the Earth's atmosphere. Asteroids and comets can both come close to the Earth's orbit. Any meteoroid released from them will be observed as a meteor should it enter the Earth's atmosphere. Since the velocity of any meteoroid released is likely to be small compared to the orbital velocity, the meteoroid will move on a heliocentric orbit that is very similar to that of the parent, and if many meteoroids are released, they will form a stream. A shower occurs when the Earth passes through such a stream. An account of the historical developments that led to this conclusions can be found in Williams (2011).

A number of different processes can change the orbit of individual meteoroids so that more and more of them have orbits that differ from the initial mean orbit. Obvious mechanisms to cause this dispersal are close planetary encounters (Hughes et al., 1981), collisions between stream meteoroids and its parent (Williams et al., 1993) and collisions with interplanetary dust particles (Trigo-Rodríguez et al., 2005). Hence, with the passage of time, the stream slowly disperses and becomes harder to identify (Williams and Ryabova, 2011). The corresponding meteor shower also becomes weaker and can be identified only as a minor shower. Eventually the complete dispersal of the stream occurs and a meteor shower can no longer be identified, though individual meteors will still be seen as part of the sporadic background before finally becoming part of the zodiacal dust cloud. It should be noted that, so far, the International Astronomical Union has not produced a definition for the difference between a minor shower and a major one. In the following discussion, we simply use the definition that a minor shower is a shower that is not generally considered to be a historic major shower.

Four meteor showers – the Leonids (#013), the Perseids (#007), the Andromedids (#018), and the April Lyrids (#006) – played a central role in the early development of the subject, helping to establish the notion that meteoroid streams originated from comets. Whipple (1951) produced a model of a comet nucleus that allowed a mathematical formulation of the ejection process of solid particles from a comet nucleus to be established. This in turn allowed the heliocentric orbit of the ejected particle (meteoroid) to be determined thereby enabling models of meteoroid streams to be formulated. Over the last fifty years there have been a multitude of such models proposed and it is outside the scope of this chapter to describe them further. Many references and descriptions can be found in Williams (2002), Jenniskens (2006), Ryabova (2006). Following the great expansion in radar observations of meteors some fifty years ago, Sekanina (1973) identified a number of new weak meteoroid streams and suggested that asteroids could be the parent, the weakness being due to far fewer meteoroids being ejected from asteroids compared to the number from comets.

Theoretical models are only of value if they can be compared to observations of real meteor showers, and this chapter is mainly concerned with interpreting and understanding these showers. One important problem that is under active consideration in this field is to determine whether the parent of a stream was a comet or an asteroid. An important aspect of this is that observations of meteors can produce information regarding both the physical and chemical properties of the meteoroid and hence, by implication, of its parent. Such information is difficult to obtain using ground-based observations because the parent may be very faint and only visible for a short time. Sporadic meteors have lost all information regarding their initial orbits and so, for them, it is impossible to identify a unique parent. All that can be done is to establish whether the parent was of asteroid or comet type. One indicator is bulk density. It is to be expected that a meteoroid that originated from a rocky asteroid would have a bulk density of the order of 3 000 kg m^{-3}, while one originating from an icy comet would have bulk densities of the order of 500 kg m^{-3}. However, the derived density depends critically on the model used for the ablation and whether or not fragmentation takes place during this phase (Moorhead et al., 2017). An early estimate of meteoroid densities was by Ceplecha (1958), who introduced a model for ablation based on the heat conduction equation through a solid body. Using this method, Ceplecha (1967) found that meteoroid densities were in the range 1 400–4 000 kg m^{-3}. Another early estimate was by Jacchia et al.

---

[1] A unique meteor shower number as assigned in IAU Meteor Data Center, http://pallas.astro.amu.edu.pl/~jopek/MDC2007/.

(1967), who obtained 260 kg m^{-3} as the typical bulk density assuming they were porous and crumbly. Verniani (1969) found that values for meteoroids in showers varied from 140 kg m^{-3} to 630 kg m^{-3}, an order of magnitude lower than previous or more recent estimates. Further discussion of meteoroid densities are beyond the scope of this chapter as they are discussed fully in Chapter 2 (Borovička et al., 2019), but the range of values obtained makes using bulk densities alone a rather unreliable tool for establishing the nature of the parent. Here we concentrate on a pairing of a parent body with a meteoroid stream based on orbital similarity. The next section deals with identifying meteor showers and determining the orbits of meteoroids in the associated streams.

9.2 Identifying Meteor Showers and Meteoroid Streams

As with many aspects of meteor astronomy, common sense can produce an acceptable starting point for a discussion. Thus, an indication that a meteor shower exists is that the number of observed meteors in a given time interval is significantly higher than the sporadic background numbers in the same interval. Changing this into a scientifically quantified statement is harder, and there is no general agreement as to exactly what significantly higher for example means. Commission 22 of the International Astronomical Union (IAU) (a predecessor of the current Commission F1) produced a report regarding definitions at its General Assembly in 1961 (see Millman, 1963). This defined a meteor shower as a number of meteors with approximately parallel trajectories and a meteoroid stream as a number of meteoroids with nearly identical orbits. Recently, some of these definitions were revisited by Commission F1. New definitions were approved by the commission's members electronically, and partly included in the triennial report that will be published in the IAU Transactions.[2] According to this report: "Meteoroid stream is a group of meteoroids which have similar orbits and a common origin. Meteor shower is a group of meteors produced by meteoroids of the same meteoroid stream". There is no agreement as to exactly what one should understand by "similar orbits" and so the new definitions are still of a general nature. Similar meteoroid trajectories imply that the meteors observed within a short time interval will appear to emanate from a fixed point on the sky called the radiant.

Further proof of the existence of a stream comes if the geocentric velocities of the meteoroids are similar. For shower meteors, the similarity of the coordinates of the radiant point can easily be detected by, for example, plotting a histogram of the numbers of meteors observed against the declination of the radiant.

An increased flux, repeated over several years, occurring at nearly the same sidereal date is also a good indicator of the existence of a shower.

Other than these general considerations, there is no quantified definition of a meteor shower or a meteoroid stream.

The heliocentric orbital elements of individual meteoroids can be calculated once precise observations of a meteor have been obtained. If the individual orbits of a set of meteoroids are similar, it is a strong indicator of the existence of a stream. With sufficient meteoroids assigned to a stream, a mean orbit can be determined (Jopek et al., 2006, 2008a). If the mean orbit is similar to the orbit of a known comet or asteroid, then an association between the two is inferred and hence a possible parent for the stream identified.

It is thus important to be able to determine whether orbits are similar, and the simplest way of doing this is to quantify the difference between two orbits. If the difference is less than a threshold (or critical) value, the orbits are regarded as being similar. The selection of the threshold value is important. Choosing too high a critical value means that a larger difference between orbits is acceptable, which can lead to false similarities being claimed. Alternatively, choosing too low a critical values implies that orbits that are really almost identical can be missed through observational errors or small perturbations to the orbital elements. Unfortunately again, other than through repeated usage of various values there is no agreement as to what value should be used. Different critical values can be employed depending on whether the search is for a broad stream or a thin filament (see e.g. Porubčan et al., 2006).

Considering its importance, it is not surprising that many so-called D-criteria have been proposed to determine whether or not orbits are similar, and these are now discussed.

9.2.1 Methods for Quantifying the Difference Between Orbits

What is required is a quantifiable measure of the distance between two orbits. Though it is not necessary to use this formulation, it is easier to visualise the problem if two orbits are associated with two points of the conceptual orbital elements space. Those generally used are eccentricity e, perihelion distance q or semi-major axis a, inclination i, argument of perihelion ω, and longitude of the ascending node Ω. Thus, the difference between two orbits has to be measured in five-dimensional phase space at least.

The first definition of the difference between two orbits A and B was by Southworth and Hawkins (1963), and is given by

$$D_{SH}^2 = [e_B - e_A]^2 + [q_B - q_A]^2 + \left[2 \cdot \sin \frac{I_{BA}}{2}\right]^2 + \left[\frac{e_B + e_A}{2}\right]^2 \left[2 \cdot \sin \frac{\pi_{BA}}{2}\right]^2. \quad (9.1)$$

Here e_A, e_B are the eccentricities and q_A, q_B the perihelion distances of the two orbits. I_{BA} is the angle between the two orbital planes and π_{BA} is the distance of the longitudes of perihelia measured from the intersection point of the orbits. Those functions are

$$\left[2 \sin \frac{I_{BA}}{2}\right]^2 = \left[2 \sin \frac{i_B - i_A}{2}\right]^2 + \sin i_B \sin i_A \left[2 \sin \frac{\Omega_B - \Omega_A}{2}\right]^2, \quad (9.2)$$

---
[2] The full text of the new definitions of notions used by meteor astronomers is available on the IAU Webpage: www.iau.org/static/science/scientific_bodies/commissions/f1/meteordefinitions_approved.pdf.

$$\pi_{BA} = (\omega_B - \omega_A) +$$
$$+ 2 \arcsin\left[\cos\frac{i_B + i_A}{2} \cdot \sin\frac{\Omega_B - \Omega_A}{2} \cdot \sec\frac{I_{BA}}{2}\right].$$

When $|\Omega_B - \Omega_A| > 180°$, the sign of arcsin should be opposite. Assuming that the Earth moves on the circular orbit of 1 [au] radius, for the meteoroid elliptic orbits the codomain of the D_{SH} function is the interval [0, 3) spanning the case of a perfect orbital similarity and dissimilarity.

Southworth and Hawkins simply had D as a function (9.1). The terminology D-criterion was introduced by Sekanina (1970a) and Lindblad (1971b), and thereafter the terminology became commonly used. However, as Kholshevnikov et al. (2016) noted, the name D-criterion for the function (9.1) itself, is regrettable, because the similarity criterion is an inequality, e.g. $D \leq const.$

Drummond (1980, 1981) proposed a variation of the distance measurement shown in Equations (9.1) and (9.2). The new D_D function utilises a particular set of weights to render each of the terms dimensionless, each linear over its range and each making an approximately equal contribution. This is given by

$$D_D^2 = \left(\frac{e_B - e_A}{e_B + e_A}\right)^2 + \left(\frac{q_B - q_A}{q_B + q_A}\right)^2 + \left(\frac{I_{BA}}{180°}\right)^2$$
$$+ \left[\left(\frac{e_B + e_A}{2}\right)\left(\frac{\theta_{BA}}{180°}\right)\right]^2, \quad (9.3)$$

where I_{BA} is the angle between the orbital planes, θ_{BA} is the angle between the lines of apses of two orbits, given by

$$I_{BA} = \arccos(\cos i_B \cos i_A + \sin i_B \sin i_A \cos(\Omega_B - \Omega_A)), \quad (9.4)$$

$$\theta_{BA} = \arccos(\sin \beta_B \sin \beta_A + \cos \beta_B \cos \beta_A \cos(\lambda_B - \lambda_A)). \quad (9.5)$$

Here λ and β are the ecliptic coordinates of the perihelion point and are given by

$\lambda = \Omega + \arctan(\cos i \tan \omega)$ (add 180° if $\cos \omega < 0$),

$\beta = \arcsin(\sin i \sin \omega).$

For the Drummond function for elliptical meteoroid orbits we have $D_D \in [0, \sqrt{3.25})$.

Jopek (1993a) showed that the two distance measurements D_{SH} and D_D are not equivalent; see Table I in the cited paper. Moreover the D_D function is biased slightly, especially when two orbits differ only by the angle of the orbital inclination. If so, the fourth term of formula (9.3) instead of being zero, makes a non-zero contribution.

As a result of analysis of the properties of D_{SH} and D_D, Jopek (1993a) proposed an alternative hybrid D_H, defined as

$$D_H^2 = [e_B - e_A]^2 + \left[\frac{q_B - q_A}{q_B + q_A}\right]^2 + \left[2 \cdot \sin\frac{I_{BA}}{2}\right]^2$$
$$+ \left[\frac{e_B + e_A}{2}\right]^2 \left[2 \cdot \sin\frac{\pi_{BA}}{2}\right]^2. \quad (9.6)$$

The codomain of the D_H function for elliptical meteoroid orbits is close to the codomain of the D_{SH} function.

Instead of using the geometrical orbital elements, the problem may be considered in terms of physical quantities. An orbit can be described by vectorial elements: the angular momentum **h** given by

$$\mathbf{h} = (h_1, h_2, h_3)^T = \mathbf{r} \times \dot{\mathbf{r}},$$

the Laplace vector **e** given by

$$\mathbf{e} = (e_1, e_2, e_3)^T = \frac{1}{k^2}\dot{\mathbf{r}} \times \mathbf{h} - \frac{\mathbf{r}}{|\mathbf{r}|},$$

and the orbital energy E given by

$$E = \frac{1}{2}\dot{\mathbf{r}}^2 - \frac{k^2}{|\mathbf{r}|}.$$

Here k is the Gauss gravity constant, and **r** and $\dot{\mathbf{r}}$ are the vectors of the heliocentric position and velocity of the meteoroid. Using these vector elements, Jopek et al. (2008b) proposed a new metric for quantifying orbital differences, namely

$$D_V^2 = w_{h1}(h_{B1} - h_{A1})^2 + w_{h2}(h_{B2} - h_{A2})^2$$
$$+ 1.5 \, w_{h3}(h_{B3} - h_{A3})^2 + w_{e1}(e_{B1} - e_{A1})^2$$
$$+ w_{e2}(e_{B2} - e_{A2})^2 + w_{e3}(e_{B3} - e_{A3})^2$$
$$+ 2 \, w_E(E_B - E_A)^2, \quad (9.7)$$

where w are weight coefficients with values that are given in Table 9.1. The distance function D_B proposed by Jenniskens (2008a) is based on three approximate dynamical invariants C_1, C_2, C_3 used by Babadzhanov (1990), and is given by

$$D_B^2 = \left(\frac{C_{A1} - C_{B1}}{0.13}\right)^2 + \left(\frac{C_{A2} - C_{B2}}{0.06}\right)^2$$
$$+ \left(\frac{C_{A3} - C_{B3}}{14.°2}\right)^2, \quad (9.8)$$

$C_1 = (1 - e^2) \cos^2 i,$

$C_2 = e^2(0.4 - \sin^2 i \sin^2 \omega),$

$C_3 = \omega + \Omega.$

Here the first invariant (C_1) corresponds to the z-component of the orbital angular momentum, the second (C_2) is taken from the secular model described by Lidov (1962) while the third invariant (C_3) is the longitude of perihelion. However, there is no information used about the dimensions of the orbit a or q in D_B, making it quite different from the D-functions described earlier (see Rożek et al. (2011)).

Though the orbital elements are easy to understand, they are not directly observed quantities. Valsecchi et al. (1999) proposed a metric, D_N, based on the geocentric parameters of the meteors, rather than those inferred for the meteoroids. D_N is formulated as follows

$$D_N^2 = [U_B - U_A]^2 + w_1[\cos \theta_B - \cos \theta_A]^2 + \Delta \xi^2, \quad (9.9)$$

Table 9.1. *Weight coefficients for Equation (9.7), (see equation (5) and Table 1, epoch 4000, in Jopek et al., 2008b).*

$w_{e1} = 2.07 \cdot 10^3$	$w_{h1} = 2.60 \cdot 10^5$	$w_E = 2.60 \cdot 10^{11}$
$w_{e2} = 9.77 \cdot 10^2$	$w_{h2} = 7.18 \cdot 10^5$	
$w_{e3} = 4.73 \cdot 10^2$	$w_{h3} = 1.49 \cdot 10^6$	

where

$$\Delta\xi^2 = \min\left[w_2\Delta\phi_I^2 + w_3\Delta\lambda_I^2, w_2\Delta\phi_{II}^2 + w_3\Delta\lambda_{II}^2\right],$$

$$\Delta\phi_I = 2\sin\frac{\phi_B - \phi_A}{2},$$

$$\Delta\phi_{II} = 2\sin\frac{180° + \phi_B - \phi_A}{2},$$

$$\Delta\lambda_I = 2\sin\frac{\lambda_B - \lambda_A}{2},$$

$$\Delta\lambda_{II} = 2\sin\frac{180° + \lambda_B - \lambda_A}{2},$$

where w_1, w_2 and w_3 are suitably defined weighting factors. Note that $\Delta\xi$ is small either if both $\phi_1 - \phi_2$ and $\lambda_B - \lambda_A$ are small, or if they are both close to $180°$. In the latter case, the two meteors would belong to the two showers corresponding to the two node crossings of essentially the same orbit, characterised by $q(1 + e) \approx 1$ and $\omega \approx 90°$ or $\omega \approx 270°$, as is the case for the Orionids and the η Aquariids.

The quantities U, θ, and ϕ are the Öpik variables described in Carusi et al. (1990). Actually, θ and ϕ are the angles that define the direction opposite to that from which the meteoroid is seen to arrive (after removal of the effect of the Earth's gravity), that is, opposite to the geocentric radiant. However, there is an important difference, the latter is defined with respect to fixed stars, whereas θ and ϕ identify the position of the antiradiant at the location where the meteor is observed, and are defined in the instantaneous geocentric reference frame in which the x-y plane coincides with the ecliptic and the x-axis coincides with the heliocentric position vector of the Earth. The third variable U is the modulus of the unperturbed geocentric velocity of the meteoroid. U and θ are the dynamical semi-invariants. The fourth variable λ is the ecliptic longitude of the Earth at the time of meteor observation.

The three quantities U, θ and ϕ can be obtained from the components of the velocity vector $\mathbf{U} = (U_x, U_y, U_z)$. These in turn can be calculated from the geocentric velocity of the meteoroid (V_G in [km/s]) and the equatorial coordinates of the meteor geocentric radiant, α_G and δ_G, all directly observed quantities. The equations are

$$\begin{pmatrix} U_x \\ U_y \\ U_z \end{pmatrix} = \hat{\mathbf{r}}(\lambda) \cdot \hat{\mathbf{p}}(\varepsilon) \cdot \frac{V_G}{29.7} \begin{pmatrix} -\cos\delta_G \cos\alpha_G \\ -\cos\delta_G \sin\alpha_G \\ -\sin\delta_G \end{pmatrix}, \quad (9.10)$$

where $\hat{\mathbf{p}}(\varepsilon)$ and $\hat{\mathbf{r}}(\lambda)$ are rotational matrices around the x- and z-axis, respectively. The angle ε denotes the inclination of the ecliptic plane to the plane of the celestial equator. In this formulation a positive rotation is assumed to be in the anticlockwise direction.

Having these components, it is straightforward to find θ and ϕ, and

$$U_x = U\sin\theta\sin\phi, \quad (9.11)$$
$$U_y = U\cos\theta,$$
$$U_z = U\sin\theta\cos\phi.$$

In applying the D_N criterion Jopek et al. (1999) put the three weighting factors in Equation (9.9) as 1 but other choices can be made; see, e.g., appendix A3 in Moorhead (2016).

The simplest quantities that come from observations are $\lambda_{\odot A}$ and $\lambda_{\odot B}$ the solar longitudes, α_A and α_B the right ascensions, δ_A and δ_B the declinations, and V_{G_A} and V_{G_B} the geocentric velocities of two meteoroids. Using these geocentric parameters, Rudawska et al. (2015) proposed a distance function D_X, defined as

$$D_X^2 = w_{\lambda_\odot}\left[2\cdot\sin\left(\frac{\lambda_{\odot A} - \lambda_{\odot B}}{2}\right)\right]^2$$
$$+ w_\alpha(|V_{G_A} - V_{G_B}| + 1)\left[2\cdot\sin\left(\frac{\alpha_A - \alpha_B}{2}\cdot\cos\delta_A\right)\right]^2$$
$$+ w_\delta(|V_{G_A} - V_{G_B}| + 1)\left[2\cdot\sin\left(\frac{\delta_A - \delta_B}{2}\right)\right]^2$$
$$+ w_V\left[\frac{V_{G_A} - V_{G_B}}{V_{G_A}}\right]^2, \quad (9.12)$$

where $w_{\lambda_\odot}$, w_α, w_δ, and w_V are suitably defined weighting factors. To normalise contributions of each term in D_X, Rudawska et al. (2015) used values: $w_{\lambda_\odot} = 0.17$, $w_\alpha = 1.20$, $w_\delta = 1.20$, and $w_v = 0.20$. It was assumed that each term of D_X should have similar contribution to the D value, and its final result would be comparable to other D-function.

The distance functions mentioned so far are not equivalent and from the mathematical point of view some of them represent pseudo metrics. Kholshevnikov et al. (2016) showed that D_{SH} and D_D do not satisfy the three axioms of the metric space. They proposed several new metrics; three of them, ρ_1, ρ_2 and ρ_5, were applied to find near-Earth objects that have a common origin. These authors concluded that the metrics ρ_1, ρ_2 and ρ_5 represent a sufficiently reliable means for determining the distance between orbits.

The first metric is defined in terms of the angular momenta $\mathbf{h}$ and Laplace vector $\mathbf{e}$ of two orbits A and B, namely

$$\rho_1^2 = \frac{(\mathbf{h}_B - \mathbf{h}_A)^2}{k^2 L} + (\mathbf{e}_B - \mathbf{e}_A)^2, \quad (9.13)$$

where k^2 is the gravitational parameter and L is a scale factor with dimension of length. Generally L is an arbitrary factor; however, for application in the Solar System it is reasonable to put $L = 1$ [au]. In term of the orbital elements the ρ_1 metric can be calculated by

$$\rho_1^2 = \frac{1}{L}(p_B + p_A - 2\sqrt{p_B p_A}\cos I_{BA}) \quad (9.14)$$
$$+ (e_B^2 + e_A^2 - 2e_B e_A \cos\theta_{BA}),$$

were I_{BA} is defined by Equation (9.4) and θ_{BA} is defined by Equation (9.5), while p_A and p_B are the semilatus rectum $p = a(1 - e^2)$.

In Kholshevnikov et al. (2016), the ρ_2 metric also has two representations, a definition by means of the orbital elements has the form

$$\rho_2^2 = (1 + e_B^2)p_B + (1 + e_A^2)p_A \quad (9.15)$$
$$- 2\sqrt{p_B p_A}(\cos I_{BA} + e_B e_A \cos\theta_{BA}).$$

In definitions of the ρ_1 and ρ_2 metrics all the p, e, ω, Ω, i orbital elements are utilised. However in cases when the longitudes of ascending nodes and the arguments of perihelions change faster than remaining elements, Kholshevnikov et al. (2016) proposed another metric, namely

$$\rho_5^2 = (1 + e_B^2)p_B + (1 + e_A^2)p_A \quad (9.16)$$
$$- 2\sqrt{(p_B p_A)}(e_B e_A + \cos(i_B - i_A)).$$

According to Kholshevnikov et al. (2016), the ρ_1, ρ_2 and ρ_5 metrics represent distances satisfying all three axioms of a metric space.

It is not easy to decide which of the distance measures represented by various D- or ρ-functions is the best to identify meteoroid streams, or indeed which of them is more reliable. Several tests limited to D-functions, carried out by Neslušan and Welch (2001); Galligan (2001) and Moorhead (2016) indicated a modest preference for the D_N function defined by formula (9.9).

For the major streams, where the number of orbits is large, there is no problem in using any of the above D-functions or ρ-metrics to identify similar orbits and calculate a mean orbit. Iterative schemes, such as proposed by Arter and Williams (1997) can be used. It is even possible to identify a sub-stream, for example Porubčan et al. (2006) found sub-streams in the Taurid complex. As the number of known orbits decreases or if the variance of the orbital separation becomes comparable to the orbital uncertainty, identification becomes problematic. This is the case for many minor streams and they are discussed in a later section.

9.2.2 Methods for Detecting Meteoroid Streams

Southworth and Hawkins (1963) proposed two definitions of a meteoroid stream. The first one is straightforward – a meteoroid stream consists of all meteoroids that have orbits concentrated around a given mean orbit. It means that one can consider a meteoroid stream as a set of points inside a hyper-sphere of constant radius D_c. Sekanina (1970a, 1976) modified this approach adding an iterative procedure during which a new mean orbit is calculated. As iteration advances, the centre of the hyper-sphere moves from one place to the other converging to the final mean orbit. The first definition and its variation are convenient for finding all members of known streams. The second definition proposed by Southworth and Hawkins is more complicated and involves using a cluster analysis algorithm based on a single neighbour linkage technique (a variate of general hierarchical clustering). It has the advantage of not requiring any prior information about the meteoroid stream orbit, and so it can be applied to new streams. Both basic definitions or some iterative variants of them have been extensively used to search for old and new streams and to identify possible genetic relationships between meteoroids, asteroids and comets (e.g.: Sekanina, 1970a,b, 1973, 1976; Arter and Williams, 1997; Jopek et al., 2003, 2010; Jopek, 2011; Rudawska and Jenniskens, 2014; Kornoš et al., 2014a).

Most minor meteoroid streams are identified following a search through databases of observed meteor parameters to find similar orbits. To help in this task, Southworth and Hawkins (1963) introduced a computer tool that consisted of all components necessary to carry out a computer cluster analysis of a given database. This involves:

- selecting a distance function (D-criterion) (for example D_{SH}),
- a rule for calculating the threshold value D_c of the orbital similarity,
- a stream searching algorithm (cluster analysis algorithm) which together with the D-function and the threshold value D_c produces an acceptable definition of a meteoroid stream.

With the passage of time, different approaches have been proposed by several authors to identify meteoroid streams. Svoreň et al. (2000), proposed a procedure called the method of indices. A set of eight geocentric and heliocentric meteoroid parameters is used and the searching algorithm is based on some kind of sorting of the meteoroid sample. Each parameter range is divided into a certain number of intervals, and indices are assigned to each meteoroid according to the interval pertinent to its parameters. The meteoroids with the same indices belong to the same stream.

The method described by Welch (2001) follows the Sekanina approach. However, it differs significantly and is more complicated. The iterative procedure starts at the point of the highest density in orbital element space. For this purpose, the data sample of discrete points is replaced by a smooth density map in orbital element space, and Welch developed a search algorithm within this map to find the core orbit of an old or a new meteoroid stream.

Galligan (2001) analysed radar orbits observed during the equinoctical year. Instead of searching among the whole $\sim 10^5$ data sample, Galligan partitioned the orbits on the basis of inclination defining five partitions. In each partition the orbits have been analysed separately. A similar approach was proposed by Moorhead (2016). However, the sporadic intrusion for each individual meteoroid stream was considered, instead for inclination bands.

In planetary astronomy, the wavelet transform method was applied for the first time by Bendjoya et al. (1991). It was applied for cluster analysis of the main belt asteroids. The wavelet transform is a kind of transformation similar to the Fourier transform. The function representing the orbital sample is given by sum of Dirac functions, and the wavelet transform turns to the computation at a given point of a wavelet coefficient. Each coefficient is an indicator of the local density at the orbital element space. For detecting meteor showers within radar data, this technique was firstly applied by Galligan and Baggaley (2002a,b), next by Brown et al. (2008, 2010a) and recently by Pokorný et al. (2017) and Schult et al. (2018).

The techniques mentioned in the preceding paragraph are more or less used automatically. However, to analyse the CAMS video data Jenniskens and Nénon (2016) decided to use interactive *CAMS Streamfinder* software to search visually for an excess number density of radiant points in $10°$ intervals of solar longitude and extract cluster using the D_H function (Equation (9.6)).

A comparison of several D-functions and cluster analysis algorithms was carried out in Rudawska and Jopek (2010). A cluster analysis algorithm based on a simplified version of the Welch algorithm and single linking technique with the D_V function generally appeared to be the most effective, whereas methods using the D_{SH} function were less effective.

Due to so many factors contributing to the final result of the meteoroid stream searching method, it seems unlikely that a generally accepted 'best' method will be found.

Selection of a *D*-function and a cluster analysis algorithm are significant components of the meteoroid stream (shower) searching. However, as mentioned earlier, two orbits are regarded as being similar if the value of the selected *D*-function does not exceed a certain threshold value D_c, and the right selection of this value is the crucial component in any stream searching. Using a four-dimensional point distribution as a model of the distribution of meteoroids orbits, Southworth and Hawkins (1963) concluded that D_c should vary inversely with the fourth root of the orbital sample size N. To find D_c they proposed the formula

$$D_c = 0.2 \cdot (360/N)^{0.25}, \qquad (9.17)$$

which was calibrated using small sample of 359 photographic meteors observed by the Super-Schmidt cameras and a list of meteoroid streams obtained earlier. The values of D_c given by (9.17), as well as its slight variation (Lindblad, 1971a) have been applied for identification of meteoroid streams, for example by Southworth and Hawkins (1963); Lindblad (1971b,a); Jopek (1986, 1993b), and recently by Holman and Jenniskens (2012).

Formula (9.17) gives the D_c value regardless of the size of the identified meteoroid group, and as result, the same threshold value has been used to identify streams of 2, 3 ... members. Also, (Jopek, 1995) has shown that for the same size N of the orbital samples, using the formula (9.17) may result in a different sporadic bias among the identified meteoroid streams.

Recently Jopek and Bronikowska (2017) have shown that, when dealing with pairs among meteoroids, the (9.17) formula overestimates the values as it does not take into account the statistical properties of the orbital samples observed by different techniques. Hence, for example, application of this formula may prove to be risky when applied to a data set consisting of meteors observed by small photographic cameras as well as meteors observed by radar systems. Therefore, Lindblad (1971b,a), Gartrell and Elford (1975) and Jopek (1986) carried out several meteoroid stream searches at different values of D_c close to the value given by the formula (9.17).

The main shortcoming of the formula (9.17) is a lack of information regarding the possibility of a false-positive identification of the group with given D_c value. Therefore, South-worth and Hawkins (1963) and Nilsson (1964) investigated the reliability of their results by searching for streams in equivalent sets of artificial data that were constructed by shuffling and re-assigning appropriate orbital elements. Due to the limitations of the computing power available at that time, however, it was rather problematic to accomplish an extensive reliability test of the detected streams. Fortunately, nowadays there is no problem in including a more detailed significance test in every stream searching process.

To find the threshold D_c, Neslušan et al. (1995, 2013b) proposed the method making use of the so-called break point on the plot of cumulative numbers of the identified meteors versus values of the *D*-function. After removing the members of the stream, the reliability of the results is estimated by comparing the density of sporadic radiants in the areas surrounding and containing the shower radiants under study – the so-called background-number-density test.

Jopek and Froeschle (1997) proposed another approach. By numerical statistical experiment, for each individual group of M members, they estimated the threshold D_c corresponding to the fixed small probability of a chance identification. However a subsequent paper, (Jopek et al., 2003) found that if a search for streams is carried out among data derived from different sources and one of the data sets has statistical properties significantly different from the others (e.g. the data set under study is seriously affected by a very specific selection effect), the use of a statistical criterion to determine thresholds does not lead to satisfactory results.

Moreover, in case of a single linkage cluster analysis technique for a large enough orbital sample size N, the threshold D_c can become arbitrarily small, and this seems to be more a bug than a feature of the method. This is related to another complication of the problem, which is to take into account the accuracy of the individual orbital catalogues.

Hence, it seems that the acceptable mutual distances between meteoroid orbits of the same stream in a given data set should be determined by physical rather than by statistical arguments. By this, we mean that the thresholds should be the result of an accurate modelling of the time evolution of a stream, taking into account all the phenomena relevant for the meteoroids of the size of interest – i.e., different physical and dynamical processes should be taken into account for photographic meteors and for radar meteors (see, e.g., Ryabova, 2016; Ishmukhametova et al., 2009). It is possible that the thresholds determined in this way would turn out to be different in different regions of the parameter space (e.g., ecliptic, low-velocity streams as opposed to high-velocity retrograde ones), and it is clear that a statistical technique such as that discussed in Jopek et al. (2003) may have problems in dealing with such a situation.

Hence, the methods applied by Galligan (2001); Welch (2001); Moorhead (2016) or based on the wavelet approaches (see Galligan and Baggaley, 2002a; Brown et al., 2008; Pokorný et al., 2017; Schult et al., 2018) will prove to be more reliable.

9.3 Analysis of Data Sets of Minor Showers

9.3.1 Definition of a Minor Shower

Historically, only the strongest showers such as the Leonids (#013) or the Perseids (#007) were recognised (Olmsted, 1834). When more accurate visual observations were carried out, more showers were identified (Cook, 1973; Jenniskens, 2006). These showers were always present but were not recognised due to poor recording methods. New and productive techniques have been developed to automatically record and analyse large amounts of meteor data produced through photography, radar and video methods. Cataloguing data on known and potential new meteor showers is carried out by the IAU Meteor Data Center (MDC) – for details see Section 9.4.3 or Jopek and Kaňuchová (2017). At the IAU General Assembly held in Prague in 2006, a Working Group was established to consider new rules and definitions relevant to meteor astronomy. It has not produced a definition for the difference between a minor shower and a major one. Obviously, a minor meteor shower has lower activity compared to a major one. Kronk (1988) uses the definition that minor showers produce less than 10 meteors per hour at the time of maximum activity. However, there is no generally accepted limit. The Meteor Data Center will implement whatever decision is eventually produced but currently makes no formal distinction between showers. In the following discussion, we simply use the definition that a minor shower is a shower that is not generally considered to be a major shower.

There is, however, a further difficulty – how to distinguish between a real minor shower and a possible random grouping of sporadic background meteors. Theoretical modelling shows (Tóth et al., 2011) that minor streams could exist that are currently under the detection level of standard statistical methods (Neslušan and Hajduková, 2017). The limit for minor meteor shower detection will be pushed by more numerous and precise orbital data as well as by more robust statistical methods. This will allow the existence of very minor showers to be confirmed in a statistical manner.

9.3.2 Sources of Minor Showers

There are three reasons for expecting minor showers to be present. The first is because the Earth is currently passing through the outer and less dense part of a meteoroid stream. If the relative geometry were different, a stronger shower might be seen. Second, there is a strong dependence on the geocentric velocity of meteoroids that allows them to produce observable meteors. A substantial number of meteoroids in a stream cannot be observed by any technique, simply because their geocentric velocity is not suitable. The final reason why only a minor shower can be observed is that the actual number of meteoroids in the stream is low. This could be either because of a lack of particles being ejected from a parent (e.g. a low activity comet, or collisions between asteroids) or because the stream is old and the originally denser stream has lost many of its meteoroids due to gravitational perturbations and non-gravitational effects.

9.3.3 Minor Showers

Some minor showers are already included in the list of established meteor showers, while some are waiting to be confirmed or better characterised. Some will eventually be excluded because they are duplicates of other showers, or a chance grouping. Several authors reported new possible showers (Jenniskens et al., 2011) but these were not confirmed when additional data became available. Removing them from the working list of showers is being considered (Jenniskens and Nénon, 2016). Statistically, however (Jenniskens et al., 2016b), about 700 meteor showers are expected to exist, of which 71% will be on Jupiter-family-type orbits. As mentioned, there are 906 meteor showers currently listed. It may be that we are close to identifying almost all the meteor showers, including the minor ones. Nevertheless, more work is needed on precise and populous datasets, especially in photographic and video ones.

It would be good if established and new photographic and video datasets could follow the CAMS example and require showers to pass the evaluation criteria of the IAU video database and be open to other workers for further research. Further, orbit calculation methods should be evaluated with consistency (Neslušan et al., 2013b) and precise orbital analyses (Hajduková et al., 2017) should be carried out.

About 10–15% of all meteor showers have known and confirmed parent bodies. For the major streams these have generally been established for some considerable time. The percentage of known parent bodies for minor streams is significantly less, and work in this field is overdue. More precise data will be very helpful to determine the mean orbital parameters for any filaments present as was shown by Spurný et al. (2017) in the case of the Southern Taurids in 2015. In this way the uncertainty in orbital phase space will be lower and similarity in phase space of metrics or orbital evolution modelling will be more significant in proving an association of the specific parent bodies with the stream. Also, the fact that near-Earth space is a chaotic regime and that, therefore, some level of uncertainty will remain when considering orbital evolution must be taken into account.

Radar observations usually cover small meteoroid particle ranges in mass and size and are suitable for continuously monitor meteor activity. Different radar systems are not limited to weather and light conditions, but have different sensitivity and observability specifications. Thus, the interpretation of radar data is not straightforward when compared with other observational techniques. Currently, there are several radar systems on both hemispheres covering the activity of both sporadic meteors and meteor shower sources. Some of the radar data from previous campaigns were analysed and we briefly mention some of them.

Galligan and Baggaley (2004) analysed about 500 000 AMOR radar orbits observed from New Zealand in 1995–1999 corresponding to limiting masses of 3×10^{-10}kg. Chau et al. (2007), using a high-power, large-aperture 50-MHz radar (HPLAR) at Jicamarca, Peru, obtained data on which interferometric modes allow the speed and trajectory of meteors to be determined. Six sporadic previously known sources were identified with speed distributions available for 17 000 meteors detected in fourteen days between 2001 and 2006. Campbell-Brown (2008) analysed 2.35 million Canadian CMOR radar high quality orbits, focusing on sporadic radiant, speed and orbital parameters distributions. The source of a ring depleted in meteor radiants at 55 degrees from the apex was attributed to shorter collisional lifetimes of meteoroids with the zodiacal cloud. Younger et al. (2009) found thirty-seven meteor showers, including nine undocumented showers, using the 33.2-MHz interferometric meteor radars located at Davis Station, Antarctica and Darwin, Australia. The data from both locations span the period of 2006–2007. The orbital elements of the streams were calculated for the thirty-one showers. Kero et al. (2012) analysed 2009–2010 MU radar (Japan) head echo observations of sporadic and shower meteors and their radiant densities and diurnal rates. They found that meteor rate variation is not symmetric with respect to the equinoxes and the north and south apex source regions fluctuated in strength. Schult et al. (2018) observed sporadic sources up to microgram masses for more than two years using the Middle Atmosphere ALOMAR Radar System (MAARSY, Norway). The large statistical amount of nearly one million meteor head detections provided an overview of the elevation, altitude, velocity and daily count rate distributions during different times of the year at polar latitudes. Schult et al. (2018) also analysed the set of MAARSY radar data using a wavelet shower search algorithm and identified thirty-three meteor showers in the data set, all of which are found in the IAU meteor shower catalogue. Only 1% of all measured head echoes are associated with meteor showers. Differences in shower duration, mass indices and enrichment of Halleyid showers by very small particles were found by comparing with other radar observations.

9.3.4 Parent Bodies

Many authors, the latest being Jenniskens (2017), have shown that some meteor showers are dynamically connected to one stream and its parent body. Some meteoroid streams might have several potential parent bodies (Porubčan et al., 2006) creating meteor complexes consisting of several showers. The situation for each meteor shower and its meteoroid stream and possible parent bodies has to be individually studied. Recently, there have been several investigations concerning the dynamical evolution of streams and their parent bodies. Neslušan et al. (2013a), Abedin et al. (2018) and Babadzhanov et al. (2017) investigated the orbital evolution of comet 96P/Machholz and the largest Marsden group comet P/1999 J6, and were able to fit eight meteor showers to that complex as had previously been shown by Babadzhanov et al. (2008).

Dumitru et al. (2017) investigated the orbital evolution of known Near-Earth Asteroids (NEAs) over 10 000 years and, using various D-criteria metrics, found that 206 NEAs could be possible parent bodies to 28 meteor showers. However, only fifty of those NEAs satisfied the condition that they are lying within the limit in at least two D-criteria. Dumitru et al. suggested the association of the Andromedids (#018) with asteroid 2000 UG11 and the October Capricornids (#233) with (4179) Toutatis. Currently, there are about 17 000 known NEOs, including cometary objects, and around 100 000 objects with a diameter larger than 100m are estimated to be in near-Earth space (Stuart and Binzel, 2004; Ye et al., 2016). Hence, more possible associations are likely to be proposed in the future. This is why detailed and cautious dynamical study is needed to avoid false or chance linking of parent bodies with meteoroid streams. Determining the orbital evolution of the parent body, as well as obtaining a precise mean orbit of the stream or filament, is essential. This might require the use of an N-body numerical integrator that includes planets and most influencing asteroids and can also include relativistic effects of the Sun and its oblatenes (e.g. Galushina et al., 2015) and non-gravitational forces, i.e. the rocket effects on the motion of comets (Marsden et al., 1973), the Yarkovsky effect on the motion of asteroids (Farinella et al., 1998) and the Poynting–Robertson (P-R) effect and the solar wind on the motion of meteoroids (Wyatt and Whipple, 1950; Whipple, 1955; Burns et al., 1979; Gustafson, 1994; Banaszkiewicz et al., 1994). Klačka (2014) and Klačka et al. (2014) found the solar wind effect is more significant than the P-R effect because the kappa distribution valid for the solar wind plays the key role. The situation is not so straightforward in some cases, e.g. Kornoš et al. (2015) pointed to a limited reliability of numerical integrations in the case of the April Lyrids (#006) and comet C/1861 G1(Thatcher).

Sato et al. (2017) compared the modelled activity of the Phoenicids in 2014 with actual visual video and radar observation and concluded that its parent comet 289P/Blanpain was less active in the early part of the twentieth century when compared with its activity in the eighteenth and nineteenth centuries. This suggests that the comet is evolving towards a dormant stage. Similarly, Koten et al. (2012) and Vaubaillon et al. (2011) compared the observed activity of the Draconids in 2011 with that predicted by modelling ejected particles from the parent comet, 21P/Giacobini-Zinner, and found that the comet was active prior its discovery early in the twentieth century. Moreover, they were able to predict jet-like ejection of particles of a specific mass range from a certain hemisphere of the comet nucleus to reproduce the observed activity. This is an illustration of how useful meteor observations and modelling might be for the broader scientific community.

Micheli et al. (2016) proposed an association of the Amor-type asteroid 2009 WN25 with the November i Draconid stream (#392), where ejection modelling of particles from a parent was used. The association of meteoroids with asteroids might be reinforced by physical observations of spectral-type or photometric parameters of asteroids. Tubiana et al. (2013, 2015) ruled out some possible asteroids previously suggested as parents in the Taurid complex (#247) by dynamical studies (Porubčan et al., 2006). Some of these asteroids appear to be S-type, which is not consistent with parameters such as strength, dynamical pressure fragmentation or spectral observations of the Taurids in the atmosphere (Matlovič et al., 2017).

9.4 Meteor Data Catalogues

Over the last ten to fifteen years an enormous increase in the volume of publicly available meteor data has taken place. The reasons behind this phenomenon are advances in technology that led to the availability of inexpensive video cameras in the marketplace and the availability of open access software over the Internet, dedicated to the reduction of the meteor observations.

At the end of the twentieth century, the only publicly available meteor data in digital form were those recorded the IAU Meteor Data Center in Lund, Sweden (IAU MDC). The archive was created by Lindblad (1987), and transformed into transmissible form (a floppy disk) by Lindblad and Steel (1994). 68 000 radiants and orbits were included in the database, of which only about 6 000 were obtained by photographic and TV techniques. The bulk of the optical orbits were obtained using photographic techniques by several stations located in North America (USA, Canada), Europe (Czechoslovakia, Ukraine) and Asia (Tajikistan, Japan).

Of the 62 000 radar meteors, about 39 000 orbits were derived by the Harvard Radar Meteor Project. The remaining radio orbits were determined by the former Soviet Union stations (Obninsk, Kharkov), the Australian station in Adelaide, and some orbits were obtained during the Soviet Equatorial Expedition to Mogadishu in Somalia. Most of these programs were carried out during the 1960s and 1970s.

In 2001 the IAU MDC was moved to the Astronomical Institute of the Slovak Academy of Sciences in Tatranska Lomnica, were it is currently maintained by Slovak astronomers (Lindblad et al., 2003; Svoren et al., 2008; Porubčan et al., 2011; Neslušan et al., 2014).[3]

---

[3] The orbital part of the IAU MDC is at the address www.astro.sk/~ne/IAUMDC/PhV2016/

9.4.1 Developments in Radio Observations

Since the 1990s several new radio surveys have started observing meteors. The Advanced Meteor Orbit Radar facility (AMOR) was developed at the Birdlings Flatfield station of the University of Canterbury, Christchurch, New Zealand (Baggaley et al., 1994; Galligan, 2001). In late 2001 the Canadian Meteor Orbit Radar (CMOR), developed by the University of Western Ontario in London, Canada, started operations (Webster et al., 2004; Jones et al., 2005). In 2008 the Southern Argentina Agile Meteor Radar (SAAMER) was installed at Rio Grande (Fritts et al., 2010; Janches et al., 2015b). In 2010, on the North Norwegian island of Andoya, the Middle Atmosphere Alomar Radar System (MAARSY) was developed (Stober et al., 2013). The main focus of this radar system is the investigation of the middle atmosphere in the polar regions. During the Geminid meteor shower in 2010, however, the system demonstrated the capabilities for extracting meteor dynamical parameters (Schult et al., 2017). Quite recently, a group from the Colorado Center for Astrodynamics Research at the University of Colorado, Boulder, installed a 36.17 MHz all-sky meteor radar at McMurdo Station, Antarctica (77.8 S, 166.7 E) that provides meteor measurements in the mesosphere and lower thermosphere.[4]

All new radar systems are extremely sensitive and provide millions of meteor echoes. Unfortunately, none of the orbital data are publicly accessible.

9.4.1.1 New Photographic and Video Techniques

At the turn of the century, new photographic and video techniques were used by several groups of professional and amateur meteor observers; these techniques are described in the following list.

- During the 1950s at the Ondřejov Astronomical Observatory (Academy of Science of Czech Republic), dedicated arrays of photographic cameras were put in operation, which later formed the European Fireball Network. Those first manually operating, low-resolution all-sky cameras have been systematically improved, and in 2001 were replaced by Autonomous Fireball Observatories (AFO) (Spurný et al., 2007). Currently, the photographic emulsion detectors of AFO have been replaced by digital versions (DAFO), (for more details, see Borovička et al., 2015).
- Observers at Western Ontario University in London, Canada pioneered TV meteor detection, which resulted in more than 500 meteor orbits being recorded (Hawkes et al., 1984; Jones and Sarma, 1985).
- The Ondřejov Observatory double-station video observation programme started in 1998. During the first 4 years of operation more than 1000 video orbits were obtained (Štork et al., 2002; Koten et al., 2003), including the first sample of 230 orbits from the Southern Hemisphere during a short campaign in New Zealand (Jopek et al., 2009, 2010).
- The International Meteor Organization (IMO) video meteor network, the main contributor to the EDMOND database, started its activity in 1999 (Molau, 2001). With about 80 cameras, between 1999 and 2017, the IMO observers recorded 3.5 million single-station meteors in over 6 400 observing nights and almost 900 000 hours of effective observing time (e.g., the year 2016 is referred to in Molau et al. (2016)).
- The Japanese meteor network SonotaCo has continued regular observations since 2007 with twenty-three observers and about eighty cameras being used (SonotaCo, 2009; SonotaCo et al., 2010). By the end of 2017 it had recorded and computed round about 255 000 meteor heliocentric orbits; see http://sonotaco.jp/doc/SNM/.
- The Fireball Recovery and InterPlanetary Observation Network (FRIPON) (Colas, 2016) is a new network equipped with about 100 all-sky cameras covering the sky over France. The main goal of the network is to recover meteorites following an observed fireballs. A few FRIPON cameras are placed in several other European countries.
- In California, USA, the Cameras for All-sky Meteor Surveillance (CAMS) started to operate in late 2010. During the first year of activity about 40 000 meteoroid orbits were obtained. Subsequently, the system expanded to include other parts of the world: BeNeLux and New Zealand (Jenniskens et al. 2011).
- The Southern sky is monitored by the Canary Island Long-Baseline Observatory system (CILBO) (Albin et al., 2017). This is operated by the Meteor Research Group of the European Space Agency and is part of the video camera system of the IMO.
- The Southern hemisphere is also observed by the Slovak group. The All-sky Meteor Orbit System (AMOS) has been installed in the Canary Islands and in the Atacama desert in Chile (Tóth et al., 2015; Tóth, 2016).
- Since 2007 a video meteor camera network has been monitoring the sky over Croatia (Croatian Meteor Network, CMN) (Korlević et al., 2013; Šegon et al., 2014). About 19 000 meteoroid orbits have been published spanning the years 2007–2013.
- In 2008 the NASA Meteoroid Environment Office established the NASA All Sky Fireball Network (Cooke and Moser, 2012). As of 2011, using 6 video cameras, this network detected 1 796 multi-station meteors.
- In Southern Ontario, Canada, the multi-instrumental Southern Ontario Meteor Network (SOMN) was developed (Weryk et al., 2008). An automated all-sky camera network has been constructed as one of its instrumental components – the ASGARD system. Instrumentation, software and early results are described in Brown et al. (2010b).

9.4.1.2 Regional Networks

In addition to those networks mentioned in the previous list, there are many regional meteor observation networks maintained by amateur and professional groups. In Europe, several local networks constitute a multi-national network — the European viDeo Meteor Observation Network (EDMONd) (Kornoš et al., 2013, 2014a,b; Koukal et al., 2014; Rudawska et al., 2015). As a result, a common orbital database was established. The local networks connected to the European viDeo MeteOr Network Database (EDMOND), are the following:

- BOAM (Base des Observateurs Amateurs de Météores), French amateur observers;

---

[4] See http://ccar.colorado.edu/meteors/, private communication from John Mario (2018).

- CEMeNt (Central European Meteor Network), a cross-border network of Czech and Slovak amateur observers;
- CMN (Croatian Meteor Network or Hrvatska Meteorska Mreža, Croatia);
- FMA (Fachgruppe Meteorastronomie, Switzerland);
- HMN (Hungarian Meteor Network/Magyar Hullócsillagok Egyesület), Hungarian amateur observers;
- IMO VMN (IMO Video Meteor Network);
- IMTN, Italian amateur observers in the Italian Meteor and TLE Networks;
- MeteorsUA (Ukraine);
- NEMETODE (Network for Meteor Triangulation and Orbit Determination, United Kingdom);
- PFN (Polish Fireball Network (Pracownia Komet i Meteorów, PKiM);
- Stjerneskud (Danish all-sky fireball cameras network, Denmark);
- SVMN (Slovak Video Meteor Network), Comenius University network;
- UKMON (UK Meteor Observation Network), a network of British amateur observers.

The EDMONd network also includes several individual observers from Bosnia and Herzegovina and Serbia, where an extensive network is being created. The CMN and IMO networks connected to EDMONd recently. More than 315 000 orbits were observed during the years 2001–2016, and these are accessible at https://fmph.uniba.sk/microsites/daa/daa/meteory/edmond/.

An open meteor database is also maintained by the Japanese SonotaCo group. About 255 000 orbits of meteoroids observed in the years 2007–2017 can be downloaded using http://sonotaco.jp/doc/SNM/.

A third open meteor database is provided by the CAMS system, http://cams.seti.org/. Currently it contains the orbits of meteors observed in California, USA and in the BeNeLux part of the network. A total of about 110 000 orbits can be downloaded.

9.4.2 Meteor Shower Searches

A search for meteor showers has been carried out on three large video databases, e.g., Jenniskens et al. (2011, 2016a); Jenniskens and Nénon (2016); Jenniskens et al. (2016b); Koseki (2014); Rudawska and Jenniskens (2014); Rudawska et al. (2015) and Hajduková et al. (2017). The data were used both for the study of individual meteor showers and also to classify sporadic meteors (Jopek and Williams, 2013). Many new meteor showers were discovered, and many old ones were confirmed. This resulted in the mean geocentric and heliocentric parameters of identified showers being submitted to the IAU MDC.

The crucial issue about the precision of the meteor orbits was raised (Spurný et al., 2017), and needs to be considered in the future meteor stream and parent bodies association. Otherwise, strong restrictive conditions on datasets need to be applied (Vereš and Toth, 2010).

9.4.3 The IAU MDC Meteor Shower Database

The shower database was formed in 2007 as a second component of the IAU Meteor Data Center – complementing the existing orbital part (Jenniskens, 2008b; Jopek and Jenniskens, 2011; Jopek and Kaňuchová, 2014, 2017). The MDC shower database is posted on the Web on two servers: one at The Institute Astronomical Observatory, Adam Mickiewicz University, Poznan, Poland (http://pallas.astro.amu.edu.pl/~jopek/MDC2007/) and the other on its mirror at the Astronomical Institute of the Slovak Academy of Sciences, Tatranska Lomnica, Slovakia (https://www.ta3.sk/IAUC22DB/MDC2007/).

The main objectives of the database are:

- to archive information on meteor showers, their geocentric and heliocentric parameters;
- to assign unique codes and names to the discovered meteor showers;
- in conjunction with the Working Group on Meteor Shower Nomenclature of IAU Commission F1, the shower database is used to formulate a descriptive list of the established meteor showers that can receive official names during the next IAU General Assembly.

9.4.4 The Naming of Meteor Showers

For the first time in the history of astronomy, meteor showers were officially named at the XXVIIth IAU General Assembly in Rio de Janeiro, Brazil, in August 2009. Sixty-four meteor showers were given official names (Jopek and Jenniskens, 2011). At the XXVIIIth General Assembly in Beijing, China, in 2012, a further thirty-one official names of meteor showers were approved (Jopek and Kaňuchová, 2014), with a further eighteen showers named officially at the XXIX General Assembly in Honolulu, USA, in 2015 (Jopek and Kaňuchová, 2017). The geocentric and heliocentric parameters of these named showers are given in Tables 9.2, 9.3 and 9.4. In Figures 9.1 and 9.2, the Hammer diagrams of their geocentric radiants are plotted using two reference systems.

At present (July 2018), 957 showers are archived and their parameters are grouped into four listings as follows: a list of all showers, a list of 112 established showers, a working list (820 showers), a list of 25 shower groups. They are supplemented by a list of ten removed showers. The content of all the lists can be displayed using a Web browser and the first four of them can be downloaded as ASCII files.

There are 373 showers that are represented by two or more sets of radiants and orbital parameters. Additional sets of parameters were collected from the literature. Some data were submitted by the MDC users. Multiple sets of shower parameters play important functions, apart from their cognitive importance; each independent set of parameters strongly confirms the existence of a particular meteor shower. The list contains 112 established showers. A shower can be moved from the working list to the list of established showers if: (1) it passes the verification procedure and (2) is given an official name, which has to be approved by the IAU F1 Commission business meeting. Occasionally an established meteor shower may be moved back to the working list. This happened at the IAU General Assembly in Honolulu when the Working Group on Meteor Shower Nomenclature proposed moving the established meteor shower No. 3 (SIA), shown in Table 9.2, back to the working list (for more details, see Jenniskens et al. (2015)). Identification of a reliable meteor shower is a complex task. The new results are compared with

Table 9.2. *Geocentric data of sixty-three showers officially named during IAU General Assembly XXVII, held in Rio de Janeiro, Brazil, in 2009. For each shower, the solar ecliptic longitude λ_S, the radiant right ascension and declination α_g, δ_g are given for J2000. The Southern iota Aquariids (#003/SIA) were removed from the original Table given in Jopek and Jenniskens (2011).*

No.	IAU No. & code		Stream name	λ_S (deg)	α_g (deg)	δ_g (deg)	V_g (km/s)
1	1	CAP	α Capricornids	127	306.6	-8.2	22.2
2	2	STA	Southern Taurids	224	49.4	13	28
3	4	GEM	Geminids	262.1	113.2	32.5	34.6
4	5	SDA	Southern δ Aquariids	125.6	342.1	-15.4	40.5
5	6	LYR	April Lyrids	32.4	272	33.3	46.6
6	7	PER	Perseids	140.2	48.3	58	59.4
7	8	ORI	Orionids	208.6	95.4	15.9	66.2
8	9	DRA	October Draconids	195.1	264.1	57.6	20.4
9	10	QUA	Quadrantids	283.3	230	49.5	41.4
10	12	KCG	κ Cygnids	145.2	284	52.7	24
11	13	LEO	Leonids	235.1	154.2	21.6	70.7
12	15	URS	Ursids	271	219.4	75.3	33
13	16	HYD	σ Hydrids	265.5	131.9	0.2	58
14	17	NTA	North. Taurids	224	58.6	21.6	28.3
15	18	AND	Andromedids	232	24.2	32.5	17.2
16	19	MON	December Monocerotids	260.9	101.8	8.1	42
17	20	COM	December Comae Berenicids	274	175.2	22.2	63.7
18	22	LMI	Leonis Minorids	209	159.5	36.7	61.9
19	27	KSE	κ Serpentids	15.7	230.6	17.8	45
20	31	ETA	η Aquariids	46.9	336.9	-1.5	65.9
21	33	NIA	North. ι Aquariids	147.7	328	-4.7	27.6
22	61	TAH	τ Herculids	72	228.5	39.8	15
23	63	COR	Corvids	94.9	192.6	-19.4	9.1
24	102	ACE	α Centaurids	319.4	210.9	-58.2	59.3
25	110	AAN	α Antliids	313.1	140	-10	42.6
26	137	PPU	π Puppids	33.6	110.4	-45.1	15
27	144	APS	Daytime April Piscids	30.3	7.6	3.3	28.9
28	145	ELY	η Lyrids	49.1	292.5	39.7	45.3
29	152	NOC	North. Daytime ω Cetids	46.7	2.3	17.8	33
30	153	OCE	South. Daytime ω Cetids	46.7	22.5	-3.6	36.6
31	156	SMA	South. Daytime May Arietids	55	33.7	9.2	28.9
32	164	NZC	North. June Aquilids	86	298.3	-7.1	36.3
33	165	SZC	South. June Aquilids	80	297.8	-33.9	33.2
34	170	JBO	June Bootids	96.3	222.9	47.9	14.1
35	171	ARI	Daytime Arietids	76.7	40.2	23.8	35.7
36	172	ZPE	Daytime ζ Perseids	78.6	64.5	27.5	25.1
37	173	BTA	Daytime β Taurids	96.7	84.9	23.5	29
38	183	PAU	Piscis Austrinids	123.7	347.9	-23.7	44.1
39	187	PCA	ψ Cassiopeiids	106	389.4	71.5	40.3
40	188	XRI	Daytime ξ Orionids	117.7	94.4	15	44
41	191	ERI	η Eridanids	137.5	45	-12.9	64
42	198	BHY	β Hydrusids	143.8	36.3	-74.5	22.8
43	206	AUR	Aurigids	158.7	89.8	38.7	65.7
44	208	SPE	September ε Perseids	170	50.2	39.4	64.5
45	212	KLE	Daytime κ Leonids	181	162.7	15.7	43.6
46	221	DSX	Daytime Sextantids	188.4	154.5	-1.5	31.2
47	233	OCC	October Capricornids	189.7	303	-10	10
48	246	AMO	α Monocerotids	239.3	117.1	0.8	63
49	250	NOO	November Orionids	245	90.6	15.7	43.7
50	254	PHO	Phoenicids	253	15.6	-44.7	11.7
51	281	OCT	October Camelopardalids	193	166	79.1	46.6
52	319	JLE	January Leonids	282.5	148.3	23.9	52.7
53	320	OSE	ω Serpentids	275.5	242.7	0.5	38.9
54	321	TCB	θ Coronae Borealids	296.5	232.3	35.8	38.66
55	322	LBO	λ Bootids	295.5	219.6	43.2	41.75
56	323	XCB	ξ Coronae Borealids	294.5	244.8	31.1	44.25
57	324	EPR	ε Perseids	95.5	58.2	37.9	44.8
58	325	DLT	Daytime λ Taurids	85.5	56.7	11.5	36.4
59	326	EPG	ε Pegasids	105.5	326.3	14.7	29.9
60	327	BEQ	β Equuleids	106.5	321.5	8.7	31.6
61	328	ALA	α Lacertids	105.5	343	49.6	38.9
62	330	SSE	σ Serpentids	275.5	242.8	-0.1	42.67
63	331	AHY	α Hydrids	285.5	127.6	-7.9	43.6

Table 9.3. *Geocentric data for thirty-one showers (streams) officially named during the IAU General Assembly XXVIII, held in Beijing, China, in 2012. The solar ecliptic longitude λ_S, the geocentric radiant right ascension and declination α_g, δ_g are given for the epoch J2000.0. (Source: Jopek and Kaňuchová, 2014).*

No.	IAU No. & code		Shower (stream) name	λ_S (deg)	α_g (deg)	δ_g (deg)	V_g (km/s)
1	11	EVI	η Virginids	354	182.1	2.6	29.2
2	23	EGE	ε Geminids	206	101.6	26.7	68.8
3	26	NDA	Northern δ Aquariids	123.4	344.7	0.4	40.5
4	100	XSA	Daytime ξ Sagittariids	304.9	284.8	−18.6	26.3
5	128	MKA	Daytime κ Aquariids	354	338.7	−7.7	33.2
6	151	EAU	ε Aquilids	59	284.9	15.6	30.8
7	175	JPE	July Pegasids	107.5	340	15	61.3
8	184	GDR	July γ Draconids	125.3	280.1	51.1	27.4
9	197	AUD	August Draconids	142	272.5	65.1	17.3
10	202	ZCA	Daytime ζ Cancrids	147	119.7	19	43.8
11	242	XDR	ξ Draconids	210.8	170.3	73.3	35.8
12	252	ALY	α Lyncids	268.9	138.8	43.8	50.4
13	257	ORS	Southern χ Orionids	260	78.7	15.7	21.5
14	333	OCU	October Ursae Majorids	202	144.8	64.5	54.1
15	334	DAD	December α Draconids	256.5	207.9	60.6	41.6
16	335	XVI	December χ Virginids	256.7	186.8	−7.9	67.8
17	336	DKD	December κ Draconids	250.2	186.0	70.1	43.4
18	337	NUE	ν Eridanids	167.9	68.70	1.1	65.9
19	338	OER	o Eridanids	234.7	60.70	−1.5	26.9
20	339	PSU	ψ Ursae Majorids	252.9	167.8	44.5	60.7
21	341	XUM	January ξ Ursae Majorids	300.6	169.0	33.0	40.2
22	346	XHE	x Herculids	352	254	48	36
23	348	ARC	April ρ Cygnids	37.0	324.5	45.9	41.8
24	372	PPS	ϕ Piscids	106.0	20.1	24.1	62.9
25	388	CTA	χ Taurids	220.0	63.2	24.7	42.1
26	390	THA	November θ Aurigids	237.0	89	34.7	33.8
27	404	GUM	γ Ursae Minorids	299.0	231.8	66.8	31.8
28	411	CAN	c Andromedids	110	32.4	48.4	59
29	427	FED	February η Draconids	315.11	239.92	62.49	35.6
30	445	KUM	κ Ursae Majorids	223.21	144.46	45.44	65.30
31	446	DPC	December ϕ Cassiopeiids	252.48	19.8	58.0	16.4

Table 9.4. *The geocentric radiant and heliocentric orbital data for eighteen meteor showers (streams) officially named at the IAU General Assembly XXIX, held in Honolulu, USA, in 2015. The solar ecliptic longitude λ_S at the time of shower maximum activity, the geocentric radiant right ascension and declination α_g, δ_g are given for the epoch J2000.0. (Source: Jopek and Kaňuchová, 2017).*

No.	IAU No. and code		Shower (stream) name	λ_S (deg)	α_g (deg)	δ_g (deg)	V_g (km/s)
1	21	AVB	α Virginids	32.0	203.5	2.9	18.8
2	69	SSG	Southern μ Sagittariids	86.0	273.2	−29.5	25.1
3	6	NCC	Northern δ Cancrids	296.0	127.6	21.5	27.2
4	97	SCC	Southern δ Cancrids	296.0	125.0	14.4	27.0
5	343	HVI	h Virginids	38.0	204.8	−11.5	17.2
6	362	JMC	June μ Cassiopeiids	74.0	17.5	53.9	43.6
7	428	DSV	December σ Virginids	262.0	200.8	5.8	66.2
8	431	JIP	June ι Pegasids	94.0	332.1	29.1	58.5
9	506	FEV	February ε Virginids	314.0	200.4	11.0	62.9
10	510	JRC	June ρ Cygnids	84	321.8	43.9	50.2
11	512	RPU	ρ Puppids	231.0	130.4	−26.3	57.8
12	524	LUM	λ Ursae Majorids	215.0	158.2	49.4	60.3
13	526	SLD	Southern λ Draconids	221.6	163	68.1	48.7
14	529	EHY	η Hydrids	256.9	132.9	2.3	62.5
15	530	ECV	η Corvids	302	192.0	−18.1	68.1
16	533	JXA	July ξ Arietids	119	40.1	10.6	69.4
17	549	FAN	49 Andromedids	112.0	20.5	46.6	60.2
18	569	OHY	o Hydrids	309	176.3	−34.1	59.1

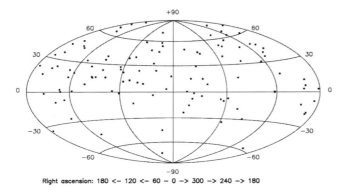

Figure 9.1. The Hammer-Aitoff diagram of 112 geocentric radiants of meteor showers listed in Tables 9.2, 9.3 and 9.4. The equatorial reference system is used. Eighty-four radiants are located on the northern celestial hemisphere.

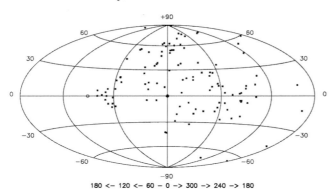

Figure 9.2. The Hammer-Aitoff diagram of 112 geocentric radiants of meteor showers listed in Tables 9.2, 9.3 and 9.4. The ecliptic Earth centered coordinates frame is used. The central point of the diagram corresponds to the position of the apex of the Earth motion. On the left the Sun position (helion) is given. Forty-one radiants are located on the helion hemisphere.

those obtained in the past. However, the complex structure of meteoroid streams and their variable dynamical evolution makes it difficult to compare data obtained at different times, hindering the task of finding a solution. Therefore, the changing of the status of a meteor shower should be considered as something not to be undertaken lightly. Fortunately, such changes happen only rarely.

In the period between two General Assemblies, a proposal can be made that a shower from the working list becomes an established shower. The nominated candidates are discussed and the list of their names is completed shortly before the IAU GA, at which they are proposed for approval.

The working list contains 691 showers that have already been published in the scientific journals or in the IMO WGN Journal, and 129 new submissions with the "pro-tempore" flag. The "pro-tempore" flag is removed when the paper describing new submission is accepted for publication.

A shower can be removed from the MDC database on the basis of the published analysis. The list of removed showers is a kind of historical archive. Ten showers are currently in this list. Showers are shifted to the list of removed showers mostly because they have proven to be duplicates.

9.4.5 Future Prospects

The rapid progress in the acquisition of meteor video data is very important from the point of view of our knowledge about the small components of the Solar System. However, some problems have been encountered. For example, Hajduková et al. (2017) compared the video meteor orbits of the Geminid meteor stream in the CAMS, SonotaCo and EDMOND catalogues, as well as Czech and Slovak observations, and concluded that the determined velocities of all video databases, except for the Ondřejov (Czech) data, are underestimated, probably as a consequence of the methods used for the astrometric and velocity measurements. The largest discrepancies were detected for the velocities recorded in the EDMOND and SonotaCo catalogues (see figure 4 in Hajduková et al. (2017)).

Some of the orbital samples have different statistical properties from those obtained by radio techniques. As an example, see figure 1 in Jopek and Bronikowska (2017). For the Harvard synoptic year sample, 19.9% of meteoroids moved on retrograde orbits; for the SonotaCo 2009 sample, 64.9% were retrograde. It is possible that the discrepancy may be explained by observational selection effects, or may be removed by more careful calibration of individual systems. Also, one should remember that the population of meteoroids observed by video and radar techniques may be different at different limiting masses.

The IAU MDC shower database is certainly not perfect – it contains some erroneous data, mistakes and various shortcomings caused by the missed recordings. However, erroneous data can be also found in the literature from which the information was derived. The meteor parameters given in the MDC are not homogeneous. They originate from different epochs and they were determined from visual, photographic, video and radar observations. Thus, the uncertainties of the parameters are different.

As far as possible, the MDC shower database will be improved. At first, the archiving of the original source meteor data found in literature will be completed. Following coments sent to the MDC by users (e.g., by Andreić et al., 2014; Koseki, 2016; Mulof, 2018), the database will be supplemented by additional information about given showers such as missing orbital elements, periods of shower activity, date of the new shower submission, etc. Also, it is projected to implement software that will make the usage of the MDC database easier.

9.5 Method for Classifying Sporadics

9.5.1 One-Parameter Classification

Sporadic meteoroids cannot be associated with a single parent body. We can classify them only as of cometary or asteroidal origin, and a number of criteria have been devised to quantify this.

In order to discriminate between the orbits of comets and asteroids (the C-A classification), Whipple (1954) proposed using an empirically founded K-criterion,

$$K = \lg\left[a(1+e)/(1-e)\right] - 1, \qquad (9.18)$$

where a is the semi-major axis in $[au]$, e the eccentricity and the logarithm is to base 10. The quantity in the square bracket is the inverse square of the orbital velocity at the aphelion point. When

$K \geq 0$, the orbit is of a cometary type, while negative values indicate an asteroidal orbit. Using this K-criterion, Whipple (1954) found that 96% of known comets and 99.8% of known asteroids were correctly classified. The criterion can also be applied to meteoroids, and Whipple classified 90% of 144 bright photographic meteors as being of cometary origin.

The K-criterion is an empirical one; it has no fundamental physical basis, and for short-period, low-eccentricity orbits it gives inconclusive results. For example, whereas for the Příbram and Neuschwanstein meteorites $K \approx 0.08$, most authors would agree that these meteorites either have originated in the main asteroid belt and were perturbed into Earth-crossing orbits or are pieces of Near-Earth asteroids.

In order to classify sporadic meteoroids, several other criteria can be used. The T-criterion is based on the Tisserand invariant and is given by

$$T = a^{-1} + 2a_J^{-1.5}\left[a(1-e^2)\right]^{(0.5)} \cos I. \qquad (9.19)$$

Here a_J and a are the semi-major axis of Jupiter's orbit and the meteoroid orbit, both in au, e is the eccentricity and I the inclination of the meteoroid orbit relative to the orbital plane of Jupiter. Kresák (1969) used the condition $T < 0.58$ to define a comet-type orbit crossing Jupiter's orbit. Kresák (1967, 1969) also proposed two additional discriminants, the P- and Q-criterion, defined by

$$P = k^2 a^{1.5} e, \qquad (9.20)$$

where k is the Gauss gravity constant. For a cometary orbit the condition $P > 2.5$ is used.

$$Q = a(1+e) \qquad (9.21)$$

with $Q > 4.6$ [au] for a cometary orbit. Q is an aphelion distance and so this condition simply requires that the orbit does not go beyond the asteroidal belt.

One additional discriminant, the E-criterion (the orbital energy E, Jopek and Williams, 2013) is given by

$$E = \frac{-k^2}{2a} \qquad (9.22)$$

with $a > 2.8$ [au] for cometary orbits.

Except for the Tisserand invariant, all the criteria discussed in this section have been derived on an empirical basis, that is, by finding a measure by means of which, in general, the known comets generate a different value from the known asteroids.

All the criteria were adjusted and tested so as to minimise the number of exceptions. At the time of Whipple the K-criterion gave a very good result. However, a half-century later, with a growing number of cometary and asteroidal samples, the efficiency of the K-criterion was clearly decreased. In Table 9.5, the result obtained by Jopek and Williams (2013) has shown that the best one-parameter C-A classifiers are the Q- and E-criteria. The orbits of sporadic meteoroids have been classified by several authors using these criteria. The results obtained were different, depending on the orbital sample and the criterion used; for details see Starczewski and Jopek (2004) or Jopek and Williams (2013). The common result was that the majority of photographic and video meteoroids were of a cometary type. For the radio meteoroids, the opposite was true. The difference was

Table 9.5. *Comets and NEAs one-parameter orbital classification. The number of exceptions (in %) not obeying the cometary and asteroidal orbital K-, P-, T-, Q-, E- criteria amongst 7830 NEAs and 780 periodic comets. (Source: Jopek and Williams (2013).*

Q [%]	E [%]	T [%]	P [%]	K [%]	
3.9	3.8	7.2	10.7	16.4	NEAs
1.0	1.5	2.2	5.8	13.8	Comets

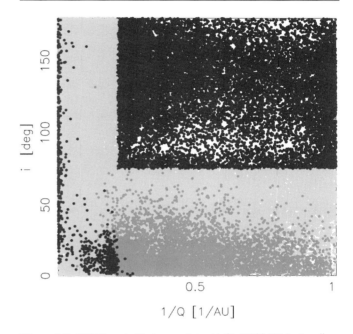

Figure 9.3. 780 Comets (dark grey dots at left), 7830 NEAs (medium grey dots and area at bottom) and 60 412 sporadic meteoroids (light grey area) on the i-$1/Q$ plane. Additionally, the set of 17 457 "asteroidal" meteoroids for which inclinations $i > 75°$ and aphelia distances $Q < 4.6$ [au] forms the grey rectangle at upper right. This rectangle contains 22% of our sporadic sample. (A black and white version of this figure will appear in some formats. For the colour version, please refer to the plate section.)

significant: 23% of all radio meteoroids and 73% of all video meteoroids moved on cometary orbits.

9.5.2 Two-Parameter Classification

However, Jopek and Williams (2013) have shown that using the one-parameter criteria can result in the fractions of meteoroids on cometary orbits being understated by up to 29%. For photographic meteoroids, the underestimation is much less, 2–5%. In the case of radio data, depending on the C-A criterion used, the underestimation can reach 15–29%. This bias is a consequence of the fact that among the sporadic meteoroids there are many asteroidal orbits ($Q < 4.6$ [au]) with high inclinations ($i > 75°$); see Figure 9.3. In this figure we produce a plot of $1/Q$ against inclination for all the sporadics in our dataset as well as relevant NEAs and comets. According to the Q-criterion, all meteoroids for which $Q < 4.6$ should be regarded as moving on asteroidal type orbits. However, in Figure 9.3, there are many "asteroidal"

Table 9.6. *The proportion of cometary orbits among 77 869 sporadic meteoroids. The percentages of meteoroids among the whole sporadic component, the radar, video and photographic meteors are given. The top and bottom sections of Table this contain the results of two and one C-A parameter classifications, respectively. More detailed results one can found in Jopek and Williams (2013).*

Q [%]	E [%]	Sample size
66.4	66.9	77 869 all meteoroids
51.3	49.9	45 539 radio meteoroids
88.5	87.2	30 899 video meteoroids
68.0	64.4	1431 photo meteoroids
44.0	41.8	77 869 all meteoroids
23.4	21.4	45 539 radio meteoroids
73.4	71.0	30 899 video meteoroids
65.5	59.6	1431 photo meteoroids

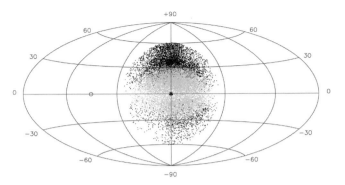

Figure 9.4. The Hammer-Aitoff diagram of 17 457 radiants of sporadic meteors plotted in the ecliptic Earth apex centred reference frame. The orbits of all the plotted meteoroids have aphelia $Q < 4.6$ [au] and inclinations $i > 75°$. The central point of the diagram corresponds to the position of the apex of the Earth motion. Two regions are seen: the north and south apex (both in the center and light grey) and toroidal concentrations (dark grey). (A black and white version of this figure will appear in some formats. For the colour version, please refer to the plate section.)

meteoroids for which $i > 75°$. But at the time of this study, only two known NEAs ('343158' and 2014PP 69) had orbital inclinations $i > 75°$. On the other hand, there is not a single comet in this region. Therefore, the source of all sporadic meteoroids plotted in the top right corner in Figure 9.3 cannot be established by the one-parameter C-A classification.

Assuming that the comet and NEAs populations are disjoint, the Q-i and E-i two-parameter criteria were proposed as more reliable tools for dynamical discrimination of the NEOs population, namely, for cometary meteoroids:

$$Q = a(1+e) > 4.6 \text{ [au]}, \quad \text{or} \quad i > 75° \quad (9.23)$$

$$E = \frac{-k^2}{2a} > -5.28 \cdot 10^{-5}, \quad \text{or} \quad i > 75° \quad (9.24)$$

where $a = 2.8$ [au].

Using those criteria, Jopek and Williams (2013) have found that among all $\sim$78 000 sporadic meteors studied, those on orbits of cometary type predominate, the fraction of such orbits being 66–67%. The fraction can differ significantly from one orbital sub-sample to the other. In Table 9.6 we get these results together and also, for comparison, the results of the C-A one-parameter classification. In Table 9.6 the values of the percentages can change slightly because of uncertainties in the meteoroid orbital elements. However, as was certified in Starczewski and Jopek (2004), the percentages given in Table 9.6 can differ at most by ±2% due to such uncertainties.

9.5.3 Cometary Origin of Sporadic Meteoroids with $Q < 4.6$ [au] and $i > 75°$

The sporadic meteoroids moving on the orbits with $Q < 4.6$ [au] and $i > 75°$ (see Figure 9.3) can be related with the group of meteor radiants clearly seen on the Hammer-Aitoff diagram made in the Earth apex rotating reference frame. Recent examples of such diagrams can be found in Galligan and Baggaley (2005); Chau et al. (2007) and Campbell-Brown (2008). On such diagrams the radiants of sporadic meteoroids clearly form six concentrations: the helion and antihelion, the north and south apex concentrations, and the north and south toroidal ones (for example, see Campbell-Brown, 2008). In Figure 9.4 we refer to such a diagram on which, one can distinguish the regions corresponding to north and south apex and toroidal concentrations occupied solely by the meteoroids for which $Q < 4.6$ [au] and $i > 75°$. All these particles occupy the upper right of Figure 9.3.

Referring to the papers by Davies (1957), Dycus and Bradford (1965), Jones et al. (2001), Wiegert et al. (2009) and Nesvorný et al. (2011) – Jopek and Williams (2013) have shown that the meteoroids lying in the toroidal and apex regions in Figure 9.4 may originate from a disruption of long-period comets or Oort Cloud comets.

This supports the statement that the two-parameter C-A criteria given by conditions (9.23) and (9.24) are sufficient to match sporadic meteoroids to the appropriate parent body population. However, due to insufficient information, we are not able to make any inferences regarding the real fractions of cometary and asteroidal sporadic meteoroids in the Solar System. Our results are concerned only with the observed samples. In general, the fractions of meteoroids on cometary orbits can differ significantly from one orbital sub-sample to another obtained by different techniques. A general trend was observed – the fraction of cometary orbits decreases with decreasing size of the observed meteors.

9.6 Sporadic Meteoroids and the Zodiacal Cloud

At distances up to a few astronomical units from the Sun, the inner part of the Solar System is filled with dust from submicrometer-sized grains to centimetre-sized meteoroids of both cometary and asteroidal origin. The dust is concentrated within a thick, near-ecliptic circumsolar disk called the zodiacal cloud. Light from the Sun in the UV, optical, and near-infrared is scattered by dust particles. It can also be absorbed and thermally emitted in the mid- to far-infrared range. The zodiacal light is the dominant source of diffuse sky brightness throughout the infrared (Reach, 1997).

The zodiacal light is visible on a clear and moonless dark night sky. Since the seventeenth century, the cause has been suspected to be an interplanetary dust cloud (Cassini, 1685). Observed from Earth, the disc appears as a strip along the zodiac, hence the name zodiacal light. It is best seen either after sunset or before sunrise as a light cone extending with a diminishing brightness with increasing elongation from the Sun along the ecliptic (e.g. Leinert, 1975). At angular distances smaller than about 20°, where the zodiacal light is not observable, it continues as the F-corona (Grotrian, 1934; Leinert, 1975; Kimura and Mann, 1998). A faint glow of oval shape in the anti-solar direction, known as the Gegenschein, is backscattered light from dust particles of the zodiacal cloud beyond the Earth's orbit (e.g. Ishiguro et al., 2013).

9.6.1 Observations of the Zodiacal Cloud

Observations of the zodiacal light have been carried out using many different methods. Scattered light observations come from the Clementine mission (Hahn et al., 2002) and the Wide-field Imager of Zodiacal light with ARray Detector (WIZARD; Usui et al., 2002). Thermal infrared observations are from the Infrared Astronomical Satellite (IRAS; Low et al., 1984), the Cosmic Background Explorer (COBE; Reach et al., 1995; Kelsall et al., 1998), the Midcourse Space Experiment (MSX; Price et al., 2003), the Infrared Space Observatory (ISO; Fixsen and Dwek, 2002; Leinert et al., 2002; Reach et al., 2003; Müller et al., 2005), the Spitzer Space Telescope (Bhattacharya and Reach, 2004; Reach et al., 2007), the Planck mission (Ade et al., 2014) and the AKARI infrared satellite (Murakami et al., 2007). Ultraviolet observations were obtained by the Faint Object Spectrograph on board the Hubble Space Telescope (Kawara et al., 2017).

Optical and infrared observations mainly cover dust particles in the size range from 1 to 100 μm. The particles mostly originate from the collisional evolution of larger meteoroid particles that were produced through cometary activity and collisions of asteroids. The lifetime of the dust particles is limited by their collisional evolution and also by the Poynting–Robertson effect, which causes them to slowly drift towards the Sun (Robertson, 1937; Burns et al., 1979). The larger particles undergo collisional disruption to become smaller fragments that are progressively more affected by Poynting–Robertson drag (Nesvorný et al., 2010). Meteoroid lifetimes were estimated by Grun et al. (1985), Steel and Elford (1986) and more early by Tokhtas'ev (1982). The estimates are based on the models of the size frequency and velocity distribution in the interplanetary environment that are constrained by in situ observations. Collisional effects are examined by considering catastrophic collisions, collision rate and generated fragments of different masses and sizes (Grun et al., 1985). Studies of particles on eccentric cometary orbits (JFCs and HTCs) (Nesvorný et al., 2011; Pokorný et al., 2014) showed that the dynamical transport of meteoroids in the millimeter size range to the vicinity of the Earth's orbit requires a longer lifetime than that predicted in Grun et al. (1985). Using the ESA's Interplanetary Meteoroid Environment Model (IMEM) (Dikarev et al., 2005), Soja et al. (2016) calculated the collisional lifetime as a function of mass for six potential parent bodies of meteoroid streams (see Figure 2 in Soja et al., 2016).

The spatial distribution of the zodiacal cloud can be deduced from the brightness data. The observed brightness produced by solar radiation scattered by the dust, as well as by thermal radiation emitted from dust particles, is derived along the line of sight for a given longitude ($\lambda - \lambda_S$) and latitude (β), where λ_S is the longitude of the Sun. The brightness results from the integrated signal at a given wavelength in a given volume element in space. Observations yield a data set of the visible zodiacal light brightness from about 15° longitude of the Sun as well as infrared observation, usually at elongation larger than 60° (Mann et al., 2006; Mann, 1998, and references therein). The observed brightness depends on the spatial distribution of the dust and on the scattering or thermal emission properties in each volume element (e.g. Dumont and Levasseur-Regourd, 1985; Kelsall et al., 1998). The data are then inverted to the spatial distribution and orbital elements distribution of the cloud under reasonable assumptions about physical properties of dust particles, size distribution, scatter function and collisional models.

Space-based observations, supplemented by some ground-based observations of the zodiacal light (in both the visible and infrared range) have revealed a complex structure of the zodiacal light, which, to a first approximation, can be considered as a very smooth brightness distribution (e.g. Leinert, 1975; Dumont and Levasseur-Regourd, 1978; Leinert et al., 1982). Under this approximation, the cloud can be considered nearly symmetric about the plane of maximum density, with an overall rotational symmetry with respect to an axis perpendicular to this plane. The plane of symmetry of this cloud is close to the ecliptic, with an inclination of about 1–3° (see, for instance, Leinert et al., 1980; James et al., 1997) while the orientation of the plane is defined by the ascending node Ω_A. Moreover, the centre of the cloud is displaced from the Sun (Dermott et al., 1995; Reach, 1997). Observations at different elongations from the Sun and different wavelengths show that the inclination of the symmetry plane and its ascending node are a function of heliocentric distance. A warped plane of symmetry was suggested by Misconi (1980) and supported by others (e.g. Gustafson and Misconi, 1986; Rowan-Robinson et al., 1990) and more recently by Stenborg and Howard (2017). These features are produced by gravitational perturbations from Jupiter. Secular changes in the eccentricitiy of particles in the disk affect the orbital decay rate, and this in turn influences the radial size distribution of material in the disk. The distribution of the pericentres of particles is biased towards the orientation in the disk, which is defined by the forced longitude of perihelion imposed by the perturbers (Jupiter, P-R drag) (Dermott et al., 1985, 1998). The distribution of inclinations of these particles' orbits is dependent on the heliocentric distance, which results in the disk appearing warped (Dermott et al., 2001). Other structures were identified in the zodiacal cloud, asteroidal dust bands (Sykes, 1990), cometary dust trails (Davies et al., 1984; Sykes and Greenberg, 1986) and resonant structures (Dermott et al., 1994; Reach et al., 1995; Jones et al., 2013).

9.6.2 Models of the Zodiacal Cloud

Models of the zodiacal cloud dust distribution have traditionally been created to fit the observations, for example, the amount of scattered light, the Doppler shifts of the solar Mg I Fraunhofer line, and to interpret thermal emission observed in various lines of sight (Nesvorný et al., 2010, and references therein). These phenomenological models can explain the size, spatial and velocity distribution of dust particles in the Solar System, but are insufficient to explain basic questions related to the origin of the zodiacal cloud, its temporal brightness variability, and the origin of dust particle influx into the Earth's atmosphere (Nesvorný et al., 2010). More helpful in this case are detailed dynamical models that consider physical properties of dust particles, such as density, geometric albedo, elemental composition, mineralogy, tensile strength, heat capacity, and so on (e.g. Levasseur-Regourd et al., 2007; Yang and Ishiguro, 2015). Individual dust particles are numerically integrated under the influence of planetary perturbations and non-gravitational forces (solar radiation and solar wind). They also consider various processes such as collision, sputtering and sublimation that change the physical properties of the dust particles during the evolution from their sources.

Nesvorný et al. (2010) generated a new model based on the dynamical evolution of dust particles originating from different sources, and investigated what fraction of the zodiacal cloud is of cometary or asteroidal origin. The source populations included in the model are asteroids, Jupiter-family comets (JFCs), Halley-type comets (HTCs) and long-period Oort-cloud comets (OCCs) (for more details about orbital distribution and lifetime of the sources, see Nesvorný et al. (2010) and references therein). A contribution from interstellar dust particles (supposed to be of diameter < 1 µm) was not considered because observations of the mid-IR spectrum of the zodiacal cloud by ISOCAM/ISO (Reach et al., 2003) suggest that a substantial part of the zodiacal cloud is produced by $\sim$ 10–100 µm particles.

The model is primarily constrained by the latitudinal (ecliptic) distribution of the IR spectrum obtained with IRAS, but is consistent also with other zodiacal observation, meteor observations, spacecraft collections of dust and with properties of recovered micrometeorites collected on Earth. Nesvorný et al. (2010) found that 85–95% of the IR emission of the zodiacal cloud is produced by dust in the inner solar system originally from JFCs; less than 10% of the IR emission comes from dust from long-period comets and less than 10% from asteroids. Dust particles coming from JFCs have masses in the range of 1–10 µg, with a typical diameter of $\sim$100 µm (Nesvorný et al., 2010). The results are consistent with those in Hahn et al. (2002) based on the Clementine spacecraft observations, and similar results are given by Ueda et al. (2017) from measurement of the Japanese infrared astronomical satellite AKARI. They conclude that zodiacal dust consists of less than 10% asteroidal dust and more than 90% cometary dust. The small fractional of asteroid contribution is also supported by the fact that whereas the mid-infrared emission from particles produced in the asteroid belt mostly does not exceed latitudes of about 30° of the ecliptic, strong thermal emission is also observed at high latitudes (Nesvorný et al., 2010).

Rowan-Robinson and May (2013) modelled the infrared emission from zodiacal dust detected by the IRAS and COBE missions. Three distinct sources of dust were considered to contribute to the model: asteroidal dust originating between 1.5 and 3.1 au, cometary dust extending from 1.5 au out to large distances, and interstellar dust extending through the whole Solar System. The authors obtained very good fits to the combined IRAS and COBE data and found the relative contributions of cometary, interstellar and asteroidal dust to the density of the smooth part of the cloud at 1.5 au are 70.4%, 7.5% and 22.2%. The cometary and asteroidal contribution are quite consistent with those of Nesvorný et al. (2010, 2011). However, the contribution of the interstellar component is quite high.

According to Nesvorný et al. (2010), dust particles coming from Jupiter Family Comets (JFC) have a typical diameter $D \sim$ 100 µm. Their orbits are circularised by the Poynting–Robertson effect before they reach 1 au. Thus, they impact Earth with relatively low speeds ($v < 20$ km s^{-1}) and low eccentricities. This is not in agreement with observations of the meteor sporadic background described by Wiegert et al. (2009). The helion/antihelion meteoroids have a measured impact speed distribution that peaks at $v \sim$ 20–30 km s^{-1}, $a \sim$1 au with a tail to 3 au, $e > 0.3$, and low inclinations (Galligan and Baggaley, 2005; Campbell-Brown, 2008). This and other arguments led Nesvorný et al. (2011) to modify their previous model. They added (1) a perihelion distance (q)-dependent meteoroid production rate, taking into account orbits with $q < 1.5$ au, (2) a continuous size frequency distribution of particles instead of the original delta functions, and (3) including a proper collisional lifetime of JFC particles and more precise data on collisional disruption in space (Nesvorný et al., 2011). These improvements resulted in a revised total mass input on Earth of 32 tons day^{-1} ($\pm$50%) representing 50%–70% of the total input. Moreover, the mass input required to keep the zodiacal cloud in a steady state is estimated to be $\sim$10 times larger than found in Nesvorný et al. (2010) because particles released closer to the Sun have shorter collisional lifetimes and have to be replenished at a faster rate (Nesvorný et al., 2011). This new estimate supports the suggestion of Nesvorný et al. (2010) that the spontaneous disruption of JFCs is the main mechanism that supplies particles into the zodiacal cloud because the observed usual cometary activity (sublimation of ices and dust release) of JFCs cannot provide the necessary amount of dust.

Nesvorný et al. (2010, 2011) conclude that the dust on near-prograde low-eccentricity orbits impact Earth at a mean geocentric speed of about 14 km s^{-1}. Therefore radars cannot detect small particles entering the atmosphere at low velocities due to the relatively low production of electrons. This is in general disagreement with decades of observations using ground-based radars (Janches et al., 2015a). Janches et al. (2014) developed a new probabilistic approach to estimate the sensitivity of the Arecibo Observatory 430 MHz radar to detect such particles in the form of meteor head echoes as a function of particle mass, incoming velocity and entry angle. The authors integrated and employed existing comprehensive models of meteoroid ablation and ionisation, including the Chemical Ablation Model (CABMOD) developed by Vondrak et al. (2008), for an accurate interpretation of radar observations. A similar approach was applied to two other High Power and Large Aperture radars (HPLA), the 440 MHz Poker Flat Incoherent Scatter Radar and the 46.5 Middle and Upper Atmosphere radar (Janches et al., 2015a). The main result is that even with the least sensitive MU radar system, the current zodiacal cloud model of Nesvorný et al. (2011) over-predicts the radar observations of detected rates and velocity distributions. Moreover, the new IR measurements by the Planck satellite (Ade et al., 2014) suggest that the particles in the zodiacal cloud could be of smaller sizes than those considered in the model of Nesvorný et al. (2011). Considering also this fact, Janches et al. (2015a) conclude that finding an agreement between the current zodiacal cloud model and the mentioned three radars (Arecibo Observatory radar and two HPLA radars) requires a re-examination of our knowledge of radar detection biases and also the physical assumptions of the zodiacal cloud model itself.

9.6.3 Dust Bands and Dust Trails in the Zodiacal Cloud

The Infrared Astronomical Satellite (IRAS) observations in 1983 revealed the existence of several zodiacal dust bands (Low et al., 1984). These bands are bright infrared strips produced by thermal emission from circumsolar rings of particles. The emission structures are located at a few roughly constant ecliptic latitudes and were later confirmed by the Cosmic Background Explorer satellite (COBE; Spiesman et al., 1995) and by a ground-based telescope (Ishiguro et al., 1999).

The dust bands can be explained as a product of asteroid collision within the Hirayama (1918) asteroid families. It is supposed that the dust bands originate from debris produced by relatively recent disruption among main belt asteroids (Sykes and Greenberg, 1986; Nesvorný et al., 2003) within about 10 million years. The collision debris is distributed along the mean inclination of the asteroid family (or collided cluster) and due to differential precession, the longitudes of ascending node of the particles are randomised within about 10^6 years, resulting in a pair of bands at inclinations $\pm i$ stretching over the all ecliptic longitudes in the infrared sky map (Dermott et al., 1984; Sykes and Greenberg, 1986). Initially, the near-ecliptic dust bands were associated with the Themis and Koronis families, and the 10° band with the Eos family (Dermott et al., 1984; Grogan et al., 1997; Reach, 1997).

Nesvorný et al. (2002, 2003, 2008) studied the connection between dust bands and young asteroid families in proper orbital elements (a_p, e_p, i_p) space. Proper elements are time-averaged orbital elements from which short and long-period perturbations are eliminated. So they are nearly constant in time (Knezevic et al., 2002, and references therein). Nesvorný et al. (2002) discovered a young compact asteroid cluster – the Karin cluster – inside the Koronis family and proposed its connection with the dust band at inclination $i_p = 2.1°$ and $a_p = 2.865$ au. They found that the Karin cluster was formed 5.8 ± 0.2 Myr ago by the collisional disruption of a ~ 25 km diameter parent body. The dust bands with the inclination 9.35° originate from the Veritas asteroid family located at 3.17 au, which comes from a more than 100 km diameter parent body disrupted at 8.3 ± 0.5 Myr ago (Nesvorný et al., 2003). In (Nesvorný et al., 2008) the discovery of sixty-five very young members of the Beagle family with mean inclination 1.37° and located at 3.157 au near the orbit of asteroid (656) Beagle is reported. However, (656) Beagle itself probably does not belong to the family (Nesvorný et al., 2003). Numerical modelling and observations of the α-band (one of the three principal bands in the zodiacal cloud) thermal emission observed by the Spitzer Space Telescope indicate that the discovered breakup creating the Beagle family is the source of the α dust band (Nesvorný et al., 2008). Additional dust bands or trails have been identified by Sykes (1988) in the IRAS data and by Reach (1997) in the COBE data. The discovery of several new, younger (< 1 Myr) asteroid clusters (Nesvorný and Vokrouhlický, 2006) provides possible new sources for these postulated dust trails. The evolution of asteroid dust trails into bands was studied in Vokrouhlický et al. (2008).

A special co-adding procedure of intensity scans applied to the IRAS data (Espy et al., 2009) led to a significant increase in the signal-to-noise ratio and revealed a much weaker dust band at 17.1° latitude. The new dust band appears at some, but not all, ecliptic longitudes. Espy et al. (2009) suppose it is the M/N band originally found by Sykes (1988) and conclude that this is a still-forming dust band evolving from a recent asteroidal disruption. Espy Kehoe et al. (2015) found that this partial dust band has been produced by the Emilkowalski cluster and estimated the amount of released dust corresponds to a regolith layer about 3 m deep on the 10 km diameter parent body.

In several cases, a coincidence of narrow dust trails with the orbits of periodic comets has been found in the IRAS data (Davies et al., 1984; Sykes et al., 1986a). A trail extends only over a part of the parent comet orbit and is close to the position of the nucleus. The particles are seen both ahead of and behind the comet. In total, eight trails are associated with known short-period comets: 67P/Churyumov-Gerasimenko, 2P/Encke, 65P/Gunn, 22P/Kopff, 7P/Pons-Winnecke, 29P/Schwassmann-Wachmann 1, 9P/Tempel 1, and 10P/Tempel 2 (Sykes et al., 1986b; Sykes and Walker, 1992). Extended dust structures were also optically observed near the orbits of three short-period comets 2P/Encke, 22P/Kopff, and 65P/Gunn (Ishiguro et al., 2007) and a long-extended trail associated with the comet 4P/Faye (Sarugaku et al., 2007).

9.7 Discussion, General Conclusions and Future Work

Since the first meteor showers were recognised more than 200 years ago, there has been a steady increase both in the number of showers and associated streams identified and in the data available regarding each individual stream. There are several reasons for this. In the early days the development of photography made the determination of a meteor trajectory much easier, while in the middle of the twentieth century radar allowed many daylight showers to be detected. More recently, a number of radar facilities regularly operate, producing a large volume of data. Further, a rapid technological advance in video cameras and in the development of the computer programming has been responsible for many networks being set up as a collaboration between amateur and professional astronomers. There is thus significantly more data available now than has been the case at any time in the past, especially for minor showers and the sporadic background. This chapter has described the work carried out and our present state of knowledge.

9.7.1 Future Work

The significant volume of work carried out in the recent past and described here has generated problems that will require further work to be carried out. The main result of this work has been a large increase in the number of known meteoroid orbits. Further, because of the advances in observational techniques this number is increasing daily. In the future, it is expected that much of the data in the large data bases will be made easily available, thus increasing the flow of scientific information even further.

This flow generates a problem of storing all the data and cataloguing it in a meaningful way that is generally available and user friendly. Simply producing a list in order of the observing date is not helpful. Doing this cataloguing places considerable strain on the Meteor Data Centre. The MDC shower database will need to be able to carry out a complete archiving of the original source meteor data found in literature. This will need to be supplemented by additional information about a given shower such as inserting missing orbital elements as they become available, periods of shower activity, date of the addition of new shower submission and so forth. There is also a need to develop and implement software that will make it easier for the user to gain maximum benefit from the Meteor Data Center.

Another problem that arises from the increased number of orbits is that of a clustering of meteor orbits coming about by chance, thereby producing an apparent meteoroid stream that is not real. Improved statistical analysis may be required here to work on models of meteor data. A problem of a similar nature is the identification of a parent body for a stream. With the large number of Near-Earth Objects now being discovered, there is a strong possibility that such an NEO can have an orbit that by chance meets our criteria for similarity with the mean orbit of a stream, but in reality be unrelated to the stream. To solve this dilemma, it may be necessary to establish that the orbital evolution of both NEO and stream have been similar for a considerable time. Doing this, however, requires considerable computational effort to numerically integrate the orbits involved over a significant time interval of several thousand years. This is important, as identifying the parent body will throw light on the age-old problem of whether the minor streams originate from asteroids and so are fairly young and formed as a oneoff event (for example, a collision), or whether their origin is a comet that formed a major stream through the normal processes and they have evolved to being minor through losing meteoroid due to various mechanisms. If the latter is the case, the study of the minor stream will throw valuable light onto the problem of the decay of major streams and in turn provide an answer to the problem of how long particles can survive in the inner Solar System.

The outstanding problem regarding sporadic meteors remains in determining the proportions that may be asteroidal or cometary in origin. The difficulty is that sporadic meteors have been moving in the Solar System for a considerable time and their orbits have evolved so that the observed orbit is no longer similar to the initial orbit. Modelling of the evolution of sets of hypothetical meteoroids on hypothetical orbits will, in the long run, produce meaningful answers to this problem, but again, this will require considerable computing resources.

Membership of the zodiacal cloud represents the final end point in the life of a meteoroid. There are well-defined structures within the clouds, but it is a long-term project to understand how these structures reflect the decay of major streams to minor streams and then the sporadic background.

Acknowledgements

The authors would like to thank Peter Brown and the anonymous reviewer for constructive comments and suggestions. We thank Galina Ryabova, David Asher, and Margaret Campbell-Brown for organizing this Meteoroids book project and for their valuable suggestions. Tadeusz J. Jopek's work was supported by project No 2016/21/B/ST9/01479 founded by the National Science Centre in Poland. J. Tóth and L. Kornoš were supported by the Slovak Research and Development Agency under contract No. APVV-16-0148, and by the Slovak Grant Agency for Science, grant No. VEGA 1/0596/18. This research has made use of NASA's Astrophysics Data System Bibliographic Services.

References

Abedin, A., Wiegert, P., Janches, D. et al. 2018. Formation and past evolution of the showers of 96P/Machholz complex. *Icarus*, **300**, 360–385.

Ade, P. A. R., Aghanim, N., Armitage-Caplan, C. et al. 2014. Planck 2013 results. XIV. Zodiacal emission. *Astronomy & Astrophysics*, **571**, A14.

Albin, T., Koschny, D., Molau, S., Srama, R., and Poppe, B. 2017. Analysis of the technical biases of meteor video cameras used in the CILBO system. *Geoscientific Instrumentation, Methods and Data Systems*, **6**, 125–140.

Andreić, Ž., Šegon, D., and Vida, D. 2014. A statistical walk through the IAU MDC database. In: Rault, J.-L., and Roggemans, P. (eds), *Proceedings of the International Meteor Conference, Giron, France, 18–21 September 2014*, pp. 126–133. Hove, Belgium: International meteor Organization, pp. 126–133

Arter, T. R., and Williams, I. P. 1997. The mean orbit of the April Lyrids. *Monthly Notices of the Royal Astronomical Society*, **289**, 721–728.

Babadzhanov, P. B. 1990. Formation of twin meteor showers. In: C.I. Lagerkvist, H. Rickman, and B.A. Lindblad (eds), *Asteroids, Comets, Meteors III, Proceedings of a Meeting (AMC 89)*, Astronomical Observatory of the Uppsala University, June 12–16, 1989, 99. 497–503. Uppsala: Universitet.

Babadzhanov, P. B., Williams, I. P., and Kokhirova, G. I. 2008. Near-Earth objects in the Taurid complex. *Monthly Notices of the Royal Astronomical Society*, **386**, 1436–1442.

Babadzhanov, P. B., Kokhirova, G. I., Williams, I. P., and Obrubov, Y. V. 2017. Investigation into the relationship between comet 96P/Machholz 1 and asteroid 2003 EH1. *Astronomy & Astrophysics*, **598**, A94.

Baggaley, W. J., Bennett, R. G. T., Steel, D. I., and Taylor, A. D. 1994. The Advanced Meteor Orbit Radar Facility: AMOR. *Quarterly Journal of the Royal Astronomical Society*, **35**, 293–320.

Banaszkiewicz, M., Fahr, H. J., and Scherer, K. 1994. Evolution of dust particle orbits under the influence of solar wind outflow asymmetries and the formation of the zodiacal dust cloud. *Icarus*, **107**, 358–374.

Bendjoya, P., Slezak, E., and Froeschle, C. 1991. The wavelet transform: a new tool for asteroid family determination. *Astronomy & Astrophysics*, **251**, 312–330.

Bhattacharya, B., and Reach, W. T. 2004. Zodiacal background: Spitzer observations vs. DIRBE-based model. In: *American Astronomical Society Meeting Abstracts*, p. 1432. Bulletin of the American Astronomical Society, vol. 36.

Borovička, J., Spurný, P., and Brown, P. 2015. Small Near-Earth Asteroids as a Source of Meteorites. In: P. Michel, F. E. DeMeo, and

W. F. Bottke (eds), *Asteroids IV*. Tucson: University of Arizona Press. pp. 257–280.

Borovička, J., Macke, R. J., Campbell-Brown, M. D. et al. 2019. Physical and Chemical Properties of Meteoroids. In: G. O. Ryabova, D. J. Asher, and M. D. Campbell-Brown (eds), *Meteoroids: Sources of Meteors on Earth and Beyond*. Cambridge, UK: Cambridge University Press, pp. 37–62.

Brown, P., Weryk, R. J., Wong, D. K., and Jones, J. 2008. A meteoroid stream survey using the Canadian Meteor Orbit Radar. I. Methodology and radiant catalogue. *Icarus*, **195**, 317–339.

Brown, P., Weryk, R. J., Kohut, S., Edwards, W. N., and Krzeminski, Z. 2010b. Development of an all-sky video meteor network in Southern Ontario, Canada: The ASGARD system. *WGN, Journal of the International Meteor Organization*, **38**, 25–30.

Brown, P., Wong, D. K., Weryk, R. J., and Wiegert, P. 2010a. A meteoroid stream survey using the Canadian Meteor Orbit Radar. II: Identification of minor showers using a 3D wavelet transform. *Icarus*, **207**, 66–81.

Burns, J. A., Lamy, P. L., and Soter, S. 1979. Radiation forces on small particles in the solar system. *Icarus*, **40**, 1–48.

Campbell-Brown, M. D. 2008. High resolution radiant distribution and orbits of sporadic radar meteoroids. *Icarus*, **196**, 144–163.

Carusi, A., Valsecchi, G. B., and Greenberg, R. 1990. Planetary close encounters: Geometry of approach and post-encounter orbital parameters. *Celestial Mechanics and Dynamical Astronomy*, **49**, 111–131.

Cassini, G. D. 1685. *Découverte de la lumière celeste qui paroist dans le zodiaque*. Impr. royale (Paris).

Ceplecha, Z. 1958. On the composition of meteors. *Bulletin of the Astronomical Institutes of Czechoslovakia*, **9**, 154–159.

Ceplecha, Z. 1967. Classification of meteor orbits. *Smithsonian Contributions to Astrophysics*, **11**, 35–60.

Chau, J. L., Woodman, R. F., and Galindo, F. 2007. Sporadic meteor sources as observed by the Jicamarca high-power large-aperture VHF radar. *Icarus*, **188**, 162–174.

Colas, F. 2016. *Official launching of FRIPON*. www.meteornews.net/2016/06/30/official-launching-fripon.

Cook, A.F. 1973. A working list of meteor streams. *Evolutionary and Physical Properties of Meteoroids, Proceedings of IAU Colloq. 13*, Albany, NY, 14–17 June 1971, pp. 183–192. National Aeronautics and Space Administration SP–319.

Cooke, W. J., and Moser, D. E. 2012. The status of the NASA All Sky Fireball Network. *Proceedings of the International Meteor Conference*, Sibiu, Romania, 15–18 September, pp. 9–12. Belgium: International Meteor Organization.

Davies, J. G. 1957. Radio Observation of Meteors. *Advances in Electronics and Electron Physics*, **9**, 95–128.

Davies, J. K., Green, S. F., Meadows, A. J., Stewart, B. C., and Aumann, H. H. 1984. The IRAS fast-moving object search. *Nature*, **309**, 315–319.

Dermott, S. F., Nicholson, P. D., Burns, J. A., and Houck, J. R. 1984. Origin of the solar system dust bands discovered by IRAS. *Nature*, **312**, 505–509.

Dermott, S. F., Nicholson, P. D., Burns, J. A., and Houck, J. R. 1985. An analysis of IRAS' solar system dust bands. In: Giese, R. H., and Lamy, P. (eds), *Properties and Interactions Of Interplanetary Dust; Proceedings of the 85th Colloquium of the International Astronomical Union*, Marseille, France, July 9–12, pp. 395–409. Dordrecht: Springer [Astrophysics and Space Science Library, vol. 119].

Dermott, S. F., Jayaraman, S., Xu, Y. L., Gustafson, B. Å. S., and Liou, J. C. 1994. A circumsolar ring of asteroidal dust in resonant lock with the Earth. *Nature*, **369**, 719–723.

Dermott, S. F., Grogan, K., Jayaraman, S., and Xu, Y. L. 1995. Rotational asymmetry of the zodiacal cloud. Page 1084 of: *American Astronomical Society Meeting Abstracts*. Bulletin of the American Astronomical Society, vol. 27.

Dermott, S. F., Grogan, K., Holmes, E. K., and Wyatt, M. C. 1998. Signatures of planets. In: Backman, D. E., Caroff, L. J., Sandford, S. A., and Wooden, D. H. (eds), *Exozodiacal Dust Workshop Conference Proceedings, pp. 59–84*. Washington DC: NASA (CP-1998-10155).

Dermott, S. F., Grogan, K., Durda, D. D. et al. 2001. Orbital evolution of interplanetary dust. In: Grün, E., Gustafson, B. A. S., Dermott, S., and Fechtig, H. (eds), *Interplanetary Dust*, pp. 569–639. Berlin, Heidelberg: Springer, 2001 (Astronomy and Astrophysics Library).

Dikarev, V., Grün, E., Baggaley, J. et al. 2005. The new ESA meteoroid model. *Advances in Space Research*, **35**, 1282–1289.

Drummond, J. D. 1980. On the meteor/comet orbital discriminant D. In: Gott, P. F., and Riherd, P. S. (eds), *Proceedings of the Southwest Regional Conference for Astronomy and Astrophysics, May 21, 1979*, vol. 5, pp. 83–86.

Drummond, J. D. 1981. A test of comet and meteor shower associations. *Icarus*, **45**, 545–553.

Dumitru, B. A., Birlan, M., Popescu, M., and Nedelcu, D. A. 2017. Association between meteor showers and asteroids using multivariate criteria. *Astronomy & Astrophysics*, **607**, A5.

Dumont, R., and Levasseur-Regourd, A. C. 1978. Zodiacal light photopolarimetry. IV – Annual variations of brightness and the symmetry plane of the zodiacal cloud: Absence of solar-cycle variations. *Astronomy & Astrophysics*, **64**, 9–16.

Dumont, R., and Levasseur-Regourd, A. C. 1985. Zodiacal light gathered along the line of sight: Retrieval of the local scattering coefficient from photometric surveys of the ecliptic plane. *Planetary and Space Science*, **33**, 1–9.

Dycus, R. D., and Bradford, D. C. 1965. Discovery of a meteor shower and its possible association with micrometeoroid showers. *The Observatory*, **85**, 88–89.

Espy, A. J., Dermott, S. F., Kehoe, T. J. J., and Jayaraman, S. 2009. Evidence from IRAS for a very young, partially formed dust band. *Planetary and Space Science*, **57**, 235–242.

Espy Kehoe, A. J., Kehoe, T. J. J., Colwell, J. E., and Dermott, S. F. 2015. Signatures of recent asteroid disruptions in the formation and evolution of Solar System dust bands. *The Astrophysical Journal*, **811**, 1–16.

Farinella, P., Vokrouhlický, D., and Hartmann, W. K. 1998. Meteorite delivery via Yarkovsky orbital drift. *Icarus*, **132**, 378–387.

Fixsen, D. J., and Dwek, E. 2002. The zodiacal emission spectrum as determined by COBE and its implications. *The Astrophysical Journal*, **578**, 1009–1014.

Fritts, D. C., Janches, D., Iimura, H. et al. 2010. Southern Argentina Agile Meteor Radar: System design and initial measurements of large-scale winds and tides. *Journal of Geophysical Research (Atmospheres)*, **115**(D14), D18112.

Galligan, D. P. 2001. Performance of the D-criteria in recovery of meteoroid stream orbits in a radar data set. *Monthly Notices of the Royal Astronomical Society*, **327**, 623–628.

Galligan, D. P., and Baggaley, W. J. 2002a. Wavelet enhancement for detecting shower structure in radar meteoroid data I. methodology. In: Green, S. F., Williams, I. P., McDonnell, J. A. M., and McBride, N. (eds), *Dust in the Solar System and Other Planetary Systems, Proceedings of the IAU Colloquium 181*, the University of Kent, Canterbury, UK, 4–10 April 2000, pp. 42–47. Oxford: Pergamon, 2002 (COSPAR colloquia series, Vol. 15).

Galligan, D. P., and Baggaley, W. J. 2002b. Wavelet enhancement for detecting shower structure in radar meteoroid data II. Application to the AMOR data. In: Green, S. F., Williams, I. P., McDonnell, J. A. M., and McBride, N. (eds), *Dust in the Solar System and Other Planetary Systems, Proceedings of the IAU Colloquium 181*, the University of Kent, Canterbury, UK, 4–10 April 2000, pp. 48–60. Oxford: Pergamon, 2002 (COSPAR colloquia series, Vol. 15).

Galligan, D. P., and Baggaley, W. J. 2004. The orbital distribution of radar-detected meteoroids of the Solar system dust cloud. *Monthly Notices of the Royal Astronomical Society*, **353**, 422–446.

Galligan, D. P., and Baggaley, W. J. 2005. The radiant distribution of AMOR radar meteors. *Monthly Notices of the Royal Astronomical Society*, **359**, 551–560.

Galushina, T. Y., Ryabova, G. O., and Skripnichenko, P. V. 2015. The force model for asteroid (3200) Phaethon. *Planetary and Space Science*, **118**, 296–301.

Gartrell, G., and Elford, W. G. 1975. Southern Hemisphere meteor stream determinations. *Australian Journal of Physics*, **28**, 591–620.

Grogan, K., Dermott, S. F., Jayaraman, S., and Xu, Y. L. 1997. Origin of the ten degree Solar System dust bands. *Planetary and Space Science*, **45**, 1657–1665.

Grotrian, W. 1934. Über das Fraunhofersche Spektrum der Sonnenkorona. Mit 10 Abbildungen. *Zeitschrift für Astrophysik*, **8**, 124–145.

Grun, E., Zook, H. A., Fechtig, H., and Giese, R. H. 1985. Collisional balance of the meteoritic complex. *Icarus*, **62**, 244–272.

Gustafson, B. Å. S. 1994. Physics of zodiacal dust. *Annual Review of Earth and Planetary Sciences*, **22**, 553–595.

Gustafson, B. A. S., and Misconi, N. Y. 1986. Interplanetary dust dynamics: I. Long-term gravitational effects of the inner planets on zodiacal dust. *Icarus*, **66**, 280–287.

Hahn, J. M., Zook, H. A., Cooper, B., and Sunkara, B. 2002. Clementine observations of the zodiacal light and the dust content of the inner Solar System. *Icarus*, **158**, 360–378.

Hajduková, Jr., M., Koten, P., Kornoš, L., and Tóth, J. 2017. Meteoroid orbits from video meteors. The case of the Geminid stream. *Planetary and Space Science*, **143**, 89–98.

Hasegawa, I. 1993. Historical records of meteor showers. In: Stohl, J., and Williams, I. P. (eds), *Meteoroids and Their Parent Bodies, Proceedings of the International Astronomical Symposium*, Smolenice, Slovakia, July 6–12, 1992, pp. 209–226. Bratislava: Astronomical Institute, Slovak Academy of Sciences.

Hawkes, R. L., Jones, J., and Ceplecha, Z. 1984. The populations and orbits of double-station TV meteors. *Bulletin of the Astronomical Institutes of Czechoslovakia*, **35**, 46–64.

Hirayama, K. 1918. Groups of asteroids probably of common origin. *The Astronomical Journal*, **31**, 185–188.

Holman, D., and Jenniskens, P. 2012. Confirmation of the Northern Delta Aquariids (NDA, IAU #26) and the Northern June Aquilids (NZC, IAU #164). *WGN, Journal of the International Meteor Organization*, **40**, 166–170.

Hughes, D. W., Williams, I. P., and Fox, K. 1981. The mass segregation and nodal retrogression of the Quadrantid meteor stream. *Monthly Notices of the Royal Astronomical Society*, **195**, 625–637.

Ishiguro, M., Nakamura, R., Fujii, Y. et al. 1999. First detection of visible zodiacal dust bands from ground-based observations. *The Astrophysical Journal*, **511**, 432–435.

Ishiguro, M., Sarugaku, Y., Ueno, M. et al. 2007. Dark red debris from three short-period comets: 2P/Encke, 22P/Kopff, and 65P/Gunn. *Icarus*, **189**, 169–183.

Ishiguro, M., Yang, H., Usui, F. et al. 2013. High-resolution imaging of the gegenschein and the geometric albedo of interplanetary Dust. *The Astrophysical Journal*, **767**, 75.

Ishmukhametova, M. G., Kondrat'eva, E. D., and Usanin, V. S. 2009. Analysis of the upper limit of the Southworth–Hawkins D criterion for the Pons–Winneckid and Perseid meteoroid streams. *Solar System Research*, **43**, 438–442.

Jacchia, L. G., Verniani, F., and Briggs, R. E. 1967. Selected results from precision-reduced Super-Schmidt meteors. *Smithsonian Contributions to Astrophysics*, **11**, 1–6.

James, J. F., Mukai, T., Watanabe, T., Ishiguro, M., and Nakamura, R. 1997. The morphology and brightness of the zodiacal light and gegenschein. *Monthly Notices of the Royal Astronomical Society*, **288**, 1022–1026.

Janches, D., Plane, J. M. C., Nesvorný, D. et al. 2014. Radar detectability studies of slow and small zodiacal dust cloud particles. I. The case of Arecibo 430 MHz meteor head echo observations. *The Astrophysical Journal*, **796**, 41.

Janches, D., Swarnalingam, N., Plane, J. M. C. et al. 2015a. Radar detectability studies of slow and small zodiacal dust cloud particles: II. A study of three radars with different sensitivity. *The Astrophysical Journal*, **807**, 13.

Janches, D., Close, S., Hormaechea, J. L. et al. 2015b. The Southern Argentina Agile MEteor Radar Orbital System (SAAMER-OS): An initial sporadic meteoroid orbital survey in the Southern sky. *The Astrophysical Journal*, **809**, 36.

Jenniskens, P. 2006. *Meteor Showers and their Parent Comets*. Cambridge, UK: Cambridge University Press.

Jenniskens, P. 2008a. Meteoroid streams that trace to candidate dormant comets. *Icarus*, **194**, 13–22.

Jenniskens, P. 2008b. The IAU meteor shower nomenclature rules. *Earth, Moon, and Planets*, **102**, 5–9.

Jenniskens, P. 2017. Meteor showers in review. *Planetary and Space Science*, **143**, 116–124.

Jenniskens, P., and Nénon, Q. 2016. CAMS verification of single-linked high-threshold D-criterion detected meteor showers. *Icarus*, **266**, 371–383.

Jenniskens, P., Gural, P. S., Dynneson, L. et al. 2011. CAMS: Cameras for Allsky Meteor Surveillance to establish minor meteor showers. *Icarus*, **216**, 40–61.

Jenniskens, P., Borovička, J., Watanabe, J. et al. 2015. Division III: Commission 22: Meteors, Meteorites and Interplanetary Dust. *Transactions of the International Astronomical Union, Series B*, **28**, 120–123.

Jenniskens, P., Nénon, Q., Gural, P. S. et al. 2016a. CAMS confirmation of previously reported meteor showers. *Icarus*, **266**, 355–370.

Jenniskens, P., Nénon, Q., Gural, P. S. et al. 2016b. CAMS newly detected meteor showers and the sporadic background. *Icarus*, **266**, 384–409.

Jones, J., and Sarma, T. 1985. Double-station observations of 454 TV meteors. II - Orbits. *Bulletin of the Astronomical Institutes of Czechoslovakia*, **36**, 103–115.

Jones, J., Campbell, M., and Nikolova, S. 2001. Modelling of the sporadic meteoroid sources. In: Warmbein, B. (ed), *Proceedings of the Meteoroids 2001 Conference*, 6–10 August 2001, Kiruna, Sweden, pp. 575–580. ESA SP-495, Noordwijk: ESA Publications Division.

Jones, J., Brown, P., Ellis, K. J. et al. 2005. The Canadian Meteor Orbit Radar: System overview and preliminary results. *Planetary and Space Science*, **53**, 413–421.

Jones, M. H., Bewsher, D., and Brown, D. S. 2013. Imaging of a circumsolar dust ring near the orbit of Venus. *Science*, **342**, 960–963.

Jopek, T. J. 1986. *Comet and meteor streams associations*. Ph.D. thesis, Astronomical Observatory, Physics Department, Adam Mickiewicz University, Poznan, Poland.

Jopek, T. J. 1993a. Remarks on the meteor orbital similarity D-criterion. *Icarus*, **106**, 603–607.

Jopek, T. J. 1993b. TV meteor streams searching. In: Stohl, J., and Williams, I. P. (eds), *Meteoroids and Their Parent Bodies, Proceedings of the International Astronomical Symposium*,

Smolenice, Slovakia, July 6–12, 1992, pp. 269–272. Bratislava: Astronomical Institute, Slovak Academy of Sciences.

Jopek, T. J. 1995. Separation of Meteor Streams from the Sporadic Background. *Earth, Moon, and Planets*, **68**, 339–346.

Jopek, T. J. 2011. Meteoroid streams and their parent bodies. *Memorie della Societa Astronomica Italiana*, **82**, 3.

Jopek, T. J., and Bronikowska, M. 2017. Probability of coincidental similarity among the orbits of small bodies – I. Pairing. *Planetary and Space Science*, **143**, 43–52.

Jopek, T. J., and Froeschle, C. 1997. A stream search among 502 TV meteor orbits. an objective approach. *Astronomy & Astrophysics*, **320**, 631–641.

Jopek, T. J., and Jenniskens, P. M. 2011. The Working Group on Meteor Showers Nomenclature: A history, current status and a call for contributions. In: Cooke, W. J., Moser, D. E., Hardin, B. F., and Janches, D. (eds), *Meteoroids: The Smallest Solar System Bodies, Proceedings of the Meteoroids 2010 Conference*, Breckenridge, Colorado, USA, May 24–28, 2010, pp. 7–13. NASA/CP-2011-216469. Huntsville, AL: NASA.

Jopek, T. J., and Kaňuchová, Z. 2014. Current status of the IAU MDC Meteor Showers Database. In: Jopek, T. J., Rietmeijer, F. J. M., Watanabe, J., and P., Williams I. (eds), *Proc. International Conf. held at A.M. University in Poznań, Poland, August 26–30, 2013, Meteoroids 2013*, pp. 353–36. Poznań: Wydanictwo Naukowe UAM.

Jopek, T. J., and Kaňuchová, Z. 2017. IAU Meteor Data Center-the shower database: A status report. *Planetary and Space Science*, **143**, 3–6.

Jopek, T. J., and Williams, I. P. 2013. Stream and sporadic meteoroids associated with near-Earth objects. *Monthly Notices of the Royal Astronomical Society*, **430**, 2377–2389.

Jopek, T. J., Valsecchi, G. B., and Froeschle, C. 1999. Meteoroid stream identification: a new approach – II. Application to 865 photographic meteor orbits. *Monthly Notices of the Royal Astronomical Society*, **304**, 751–758.

Jopek, T. J., Valsecchi, G. B., and Froeschlé, C. 2003. Meteor stream identification: a new approach – III. The limitations of statistics. *Monthly Notices of the Royal Astronomical Society*, **344**, 665–672.

Jopek, T. J., Rudawska, R., and Pretka-Ziomek, H. 2006. Calculation of the mean orbit of a meteoroid stream. *Monthly Notices of the Royal Astronomical Society*, **371**, 1367–1372.

Jopek, T. J., Rudawska, R., and Pretka-Ziomek, H. 2008a. Erratum: Calculation of the mean orbit of a meteoroid stream. *Monthly Notices of the Royal Astronomical Society*, **384**, 1741–1741.

Jopek, T. J., Rudawska, R., and Bartczak, P. 2008b. Meteoroid stream searching: The use of the vectorial elements. *Earth, Moon, and Planets*, **102**, 73–78.

Jopek, T. J., Koten, P., and Pecina, P. 2009. Meteoroid streams identification amongst 231 Southern Hemisphere video meteors. *European Planetary Science Congress*, **4**, EPSC2009–747. Available at: http://meetingorganizer.copernicus.org/EPSC2009/EPSC2009-747.pdf.

Jopek, T. J., Koten, P., and Pecina, P. 2010. Meteoroid streams identification amongst 231 Southern hemisphere video meteors. *Monthly Notices of the Royal Astronomical Society*, **404**, 867–875.

Kawara, K., Matsuoka, Y., Sano, K. et al. 2017. Ultraviolet to optical diffuse sky emission as seen by the Hubble Space Telescope Faint Object Spectrograph. *Publications of the Astronomical Society of Japan*, **69**, 31.

Kelsall, T., Weiland, J. L., Franz, B. A. et al. 1998. The COBE diffuse infrared background experiment search for the cosmic infrared background. II. Model of the interplanetary dust cloud. *The Astrophysical Journal*, **508**, 44–73.

Kero, J., Szasz, C., Nakamura, T. et al. 2012. The 2009–2010 MU radar head echo observation programme for sporadic and shower meteors: Radiant densities and diurnal rates. *Monthly Notices of the Royal Astronomical Society*, **425**, 135–146.

Kholshevnikov, K. V., Kokhirova, G. I., Babadzhanov, P. B., and Khamroev, U. H. 2016. Metrics in the space of orbits and their application to searching for celestial objects of common origin. *Monthly Notices of the Royal Astronomical Society*, **462**, 2275–2283.

Kimura, H., and Mann, I. 1998. Brightness of the solar F-corona. *Earth, Planets, and Space*, **50**, 493–499.

Klačka, J. 2014. Solar wind dominance over the Poynting–Robertson effect in secular orbital evolution of dust particles. *Monthly Notices of the Royal Astronomical Society*, **443**, 213–229.

Klačka, J., Petržala, J., Pástor, P., and Kómar, L. 2014. The Poynting–Robertson effect: A critical perspective. *Icarus*, **232**, 249–262.

Knezevic, Z., Lemaître, A., and Milani, A. 2002. The determination of asteroid proper elements. In: W. F. Bottke Jr., A. Cellino, P. Paolicchi, and R. P. Binzel (eds), *Asteroids III*, pp. 603–612. Tucson: University of Arizona Press.

Korlević, K., Šegon, D., Andreić, Ž. et al. 2013. Croatian meteor network catalogues of orbits for 2008 and 2009. *WGN, Journal of the International Meteor Organization*, **41**, 48–51.

Kornoš, L., Koukal, J., Piffl, R., and Tóth, J. 2013. Database of meteoroid orbits from several European video networks. In: Gyssens, M., and Roggemans, P. (eds), *Proceedings of the International Meteor Conference, 31st IMC, La Palma, Canary Islands, Spain, 2012*, pp. 21–25. Hove, Belgium: International Meteor Organization.

Kornoš, L., Matlovič, P., Rudawska, R. et al. 2014a. Confirmation and characterization of IAU temporary meteor showers in EDMOND database. In: Jopek, T. J., Rietmeijer, F. J. M., Watanabe, J., and P., Williams I. (eds), *Proc. International Conf. held at A.M. University in Poznań, Poland, August 26–30, 2013, Meteoroids 2013*, pp. 225–233. Poznań: Wydanictwo Naukowe UAM.

Kornoš, L., Koukal, J., Piffl, R., and Tóth, J. 2014b. EDMOND Meteor Database. In: Gyssens, M., Roggemans, P., and Zoladek, P. (eds), *Proceedings of the International Meteor Conference, Poznan, Poland, 22–25 August 2013*, pp. 23–25. Hove, Belgium: International Meteor Organization.

Kornoš, L., Tóth, J., Porubčan, V. et al. 2015. On the orbital evolution of the Lyrid meteoroid stream. *Planetary and Space Science*, **118**, 48–53.

Koseki, M. 2014. Various meteor scenes II: Cygnid-Draconid Complex (κ-Cygnids). *WGN, Journal of the International Meteor Organization*, **42**, 181–197.

Koseki, M. 2016. Research on the IAU meteor shower database. *WGN, Journal of the International Meteor Organization*, **44**, 151–169.

Koten, P., Spurný, P., Borovička, J., and Stork, R. 2003. Catalogue of video meteor orbits. Part 1. *Publications of the Astronomical Institute of the Czechoslovak Academy of Sciences*, **91**, 1–32.

Koten, P., Vaubaillon, J., Toth, J. et al. 2012. Activity of Draconid meteor shower during outburst on 8th October 2011. *Asteroids, Comets, Meteors 2012*, May 16–20, 2012, Niigata, Japan. LPI Contribution No. 1667. Available at: http://adsabs.harvard.edu/abs/2012LPICo1667.6225K.

Koukal, J., Tóth, J., Piffl, R., and Kornoš, L. 2014. Some interesting meteor showers in EDMOND database. *WGN, Journal of the International Meteor Organization*, **42**, 7–13.

Kresák, Ľ. 1967. Relation of meteor orbits to the orbits of comets and asteroids. *Smithsonian Contributions to Astrophysics*, **11**, 9–33.

Kresák, L. 1969. The discrimination between cometary and asteroidal meteors. I. The orbital criteria. *Bulletin of the Astronomical Institutes of Czechoslovakia*, **20**, 177–187.

Kronk, G. W. 1988. *Meteor showers. A Descriptive Catalog*. Hillside, NJ: Enslow Publishers

Leinert, C. 1975. Zodiacal light – A measure of the interplanetary environment. *Space Science Reviews*, **18**, 281–339.

Leinert, C., Richter, I., Pitz, E., and Hanner, M. 1980. The plane of symmetry of interplanetary dust in the inner solar system. *Astronomy & Astrophysics*, **82**, 328–336.

Leinert, C., Richter, I., Pitz, E., and Hanner, M. 1982. HELIOS zodiacal light measurements – A tabulated summary. *Astronomy & Astrophysics*, **110**, 355–357.

Leinert, C., Ábrahám, P., Acosta-Pulido, J., Lemke, D., and Siebenmorgen, R. 2002. Mid-infrared spectrum of the zodiacal light observed with ISOPHOT. *Astronomy & Astrophysics*, **393**, 1073–1079.

Levasseur-Regourd, A. C., Mukai, T., Lasue, J., and Okada, Y. 2007. Physical properties of cometary and interplanetary dust. *Planetary and Space Science*, **55**, 1010–1020.

Lidov, M. L. 1962. The evolution of orbits of artificial satellites of planets under the action of gravitational perturbations of external bodies. *Planetary and Space Science*, **9**, 719–759.

Lindblad, B. A. 1971a. A computerized stream search among 2401 photographic meteor orbits. *Smithsonian Contributions to Astrophysics*, **12**, 14–24.

Lindblad, B. A. 1971b. A stream search among 865 precise photographic meteor orbits. *Smithsonian Contributions to Astrophysics*, **12**, 1–13.

Lindblad, B. A. 1987. The IAU Meteor Data Center in Lund. *Publications of the Astronomical Institute of the Czechoslovak Academy of Sciences*, **67**, 201–204.

Lindblad, B. A., and Steel, D. I. 1994. Meteoroid orbits available from the IAU Meteor Data Center. In: Milani, A., di Martino, M., and Cellino, A. (eds), *Asteroids, Comets, Meteors 1993. Proceedings of the 160th International Astronomical Union Symposium*, Belgirate, Italy, June 14–18, 1993, pp. 497–500. Dordrecht, The Netherlands: Kluwer Academic Publishers.

Lindblad, B. A., Neslušan, L., Porubčan, V., and Svoreň, J. 2003. IAU Meteor Database of photographic orbits version 2003. *Earth, Moon, and Planets*, **93**, 249–260.

Low, F. J., Young, E., Beintema, D. A. et al. 1984. Infrared cirrus: New components of the extended infrared emission. *The Astrophysical Journal*, **278**, L19–L22.

Mann, I. 1998. Zodiacal cloud complexes. *Earth, Planets, and Space*, **50**, 465–471.

Mann, I., Köhler, M., Kimura, H., Cechowski, A., and Minato, T. 2006. Dust in the solar system and in extra-solar planetary systems. *Astronomy & Astrophysics Review*, **13**, 159–228.

Marsden, B. G., Sekanina, Z., and Yeomans, D. K. 1973. Comets and nongravitational forces. V. *The Astronomical Journal*, **78**, 211.

Matlovič, P., Tóth, J., Rudawska, R., and Kornoš, L. 2017. Spectra and physical properties of Taurid meteoroids. *Planetary and Space Science*, **143**, 104–115.

Micheli, M., Tholen, D. J., and Jenniskens, P. 2016. Evidence for 2009 WN$_{25}$ being the parent body of the November i-Draconids (NID). *Icarus*, **267**, 64–67.

Millman, P. M. 1963. Terminology in Meteoritic Astronomy. *Meteoritics*, **2**, 7–10.

Misconi, N. Y. 1980. The symmetry plane of the zodiacal cloud near 1 AU. In: Halliday, I., and McIntosh, B. A. (eds), *Solid Particles in the Solar System. Proceedings of the IAU Symposium 90*, Ottawa, Canada, August 27–30, 1979, pp. 49–52. Dordrecht: D. Reidel Publishing Co.

Molau, S. 2001. The AKM video meteor network. Pages 315–318 of: Warmbein, B. (ed), *Meteoroids 2001 Conference*. Hillside, NJ: Enslow Publishers.

Molau, S., Crivello, S., Goncalves, R. et al. 2016. Results of the IMO Video Meteor Network - June 2016, and photometry algorithms. *WGN, Journal of the International Meteor Organization*, **44**, 198–204.

Moorhead, A. V. 2016. Performance of D-criteria in isolating meteor showers from the sporadic background in an optical data set. *Monthly Notices of the Royal Astronomical Society*, **455**, 4329–4338.

Moorhead, A. V., Blaauw, R. C., Moser, D. E. et al. 2017. A two-population sporadic meteoroid bulk density distribution and its implications for environment models. *Monthly Notices of the Royal Astronomical Society*, **472**, 3833–3841.

Müller, T. G., Ábrahám, P., and Crovisier, J. 2005. Comets, asteroids and zodiacal light as seen by Iso. *Space Science Reviews*, **119**, 141–155.

Mulof, A. 2018. Private Comunication.

Murakami, H., Baba, H., Barthel, P. et al. 2007. The Infrared Astronomical Mission AKARI*. *Publications of the Astronomical Society of Japan*, **59**, S369–S376.

Neslušan, L., and Hajduková, M. 2017. Separation and confirmation of showers. *Astronomy & Astrophysics*, **598**, A40.

Neslušan, L., and Welch, P. G. 2001. Comparison among the Keplerian-orbit-diversity criteria in major-meteor-shower separation. In: Warmbein, B. (ed), *Proceedings of the Meteoroids 2001 Conference*, Kiruna, Sweden, August 6–10, 2001, pp. 113–118, ESA SP-495. Noordwijk: ESA Publications Division.

Neslušan, L., Svoreň, J., and Porubčan, V. 1995. A procedure of selection of meteors from major streams for determination of mean orbits. *Earth, Moon, and Planets*, **68**, 427–433.

Neslušan, L., Kaňuchová, Z., and Tomko, D. 2013a. The meteor-shower complex of 96P/Machholz revisited. *Astronomy & Astrophysics*, **551**, A87.

Neslušan, L., Svoreň, J., and Porubčan, V. 2013b. The method of selection of major-shower meteors revisited. *Earth, Moon, and Planets*, **110**, 41–66.

Neslušan, L., Porubčan, V., and Svoreň, J. 2014. IAU MDC Photographic Meteor Orbits Database: Version 2013. *Earth, Moon, and Planets*, **111**, 105–114.

Nesvorný, D., and Vokrouhlický, D. 2006. New candidates for recent asteroid breakups. *The Astronomical Journal*, **132**, 1950–1958.

Nesvorný, D., Bottke, Jr., W. F., Dones, L., and Levison, H. F. 2002. The recent breakup of an asteroid in the main-belt region. *Nature*, **417**, 720–771.

Nesvorný, D., Bottke, W. F., Levison, H. F., and Dones, L. 2003. Recent origin of the Solar System dust bands. *The Astrophysical Journal*, **591**, 486–497.

Nesvorný, D., Bottke, W. F., Vokrouhlický, D. et al. 2008. Origin of the near-ecliptic circumsolar dust band. *The Astrophysical Journal*, **679**, L143.

Nesvorný, D., Jenniskens, P., Levison, H. F. et al. 2010. Cometary origin of the zodiacal cloud and carbonaceous micrometeorites. Implications for hot debris disks. *The Astrophysical Journal*, **713**, 816–836.

Nesvorný, D., Janches, D., Vokrouhlický, D. et al. 2011. Dynamical model for the zodiacal cloud and sporadic meteors. *The Astrophysical Journal*, **743**, 129.

Nilsson, C. S. 1964. A southern hemisphere radio survey of meteor streams. *Australian Journal of Physics*, **17**, 205.

Olmsted, D. 1834. Observations of the meteors of November 13, 1833. *American Journal of Science*, **25**, 354–411.

Pokorný, P., Vokrouhlický, D., Nesvorný, D., Campbell-Brown, M., and Brown, P. 2014. Dynamical model for the toroidal sporadic meteors. *The Astrophysical Journal*, **789**, 25.

Pokorný, P., Janches, D., Brown, P. G., and Hormaechea, J. L. 2017. An orbital meteoroid stream survey using the Southern Argentina Agile MEteor Radar (SAAMER) based on a wavelet approach. *Icarus*, **290**, 162–182.

Porubčan, V., Kornoš, L., and Williams, I. P. 2006. The Taurid complex meteor showers and asteroids. *Contributions of the Astronomical Observatory Skalnate Pleso*, **36**, 103–117.

Porubčan, V., Svoreň, J., Neslušan, L., and Schunova, E. 2011. The updated IAU MDC catalogue of photographic meteor orbits. In: Cooke, W. J., Moser, D. E., Hardin, B. F., and Janches, D. (eds), *Meteoroids: The Smallest Solar System Bodies, Proceedings of the Meteoroids 2010 Conference*, Breckenridge, CO, May 24–28, 2010, pp. 338–343, NASA/CP-2011-216469. Huntsville, AL, NASA.

Price, S. D., Noah, P. V., Mizuno, D., Walker, R. G., and Jayaraman, S. 2003. Midcourse space experiment mid-infrared measurements of the thermal emission from the zodiacal dust cloud. *The Astronomical Journal*, **125**, 962–983.

Reach, W. T. 1997. The structured zodiacal light: IRAS, COBE, and ISO observations. In: Okuda, H., Matsumoto, T., and Rollig, T. (eds), *Diffuse Infrared Radiation and the IRTS*, San Francisco, CA, vol. 124, pp. 33–40. Astronomical Society of the Pacific Conference Series.

Reach, W. T., Franz, B. A., Weiland, J. L. et al. 1995. Observational confirmation of a circumsolar dust ring by the COBE satellite. *Nature*, **374**, 521–523.

Reach, W. T., Morris, P., Boulanger, F., and Okumura, K. 2003. The mid-infrared spectrum of the zodiacal and exozodiacal light. *Icarus*, **164**, 384–403.

Reach, W. T., Kelley, M. S., and Sykes, M. V. 2007. A survey of debris trails from short-period comets. *Icarus*, **191**, 298–322.

Robertson, H. P. 1937. Dynamical effects of radiation in the solar system. *Monthly Notices of the Royal Astronomical Society*, **97**, 423–438.

Rowan-Robinson, M., and May, B. 2013. An improved model for the infrared emission from the zodiacal dust cloud: Cometary, asteroidal and interstellar dust. *Monthly Notices of the Royal Astronomical Society*, **429**, 2894–2902.

Rowan-Robinson, M., Hughes, J., Vedi, K., and Walker, D. W. 1990. Modelling the IRAS zodiacal emission. *Monthly Notices of the Royal Astronomical Society*, **246**, 273–277.

Rożek, A., Breiter, S., and Jopek, T. J. 2011. Orbital similarity functions – Application to asteroid pairs. *Monthly Notices of the Royal Astronomical Society*, **412**, 987–994.

Rudawska, R., and Jenniskens, P. 2014. New meteor showers identifed in the CAMS and SonotaCo meteoroid orbit surveys. In: Jopek, T. J., Rietmeijer, F. J. M., Watanabe, J., and P., Williams I. (eds), *Proc. International Conf. held at A.M. University in Poznań, Poland, August 26–30, 2013, Meteoroids 2013*, pp. 217–224. Poznań: Wydanictwo Naukowe UAM.

Rudawska, R., and Jopek, T. J. 2010. Study of meteoroid stream identification methods. In: Fernandez, J. A., Lazzaro, D., Prialnik, D., and Schulz, R. (eds), *Icy Bodies of the Solar System Proceedings*, IAU Symposium No. 263, 2009, pp. 253–256. Cambridge: Cambridge University Press.

Rudawska, R., Matlovič, P., Tóth, J., and Kornoš, L. 2015. Independent identification of meteor showers in EDMOND database. *Planetary and Space Science*, **118**, 38–47.

Ryabova, G. O. 2006. Meteoroid streams: mathematical modelling and observations. In: Lazzaro, D., Ferraz-Mello, S., and Fernández, J.A. (eds), *Asteroids, Comets, Meteors, Proceedings of the 229th Symposium of the International Astronomical Union*, Búzios, Rio de Janeiro, Brasil, August 7–12, 2005, IAU Symposium, vol. 229, pp. 229–247. Cambridge: Cambridge University Press.

Ryabova, G. O. 2016. A preliminary numerical model of the Geminid meteoroid stream. *Monthly Notices of the Royal Astronomical Society*, **456**, 78–84.

Sarugaku, Y., Ishiguro, M., Pyo, J. et al. 2007. Detection of a Long-Extended Dust Trail Associated with Short-Period Comet 4P/Faye in 2006 Return. *Publications of the Astronomical Society of Japan*, **59**, L25–L28.

Sato, M., Watanabe, J.-i., Tsuchiya, C. et al. 2017. Detection of the Phoenicids meteor shower in 2014. *Planetary and Space Science*, **143**, 132–137.

Schult, C., Stober, G., Janches, D., and Chau, J. L. 2017. Results of the first continuous meteor head echo survey at polar latitudes. *Icarus*, **297**, 1–13.

Schult, C., Brown, P., Pokorný, P., Stober, G., and Chau, J. L. 2018. A meteoroid stream survey using meteor head echo observations from the Middle Atmosphere ALOMAR Radar System (MAARSY). *Icarus*, **309**, 177–186.

Šegon, D., Gural, P., Andreić, Ž. et al. 2014. A parent body search across several video meteor data bases. In: Jopek, T. J., Rietmeijer, F. J. M., Watanabe, J., and P., Williams I. (eds), *Proc. International Conf. held at A.M. University in Poznań, Poland, August 26–30, 2013, Meteoroids 2013*. Poznań: Wydanictwo Naukowe UAM.

Sekanina, Z. 1970a. Statistical model of meteor streams. I. Analysis of the Model. *Icarus*, **13**, 459–474.

Sekanina, Z. 1970b. Statistical Model of meteor streams. II. Major Showers. *Icarus*, **13**, 475–493.

Sekanina, Z. 1973. Statistical Model of meteor streams. III. Stream search among 19303 radio meteors. *Icarus*, **18**, 253–284.

Sekanina, Z. 1976. Statistical model of meteor streams. IV. A study of radio streams from the synoptic year. *Icarus*, **27**, 265–321.

Soja, R. H., Schwarzkopf, G. J., Sommer, M. et al. 2016. Collisional lifetimes of meteoroids. In: Roggemans, A., and Roggemans, P. (eds), *International Meteor Conference Egmond, the Netherlands, 2–5 June 2016*, pp. 284–286. Hove, Belgium: IMO

SonotaCo. 2009. A meteor shower catalog based on video observations in 2007–2008. *WGN, Journal of the International Meteor Organization*, **37**, 55–62.

SonotaCo, T., Molau, S., and Koschny, D. 2010. Amateur contributions to meteor astronomy. *European Planetary Science Congress*, **5**, EPSC2010-798. Available at: http://meetingorganizer.copernicus.org/EPSC2010/EPSC2010-798.pdf.

Southworth, R. B., and Hawkins, G. S. 1963. Statistics of meteor streams. *Smithsonian Contributions to Astrophysics*, **7**, 261–285.

Spiesman, W. J., Hauser, M. G., Kelsall, T. et al. 1995. Near- and far-infrared observations of interplanetary dust bands from the COBE diffuse infrared background experiment. *The Astrophysical Journal*, **442**, 662–667.

Spurný, P., Borovička, J., and Shrbený, L. 2007. Automation of the Czech part of the European fireball network: Equipment, methods and first results. In: Valsecchi, G. B., Vokrouhlický, D., and Milani, A. (eds), *Near Earth Objects, our Celestial Neighbors: Opportunity and Risk*, Proceedings of IAU Symposium 236, pp. 121–130. Cambridge: Cambridge University Press, 2007

Spurný, P., Borovička, J., Mucke, H., and Svoreň, J. 2017. Discovery of a new branch of the Taurid meteoroid stream as a real source of potentially hazardous bodies. *Astronomy & Astrophysics*, **605**, A68.

Starczewski, S., and Jopek, T. J. 2004. Dynamical relation of Meteoroids to comets and asteroids. *Earth, Moon, and Planets*, **95**, 41–47.

Steel, D. I., and Elford, W. G. 1986. Collisions in the solar system. III - Meteoroid survival times. *Monthly Notices of the Royal Astronomical Society*, **218**, 185–199.

Stenborg, G., and Howard, R. A. 2017. The evolution of the surface of symmetry of the interplanetary dust from 24 deg to 5 deg elongation. *The Astrophysical Journal*, **848**, 57.

Stober, G., Schult, C., Baumann, C., Latteck, R., and Rapp, M. 2013. The Geminid meteor shower during the ECOMA sounding rocket campaign: Specular and head echo radar observations. *Annales Geophysicae*, **31**, 473–487.

Štork, R., Koten, P., Borovička, J., and Spurný. 2002. Double station meteors recorded in Ondřejov's programme. In: Warmbein, B. (ed), *Proceedings of Asteroids, Comets, Meteors – ACM 2002 International Conference*, 29 July–2 August 2002, Berlin, Germany, pp. 182–192. ESA SP-500. Noordwijk, Netherlands: ESA Publications Division, ISBN 92-9092-810-7.

Stuart, J. S., and Binzel, R. P. 2004. Bias-corrected population, size distribution, and impact hazard for the near-Earth objects. *Icarus*, **170**, 295–311.

Svoren, J., Porubcan, V., and Neslusan, L. 2008. Current status of the Photographic Meteoroid Orbits Database and a call for contributions to a new version. *Earth, Moon, and Planets*, **102**, 11–14.

Svoreň, J., Neslušan, L., and Porubčan, V. 2000. A search for streams and associations in meteor databases. Method of Indices. *Planetary and Space Science*, **48**, 933–937.

Sykes, M. V. 1988. IRAS observations of extended zodiacal structures. *The Astrophysical Journal*, **334**, L55–L58.

Sykes, M. V. 1990. Zodiacal dust bands: Their relation to asteroid families. *Icarus*, **85**, 267–289.

Sykes, M. V., and Greenberg, R. 1986. The formation and origin of the IRAS zodiacal dust bands as a consequence of single collisions between asteroids. *Icarus*, **65**, 51–69.

Sykes, M. V., and Walker, R. G. 1992. Cometary dust trails: I. Survey. *Icarus*, **95**, 180–210.

Sykes, M. V., Hunten, D. M., and Low, F. J. 1986a. Preliminary analysis of cometary dust trails. *Advances in Space Research*, **6**, 67–78.

Sykes, M. V., Lebofsky, L. A., Hunten, D. M., and Low, F. 1986b. The discovery of dust trails in the orbits of periodic comets. *Science*, **232**, 1115–1117.

Tokhtas'ev, V. S. 1982. Influence of collisions of meteor bodies on the evolution rate of meteoroid streams. (in Russian). In: Bel'kovich, O. I., Babadzhanov, P. B., Bronshten, V. A., and Sulejmanov, N. I. (eds), *Meteor Matter in the Interplanetary Space, Proceedings of the All-USSR Symposium* held, Kazan, USSR, Sept. 9–11, 1980, pp. 162–174. Moscow-Kazan.

Tóth, J. 2016. *AMOS in Chile*. Available at: www.meteornews.net/2016/04/09/amos-in-chile/.

Tóth, J., Vereš, P., and Kornoš, L. 2011. Tidal disruption of NEAs – a case of Příbram meteorite. *Monthly Notices of the Royal Astronomical Society*, **415**, 1527–1533.

Tóth, J., Zigo, P., Kalmančok, D. et al. 2015. 5 months of AMOS on the Canary Islands. In: Rault, J.-L., and Roggemans, P. (eds), *International Meteor Conference Mistelbach, Austria*, pp. 63–65. Hove, Belgium: IMO.

Trigo-Rodríguez, J. M., Betlem, H., and Lyytinen, E. 2005. Leonid Meteoroid Orbits Perturbed by Collisions with Interplanetary Dust. *The Astrophysical Journal*, **621**, 1146–1152.

Tubiana, C., Snodgrass, C., Michelsen, R. et al. 2013. Are 2P/Encke, the Taurid complex NEOs and CM chondrites related? *European Planetary Science Congress*, **8**, 773–774. Available at: http://meetingorganizer.copernicus.org/EPSC2013/EPSC2013-773.pdf.

Tubiana, C., Snodgrass, C., Michelsen, R. et al. 2015. 2P/Encke, the Taurid complex NEOs and the Maribo and Sutter's Mill meteorites. *Astronomy & Astrophysics*, **584**, A97.

Ueda, T., Kobayashi, H., Takeuchi, T. et al. 2017. Size dependence of dust distribution around the Earth orbit. *The Astronomical Journal*, **153**, 232.

Usui, F., Ishiguro, M., Kwon, S. M. et al. 2002. WIZARD: A New System for Observing Zodiacal Light. In: Ikeuchi, S., Hearnshaw, J., and Hanawa, T. (eds), *8th Asian-Pacific Regional Meeting, Volume II*, pp. 231–232.

Valsecchi, G. B., Jopek, T. J., and Froeschle, C. 1999. Meteoroid stream identification: A new approach – I. Theory. *Monthly Notices of the Royal Astronomical Society*, **304**, 743–750.

Vaubaillon, J., Watanabe, J., Sato, M., Horii, S., and Koten, P. 2011. The coming 2011 Draconids meteor shower. *WGN, Journal of the International Meteor Organization*, **39**, 59–63.

Vereš, P., and Toth, J. 2010. Analysis of the SonotaCo video meteoroid orbits. *WGN, Journal of the International Meteor Organization*, **38**, 54–57.

Verniani, F. 1969. Structure and fragmentation of meteoroids. *Space Science Reviews*, **10**, 230–261.

Vokrouhlický, D., Nesvorný, D., and Bottke, W. F. 2008. Evolution of dust trails into bands. *The Astrophysical Journal*, **672**, 696–712.

Vondrak, T., Plane, J. M. C., Broadley, S., and Janches, D. 2008. A chemical model of meteoric ablation. *Atmospheric Chemistry & Physics Discussions*, **8**, 14557–14606.

Webster, A. R., Brown, P. G., Jones, J., Ellis, K. J., and Campbell-Brown, M. 2004. Canadian Meteor Orbit Radar (CMOR). *Atmospheric Chemistry & Physics*, **4**, 679–684.

Welch, P. G. 2001. A new search method for streams in meteor data bases and its application. *Monthly Notices of the Royal Astronomical Society*, **328**, 101–111.

Weryk, R. J., Brown, P. G., Domokos, A. et al. 2008. The Southern Ontario All-sky meteor camera network. *Earth, Moon, and Planets*, **102**, 241–246.

Whipple, F. L. 1951. A comet model. II. Physical relations for comets and meteors. *The Astrophysical Journal*, **113**, 464–473.

Whipple, F. L. 1954. Photographic meteor orbits and their distribution in space. *The Astronomical Journal*, **59**, 201–217.

Whipple, F. L. 1955. A comet model. III. The zodiacal light. *The Astrophysical Journal*, **121**, 750.

Wiegert, P., Vaubaillon, J., and Campbell-Brown, M. 2009. A dynamical model of the sporadic meteoroid complex. *Icarus*, **201**, 295–310.

Williams, I. P. 2002. The Evolution of Meteoroid Streams. In: E. Murad and I. P. Williams (eds), *Meteors in the Earth's Atmosphere: Meteoroids and Cosmic Dust and Their Interactions with the Earth's Upper Atmosphere*. Cambridge, UK: Cambridge University Press, pp. 13–32.

Williams, I. P. 2011. The origin and evolution of meteor showers and meteoroid streams. *Astronomy and Geophysics*, **52**(2), 2.20–2.26.

Williams, I. P., and Ryabova, G. O. 2011. Meteor shower features: are they governed by the initial formation process or by subsequent gravitational perturbations? *Monthly Notices of the Royal Astronomical Society*, **415**, 3914–3920.

Williams, I. P., Hughes, D. W., McBride, N., and Wu, Z. 1993. Collisions between the nucleus of Comet Halley and dust from its own meteoroid stream. *Monthly Notices of the Royal Astronomical Society*, **260**, 43–48.

Wyatt, S. P., and Whipple, F. L. 1950. The Poynting-Robertson effect on meteor orbits. *The Astrophysical Journal*, **111**, 134–141.

Yang, H., and Ishiguro, M. 2015. Origin of Interplanetary Dust through Optical Properties of Zodiacal Light. *The Astrophysical Journal*, **813**, 87.

Ye, Q.-Z., Brown, P. G., and Pokorný, P. 2016. Dormant comets among the near-Earth object population: A meteor-based survey. *Monthly Notices of the Royal Astronomical Society*, **462**, 3511–3527.

Younger, J. P., Reid, I. M., Vincent, R. A., Holdsworth, D. A., and Murphy, D. J. 2009. A southern hemisphere survey of meteor shower radiants and associated stream orbits using single station radar observations. *Monthly Notices of the Royal Astronomical Society*, **398**, 350–356.

10

Interstellar Meteoroids

Mária Hajduková Jr., Veerle Sterken and Paul Wiegert

10.1 Introduction

Meteor observations provide us with primary information about the meteoroid population in our Solar System, and address the important question of the origin of these meteoroids. Dust particles, larger grains, and fragments produced by collisions and disintegration processes are subject to gravitational and non-gravitational forces, which cause their dynamical and physical evolution and which can obscure their place and time of origin. This chapter is dedicated to those particles which arrive from outside the Solar System and to the problem of distinguishing them from local meteoroids.

The question of the origin of meteoroids, and in particular, the fraction of interstellar particles in the Solar System, has been a matter of long debate. The reason for this has been the abundance of hyperbolic orbits among detected meteors, as a hyperbolic excess above the escape velocity with respect to the Sun indicates a possible interstellar origin. In fact, a hyperbolic orbit is the only easily measurable property of a meteoroid that might indicate an interstellar origin, and so speed measurements are central to this discussion.

How much of the dust population originates locally in the Solar System and how much comes from beyond? How big is the probability that an interstellar particle passing through the Solar System will hit the Earth? How are the speed measurements by which interstellar meteoroids are identified affected by the uncertainties of the statistical treatment or the measurement errors?

In the first part of this chapter, we introduce interstellar particles from a theoretical point of view, describing their dynamical behaviour, orbital characteristics and their expected velocities measured at Earth. We examine the possible sources of the hyperbolicity of the meteoroid's orbit and demonstrate the problem of distinguishing the particles of interstellar origin from those produced in the Solar System or caused by the process of measuring. The second part of this chapter deals with experimental results based on Earth-based and space-borne observations. These have given rise to many searches within the last twenty-five years, the main results of which are summarised here and linked to the historical context.

10.1.1 Historical Background

During the first half of the twentieth century, the general opinion, influenced mostly by the work of Hoffmeister and Öpik, was that the majority of meteors are produced by interstellar particles. This finding was based on the results of visual meteor observations and estimates of the meteor angular velocities. In the 1925 catalogue of visual fireballs measurements (Von Nießl and Hoffmeister, 1925), 79 percent of the meteors listed had hyperbolic orbits with respect to the Sun.

In the years 1931–1933, the Harvard Observatory organised the Arizona Expedition, the results of which supported this attitude. From rocking-mirror visual observations of angular velocities, Öpik (1940) concluded that about 60% of sporadic meteors were moving in hyperbolic orbits, with heliocentric velocities in excess of the parabolic limit, and defended their interstellar origin on a long-term basis (Öpik, 1950). Not until nearly thirty years later (Öpik, 1969) did he concede that the analysis of the velocities (not their observations) in the Arizona results was faulty and declared that the high fraction of hyperbolic meteors found by Öpik (1940) did not exist, proposing the true fraction of hyperbolic orbits was under 1%.

The theory of the predominance of the interstellar meteoroids over local interplanetary ones became indefensible with the progress of observational techniques and consequent improvements in velocity determination; mainly with double-camera photography and direct velocity measurements using the rotating shutter method.

Based on the first photographic observations, Whipple (1940) found that Hoffmeister's most significant interstellar stream was in fact associated with the comet Encke, a definitely local source with one of the shortest periods of all known comets and an aphelion in the asteroid belt.

Porter (1944), analysing British visual meteor data, found no direct evidence for the existence of an excess of hyperbolic velocities. He declared that, with a few doubtful exceptions, all meteors in his sample were members of the Solar System and pointed out the lack of any systematic analysis of the meteor data of the Arizona observations.

The new radio-echo techniques for velocity measurements, developed after 1945, brought the following results in the study of hyperbolic meteors: McKinley (1951), analysing thousands of meteor velocities in the records from the Ottawa radar, did not find one that he could definitely assert to be a meteor from interstellar space; radar meteor observations from Jodrell Bank (Almond et al., 1953) showed no evidence of a significant hyperbolic velocity component either.

Summarising the results obtained by different techniques, Lovell (1954) sceptically suggested that if any of the observed hyperbolic meteors were real, their possible origin was in planetary perturbations.

The final breakthrough was brought about by precise photographic observations at the Smithsonian Astrophysical

Observatory (the Harvard Super-Schmidt photographic program), the results of which were published by Jacchia and Whipple (1961). Their data contained so few hyperbolic orbits that they announced that there were statistically hardly any hyperbolic meteors.

In the following years, research into hyperbolic meteors, and especially interstellar meteoroids, focused on the determination of their ratio in the population of observed meteors. Štohl (1970) studied the hyperbolicity of meteors and also investigated the effect of planetary perturbations on meteoroid orbits. Comparing the results of the photographic catalogues collected until 1970, he found large differences in the percentage of hyperbolic orbits (from less than 2% to more than 20%) and showed the percentage clearly depended on the accuracy of the speed measurement.

Analyses of the Kharkov radar data (Tkachuk and Kolomiets, 1985; Andreev et al., 1987) yielded 1–2% hyperbolic orbits among more precise radar orbits and 5% for orbits of lower accuracy. The authors pointed out a lack of criteria for distinguishing real hyperbolic orbits.

Sarma and Jones (1985) reported 19% apparent hyperbolic orbits in their double-station video meteor data, of which 26 (6%) were hyperbolic on the basis of 95% confidence limits and only 8 (1.7%) were regarded as truly hyperbolic after considering the uncertainties arising from difficulties with the velocity determination.

Hajduková (1993), analysing photographic orbits collected in the IAU Meteor Data Centre, determined that the ratio of hyperbolic orbits in the data shrank considerably from 12% to 0.02% after a detailed error analysis. She concluded that there is no evidence for meteors of interstellar origin in the photographic surveys and that the hyperbolic velocities observed at Earth are most likely the consequence of observational and measurement errors, mostly in the velocity.

A new period in the research of interstellar meteoroids started with the first in situ measurements of interstellar dust by detectors, mainly on board the Ulysses and Galileo spacecraft. Research moved towards particles of smaller sizes, typically less than 1 μm, and too small to be detected as meteors at the Earth, but measurable with space-borne dust detectors. Grün and his colleagues (Grün et al., 1993) confidently reported the identification of interstellar dust (ISD) grains flowing through the heliosphere, originating from the Local Interstellar Cloud (LIC). They reported that small dust impacts can be clearly distinguished from noise for most of the events due to the multi-coincidence characteristics of the instrument of the detector (Baguhl et al., 1993).

Around the same time, Baggaley et al. (1993a), on the basis of observations from the Advanced Meteor Orbit Radar, dealt with the influx of meteoroids with extremely high velocities. Their work and other related measurements are discussed more fully in Section 10.5.2.1.

From this point, monitoring of the influx of interstellar particles (ISP) by various techniques in the following years enabled tens of papers about their detection to be published. These have brought valuable information about the number of interstellar particles registered by various techniques, which depend on their distance from the Sun, their location (near ecliptic planetary regions or in high ecliptic latitudes) and on their size.

10.1.2 Significance of Interstellar Meteoroid Detections

This topic of debate is important because a conclusive detection of interstellar meteoroids from radar measurements would have significant consequences for astronomy. Although the size distribution of the interstellar dust particles declines exponentially with the particle size, the mass distribution is dominated by large particles (Landgraf et al., 2000), e.g., larger than about 1 μm. Therefore, a reliable detection of a large-ISD-particle flux would considerably change the interstellar gas-to-dust mass ratio. The most recent gas-to-dust mass ratio – derived from in situ dust measurements of particles from the LIC by the Ulysses spacecraft – was determined to be $R_{g/d} = 193^{+85}_{-57}$ (Krüger et al., 2015). For this number, a dust inflow velocity of $v_\infty = 23.2$ km s^{-1} was assumed. The gas-to-dust mass ratio is about 20% higher when assuming $v_\infty = 26$ km s^{-1} (Krüger et al., 2015).

Moreover, the presence of large interstellar particles and their characterisation would provide more information on the processes taking place in interstellar space like collisions, shocks and mixing. Several consequences, like a non-homogeneous distribution of the dust in the interstellar medium, are discussed in Grün and Landgraf (2000).

Finally, detection and analysis of large interstellar dust fluxes or interstellar meteors may, for the first time, permit the study of debris disks from other stars from observations of dust in the Solar System, complementing astronomical observations. This is not possible using smaller (micron-sized) interstellar dust particles from spacecraft in situ data since they do not directly represent their source region (see Section 10.2). Interstellar meteoroids could thus provide unique information, because in situ measurements, sample return missions and astronomical observations often have limitations in their ability to detect these large grains.

10.2 Dynamics of Extra-Solar System Particles

Interstellar dust in the heliosphere may come from different sources in our local galactic neighbourhood via different ejection mechanisms (see Section 10.3), but one confirmed source of interstellar dust is the Local Interstellar Cloud (LIC) through which the Sun currently moves with a relative speed of about 26 km s^{-1}, in the direction of the neighbouring G-cloud. The Local Interstellar Cloud is a dense (0.3 H cm^{-3}; Frisch et al., 1999), warm, partially ionised cloud consisting of H and He gas and about 0.5–1% dust by mass. The LIC is embedded in a low density region of space called the *Local Bubble* that was likely excavated by a few supernovae. The dust in the LIC is coupled to its magnetic fields on distance scales of ∼ 0.06 pc (for a 1 μm particle radius) to ∼ 6 pc for a 10 μm particle radius (assuming particle material density $\rho = 1000$ kg m^{-3}), because they are charged and move through the Local Interstellar Magnetic Field (Grün and Svestka, 1996)[1]. Their surface equilibrium potential is estimated to be about +5 or +12 V for

---
[1] The LIC is about 9 pc in size. Table 10.1 shows the gyroradii in the LIC for various particle sizes.

Table 10.1. *Particle masses (in kg) with corresponding radii for different particle bulk material densities (ρ in g cm^{-3}). β (ratio of solar radiation pressure to gravity for a dust particle) and Q/m (charge-to-mass ratio of a dust particle) are calculated assuming the "adapted astronomical silicates β-curve" and a bulk material density of 2.5 g cm^{-3}. The Gyroradius in the Local Interstellar Cloud is given for particle surface potential U = +0.5 V, interstellar magnetic field strength B = 0.5 nT and particle bulk material density $\rho = 2.5$ g cm^{-3}*

Mass [kg]	$\rho = 3.3$ Radius	$\rho = 2.5$ Radius	$\rho = 0.7$ Radius	β[–]	Q/m [C kg^{-1}]	Gyroradius
	[mm]	[mm]	[mm]			[pc]
1.	41.67	45.71	69.87	0.	2.5×10^{-11}	8.5×10^7
10^{-1}	19.34	21.21	32.42	0.	1.2×10^{-10}	1.8×10^7
10^{-2}	8.98	9.85	15.05	0.	5.5×10^{-10}	3.9×10^6
10^{-3}	4.16	4.57	6.99	0.	2.5×10^{-09}	8.5×10^5
10^{-4}	1.93	2.12	3.24	0.	1.2×10^{-08}	1.8×10^5
	[µm]	[µm]	[µm]			
10^{-5}	898	985	1505	0.	5.5×10^{-08}	3.9×10^4
10^{-6}	417	458	699	0.	2.5×10^{-07}	8.5×10^3
10^{-7}	193	212	324	0.	1.2×10^{-06}	1.8×10^3
10^{-8}	89.8	98.5	150.5	0.	5.5×10^{-06}	395
10^{-9}	41.7	45.7	69.9	0.006	2.5×10^{-05}	82
10^{-10}	19.3	21.2	32.4	0.01	1.2×10^{-05}	18
10^{-11}	8.98	9.85	15.05	0.03	5.5×10^{-04}	3.9
10^{-12}	4.17	4.57	6.97	0.07	2.5×10^{-03}	0.82
						[AU]
10^{-13}	1.93	2.12	3.24	0.15	1.2×10^{-02}	3.7×10^4
10^{-14}	0.89	0.98	1.60	0.36	5.5×10^{-02}	8.1×10^3
10^{-15}	0.42	0.46	0.70	0.88	0.25	1.8×10^3
10^{-16}	0.19	0.21	0.32	1.54	1.2	371
10^{-17}	0.09	0.10	0.15	1.29	5.5	81
10^{-18}	0.04	0.05	0.07	0.50	25.4	18

silicate and graphite grains respectively in the Local Bubble, and +0.5 to +1 V in the local interstellar cloud (Grün and Svestka, 1996). The surface equilibrium potential increases to +2–+3 V (Alexashov et al., 2016) or +8–+10 V (Slavin et al., 2012) in the outer boundary regions of the heliosphere, with the consequence that large particles can penetrate freely into the heliosphere, while smaller ones are prevented from entering the heliosphere and are deflected around it instead. Linde and Gombosi (2000) calculated the filtering of dust at the heliospheric boundary and concluded that particles between 0.1 and 0.2 µm are filtered out. This seemed somewhat pessimistic and was based on a defocusing configuration of the heliospheric magnetic fields. Slavin et al. (2012) found that particles smaller than 0.01 µm are completely excluded from the heliosphere.

Once the interstellar dust moves inside the heliosphere, it is also subject to forces that influence its trajectories and that may even prevent it from reaching the Solar System. The most important of these forces are solar gravity, the solar radiation pressure force and the Lorentz force due to the motion of the charged dust particles in the Interplanetary Magnetic Field (IMF). Other forces like solar wind drag, Coulomb drag, or Poynting–Robertson drag and the Yarkovsky effect play a negligible role for ISD (Altobelli, 2004). We assume that there is no mass loss for the particles.

The solar radiation pressure and gravity both depend on the square distance to the Sun, and hence, can be written in one term $\beta = F_{SRP}/F_{grav}$ that depends on the grain composition, morphology, particle size, density and on the solar radiation. The relation between the particle size and β-value is visualised in a so-called β-curve. Figure 10.1 shows such curves for different materials. "Astronomical silicates" (see Figure 10.1) are hypothetical silicates, with for instance inclusions or mantles, that have optical constants consistent with the astronomical observations of interstellar silicates (Draine and Lee, 1984).

ISD particles with $\beta = 1$ move straight through the Solar System without being perturbed by solar gravity or the radiation pressure force. Particles with $\beta > 1$ decelerate when approaching the Sun, and then accelerate again on their hyperbolic orbit into interstellar space. Such particles typically have diameters between about 0.2 and 0.6 µm for silicates (see Figure 10.1, around $\sim 8 \times 10^{-18}$ kg and 2×10^{-16} kg). On the contrary, particles with $\beta < 1$ (diameters larger than about one micron) accelerate near the Sun and then slow down again to continue on their hyperbolic trajectories into interstellar space (Sterken et al., 2012, Fig. 1–3). Particles with $\beta < 1$ can reach speeds of up to 40 km s^{-1} at Earth's distance from the Sun, instead of the undisturbed 26 km s^{-1} for particles from the LIC of about 1.5 µm diameter (Strub et al., 2019). The largest particles reach speeds up to 49 km s^{-1} (heliocentric velocity). The gravitational focusing from the Sun causes a region of higher interstellar dust number density to be located "downstream" from the Sun (Landgraf, 2000). The number density there increases up to a factor

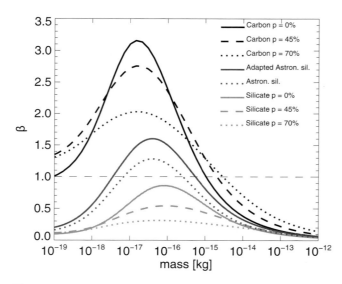

Figure 10.1. The β-mass relation for several particle sizes and material properties. Porosity p is the ratio of the volume of void space to the total (bulk) material space of the particle. Porosity $p = 0\%$ corresponds to compact material. Credit: adapted from Kimura and Mann (1999) and Sterken et al. (2012), reproduced with permission © ESO and the Astronomical Institute of the Slovak Academy of Sciences. (A black and white version of this figure will appear in some formats. For the colour version, please refer to the plate section.)

of 5 (for LIC particles) at Earth's distance from the Sun, and depending on particle size. The Earth passes this enhanced-density region of (large) dust from the LIC every year, around December 13 (Strub et al., 2019).

On the contrary, particles with $\beta > 1$ create a void region around the Sun, called the "β-cone" where these particles with $\beta > \beta_{cone}$ cannot enter. Since the Earth is within the β-cone for $\beta = 1.4$, interstellar dust with $\beta > 1.4$ cannot be detected from 1 AU (Strub et al., 2019). Only particles smaller than about 0.3 µm in diameter or larger than about 0.6 µm in diameter (assuming the adapted astronomical silicates β-curve; see Figure 10.1) can be detected locally near the Earth (Strub et al., 2019). Particles between 0.3 and 0.6 µm in diameter are in the size range of the bulk of the particle sizes as measured by Ulysses for ISD from the LIC.

As the interstellar dust particles move through the heliosphere plasma, they emit electrons by photoionisation from the solar UV radiation. They also pick up electrons and protons from the solar wind and they emit electrons by secondary electron emission from energetic electrons and ions. These processes are in equilibrium and as a result, the particles have a more or less constant positive potential of typically $U = +5$ V in interplanetary space because of the dominance of photoionisation (Grün and Landgraf, 2001) and because the plasma density as well as UV radiation intensity decline with the square of the distance to the Sun. In changing plasma environments, like close to the Sun (< 1 AU), in planetary magnetospheres, or near the heliospheric boundary regions, the particles have different surface potentials and can have negative, or time-varying potentials (Slavin et al., 2012, Fig. 2). The smaller the particle, the larger its charge-to-mass ratio (Q/m), and the more it is influenced by the Lorentz force. These particles are moving through the interplanetary magnetic field with alternating sectors of positive and negative polarity, and hence, they are deflected by the Lorentz force in an alternating manner. The polarity of the interplanetary magnetic field depends on the solar cycle, and as a result the submicron-sized ISD particles are deflected away from the solar equatorial plane for a large part of the eleven-year solar cycle, and focused near the solar equatorial plane in the subsequent solar cycle.

Very small particles (smaller than about 0.01 micron) have a large charge-to-mass ratio and thus couple tightly to local magnetic fields. Therefore, they do not enter the heliosphere but are deflected at the heliosphere boundary (Slavin et al., 2012; Alexashov et al., 2016). However, these are not relevant for radar measurements. Particles larger than that can enter the heliosphere and may reach the Solar System. They are deflected towards or away from the equatorial plane by the Lorentz force, depending on the polarity of the local interplanetary magnetic field and their size (and hence, charge-to-mass ratio). Particles entering at low ecliptic latitudes, for instance those coming from the LIC, are more likely to reach the Solar System because they pass more alternating magnetic field polarities (assuming a Parker interplanetary magnetic field from Parker, 1958). If large particles are observed from higher latitudes than those which come from the LIC (for instance, as claimed by Taylor et al., 1994), there are not necessarily also smaller particles from these latitudes, because they may be deflected away by the Lorentz force as they pass through fewer alternating polarities of the magnetic field. Moreover, the Lorentz force would likely scatter their direction of arrival.

To summarise, the smallest particles' trajectories are mostly influenced by the Lorentz force, while the largest (a few µm in radius) are mostly influenced by solar gravity. For the flux of large ISD, the relative velocity of the Earth with respect to the ISD inflow vector plays a major role, in addition to the gravitational focusing[2]. This flux modulation leads to variations in relative ISD speeds (with respect to Earth) from a few km s^{-1} to up to 60 km s^{-1} for typical "big" Ulysses-type LIC-particles (1 µm diameter), and the relative flow direction can vary over a full 360° (Strub et al., 2019). Larger particles with β close to zero (e.g., 10 µm diameter) would have speeds relative to Earth from 19 to 79 km s^{-1}. The fluctuations in total mass flux of ISD onto Earth are dominated by the biggest particles because these contain most of the total dust mass. They are mainly caused by the Earth's relative motion with respect to the ISD, and by the yearly increase in local ISD number density when the Earth passes near the gravitational focusing region of the dust particles, downstream from the Sun with respect to the inflow direction (Strub et al., 2019).

For large particles (interstellar meteoroids, a few tens of microns) that are not coupled to the LIC and that come from other star systems, the velocity at which the particle (ejected from the other star system) arrives at the Earth is a superposition of its ejection velocity and the star's velocity relative to the Sun. The local stellar velocity distribution ranges from 15–40 km s^{-1} depending on the star's spectral type (Dehnen and Binney, 1998). The particle ejection velocity depends on the source: Murray et al. (2004) found speeds of ~ 1 km s^{-1} for debris

---

[2] Of course, this is also important for small ISD, in addition to velocity effects from the Lorentz and radiation pressure forces.

disks containing gas giants, ~ 10 km s^{-1} for Asymptotic Giant Branch (AGB) star dust ejection, and perhaps ~ 100 km s^{-1} in collimated bipolar outflows from young stellar objects. In any case, the excess speed v_a at which the interstellar particle arrives at the edge of the Solar System might be tens of km s^{-1}.

The speed measured by a detector at Earth has the additional component resulting from the fall into the Sun's potential well. As a result, at the Earth a particle arriving with speed v_a will have a measured heliocentric speed $v_H = \sqrt{42^2 + v_a^2}$, which is 49 km s^{-1} (or 7 km s^{-1} above the parabolic velocity) for an arrival speed of 25 km s^{-1}. Interstellar meteors arriving from behind the Sun's motion with respect to the local standard of rest may of course arrive with almost zero excess velocity; typically, however, we expect them to arrive at Earth with speeds exceeding the Sun's escape speed by a few km s^{-1}. It is worth noting that meteoroids with hyperbolic speeds relative to the Sun will not necessarily have high speeds relative to the Earth. A particle on a hyperbolic orbit could arrive from behind our planet at a relative speed as low as 12 km s^{-1}, and in fact any sample of measured interstellar meteors should show a range of in-atmosphere speeds, not only high ones.

10.3 Hyperbolic Meteoroids Produced in the Solar System

It is possible for material on a bound orbit around our Sun to be transferred to a hyperbolic orbit. If this material should intersect with the Earth as it leaves the Solar System on its now-unbound orbit, it will appear as a hyperbolic meteor. However it is not of interstellar origin, and so these ejection processes must be understood in order to recognise true interstellar meteors.

Planetary perturbations can generate hyperbolic orbits from asteroidal or cometary particles through the slingshot effect (Lovell, 1954, Chapter XII; Öpik, 1969). These can occur when a planet encounters either sporadic meteoroids or a meteoroid stream. The later is relatively uncommon; however, Comet C/1995 O1 Hale–Bopp had a pre-perihelion encounter with Jupiter that could have scattered particles onto orbits reaching our planet at speeds almost 1 km s^{-1} above the parabolic velocity at Earth (Wiegert, 2014). Sporadic scattering is more common, but still relatively rare. Only one in 10 000 sporadics are expected to have been scattered to hyperbolic velocity by a planet (Wiegert, 2014). These originate mostly from Mercury and Venus, because of the higher dust densities and the abundance of meteoroids on nearly-unbound orbits there. However, these hyperbolic meteoroids are usually travelling only a few hundreds of meters per second above the heliocentric escape speed at Earth, and so such meteors are unlikely to be confused with interstellar meteors, as these are expected to have much higher heliocentric speeds.

Hyperbolic meteors can also be generated locally by solar radiation pressure. Particles with sufficiently large area-to-mass ratios find themselves in a state where the outward acceleration due to solar radiation pressure exceeds gravity, and they accelerate outwards. These are the "beta meteoroids" (Zook and Berg, 1975), typically less than 1 micron across, but their small sizes make them unlikely to be observed with traditional meteor techniques.

Comets approaching the Sun on highly eccentric orbits may release material onto hyperbolic orbits through outgassing processes. Though there is not one universally-accepted model for cometary ejection speeds (Whipple, 1950; Jones, 1995; Crifo, 1995; Ryabova, 2013) they are rarely expected to reach above 1 km s^{-1}, even near the Sun for comets with very low perihelia. Such meteoroids that reach the Earth will have slowed as they climb out of the Sun's gravitational well and are not likely to be misidentified as interstellar in origin.

Collisions between asteroids and impacts by meteoroids onto asteroid surfaces can release material onto unbound orbits. However, ejecta is usually released at speeds significantly below that of the impactor, making the process relatively inefficient. The ejecta velocity distribution is sensitive to the mechanical properties of the target and impactor (Waza et al., 1985), but ejection velocities are not observed above the impactor velocity. Random velocities in the asteroid belt are 4–6 km s^{-1} (Farinella and Davis, 1992; Bottke et al., 1994) while a Δv of 7 km s^{-1} is needed to put an asteroid on a near-circular orbit at 3 AU on a hyperbolic one; so asteroid collisions may in principle produce hyperbolic ejecta, but such events are rare.

Impacts by meteoroids into asteroid targets can occur at much higher relative velocities than inter-asteroid collisions, as the meteoroid environment contains a significant retrograde population, the apex meteoroids (Jones and Brown, 1993) thought to be derived from retrograde comets (Wiegert et al., 2009; Nesvorný et al., 2011). However, experiments show that meteoroid impacts typically produce relatively low ejection velocities, typically $\sim 1\%$ of the impactor's (Braslau, 1970; Hartmann, 1985; Housen and Holsapple, 2011; Wiegert, 2015).

Planetary magnetic fields are known to accelerate charged particles. High-speed sub-micron grains were measured at Jupiter by the Ulysses spacecraft (Grün et al., 1993; Kempf et al., 2005) and were inferred to have originated from Io, become electrically charged in the Jovian plasma environment and accelerated by its magnetic field. Similar particles detected by the Cassini spacecraft at Saturn (Kempf et al., 2005) are thought to originate in Saturn's outer main ring. These small particles can be accelerated to very high speeds (larger than 100 km s^{-1}) (Horanyi et al., 1993; Flandes et al., 2011).

10.4 Effect of Measurement Errors on the Resulting Orbit

Meteor observations, especially in the last two decades, are a rich data resource, extremely important for statistical evidence on the nature of the meteoroid orbits in the Solar System. However, the use of the orbits on an individual basis is very problematic and requires high-accuracy data to ensure that the resulting analyses are not biased by effects of measurement and determination errors. Discriminating between orbits of different natures for individual meteoroids is demanding, even for the most accurate photographic meteors (Kresák and Kresáková, 1976; Steel, 1996; Hughes and Williams, 2000).

In optical surveys, meteors are observed through visible light emitted during the ablation process, when meteoroids pass through the atmosphere. Measured parameters of a meteor (position and speed) are used to determine the orbit of the meteoroid around the Sun. The orbital elements derived indicate the particles' origin. However, the semi-major axis a, which defines the type of the orbit, strongly depends upon the derived heliocentric velocity v_H, eventuating from the fundamental equation of the motion of a particle in the Sun's gravitational field:

$$v_H^2 = GM_S \left(\frac{2}{r} - \frac{1}{a} \right), \qquad (10.1)$$

where r is the heliocentric distance (for the meteor observations in the Earth's atmosphere $r \approx 1$ AU), M_S is the solar mass and G is the gravitational constant.

The determination of the heliocentric velocity v_H from photographic meteor observations proceeds in several steps involving various corrections (for atmospheric deceleration, diurnal aberration, acceleration by the Earth's gravitational field, and vector addition with the Earth's motion) (e.g. Kresák, 1992). Hence, the resulting value of v_H, determined, consequentially, from the measured speed v, the non-atmospheric velocity v_∞ and geocentric velocity v_G, is influenced by various errors: incorrect determination of the radiant, leading to incorrect heights and changing of the elongation of the radiant from the apex; errors in timing affecting the right ascension; difficulties with short meteor trail data reduction; changes in the rotation velocity of a sector or other errors arising from the equipment used. All these sources of uncertainty vary in importance, and cannot be easily separated from one another, but each of them tends to increase the inaccuracy of the determined values. At the end, the error of v_H can easily exceed 1 km s^{-1}. Such errors can transfer near-parabolic orbits over the parabolic limit and create an artificial population of hyperbolic meteors (Hajduková, 1994). The higher the velocity v_H, the smaller the error needed for this change.

This effect can be demonstrated by a diagram showing the correlation between the non-atmospheric velocity v_∞ (or geocentric velocity v_G) and the angular elongation of the apparent radiant from the apex, ε_A. Based on Kresák and Kresáková (1976), we constructed, for different values of semi-major axis a, a diagram (Figure 10.2) showing the relation between these two quantities:

$$v_H^2 = v_G^2 + v_0^2 - 2v_G v_0 \cos \varepsilon_A, \qquad (10.2)$$

where v_0 is the mean heliocentric velocity of the Earth.

The diagram allows us to estimate the required measurement accuracy in speed and radiant position needed to discriminate between bound and unbound orbits. The required accuracy is different for different regions in the diagram, determined by the resolution within the particular area between individual orbits. The curves represent orbits with different values of semi-major axes (marked in the graph).

For the plot, we used rough data from the photographic catalogues of the Meteor Data Centre of the International Astronomical Union (IAU MDC), version 2003 (Lindblad et al., 2003). The database is created from more than fifteen different catalogues, each containing data obtained with different equipment and different determination software. Each of them differs in the proportion of hyperbolic orbits they contain, which raises concerns about the accuracy of their speed determinations. However, to reveal the sources of the inconsistencies is problematical.

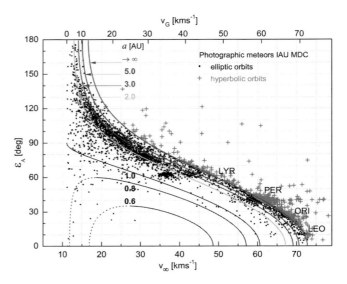

Figure 10.2. The effect of the measurement errors in radiant position and velocity on the resulting semi-major axis of the orbit. The angular elongation of the apparent radiant from the apex ε_A is plotted against the non-atmospheric velocity of meteors v_∞, using rough photographic data of the IAU MDC (Lindblad et al., 2003) in which 11.5% of orbits were determined to be hyperbolic (crosses). The curves, representing the relation between ε_A and v_G according to Equation (10.2), are constructed for different values of semi-major axes a. (A black and white version of this figure will appear in some formats. For the colour version, please refer to the plate section.)

Of the 4581 meteors, 4054 meteors are on elliptic orbits (black circles) and 527 (11.5%) were determined by authors of the catalogues to be hyperbolic (crosses). These are beyond the parabolic limit in the right part of the diagram, in contrast to Kresák and Kresáková (1976), who used a sample of the most precise 413 Super-Schmidt meteor data (Jacchia and Whipple, 1961), leaving this part of the graph empty.

Due to the strong correlation between velocity and apex elongation, meteors are distributed in a very narrow zone of the diagram, where discriminating between orbits of different semi-major axes is most challenging (due to the resolution within this zone). It is clearly seen that for large a, the value of the semi-major axis derived is strongly affected by small errors in the measured speed or radiant position, especially for $\varepsilon_A < 60$ deg. Following the example of Kresák and Kresáková (1976), discrimination between a long-period orbit and an orbit with its aphelion near Jupiter would demand a resolution better than ± 3 km s^{-1} in speed and ± 5 deg in radiant coordinates. The only regions of the diagram where a lower measuring accuracy would be sufficient is the lower left or upper right. The first area corresponds to orbits with aphelia near the Earth (in Figure 10.2, shown for semi-major axes of 0.6 and 0.8 AU). The dotted lines correspond to particles overtaken by the Earth and solid lines to the retrograde particles encountering the Earth head-on. The number of meteors in this area is low. The upper right region of the diagram corresponds to highly hyperbolic orbits, which are absent among the data investigated. There are few cases of high hyperbolic excesses in v_H; however, they belong exclusively to the catalogues of lowest accuracy. The vast majority of hyperbolic orbits in the diagram are concentrated in the line along the parabolic limit, where conditions are most demanding.

A detailed analysis of the same sample of the IAU MDC data (Hajduková, 2008) showed that the number of hyperbolic meteors in this sample does not in any case represent the number of interstellar meteors; the hyperbolicity of the vast majority of them is only apparent, caused by various errors. The upper limit of possible interstellar meteors in the data was statistically determined to be of the order of 10^{-3}. The individual cases, however, cannot be identified as they are hidden within the error bars.

The first clear evidence of the influence of measurement errors, seen in Figure 10.2, is the concentration of shower meteors (the Perseids, Orionids, Lyrids and Leonids) among the hyperbolic orbits. These showers have known local sources, and so the presence of apparently hyperbolic orbits among them points to incorrect or insufficiently precise speed determination. To follow the influence of the errors on the sample of hyperbolic orbits, diagrams showing the position of radiants of orbits for the selected intervals of values of $1/a$ close to the parabolic limit and beyond were constructed from various photographic and video data (Hajduková, 2008, 2011). In the case of true hyperbolic meteors, a gradual decrease in the concentration of shower radiants with decreasing values of $1/a$ would be expected, but actually the opposite was found. Their concentration is higher among the orbits with the highest hyperbolic excesses, reaching a proportion of 1:1.

Moreover, analysis of hyperbolic shower meteors observed by different techniques showed that the proportion of hyperbolic orbits is different in different meteor showers. A clear dependence of the contribution of hyperbolic meteors in meteor showers on the mean heliocentric velocity of a particular shower ($N_{e>1}/N = f(v_H)$) was found, which was true for the radio, photographic, and video data investigated (Hajduková, 2008, 2011). At any rate, these results and the assumption that the shower meteor orbits were determined with the same precision as non-shower data within the same catalogue, led Hajdukova to conclude that there is a lack of statistical argument for the presence of meteors produced by interstellar particles observed in the Earth's atmosphere.

The disputability of hyperbolic shower meteors was noticed early by Fisher (1928) and Watson (1939), who pointed out the problem of over-estimation of the measured velocities leading to the artificial interstellar population. However, influenced by results of Hoffmeister and Öpik, the existence of interstellar meteor showers was seriously considered at that time.

Concerning hyperbolic shower meteors from radar surveys, Taylor et al. (1994) tested the very high velocity meteors observed by AMOR on the Eta Aquariid meteor shower. Approximately half of about 500 Eta Aquariids in their sample had hyperbolic orbits with $a < 0$ and $e > 1$, with very extreme values of $e = 1.5$. This is a large fraction, yet perhaps not surprising, seeing that radio data require even more complicated treatment to yield the heliocentric orbital parameters (e.g. Šimek, 1966; Sekanina and Southworth, 1975; Taylor and Elford, 1998; Galligan and Baggaley, 2004) than is the case with photographic data reduction.

Meteor radars detect echoes from the ionisation produced by meteoroids entering the atmosphere and suffer from a number of instrumental detection biases (see Chapter 3, Kero et al. 2019), such as the initial trail radius effect, the finite velocity effect, the pulse repetition factor, and Faraday rotation, described in detail in Moorhead et al. (2017). The increase in ionisation that occurs with increasing meteor speed is itself an observation bias that requires the largest correction to the speed distribution. Corrections to the observed radar speed distribution have been gradually improved by several authors, e.g. Taylor (1995) using the Harvard Radio Meteor Project (HRMP) data; Brown et al. (2004) using Canadian Meteor Orbit Radar (CMOR) data; Hunt et al. (2004), Close et al. (2007) analysing ALTAIR radar data; and Janches et al. (2014) using Arecibo observations. Recently, Moorhead et al. (2017) presented a method for correcting all the radar observation biases and applied it to the CMOR data. In spite of that, an unrealistically high fraction of seemingly hyperbolic meteors remained in their sample. They concluded that the uncertainty smoothing from the observed speed distribution has not yet been successfully deconvolved and that a thorough exploration of all sources of uncertainty and comparison of the results with other datasets and dynamical models is necessary (Moorhead et al., 2017).

In summary, we can conclude that to distinguish the detected interstellar particles from interplanetary ones is a significant challenge. Moreover, the expected hyperbolic excesses of a meteor's heliocentric velocity (see Section 10.2) are of the same order as the uncertainty in the velocity determination. This requires high-accuracy meteors and a proper error examination, failing which, each analysis using the velocity data will be seriously affected by measurement errors.

The following sections will show a review of studies related to interstellar meteoroids, each of which dealt with this problem to a greater or lesser degree.

10.5 Meteor Observations of Interstellar Particles

Searches for meteoroids of interstellar origin have been carried out using different observational techniques (photographic, image-intensified video and radar) as well as by data-mining for hyperbolic orbits in various meteor catalogues. Their detection, however, remains controversial. Here we summarise a portion of the relevant literature on the subject.

10.5.1 Optical Surveys

10.5.1.1 Photographic

Photographic orbits from different catalogues of the IAU Meteor Data Centre (Lindblad, 1987) were analysed by Hajduková (1994), who determined an upper limit for possible interstellar particles of 0.02% for masses $> 10^{-3}$ kg. Analysing the updated catalogues of the IAU MDC (Lindblad et al., 2001), Hajduková and Paulech (2002) concluded that the proportion of possible interstellar meteoroids in the mass range 10^{-4}–10^0 kg does not exceed 2.5×10^{-4} of the total number of meteors. This value was used to set an upper limit for the flux of interstellar particles for the photographic mass interval, $f_{ISP} < 10^{-18} \mathrm{m}^{-2} \mathrm{s}^{-1}$, derived by the authors using the interplanetary particle distribution given by Divine et al. (1993).

10.5.1.2 Video

Hyperbolic meteors observed with video detectors were examined in several studies by Hawkes, Woodworth and their colleagues, who determined the contribution of interstellar meteoroids to the meteoroid population to be 1–2 % for masses $> 10^{-8}$ kg (Hawkes et al., 1999). The proportion of hyperbolic orbits in their data decreased from 6.5 %, as reported in the first study (Hawkes et al., 1984), to 1.3 %, when applying the 95 % confidence limit (Hawkes and Woodworth, 1997a). After a comprehensive error analysis of 160 meteors, observed up to +8 magnitude by image-intensified video detectors, the authors announced the detection of two meteoroids, which arrived on high inclination orbits from interstellar space (Hawkes and Woodworth, 1997b). The two observed events were caused by particles of masses of the order of 10^{-7} kg, with heliocentric velocities of 49.9 ± 1.7 km s^{-1} and 48.4 ± 0.6 km s^{-1}, respectively. Based on their analyses, the authors concluded that approximately 2% of the total meteoroid flux is interstellar in origin, and determined the flux of heliocentric hyperbolic meteoroids (at 1 AU) to be $f_{ISP} = 1.3 \times 10^{-3}$ meteors, of mass greater than 5×10^{-8} kg, impacting a 100 m^2 detector per year (Hawkes et al., 1999).

An analysis based on a one-year survey for interstellar meteoroids using the Canadian Automated Meteor Observatory (CAMO), was reported by Musci et al. (2012). The two station automated electro-optical system detects meteors up to +5 magnitude with an average uncertainty of 1.5% in speed and about 0.4 deg in radiant direction. From a total of 1739 meteor orbits, the authors found 22 potential hyperbolic meteors; none of them had a heliocentric velocity greater than 45 km s^{-1}. After a detailed examination and checking for close encounters with planets, the authors concluded that their few identified hyperbolic events are most likely the result of measurement errors. They determined an upper limit of the flux of interstellar meteoroids at Earth as $f_{ISP} < 6 \times 10^{-14}$ m^{-2}s^{-1} and declared that they found no clear evidence of interstellar meteoroids with a limiting mass of $m > 2 \times 10^{-7}$ kg.

Hyperbolic orbits from the SonotaCo catalogue (SonotaCo, 2009) and the European Video Meteor Network Database (EDMOND) (Kornoš et al., 2014) were analysed by Hajduková et al. (2014a,b); the hyperbolic excesses were in all cases very low and within the error bars. The flux of interstellar particles with a limiting mass $m > 10^{-4}$ kg was estimated to be approximately $f_{ISP} \sim 10^{-16}$ m^{-2}s^{-1}.

10.5.2 Radar Surveys

10.5.2.1 The AMOR Results

Significant contributions to the problem of hyperbolic meteors were made near the turn of the century in the work of Baggaley, Taylor, Bennett and others, based on observations from the Advanced Meteor Orbit Radar (Baggaley et al., 1993a; Baggaley and Bennett, 1996). This high-power radar system was located near Christchurch in New Zealand, and operated between 1990 and 2000, carrying out a survey of southern hemisphere radiants in order to determine the heliocentric orbits of the interplanetary dust and meteoroids that impact the Earth. The system sensitivity was +13 magnitude, corresponding to about 100 μm-sized particles, or to a mass limit of about 10^{-9} kg (Baggaley et al., 1994). The fan-shaped radar antenna beam, narrow in azimuth yet broad in elevation, permitted a large declination response and hence a large ecliptic latitude sampling of the celestial sphere. Measuring a few 10^3 orbits daily, their data doubled the existing total database from all other orbit-measuring techniques at that time. For the first time, meteoroid orbits were determined using digitised raw data (Baggaley et al., 1993b). Orbital uncertainties (dependent on the meteors' atmospheric speed) were determined to be about 2 degrees in angular elements and about 5% in size elements (Baggaley, 2000).

The AMOR analyses yielded about 1 % extremely hyperbolic orbits of meteoroids in the mass range of 10^{-7}–10^{-10} kg, with very high geocentric velocities in excess of 100 km s^{-1} (Baggaley et al., 1993a; Taylor et al., 1994). This arbitrary limit was imposed to select values which are about five times the standard error above the parabolic limit and thus, well beyond the 3 sigma error. The velocities were derived from the time-lag values and from Verniani's ionisation relation (Verniani, 1973). The authors write about determined uncertainties of about 6 km s^{-1} for a speed of $v_G = 72$ km s^{-1} (Baggaley et al., 1993a). The reality of the high-speed meteors was supported by their height distribution, which showed a peak 6–8 km higher than that for the full sample. Baggaley and his colleagues explained these very-high-velocity meteors as being produced by meteoroids on unbound orbits, and thus, they reported the detection of interstellar meteoroids in the Earth's atmosphere and presented a large sample of interstellar particles (Taylor et al., 1996; Baggaley, 1999). However, the authors did not provide any estimates of their flux.

One of the indicators of the meteors' interstellar origin were the directions of their heliocentric influx (Baggaley, 1999). Mapping the influx directions showed, other than a broad interstellar inflow (a high southern latitude open orbit component), the presence of discrete sources in the vicinity of the Sun (Baggaley and Galligan, 2001). Measuring a general background influx of extra-Solar System particles from southern ecliptic latitudes, the authors found enhanced areas. From intra-annual variations in the influx, they inferred the existence of discrete sources, one associated with the Sun's motion about the Galactic centre, and the other with the motion of the Solar System relative to nearby A-type stars (Taylor et al., 1996). The dominant compact directional inflow was believed to appear from the direction of the main-sequence debris-disk star β Pictoris (Baggaley, 2000). This was, however, questioned in a theoretical approach by Murray et al. (2004), who estimated that the observed particle flux coming from the discrete source observed by Baggaley (2000) is several orders of magnitude larger than would be expected from a debris disk at 20 pc. Moreover, the authors showed that the location of the source on the sky is inconsistent with particles ejected from β Pictoris.

The other indicator showing that the meteoroids are interstellar was the distribution of their orbital inclinations, which was found to be a clear function of the heliocentric velocity (Baggaley, 1999). For the closed orbit population, retrograde motion contributed about 5% of the meteors, whereas for the open orbits, this value rose to 41%. It has to be noted that the inclinations were derived using several corrections steps. Taking into account the observational bias effects, Galligan and Baggaley (2002) described their analysis procedures and demonstrated the inclination distribution at successive levels of applied correction. They showed that about half of the directly observed inclination distributions was made up of progrades. This fraction was substantially over-represented due to ionisation efficiency, which increases with speed.

In any case, the analysis of the AMOR orbital data showed that the discrete features were present in all seasons within the year examined (Baggaley, 1999). The inclination distribution of those meteoroids that have a source direction exhibited clear seasonal changes, which, according to the author, can only be explained as being due to sources external to the Solar System. This was supported by a theoretical study by Baggaley and Neslušan (2002). Considering an influx of a collinear stream of interstellar particles (of sizes greater than 20 μm) into the Solar System, the authors modelled its evolution. The expected meteoroids at the Earth exhibited cyclic (seasonal) changes in the observed orbital elements. All elements, except for the inclination, depended on the dispersion of their original (far-Sun) speed; therefore, the variations in the inclination were most clearly discernible.

The reliability of the AMOR interstellar detections stands or falls on the reality of their extremely high velocities. The mean speed for the hyperbolic sample analysed in the paper (Taylor et al., 1996) was determined to be 164 km s^{-1}, with an uncertainty of ±30 km s^{-1}. In this paper, the authors outlined their statistical approach for confirmation of the high velocities, which was based on a comparison of the Fresnel and the time-lag determination (for the sample of meteors with lower velocities for which both methods were applied). Hajduk (2001) re-examined these methods and called attention to several contradictions; he concluded that the sets of highly hyperbolic velocities (in the range of 100–500 km s^{-1}) cannot be accepted unless an independent confirmation of the AMOR observational data is made. Recognising the limitations of the data is of high importance, since observations and orbital distributions are used as inputs in the modelling of the interplanetary dust population. The question whether these very high velocities are a consequence of a systematic error in the processing of the observations or not has not been proven as yet, although these high speeds have not been confirmed by other radar programs and observations.

The problem of meteors with high geocentric velocity and their connection to faint optical meteors at great heights was discussed by Woodworth and Hawkes (1996). The authors found no meteors with beginning heights greater than approximately 120 km, but argued that, because of the small field of view and the optimum intersection height assumed, they had an observational bias against high (and low) meteors. To reconcile their results with AMOR observations, the authors suggested several explanations: there could exist a flux of very fast and high meteors only at masses smaller than the limiting mass of their observations; or they could come from a radiant which strongly prefers southern hemisphere observers; or some constituent of the meteoroids could ablate at very great heights, producing ionisation but not luminosity. Or, on the other hand, the ionising efficiency, which might increase with velocity far more rapidly than the luminous efficiency factor, could bias radar results in favour of very fast meteors (Woodworth and Hawkes, 1996).

In two other works, Rogers et al. (2005), and extended in Hill et al. (2005), the authors modelled the heights, trail lengths and luminosities expected from high geocentric velocity meteor ablation in the Earth's atmosphere and suggested that very high velocity meteors (with entry masses larger than about 10^{-8} kg) could, if they exist, be observed with current electro-optical technology, although there may be observational biases against their detection.

Moreover, the interstellar meteoroids detected by AMOR are also expected to be accompanied by a large number of even smaller grains, which, however, have not been detected by the space detectors on board the Ulysses and Galileo spacecrafts (Landgraf et al., 2000) which are sensitive in this size range, as Earth-bound detectors are not. Unless, as the authors speculated, the small grains have been stopped on their way to the Solar System by an interstellar cloud.

In spite of the controversy surrounding their highest velocity results and their applications, AMOR provided valuable information about the meteoroid population and contributed enormously to our understanding of the evolutionary processes of the small body dynamics.

10.5.2.2 Other Radar Surveys

Using the Arecibo Observatory UHF Radar, Mathews et al. (1999) analysed thirty-two registered micrometeors (mass range of 10^{-9}–10^{-12} kg), fifteen of which were suitable for velocity determination. The velocities, determined by the direct Doppler method, were greater than the escape velocity for eleven of fifteen particles. In one case, the authors suggested it was of interstellar origin. The errors in the orbital elements were estimated by the authors (with several assumptions) to be 10% to 20%. (Janches et al., 2000). In their next study (Meisel et al., 2002a), 143 hyperbolic micrometeors were found in a set of 3000 particles showing measurable deceleration. The sample of hyperbolic orbits was analysed in detail and a final list of 108 interstellar particle candidates was presented. The authors, trying to establish the ISP origins, suggested they had arrived from the direction of the vicinity of the Geminga pulsar (Meisel et al., 2002b). However, according to Musci et al. (2012), they had assumed that all meteors came down the main beam and were not in one of the sidelobes. Moreover, meteors crossing the main beam come in at an angle that cannot be measured, meaning that only the radial velocity is truly known. This leads to large uncertainties in velocity. Thus, these Arecibo radar results remain disputed (Musci et al., 2012).

Radio data on meteoroids with a limiting mass of 10^{-7} kg, collected in the IAU Meteor Data Centre (Lindblad, 2003), were analysed by Hajduková (2008), with special emphasis on the 39145 orbits of the Harvard Radio Meteor Project. The examination indicated large errors in the velocity, reaching as much as 10 km s^{-1}. The author determined that only 10^{-3} of the total number of Harvard radar orbits could be of interstellar origin. Other older radar meteors, observed by Meteor Automatic Radar System (MARS) in Kharkiv, were reanalysed by Kolomiyets (2015), who did not refute the reality of the MARS hyperbolic meteoroid orbits reported earlier (Andreev et al., 1987).

The Canadian Meteor Orbit Radar (CMOR) was also used in a search for interstellar meteors, conducted over 2.5 years by

Weryk and Brown (2004). CMOR is a interferometric HF/VHF meteor radar with a radio magnitude limit of +8, which corresponds to an effective limiting mass of 4×10^{-8} kg at typical interplanetary meteoroid encounter speeds (Jones et al., 2005). The system is used for time-of-flight velocity measurements and provides individual error estimates on all measured and derived quantities for each echo. Thus, the authors examined the data on a case-by-case basis for interstellar meteoroids (Weryk and Brown, 2004) and, when measurement errors were taken into account, they found only 12 meteors of possible interstellar origin among more than 1.5 million measured orbits. This corresponded to only 0.0008% of all the observed radar echoes. The authors determined a lower limit on the ISP flux at the Earth to be 6×10^{-15} meteoroids m^{-2} s^{-2} for the 2σ criteria.

Meteor observations with the European Incoherent Scatter Scientific Association (EISCAT) radars were reported by Brosch et al. (2013). Examination of the statistical properties of the radar echoes and their Doppler velocity, derived from the VHF data, showed a lack of extreme velocity meteors. Considering only the meteors that encounter the Earth head-on, the authors reported no incoming velocities measured above 72 km s^{-1} and indicated that interstellar meteors are extremely rare. This confirmed previous observations by the EISCAT UHF (Szasz et al., 2008), in which, after all corrections have been made, only 4 of the 410 detected meteors had heliocentric velocities slightly exceeding the Solar System escape velocity. The authors did not claim their interstellar origin because the hyperbolic excesses were smaller than the uncertainties of the data treatment.

10.5.3 1I/'Oumuamua

The discovery of the first macroscopic interstellar object 1I/'Oumuamua (Meech et al., 2017) opens the door to the study of interstellar fireballs in the Earth's atmosphere, though none have been reported to date. An order of magnitude estimate based on the single detected object 1I/'Oumuamua can be obtained, though its reliability is perhaps not very high. 1I was detected at 0.22 AU from Earth: so if we assume that one interstellar object of size 100 m (approximate average diameter of 1I, Knight et al. (2017)) passes through this Earth-centred cross-section ($\sim 10^{15}$ km^2) every 10 years (the time PanSTARRS has been observing) then there should cumulatively be $N(D > 10 \text{ cm}) \sim \left(\frac{0.1 \text{ m}}{100 \text{ m}}\right)^{-p+1}$ times more objects at sizes down to 10 cm, where p is the assumed slope of the differential size distribution. Adopting $p = 3.5$ from the canonical paper of Dohnanyi (1969) gives ~ 0.1 interstellars of 10 cm striking the Earth per year. If 50% strike during the day, and 70% over the oceans, we might expect only one at night over land per several decades. This low rate can likely be accommodated within current observational constraints from all-sky meteor networks but needs much more careful analysis.

10.6 In-Situ Measurements of Interstellar Dust

10.6.1 In-Situ Measurements

Predictions of interstellar dust entering the Solar System were made by Bertaux and Blamont (1976). Levy and Jokipii (1976) recognised the role of the Lorentz force herein. Interstellar dust trajectories in the heliosphere were studied by Gustafson and Misconi (1979) and Morfill and Grün (1979) with a focus on the effects of the Lorentz force. Finally, in 1992, the first interstellar dust impacts were detected in situ on a dust instrument on board the Ulysses mission (Grün et al., 1993). Ulysses flew out of the ecliptic plane, which facilitated the reliable detection of interstellar dust particles amongst the interplanetary dust population (see Section 10.6.3). The mission lasted from 1990 until 2009 and the ISD data cover a time-span from 1992 to 2008.

A similar (twin) instrument flew on Galileo from 1989 until 2003 towards and around Jupiter. It confirmed the measurements of interstellar dust made by Ulysses, but *in* the ecliptic plane and during its cruise phase (Baguhl et al., 1996). Reliable identification of these particles was only possible outside of the asteroid belt because of the geometry of the Galileo spacecraft trajectory.

Landgraf et al. (2000) performed Monte Carlo simulations of interstellar dust trajectories in the heliosphere (without boundary regions) and compared these with the Ulysses measurements available then. The depletion seen for the smallest particles could not be justified by instrument sensitivity alone, but was best explained by the effect of the Lorentz force, using the simulations. For the radiation pressure-mass relation, the β-curve for "astronomical silicates" was used from Gustafson (1994) with corresponding dust bulk material density of 2.5 g cm^{-3}. With these assumptions, a best fit was found between data and simulations for ISD particles with size 0.3 μm (bulk population), 0.4 μm, and to a lesser extent 0.2 μm (Landgraf et al., 2003). The Ulysses data, supported by these simulations, thus confirmed the early predictions from studies like Morfill and Grün (1979). Landgraf et al. (1999) also constrained the optical dust properties (β-values) from the Ulysses ISD measurements by analyzing the spatial distribution of the particles with respect to their mass. This is called *"interstellar dust β-spectroscopy"* (Altobelli et al., 2005). Landgraf et al. (1999) concluded that the maximum β-value of the dust particles as measured by Ulysses is between 1.4 and 1.8. Follow-up simulation studies (e.g. Sterken et al. 2012) used this maximum value in the so-called "adapted astronomical silicate" β-curve. This is the original curve from Gustafson (1994) multiplied to reach a maximum of $\beta = 1.6$.

Cassini carried an impact ionisation dust detector (the *Cassini Cosmic Dust Analyser*) that was equipped with a high rate detector and a time-of-flight (TOF) mass spectrometer (Srama et al., 2004). During the cruise phase to Saturn, interstellar dust was identified inside the Earth's orbit (Altobelli et al., 2003) for the first time. The fluxes were consistent with simultaneous ISD measurements at Ulysses. The measured mass range was between 5×10^{-17} kg and 10^{-15} kg, proving that both small and larger (about 1 μm diameter) interstellar dust can also reach the inner Solar System. Altobelli et al. (2005) expanded the earlier Galileo data analysis of Baguhl et al. (1996) by using new selection criteria, which allowed studying ISD impacts down to 0.7 AU, including "ISD β-spectroscopy".

Helios data allowed further analysis of ISD near the Earth between 0.3 and 1 AU from the Sun (Altobelli et al., 2006) and provided the first mass spectra of the particles using a time-of-flight mass spectrometer (mass resolution: $\frac{M}{\Delta m} \approx 5$ Altobelli et al., 2006). The results indicated compositions compatible with silicates, and a mix of various minerals were present in the data.

Krüger et al. (2007) reported on a shift in latitude of the direction from which the interstellar dust particles came, as measured by Ulysses in 2005. The cause of this was unknown.

The Stardust mission was launched in 1999 and captured cometary dust on one side of a collector consisting of aerogel tiles, while the other side was used to capture interstellar dust particles. This was done at speeds as low as possible (a few km s^{-1}) in order to capture the particles as intact as possible. The interstellar dust was collected between approximately 1 and 2 AU, during two specific periods in 2000 and 2002, totalling 195 days (Westphal et al., 2014). The collector was brought back to Earth in 2006, after which the tiles were cut and then scanned under an automatic microscope. These images were made available on the Internet for lay people to search for the particles after some initial online training. This was a successful example of citizen–science. Three interstellar dust candidates were found in the aerogel tiles, and another four craters from presumably interstellar impacts (with remnants) were found in the aluminum foils that surround the tiles. The particles were very diverse in (crystal) structure, composition and size, and particularly surprising was the low density of the candidates in the aerogel (Westphal et al., 2014).

The complete Ulysses dataset from 1992 until 2008 was finally analysed for ISD by Krüger et al. (2015) for the mass distribution, by Strub et al. (2015) for the flux and variability, and by Sterken et al. (2015) for the interpretation of the data using Monte Carlo computer simulations. This dataset totalled more than 900 selected ISD particles spanning three quarters of a Hale cycle. The flux of ISD in the Solar System as measured during these sixteen years of operations was on average 7×10^{-5} m^{-2} s^{-1}, with fluctuations from about 1×10^{-5} m^{-2} s^{-1} to almost 2×10^{-4} m^{-2} s^{-1} (Strub et al., 2015). The bulk of the particles had masses between 10^{-17} and 10^{-16} kg (Krüger et al., 2015) with minima and maxima of about 10^{-18} kg and 1.7×10^{-14} kg (equivalent to 0.1 and 2.3 μm diameter assuming bulk material densities of 2.5 g cm^{-3}).

The reanalysis of the Ulysses ISD data from 1992 until about 2000 by Krüger et al. (2015); Strub et al. (2015); Sterken et al. (2015) confirmed the earlier results from studies by Landgraf et al. (2000). The shift in dust flow direction as shown by Krüger et al. (2007) could possibly be explained with the simulations using dust properties of charge-to-mass ratio $Q/m \approx 4$ C/kg, but if these would be the bulk of the dust particles, then the fluxes in the first period of Ulysses observations would have been larger, which was not the case. It is likely that the shift in dust flow direction (latitude) for these particles is due to Lorentz forces and not dynamics outside of the heliosphere, but a single fit of the simulations to all the data at once (1992–2008) remains to be found in order to prove and confirm this proposed mechanism. This also would imply that the "big" particles (> 0.2 μm) in the Ulysses data are fluffy rather than compact, because for these masses a lower Q/m would have been expected (Sterken et al., 2015).

Since one single dominant dust size or population (characterised by one set of β and Q/m) cannot explain the full sixteen years of data, Sterken et al. (2015) speculated that the heliosphere boundary plays an extra role in the filtering of ISD on its way to the Solar System. Full simulations, including boundary regions of the heliosphere, are probably necessary to find one fit for all the data at once. The study suggests that big particles are possibly fluffy while smaller ones may be more compact.

The Cassini Dust Analyser TOF mass spectrometer detected interstellar particles during Cassini's mission around Saturn. In a timespan from mid-2004 until 2013, thirty-six particles were identified as interstellar. Due to sensitivity limits of the instrument, only particles from 10^{-19} to 5×10^{-16} kg were in the range of masses that could be analysed. The dust composition was rather homogenous, implying repeated processing of the dust in the interstellar medium (Altobelli et al., 2016). The dust consisted of magnesium-rich silicate particles, some with iron inclusions (Altobelli et al., 2016). The derived β-curve was rather compatible with compact particles, unlike their larger counterparts from the Stardust and Ulysses missions.

Several spacecraft (e.g. Voyager, Gurnett et al., 1983) have detected dust impacts using plasma wave instruments (antennas) that pick up a signal from the small plasma cloud that arises temporarily after a dust impact on the spacecraft body. So far, two missions have detected ~ 1 μm impacts of dust with a yearly variation in the flux that is compatible with the direction of the flow of interstellar dust from the LIC in the Solar System. These are the STEREO and WIND missions (Zaslavsky et al., 2012; Malaspina et al., 2014), both in the vicinity of the Earth (1 AU). The WIND mission's twenty-two years of interstellar dust measurements (1994–2016) have been archived in a database for future use (Malaspina and Wilson, 2016). The data partially overlap the Ulysses data (1994–2008) and span a full Hale cycle.

10.6.2 Impact Ionisation Instruments, Laboratory Calibrations and Instrument Limitations

The Ulysses Dust Analyser instrument is a typical impact ionisation detector with a maximum sensitive area of 0.1 m^2 and an effective solid angle of 1.45 sr[3] (corresponding to a 140° opening angle, Baguhl et al., 1996). It is sensitive to dust particles of masses between about 10^{-19} and 10^{-9} kg, but this sensitivity is speed-dependent (see, e.g., figure 8 in Grün et al. 1992). For impact speeds of 40 km s^{-1}, the measurement limits are 10^{-19} kg $< m < 10^{-13}$ kg and for impact speeds of 5 km s^{-1}, the limits are 10^{-16} kg $< m < 10^{-10}$ kg (Grün et al., 1992). Above this mass, the instrument works as a threshold detector (Grün et al., 1992) and is saturated, while below this mass the instrument sensitivity is not high enough to reliably detect an impact. This means that such an instrument in an orbit near the Earth would be sensitive to different parts of the interstellar dust mass distribution depending on the time of the year. Instrument calibrations were performed using iron, carbon and silicates (Goeller and Grün, 1989). The Cassini instrument has similar characteristics, and calibrations have been performed for a larger variety of materials. Fielding et al. (2015) present a comprehensive review of these materials (e.g. Polypyrrole-coated mineral grains).

---

[3] A flat plate has an "effective solid angle" of π sr (Goeller and Grün, 1989).

The measurement principle is based on impact ionisation: when dust particles with velocities larger than about 1 km s^{-1} hit the gold-coated spherical target of the instrument, the dust particle and part of the target vaporises and ionises. The ions and electrons are separated by the electric field applied between the target (0 V for Ulysses) and the ion collector (-350 V for Ulysses). The speed of the impacting particle is estimated using the rise-time of the signal while the mass is estimated based on the velocity and the total charge in the plasma cloud after impact: $Q \approx m^\alpha v^\gamma$ with $\alpha \approx 1$ and $\gamma \approx 1.5$–5.5 depending on the speed range. Krüger et al. (2015) used calibration curves from Grün et al. (1995) for the reduction of the sixteen years of Ulysses ISD data. The speed is determined with an accuracy of typically a factor of 2 (using only one channel) resulting in an accuracy of a factor of 10 for the mass (Grün et al., 1992). Because of the large uncertainty in the speed determination, Landgraf et al. (2000) and Krüger et al. (2015) have obtained more reliable results for the ISD mass distributions using simulated velocities instead of measured in order to determine the measured masses. Apart from dust impact velocity and mass, the instrument also measures impact direction as inferred from its pointing direction on the spinning spacecraft. The electric charge of the largest particles is also measured through the induction on an entrance grid, but only for very large particles (Grün et al., 1992; Krüger et al., 2015).

10.6.3 ISD Selection Criteria

The orbit of Ulysses largely facilitated the distinction of interstellar dust from interplanetary dust because it was more or less perpendicular to the inflow direction of the ISD from the LIC: only at perihelion would the ISD come from the same (prograde) direction as the IDPs, while during the rest of the orbit, ISD would rather come from the retrograde direction with respect to Ulysses, and opposite to most of the IDPs. In addition, Ulysses flew out of the ecliptic plane, where IDPs are much less present than in the ecliptic plane where distinguishing IDPs from ISD is a lot more challenging (e.g. during the Galileo mission). While Krüger et al. (2015) focused on the mass distribution of ISD and therefore did not select on impact charge signal, Strub et al. (2015) avoided biases in the directionality of the dust flow and selected particles based on impact velocity. The following selection criteria were used for the analysis of sixteen years of Ulysses data:

- Impacts were distinguished from noise events using instrument-specific procedures (see Krüger et al., 2015).
- Both Krüger et al. (2015) and Strub et al. (2015) excluded all data during intervals of Jupiter dust stream periods as well as around perihelion (latitudes $|b| < 60°$).
- Krüger et al. (2015) excluded data at ecliptic latitudes larger than 60° near perihelion where the ion charge signal $Q_I \leq 10^{-13}$ C. These were considered to be β-meteoroids.
- Krüger et al. (2015) required the instrument to point within $\pm 90°$ from the interstellar Helium upstream direction, except in 2005/2006, where this requirement was relaxed with another 40° (in one direction) to account for the shift in dust flow direction that was observed.
- Strub et al. (2015) ignored all impacts with $Q_I \leq 1 \times 10^{-13}$ C (Jupiter dust streams, in addition to the periods mentioned in previous points).
- Strub et al. (2015) ignored all impacts with speeds $V_{imp} \leq 11.6$ km s^{-1} to exclude interplanetary dust that has lower velocities than interstellar dust.

The selection of Krüger et al. (2015) resulted in a dataset of 987 particles, while the selection of Strub et al. (2015) resulted in a dataset of 580 particles.

10.7 Summary

The Solar System is not an isolated system. Its interaction with the interstellar medium, due to the Sun's motion relative to the local interstellar cloud (LIC), should lead to the presence of interstellar particles (ISP), or, at least, to interstellar dust (ISD). Interstellar grains passing through the Solar System have been observed by dust detectors. Cassini, Ulysses, Helios and Galileo all had impact ionisation detectors that provided coherent numbers for the ISD flux throughout the Solar System and throughout time for a mass range of about 10^{-19}–10^{-13} kg. The upper limit of the mass range is limited by instrument saturation, and by the smaller number of large particles in the size distribution that could hit an instrument surface area of only 0.1 m^2. The lowest-mass-particle measurements are limited because of instrument sensitivity (speed-dependent) and the filtering of small particles due to the Lorentz force. Impact measurements on spacecraft with antennas like WIND and STEREO provide useful long-term monitoring possibilities for ISD particles in addition to dedicated cosmic dust detectors.

The situation with the interstellar meteoroids observed in the Earth's atmosphere is not so clear. On a theoretical basis, it was shown that particles larger than about 10 μm can travel tens of parsecs through the interstellar medium (Murray et al., 2004), theoretically creating the possibility of their detection by ground-based radar systems. Measurements of them have not yet been satisfactorily convincing.

Owing to the limitations of the accuracy of current meteor measurements, distinguishing interstellar particles from local meteoroids is very challenging. Hence, their proportion among interplanetary meteors has not yet been unambiguously determined, though it seems to increase strongly with decreasing particle mass. The reported derived fluxes of ISPs observed by different techniques (Sections 10.5.1, 10.5.2 and 10.6.1) are synthetised in several studies, and summarised in this section. Determination of the fluxes of interstellar particles obtained from different surveys allows the mass indices for a broad scale of masses to be derived and provides a comprehensive perspective. The reported ISP fluxes from various papers were recently compiled by Musci et al. (2012). Since no new studies reporting the detection of interstellar meteoroids have been published since that time, we use their figure to give an overview of interstellar particles detected to date. Their examination of the literature is summarised in Figure 10.3.

Figure 10.3 shows the fluxes for ISPs in a broad mass range, exceeding 20 orders of magnitude, with scales showing the approximate sensitivity for different detectors. The reported

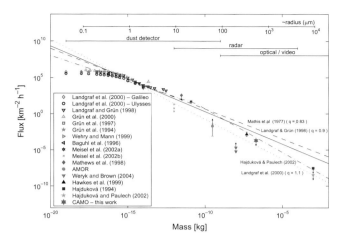

Figure 10.3. A summary of the interstellar meteoroid fluxes from the literature, from Musci et al. (2012). Different symbols designate primary sources of interstellar data. The lines in different styles represent the modelled mass distributions or those fitted by power-law functions determined by different authors. (A black and white version of this figure will appear in some formats. For the colour version, please refer to the plate section.)

detections of ISPs are shown with different symbols designating primary data sources.

The fluxes of the ISD data used in Figure 10.3 are based on Landgraf and Grün (1998) and Landgraf et al. (2000), who analysed the Ulysses and Galileo data. In contradiction to their modelled mass distributions (solid and dotted lines), the measured flux of the smallest grains (for masses $< 10^{-16}$ kg) showed a deficiency. This was explained by defocusing Lorentz forces, which kept the smaller interstellar particles out of the heliosphere (Baguhl et al., 1996). On the other side, the flux of big interstellar grains from dust detectors was overestimated when using the measured impact velocities. The authors attributed it to possible mis-identification of interplanetary particles (the interstellar data set may be contaminated by interplanetary grains) or to systematic deviation of the measured impact velocity to lower values. Therefore, for the mass distribution determination, calculated values were used instead of measured velocities. This resulted in an increase in the number of small grains ($m < 10^{-16}$ kg) and a decrease in the number of large grains ($m > 10^{-16}$ kg).

Landgraf et al. (2000) extrapolated the dust data to larger masses by fitting a power law function (with the exponent $q = 1.1$, dotted line in Figure 10.3) and compared them with the fluxes derived from the AMOR radar measurements (for meteoroids with $m > 10^{-10}$ kg). The authors concluded that these two samples are not kinematically related to each other. From the radar data, only one source (interstellar particles coming from the same direction as the dust detectors data) was compatible with the extrapolation of the in situ data; the fluxes of the other two directions examined (from the southern ecliptic hemisphere and from a discrete source) were one or two orders of magnitude larger.

Hajduková and Paulech (2002) summarised ISP fluxes obtained from observations with different techniques for particles masses from 10^{-20} to 10^{-2} kg and compared the resulting ISP flux (dash-dotted line in Figure 10.3) with the flux model of interplanetary particles suggested by Fechtig (1973). When determining the mass distributions, the authors obtained different mass indices for two different mass intervals, matching the break at masses between 10^{-10} and 10^{-11} suggested by Fechtig (1973). The distribution was steeper for larger particles (with an exponent in the power-law fit of $q \sim 1.2$, as estimated by Musci et al. (2012)), and shallower for smaller particles ($q \sim 0.7$).

Comparison of the ISP flux with another flux model derived by Grun et al. (1985), which gives slightly higher fluxes than derived by Fechtig (1973), also implies an increasing proportion of the ISPs toward smaller masses.

Hajduková and Hajduk (2006), following the summarised experimental data, obtained a smooth change of the curve of the mass distribution for ISPs. The substantial change in the integrated mass distribution indices was, however, also in the mass interval corresponding to the break seen in previous studies. Comparison of their resulting ISP flux with the interplanetary flux model by Divine (1993) showed that the ISP flux is more than two orders of magnitude lower than the flux of interplanetary meteoroids in the mass range of large particles, but strongly increases towards fainter particles.

To match spacecraft dust measurements, new models were developed that consider the population of hyperbolic orbits of dust particles, which is predicted to disappear for larger masses (Staubach et al., 1997; Dikarev et al., 2002).

In Figure 10.3, no comparison of the mass distributions of the ISPs with interplanetary flux models is shown. But they are compared with the size distribution of interstellar grains from the observed interstellar extinction, derived for masses below 10^{-15} kg by Mathis et al. (1977). Musci et al. (2012) extended it to larger particles (dashed line in Figure 10.3) and found that the resulting flux derived from CAMO data (and other optical and radar data in this mass range) is clearly below it, which means that the slope for larger masses has to be much steeper.

We can conclude that in spite of the relatively large number of hyperbolic orbits determined from the meteor observations by Earth-based systems, at most only a fraction could be expected to be produced by interstellar particles. However, none of them have yet been convincingly demonstrated and their presence among detected meteors lacks statistical significance. The only dependable measured interstellar particles in our Solar System which we have to date are the detections of the dust instruments, mainly on board Ulysses and Galileo spacecraft.

10.8 Future Work

Research into interstellar particles has, for decades, brought controversial results, mostly due to meteor observations being misinterpreted as indicating the detection of interstellar meteoroids. This paper shows that the identification of meteors produced by interstellar particles requires an understanding of their dynamical behaviour, highly accurate measurements, and a thorough error examination. Hence, without any improvements to the velocity measurements, which are crucial for orbit determination, the problem of distinguishing interstellar meteors from interplanetary ones remains.

Earth-based and space-borne observations use different techniques with different instrumental sensitivities, and the data processing is performed using different software. In addition, there are obvious constraints related to the heliocentric distance of the measurements, and their latitude and speed. The instruments for meteor observations in the Earth's atmosphere move at a heliocentric velocity of about 29.8 km s^{-1}, in comparison with the dust detectors, where this value varies for individual spacecraft.

Thus, each individual observation gives us some information while still leaving gaps in the data between different techniques, and together they gradually reveal the population of interstellar particles in the Solar System.

Missions with detectors out of the ecliptic plane, long-term monitoring, as well as missions far from the Sun (e.g., 10 AU for reaching particles with $\beta \approx 5$) are important tools to constrain interstellar dust properties. Space-borne instruments with trajectory sensors would allow better orbit determination of the ISD. The most interesting would be to be able to bridge the gap between particle sizes observed by in-situ instruments (up to about 5 μm) and those observable by radar and optical technologies (typically as from a few tens of μm). Large surface areas, long exposure times, high-precision directional information, high-precision velocity determination, and sensitivity to the largest/smallest particles (for space/Earth instruments respectively) would be characteristics needed for an instrument[4] targeted at detecting these intermediate-range interstellar dust particles. Current selection criteria may also exclude interstellar dust particles that come from different directions than from the solar apex direction, and thus dust from sources other than the LIC.

Improvements in ground-based radar systems should allow the detection of interstellar meteoroids in the 10–100-micrometer range (Murray et al., 2004). This would make it possible to register interstellar particles ejected from other star systems. The forthcoming radar system EISCAT 3D (Pellinen-Wannberg et al., 2016) seems promising for registering even smaller particles (with a limiting mass of 10^{-12} kg).

To prove the existence of meteoroids with very high geocentric velocities, search strategies need also to be optimised. Modelling the atmospheric ablation and the luminosity of meteoroids with geocentric velocities, Rogers et al. (2005) suggested improvements to electro-optical technology: using several multi-station observing systems with slightly overlapping fields to create a large net field of view; and, to avoid the observational bias due to the optimum intersection, the system configuration of an altitude of 120 km is required.

Experimental approaches, as made by Thomas et al. (2016) measuring the ionisation efficiency of meteors (up to 70 km s^{-1}) and high-altitude ablation in the laboratory, could also cast light on this problem.

So far, the focus of interstellar dust trajectory modelling has been mainly on the effects of the dynamic IMF, solar radiation pressure and gravity on the particle trajectories within 50 AU from the Sun. Static models were also used of the whole heliosphere (Slavin et al., 2012), but they do not yet represent the ISD distribution in the Solar System, because the ISD travels for more than 20 years (2 solar cycles) through a dynamic heliosphere. Alexashov et al. (2016) focused on the distribution of the dust outside of the Heliopause, based on a 3D kinetic-magnetohydrodynamics model of the solar wind-Local Interstellar Medium (LISM) interaction. Understanding the trajectories of ISD particles smaller than half a micron in diameter ("interstellar dust") requires the inclusion of the (dynamic) outer heliosphere regions in the currently existing dynamic dust-heliosphere models of interplanetary space (Sterken et al., 2015). Data of the ISD flux and flow direction, taken by Ulysses between 1992 and 2008, can constrain such models.

In 2017, the first interstellar asteroid 1I/'Oumuamua was discovered with a hyperbolic trajectory around the Sun. An abundance of at least 6.0×10^{-3} AU^{-3} was estimated for such objects in interstellar space (Feng and Jones, 2018). A continuous mass distribution of interstellar objects (meteoroids, dust particles) is thus very likely present in the interstellar medium. Backtracking is difficult for individual objects due to gravitational scattering from random stellar encounters (Zhang, 2018) and because of the limitations of current astrometry, but for younger objects, this may be possible. The discovery of this interstellar asteroid sheds new light on the possible existence of interstellar meteoroids and shows the importance of closing the gap between measurements of interstellar dust particles with in-situ data, and the ground-based methods for their larger counterparts.

Acknowledgements

This work was supported by the Natural Sciences and Engineering Research Council of Canada, grant No. RGPIN-2018-05659, by the Slovak Scientific Grant Agency, grant No. VEGA 2/0037/18, and by the Slovak Research and Development Agency, contract No. APVV-16-0148.

References

Alexashov, D. B., Katushkina, O. A., Izmodenov, V. V., and Akaev, P. S. 2016. Interstellar dust distribution outside the heliopause: Deflection at the heliospheric interface. *Monthly Notices of the Royal Astronomical Society*, **458**, 2553–2564.

Almond, M., Davies, J. G., and Lovell, A. C. B. 1953. The velocity distribution of sporadic meteors. IV. Extension to magnitude + 8, and final conclusions. *Monthly Notices of the Royal Astronomical Society*, **113**, 411–427.

Altobelli, N. 2004 (May). *Monitoring of the Interstellar Dust Stream in the Inner Solar System Using Data of Different Spacecraft*. Ph.D. thesis, Ruprecht-Karls-Universität Heidelberg.

Altobelli, N., Kempf, S., Landgraf, M. et al. 2003. Cassini between Venus and Earth: Detection of interstellar dust. *Journal of Geophysical Research (Space Physics)*, **108**, 8032.

Altobelli, N., Kempf, S., Krüger, H. et al. 2005. Interstellar dust flux measurements by the Galileo dust instrument between the orbits of Venus and Mars. *Journal of Geophysical Research (Space Physics)*, **110**, A07102.

Altobelli, N., Grün, E., and Landgraf, M. 2006. A new look into the Helios dust experiment data: presence of interstellar dust inside the Earth's orbit. *Astronomy & Astrophysics*, **448**, 243–252.

---

[4] Also called a Swiss army knife.

Altobelli, N., Postberg, F., Fiege, K. et al. 2016. Flux and composition of interstellar dust at Saturn from Cassini's Cosmic Dust Analyzer. *Science*, **352**, 312–318.

Andreev, G. V., Kashcheev, B. L., and Kolomiets, S. V. 1987. Contradictions of the problem of hyperbolic meteors. *Meteornye Issledovaniia*, **13**, 93–104.

Baggaley, J., and Galligan, D. 2001. Monitoring the near-space meteoroid population using the AMOR Radar Facility. In: H. Sawaya-Lacoste (ed.), *Proceedings of the Third European Conference on Space Debris, 19 - 21 March 2001, Darmstadt, Germany*, pp. 219–221. ESA SP-473, Vol. 1, Noordwijk, Netherlands: ESA Publications Division, ISBN 92-9092-733-X, 2001.

Baggaley, J. W., and Bennett, R. G. T. 1996. The Meteoroid Orbit Facility Amor: Recent Developments. Pages 65–70 of: Gustafson, B. Å. S., and Hanner, M. S. (eds), *IAU Colloq. 150: Physics, Chemistry, and Dynamics of Interplanetary Dust*, Gainesville, FL, 14–18 August 1995. [Astronomical Society of the Pacific Conference Series, vol. 104]. San Francisco: Astronomical Society of the Pacific.

Baggaley, W. J. 1999. The interstellar particle component measured by AMOR. Pages 265–273 of: Baggaley, W. J., and Porubcan, V. (eds), *Meteoroids 1998*. Bratislava: Astronomical Institute of the Slovak Academy of Sciences

Baggaley, W. J. 2000. Advanced Meteor Orbit Radar observations of interstellar meteoroids. *Journal of Geophysical Research*, **105**, 10353–10362.

Baggaley, W. J., and Neslušan, L. 2002. A model of the heliocentric orbits of a stream of Earth-impacting interstellar meteoroids. *Astronomy & Astrophysics*, **382**, 1118–1124.

Baggaley, W. J., Taylor, A. D., and Steel, D. I. 1993a. The influx of meteoroids with hyperbolic heliocentric orbits. Pages 53–56 of: Stohl, J., and Williams, I. P. (eds), *Meteoroids and their Parent Bodies*, Smolenice, Slovakia, July 6–12, 1992. Bratislava: Astronomical Institute, Slovak Academy of Sciences.

Baggaley, W. J., Taylor, A. D., and Steel, D. I. 1993b. The southern hemisphere meteor orbit radar facility: AMOR. Pages 245–248 of: Stohl, J., and Williams, I. P. (eds), *Meteoroids and their Parent Bodies*, Smolenice, Slovakia, July 6–12, 1992. Bratislava: Astronomical Institude, Slovak Academy of Sciences.

Baggaley, W. J., Bennett, R. G. T., Steel, D. I., and Taylor, A. D. 1994. The Advanced Meteor Orbit Radar Facility: AMOR. *Quarterly Journal of the Royal Astronomical Society*, **35**, 293–320.

Baguhl, M., Grün, E., Linkert, G., Linkert, D., and Siddique, N. 1993. Identification of "small" dust impacts in the Ulysses dust detector data: Relevance to Jupiter dust streams Page 1036 of: *AAS/Division for Planetary Sciences Meeting Abstracts #25*. Bulletin of the American Astronomical Society, vol. 25.

Baguhl, M., Grün, E., and Landgraf, M. 1996. In situ measurements of interstellar dust with the ULYSSES and Galileo spaceprobes. *Space Science Reviews*, **78**, 165–172.

Bertaux, J. L., and Blamont, J. E. 1976. Possible evidence for penetration of interstellar dust into the solar system. *Nature*, **262**, 263–266.

Bottke, W. F., Nolan, M. C., Greenberg, R., and Kolvoord, R. A. 1994. Velocity distributions among colliding asteroids. *Icarus*, **107**, 255–268.

Braslau, D. 1970. Partitioning of energy in hypervelocity impact against loose sand targets. *Journal of Geophysical Research*, **75**, 3987–3999.

Brosch, N., Häggström, I., and Pellinen-Wannberg, A. 2013. EISCAT observations of meteors from the sporadic complex. *Monthly Notices of the Royal Astronomical Society*, **434**, 2907–2921.

Brown, P., Jones, J., Weryk, R. J., and Campbell-Brown, M. D. 2004. The velocity distribution of meteoroids at the Earth as measured by the Canadian Meteor Orbit Radar (CMOR). *Earth, Moon, and Planets*, **95**, 617–626.

Close, S., Brown, P., Campbell-Brown, M., Oppenheim, M., and Colestock, P. 2007. Meteor head echo radar data: Mass-velocity selection effects. *Icarus*, **186**, 547–556.

Crifo, J. F. 1995. A general physicochemical model of the inner coma of active comets. 1: Implications of spatially distributed gas and dust production. *The Astrophysical Journal*, **445**, 470–488.

Dehnen, W., and Binney, J. J. 1998. Local stellar kinematics from HIPPARCOS data. *Monthly Notices of the Royal Astronomical Society*, **298**, 387–394.

Dikarev, V., Jehn, R., and Grün, E. 2002. Towards a new model of the interplanetary meteoroid environment. *Advances in Space Research*, **29**, 1171–1175.

Divine, N. 1993. Five populations of interplanetary meteoroids. *Journal of Geophysical Research*, **98**, 17029–17048.

Dohnanyi, J. S. 1969. Collisional model of asteroids and their debris. *Journal of Geophysical Research*, **74**, 2531–2554.

Draine, B. T., and Lee, H. M. 1984. Optical properties of interstellar graphite and silicate grains. *The Astrophysical Journal*, **285**, 89–108.

Farinella, P., and Davis, D. R. 1992. Collision rates and impact velocities in the Main Asteroid Belt. *Icarus*, **97**, 111–123.

Fechtig, H. 1973. Cosmic dust in the atmosphere and in the interplanetary space at 1 AU today and in the early Solar System Pages 209–221 of: Hemenway, C. L., Millman, P. M., and Cook, A. F. (eds), *Evolutionary and Physical Properties of Meteoroids*. National Aeronautics and Space Administration SP, vol. 319. State University of New York, Albany, NY, June 14–17. Washington: NASA

Feng, F., and Jones, H. R. A. 2018. Oumuamua as a messenger from the local association. *The Astrophysical Journal*, **852**, L27.

Fielding, Lee A., Hillier, Jon K., Burchell, Mark J., and Armes, Steven P. 2015. Space science applications for conducting polymer particles: Synthetic mimics for cosmic dust and micrometeorites. *Chemical Communications*, **51**, 16886–16899.

Fisher, W. J. 1928. Remarks on the Fireball Catalogue of von Niessl and Hoffmeister. *Harvard College Observatory Circular*, **331**, 1–8.

Flandes, A., Krüger, H., Hamilton, D. P. et al. 2011. Magnetic field modulated dust streams from Jupiter in interplanetary space. *Planetary and Space Science*, **59**, 1455–1471.

Frisch, P. C., Dorschner, J. M., Geiss, J. et al. 1999. Dust in the local interstellar wind. *The Astrophysical Journal*, **525**, 492–516.

Galligan, D. P., and Baggaley, W. J. 2002. The meteoroid orbital distribution at 1 AU determined by Amor. Pages 225–228 of: Warmbein, B. (ed), *Asteroids, Comets, and Meteors – ACM 2002 International Conference*, 29 July–2 August 2002, Berlin, Germany. Ed. Barbara Warmbein. ESA SP–500. Noordwijk, Netherlands: ESA Publications Division, ISBN 92-9092-810-7.

Galligan, D. P., and Baggaley, W. J. 2004. The orbital distribution of radar-detected meteoroids of the Solar system dust cloud. *Monthly Notices of the Royal Astronomical Society*, **353**, 422–446.

Goeller, J. R., and Grün, E. 1989. Calibration of the Galileo/Ulysses dust detectors with different projectile materials and at varying impact angles. *Planetary and Space Science*, **37**, 1197–1206.

Grün, E., and Landgraf, M. 2000. Collisional consequences of big interstellar grains. *Journal of Geophysical Research*, **105**, 10291–10298.

Grün, E., and Landgraf, M. 2001. Fast dust in the heliosphere. *Space Science Reviews*, **99**, 151–164.

Grün, E., and Svestka, J. 1996. Physics of interplanetary and interstellar Dust. *Space Science Reviews*, **78**, 347–360.

Grun, E., Zook, H. A., Fechtig, H., and Giese, R. H. 1985. Collisional balance of the meteoritic complex. *Icarus*, **62**, 244–272.

Grün, E., Fechtig, H., Kissel, J. et al. 1992. The ULYSSES dust experiment. *Astronomy & Astrophysics Supplement*, **92**, 411–423.

Grün, E., Zook, H. A., Baguhl, M. et al. 1993. Discovery of Jovian dust streams and interstellar grains by the ULYSSES spacecraft. *Nature*, **362**, 428–430.

Grün, E., Baguhl, M., Hamilton, D. P. et al. 1995. Reduction of Galileo and Ulysses dust data. *Planetary and Space Science*, **43**, 941–951.

Gurnett, D. A., Grun, E., Gallagher, D., Kurth, W. S., and Scarf, F. L. 1983. Micron-sized particles detected near Saturn by the Voyager plasma wave instrument. *Icarus*, **53**, 236–254.

Gustafson, B. A. S. 1994. Physics of Zodiacal Dust. *Annual Review of Earth and Planetary Sciences*, **22**, 553–595.

Gustafson, B. A. S., and Misconi, N. Y. 1979. Streaming of interstellar grains in the solar system. *Nature*, **282**, 276–278.

Hajduk, A. 2001. On the very high velocity meteors. Pages 557–559 of: Warmbein, B. (ed), *Meteoroids 2001 Conference*. ESA Special Publication, vol. 495, 6–10 August 2001, Swedish Institute for Space Physics, Kiruna, Sweden. ESTEC, Noordwijk, the Netherlands.

Hajduková, M. Jr. 1993. On the hyperbolic and interstellar meteor orbits. Pages 61–64 of: Stohl, J., and Williams, I. P. (eds), *Meteoroids and their Parent Bodies*. Bratislava: Astronomical Institute of the Slovak Academy of Sciences.

Hajduková, M. Jr. 1994. On the frequency of interstellar meteoroids. *Astronomy & Astrophysics*, **288**, 330–334.

Hajduková, M. 2008. Meteors in the IAU Meteor Data Center on Hyperbolic Orbits. *Earth, Moon, and Planets*, **102**, 67–71.

Hajduková, M. Jr. 2011. Interstellar meteoroids in the Japanese tv catalogue. *Publications of the Astronomical Society of Japan*, **63**, 481–487.

Hajduková, M. Jr. and Hajduk, A. 2006. Mass distribution of interstellar and interplanetary particles. *Contributions of the Astronomical Observatory Skalnate Pleso*, **36**, 15–25.

Hajduková, M. Jr. and Paulech, T. 2002. Interstellar and interplanetary meteoroid flux from updated IAU MDC data. Pages 173–176 of: Warmbein, B. (ed), *Asteroids, Comets, and Meteors – ACM 2002 International Conference*, 29 July – 2 August 2002, Berlin, Germany. ESA SP–500. Noordwijk, Netherlands: ESA Publications Division, ISBN 92-9092-810-7.

Hajduková, M., Kornoš, L., and Tóth, J. 2014a. Frequency of hyperbolic and interstellar meteoroids. *Meteoritics and Planetary Science*, **49**, 63–68.

Hajduková, M. Jr., Kornoš, L., and Tóth, J. 2014b. Hyperbolic Orbits in the EDMOND. Pages 289–295 of: Jopek, T.J., Rietmeijer, F.J., Watanabe, M. J., and Williams, I.P. (eds), *Proceedings of the Meteoroids 2013*. Poznan: Adam Mickiewicz University Press.

Hartmann, W. K. 1985. Impact experiments. I - Ejecta velocity distributions and related results from regolith targets. *Icarus*, **63**, 69–98.

Hawkes, R. L., and Woodworth, S. C. 1997a. Do some meteorites come from interstellar space? *Journal of the Royal Astronomical Society of Canada*, **91**, 68–73.

Hawkes, R. L., and Woodworth, S. C. 1997b. Optical detection of two meteoroids from interstellar space. *Journal of the Royal Astronomical Society of Canada*, **91**(Oct.), 218–219.

Hawkes, R. L., Jones, J., and Ceplecha, Z. 1984. The populations and orbits of double-station TV meteors. *Bulletin of the Astronomical Institutes of Czechoslovakia*, **35**, 46–64.

Hawkes, R. L., Close, T., and Woodworth, S. 1999. Meteoroids from outside the Solar System. Pages 257–264 of: Baggaley, W. J., and Porubcan, V. (eds), *Meteoroids 1998*. Bratislava: Astronomical Institute of the Slovak Academy of Sciences.

Hill, K. A., Rogers, L. A., and Hawkes, R. L. 2005. High geocentric velocity meteor ablation. *Astronomy & Astrophysics*, **444**, 615–624.

Horanyi, M., Morfill, G., and Grun, E. 1993. Mechanism for the acceleration and ejection of dust grains from Jupiter's magnetosphere. *Nature*, **363**, 144–146.

Housen, K. R., and Holsapple, K. A. 2011. Ejecta from impact craters. *Icarus*, **211**, 856–875.

Hughes, D. W., and Williams, I. P. 2000. The velocity distributions of periodic comets and stream meteoroids. *Monthly Notices of the Royal Astronomical Society*, **315**, 629–634.

Hunt, S. M., Oppenheim, M., Close, S. et al. 2004. Determination of the meteoroid velocity distribution at the Earth using high-gain radar. *Icarus*, **168**, 34–42.

Jacchia, L. G., and Whipple, F. L. 1961. Precision orbits of 413 photographic meteors. *Smithsonian Contributions to Astrophysics*, **4**, 97–129.

Janches, D., Mathews, J. D., Meisel, D. D., Getman, V. S., and Zhou, Q.-H. 2000. Doppler studies of Near-Antapex UHF radar micrometeors. *Icarus*, **143**, 347–353.

Janches, D., Plane, J. M. C., Nesvorný, D. et al. 2014. Radar detectability studies of slow and small zodiacal dust cloud particles. I. The case of Arecibo 430 MHz meteor head echo observations. *The Astrophysical Journal*, **796**, 41.

Jones, J. 1995. The ejection of meteoroids from comets. *Monthly Notices of the Royal Astronomical Society*, **275**, 773–780.

Jones, J., and Brown, P. 1993. Sporadic meteor radiant distributions - Orbital survey results. *Monthly Notices of the Royal Astronomical Society*, **265**, 524–532.

Jones, J., Brown, P., Ellis, K. J. et al. 2005. The Canadian Meteor Orbit Radar: System overview and preliminary results. *Planetary and Space Science*, **53**, 413–421.

Kempf, S., Srama, R., Horányi, M. et al. 2005. High-velocity streams of dust originating from Saturn. *Nature*, **433**, 289–291.

Kero, J., Campbell-Brown, M., Stober, G. et al. 2019. Radar observations of meteors. In: G. O. Ryabova, D. J. Asher, and M. D. Campbell-Brown (eds), *Meteoroids: Sources of Meteors on Earth and Beyond*. Cambridge, UK: Cambridge University Press, pp. 65--89.

Kimura, H., and Mann, I. 1999. Radiation pressure on porous micrometeoroids. Pages 283–286 of: Baggaley, W. J., and Porubcan, V. (eds), *Meteoroids 1998*. Bratislava: Astronomical Institute of the Slovak Academy of Sciences.

Knight, M. M., Protopapa, S., Kelley, M. S. P. et al. 2017. On the rotation period and shape of the hyperbolic asteroid I1/'Oumuamua (2017 U1) from its lightcurve. *The Astrophysical Journal*, **851**, L31.

Kolomiyets, S. V. 2015. Uncertainties in MARS meteor Orbit radar data. *Journal of Atmospheric and Solar-Terrestrial Physics*, **124**, 21–29.

Kornoš, L., Koukal, J., Piffl, R., and Tóth, J. 2014. EDMOND Meteor Database. Pages 23–25 of: Gyssens, M., Roggemans, P., and Zoladek, P. (eds), *Proceedings of the International Meteor Conference, Poznan, Poland, 22–25 August 2013*. Hove, Belgium; IMO.

Kresák, L. 1992. On the ejection and dispersion velocities of meteor particles. *Contributions of the Astronomical Observatory Skalnate Pleso*, **22**, 123–130.

Kresák, L., and Kresáková, M. 1976. A note on meteor and micrometeoroid orbits determined from rough velocity data. *Bulletin of the Astronomical Institutes of Czechoslovakia*, **27**, 106–109.

Krüger, H., Landgraf, M., Altobelli, N., and Grün, E. 2007. Interstellar dust in the Solar System. *Space Science Reviews*, **130**, 401–408.

Krüger, H., Strub, P., Grün, E., and Sterken, V. J. 2015. Sixteen years of Ulysses interstellar dust measurements in the Solar System. I. Mass distribution and gas-to-dust mass ratio. *The Astrophysical Journal*, **812**, 139.

Landgraf, M. 2000. Modeling the motion and distribution of interstellar dust inside the heliosphere. *Journal of Geophysical Research*, **105**, 10303–10316.

Landgraf, M., and Grün, E. 1998. In Situ Measurements of Interstellar Dust. Pages 381–384 of: Breitschwerdt, D., Freyberg, M. J., and Truemper, J. (eds), *IAU Colloq. 166: The Local Bubble and Beyond*. Lecture Notes in Physics, Berlin: Springer Verlag, vol. 506.

Landgraf, M., Augustsson, K., Grün, E., and Gustafson, B. Å. S. 1999. Deflection of the local interstellar dust flow by solar radiation pressure. *Science*, **286**, 2319–2322.

Landgraf, M., Baggaley, W. J., Grün, E., Krüger, H., and Linkert, G. 2000. Aspects of the mass distribution of interstellar dust grains in the solar system from in situ measurements. *Journal of Geophysical Research*, **105**, 10343–10352.

Landgraf, M., Krüger, H., Altobelli, N., and Grün, E. 2003. Penetration of the heliosphere by the interstellar dust stream during solar maximum. *Journal of Geophysical Research (Space Physics)*, **108**, 8030.

Levy, E. H., and Jokipii, J. R. 1976. Penetration of interstellar dust into the solar system. *Nature*, **264**, 423–424.

Lindblad, B. A. 1987. The IAU Meteor Data Center in Lund. *Publications of the Astronomical Institute of the Czechoslovak Academy of Sciences*, **67**, 201–204.

Lindblad, B. A. 2003. Private Communication.

Lindblad, B. A., Neslušan, L., Svoreň, J., and Porubčan, V. 2001. The updated version of the IAU MDC database of photographic meteor orbits. Pages 73–75 of: Warmbein, B. (ed), *Meteoroids 2001 Conference*. ESA Special Publication, vol. 495 , 6–10 August 2001, Swedish Institute for Space Physics, Kiruna, Sweden. ESTEC, Noordwijk, the Netherlands.

Lindblad, B. A., Neslušan, L., Porubčan, V., and Svoreň, J. 2003. IAU Meteor Database of photographic orbits version 2003. *Earth, Moon, and Planets*, **93**, 249–260.

Linde, T. J., and Gombosi, T. I. 2000. Interstellar dust filtration at the heliospheric interface. *Journal of Geophysical Research*, **105**, 10411–10418.

Lovell, A. C. B. 1954. *Meteor Astronomy*. Oxford: Clarendon Press.

Malaspina, D. M., and Wilson, L. B. 2016. A database of interplanetary and interstellar dust detected by the Wind spacecraft. *Journal of Geophysical Research (Space Physics)*, **121**, 9369–9377.

Malaspina, D. M., Horányi, M., Zaslavsky, A. et al. 2014. Interplanetary and interstellar dust observed by the Wind/WAVES electric field instrument. *Geophysical Research Letters*, **41**, 266–272.

Mathews, J. D., Meisel, D. D., Janches, D., Getman, V. S., and Zhou, Q.-H. 1999. Possible origins of low inclination antapex micrometeors observed using the Arecibo UHF radar. Pages 79–82 of: Baggaley, W. J., and Porubcan, V. (eds), *Meteoroids 1998*. Bratislava: Astronomical Institute of the Slovak Academy of Sciences.

Mathis, J. S., Rumpl, W., and Nordsieck, K. H. 1977. The size distribution of interstellar grains. *The Astrophysical Journal*, **217**, 425–433.

McKinley, D. W. R. 1951. Meteor velocities determined by radio observations. *The Astrophysical Journal*, **113**, 225–267.

Meech, K. J., Weryk, R., Micheli, M. et al. 2017. A brief visit from a red and extremely elongated interstellar asteroid. *Nature*, **552**, 378–381.

Meisel, D. D., Janches, D., and Mathews, J. D. 2002a. Extrasolar micrometeors radiating from the vicinity of the local interstellar bubble. *The Astrophysical Journal*, **567**, 323–341.

Meisel, D. D., Janches, D., and Mathews, J. D. 2002b. The size distribution of arecibo interstellar particles and its implications. *The Astrophysical Journal*, **579**, 895–904.

Moorhead, A. V., Brown, P. G., Campbell-Brown, M. D., Heynen, D., and Cooke, W. J. 2017. Fully correcting the meteor speed distribution for radar observing biases. *Planetary and Space Science*, **143**, 209–217.

Morfill, G. E., and Grün, E. 1979. The motion of charged dust particles in interplanetary space - II. Interstellar grains. *Planetary and Space Science*, **27**, 1283–1292.

Murray, N., Weingartner, J. C., and Capobianco, C. 2004. On the flux of extrasolar dust in Earth's atmosphere. *The Astrophysical Journal*, **600**, 804–827.

Musci, R., Weryk, R. J., Brown, P., Campbell-Brown, M. D., and Wiegert, P. A. 2012. An optical survey for millimeter-sized interstellar meteoroids. *The Astrophysical Journal*, **745**, 161.

Nesvorný, D., Vokrouhlický, D., Pokorný, P., and Janches, D. 2011. Dynamics of dust particles released from Oort Cloud comets and their contribution to radar meteors. *The Astrophysical Journal*, **743**, 37.

Öpik, E. 1940. Meteors (Council report on the progress of astronomy). *Monthly Notices of the Royal Astronomical Society*, **100**, 315–326.

Öpik, E. J. 1950. Interstellar meteors and related problems. *Irish Astronomical Journal*, **1**, 80–96.

Öpik, E. J. 1969. NEWS AND COMMENTS- Interstellar Meteors. *Irish Astronomical Journal*, **9**, 156–159.

Parker, E. N. 1958. Dynamics of the interplanetary gas and magnetic fields. *The Astrophysical Journal*, **128**, 664–676.

Pellinen-Wannberg, A., Kero, J., Häggström, I., Mann, I., and Tjulin, A. 2016. The forthcoming EISCAT_3D as an extra-terrestrial matter monitor. *Planetary and Space Science*, **123**, 33–40.

Porter, J. G. 1944. An analysis of British meteor data: Part 2, Analysis. *Monthly Notices of the Royal Astronomical Society*, **104**, 257–271.

Rogers, L. A., Hill, K. A., and Hawkes, R. L. 2005. Mass loss due to sputtering and thermal processes in meteoroid ablation. *Planetary and Space Science*, **53**, 1341–1354.

Ryabova, G. O. 2013. Modeling of meteoroid streams: The velocity of ejection of meteoroids from comets (a review). *Solar System Research*, **47**, 219–238.

Sarma, T., and Jones, J. 1985. Double-station observations of 454 TV meteors. I - Trajectories. *Bulletin of the Astronomical Institutes of Czechoslovakia*, **36**, 9–24.

Sekanina, Z., and Southworth, R. B. 1975. *Physical and dynamical studies of meteors. Meteor-fragmentation and stream-distribution studies*. NASA CR 2615. Cambridge, MA: Smithsonian Institution.

Šimek, M. 1966. Some errors in determining meteor velocities by the diffraction method. *Bulletin of the Astronomical Institutes of Czechoslovakia*, **17**, 90–92.

Slavin, J. D., Frisch, P. C., Müller, H.-R. et al. 2012. Trajectories and distribution of interstellar dust grains in the heliosphere. *The Astrophysical Journal*, **760**, 46.

SonotaCo. 2009. A meteor shower catalog based on video observations in 2007–2008. *WGN, Journal of the International Meteor Organization*, **37**, 55–62.

Srama, R., Ahrens, T. J., Altobelli, N. et al. 2004. The Cassini Cosmic Dust Analyzer. *Space Science Reviews*, **114**, 465–518.

Staubach, P., Grün, E., and Jehn, R. 1997. The meteoroid environment near Earth. *Advances in Space Research*, **19**, 301–308.

Steel, D. 1996. Meteoroid orbits. *Space Science Reviews*, **78**, 507–553.

Sterken, V. J., Altobelli, N., Kempf, S. et al. 2012. The flow of interstellar dust into the solar system. *Astronomy & Astrophysics*, **538**, A102.

Sterken, V. J., Strub, P., Krüger, H., von Steiger, R., and Frisch, P. 2015. Sixteen years of Ulysses interstellar dust measurements in the Solar System. III. Simulations and data unveil new insights into local interstellar dust. *The Astrophysical Journal*, **812**, 141.

Štohl, J. 1970. On the problem of hyperbolic meteors. *Bulletin of the Astronomical Institutes of Czechoslovakia*, **21**, 10–17.

Strub, P., Krüger, H., and Sterken, V. J. 2015. Sixteen years of Ulysses interstellar dust measurements in the Solar System. II. Fluctuations in the dust flow from the data. *The Astrophysical Journal*, **812**, 140.

Strub, P., Sterken, V.J., Soja, R. et al. 2019. Heliospheric modulation of the interstellar dust flow on to Earth. *Astronomy & Astrophysics*, **621**, A54.

Szasz, C., Kero, J., Meisel, D. D. et al. 2008. Orbit characteristics of the tristatic EISCAT UHF meteors. *Monthly Notices of the Royal Astronomical Society*, **388**, 15–25.

Taylor, A. D. 1995. The Harvard Radio Meteor Project meteor velocity distribution reappraised. *Icarus*, **116**, 154–158.

Taylor, A. D., and Elford, W. G. 1998. Meteoroid orbital element distributions at 1 AU deduced from the Harvard Radio Meteor Project observations. *Earth, Planets, and Space*, **50**, 569–575.

Taylor, A. D., Baggaley, W. J., Bennett, R. G. T., and Steel, D. I. 1994. Radar measurements of very high velocity meteors with AMOR. *Planetary and Space Science*, **42**, 135–140.

Taylor, A. D., Baggaley, W. J., and Steel, D. I. 1996. Discovery of interstellar dust entering the Earth's atmosphere. *Nature*, **380**, 323–325.

Thomas, E., Horányi, M., Janches, D. et al. 2016. Measurements of the ionization coefficient of simulated iron micrometeoroids. *Geophysical Research Letters*, **43**, 3645–3652.

Tkachuk, A. A., and Kolomiets, S. V. 1985. The distribution of angular elements in the near-parabolic and near-hyperbolic orbits of meteoric bodies. *Meteornye Issledovaniia*, **10**, 67–74.

Verniani, F. 1973. An analysis of the physical parameters of 5759 Faint Radio Meteors. *Journal of Geophysical Research*, **78**, 8429–8462.

Von Nießl, G., and Hoffmeister, C. 1925. *Katalog der Bestimmungsgrößen für 611 Bahnen großer Meteore*. Vol. 100. Wien: Denkschriften der Akademie der Wissenschaft. Mathematisch-Naturwissenschaftliche Klasse.

Watson, Fletcher. 1939. A study of fireball radiant positions. *Proceedings of the American Philosophical Society*, **81**(4), 473–479.

Waza, T., Matsui, T., and Kani, K. 1985. Laboratory simulation of planetesimal collision: 2. Ejecta velocity distribution. *Journal of Geophysical Research: Solid Earth*, **90**(B2), 1995–2011.

Weryk, R. J., and Brown, P. 2004. A search for interstellar meteoroids using the Canadian Meteor Orbit Radar (CMOR). *Earth, Moon, and Planets*, **95**, 221–227.

Westphal, A. J., Stroud, R. M., Bechtel, H. A. et al. 2014. Evidence for interstellar origin of seven dust particles collected by the Stardust spacecraft. *Science*, **345**, 786–791.

Whipple, F. L. 1940. Photographics meteor studies. III. The Taurid shower. *Proceedings of the American Philosophical Society*, **83**, 711–745.

Whipple, F. L. 1950. A comet model. I. The acceleration of comet Encke. *The Astrophysical Journal*, **111**, 375–394.

Wiegert, P., Vaubaillon, J., and Campbell-Brown, M. 2009. A dynamical model of the sporadic meteoroid complex. *Icarus*, **201**, 295–310.

Wiegert, P. A. 2014. Hyperbolic meteors: Interstellar or generated locally via the gravitational slingshot effect? *Icarus*, **242**, 112–121.

Wiegert, P. A. 2015. Meteoroid impacts onto asteroids: A competitor for Yarkovsky and YORP. *Icarus*, **252**, 22–31.

Woodworth, S. C., and Hawkes, R. L. 1996. Optical Search for High Meteors in Hyperbolic Orbits. Pages 83–86 of: Gustafson, B. Å. S., and Hanner, M. S. (eds), *Physics, Chemistry, and Dynamics of Interplanetary Dust*. Proceedings of the 150th colloquium of the International Astronomical Union held in Gainesville, Florida, USA 14–18 August 1995. [Astronomical Society of the Pacific Conference Series, vol. 104]. San Francisco: Astronomical Society of the Pacific.

Zaslavsky, A., Meyer-Vernet, N., Mann, I. et al. 2012. Interplanetary dust detection by radio antennas: Mass calibration and fluxes measured by STEREO/WAVES. *Journal of Geophysical Research (Space Physics)*, **117**, A05102.

Zhang, Q. 2018. Prospects for backtracing 1I/Oumuamua and future interstellar objects. *The Astrophysical Journal*, **852**, L13.

Zook, H. A., and Berg, O. E. 1975. A source for hyperbolic cosmic dust particles. *Planetary and Space Science*, **23**, 183–203.

Part V

Hazard

11

The Meteoroid Impact Hazard for Spacecraft

Gerhard Drolshagen and Althea V. Moorhead

11.1 Introduction

The outer surfaces of spacecraft in all orbits are exposed to impacts from meteoroids. Impact velocities in Earth orbit range from 11 to nearly 74 km s^{-1}. This high speed can lead to impact damage from even very small particles. This is a risk that has to be considered for every spacecraft design and operation. The analysis of retrieved hardware has provided a wealth of information on the meteoroid environment near Earth which is used for the development and validation of meteoroid flux models. The severity of potential damage from a meteoroid impact depends on the parameters of the impacting particle (like speed, size and impact direction) and on the type of effect. Such effects can be of a transient nature, such as interference with measurements or the operation of the spacecraft, or they may cause permanent degradation or even destruction of hardware.

Reviews of impacts on spacecraft and resulting effects are given in Foschini (2002) and Inter-Agency Space Debris Coordination Committee (2014). The risk from hypervelocity impacts is now regularly assessed for most spacecraft and, for some interplanetary missions, shielding against meteoroids is a main design driver. The list of observed or recognised impact effects is steadily increasing. It includes damage on exposed x-ray sensors, attitude changes of spacecraft requiring high pointing stability, electrical effects from the impact plasma, punctures of heat pipes, fluid containers or pressure vessels, and the cutting of tethers and cables and other structural damages. These effects have to be considered anywhere in space outside of "sufficiently thick" atmospheres.

Meteoroid engineering models can be used to quantitatively assess the impact risk; these models are developed for Earth orbit (Divine, 1993; Smith, 1994; Staubach et al., 1997; McNamara et al., 2004; Dikarev et al., 2005), interplanetary space (ibid.), special locations such as within the Jovian system (Liu et al., 2016), and special events such as meteor storms (Cooke and Moser, 2010) or a close encounter with a cometary coma (Moorhead et al., 2014). The meteoroid environment encountered by spacecraft in orbit is described in the next section. It is followed by an overview of the main features of a hypervelocity impact and the resulting effects, from structural damage to electrical effects caused by the impact-generated plasma. Observed spacecraft anomalies attributed to impacts are presented next. The standard impact risk assessment procedure and potential operational measures are outlined thereafter. Besides posing a threat to spacecraft, impacts can be used to measure the meteoroid flux, often using dedicated impact detectors on spacecraft. A brief overview will be given on the main technologies used for *in situ* impact detectors. The chapter concludes by discussing suggested future work.

11.2 The Meteoroid Environment Encountered by Spacecraft

The impact risk and resulting effects on spacecraft depend on several variables. The meteoroid flux determines the general risk or probability of being hit by a meteoroid; this flux is a function of meteoroid size. For an individual impact the size and impact velocity of the impacting meteoroid are most important. Other relevant parameters are the shape, material density and incident angle of the impactor. Whether an impact has any noticeable effect or even causes damage depends on the sensitivity of a spacecraft subsystem. Specific impact effects have different dependencies on mass and, to an even larger extent, velocity of an impacting meteoroid. Impact energy tends to determine the extent of the physical damage caused, the momentum of the impactor determines momentum transfer, and electrical effects may be a steep function of impact velocity.

11.2.1 Impact Velocity

Meteoroid impact velocities relative to the Earth's atmosphere can range from about 11 to nearly 74 km s^{-1}. The lower limit of this range is determined by the gravitational attraction of Earth; particles on interplanetary trajectories necessarily impact the Earth's atmosphere with speeds exceeding the escape velocity (approximately 11.1 km s^{-1} at a typical meteor altitude of 100 km). The upper limit of nearly 73.6 km s^{-1} is a combination of the Earth's maximum orbital velocity, the speed of an object on a parabolic heliocentric orbit at the corresponding point in the Earth's orbit, and the Earth's gravitational pull (this again corresponds to an altitude of 100 km). Only particles impacting the Earth in a "head-on" direction can achieve such speeds. The impact speed must also take into account the spacecraft's state; for instance, a spacecraft in a typical low Earth orbit may contribute (or remove) up to 7.8 km s^{-1} from the relative impact speed. This spacecraft motion can extend the range of impact speeds to Earth-orbiting spacecraft from about 4 to 81 km s^{-1}.

In interplanetary space, there is no minimum impact velocity. Although impact velocity is best described by a distribution of speeds, an average or characteristic impact velocity around 20 km s^{-1} is sometimes quoted for 1 au (e.g., Grün et al., 1985); the actual characteristic speed will depend on both the speed distribution and the application (average for momentum,

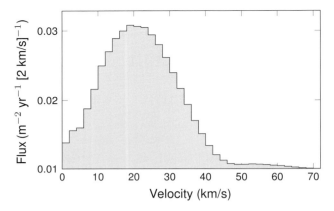

Figure 11.1. Meteoroid velocity distribution as modeled by the Meteoroid Engineering Model (McNamara et al., 2004) Release 2.0.5. Velocities are relative to a spacecraft orbiting the Sun at 1 au (i.e., the spacecraft is on an Earth-like orbit without considering planetary effects such as gravitational focussing and planetary shielding).

root-mean-square for kinetic energy, etc.). A distribution of meteoroid velocities relative to a spacecraft orbiting the Sun at 1 au is depicted in Figure 11.1. This velocity distribution was generated using the Meteoroid Engineering Model (McNamara et al., 2004); alternative models (Taylor, 1995) have different speed distributions (Drolshagen et al., 2008). Impact velocities will tend to be higher near the Sun and lower at larger distances from the Sun.

Compared to the "typical" speed of 20 km s^{-1} (at 1 au), stream meteoroids typically have higher velocities. For instance, the Perseids intercept the Earth at fixed speeds of nearly 60 km s^{-1}, the Leonids at 71 km s^{-1}, and Geminids at 35 km s^{-1} (Jenniskens, 1994; SonotaCo, 2009; Brown et al., 2010; Jenniskens et al., 2016). The stream velocity can differ by 1–2 km s^{-1} depending on the reference used.

Near some planets the meteoroid environment can be dominated by a locally created population. This is the case for the Jovian and Saturnian systems, where ejecta from impacts on the moons and collisions of ring particles create a local meteoroid population (Liu et al., 2016). Spacecraft impacts from such local meteoroids will typically be at velocities that are less than the local escape velocity; within the Jovian system, for example, impacts on a spacecraft orbiting in the same direction as the moons at low inclination and eccentricity will tend to be at speeds between 5 and 10 km s^{-1}. This is a result of the distribution of the local meteoroid population which peaks at ca 30° in inclination and at an eccentricity of 0.4 (Liu et al., 2016).

11.2.2 Impact Flux and Impactor Masses

According to a definition adopted by the International Astronomical Union in 2017, a meteoroid is defined as a natural object in space which is between 30 microns and 1 metre in size (Commission F1 of the International Astronomical Union, 2017). Smaller objects are considered dust and larger minor bodies are asteroids or comets. In this chapter, interplanetary dust will be sometimes implicitly included when small meteoroids are addressed.

Spacecraft are exposed to meteoroids of all sizes. Figure 11.2 shows the meteoroid flux onto the Earth as a function of limiting mass for the mass range 10^{-21}–10^{12} kg. The largest object that impacts Earth on average on a daily basis is about 0.5 m in size. Spherical shape and a material density of 2.5 g cm^{-3} are assumed for the conversion from mass to size. The impact flux increases strongly with decreasing particle size. A spacecraft with an exposed surface area of a few square metres will be hit daily by several micron-sized particles and every few years by a meteoroid of at least 0.2 mm or larger. If the cross-sectional area of the target is considered, the fluxes are four times larger than the surface fluxes (for convex objects).

Figure 11.2 is based on a variety of data sources, including spacecraft impacts (Grün et al., 1985; Drolshagen et al., 2017), lunar impact flashes (Suggs et al., 2014), optical meteor observations (Halliday et al., 1996; Brown et al., 2002; Drolshagen et al., 2017), and infrasound measurements (Silber et al., 2009). In order to derive a mass-limited flux measurement from each data set, the respective authors made some assumptions about the meteoroid speeds. For instance, Grün et al. (1985) assumed a characteristic impact speed of 20 km s^{-1}, while Koschny et al. (2017) assumed a speed distribution in order to derive the flux data presented in Drolshagen et al. (2017). A different choice of characteristic speed or speed distribution will result in a different flux; for instance, the relatively low speed of 17 km s^{-1} assumed by Love and Brownlee (1993) yielded a mass influx measurement that was higher than that of previous studies. A shift in the assumed speed will also correspond in a shift in the mass at which the mass influx peaks; according to Love and Brownlee (1993), this lies between 10^{-9} and 10^{-8} kg. It should be noted that the correct velocity distribution of meteoroids in space still has considerable uncertainties. Meteoroid detection methods are usually based on monitoring their effects, like meteors in the atmosphere or impact craters. Careful de-biasing is required, and far from trivial, to obtain the correct un-biased velocity distribution.

The log-log plot in Figure 11.2 suggests good agreement between the various measurements and models. However, there are still considerable differences and uncertainties, especially for the mass range 10^{-6}–10^4 kg. The best fit to the curves in Figure 11.2 gives a total daily mass influx to Earth of 54 t (for an upper size limit of 1 km; Drolshagen et al., 2017). Other models give daily accumulation masses of Earth between 5 and 300 t. At least values between 30 and 150 t are still within the range of uncertainty of the latest measurements and models. (For a more detailed discussion, see Drolshagen et al., 2017).

Impact fluxes are also modified by the gravitational attraction and shielding of planets and moons. A massive body's gravitational attraction will enhance the local meteoroid flux and accelerate meteoroids to higher speeds (Öpik, 1951; Kessler, 1972; Jones and Poole, 2007). Globally, the flux enhancement takes the form

$$\eta_g = 1 + \frac{v_{\text{esc},h}^2}{v_g^2} = \frac{v_h^2}{v_g^2}, \quad (11.1)$$

where η_g is the factor by which the flux is enhanced at a given altitude over a massive body, averaged over all angles; $v_{\text{esc},h}$ is the escape velocity at the given altitude; v_g is the velocity of the meteoroid relative to the massive body before entering the

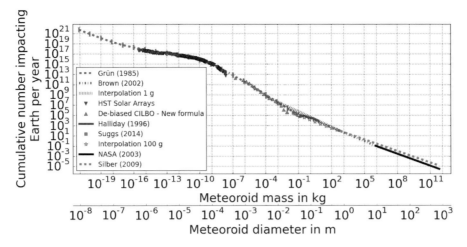

Figure 11.2. Cumulative number of impacts of natural particles onto the Earth per year (from Drolshagen et al., 2017). (A black and white version of this figure will appear in some formats. For the colour version, please refer to the plate section.)

body's sphere of gravitational influence; and v_h is the velocity of the meteoroid relative to the massive body at a given altitude h, where $v_h = \sqrt{v_g^2 + v_{esc,h}^2}$. A large body can also physically block some meteoroids from reaching the spacecraft in a phenomenon that we call shielding. This shielding factor is

$$\eta_s = \frac{1 + \cos \psi}{2}, \text{ where} \quad (11.2)$$

$$\sin \psi = \frac{v_H}{v_h} \frac{R_b + H}{R_b + h}. \quad (11.3)$$

Here, v_H is speed of the meteoroid at the surface of the massive body or at the top of its atmosphere ($v_H = \sqrt{v_g^2 + v_{esc,H}^2}$), R_b is the radius of the massive body, and H is the effective thickness of its atmosphere for the purposes of blocking meteoroids ($H = 0$ for airless bodies such as the Moon).

Equations (11.1) and (11.2) are global averages; the local gravitational focusing and shielding factors will depend on the relative position of the spacecraft, planet or moon, and meteoroid radiant (see Jones and Poole, 2007, for a full directional treatment). Note also that, because gravitational focusing is a function of speed, the meteoroid speed distribution will vary with altitude.

Finally, a spacecraft's velocity will affect the meteoroid flux onto its surface. The flux from meteoroid radiants anti-aligned with the spacecraft's ram direction will be enhanced, and the flux from the opposite direction diminished (Kessler, 1972).

11.3 Hypervelocity Impacts

If the impact velocity exceeds the speed of sound in the target material (about 5 km s^{-1} for aluminium) it is called a hypervelocity impact (HVI). In this case the impact is more similar to an explosion than it is to a forward-directed "bullet shot" (at 2.9 km s^{-1} the kinetic energy of a particle is equivalent to the chemical energy of the same mass of TNT). This is reflected in the effects on spacecraft resulting from such an HVI.

Hypervelocity impacts create a shock wave in the target material and lead to very high pressures (>100 GPa) and temperatures (>10 000 K). The first impact signal is usually a flash of light. The maximum pressures and temperatures are a strong function of the impact velocity. The high pressures and temperatures lead to structural failure of a certain target volume that is ejected from the impact site, leaving a crater surrounded by an area of damaged target material.

If the target is penetrated, material is ejected to both sides of the hole: both backward (i.e., in the direction where the object came from) and forward (in the flight direction). Figure 11.3 shows a combination of photographs that demonstrate some of the main effects of a hypervelocity impact. In this experiment, an aluminium projectile of 0.32 g mass is impacting a 1 mm thick aluminium wall under normal incidence with $v = 5.5$ km s^{-1} (Schimmerohn et al., 2010). The light flash and both uprange and downrange ejecta are clearly visible.

The main impact process lasts only a few microseconds. The impacting object and the target material are fragmented, melted, or vaporised. The ratio depends on the impact velocity and materials. For impacts with velocities up to about 5 km s^{-1}, most of the ejected material takes the form of solid fragments. Above 20–25 km s^{-1} the ejecta is completely vaporised. Ejecta from impacts in the velocity range 5–20 km s^{-1} are a mixture of solid fragments, molten droplets and vapour.

A small fraction of the ejected material is ionised. This fraction is a strong function of the impact velocity but does not exceed about 1% for even the fastest meteoroids. This fraction can be deduced from the measured charge yield of a hypervelocity impact test (see e.g. Collette et al., 2014) and also from Equation (11.5), considering that the ejected mass is at least a factor of 10 larger than the impactor mass). The impact plasma can interact with charged spacecraft surfaces, and interfere with the spacecraft electronics and signals.

The ejected mass can be much larger (by a factor of 10–1000) than the mass of the original impactor (see, e.g., McDonnell, 2005; Inter-Agency Space Debris Coordination Committee, 2014). Most of the impact energy ends up in the ejecta. Ejecta can be very numerous. For solid ejecta, the largest particles are comparable to the size of the impactor and the maximum ejection velocity is similar to the impact velocity. A small portion can even be ejected at higher velocities (jetting effect).

Figure 11.3. Effects of hypervelocity impacts (from Schimmerohn et al., 2010). Image credit: Fraunhofer EMI.

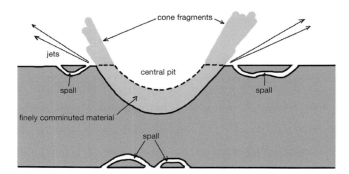

Figure 11.4. Sketch of main impact crater features (adapted from McDonnell and Gardner, 1998).

For brittle materials (e.g., glass) larger spall fragments can be ejected at lower velocities around the main crater. The damaged area of the main impact feature can be ten to twenty times the size of the impactor. For laminate structures even larger fragments can break away, as observed on several craters of the retrieved Hubble Space Telescope (HST) solar arrays (McDonnell, 2005). A schematic view of the main cratering and ejection features resulting from a hypervelocity impact is presented in Figure 11.4.

11.3.1 Crater Shapes and Penetration Capabilities

Most craters are roughly hemispherical. For ductile materials, only impacts under an angle of more than approximately 60° from the surface normal cause craters with an elongated shape. In other materials, like compounds with fibres and laminate structures, the impact features can have more anisotropic shapes determined by the target composition. In ductile materials, like metals, crater diameters and penetration depth are typically two to five times larger than the diameter of the impactor. These values are based on empirically derived damage equations as presented, e.g., in the IADC protection manual (Inter-Agency Space Debris Coordination Committee, 2014). In brittle material a zone of shattered material often surrounds a central crater. The visible crater size can be ten to twenty times the size of the impactor (McDonnell, 2005). If the target is thin relative to the impactor (foils), the hole size approaches the size of the impactor. Figure 11.5 shows a typical impact crater on a retrieved solar cell of the Hubble Space Telescope.

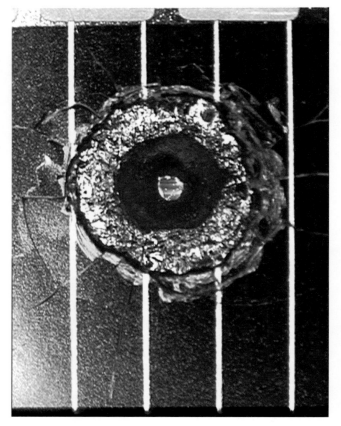

Figure 11.5. Impact crater on Hubble Space Telescope solar cell retrieved in 2002. The crater diameter is approximately 5 mm (McDonnell, 2005). (A black and white version of this figure will appear in some formats. For the colour version, please refer to the plate section.)

If a target is penetrated by more than about 50% of its thickness, spall will detach from its rear side and/or a complete hole will form (Hayashida and Robinson, 1991). Long thin structures like tethers, cables or booms can be cut by meteoroids which are only about one-third the size of the target diameter (McBride and Taylor, 1997; Sabath and Paul, 1997; Inter-Agency Space Debris Coordination Committee, 2014). A pressure vessel like a tank or pipe can burst if the penetration depth exceeds one-third to one-quarter of the vessel's wall thickness. It should be pointed out that most of the numbers given in the section above for crater sizes and thresholds are general indications only. They are intentionally kept vague as they will differ in each specific case depending on the parameters of the impacting particle and on the target properties. For some commonly used targets and shielding configurations, specific "damage equations" have been derived that relate impact features to impactor properties (see e.g. Miller, 2017; Inter-Agency Space Debris Coordination Committee, 2014). These are empirically derived and often based on a rather limited number of impact calibration tests. For high-value spacecraft where more accurate and quantitative knowledge of potential impact damage is needed, dedicated impact test campaigns are often required.

If the target is composed of several layers with a spacing in between, this configuration can act as a so-called Whipple shield. In a Whipple shield the outer layer breaks up and melts and vaporises the impactor. The debris cloud spreads within the spacing and when impacting the inner wall the impact load is spread out upon a larger area. Such a Whipple shield is a much more effective impact protector than a single wall consisting of the same total amount of material. It requires a minimum spacing between the outer "bumper" and the inner wall of 1–2 cm. This minimum distance is required to allow some spreading of the debris cloud after penetration of the bumper and to prevent spall from the inner side of the bumper hole from connecting with the inner wall. A general rule is that the minimum spacing of a Whipple shield should be fifteen times the maximum particle size to be defeated. It is most effective in the hypervelocity regime above approximately 7 km s^{-1} where vaporisation and melting of the impactor is most complete, and its protection increases with increasing separation distance of the walls. Some other parameters, like the shape and material density of the impactor, will be important for the resulting shape of the impact feature and for the penetration capability and resulting damage. Most impact tests have been carried out with spherical metal objects. The impact effects of, e.g., tethers, plates or rods will be different from spheres and depend on the orientation during the impact as well. Efforts are ongoing to study these parameters as well. More detailed information on structural effects and protective impact shields can be found in the Protection Manual of the IADC (Inter-Agency Space Debris Coordination Committee, 2014).

11.3.2 Impacts from Man-Made Debris

Impact effects from meteoroids and man-made space debris (also called orbiting debris) are very similar. Typical impact velocities to spacecraft in low Earth orbit are 20 km s^{-1} for meteoroids and 10 km s^{-1} for space debris. That puts all of them in the hypervelocity regime. The crater shape alone does not allow a distinction between meteoroids and space debris as the source. Information on the type of source could be obtained by the impactor trajectory, but that data is usually not available. A chemical residue analysis is one possible method of distinction. Typical elements indicating natural meteoroids include: Ca, Mg, Fe, Ni, Si, S, O and C. Man-made space debris often contains: Cr, Zn, Cu, Mn, Al, Ti, Na, K, Si, O, C and Fe. Several elements are found in both types of impactor (e.g., O, C, Fe, Si). Only a combination of several typical elements found in individual residue fragments should be considered as indication of the type of source. Some personal interpretations in the identification of the impactor are unavoidable but minimised by a systematic 2-D scan around an impact crater for different elements. For retrieved HST solar cells, a detailed analysis of impact residues allowed the identification of the type of impactor in 75% of the analysed craters (McDonnell, 2005). The results for this specific case (HST was at 600 km altitude) indicated that impactors smaller than 10 microns were mainly made of space debris, while for the larger objects meteoroids dominated the impact flux. According to flux models, space debris should again dominate for fluxes on HST larger than 1 mm, but there were not enough craters in the resulting size range for statistically meaningful results (McDonnell, 2005). It should be kept in mind that as of today space debris is only relevant for Earth-orbiting spacecraft, while meteoroids are present throughout the Solar System and beyond.

11.4 Impact Effects

The consequences of impacts depend on the parameters of the impacting particle, the characteristics of the target, and the effects under consideration. In this section, we first address mechanical and structural effects in order of increasing impact energy or particle size. Effects of the impact plasma will be presented thereafter.

11.4.1 Impacts on Exposed Sensors

Micron-sized particles can damage exposed CCD sensors if they can reach the sensitive detector. Individual pixels are only a few to a few tens of microns in size. An impact by a micron-sized particle can lead to permanently bright pixels. This was noticed with some surprise on ESA's XMM-Newton satellite (a scientific mission to observe sky sources in x-rays). Its mirrors were made of very smooth concentric shells to direct x-rays entering at very shallow angles onto the CCD sensors. The mirror aperture was open and two reflections were required from the aperture to the sensor, which was about 10 m away. The sensors were only protected by a micron-sized foil. During normal operations, thirty-five bright pixels suddenly appeared, and these remained permanently bright thereafter (Strüder et al., 2001). This required a re-calibration of the image processing software to account for the damaged pixels (Strüder et al., 2001). A study program was initiated to analyse the cause of the anomaly. Experimental tests with the dust accelerator in Heidelberg on a representative set-up, including mirror samples and a working CCD, revealed the process. Under very grazing angles (up to 2–3 degrees from the surface plane), particles can be scattered on very smooth surfaces without breaking up. These conditions were fulfilled by the x-ray mirrors. It was even found that the incidence angle was further reduced by conversion of the energy component normal to flight to internal energy. The scattered dust then left the mirror almost parallel to the surface and could be scattered again on the second mirror, which focused the particle onto the sensor. The thin foil in front of the sensor was penetrated by the particle, creating more fragments, which were still big enough to damage several pixels within the spray area.

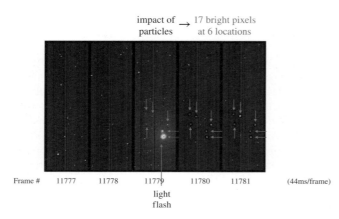

Figure 11.6. Impacts on exposed CCD of the type used by the XMM-Newton x-ray telescope (from Meidinger et al., 2003). Bright pixels on CCD following as a result of an impact test by a micron sized particle. (A black and white version of this figure will appear in some formats. For the colour version, please refer to the plate section.)

Figure 11.6 shows the results of the laboratory impact test on a real representative CCD (Meidinger et al., 2003). The bright pixels (seventeen in this case) are clearly visible. In parallel with the experimental test, numerical hydrocode simulations were carried out to analyse the scattering process on smooth surfaces. These numerical studies confirmed the observations and test results. Particles under very grazing incidence can be scattered intact on smooth surfaces and even become focused by such mirrors (Palmieri et al., 2003). Similar events were later observed again on XMM-Newton and also reported by the Swift x-ray telescope (Carpenter et al., 2008). Fluxes of 1-micron-sized particles from the standard meteoroid reference model for 1 au (Grün et al., 1985) agreed with the observations if an opening angle for scattering of about 2° from the mirror plane is assumed. This was a case where in-orbit observations, ground-based impact tests and numerical observations were all in agreement.

Later during the XMM-Newton mission, another event, likely resulting from an impact, permanently damaged a complete section of one CCD. The exposure maps before and after the event are shown in Figure 11.7.

11.4.2 Momentum Transfer by Impacts

Spacecraft with very accurate and stable pointing requirements are susceptible to an attitude change by the momentum transfer of an impact. Astrometry missions like Hipparcos and Gaia are prime examples of such missions. Other missions requiring high positional and attitude precision include gravity wave missions, like the planned LISA, or other formation flying systems. The momentum transfer to any satellite depends on the incident momentum and a potential enhancement of the momentum transfer by ejecta. A momentum transfer is most noticeable if it leads to a change in the attitude and pointing direction of the spacecraft.

ESA's Hipparcos satellite detected 157 impact-induced attitude changes during 682 days of operation (see Figure 11.8). For thirty-eight of these events it was even possible to deduce the x, y and z components of the momentum transfer (Ernst, 2007). When the software of the spacecraft noticed the attitude

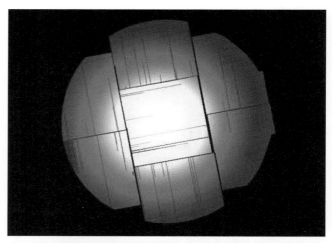

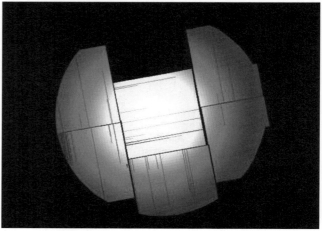

Figure 11.7. Exposure maps before (top) and after (bottom) the loss of a section of the MOS1 CCD sensor of the XMM-Newton telescope following an impact. Image credit: ESA. (A black and white version of this figure will appear in some formats. For the colour version, please refer to the plate section.)

drift, automatic corrective action was taken. The attitude change resulting from an impact depends on the transferred momentum and on the impact location relative to the moments of inertia of the satellite. For any individual momentum transfer, the exact impact location is not known. Some information is obtained when several directional components are measured. For Hipparcos the momentum transfer detection threshold could be reached by an impactor as small as 10 microns in size (Ernst, 2007). Hipparcos was operating in a GTO orbit of 500 km by 35900 km. In this orbit it could also experience hits from space debris particles. However, most of the time the satellite was at altitudes where little space debris is expected. The majority of the recorded impacts certainly result from natural meteoroids.

Attitude changes by small impactors were also observed by Gaia (T. Prusti, private communication, and European Space Agency and Gaia Data Processing and Analysis Consortium, 2017) and by the LISA Pathfinder mission (Thorpe et al., 2016), a precursor to the gravity wave mission LISA. Gaia is heavier than Hipparcos, but it is also more sensitive to even smaller attitude changes because of its improved pointing accuracy. As expected, a few impacts per day lead to a noticeable attitude change because of the momentum transfer. As for Hipparcos,

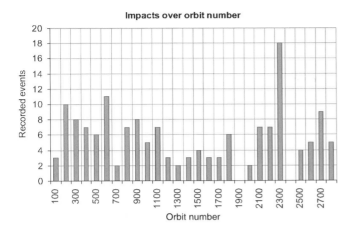

Figure 11.8. Recorded attitude changes by the Hipparcos astrometry satellite (from Ernst, 2007).

Figure 11.9. Impacts on front side of HST solar array. Each solar cell is 20 mm by 40 mm in size. The bright grid fingers are spaced by 1.25 mm. Image credit: ESA. (A black and white version of this figure will appear in some formats. For the colour version, please refer to the plate section.)

such attitude changes were automatically corrected by the on-board attitude and orbital control system.

Impacts from larger meteoroids at the same velocity will lead to larger momentum transfers which can also be noticed by less sensitive spacecraft. Such events are less frequent but have been observed as well, sometimes resulting in a noticeable orbital change or structural damage.

11.4.3 Impact Craters

Numerous impact craters have been observed in each case when hardware was retrieved after space exposure. Prime examples are the Long Duration Exposure Facility (Love et al., 1995), The European Retrievable Carrier (Drolshagen et al., 1996) and numerous impacts on Shuttle windows and the HST solar arrays retrieved in 1993 and 2002, which presented thousands of craters visible to the naked eye (McDonnell, 2005). Figures 11.9–11.13 show typical impact craters on HST solar cells. During the post-flight impact surveys of retrieved HST solar arrays more than 1000 high-resolution colour images were taken. Some representative examples are reproduced in McDonnell (2005). The full set is made available at ESA's Meteoroids and Debris Website (MADWEB).[1] The total solar cell compound was only 0.7 mm

[1] http://space-env.esa.int/madweb/

Figure 11.10. Impact features on front side of HST solar array caused by impact on back side. Crater size is 3.5 mm. Image credit: ESA. (A black and white version of this figure will appear in some formats. For the colour version, please refer to the plate section.)

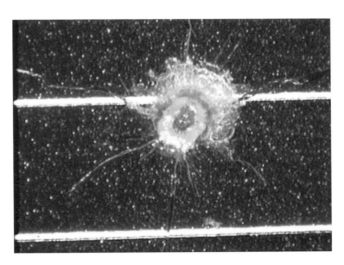

Figure 11.11. Small impact feature (about 0.5 mm crater size) on front side of HST solar array. The "butterfly" shape indicates a grazing incidence (> 70° from surface normal). Image credit: ESA. (A black and white version of this figure will appear in some formats. For the colour version, please refer to the plate section.)

thick. As a consequence, features from impacts on one side could be seen on the other side as well, and the solar array could be completely punctured. One method of distinguishing whether impacts are from meteoroids or man-made debris is a chemical analysis of residues found in impact craters. The most successful method proved to be 2-D scanning of the crater area and its immediate surroundings for various elements. Many elements are found in both meteoroids and debris, but a combination of elements in detected grains makes it possible to identify the impactor with a certain confidence. Nevertheless, it is clear that this procedure still leaves room for some personal interpretation of the results. Impact residues could be found in about 75% of the craters analysed (McDonnell, 2005). The chemical analysis of crater residues showed that most of the craters visible to

Figure 11.12. Front view of an impact hole in HST solar cell. The hole size is 3 mm. Image credit: ESA. (A black and white version of this figure will appear in some formats. For the colour version, please refer to the plate section.)

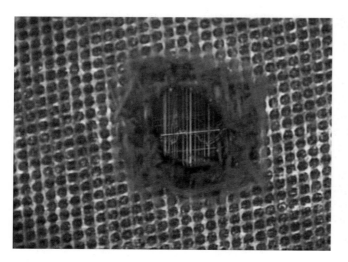

Figure 11.13. Rear view of the impact hole shown in Figure 11.12. The hole size is 3 mm. Image credit: ESA. (A black and white version of this figure will appear in some formats. For the colour version, please refer to the plate section.)

the naked eye were caused by natural meteoroids and not by man-made space debris (McDonnell, 2005). Examples of impact craters on the windows of the Space Shuttle are shown in Figure 11.14. Figure 11.15 shows the solar array of Sentinel 1A, in space, which suffered from an impact on 2016 August 26. It was first noticed by a change in the attitude and orbit and a power loss of 5%. A small camera which was on-board to confirm the proper deployment of the arrays was activated and showed an impact feature approximately 40 cm in diameter. Most likely it is the result of the impact of a 1-cm particle hitting the back side. In addition, the impact resulted in seven trackable fragments (Krag, personal communication). The spacecraft has remained operable despite this damage.

11.4.4 Severing of Long Wires

Experimental tests have shown that long, thin structures like wires, tethers, cables and booms can be cut by particles that are only about 30–40% the size of the wire thickness (McBride and Taylor, 1997; Sabath and Paul, 1997; Inter-Agency Space Debris Coordination Committee, 2014). The total exposed area of long tethers or harnesses can be several m^2. A 1-mm-thick tether of 10 km length has an exposed area of 31 m^2. If it can be cut by a meteoroid one-third of its thickness – i.e., by a 0.333-mm particle – the average lifetime will only be a few weeks. This prohibits the use of long single-strand tethers and requires a different design, such as a broader flat band or a type of ladder with redundant interconnections.

The high risk of thin structures being cut has been verified by observations. For example, the 20-km-long tether deployed in 1994 by the Small Expendable Deployer System, SEDS II, experiment was cut after 3.7 days (Chen et al., 2013). A 4000-m tether of the Tether Physics and Survivability Experiment (TiPS) was deployed in 1996 and cut in 2006 (Chen et al., 2013). The short lifetime of the SEDS II tether might have been an "unlucky" event, but single-strand tethers always have a high risk of being severed.

Retrieved hardware also showed examples of cut wires and springs. The HST solar arrays were held expanded by 0.5-mm thin wire springs. One of the springs on the arrays retrieved in 2002 was cut by an impact. The broken spring is shown in Figures 11.16 and 11.17. A wire of the impact detector DEBIE-2, deployed on the International Space Station (ISS) and later retrieved, was also cut by an impact; see Figure 11.18 (Menicucci et al., 2012).

11.4.5 Other Structural Damage

In addition to producing a primary crater, particle impacts can produce ejecta that generate secondary impacts and lead to heavily cratered areas. Smaller impactors may not individually produce sizable craters, but, due to their high numbers, their cumulative effect is analogous to sandblasting. Numerous tiny impacts can lead to a change of surface properties, and can erode or otherwise degrade mirrors and sensors.

On the other end of the size spectrum, particles are much less numerous but individually more hazardous. Sufficiently large particles may puncture pipes or pressure vessels. Pressure vessels on spacecraft in orbit are always considered as high-risk objects under impacts. HVI tests have shown that pressurised tanks can rupture or burst following an impact well before the tank wall is actually penetrated. The eventual reaction of the tank under impact depends on many parameters like the tank geometry, size, fluid or gas type and pressure, and impact location, in addition to the characteristics of the impacting particle. Overviews of impact tests and results are given in Putzar et al. (2008) and Inter-Agency Space Debris Coordination Committee (2014). For protection, pressure vessels are often located well inside a spacecraft or have extra shielding.

Even larger impactors can lead to severe structural damage or even complete destruction. Hypervelocity impact tests have demonstrated that a target is usually completely destroyed if the impact energy exceeds 40 kJ per kg of target mass (Johnson

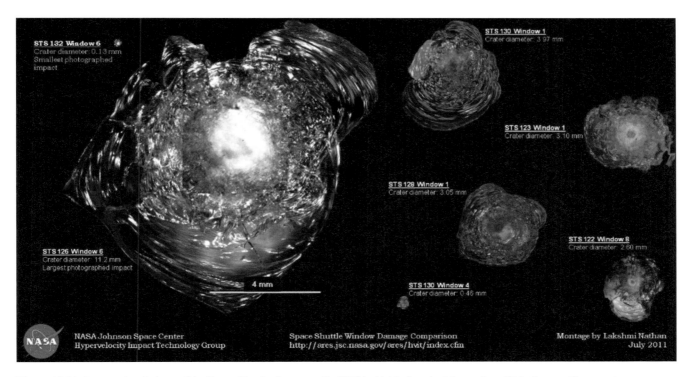

Figure 11.14. Impacts in windows of the Space Shuttle. Image credit: NASA. (A black and white version of this figure will appear in some formats. For the colour version, please refer to the plate section.)

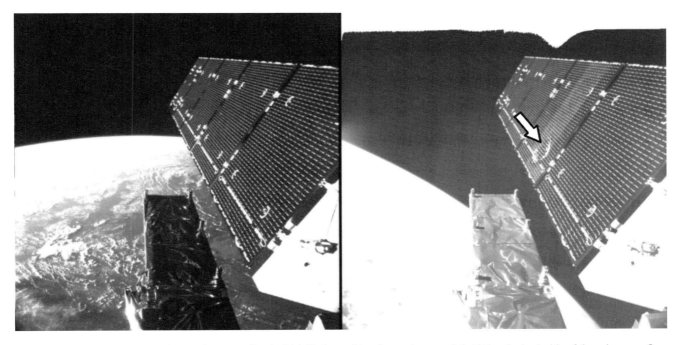

Figure 11.15. A 40 cm diameter impact feature on Sentinel 1A likely resulting from a 1-cm particle hitting the back side of the solar array. Image credit: ESA.

et al., 2001). The DebriSat experiment, in which a model satellite was subjected to a hypervelocity impact test in a laboratory setting, achieved an impact energy of 235 kJ per kg of target mass and resulted in complete destruction of the target, generating some 126 000 fragments (Ausay et al., 2017).

11.4.6 Impact Plasma

The extreme conditions during a hypervelocity impact partially ionise the projectile and target material and create plasma. Near the impact site this plasma can be very dense ($>10^{12}$ ions per cm^3). The initial plasma expansion velocity is similar to the bulk

Figure 11.16. A broken spring on HST solar array. The wire thickness is 0.5 mm. Image credit: ESA. (A black and white version of this figure will appear in some formats. For the colour version, please refer to the plate section.)

Figure 11.17. The other part of the broken spring on the HST solar array. The wire thickness is 0.5 mm. Image credit: ESA. (A black and white version of this figure will appear in some formats. For the colour version, please refer to the plate section.)

ejection velocity, corresponding to a plasma particle energy of a few eV. The rise time of a plasma signal is a few microseconds over some tens of cm distance. Plasma generation during impact depends strongly on the impact parameters, especially on the impact velocity (see Equation (11.4)). A sketch of the impact plasma generation is shown in Close et al. (2010). Figure 11.19 shows a sequence of a hypervelocity impact test from the Ernst Mach Institute in Freiburg, Germany to study the creation of the light flash and plasma cloud (Schimmerohn, 2012). This test was for a 1.6-mm aluminium sphere impacting at 7.4 km s^{-1} at an angle of 45° from normal. The particle size and location 4 s before impact is shown in the left picture. The initial electron density was about 10^{19} cm^{-3}, and the electron temperature $\approx$ 10 eV. In Frame 4, both parameters are lower by a factor of 10.

Figure 11.18. A broken wire on DEBIE-2 impact detector. The wire thickness is 0.075 mm (from Menicucci et al., 2012). Image credit: ESA. (A black and white version of this figure will appear in some formats. For the colour version, please refer to the plate section.)

Because of the high dependence on velocity, an impact by a 150–200 micron particle at 50 km s^{-1} could produce about the same charge, but with higher initial charge density.

The impact-generated plasma charge (Q) is mainly a function of the mass (m) and velocity (v) of the projectile (McBride and McDonnell, 1999):

$$Q \approx m^\alpha \cdot v^\beta, \tag{11.4}$$

where α is close to unity and derived values for β are typically between 2.5 and 7. McBride and McDonnell (1999) provide the following empirical relation for impact-generated plasma on the DEBIE detector (Schwanethal et al., 2005):

$$Q = 0.1\, m \left(\frac{m}{10^{-11}\,\text{g}}\right)^{0.02} \left(\frac{v}{5\,\text{km s}^{-1}}\right)^{3.48} \tag{11.5}$$

where Q is the charge in Coulomb, m the projectile mass in g and v the projectile velocity in km s^{-1}. To obtain formally correct units, a factor of C g^{-1} must be applied to Equation (11.5) to convert its output from grams to Coulombs. Note the steep dependence on impactor velocity. This equation was obtained during extensive calibration tests of the DEBIE impact ionisation detector. Projectile diameters ranged from sub-micron to a few microns and impact velocities ranged from a few km s^{-1} up to around 70 km s^{-1}. It is believed that this equation remains valid for larger masses as well. According to Equation (11.5), an impact from a Leonid meteoroid stream particle with mass of 1 microgram and an impact speed of 72 km s^{-1} will generate a plasma charge of 1.35×10^{-3} Coulomb.

The percentage of ionisation in the impact ejecta cloud depends on the material of the impactor and target and most of all on the impact velocity. A systematic study of the velocity dependence of the ionisation for different projectiles was performed by Stübig (2002) using the Van de Graaff accelerator at the Max Planck Institut für Kernforschung in Heidelberg. For sub-micron- and micron-sized particles, this electrostatic dust accelerator can reach velocities up to several tens of km s^{-1}, covering the full velocity range of meteoroids encountered in space. The experimental tests were performed for the calibration

Table 11.1. *Measured charge yield as a function of velocity for different projectile materials.*

Material	Impact speed range (km s^{-1})	Charge/mass (C kg^{-1}) Göller and Grün (1989)	Charge/mass (C kg^{-1}) Stübig (2002)
Iron	$v < 6$	1.0e-01 $v^{2.9}$	7.0e-02 $v^{3.4}$
	$6 < v < 12$	1.1e-00 $v^{1.5}$	1.9e-00 $v^{1.4}$
	$12 < v < 60$	1.5e-04 $v^{5.2}$	2.5e-04 $v^{5.0}$
Carbon	$v < 12$	8.0e-02 $v^{3.1}$	
	$12 < v < 20$	2.3e-00 $v^{1.8}$	5.0e-01 $v^{1.9}$
	$20 < v$	6.0e-05 $v^{5.3}$	
Silicate	$v < 6$	3.3e-02 $v^{4.6}$	
	$6 < v < 15$	5.5e-00 $v^{1.7}$	
	$15 < v$	5.1e-03 $v^{4.2}$	

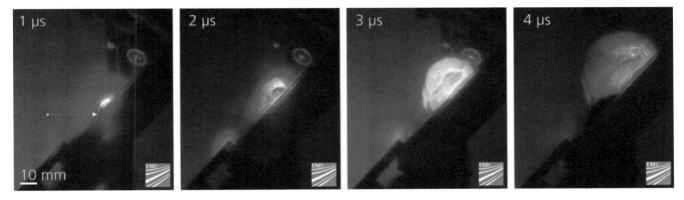

Figure 11.19. Impact light flash from a hypervelocity impact test of a 1.6-mm Al sphere impacting a solar generator sample at 7.4 km s^{-1} and an incidence angle of 45°. The tests were performed by EMI. Image source: Fraunhofer Ernst-Mach-Institut. (A black and white version of this figure will appear in some formats. For the colour version, please refer to the plate section.)

of the Cosmic Dust Analyser (CDA) flown on the Cassini mission. The CDA is an impact ionisation detector. Some results for different projectiles are reproduced in Table 11.1 (from Stübig, 2002). They are, strictly speaking, only valid for the CDA instrument. A comparison is made with earlier results (Göller and Grün, 1989). The results show that the plasma yield is more steeply dependent on velocity at high and low velocities than at intermediate velocities (with this intermediate range varying from 6–12 km s^{-1} to 12–20 km s^{-1}, depending on the material; see Table 11.1). The velocity dependence of the charge yield is much weaker for intermediate velocities. An offered explanation is that at low velocities a combination of secondary ejecta and surface contaminants with low ionisation potential dominates the plasma yield. Above a certain energy the secondary ion production is no longer rising and primary ions from the impact site will dominate the total ion yield. These ions are produced from a volume leading eventually to the steepest yield function (Göller and Grün, 1989; Stübig, 2002). Recent results on the micrometeoroid impact charge yield were presented by Collette et al. (2014). A comparison of the results from Göller and Grün (1989) and Stübig (2002) for iron and carbon shows good agreement for iron, while differences for carbon can reach almost an order of magnitude. A discussion and graphical representation of the differences are given in Stübig (2002).

11.4.7 Plasma Effects

The impact-generated plasma will normally recombine quickly. If some electrical fields are present, the positive and negative charges can be separated and a current pulse into neighbouring spacecraft wires or components could be generated. Every hypervelocity impact will also produce a flash of light. The light intensity or luminosity is believed to be a strong function of the impact velocity.

The plasma can lead to electromagnetic interference with spacecraft sub-systems and noise in scientific instruments. It is suspected that plasma oscillations of the dense and expanding impact plasma can lead to radiated interference with sensitive equipment (Foschini, 1998).

High-density impact plasma ($> 10^{20}$ electrons cm^{-3}) and its evolution (expansion, charge separation) can cause:

- Voltage spikes that could couple into cables
- Current spikes (Up to ampere level for 100 μm particles and high impact speeds and depending on the electrical configuration)
- Light flashes
- Radio emission from plasma oscillations
- Transient events in electronics
- Upsets/anomalies in computers and instruments (requiring restart)

- Permanent damage in computers/electronics
- Electrostatic discharges on pre-charged surfaces
- Triggering of arc discharges on operating solar arrays

The impact plasma can trigger a discharge of electrostatically charged dielectric surfaces or bulk material, releasing a current that is much larger than would be possible by the impact alone. This discharge current can reach ampere levels. Such a triggering of discharges by impact-generated plasma has been experimentally verified (Levy et al., 1997). The threshold (projectile size, generated plasma quantity, pre-impact charging level, etc.) for the triggering of discharges is largely still unknown and requires further investigation. However, it is already clear that even relatively small particles in the sub-millimetre size range could trigger discharges. These small-size particles are much more abundant than bigger particles, which could inflict structural damage.

Recent studies addressed a potential electromagnetic pulse generated by a hypervelocity impact (Close et al., 2010) and studied the freely expanding plasma from HVI (Lee et al., 2012). As plasma effects depend sensitively on electrical material properties and electrical parameters and circuits, like grounding, connections and pre-charge conditions, more work is needed for a quantitative understanding of impact induced plasma effects and synergies with other parameters of the space environment. An example is an impact-triggered electrostatic discharge of a pre-charged surface. Dielectric surfaces or materials can accumulate large amounts of negative charge. This can happen over hours or days for surfaces. Thicker internal insulators can collect electrons over weeks. Such pre-charged surfaces can suddenly release their charge as an electrostatic discharge, either spontaneously by exceeding an electric field threshold or by a triggering event. The shock wave and/or ionisation cloud can trigger such a discharge. This synergetic effect requires the presence of energetic electrons to charge the spacecraft and subsequently an impact, which can be from a rather small particle. Individually, the charging and impact might not have shown any effects.

11.5 Anomalies and Operational Measures

As is clear from the preceding sections of this chapter, a variety of possible meteoroid impact effects are capable of producing spacecraft anomalies or even failures. This section presents selected examples of suspected meteoroid-induced anomalies and discusses potential methods for reducing the risk of impact-related failures.

11.5.1 Example Anomalies

Missions experience anomalies and failures that, in many cases, cannot be definitively tied to a single cause. Even in cases where the cause of the anomaly may be known to a spacecraft's operators, that information may not be broadly shared. Therefore, it is impossible to know how often spacecraft experience meteoroid-related anomalies. Nevertheless, there are a handful of cases in which meteoroid impacts are either known or suspected. This section presents a few such examples.

11.5.1.1 Mariner IV, 1967

Mariner IV performed the first successful flyby of the planet Mars in 1965. Its suite of instruments included a 484 cm^2 impact detector known as the Cosmic Dust Detector. After completing the flyby, Mariner IV remained on an eccentric heliocentric orbit. Overall, Mariner's measurements of the interplanetary flux were within an order of magnitude of existing models (Naumann, 1966). However, Mariner's dust detector recorded two bursts of activity on September 14 and December 10–11 of 1967.

The first set of impacts on September 14 coincided with an attitude change and a one-degree drop in temperature in the interior of the spacecraft (Science and Technology Division, Library of Congress, 1968). The Cosmic Dust Detector recorded seventeen impacts within a period that is variously recorded as 15 (Science and Technology Division, Library of Congress, 1968) or 45 minutes (NASA, 1971), which would imply that the spacecraft was hit by thousands of particles of a similar size. This intense level of activity exceeds even that of meteoroid showers, and it has been hypothesised that the spacecraft passed near the nucleus of an unknown comet.

Mariner IV survived the September 14 onslaught and continued to operate normally. However, it experienced a second set of impacts in December of that year. This period of activity was less acute: five times as many impacts occurred, but over a period of about a day. It caused attitude disturbances that could not be corrected, as the gas supply had run out several days earlier. The mission was terminated on 1967 December 21.

11.5.1.2 ISEE-3, Impact in Period 1978–1982

The International Sun-Earth Explorer-3 (ISEE-3) spacecraft was a joint NASA-ESRO mission designed primarily to study the solar wind and its interaction with the Earth's magnetic field at the outer edges of the Earth's magnetosphere. It was the first spacecraft to be placed in a halo orbit about the L1 Lagrangian point. Thus, it inhabited a "pristine" region of space from an orbital debris point of view, and any impacts could be assumed to be natural rather than man-made.

It was reported that an instrument conducting a low energy cosmic ray experiment on the spacecraft was punctured by a meteoroid (Bloomquist and Graham, 1983), although the exact date of the anomaly was not provided. Bedingfield et al. (1996) claim that the impact negatively affected the mission in the form of a 25% data loss, although they identify the mission as ISEE-1, raising doubts as to whether these two works reference the same event and spacecraft.

ISEE-3 appears to have suffered no long-term effects from the impact, and was later used to visit comet 21P/Giacobini-Zinner as the International Cometary Explorer (ICE).

11.5.1.3 Olympus 1, 1993

Olympus 1 was an experimental communications satellite built for the European Space Agency (ESA). It was launched in 1989, but in 1991 it began to experience a series of problems, starting with a loss of tracking in one of its solar arrays in January of that year (Lenorovitz, 1991). Attitude control was lost in May 1991 due to a combination of sensor problems and untested command sequences, and seventy-two days elapsed before the satellite was recovered (Lenorovitz, 1991; Caswell et al., 1995).

In August 1993, Olympus 1 again experienced a failure of attitude control and entered an uncontrolled spin from which it could not be recovered (Caswell et al., 1995). The date of the anomaly – 1993 August 11 – coincided with the peak of the Perseid meteor shower. Unusually high activity had been predicted for the 1993 Perseids, and NASA had even delayed a shuttle launch to avoid the shower (an outburst of at least 250 meteors per hour did occur; Rendtel, 1993). In this context, the anomaly seemed to some likelihood to be a Perseid impact.

ESA performed an investigation of the potential causes of the failure. The anomaly could not be definitively linked to a Perseid impact, but it was noted that the spacecraft had a significant surface area (8.5 m^2) exposed to the Perseid radiant at the time of the impact, that charge from a meteoroid impact plasma could have been conducted to the gyro through the spacecraft umbilical, and that the Perseids were five times more likely to produce an impact plasma than a sporadic meteoroid impact (Caswell et al., 1995). The relative risk from stream meteoroids is further discussed in Section 11.5.2.4.

11.5.1.4 Chandra, 2003

Chandra is a space-based X-ray observatory launched and operated by NASA. Its geocentric distance ranges from 16 000 to 133 000 km, placing it well out of range of the man-made debris environment in low Earth orbit. On 2003 November 15, the Chandra spacecraft exhibited a sudden attitude change of about 13 arcseconds (Bucher and Martin, 2004).

Initially, a Leonid impact was hypothesised due to the timing of the event during the annual Leonid meteor shower. At typical Leonid speeds, the impact would correspond to a particle of approximately 1 mm in diameter. However, a subsequent analysis indicated that an impact by a sporadic meteoroid was five times more likely (Cooke, personal communication). This case illustrates the tendency to assign impacts to meteor showers despite their low contribution to meteoroid impact risk (see Section 11.5.2.4).

11.5.2 Risk Assessment and Mitigation

The risk of meteoroid impacts can be mitigated through spacecraft design or by operational measures. Design-based measures are generally required for mitigating the sporadic complex, while operational measures can be used to temporarily reduce exposure to short-term, directional enhancements of the meteoroid environment such as shower outbursts. Beyond this general rule, the exact mitigation measures will depend on the particular needs of the mission. For example, manned missions may prioritise protection against spacecraft punctures in order to avoid loss of life. In contrast, meteoroid momentum and energy transfer can pose the greatest operation risk to sensitive space telescopes such as Gaia (Risquez et al., 2012). In some cases, it can be difficult to determine the relative importance of the different modes of meteoroid-induced failure; the limiting mass for a damaging electrostatic discharge, for instance, is uncertain.

11.5.2.1 Evaluation of Risk

The first step in any impact risk assessment is an identification of the potential problem or failure mode that could result from an impact. All effects described previously in this chapter should be considered. Potential problems that are not direct failures of a system or sub-system include measurement disturbances, e.g., through an impact-induced attitude change or through electromagnetic interference with impact plasma. Other problems could be degradation of surfaces, solar cells and sensors. Direct failures of a subsystem include the cutting of tethers, harnesses or wires and the penetration of pressurised volumes, pipes or tanks. The penetration of the outer wall of an electronics box is usually considered as a failure of the corresponding subsystem as well.

The second step is an identification of mission requirements which could be threatened by impacts. Typical relevant requirements can address the performance or operational availability of a spacecraft or the structural integrity of a space system or sub-system. Examples of specific requirements are the minimum performance of solar arrays at the end of life, or requirements for the spacecraft software to correct recorded attitude changes within a certain period. Requirements on structural integrity are usually given in the form of a probability of no penetration for a system or sub-system under consideration.

The third step of the impact risk assessment is an identification of the affected area for a given failure mode, its exposure to the meteoroid environment and the type and thickness of shielding, if any. The affected target area could be the whole spacecraft, as in the case of an attitude change, or just a small part of the harness which is exposed.

In the next step, the critical particle size for an identified problem or failure has to be evaluated. In some cases, any impacting particle has to be considered and the potential problem results from an accumulation of effects (e.g. degradation of surfaces, mirrors or solar cells). In other cases, all objects above a given threshold in impact momentum or energy are of concern (e.g., for attitude changes or the triggering of electrostatic discharges). A frequently-used failure criterion is the cutting of a narrow structure, like a tether or cable or the penetration of an outer wall. In each case where the impactor needs to exceed a certain threshold to cause damage, the critical minimum particle size for this damage must be found. For a momentum change it will depend on the sensitivity of the spacecraft, its moments of inertia, on the impact location, and of course on the transferred momentum. This momentum transfer depends not only on the momentum of the impactor but also on potential impact ejecta which can enhance the momentum transfer by factors of 2–5 or more (Jutzi et al., 2015), depending on the impact velocity and target properties.

11.5.2.2 Damage Equations and Failure Probabilities

The critical particle size for penetration of a structure is usually obtained by so-called ballistic limit or damage equations. These damage equations give the critical particle size as a function of the parameters of the impacting particle (like impact velocity and direction and the material density) and properties of the target (like the wall thickness, material and density).

Table 11.2. *Values for failure mode constant K_f of Equation (11.6).*

Failure Mode Constant	Effect
$3.0 \leq K_f$	crater generation only (in this case, Equation (11.6) gives the critical particle size for penetration depth t into the target)
$2.2 \leq K_f < 3.0$	spallation of the plate
$1.8 \leq K_f < 2.2$	spall detaches from plate
$K_f < 1.8$	perforation of the plate

An example is given in Equation (11.6), which presents the critical particle size for the crater depth or penetration for impacts on a single Al plate (Miller, 2017):

$$d_c = \left[\frac{t}{K_f K_m \rho^{0.52} (v \cos\theta)^{0.667}} \right]^{0.947}, \quad (11.6)$$

where d_c is the threshold particle diameter for penetration in cm, t is the wall thickness in cm, K_f is the failure mode constant (see Table 11.2), K_m the material constant (0.33 for Al), ρ the material density of the projectile in g cm^{-3}, v the impact velocity in km s^{-1}, and θ is the impact angle (0° when the impactor velocity is perpendicular to the surface).

Ballistic limit equations (BLEs) have been derived empirically from hypervelocity gun tests of man-made materials (see, for example, Hayashida and Robinson, 1991; Inter-Agency Space Debris Coordination Committee, 2014). As these equations are not derived from basic physical principles, numerous damage equations exist for different types of shielding configurations. It is not possible to derive a dedicated equation for each specific type of shielding. To give just some examples, the outer shielding could consist of single walls made of metal, carbon fibres or compounds, multiple walls with spacings and with or without potential internal layers or honeycomb structures. And all could have thermal foils on top or not. In practice, one often tries to represent the real compound by a simplified shield with equivalent thicknesses for which a damage equation exists. Obviously, this introduces uncertainties in the impact risk assessment.

After these steps, the critical particle size for the problem or failure under consideration can now be determined. Important impact parameters for a damage assessment are the impact velocity and direction (see also Equation (11.6)). For a first impression of the impact risk a simplified calculation is often performed with fixed impact direction (usually 45°) and impact velocity (20 km s^{-1} for meteoroids) and projectile material density of 2.5 g cm^{-3}. With these fixed parameters and the selected damage equation the critical particle size is easily obtained, in most cases by a very simple calculation. It is assumed that all particles larger than the critical size will cause the failure as well.

Next, for a quantitative evaluation, the fluxes encountered by the mission for the determined critical particle size have to be obtained. The mission fluence is then the time integrated flux.

The damage fluence is a function of the spacecraft's trajectory, mission duration, and the exposed surface area and orientation of the sub-system under study. Models such as NASA's Meteoroid Engineering Model (MEM; McNamara et al., 2004), the interplanetary model by Grün et al. (1985), and ESA's Interplanetary Meteoroid Engineering Model (IMEM; Dikarev et al., 2005) provide the meteoroid flux encountered along a given spacecraft trajectory. For the Jovian local meteoroid environment, a dedicated model, JMEM (Liu et al., 2016), has recently been developed. The fluxes on a randomly oriented plate are often used for an initial, simplified assessment.

For a given damage equation (or shielding type), we use f to denote the flux of meteoroids larger than the calculated critical particle size d_c. The average number of failures on an exposed surface of area A during a mission duration t will then be $N_f = f A t$. Shadowing from other spacecraft surfaces has to be considered to determine the exposed surface area. Note that the number of failures increases linearly with exposed area and exposure time.

Once N_f is obtained, the probability of exactly n failures occurring in the considered time interval t is given by Poisson statistics:

$$P_n = \frac{N_f^n}{n!} e^{-N_f}. \quad (11.7)$$

The probability of no failures, P_0, is thus given by:

$$P_0 = e^{-N_f}. \quad (11.8)$$

For values of $N_f \ll 1$, the probability of at least one failure, Q, is approximately equal to N_f:

$$Q = 1 - P_0 = 1 - e^{-N_f} \quad (11.9)$$
$$\approx 1 - (1 - N_f) = N_f \text{ for small values of } N_f. \quad (11.10)$$

If the requirement is met with a margin of at least a factor of 3–4, the risk assessment is completed with this simplified analysis and the impact risk is acceptable. This factor is based on practical experience with comparisons of simplified and fully 3-D calculations. It is an indication only and not a written guideline or specification. If the risk is close to or higher than the requirement, a more detailed and fully 3-D analysis should be performed. This will consider the full spacecraft geometry and flight attitude, shielding types and thicknesses, shielding from other spacecraft parts and the full velocity and directional distributions of the impactors. Such a calculation requires a numerical tool, like ESABASE/DEBRIS (Miller, 2017) or BUMPER II (Hyde et al., 2006). These tools also include flux models and damage equations and they will give the failure rate, N_f, and the probability of no failures, P_0. An example of ESABASE/DEBRIS results of such a fully 3-D impact risk assessment is shown in Figure 11.20, where the number of predicted impacts per m^2 from meteoroids with diameter >10 microns is displayed on a model of the Hubble Space Telescope for a mission duration of 3.62 years (the time between deployment and first replacement of the solar arrays).

11.5.2.3 Spacecraft Design and Shielding

Mitigation through spacecraft design is necessary for protection against the impact risk from meteoroids. In some cases, shielding is not possible and mitigation measures consist of

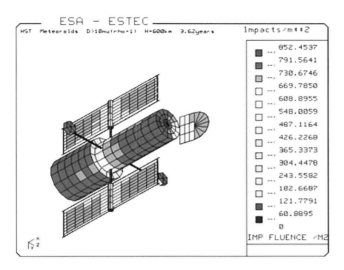

Figure 11.20. Predicted number of impacts from meteoroids with diameter > 10 microns during the first 3.62 years of the HST mission. (A black and white version of this figure will appear in some formats. For the colour version, please refer to the plate section.)

Figure 11.21. Two Whipple shield configurations that have been subjected to impact testing. Image credit: NASA. (A black and white version of this figure will appear in some formats. For the colour version, please refer to the plate section.)

redundancy (e.g. for wires), overdesign to account for degradation (e.g. solar arrays) or software and control measures (e.g. countering attitude changes). The impact risk from individual particles could be reduced by re-locating sensitive items or by having extra redundancy.

In other cases, enhanced shielding protection might be required to fulfil safety criteria. Thicker walls are one option. More efficient protection is provided by a so-called Whipple shield as described in Section 11.3.1. The outer wall or bumper may be as simple as a single aluminium plate offset from the spacecraft's surface, or a complex structure containing multiple layers of Kevlar, Nextel or other materials. When a particle strikes a layer with a thickness of at least 1/6 of its own diameter, it will be vaporised or significantly disrupted (Whipple, 1947). A Whipple shield is most effective in the hypervelocity regime above approximately 7 km s^{-1}. Figure 11.21 shows examples of Whipple shields with double and multiple walls. A single plate is often sufficient to disrupt a meteoroid, although additional layers may be needed for slower-moving particles (Christiansen et al., 1995). The resulting cloud of impact debris spreads and, when hitting the second wall, its load is distributed over a larger area. In this way a relatively thin second wall can prevent a penetration. Spacing between the walls is important, and increased distances provide better protection for the same wall thicknesses. Multi-layer insulation (MLI) can itself act as a bumper if it is spaced at least a few cm from the main wall and depending on the parameters v, d, etc. of the impactor.

Besides shielding, other design measures can be used to minimise risk. The use of redundant wiring can allow the mission to tolerate the occasional severing of wires. Sensitive components may be placed behind or inside less sensitive components. Spacecraft designed to minimise differential charging may also be less susceptible to electrical damage caused by meteoroid impact plasmas. Finally, impact-induced attitude changes should be considered when determining the amount of fuel necessary for the mission.

11.5.2.4 Operational Measures

The sporadic meteoroid complex constitutes the bulk of the risk to spacecraft; for a limiting kinetic energy equivalent to a 1 mm meteoroid moving at 20 km s^{-1}, sporadics constitute an estimated 98% of predicted impacts. Sporadic meteoroids are present throughout the year and are not limited to specific radiants (although they are concentrated in broad groups termed "sources"; e.g., Weiss and Smith, 1960). Individual millimetre-sized meteoroids cannot be detected and avoided, and thus operational measures are not useful against the sporadic complex.

In contrast, a meteor shower can produce a short-term elevation of risk (Moorhead et al., 2017). In some cases, the shower may be extremely brief; the Ursids, for example, last for less than one day. Such brief enhancements may be mitigated operationally. These operational measures can consist of delaying operations, altering the spacecraft's trajectory, reorienting the spacecraft, or powering off components.

Steel (1998) reported that shuttle launches were delayed due to Perseid shower activity; later, at least one launch was delayed due to the Leonids (O'Brien, 1999). The Leonid meteor shower does not pose any great risk to spacecraft in a typical year, but occasionally exhibits massive outbursts that can be of great concern. In November 1999, a predicted Leonid outburst prompted a delay in the shuttle launch schedule in an effort to avoid the storm.

The International Space Station is itself well-shielded against particle impacts. However, astronauts occasionally venture outside the station on planned "extravehicular activities" (EVAs). The planning stages of each EVA involves an assessment of the meteoroid shower risk at the time of the EVA to ensure that the hazards posed to the astronauts are within acceptable limits.

Besides delaying launch or operations, a mission may also alter its trajectory to avoid a brief or localised meteoroid flux enhancement. For example, the New Horizons mission developed (but did not use) several alternative trajectories for its Pluto fly-by (see Figure 11.22) due to concerns that a substantial debris population could be present around Pluto (see Lauer et al.,

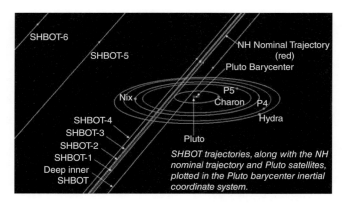

Figure 11.22. Nominal and alternate trajectories developed for the New Horizons Pluto fly-by, the latter of which are termed "Safe Haven by Other Trajectory", or SHBOT. These trajectories were designed to avoid any late-detected debris concentrations, but were not ultimately used. Image credit: NASA/JHUAPL. (A black and white version of this figure will appear in some formats. For the colour version, please refer to the plate section.)

2018, for a summary of the dust survey). In 2014, the Mars Reconnaissance Orbiter (MRO), Mars Atmosphere and Volatile Evolution mission (MAVEN) and Mars Odyssey spacecraft performed maneuvers to phase their orbit such that the planet Mars would shield them from the coma of comet C/2013 A1 (Siding Spring).

All meteoroids belonging to a shower or stream essentially move in parallel; thus, shower-related impact risk can also be mitigated through reorientation. The spacecraft can be turned to either minimise the cross-sectional area presented to the shower, or to expose a less sensitive side. For instance, the Hubble Space Telescope has been reoriented to reduce Perseid impacts on the telescope (Steel, 1998).

11.6 Impact Detectors

Several of the impact effects described in this chapter can can be used to design *in situ* impact detectors which use the effects of an impact as a detection method. Such impact detectors can be passive or active and several detection principles can be combined.

11.6.1 Passive Impact Detectors

Passive detectors are surfaces or specific material exposed to space without any active signal processing. They need to be retrieved for analysis on the ground. Examples include simple surfaces of a plain material or special designs aimed at a more or less intact capture of impactors. These capture cells can be made, e.g., of several foils or a material of very low density, like aerogel, which should gradually slow down and capture the particle (Burchell et al., 2008). Such capture cells were successfully used to retrieve cometary and interplanetary particles during the Stardust mission (Tsou et al., 2003). The advantage of passive detectors is their simplicity and the possibility to deploy relatively large surfaces. The disadvantage is the need for retrieval, which limits the type of orbit or requires a dedicated re-entry module. In addition, passive cells usually do not provide any time information on the impact event.

Retrieved space hardware can also be considered as a sort of passive impact detector if it is made available for a post-flight impact analysis.

11.6.2 Active Impact Detectors

Active *in situ* impact detectors acquire and send data to ground. They require all the resources of a space instrument such as power and telemetry. Their big advantage is that they do not need to be retrieved, making them the only option for most missions.

Almost all types of impact effects have been used for *in situ* impact detections. Impact ionisation detectors are considered to be the most sensitive. The electrons and ions of the impact plasma are collected and measured. The rise time of the signal is linearly related to the impact velocity. The signal strength is a strong function of the velocity (see Equations (11.4) and (11.5)), making this type of detector most suited for high-velocity impacts. Impact ionisation detectors can record impacts from sub-micron sized particles. Such detectors have flown on interplanetary spacecraft like Ulysses (to Jupiter and the solar poles; Grün et al., 1992b; Strub et al., 2015), Galileo (to Jupiter; Grün et al., 1992a; Krüger et al., 2005) and LADEE (to the moon; Elphic et al., 2014; Horányi et al., 2015). An enhanced version is the Cosmic Dust Analyzer (CDA; Altobelli et al., 2016) which was a main instrument of the Cassini mission to Saturn (see Figure 11.23). It included a mass spectrometer, which allowed the identification of the composition of some impactors.

Impact ionisation detectors have also flown in Earth orbit, like GORID (Geostationary Orbit Impact Detector, identical to the Ulysses detector) in geostationary orbit (Svedhem et al., 2000) and DEBIE (DEBris In-orbit Evaluator) in polar orbits and on the ISS (Schwanethal et al., 2005). Because of their high sensitivity, impact ionisation detectors are susceptible to interference from ambient plasma.

Another type of active detector measures the momentum or energy of an impactor. Examples include polyvinylidene fluoride (PVDF) detectors and piezo-electric transducers (PZT). PVDF detectors consist of an electrostatically charged material between two conducting surfaces. An impact releases the stored charge within a certain volume and the resulting current can be measured. The recorded signal is proportional to the impact energy (Simpson and Tuzzolino, 1985). PZT or other types of acoustic sensors measure the shock wave or acoustic waves created by the impact. The signal strength is usually proportional to the transferred momentum of the impactor (Kobayashi et al., 2013). Such momentum detectors are less sensitive than impact ionisation detectors, but they can still detect particles a few microns in size. In addition, they are rather simple and robust. PVDF or PZT detectors can be used as stand-alone detectors but they are also often combined with other types of detection either as a counter for high flux rates (like on Cassini) or to have a coincidence signal (like for DEBIE). The DEBIE detector combined an impact ionisation and PZT momentum detector. The smallest impacts were only detected by the more sensitive ionisation sensor, but above a certain threshold (of the momentum of the impactor) a signal was recorded simultaneously in the

Figure 11.23. The Cosmic Dust Analyzer on Cassini. Image credit: Lossen-Fotografie/MPIK Heidelberg. (A black and white version of this figure will appear in some formats. For the colour version, please refer to the plate section.)

ionisation and the momentum sensor. This coincidence detection improved the confidence that the signal was from a real impact (Schwanethal et al., 2005).

The penetrating capability of hypervelocity impacts can itself be used as a detection method. In fact, some of the first *in situ* detectors flown on Helios and the Pioneer spacecraft consisted of arrays of pressurised cans. A sudden pressure loss of these cans could be attributed to an impact large enough to penetrate the can wall, and the rate of the pressure loss could be taken as a crude indicator of the impactor energy. This type of threshold detector provided some of the very first measurements of meteoroid fluxes in interplanetary space. Alternatively, one could construct a detector out of a thin wire mesh and measure how frequently the wires are severed. Several such designs have been proposed by NASA (Hamilton et al., 2017) and DLR (Bauer et al., 2015).

Laser light curtains provide yet another method for detecting particles. In these detectors, laser diodes are arranged to form a thin "curtain" of light; particles passing through this curtain reflect and scatter light, which is in turn recorded by detectors such as photodiodes. In several cases, a time-of-flight measurement is possible, which allows the trajectory of the impactor to be deduced. Rosetta carried one such light curtain detector in the form of GIADA (Grain Impact Analyser and Dust Accumulator), which was designed to detect relatively slow particles drifting away from the nucleus of comet 67P (Colangeli et al., 2007).

The set of impact detection methods discussed here is far from a complete list. Other effects associated with impacts such as light flashes, attitude changes, and electrical interference (e.g., with radio signals) can also be used as detection methods. Basically, all known impact effects have been proposed, realised, or considered for an impact detector. Furthermore, detectors frequently incorporate multiple detection units or methods in order to better characterise impacts and impact rates by collecting more detailed information on the impact location or impactor trajectory.

11.7 Future Work

When meteoroid impacts on spacecraft are addressed, the main emphasis is usually on the potential damage and risk to the mission. For a quantitative and reliable assessment of the impact risk, both the impact effects and the fluxes of the impacting meteoroids must be known. This includes secondary effects such as the triggering of electrostatic discharges and of the impact ejecta on neighbouring surfaces. In addition to flux, it is also necessary to know the distribution of impact velocities and directions as well as the shape and densities of the impactors. At present most of these parameters and distributions are not well known.

The understanding of impact effects is far from satisfactory. Ballistic limit or damage equations are only available for a limited number of shielding types and materials. The shielding protection provided by more complex structures and modern composites is poorly understood. Present damage equations are empirically derived from a limited number of test shots, and it has never been possible to sample the full parameter space of impacting particles. Particles larger than a few hundred microns are usually fired by light gas guns, which are limited to speeds below 10 km s^{-1}. This is well below the typical impact velocity of meteoroids in space. Electrostatic accelerators can reach several tens of km s^{-1} but only for micron-sized particles.

Some impact effects, like the triggering of electric discharges or other plasma effects, are inherently difficult to study. Slight modifications of the materials, the circuitry or the charge conditions prior to the impact can lead to very different results.

More hypervelocity impact tests will improve our knowledge and lead to better damage equations. Such systematic tests are certainly needed. One alternative solution to close the knowledge gap could be numerical hydrocode simulations. In principle, such calculations can simulate all aspects of a hypervelocity impact from the conditions of the impact plasma, gas and fragments to the resulting crater size or penetration capability. Therefore hydrocode is currently "trained" on test data and used to interpolate between test points. An extrapolation by hydrocode to the high velocity regime is considered to have large uncertainties. For the hydrocodes to work (or be written) without such fits, first we need to understand all the physics. The available computing capabilities are sufficient for realistic large-scale simulations. In practice, it has also proven very difficult to find the proper material parameters needed for realistic simulations, but we expect to see progress in this field in the future.

Another area requiring more study is the development of reliable meteoroid flux models. Even in near-Earth space, meteoroid fluxes have estimated uncertainties of at least a factor of 3. This might come as some surprise, as meteors have been observed for thousands of years, but the entire suite of measurements and observations used to derive flux models probe meteoroid effects. This includes both the analysis of impact craters and radar and optical measurements of meteors. These methods all suffer from biases. For instance, an impact crater of a given size may be the product of a larger, slower particle or a smaller, faster particle (impact angle and particle density are also entangled in crater depth). And it is well known that meteor measurements are heavily biased towards faster particles, but the exact dependence of the luminosity and ionisation on the meteoroid velocity is unknown. Space-based *in situ* measurements are only possible for relatively small sensor areas limiting the sampled population to relatively small sizes. In addition, they are also based on effects like the momentum or energy transferred during impact. The uncertainty of the meteoroid population away from Earth in interplanetary space is much larger still. For heliocentric distances between Mercury and Saturn, the uncertainty in meteoroid fluxes for a given size is at least a factor of 10. To date, it has not been possible to fit fluxes derived from meteor observations and from infrared measurements of the zodiacal light simultaneously.

From a risk point of view, the meteoroid population in the size range between a few hundred microns and a few millimetres is most important. These objects are large enough to cause direct damage on spacecraft, and they are fairly abundant. Larger meteoroids will cause even more damage, but there are far fewer of them. This size range is sampled by meteor observations. A thorough understanding of the underlying physics and biases of meteors could help to derive accurate meteoroid population models. Space-based *in situ* measurements with detectors of a few square metres in size could also sample this population and provide complementary data. In some sense impacts on spacecraft will also help to better understand the meteoroid population.

The final step would be the derivation of a reference meteoroid population model that starts from the sources of meteoroids, i.e. comets and to a lesser extent asteroids, and follows their evolution for a sufficiently long term to obtain an equilibrium population. When this is done correctly, including all sources, interactions, and sinks, the model should describe reality and be able to fit existing measurements consistently. Efforts in this direction are ongoing. Such a model assumes that the meteoroid population is in a steady state at present. Whether that is true is another open question. Although a reliable meteoroid flux model together with an accurate understanding of the impact effects will not prevent any impacts on spacecraft, it will permit a proper consideration and mitigation of this risk.

References

Altobelli, N., Postberg, F., Fiege, K. et al. 2016. Flux and composition of interstellar dust at Saturn from Cassini's Cosmic Dust Analyzer. *Science*, **352**, 312–318.

Ausay, E., Cornejo, A., Horn, E. et al. 2017. A comparison of the SOCIT and DebriSat Experiments. In: T. Flohrer and F. Schmitz (eds.), *Proceedings of the 7th European Conference on Space Debris, 18–21 Apr. 2017, Darmstadt, Germany*, 8 pages. Darmstadt, Germany: ESA Space Debris Office. https://conference.sdo.esoc.esa.int/proceedings/sdc7/paper/729/SDC7-paper729.pdf.

Bauer, W., Romberg, O., and Putzar, R. 2015. Experimental verification of an innovative debris detector. *Acta Astronautica*, **117**, 49–54.

Bedingfield, K. L., Leach, R. D., and Alexander, M. B. 1996. Spacecraft system failures and anomalies attributed to the natural space environment. *NASA/RP-1390*, Aug. Alabama: National Aeronautics and Space Administration, Marshall Space Flight Center, MSFC, https://ntrs.nasa.gov/search.jsp?R=19960050463.

Bloomquist, C., and Graham, W. 1983. Analysis of spacecraft on-orbit anomalies and lifetimes. *NASA/CR-170565*, Feb, NASA Contractor Report 170565. Los Angeles, CA: PRC Systems Services. https://ntrs.nasa.gov/search.jsp?R=19830019777.

Brown, P., Spalding, R. E., ReVelle, D. O., Tagliaferri, E., and Worden, S. P. 2002. The flux of small near-Earth objects colliding with the Earth. *Nature*, **420**, 294–296.

Brown, P., Wong, D. K., Weryk, R. J., and Wiegert, P. 2010. A meteoroid stream survey using the Canadian Meteor Orbit Radar. II: Identification of minor showers using a 3D wavelet transform. *Icarus*, **207**, 66–81.

Bucher, S., and Martin, E. 2004. Chandra Flight Note 439: 2003:319 Attitude Disturbance. *Internal memo from Chandra project: June 15*.

Burchell, M. J., Fairey, S. A. J., Wozniakiewicz, P. et al. 2008. Characteristics of cometary dust tracks in Stardust aerogel and laboratory calibrations. *Meteoritics and Planetary Science*, **43**(Feb.), 23–40.

Carpenter, J. D., Wells, A., Abbey, A. F., and Ambrosi, R. M. 2008. Meteoroid and space debris impacts in grazing-incidence telescopes. *Astronomy and Astrophysics*, **483**, 941–947.

Caswell, R. D., McBride, N., and Taylor, A. D. 1995. Olympus end of life anomaly – A Perseid meteoroid event? *International Journal of Impact Engineering*, **17**, 139–150.

Chen, Y., Huang, R., Ren, X., He, L., and He, Y. 2013. History of the Tether Concept and Tether Missions: A Review. *ISRN Astronomy and Astrophysics*, **2013**, 502973, 7 pages.

Christiansen, E. L., Crews, J. L., Kerr, J. H., Cour-Palais, B. G., and Cykowski, E. 1995. Testing the validity of cadmium scaling. *International Journal of Impact Engineering*, **17**, 205–215.

Close, S., Colestock, P., Cox, L., Kelley, M., and Lee, N. 2010. Electromagnetic pulses generated by meteoroid impacts on spacecraft. *Journal of Geophysical Research (Space Physics)*, **115**, A12328.

Colangeli, L., Lopez Moreno, J. J., Palumbo, P. et al. 2007. GIADA: The Grain Impact Analyser and Dust Accumulator for the Rosetta space mission. *Advances in Space Research*, **39**, 446–450.

Collette, A., Grün, E., Malaspina, D., and Sternovsky, Z. 2014. Micrometeoroid impact charge yield for common spacecraft materials. *Journal of Geophysical Research (Space Physics)*, **119**, 6019–6026.

Commission F1 of the International Astronomical Union. 2017. Definitions of terms in meteor astronomy. IAU, www.iau.org/static/science/scientific_bodies/commissions/f1/meteordefinitions_approved.pdf.

Cooke, W. J., and Moser, D. E. 2010. The 2011 Draconid Shower Risk to Earth-Orbiting Satellites. *NASA Technical Reports Server Document 20100024125*. https://ntrs.nasa.gov/search.jsp?R=20100024125.

Dikarev, V., Grün, E., Baggaley, J. et al. 2005. The new ESA meteoroid model. *Advances in Space Research*, **35**, 1282–1289.

Divine, N. 1993. Five populations of interplanetary meteoroids. *Journal of Geophysical Research*, **98**, 17029–17048.

Drolshagen, G., McDonnell, J. A. M., Stevenson, T. J. et al. 1996. Optical survey of micrometeoroid and space debris impact features on EURECA. *Planetary and Space Science*, **44**, 317–340.

Drolshagen, G., Dikarev, V., Landgraf, M., Krag, H., and Kuiper, W. 2008. Comparison of meteoroid flux models for near Earth space. *Earth, Moon, and Planets*, **102**, 191–197.

Drolshagen, G., Koschny, D., Drolshagen, S., Kretschmer, J., and Poppe, B. 2017. Mass accumulation of earth from interplanetary dust, meteoroids, asteroids and comets. *Planetary and Space Science*, **143**, 21–27.

Elphic, R. C., Hine, B., Delory, G. T. et al. 2014. The lunar atmosphere and dust environment explorer (LADEE): Initial science results. *45th Lunar and Planetary Science Conference. The Woodlands, Texas. March 17–21, 2014.* Houston, TX: Lunar and Planetary Institute, LPI Contribution No. 1777, abstract 2677.

Ernst, R. 2007. Momentum transfer to orbiting satellites by micrometeoroid impacts. *Final Report of stage at ESTEC*.

European Space Agency and Gaia Data Processing and Analysis Consortium. 2017. https://gea.esac.esa.int/archive/documentation/GDR1/index.html.

Foschini, L. 1998. Electromagnetic interference from plasmas generated in meteoroids impacts. *Europhysics Letters (EPL)*, **43**, 226–229.

Foschini, L. 2002. Meteoroid Impacts on Spacecraft. In: E. Murad and I. P. Williams (eds), *Meteors in the Earth's Atmosphere*. Cambridge, UK: Cambridge University Press, pp. 249–263.

Göller, J. R., and Grün, E. 1989. Calibration of the Galileo/Ulysses dust detectors with different projectile materials and at varying impact angles. *Planetary and Space Science*, **37**, 1197–1206.

Grün, E., Zook, H. A., Fechtig, H., and Giese, R. H. 1985. Collisional balance of the meteoritic complex. *Icarus*, **62**, 244–272.

Grün, E., Fechtig, H., Hanner, M. S. et al. 1992a. The Galileo Dust Detector. *Space Science Reviews*, **60**, 317–340.

Grün, E., Fechtig, H., Kissel, J. et al. 1992b. The ULYSSES dust experiment. *Astronomy and Astrophysics Supplement Series*, **92**, 411–423.

Halliday, I., Griffin, A. A., and Blackwell, A. T. 1996. Detailed data for 259 fireballs from the Canadian camera network and inferences concerning the influx of large meteoroids. *Meteoritics and Planetary Science*, **31**, 185–217.

Hamilton, H., Liou, J.-C., Anz-Meador, P. D. et al. 2017. Development of the Space Debris Sensor. In: T. Flohrer, F. Schmitz (eds), *Proceedings of the 7th European Conference on Space Debris*, 11 pages. Darmstadt, Germany: ESA Space Debris Office. https://conference.sdo.esoc.esa.int/proceedings/sdc7/paper/965/SDC7-paper965.pdf.

Hayashida, K. B., and Robinson, J. H. 1991. Single wall penetration equations. *NASA/TM-103565*, Dec. Alabama: National Aeronautics and Space Administration, Marshall Space Flight Center, MSFC. NASA Technical Memorandum 103565. https://ntrs.nasa.gov/search.jsp?R=19920007464.

Horányi, M., Szalay, J. R., Kempf, S. et al. 2015. A permanent, asymmetric dust cloud around the Moon. *Nature*, **522**, 324–326.

Hyde, J. L., Christiansen, E. L., Lear, D. M., and Prior, T. G. 2006. Overview of recent enhancements to the Bumper-II meteoroid and orbital debris risk assessment tool. *57th International Astronautical Congress, 2006, IAC-06-B6.3.03*, 1–7. International Astronautical Federation/AIAA (IAF). Available at: https://ntrs.nasa.gov/archive/nasa/casi.ntrs.nasa.gov/20060047566.pdf.

Inter-Agency Space Debris Coordination Committee. 2014. Protection Manual, Version 7.0. *IADC-04-03*, Sept. www.iadc-online.org/Documents/IADC-04-03_Protection_Manual_v7.pdf.

Jenniskens, P. 1994. Meteor stream activity I. The annual streams. *Astronomy and Astrophysics*, **287**, 990–1013.

Jenniskens, P., Nénon, Q., Albers, J. et al. 2016. The established meteor showers as observed by CAMS. *Icarus*, **266**, 331–354.

Johnson, N. L., Krisko, P. H., Liou, J.-C., and Anz-Meador, P. D. 2001. NASA's new breakup model of evolve 4.0. *Advances in Space Research*, **28**, 1377–1384.

Jones, J., and Poole, L. M. G. 2007. Gravitational focusing and shielding of meteoroid streams. *Monthly Notices of the Royal Astronomical Society*, **375**, 925–930.

Jutzi, M., Holsapple, K., Wünneman, K., and Michel, P. 2015. Modeling Asteroid Collisions and Impact Processes. In: P. Michel, F. E. DeMeo, and W. F. Bottke (eds), *Asteroids IV*. Tucson: University of Arizona Press, pp. 679–700.

Kessler, D. J. 1972. A Guide to Using meteoroid-environment models for experiment and spacecraft design applications. *NASA/TN-D-6596*, Mar. Washington, DC: National Aeronautics and Space Administration. https://ntrs.nasa.gov/search.jsp?R=19720012228.

Kobayashi, M., Miyachi, T., Hattori, M. et al. 2013. Dust detector using piezoelectric lead zirconate titanate with current-to-voltage converting amplifier for functional advancement. *Earth, Planets, and Space*, **65**, 167–173.

Koschny, D., Drolshagen, E., Drolshagen, S. et al. 2017. Flux densities of meteoroids derived from optical double-station observations. *Planetary and Space Science*, **143**, 230–237.

Krüger, H., Linkert, G., Linkert, D., Moissl, R., and Grün, E. 2005. Galileo long-term dust monitoring in the Jovian magnetosphere. *Planetary and Space Science*, **53**, 1109–1120.

Lauer, T. R., Throop, H. B., Showalter, M. R. et al. 2018. The New Horizons and Hubble Space Telescope search for rings, dust, and debris in the Pluto-Charon system. *Icarus*, **301**, 155–172.

Lee, N., Close, S., Lauben, D. et al. 2012. Measurements of freely-expanding plasma from hypervelocity impacts. *International Journal of Impact Engineering*, **44**, 40–49.

Lenorovitz, J. M. 1991. Natural orbital environment guidelines for use in aerospace vehicle development. *Aviation Week & Space Technology*, **135**, 60–61.

Levy, L., Mandeville, J.C., Siguler, J.M. et al. 1997. Simulation of in-flight ESD anomalies triggered by photoemission, micrometeoroid impact and pressure pulse. *IEEE Transactions on Nuclear Science*, **44**, 2201–2208.

Liu, X., Sachse, M., Spahn, F., and Schmidt, J. 2016. Dynamics and distribution of Jovian dust ejected from the Galilean satellites. *Journal of Geophysical Research (Planets)*, **121**, 1141–1173.

Love, S. G., and Brownlee, D. E. 1993. A direct measurement of the terrestrial mass accretion rate of cosmic dust. *Science*, **262**, 550–553.

Love, S. G., Brownlee, D. E., King, N. L., and Horz, F. 1995. Morphology of meteoroid and debris impact craters formed in soft metal targets on the LDEF satellite. *International Journal of Impact Engineering*, **16**, 405–418.

McBride, N., and McDonnell, J. A. M. 1999. Meteoroid impacts on spacecraft: Sporadics, streams, and the 1999 Leonids. *Planetary and Space Science*, **47**, 1005–1013.

McBride, N., and Taylor, E. 1997. The risk to satellite tethers from meteoroid and debris impacts. In: Kaldeich-Schuermann, B. (ed), *Second European Conference on Space Debris: ESOC, Darmstadt, Germany, 17–19 March 1997*. ESA Special Publication, vol. 393, Pages 643–648.

McDonnell, J. A. M., ed. 2005. Post-Flight Impact Analysis of HST Solar Arrays – 2002 Retrieval, Final Report of ESA Contract No.16283/02/NL/LvH, UK, 2002.

McDonnell, J. A. M., and Gardner, D. J. 1998. Meteoroid morphology and densities: Decoding satellite impact data. *Icarus*, **133**, 25–35.

McNamara, H., Jones, J., Kauffman, B. et al. 2004. Meteoroid engineering model (MEM): A meteoroid model for the inner Solar System. *Earth, Moon, and Planets*, **95**, 123–139.

Meidinger, N., Aschenbach, B., Braeuninger, H. W. et al. 2003. Experimental verification of a micrometeoroid damage in the pn-CCD camera system aboard XMM-Newton. Pages 243–254 of: Truemper, J. E., and Tananbaum, H. D. (eds), *X-Ray and Gamma-Ray Telescopes and Instruments for Astronomy*, vol. 4851. SPIE Proceedings, Bellingham, WA.

Menicucci, A., Drolshagen, G., Mooney, C., Butenko, Y., and Kuitunen, J. 2012. DEBIE (Debris-In-Orbit-Evaluator) on-board of ISS: Results from impact data and post-flight analysis. *63rd International Astronautical Congress, Naples, 2012*, **IAC-12**, A6.1.11, 2232–2237

Miller, A. 2017, ESABASE2/Debris Release 10.0, Technical Description, Ref. R077-231rep_01_07_Debris_Technical_Description. Available at: https://esabase2.net/wp-content/uploads/2019/01/ESABASE2-Debris-Technical-Description.pdf.

Moorhead, A. V., Wiegert, P. A., and Cooke, W. J. 2014. The meteoroid fluence at Mars due to Comet C/2013 A1 (Siding Spring). *Icarus*, **231**, 13–21.

Moorhead, A. V., Cooke, W. J., and Campbell-Brown, M. D. 2017. Meteor shower forecasting for spacecraft operations. In: T. Flohrer and F. Schmitz (eds), *Proceedings of the 7th European Conference on Space Debris*, 11 pages. Darmstadt, Germany: ESA Space Debris Office. https://conference.sdo.esoc.esa.int/proceedings/sdc7/paper/77/SDC7-paper77.pdf.

NASA. 1971. Mariner-Venus 1967. *NASA/SP-190*, Jan. Washington, DC: National Aeronautics and Space Administration.

Naumann, R. J. 1966. The near-Earth meteoroid environment. *NASA-TN-D-3717*, Nov., 1–45. Washington, DC: National Aeronautics and Space Administration. https://ntrs.nasa.gov/search.jsp?R=19670001471.

O'Brien, M. 1999. The shuttle shuffle: Not for the faint of heart. *CNN.com*, Sept. www.cnn.com/TECH/space/9909/08/downlinks/.

Öpik, E. J. 1951. Collision probability with the planets and the distribution of interplanetary matter. *Proceedings of the Royal Irish Academy, Section A*, **54**, 165–199.

Palmieri, D., Drolshagen, G., and Lambert, M. 2003. Numerical simulation of grazing impacts from micron sized particles on the XMM-Newton mirrors. *International Journal of Impact Engineering*, **29**, 527–536.

Putzar, R., Schaefer, F., and Lambert, M. 2008. Vulnerability of spacecraft harnesses to hypervelocity impacts. *International Journal of Impact Engineering*, **35**, 1728–1734.

Rendtel, J. 1993. Perseids 1993: A first analysis of global data. *WGN, Journal of the International Meteor Organization*, **21**, 235–239.

Risquez, D., van Leeuwen, F., and Brown, A. G. A. 2012. Dynamical attitude model for Gaia. *Experimental Astronomy*, **34**, 669–703.

Sabath, D, and Paul, K.G. 1997. Hypervelocity impact experiments on tether materials. *Advances in Space Research*, **20**, 1433–1436.

Schimmerohn, M. 2012. Susceptibility of solar arrays to micrometeorite impact. *Final report of ESA contr. 22462/09/NL/GLC, Fraunhofer Ernst-Mach-Institut report I-40/12*.

Schimmerohn, M., Osterholz, J., Rott, M. et al. 2010. Understanding the mechanisms of hypervelocity impact induced failures on space solar arrays. In: *Proc. 25th European Conference on Photovoltaic Solar Energy conversion*, pp. 98–105. München: WIP Renewable Energies. www.tib.eu/en/search/id/TIBKAT%3A642613710/Proceedings-25th-European-Photovoltaic-Solar-Energy/.

Schwanethal, J. P., McBride, N., Green, S. F., McDonnell, J. A. M., and Drolshagen, G. 2005. Analysis of impact data from the DEBIE (debris In-Orbit Evaluator) sensor in polar low earth orbit. Pages 177–182 of: Danesy, D. (ed), *4th European Conference on Space Debris*. ESA Special Publication, vol. 587. Noordwijk, The Netherlands: ESA Publication Division, ESTEC.

Science and Technology Division, Library of Congress. 1968. Astronautics and Aeronautics, 1967: Chronology on Science, Technology, and Policy. *NASA Special Publication*, **4008**, 270–271. Washington, DC: National Aeronautics and Space Administration. https://ntrs.nasa.gov/search.jsp?R=19690016269.

Silber, E. A., ReVelle, D. O., Brown, P. G., and Edwards, W. N. 2009. An estimate of the terrestrial influx of large meteoroids from infrasonic measurements. *Journal of Geophysical Research (Planets)*, **114**, 8.

Simpson, J.A., and Tuzzolino, A.J. 1985. Polarized polymer films as electronic pulse detectors of cosmic dust particles. *Nuclear Instruments and Methods in Physics Research Section A: Accelerators, Spectrometers, Detectors and Associated Equipment*, **236**, 187–202.

Smith, R. E. 1994. Natural orbital environment guidelines for use in aerospace vehicle development. *NASA/TM-4527*, June. Alabama: National Aeronautics and Space Administration, Marshall Space Flight Center, MSFC. https://ntrs.nasa.gov/search.jsp?R=19940031668.

SonotaCo. 2009. A meteor shower catalog based on video observations in 2007–2008. *WGN, Journal of the International Meteor Organization*, **37**, 55–62.

Staubach, P., Grün, E., and Jehn, R. 1997. The meteoroid environment near Earth. *Advances in Space Research*, **19**, 301–308.

Steel, D. 1998. The Leonid Meteors: Compositions and consequences. *Astronomy and Geophysics*, **39**, 24–26.

Strub, P., Krüger, H., and Sterken, V. J. 2015. Sixteen years of Ulysses interstellar dust measurements in the solar system. II. Fluctuations in the dust flow from the data. *The Astrophysical Journal*, **812**, 140.

Strüder, L., Aschenbach, B., Bräuninger, H. et al. 2001. Evidence for micrometeoroid damage in the pn-CCD camera system aboard XMM-Newton. *Astronomy and Astrophysics*, **375**, L5–L8.

Stübig, M. 2002. New insights in impact ionization and in time-of-flight mass spectroscopy with micrometeoroid detectors by improved impact simulations in the laboratory. Dissertation, Ruprecht-Karls-Universität Heidelberg.

Suggs, R. M., Moser, D. E., Cooke, W. J., and Suggs, R. J. 2014. The flux of kilogram-sized meteoroids from lunar impact monitoring. *Icarus*, **238**, 23–36.

Svedhem, H., Drolshagen, G., Grün, E., Grafodatsky, O., and Prokopiev, U. 2000. New results from in situ measurements of Cosmic Dust — Data from the GORID experiment. *Advances in Space Research*, **25**, 309–314.

Taylor, A. D. 1995. The Harvard Radio Meteor Project meteor velocity distribution reappraised. *Icarus*, **116**, 154–158.

Thorpe, J. I., Parvini, C., and Trigo-Rodríguez, J. M. 2016. Detection and measurement of micrometeoroids with LISA Pathfinder. *Astronomy and Astrophysics*, **586**, A107.

Tsou, P., Brownlee, D. E., Sandford, S. A., Hörz, F., and Zolensky, M. E. 2003. Wild 2 and interstellar sample collection and Earth return. *Journal of Geophysical Research (Planets)*, **108**, 8113, doi:10.1029/2003JE002109.

Weiss, A. A., and Smith, J. W. 1960. A southern hemisphere survey of the radiants of sporadic meteors. *Monthly Notices of the Royal Astronomical Society*, **121**, 5–16.

Whipple, F. L. 1947. Meteorites and space travel. *The Astronomical Journal*, **52**, 131.

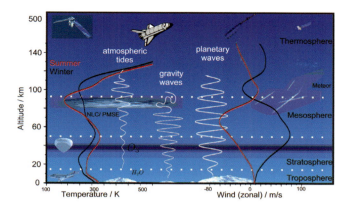

Figure 3.4. Schematic view of the vertical structure of the atmosphere at mid- and high latitudes.

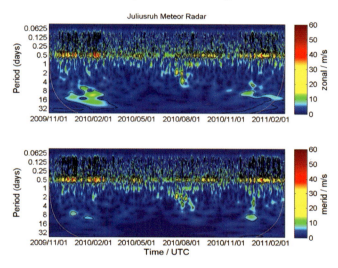

Figure 3.5. Wavelet spectra of zonal and meridional wind measurements for a mid-latitude station in Juliusruh.

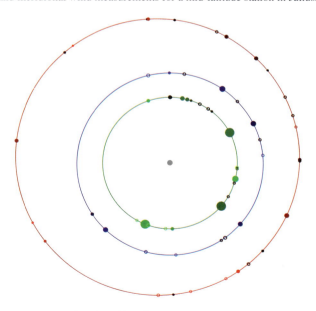

Figure 5.2. Locations of meteoroid streams encountering the orbits of Earth (middle circle), Venus (inner circle) and Mars (outer circle) based on data from Christou (2010). The First Point of Aries is towards the right. Open symbols correspond to Encke-type Comets, filled symbols to Intermediate Long Period and Halley-Type Comets. The size of each symbol is proportional to the encounter speed while the brightness indicates the solar elongation of the radiant.

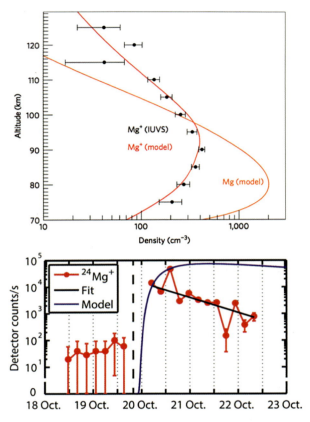

Figure 5.5. Top: Mg⁺ altitude profiles derived from *MAVEN* IUVS at orbit 3040 compared with the model prediction. Note that Mg is not detected despite large predicted concentrations. From Crismani et al. (2017). Bottom: Temporal evolution of the abundances of Mg⁺ measured by *MAVEN*/NGIMS at periapsis from 2014 October 18 to 2014 October 23. The exponential decay can be fitted by a time constant of 1.8 days. The dashed line marks the predicted time of maximum flux of C/2013 A1 dust. Predicted signal levels were derived by a 1-D model. Error bars reflect 3 × standard deviation of the sampled data due to counting statistics. Instrument background is 10–100 counts/s. From Benna et al. (2015).

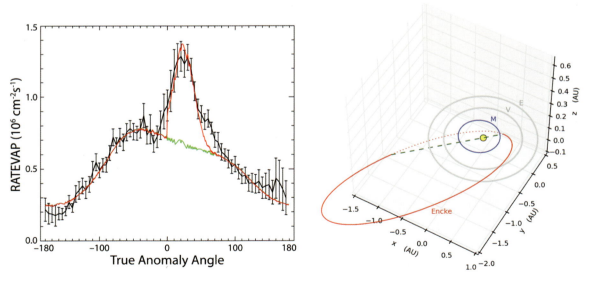

Figure 5.7. Left panel: Total planetary calcium source rate to Mercury's exosphere is a periodic function of the planet's TAA. Mercury is at perihelion when TAA = 0° and at aphelion when TAA = ±180°. The black curve is this total rate summed over the planet at each TAA, derived from observations obtained by the *MESSENGER* MASCS spectrometer 2011 March – 2013 March (from Burger et al., 2014). The red line is the modeled contribution from a cometary dust stream with peak density at TAA = 25° plus that due to an interplanetary dust-disk. The green line is the contribution from the disk. Adapted from Killen and Hahn (2015). Right panel: Encke's orbit (red) along with those of Mercury (blue circle M), Venus and Earth (grey circles V and E). After Killen and Hahn (2015).

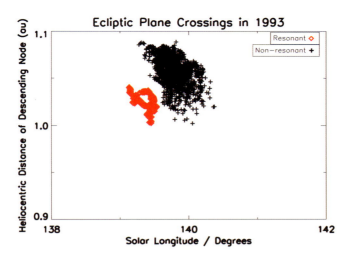

Figure 7.5. Ecliptic plane crossings of Perseids: heliocentric distance of descending node versus solar longitude (J2000) in AD 1993, for three-body resonant meteoroids (2000 red diamonds, showing dense structures) and non-resonant (2000 black crosses, showing significant dispersion) ones. Integration started at the 69 BC perihelion time of parent comet 109P/Swift-Tuttle (figure 6b from Sekhar et al., 2016).

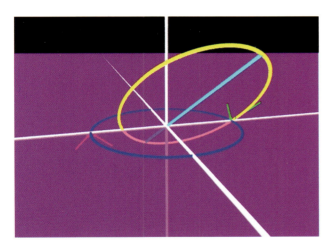

Figure 7.6. Circulation of the argument of pericenter causing four meteor showers from a single stream (personal communication from Paul Wiegert, University of Western Ontario).

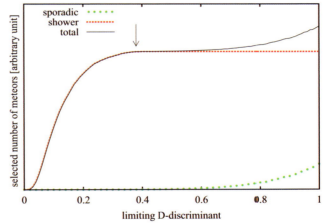

Figure 7.8. The dependence of the number of meteors selected from a database on the threshold value of the D-discriminant. The black solid curve shows the dependence for the selected number from a real database. This dependence is the superposition of the selected shower meteors (red dashed curve) and sporadic background meteors (green dotted curve). Taken from Neslušan et al. (2013a).

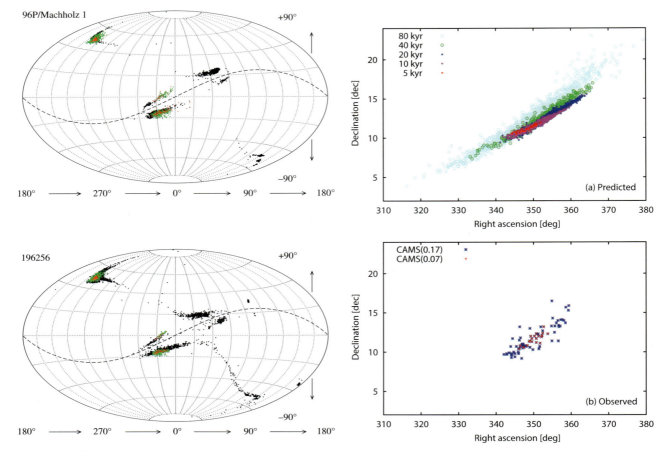

Figure 7.10. The distribution of the predicted and observed radiants of the meteoroid complex generated by 96P/Machholz 1 (upper plot) and asteroid 196 256 (lower plot). The black dots illustrate the radiants of the test particles approaching the Earth's orbit within 0.05 AU at the end of their followed orbital evolution, and considered as causing meteor showers. The red (green) dots show the radiants of real meteors of three of the 96P-complex showers, Quadrantids, Northern δ-Aquariids, and Southern δ-Aquariids, identified in the IAU MDC photographic (SonotaCo video) databases (taken from Neslušan et al., 2014b).

Figure 7.12. The positions of geocentric radiants of theoretical particles (upper plot) corresponding to the July Pegasids, IAU #175, and real July Pegasids (lower plot) separated from the CAMS video database (blue asterisks for $D_{lim} = 0.17$ and red crosses for $D_{lim} = 0.07$). The theoretical radiants were obtained for $\beta = 10^{-5}$ and $t_{ev} = 5, 10, 20, 40,$ and 80 kyr. The positions for the different t_{ev} are distinguished using the different marks and colors (taken from Hajduková and Neslušan, 2017, reproduced with permission © ESO).

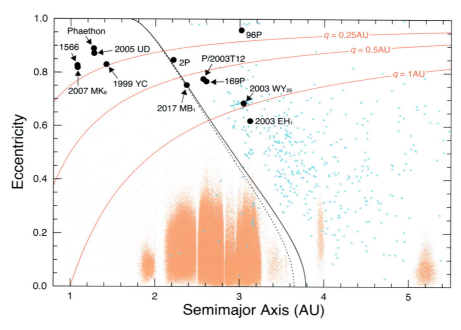

Figure 8.1. Distribution of the parent bodies (black circles) in the semimajor axis versus orbital eccentricity plane (cf. Table 8.1). The distributions of asteroids (brown dots) and comets (light-blue dots) are shown for reference. The lines for $T_J = 3.08$ with $i = 0°$ (solid black curve) and with $i = 9°$ (dotted black curve) broadly separate asteroids and comets. Perihelion distances $q = 0.25, 0.5$ and 1 AU are shown as solid red curves.

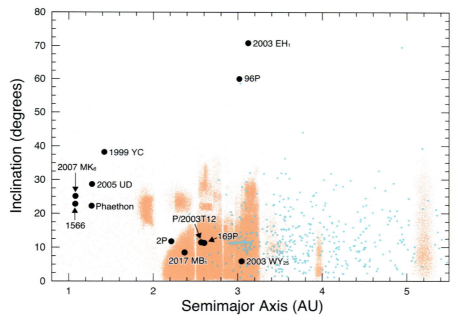

Figure 8.2. Same as Figure 8.1 but semimajor axis versus inclination.

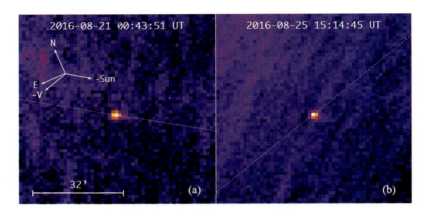

Figure 8.7. STEREO-A image taken on 2016 August 20 of (a) (3200) Phaethon and tail aligned on the position angles of the anti-solar direction and (b) the whole image sequence on the negative heliocentric velocity vector projected on the sky plane. Geometric parameters are $R \sim 0.14$ AU, Phaethon-STEREO distance ~ 0.9 AU, and the phase angle $\sim 120°$. The projected Sun-Phaethon line (the narrow white line across the left panel) is drawn to better illustrate that the direction of the tail was anti-solar, while the white dotted line across the right panel is the orientation. From Hui and Li (2017).

Figure 8.10. The 24 μm infrared image of comet 2P/Encke obtained in 2004 June by the Spitzer Space Telescope. The image field of view is about 6′ and centered on the nucleus. The near horizontal emission is produced by recent cometary activity, and the diagonal emission across the image is the meteoroid stream along the orbit (Kelley et al.,2006; see also Reach et al., 2007). Courtesy NASA/JPL-Caltech/M. Kelley (Univ. of Minnesota).

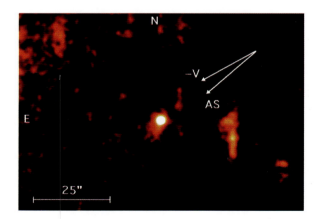

Figure 8.12. 2003 WY$_{25}$ (=289P/Blanpain) imaged in R-band, 500 sec integration at $R = 1.6$ AU, $\Delta = 0.7$ AU and $\alpha = 20.7°$ using UH2.2 on UT 2004 March 20. A faint coma is apparent, extending to the southeast. Arrows show the directions of the negative heliocentric velocity vector (marked "–V") and the anti-solar direction ("AS"). The estimated radius ~ 160 m is the smallest active cometary parent ever observed. From Jewitt (2006).

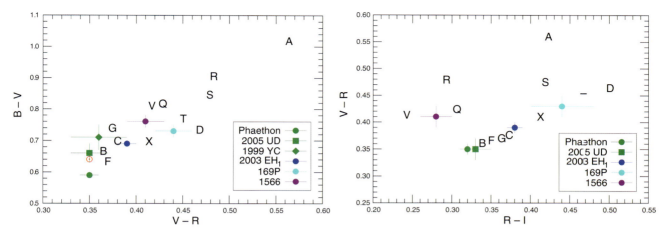

Figure 8.13. Color plots of $V-R$ vs. $B-V$ showing parent bodies (filled symbols) from Table 8.2, and Tholen taxonomic classifications (Tholen, 1984), as tabulated by Dandy et al. (2003). The color of the Sun (red circle) is also plotted.

Figure 8.14. The same as Figure 8.13 but in the $R-I$ vs. $V-R$ color plane. The color of the Sun is exactly coincident with that of 2005 UD.

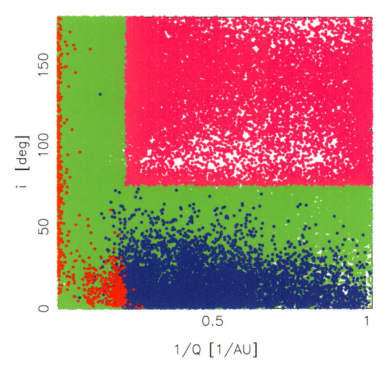

Figure 9.3. 780 Comets (red), 7 830 NEAs (blue) and 60 412 sporadic meteoroids (green) on the i-$1/Q$ plane. Additionally, the set of 17 457 "asteroidal" meteoroids for which inclinations $i > 75°$ and aphelia distances $Q < 4.6$ [au] forms a magenta rectangle. This rectangle contains 22% of our sporadic sample.

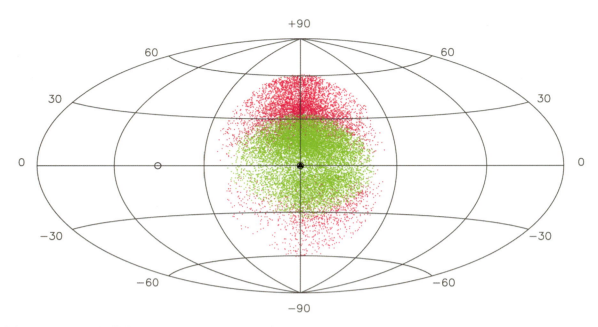

Figure 9.4. The Hammer-Aitoff diagram of 17 457 radiants of sporadic meteors plotted in the ecliptic Earth apex centred reference frame. The orbits of all the plotted meteoroids have aphelia $Q < 4.6$ [au] and inclinations $i > 75°$. The central point of the diagram corresponds to the position of the apex of the Earth motion. Two regions are seen: the north and south apex (both in green) and toroidal concentrations (magenta).

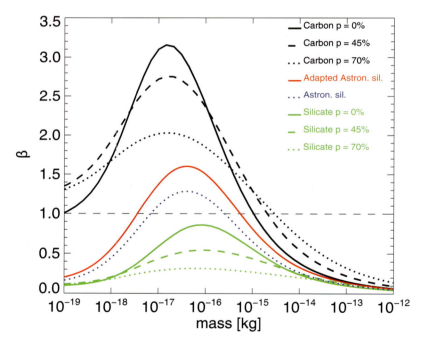

Figure 10.1. The β-mass relation for several particle sizes and material properties. Porosity p is the ratio of the volume of void space to the total (bulk) material space of the particle. Porosity $p = 0$ % corresponds to compact material. Credit: adapted from Kimura and Mann (1999) and Sterken et al. (2012), reproduced with permission © ESO and the Astronomical Institute of the Slovak Academy of Sciences.

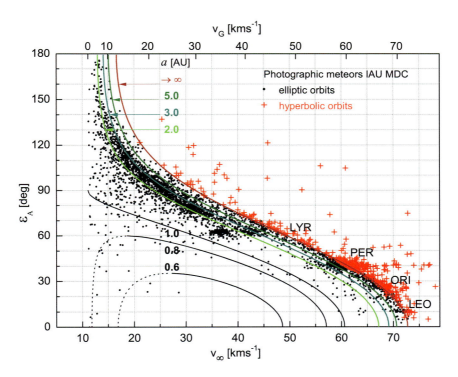

Figure 10.2. The effect of the measurement errors in radiant position and velocity on the resulting semi-major axis of the orbit. The angular elongation of the apparent radiant from the apex ε_A is plotted against the non-atmospheric velocity of meteors v_∞, using rough photographic data of the IAU MDC (Lindblad et al., 2003) in which 11.5% of orbits were determined to be hyperbolic (crosses). The curves, representing the relation between ε_A and v_G according to Equation (10.2), are constructed for different values of semi-major axes a.

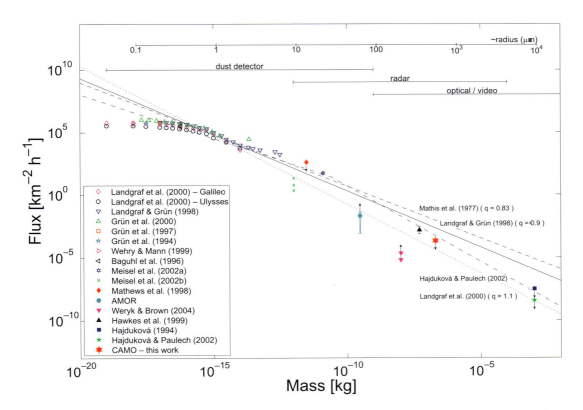

Figure 10.3. A summary of the interstellar meteoroid fluxes from the literature, from Musci et al. (2012). Different symbols designate primary sources of interstellar data. The lines in different styles represent the modelled mass distributions or those fitted by power-law functions determined by different authors.

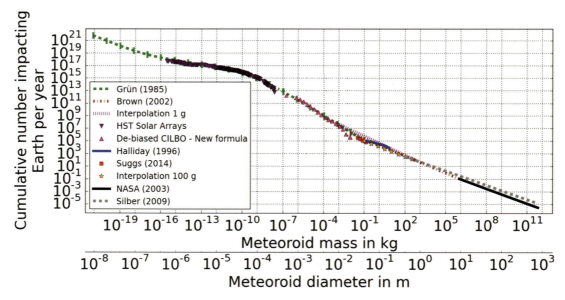

Figure 11.2. Cumulative number of impacts of natural particles onto the Earth per year (from Drolshagen et al., 2017).

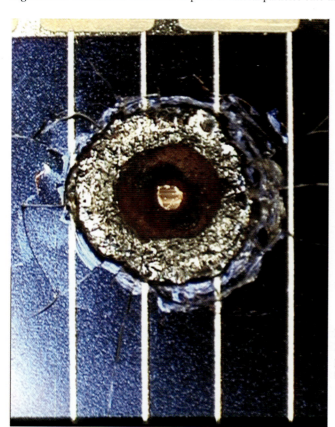

Figure 11.5. Impact crater on Hubble Space Telescope solar cell retrieved in 2002. The crater diameter is approximately 5 mm (McDonnell, 2005).

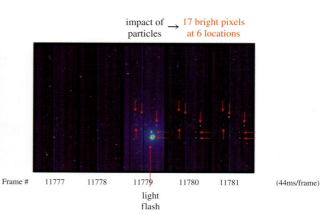

Figure 11.6. Impacts on exposed CCD of the type used by the XMM-Newton x-ray telescope (from Meidinger et al., 2003). Bright pixels on CCD following as a result of an impact test by a micron sized particle.

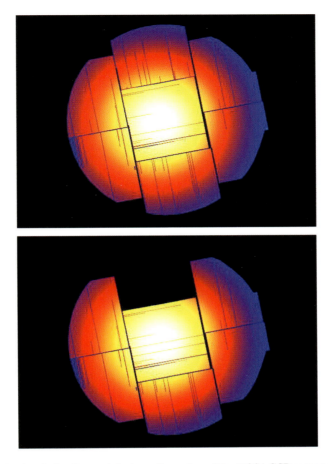

Figure 11.7. Exposure maps before (top) and after (bottom) the loss of a section of the MOS1 CCD sensor of the XMM-Newton telescope following an impact. Image credit: ESA.

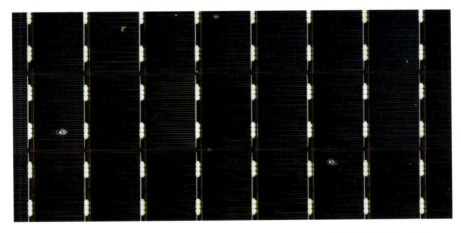

Figure 11.9. Impacts on front side of HST solar array. Each solar cell is 20 mm by 40 mm in size. The bright grid fingers are spaced by 1.25 mm. Image credit: ESA.

Figure 11.10. Impact features on front side of HST solar array caused by impact on back side. Crater size is 3.5 mm. Image credit: ESA.

Figure 11.11. Small impact feature (about 0.5 mm crater size) on front side of HST solar array. The "butterfly" shape indicates a grazing incidence (> 70° from surface normal). Image credit: ESA.

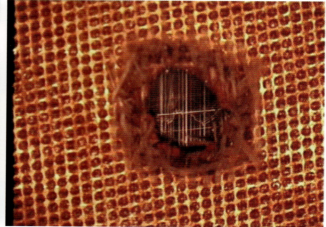

Figure 11.12. Front view of an impact hole in HST solar cell. The hole size is 3 mm. Image credit: ESA.

Figure 11.13. Rear view of the impact hole shown in Figure 11.12. The hole size is 3 mm. Image credit: ESA.

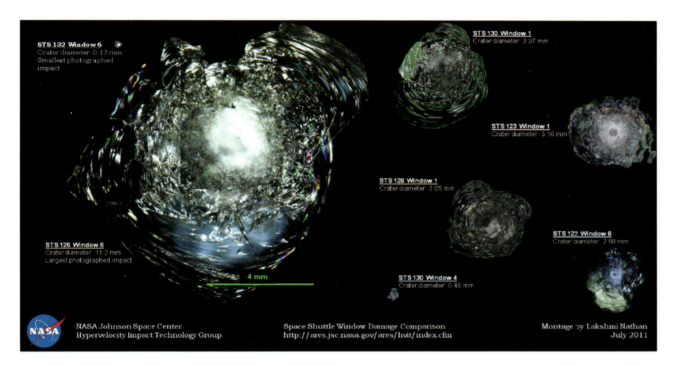

Figure 11.14. Impacts in windows of the Space Shuttle. Image credit: NASA.

Figure 11.16. A broken spring on HST solar array. The wire thickness is 0.5 mm. Image credit: ESA.

Figure 11.17. The other part of the broken spring on the HST solar array. The wire thickness is 0.5 mm. Image credit: ESA.

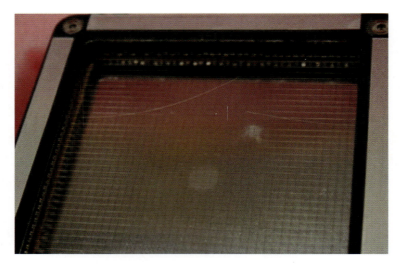

Figure 11.18. A broken wire on DEBIE-2 impact detector. The wire thickness is 0.075 mm (from Menicucci et al., 2012). Image credit: ESA.

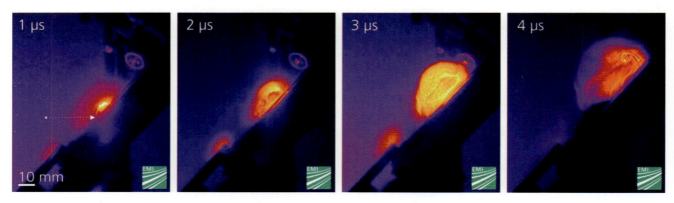

Figure 11.19. Impact light flash from a hypervelocity impact test of a 1.6-mm Al sphere impacting a solar generator sample at 7.4 km s^{-1} and an incidence angle of 45°. The tests were performed by EMI. Image source: Fraunhofer Ernst-Mach-Institut.

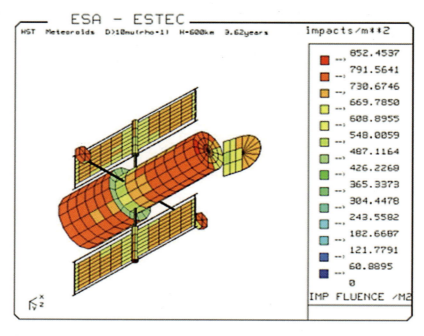

Figure 11.20. Predicted number of impacts from meteoroids with diameter > 10 microns during the first 3.62 years of the HST mission.

Figure 11.21. Two Whipple shield configurations that have been subjected to impact testing. Image credit: NASA.

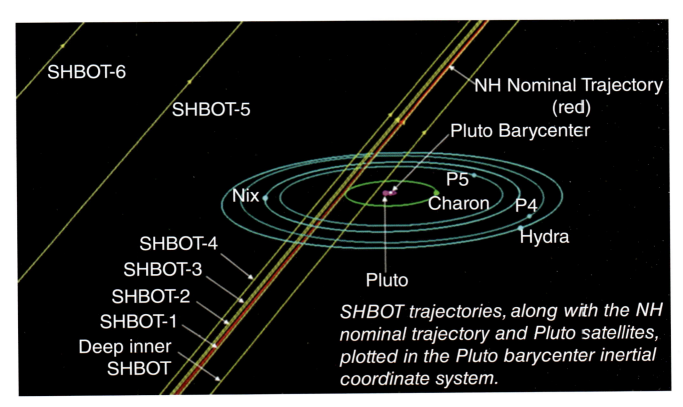

Figure 11.22. Nominal and alternate trajectories developed for the New Horizons Pluto fly-by, the latter of which are termed "Safe Haven by Other Trajectory", or SHBOT. These trajectories were designed to avoid any late-detected debris concentrations, but were not ultimately used. Image credit: NASA/JHUAPL.

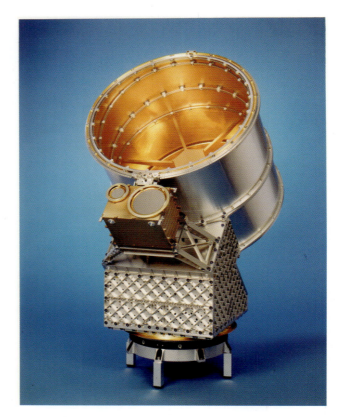

Figure 11.23. The Cosmic Dust Analyzer on Cassini. Image credit: Lossen-Fotografie/MPIK Heidelberg.

Figure 12.11. A map of approximate locations of injuries due to the Tunguska event. The area of fallen trees is shown near the epicenter (Longo et al., 2005). Locations with reported injuries are marked by circles. The numbers at the circles indicate reported stress/panic cases. Diamonds mark damaged windows and roofs. Oval contours correspond to the approximation model for rapid estimates of hazardous effects (Glazachev et al., 2019); the numbers correspond to the relative overpressure levels.

12

Impact Hazard of Large Meteoroids and Small Asteroids

Vladimir Svetsov, Valery Shuvalov, Gareth Collins, and Olga Popova

12.1 Introduction

At first glance, asteroids and comets measuring from a meter to tens of meters do not pose a significant danger to people, but, owing to their high speed, they carry a lot of energy. For example, a stone asteroid 6 m in diameter with a density of 3.3 g cm^{-3} and with a speed of 20 km s^{-1} has a kinetic energy equivalent to 17 kt of TNT – that is, the same energy as the atomic bomb dropped on Hiroshima (1 kt is the energy equivalent to 1000 metric tons of TNT, which is 4.2×10^{12} J). Even a 1-m-diameter meteoroid with the same parameters has an energy equivalent to 80 tons of TNT. After the release of approximately the same energy in the accidental explosion that occurred in 1896 at Braamfontein station (a suburb of Johannesburg) a crater with a diameter of about 50 m was formed, over 70 people were killed and more than 200 were injured (Shorten, 1970). An explosion of somewhat less power at Iri station, Jeollabuk-do, South Korea, on 1977 November 11 killed 56 people and seriously injured 185 others (Korea Dynamite Explosion November 11, 1979). Many non-nuclear ground explosions are described, for example, in White (1994); Mannan (2012).

The frequency of collisions of large meteoroids and small asteroids with Earth has been obtained from observations of bolides (bright fireballs) with ground-based cameras (Halliday et al., 1996) and satellite records (Nemtchinov and Popova, 1997; Brown et al., 2002), from the size-frequency distribution of lunar craters translated to Earth (Ivanov, 2008), and from astronomical observations of near-Earth asteroids (NEAs, Rabinowitz et al., 2000; Harris and D'Abramo, 2015; Schunová-Lilly et al., 2017; Tricarico, 2017; Trilling et al., 2017). The probability of entry into the atmosphere of bodies with a diameter greater than D_P, derived from lunar craters, can be approximately described by the relation $P(>D_P) \approx 8 \times 10^{-8} D_P^{-2.95}$, where D_P is in kilometers and P is in impacts per year (Ivanov and Hartmann, 2007; Ivanov, 2008). The observational data on near-Earth asteroids gives a somewhat higher frequency of impacts. According to the estimates of Harris and D'Abramo (2015), impacts of bodies 20 meters in size (such as the recent bolide over Chelyabinsk, Russia) occur every 50 years, and impacts of bodies 40 meters in diameter (the approximate minimum size of the Tunguska impactor) occur every 500 years. Stony bodies 10 m in size encounter Earth about once every 10 years, and three-meter bodies more often than once a year. However, these impact frequencies are rather uncertain. The frequencies obtained from different telescopic observations can differ by a factor of three in the decameter-sized range. For example, the frequency for bodies with a diameter of 10 m, derived from the observations of Schunová-Lilly et al. (2017), is one impact in 30 years, and the average time between impacts of asteroids with a diameter of 50 m varies from ∼1000 years (Harris and D'Abramo, 2015; Schunová-Lilly et al., 2017) to ∼500 years (Trilling et al., 2017) and ∼300 years (Tricarico, 2017). Bolide observations provide similar estimates, within the limits of their errors (Brown et al., 2002, 2013). There appears to be an abrupt decrease in impact frequency in the range of asteroid diameters from about 50 m to roughly 500 m, or half a kilometer. A possible explanation for this change is that it corresponds to a transition in internal structure of asteroids from strong monolithic bodies, with long dynamical lifetimes, to weak rubble piles held together mainly by gravity with much shorter lifespans (Harris and D'Abramo, 2015). Note, however, that for sizes over 50 m, the discrepancy in the impact rate between the different observations is more than threefold, and some of the telescopic observations (Tricarico, 2017) do not show the decrease in the frequency of impacts.

Earth's atmosphere has a thickness equivalent in mass to a ten-meter layer of water or a layer of stone about three meters thick, and therefore the atmosphere protects Earth's surface from impacts of meteoroids of approximately this scale. But the deceleration of a meteoroid and the transfer of its kinetic energy to the atmosphere, even at high altitude, creates an air shock wave that reaches Earth's surface and can cause significant damage, similar to blastwaves caused by high-altitude explosions. This was demonstrated by the Chelyabinsk asteroid, which released the bulk of its energy at an altitude of about 30 km and injured more than 1600 people (Popova et al., 2013). As a result, asteroids of more than ten meters in size are now considered dangerous (Shustov et al., 2017). However, at present, only about 0.1% of all near-Earth objects with a diameter of more than ten meters have been detected (Emelyanenko and Naroenkov, 2015). In addition, the diameters of most small asteroids are not well determined, as they are usually estimated based on asteroid brightness and distance, and albedo is often poorly constrained. The majority of near-Earth objects are stony ordinary chondrites and have LL-mineralogy (Dunn et al., 2013). Although iron meteoroids and small asteroids encounter Earth much less frequently, the number that reach Earth's surface is disproportionately higher than stony meteoroids due to their greater density and strength.

In assessing the hazards of meteoroids and small asteroids, we determine the consequences that cause damage. These consequences can be local or regional, and their importance depends on the frequency of impacts. It is customary to consider a risk, which is defined as the consequence of an occurring event weighted by the probability of the event and which can be measured in, for example, average number of casualties per year. The event probability includes parameters of the impacting asteroid or comet and the chance of striking a certain location with a certain population and infrastructure. The distributions of several meteoroid parameters, such as size, velocity and angle of entry into the atmosphere, are quite well constrained from observations and calculations. The fraction of bodies with a certain composition, density and (especially) strength can be predicted with a lesser degree of certainty.

The ensemble risk has been assessed in several studies in which the consequences of impacts were determined using some approximate models of asteroid's entry, breakup, and ground damage and estimates of population affected by a certain impact (Reinhardt et al., 2016; Shuvalov et al., 2017; Rumpf et al., 2017a; Mathias et al., 2017; Stokes et al., 2017). In these risk assessments, approximate models that give fast predictions were used to determine the impact effects, but in a real situation (for example, when a dangerous object is detected) it is desirable to predict the consequences as accurately as possible. The main effects of the impacts of meteoroids and small asteroids, which deposit most of their energy in the atmosphere, consist in the action of an air shock wave, thermal radiation and the seismic wave generated by this shock wave. Cratering, ejecta deposition and tsunamis are not important at this scale of impactors, because they seldom strike the ground with high speed. In this chapter, we focus on these important processes and summarize recently developed models of their effects. In addition, we consider the influence of the impacts on the ionosphere, which may be important in connection with the development of various means of communication. Since small iron bodies can still reach the surface, we also consider such events. From the point of view of the asteroid danger, it would be ideal to have a tool that could quickly predict all the consequences of the impact given the known parameters of an asteroid and the place of a fall. We describe existing calculators and suggest possibilities for how they might be improved.

12.2 Energy Deposition in the Atmosphere

Asteroids have different compositions, densities, strengths, and shapes and can enter the atmosphere at different angles and speeds. However, we can consider the most probable parameters and estimate how the impact effects depend on the size of the body. In these estimates, we will assume that the asteroids are stony with a density characteristic of ordinary chondrites, which constitute the bulk of meteorite falls, low strength, which is derived from meteoric observations, and have a compact shape. Entry velocity can vary from the escape velocity (about 11 km s^{-1}) to 72 km s^{-1}; however, as the height of deceleration in the atmosphere only weakly depends on velocity (see Section 12.2), we will consider the average entry velocity, which is about 20 km s^{-1}. The probability density of entry at different angles is equal to the sine of twice the entry angle, therefore the most probable angle of entry is 45°, and impacts at very sharp angles are unlikely (only about 7% of impacts occur at angles less than 15°).

As a rule, meteoroids smaller than ten meters in size deposit their energy at altitudes of twenty to fifty kilometers (Brown et al., 2016) and shock waves reach the ground as a weak shock, with a relatively small overpressure, the pressure at the shock front in excess of ambient atmospheric pressure (Gi et al., 2018). Bodies from 20 to 100 m in size are decelerated at heights ranging from several km to 20–30 km. In this case, the shock waves that reach the ground are of a much larger amplitude, which can cause considerable damage (such as, for example, the Tunguska event in 1908). Bodies larger than about 100 m in size not only create shock waves in the atmosphere, but reach the ground on average with a rather high speed. The impact generates shock waves in the ground and leads to the formation of an impact crater with a diameter exceeding the body size by more than ten times (Melosh, 1989). The propagation of seismic waves in the ground can lead to earthquakes, landslides, and rockfalls at considerable distances; and impact into water can cause the formation of tsunami waves that can propagate over long distances without significant attenuation. Note that the boundaries between different scenarios are approximate; they depend on the nature of a body entering the atmosphere and the slope of its trajectory. The aforementioned threshold sizes are typical for stony bodies; they are larger for cometary bodies and smaller for iron bodies.

Since meteoroids and small asteroids with a size less than 100 m in most cases are destroyed and strongly decelerated in Earth's atmosphere, they transfer their kinetic energy to the surrounding air. In this process, most of the energy is released along a small portion of the trajectory (of the order of the atmospheric scale height) and, therefore, at large distances from the entry point (for example, at Earth's surface), the shock wave is similar to one generated by a point explosion. For this reason, the process of meteoroid disruption and deceleration in the atmosphere is often called a meteor explosion or airburst, and its effects are estimated using formulas derived from the results of chemical or nuclear explosion tests. However, a thermal explosion does not occur when a cosmic body moves in the atmosphere; the only mechanism for energy release is aerodynamic braking of the meteoroid itself, as well as the deceleration of its vapors and fragments.

The motion and deceleration in the atmosphere is accompanied by the formation of a shock wave, which is perceived as a blast wave at large distances. Shock waves are also accompanied by thermal radiation emitted by heated air and vapor. A vast database on the effects of shock waves on different objects was collected after nuclear tests in the 1950s and 1960s (Glasstone and Dolan, 1977) and has been widely used in assessing effects of the impacts of cosmic bodies. However, the analogy between concentrated explosions (chemical or nuclear) and airbursts is not perfect. This was demonstrated, for example, by numerical simulations of shock waves produced in the atmosphere by the Chelyabinsk meteorite (Popova et al., 2013), which led to the butterfly pattern of overpressure on the ground. Numerical assessment by Collins et al. (2017) suggests that overpressures from an oblique impact and a point explosion are broadly consistent at radial distances from ground zero that exceed three times the burst height. However, in the Tunguska event, whose energy was released at altitudes of 5–10 km, the trees were felled over a butterfly-shaped area of about 2000 km^2

(Vasilyev, 1998), that is, at larger radial distances. These examples show that numerical simulations of aerial shock waves are necessary for the precise determination of the shock wave effects on the ground.

The area on the ground where destructive effects of a shock wave and thermal radiation are observed is determined primarily by the released energy, and the shape of this area depends on the details of energy release during the flight accompanying the disruption of the body. If the object enters the atmosphere vertically, the shape of this area is circular, which is also typical for a concentrated explosion (Popova et al., 2013). Such a trajectory is extremely rare: in most cases, the object enters at an oblique angle to the surface, which results in an asymmetric blast radius. However, for a quick approximate evaluation of the effects caused by the fall of a cosmic body, the analogy with explosions is very useful. To use this analogy with explosions, one needs to know the height of an explosion, which can be calculated using various models.

There are many analytical and semi-analytical models of various levels of complexity that describe the destruction and deceleration of meteoroids in the atmosphere, based on the solutions of single-body flight equations of meteor physics: from the simple pancake models for energy deposition in the atmosphere (e.g., Hills and Goda, 1993; Chyba et al., 1993) to more complex models that take into account fragmentation into separate fragments, such as the recent fragment-cloud model (Wheeler et al., 2017). Ideally, the mass movement and the duration of energy release should be taken into account (Aftosmis et al., 2016; Stokes et al., 2017). These models are able to reproduce energy deposition as a function of altitude based on observed light curves, for example, the Chelyabinsk event, by selecting appropriate parameters used in the models. But in general, the results depend on poorly known parameters (different in different models). For example, according to the fragment-cloud model, when a 2.5-g cm^{-3} stony body with a diameter of 100 m and a velocity of 20 km s^{-1} falls at an angle of 45°, the peak of energy release is at an altitude of 19 km, and full deceleration occurs at an altitude of 9 km (Wheeler et al., 2017; Mathias et al., 2017). For the same meteoroid parameters, the model of Chyba et al. (1993) predicts that the mass of the body hits the surface at a speed of 5 km s^{-1} and produces a crater with a diameter of about 1 km (Collins et al., 2005). And a hydrodynamic model (see Section 12.2) shows that a vaporized mass reaches the ground at a speed of about 10 km s^{-1}. Let us consider in more detail this latter approach, based on direct gasdynamic modeling of the motion and deformation of a meteoroid in the atmosphere and first developed to predict the interaction of fragments of the comet Shoemaker-Levy 9 with Jupiter (e.g., Boslough et al., 1995).

The disruption and deceleration of a meteoroid in the atmosphere and the subsequent propagation of the shock wave over long distances were simulated using a two-step model described by Shuvalov et al. (2013). In the first step, the movement of the meteoroid in the atmosphere was modeled taking into account its deformation, deceleration, destruction and evaporation. The numerical simulation included the model, equations and numerical scheme described by Shuvalov and Artemieva (2002) and Shuvalov and Trubetskaya (2007). The model was used to simulate the motion of meteoroids at altitudes where the aerodynamic load significantly exceeds its strength; for this reason it was assumed that the meteoroid has already been fragmented and its behavior can be described by the strengthless, hydrodynamic approximation. The mathematical problem was solved in a coordinate system associated with the falling body blown by air, whose density changed depending on the stratification of the atmosphere, and the velocity of the oncoming flow was equal to the velocity of the meteoroid. The calculations in this first step were terminated when the meteoroid was destroyed and almost completely slowed down (its velocity in Earth's reference frame decreased fivefold, i.e., when its further fall did not affect a determined altitude of an explosion), or when the meteoroid reached the surface of the Earth. The distributions of gasdynamic and thermodynamic parameters in the atmosphere were used as initial data for the second stage of simulation.

In the second step of the model, the propagation of an air shock wave over long distances was simulated in a coordinate system associated with Earth's surface. Both calculation steps were implemented using the numerical method SOVA (Shuvalov, 1999). The same method was also used for calculating the propagation of the shock wave from point explosions. The model used tabular equations of state and radiation absorption coefficients of air (Kuznetsov, 1965; Avilova et al., 1970)) and vapor of H chondrite (Kosarev, 1999) and cometary material (Kosarev et al., 1996). For condensed stony material, a tabular equation of state obtained using the ANEOS program (Thompson and Lauson, 1972) with input data for granite (Pierazzo et al., 1997) was used. Tables for the equation of state are available only for a very limited set of rocks, and granite was chosen as an approximation for a stony body. The comet nuclei were considered to consist of ice, for which the Tillotson equation of state was used (Tillotson, 1962). The simulations were carried out for the impacts of spherical stony bodies from 20 to 300 m in diameter with a density of 2.65 g cm^{-3} at various entry angles. Meteoroids of the considered dimensions are noticeably deformed and crushed into pieces at altitudes where the aerodynamic load is much higher than the plastic limit, which makes it possible to neglect the strength of the body.

Figure 12.1 shows the flow around a 40-meter asteroid that enters the atmosphere vertically with a speed of 20 km s^{-1}, at different altitudes. At an altitude of about 30 km, the meteoroid begins to deform, and wave-like perturbations appear on its surface due to the development of Rayleigh-Taylor and Kelvin-Helmholtz instabilities. The increase in aerodynamic loading flattens the meteoroid. At heights of 20–25 km (along a trajectory section of about 10 km) as a result of transverse expansion, the meteoroid turns into a pancake-like structure in qualitative agreement with approximate analytical models (Grigorian, 1979; Chyba et al., 1993; Hills and Goda, 1993). The growth of instabilities causes the fragmentation of the meteoroid; at heights below 15 km, it turns into a jet consisting of vaporized material, air heated in the bow shock wave, and fragments of the falling body. At its onset, the velocity of this jet is about 18 km s^{-1}; i.e., slightly different from the initial velocity of the meteoroid. Thus, destruction and fragmentation of the body under consideration occur before it begins to slow down significantly. For this reason, the deceleration and loss of energy of such bodies cannot be modeled accurately using the equations of the physical theory of meteors, since they do not describe the motion of a gas jet. The fragmentation leads to

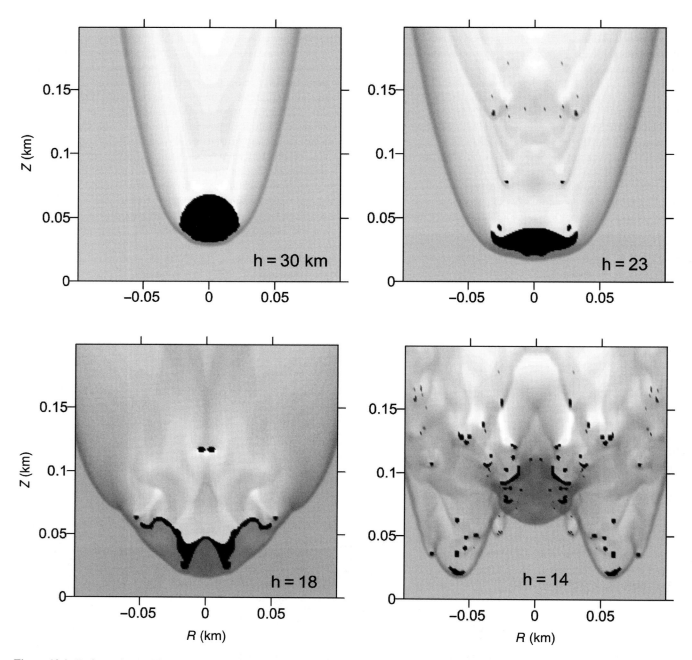

Figure 12.1. Deformation and fragmentation of an asteroid with diameter of 40 m in the coordinate system R, Z associated with a moving object. h is the altitude of the flight, the impact velocity is 20 km s^{-1}, the impact angle is 90°. Color of different intensity shows relative density distributions $\rho/\rho_0(z)$ (the darker the denser).

an increase in the evaporated surface and, consequently, to an increase in the rate of ablation. At an altitude of about 10 km, the fragments of meteoroids completely vaporize, and the jet turns into a gas jet (consisting of air and vapor), which is almost completely decelerated at an altitude of about 6 km.

The height at which the velocity of the body is reduced by one half (which can be considered as the effective height of the explosion or burst altitude) for each simulation is presented in Tables 12.1 and 12.2. The simulations for various speeds show that the burst altitude depends only weakly on the entry velocity of the body, in accordance with the approximate models (Grigorian, 1979; Grigoryan et al., 2013; Hills and Goda, 1993),

since the increase in the cross-section of the body and the deceleration efficiency depends mostly on the distance traveled.

The data presented in Tables 12.1 and 12.2 can be approximated by a simple function:

$$\frac{h_t}{H} = -1.3 \cdot \ln\left(\frac{D}{H}\left(\frac{\rho_m}{\rho_0}\right)^{2/3} \sin\alpha\right) + 1 \quad (12.1)$$

where H is the atmospheric scale height, ρ_0 and ρ_m are the density of air at sea level and the density of the meteoroid, respectively, D is the meteoroid diameter, h_t is the burst altitude and α is the trajectory angle at the top of the atmosphere.

The results of individual calculations often differ arbitrarily from this approximate dependence by 2–3 km. Such a difference seems quite natural, since even the results of several calculations with the same initial conditions may differ by the same 2–3 km from each other (Shuvalov and Trubetskaya, 2007). This is because the process of deformation and fragmentation of a quasi-liquid meteoroid is accompanied by the development of Rayleigh-Taylor and Kelvin-Helmholtz hydrodynamic instabilities on the body surface (Svetsov et al., 1995), which is random in nature. Therefore, deformation, fragmentation, and deceleration processes develop differently in similar calculations with the same initial data, which inevitably leads to different effective burst altitudes.

The results of the calculations show that the atmosphere hinders the impact on the ground even if the body is quite large (but does not have high strength). Even a stony asteroid 100 m in diameter with an angle of entry into the atmosphere of 45° loses most of its energy in the air. The main hazard of meteoroids and small asteroids, therefore, is the action of the air shock wave.

Table 12.1. *Effective heights of explosions (in km) for stony asteroids. D is asteroid diameter, α is an impact angle.*

	impact angle α, degrees						
D, m	90	60	45	30	15	10	5
20	16	18	19	23	29	32	35
40	10	13	14.5	16	21	25	29
70	3	6	8	13	19	23	26
100			0.2	8	13	16	21
200				1	6	12	18
500						0	7.5

Table 12.2. *Effective heights of explosions (in km) for icy bodies. D is comet diameter, α is an impact angle.*

	impact angle α, degrees						
D, m	90	60	45	30	15	10	5
40	15	18	21	25	31	32	36
100		6	8	14	18	27	28
200				5.3	10	18	21
500						6	12.5

12.3 Shock Waves on the Ground

The model described in the previous section for objects with low strength was also used to calculate the overpressure and wind speed behind the shock wave as it propagated over the ground surface after the vertical falls of spherical stony asteroids with a speed of 20 km s^{-1} (Shuvalov et al., 2013, 2016). Figure 12.2 shows the aerial shock wave created by the deceleration of an asteroid with a diameter of 40 m (the same scenarios as shown in Figure 12.1). This is a typical aerial burst produced by an asteroid or comet. As mentioned in the previous section, after meteoroid disruption and total vaporization at an altitude of 10 km, a high-velocity gas jet consisting of hot vapor and air is decelerated at an altitude of about 6 km. The shock wave reaches the ground with an amplitude corresponding to a blast overpressure Δp equal to about 20 kPa, then reflects from the surface

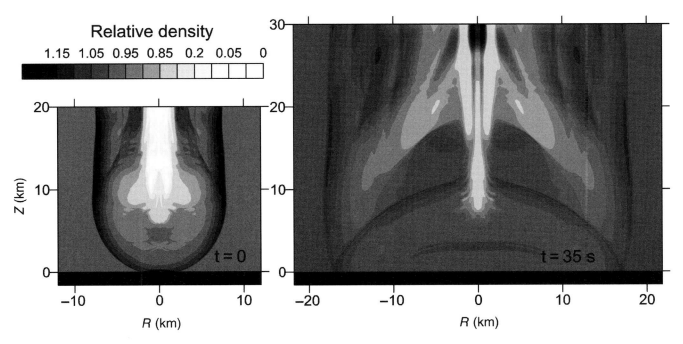

Figure 12.2. Distributions of relative density $\rho/\rho_0(z)$ for a vertical impact of an asteroid with a diameter of 40 m (the same as in Figure 12.1, but for a later time). The time is counted from the moment when the shock wave contacts the Earth's surface.

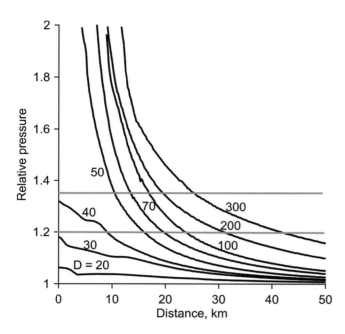

Figure 12.3. Maximum relative pressure p/p_0 reached on the surface at different distances from the impact point upon impacts of asteroids with different dimensions. The numbers near each curve show the diameter of the falling body in meters.

Table 12.3. *Radii of areas where the overpressure exceeds 35 and 20 kPa and damage efficiencies ξ_{20} and ξ_{35} showing damaged areas per unit of the mass of the impacting asteroid with a diameter D. The asteroid density is 2.65 g cm^{-3}. The impacts are vertical.*

D, m	30	40	50	70	100	200	300
R_{35}, km	–	4.5	10.5	13.3	16.5	20	30
R_{20}, km	2	9	16	20	24	31	42
ξ_{35}, $\frac{m^2}{kg}$	–	0.73	2.0	1.2	0.62	0.11	0.07
ξ_{20}, $\frac{m^2}{kg}$	0.33	2.87	4.6	2.6	1.3	0.27	0.14

and interacts with the descending wave; at large distances (about 10 km), this leads to the formation of a wavefront propagating along the surface (i.e., Mach reflection occurs).

The destructive effects of a shock wave are usually associated with the maximum overpressure behind the shock wave front. The results of nuclear tests (Glasstone and Dolan, 1977) show that destruction of brick walls with a thickness of 24–36 cm begins at an overpressure of 20 kPa (0.2 bar, 3 psi, pound-force per square inch); walls of concrete blocks with a thickness of 24–36 cm are completely destroyed at an overpressure of 35 kPa (0.35 bar, 5 psi).

Some results of numerical simulations of vertical impacts with a speed of 20 km s^{-1} are presented in Figure 12.3 where the maximum overpressure is shown as a function of distance from ground zero – the point of the first contact of the shock with the ground. The graph shows the effects of asteroids of different sizes, including relatively large asteroids that strike the ground and produce craters. Table 12.3 presents the dimensions of areas in which the overpressure exceeds the critical values $\Delta p = 35$ and 20 kPa for different airburst scenarios. The results show that asteroids with a diameter less than 20 m do not damage buildings severely (Figure 12.3). The maximum overpressure resulting from the fall of a 10-m body does not exceed 1–2 kPa. On the other hand, building damage is very likely if the falling body is larger than 30 m and the airburst occurs over inhabited regions. Note that these results are for stony asteroids with low strength, entering the atmosphere vertically. Heavier and stronger iron meteoroids can reach Earth's surface and form craters or crater fields even when their dimensions are of several meters. Lighter and weaker cometary bodies with the same dimensions are more easily destroyed and are vaporized in the atmosphere at higher altitudes (Shuvalov and Trubetskaya, 2007). The critical size of a comet that results in damage on the ground, is larger than the corresponding size of asteroids by approximately a factor of 1.5.

A destruction efficiency ξ can be defined as the ratio of the area of Earth's surface over which the critical overpressure Δp is reached to the mass of the impactor:

$$\xi = \frac{\pi R^2}{\frac{1}{6}\pi D^3 \rho_m} \qquad (12.2)$$

where R is the radius of the destruction zone, D is the diameter of the impactor, and ρ_m is the density of the impactor. The parameter ξ has dimensions of m^2 kg^{-1}. The values ξ_{20} and ξ_{35} for $\Delta p = 20$ kPa $= 0.2 p_0$ and $\Delta p = 35$ kPa $= 0.35 p_0$ (where p_0 is the normal pressure near Earth's surface), respectively, are presented in Table 12.3. Parameters ξ show the destroyed area (in square meters) per unit of mass (kilogram) of the falling body. The results in Table 12.3 show that a maximum destruction efficiency occurs for asteroids at a size of about 50 m (for vertical impact).

Another damage factor is the dynamic pressure associated with the strong wind arising after the shock wave front. The spatial distribution of maximum wind velocities is very similar to the distribution of overpressures. As in the consideration of overpressure, there is a maximum wind efficiency ξ_u (also defined using equation 12.2 where R is now the radius of the area where wind velocity exceeds 30 m s^{-1}) which relates to the same asteroid size of about 50 m.

To assess the risks associated with the blastwave damage, it is customary to take into account the probability of the catastrophic event multiplied by a value characterizing the damage caused by the event. In this case, it is natural to consider the product F of the probability of the impact of a cosmic body of a given size and the area of the region of damage caused by this impact or, equivalently, the area of the damaged region divided by the time interval between impacts. The physical meaning of the parameter F is the average area of damage per unit of time (e.g., per year). Table 12.4 presents the results of calculations of risks using the damage areas obtained by Shuvalov et al. (2013) and the frequency of asteroid falls taken from the work of Harris and D'Abramo (2015). It is also taken into account that the land is one third of the Earth's surface. The results show that, on average, about half a square kilometer of Earth's land surface is damaged in a year by asteroids smaller than 100 m.

Table 12.4. *Risks F associated with the fall of asteroids of different sizes, with different criteria for estimating the damaged area.*

D, m	30	40	50	70	100	200	300
F_{35}, km²/year	–	0.05	0.12	0.08	0.03	0.01	0.01
F_{20}, km²/year	0.02	0.21	0.28	0.18	0.07	0.03	0.03
F_u, km²/year	0.15	0.44	0.48	0.31	0.11	0.06	0.05

The maximum annual damage is achieved by the impacts of asteroids with dimensions of about 50 m. However, it should be kept in mind that the average recurrence interval of asteroids 50 m in diameter is from 300 to 1000 years.

Mathias et al. (2017) developed a probabilistic asteroid impact risk model for quantifying the hazard posed by potential asteroid impacts on the Earth's population. The probabilistic model consists of a Monte Carlo framework that simulates a large set of impact scenarios selected from distributions of asteroid and entry characteristics and the fragment-cloud model (Wheeler et al., 2017) combined with scaling relations derived from nuclear sources for calculations of damage areas. The real population distribution on the Earth was taken into account. It was estimated that, statistically, the impacts of 100-m-diameter or smaller asteroids result in one casualty per year. A rough estimate based on the data of Table 12.4 (for vertical impacts of stony asteroids with a velocity of 20 km s^{-1}) and assuming that the damage area is a region with overpressure of more than 4 psi and that the average population density is 50 people per km² gives approximately 25 casualties per year. On the one hand, this discrepancy tells us that the fragment-cloud model gives smaller areas of damage, especially in comparison with the vertical impacts, and on the other hand, it shows that it is necessary to take into account the strength, the velocity distribution of impacts, the density distribution of asteroids and the distribution of population.

The results presented in Tables 12.3 and 12.4 concern vertical impacts only. Shuvalov et al. (2017) varied the entry angle of a Chelyabinsk-like meteoroid to test the idea that a steeply descending asteroid bursts at much lower altitude and its downward momentum brings it even closer to the ground, causing much more severe damage (Kring and Boslough, 2014). Peak overpressures on the ground obtained in the simulations for different impact angles are shown in Figure 12.4. The region with high overpressure (> 1 kPa) covers roughly the same area at all the considered entry angles, but the shapes of these regions are different. The elongation across trajectory (cylindrical component) decreases with increasing entry angle to the horizontal. The maximum value of peak overpressure Δp increases with trajectory angle from 3.3 kPa at 18° up to 5.2 kPa at 90°. The relatively small growth in the maximum overpressure with angle, despite the lower altitude of the airburst, is probably a consequence of upward flow of hot air and vapor along the wake, which results in the attenuation of the shock wave.

Similar results for the Chelyabinsk event were obtained in the work of Aftosmis et al. (2016), where the depositions of mass, momentum, and energy along the trajectory derived from the observational data were used as initial data for a three-dimensional hydrodynamic simulation of the propagation of a shock wave in the atmosphere. These results, like those described above for an angle of 18°, are consistent with observational data on broken window glass In addition, this paper shows that both static and time-varying energy sources give similar results for overpressure on the ground. For a vertical entry, a linear source and a point-like explosion also gave almost identical results. For the prediction of the consequences of an impact with given parameters, this approach requires the energy deposition along the trajectory. However, with an increase in the size of a body, the very concept of the released energy becomes ambiguous. In models based on the single-body flight equations, energy release ends at the time of complete vaporization. According to the hydrodynamic simulations described in Section 12.2, during the fall of bodies of 40–100 km in size, a jet of vapor and air forms at the time of complete vaporization. This jet has a velocity close to the initial velocity of the body and is decelerated and releases its energy 3–5 km lower than the altitude of complete vaporization.

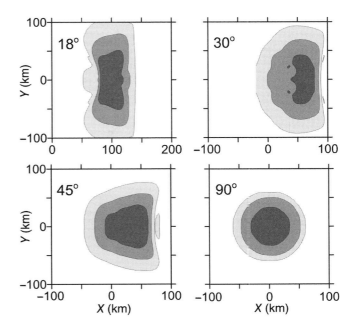

Figure 12.4. Surface peak overpressures for four entry angles (indicated on the panels). Light gray is overpressure $\Delta p > 0.5$ kPa (relative overpressure > 0.005), gray corresponds to $\Delta p > 1$ kPa (relative overpressure > 0.01), and dark gray corresponds to $\Delta p > 2$ kPa (relative overpressure > 0.02). The horizontal axis corresponds to a distance along trajectory projection, it is counted from the point where initial trajectory (without deceleration) would intersect the ground. The vertical axis shows a distance perpendicular to trajectory projection.

We described the results of hydrodynamic simulations at various asteroid diameters for vertical impacts. Similar calculations can be carried out for oblique impacts, as was done for the case of the Chelyabinsk meteorite. So far such calculations have not been published, and for the estimates of the effect of shock waves on the ground, one can use data from nuclear explosions (Glasstone and Dolan, 1977) and more powerful point explosions simulated by Aftosmis et al. (2018), and using appropriate explosion heights, for example, from Equation (12.1). Note that the overpressure at the 4-psi level, obtained for vertical impacts of asteroids from 40 to 70 m in diameter, is in good agreement with the results for atmospheric explosions given in Aftosmis et al. (2018) and Stokes et al. (2017).

12.4 Perturbations in the Ionosphere

Ionospheric disturbances are considered one of the dangerous consequences of meteoroid impacts (Nemtchinov et al., 2008). With a steady increase in the use of the Global Navigation Satellite System (GNSS), which includes the United States' Global Positioning System (GPS), reliable navigation is becoming increasingly important. Ionospheric disturbances are one of the most challenging problems in GNSS navigation. The influence of geomagnetic storms on GPS positioning is widely studied (e.g., Bergeot et al., 2011; Afraimovich et al., 2013; Jacobsen and Andalsvik, 2016; Luo et al., 2018). Under geomagnetic storm conditions, the electron columnar number density is several times higher than the values in the undisturbed ionosphere (Luo et al., 2018). It is known that geomagnetic

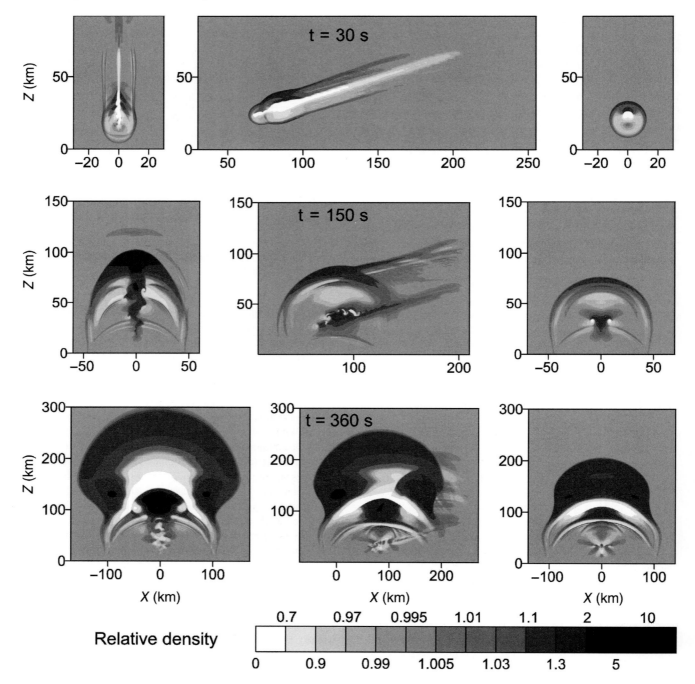

Figure 12.5. Distributions of relative density $\rho/\rho_0(z)$ at 30, 150, and 360 s after the Chelyabinsk class impact. Left images – vertical impact; right images – point explosion at an altitude of 20 km; central images – oblique impact.

disturbances registered after the Tunguska event of 1908 lasted about four hours (Vasilyev, 1998).

Atmospheric disturbances generated by the Tunguska and Chelyabinsk impacts were studied by Shuvalov and Khazins (2018) using numerical simulations based on the SOVA hydrocode (Shuvalov, 1999). Figure 12.5 demonstrates the initial stage of plume formation for three models of the impact similar in energy to the Chelyabinsk event. In the first simulation a 19-m-diameter dunite asteroid ($\rho = 3,300$ kg m^{-3}) penetrated Earth's atmosphere at a velocity of 20 km s^{-1} and angle of 19°. The second simulation considered a vertical impact of the same meteoroid. The third simulation considered a point explosion with an energy of 500 kt TNT (the same as the energy of the meteoroid) at an altitude of 20 km. By 30 s after the impact/explosion, flow fields differ considerably from each other. The difference is still clearly seen at 150 s after impact. However, by 360 s the flow fields look very similar; the shape of the energy release does not greatly affect the late stage of the evolution of atmospheric disturbances.

To estimate the degree of atmospheric disturbances at each point in space Shuvalov and Khazins (2018) used a non-dimensional value $\varepsilon = \max(\text{abs}(\rho/\rho_0 - 1))$, where $\rho_0(z)$ is a normal air density at an altitude z, i.e., a maximum deviation of local density from its equilibrium value. Figure 12.6 (a–c) shows ε-distributions for all three runs. The distributions do not differ much from each other. The size of a region where $\varepsilon > 0.05$ is about 3000–4000 km, at distances up to 500 km the density disturbances exceed 20%. The slope of the trajectory leads to a slight shift and a decrease in the perturbed region.

The atmospheric flow field resulting from a Tunguska-like impact differs considerably from the Chelyabinsk case. The difference is explained by the strong influence of the meteor wake (the hot rarefied channel remaining in the atmosphere behind the decelerated body). Since hydrostatic equilibrium is violated in the wake, the mixture of hot air and vapor is accelerated upward. This leads to the formation of an atmospheric plume that rises along the wake and carries dense air from a lower part of the atmosphere to great heights, up to several thousands of kilometers. Figure 12.7 illustrates some results of numerical simulations (Shuvalov and Khazins, 2018) of the impact of a 80-m-diameter comet. Impact angle is 45°, impact velocity is 30 km s^{-1}. At altitudes of about 1500 km, the plume is decelerated by gravity and begins to fall back, which leads to large-scale oscillations.

The plume oscillations generate shock waves expanding up and down, which turns the energy of the plume into heat. The heated region of the atmosphere expands in the transverse direction, generating a horizontally expanding shock wave. In both (Chelyabinsk and Tunguska-like) cases the atmospheric disturbances are concentrated at altitudes of more than 100 km; below 100 km the disturbances strongly attenuate due to sharply increasing air density (due to the small atmospheric scale). Unlike the case of the Chelyabinsk event, atmospheric disturbances caused by the Tunguska-like impact are much stronger than the disturbances caused by a spherical explosion of the same energy (see Figure 12.6(d–e)). This is due to the influence of the meteor wake.

Hydrodynamic modeling made it possible to determine the amount and distribution of dense air injected into the ionosphere after the impacts. This air instantaneously begins to ionize by the hard electromagnetic radiation of the Sun and energetic particles with energies from tens of keV to MeV, constantly present in the near-Earth space. The electron concentrations can exceed by several orders of magnitude the background electron concentration of the ionosphere. Possible plasmodynamic effects associated with the ionization of ejected air, with the motion of the overlying layers of the ionosphere and the plasmasphere were not considered. To account for these effects requires the development of another class of models, which take into account processes that, to date, are poorly understood. But already we can say something about the danger associated with the ejection of matter into the ionosphere.

When satellite signals pass through ionospheric irregularities, they encounter signal attenuation and even sometimes complete loss of lock. For example, Swarm satellites repeatedly encountered a total loss of signal from all GPS satellites, which made the precise orbit determination for Swarm impossible (Xiong et al., 2016). The interruption of GPS signal correlates with strong electron density variations (around 10^{10}–10^{11} m^{-3}) associated with large electron density gradients.

The modeling results of atmosphere density variation due to a Tunguska-like impact demonstrate that the neutral ionosphere density increases by more than an order of magnitude at altitudes of 250–300 km (Figure 12.6). Assuming the electron density increases correspondingly, the electron density variations in daytime will exceed 10^{11}–10^{12} m^{-3} and will cause serious disturbance or failure of GPS positioning. This can be dangerous, for example, for aircraft navigation or in military operations, or create difficulties for helping people in an emergency.

12.5 Seismic Effects

The impacts of cosmic objects can cause seismic effects when the body reaches the ground and when it is decelerated in the atmosphere. Meteoroids rarely retain a sufficient fraction of their original velocity at the Earth's surface with enough energy to produce impact craters and noticeable seismic effects, and the seismic efficiency of impact coupling is poorly known (Edwards et al., 2008).

When a body is decelerated at some altitude, the shock wave propagating in the atmosphere reaches the ground and generates seismic waves similar to a concentrated explosion in the atmosphere. Rayleigh seismic waves (Ben-Menahem and Singh, 1981) were frequently recorded after the falls of cosmic objects (Edwards et al., 2008). Rayleigh surface waves caused by the action of an air shock wave after the fall of the Tunguska cosmic body in 1908 in Central Siberia were recorded at stations in Irkutsk, Tashkent, Tiflis and Jena, and the magnitude of the seismic source was estimated to be in the range from 4.8 to 5.2 (Ben-Menahem, 1975; Pasechnik, 1976). The Chulymsky bolide, that fell on 1984 February 26 in the basin of the river Chulym, Tomsk region, Siberia, caused a seismic event of magnitude 3.4 ± 0.3 (Ovchinnikov and Pasechnik, 1988). After the fall of the Chelyabinsk meteorite in 2013, seismic oscillations were recorded by a large number of stations at distances of hundreds to thousands of kilometers. The magnitude of the seismic event from the Rayleigh wave was estimated from 3.7 ± 0.3 (Tauzin

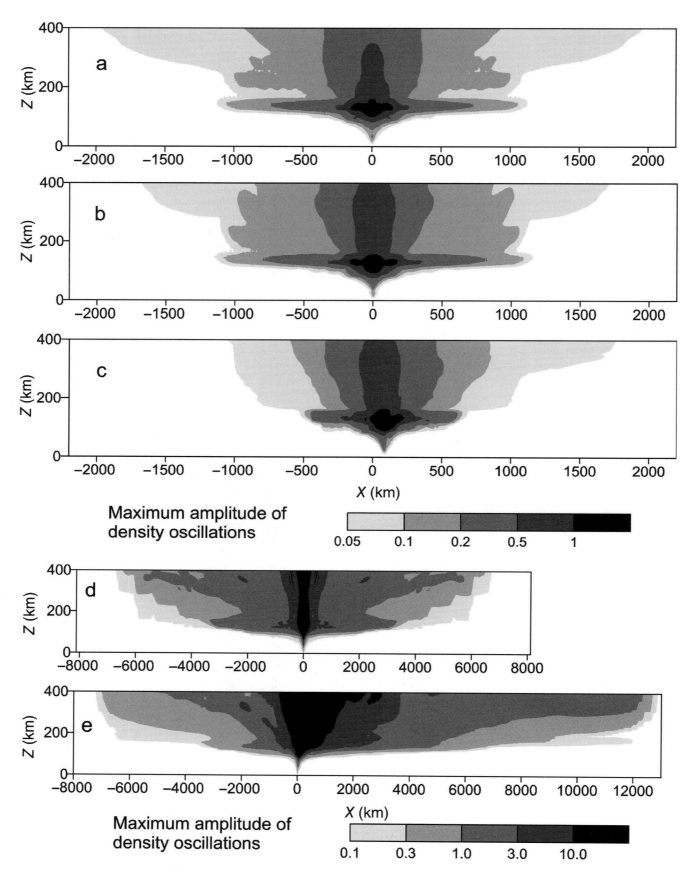

Figure 12.6. ε-distributions for different Chelyabinsk-like (a – spherical explosion; b – vertical impact; c – 19 degree impact) and Tunguska-like (d – spherical explosion; e – 45 degree impact) models.

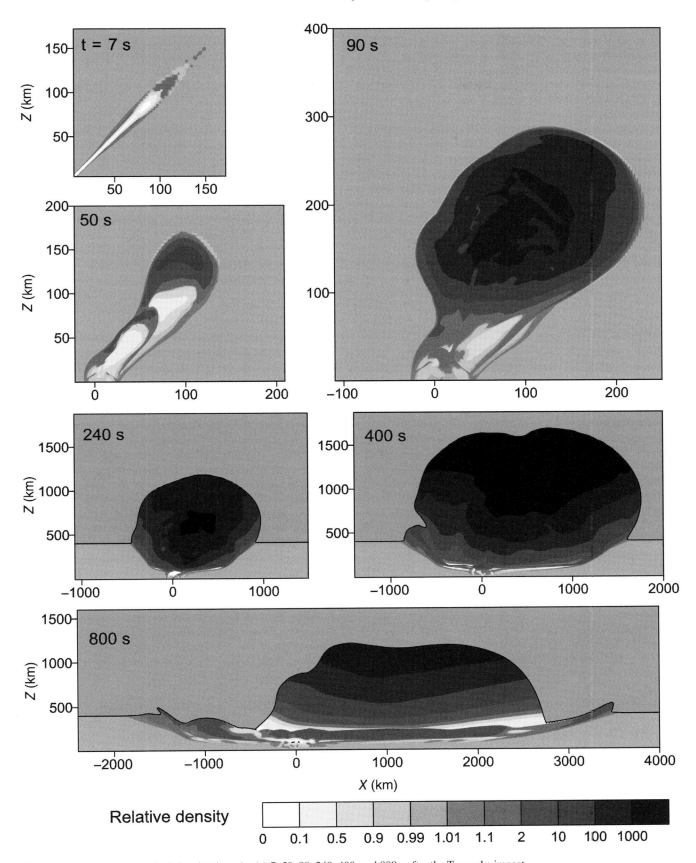

Figure 12.7. Distributions of relative density $\rho/\rho_0(z)$ 7, 50, 90, 240, 400, and 800 s after the Tunguska impact.

et al., 2013) to 4.0 ± 0.3 (Brown et al., 2013). The energy of the corresponding explosion was determined by Brown et al. (2013) according to the theory of generation of surface seismic waves by high-altitude nuclear explosions (Harkrider et al., 1974) as 420 ± 200 kt TNT, which is consistent with energy estimates from atmospheric infrasound waves of 570 ± 150 kt and from records of thermal radiation of 590 ± 50 kt (Popova et al., 2013).

Knowing the pressure distribution on the ground surface $P(x, y, t)$ caused by an aerial shock wave, it is possible to estimate the magnitude of the seismic event. Let us consider the efficiency of the coupling of the air wave into surface seismic waves. To this end, the following method was proposed by Svetsov (2007), based on the solution of the Lamb problem for an elastic half-space. First, the pressure spectrum is calculated

$$p(\omega, k_x, k_y) = \frac{1}{2\pi^2} \int_{-\infty}^{\infty} \int_{-\infty}^{\infty} \int_{-\infty}^{\infty} \times (P(x, y, t) - P_0) e^{-i(k_x x + k_y y + \omega t)} dx dy dt \quad (12.3)$$

where P_0 is the atmospheric pressure at the surface, and k_x and k_y are the wave numbers. Then the following quantity A_R is calculated:

$$A_R = \frac{1}{P_0(\omega_2 - \omega_1)} \int_{\omega_1}^{\omega_2} \left| p\left(\omega, \frac{\omega}{c_R \sqrt{2}}, \frac{\omega}{c_R \sqrt{2}}\right) \right| \omega d\omega \quad (12.4)$$

where integration limits $\omega_1 = 0.25$ s^{-1} and $\omega_2 = 0.42$ s^{-1} (corresponding to periods 25 and 15 s) are selected in accordance with the range of recorded frequencies of surface waves in air explosions, and c_R is the phase velocity of the Rayleigh wave (3.5 km s^{-1}). This interval of periods lies within the basic mode of the Rayleigh wave (Pasechnik, 1970) and includes the period of maximum sensitivity of seismographs used to monitor nuclear tests (Ben-Menahem and Toksöz, 1964; Crampin, 1966). At the same time, four seismographs that recorded the Tunguska event, according to Pasechnik (1970), recorded periods of oscillations ranging from 15 to 25 seconds. The phase velocity of the surface wave c_R was taken from Aki and Richards (1980) using the standard Gutenberg model, but variation of this velocity has little effect on the result. The physical meaning of A_R is the force acting on the target, but divided by P_0, and therefore it has a dimension of area.

Svetsov (2007) calculated the quantity A_R from numerical simulations of a series of the most powerful atmospheric explosions performed in 1961–1962 over Novaya Zemlya, for which the necessary input data have been published. A comparison of the computational results with the measured experimental seismic magnitudes M_S shows that the following relation is valid with a good accuracy:

$$M_S = \log_{10} A_R + 4.92 \quad (12.5)$$

where A_R is measured in km^2. This relation can also be used for the falls of small asteroids onto Earth.

Numerical simulation of the processes accompanying the fall of the Chelyabinsk asteroid in the atmosphere (Shuvalov et al., 2017) made it possible to determine the pressure on Earth's surface as a function of space and time. Two airburst scenarios with an initial asteroid kinetic energy equivalent to 300 and 500 kt were simulated. The magnitudes calculated using Equations (12.3), (12.4) and (12.5) were 3.85 for an initial meteoroid energy of 300 kt and 4.0 for an energy of 500 kt. Both values of M_S are close to the magnitude $M_S = 4.0$ obtained by Brown et al. (2013) (supplementary information) by averaging over thirty-three registrations of different stations.

Numerical models of explosions at various altitudes show that the pressure magnitudes are rather weakly dependent on the height of the explosion. It is obvious that when the height of an explosion increases, the amplitude of the shock wave incident on the ground decreases, but at the same time the area over which pressure exceeds normal pressure increases. Thus, the overall force can be greater; the increased area compensates for the pressure decrease. We calculate the pressures over a certain interval of periods from 15 to 25 s, and therefore the main contribution in the magnitude give regions with a size of the order of the wavelength of 50–100 km. When an explosion of low power occurs at a low altitude, the region of action is small because of the increase in the curvature of the front of the shock wave, which reduces the contribution to the integral in the chosen frequency interval.

The magnitudes of the seismic waves generated by airbursts were derived from models of impacts of stone and ice bodies with diameters of 30, 50 and 100 m with speeds from 11 to 70 km s^{-1} and angles of entry into the atmosphere from 15° to 90° (Svetsov et al., 2017). For oblique impacts the region influenced by the shock wave is close in shape to an ellipse, and the maximum pressure is higher in the direction opposite to the movement of the impactor (uprange).

In Figure 12.8, the dependence of the magnitudes of seismic events on the kinetic energy of the impactor E is plotted for all model scenarios. For a given energy of the body, the seismic magnitude weakly (and rather chaotically) depends on the velocity, size and density of the impactor, since the magnitude weakly depends on the height of the explosion. Randomness in the results is because the destruction of the body during its braking in the atmosphere is unstable (see Section 12.2), and the energy release in the final section of the trajectory can differ even for scenarios with the same parameters (Shuvalov and Trubetskaya, 2007; Shuvalov et al., 2016). At the same time, the magnitude decreases for vertical impacts due to energy losses caused by the movement of vapor and air upward along the trail remaining in the atmosphere after the deceleration of the body.

If we exclude vertical impacts, the dependence $M_S(E)$ can be approximated by the formula

$$M_S = \frac{2}{3} \log_{10} E - 6.27 \pm 0.3, \quad (12.6)$$

where E is measured in Joules.

Airbursts with a vertical impact trajectory are approximated with greater accuracy by the dependence

$$M_S = \frac{2}{3} \log_{10} E - 6.53. \quad (12.7)$$

If we use the classical expression for seismic energy E_S

$$M_S = \frac{2}{3}(\log_{10} E_S - 4.8), \quad (12.8)$$

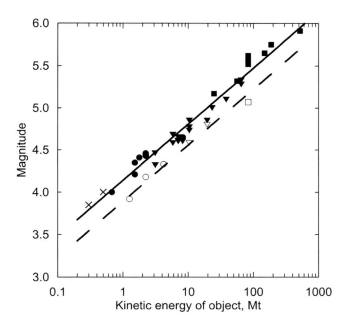

Figure 12.8. Magnitudes of seismic events caused by the falls of cosmic bodies of different sizes with different velocities and angles of entry into the atmosphere. The circles indicate bodies 30 m in size, the triangles correspond to bodies 50 m in size, and the squares correspond to 100 m in size. Empty symbols correspond to entry angles of 90° (vertical impacts). The crosses show magnitudes calculated for the fall of the Chelyabinsk meteorite under the assumption that its energy was 300 and 500 kt. The solid line is the approximation (12.6), the dashed line is the approximation (12.7) for vertical impacts.

then seismic efficiency E_S/E for airbursts from oblique impacts is

$$\log_{10} \frac{E_S}{E} = -4.6 \pm 0.3. \qquad (12.9)$$

It follows that the seismo-acoustic coupling efficiency of airbursts is equal to $E_S/E = 2.5 \times 10^{-5}$ to within a factor of 2. For vertical impacts, seismic efficiency is about 1×10^{-5}, which is close to the seismic efficiency estimated for airbursts (Griggs and Press, 1961). However, the observations of meteors give a very wide variation of acoustic coupling efficiency from 10^{-2} to 10^{-7} (Edwards et al., 2008).

There is no explicit dependence of E_S/E on the size and density of the body, and on the angle of entry up to 60°. Only for very steep trajectories of $\sim 90°$ does E_S/E decrease. The average value $E_S/E = 2.5 \times 10^{-5}$ is 4 times smaller than the magnitude of the seismic efficiency of impact coupling usually assumed for crater-forming impacts of large cosmic bodies (Collins et al., 2005), although this value is itself highly uncertain. Up to a factor of 2, the average acoustic coupling efficiency E_S/E is consistent with the seismic efficiency of experimental and calculated explosions in the atmosphere.

Thus, according to Equations (12.6) and (12.7), it is easy to determine the magnitude of seismic events during the impacts of meteoroids and small asteroids. The damage on the surface caused by the seismic event is obviously much smaller than that caused by the shock wave itself. However, the seismic effect can be dangerous for objects that are underground.

12.6 Effects of Thermal Radiation

Thermal radiation from fireballs can be strong enough to be dangerous to people and can ignite fires. The 1908 Tunguska airburst, caused by the entry of an object about 50 m in diameter, generated a forest fire within a radius of 10–15 km (e.g., Vasilyev, 1998; Svetsov, 2002). Effects of thermal radiation can be estimated if one assumes that the fireball is similar to a point source with the amount of isotropically radiated energy equal to some fraction (luminous efficiency) of impactor kinetic energy (Collins et al., 2005). However, fireballs change their shape, and can differ significantly from a sphere, especially in the case of airbursts. Moreover, the luminous efficiency is not well known. An average value of this coefficient for large meteoroids 1–10 m in size entering Earth's atmosphere is about 5–10%, increasing with meteoroid size and entry velocity (Nemtchinov and Popova, 1997). The luminous efficiency of the Chelyabinsk fireball was estimated from observational data to be 14–17% (Popova et al., 2013) or 17% (Brown et al., 2013). On the other hand, the luminous efficiency of nuclear explosions is greater than 35% and can reach $\sim 60\%$ for high-altitude air bursts (Glasstone and Dolan, 1977; Svetsov, 1994). For crater-forming impacts, Collins et al. (2005) assumed a luminous efficiency of 0.3%, substantially less than estimates from fireballs.

If the energy converted to radiation is not small, it is necessary to take into account radiation transfer in the course of hydrodynamic simulations of impacts. However, direct numerical solution of the equation of radiation transfer at every time step dictated by stability conditions of discrete hydrodynamic equations requires huge computer resources because the spectral intensity depends on two angles and photon energies as well as three spatial coordinates. If a heated volume is optically thin (each radiating point source can be freely seen from the cold surroundings), radiation cooling can be easily added to the hydrodynamic equations.

Recently, numerical simulations of impacts of 300-m-diameter asteroids (similar to the asteroid Apophis) were carried out, in which the equations of radiative transfer, added to the equations of gas dynamics, were used either in the approximation of radiative heat diffusion, if the radiating volume is optically thick, or in the optically thin approximation for small optical depths (Shuvalov et al., 2017). The impactor was treated as a strengthless quasi-liquid object. Both stages of flight through the atmosphere and impact on the ground were considered. Radiation energy that falls on the ground ranged from 0.6% to 4.5% for entry angles of 90 and 45°.

The same method was used to calculate radiation fluxes on the ground during the fall of the Chelyabinsk bolide (Shuvalov et al., 2017; Svetsov et al., 2018). The luminous efficiency obtained in these calculations was 17–18%, which is close to the estimates based on the observational data. A thermal effect on the ground depends on the radiation flux density and on the integral of this quantity over time – the surface energy density or radiant exposure. The calculated maximum of this value of 3.3 J cm^{-2} is achieved in the central region located below the point of the rectilinear trajectory of the body at an altitude of 31.5 km. This thermal exposure is not dangerous; even for sensitive skin, first-degree burns require an exposure of approximately 6 J cm^{-2} (Glasstone and Dolan, 1977).

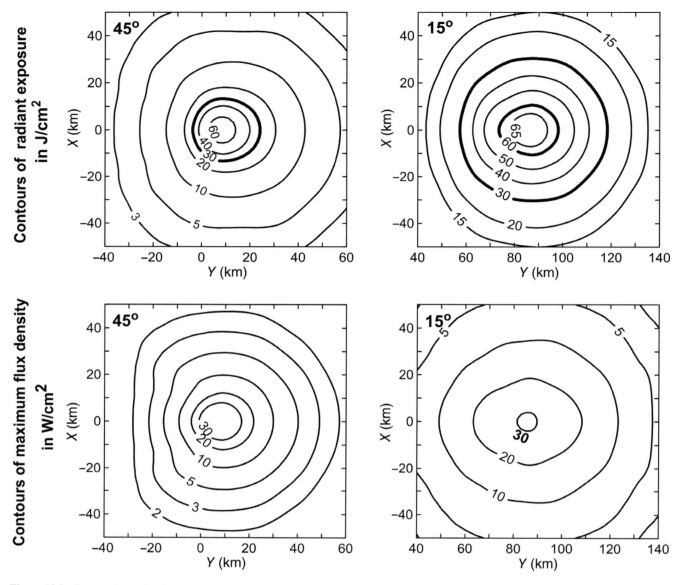

Figure 12.9. Contour lines of radiant exposure and maximum radiation flux densities on the Earth's surface calculated for the impact of a 50-m-diameter stony asteroid with a speed of 20 km s^{-1} at angles 45° and 15°. It was assumed that atmospheric visibility is 30 km.

Quite recently, numerical simulations were performed for impacts of strengthless spherical stony asteroids with a density of 3.3 g cm^{-3} (Svetsov and Shuvalov, 2018). The diameters of the bodies were 30, 50 and 100 m, the impact velocity was 20 km s^{-1}, and entry angles ranged from 15 to 90° degrees to the horizontal. Figure 12.9 shows contours of radiant exposure and maximum (in time) flux density on the ground, after the entry of a 50-m-diameter stony asteroid at angles of 45° and 15°. Both quantities, radiant exposure and maximum flux density, were calculated for an area oriented to the fireball (that is, they represent the maximum values for the orientation of the area). Such an asteroid has a kinetic energy equivalent to about 10 Mt TNT, which is a typical estimate of the energy of the Tunguska event.

As follows from the nuclear tests (Glasstone and Dolan, 1977), dry rotted wood, leaves and grass can be ignited if radiant exposure E exceeds 20–30 J cm^{-2}, and fir plywood can be ignited if $E > 60$–80 J cm^{-2}. A radiant exposure of 10 J cm^{-2} is the lowest limit for first-degree skin burns. Isolines with these values are shown in Figure 12.9. The scenario of an asteroid 50 m in diameter entering the atmosphere at an angle of 45° approximates the Tunguska event quite well, since the area of the fire in this case had an average radius of 10–15 km.

Figure 12.10 shows the luminous efficiency of bolides produced by these asteroids and average radii of areas on the ground where radiant exposure exceeds 10, 30, and 70 J cm^{-2}. Stony asteroids having low strength with a diameter of 100 m do not produce craters if they enter the atmosphere at angles smaller than 45°. For steeper entry angles the impact creates a crater; however, the majority of the energy is released in the atmosphere before the impact on the ground if the entry angle is smaller than 60°. From the point of view of thermal radiation damage, thirty-meter bodies are, on average, safe for humans, although they can cause temporary blindness, since the flux density of radiation

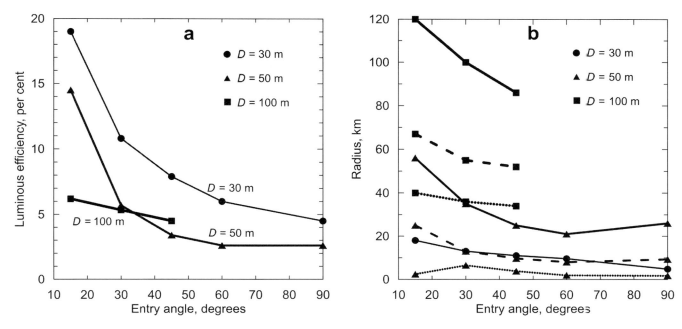

Figure 12.10. (a) Bolometric luminous efficiency (fraction of impactor kinetic energy emitted as thermal radiation) as a function of entry angles for asteroids with various diameters D; entry velocity is 20 km s^{-1}. (b) Average radius of areas where radiant exposure is higher than 10 J cm^{-2} (solid lines), 30 J cm^{-2} (dashed lines) and 70 J cm^{-2} (dotted lines).

may exceed the maximum solar flux density (at the zenith on a clear day) by about a hundred times. Asteroids 50 m in size with properties considered here can initiate forest fires within a radius of 10–20 km, and bodies with a diameter of 100 m can ignite fires in an area with a radius of more than 50 km. In these cases, the area of fire generation and thermal damage may be comparable or even exceed the area of serious damage from the shock wave.

12.7 Impact Craters Produced by Small Asteroids

As mentioned in Section 12.2, stony meteoroids smaller than 100 m in size usually decelerate at altitudes from a few kilometers to 20–50 km. These objects deposit most of their initial kinetic energy in the atmosphere, and only a relatively small portion of their initial mass and energy can reach the ground. Nevertheless, in some cases, this energy is enough to leave traces on the ground or even produce impact craters. Besides, special attention should be given to iron meteoroids, whose strength is higher than that of stones, as they penetrate deeper into the atmosphere without significant deceleration. Because of the greater density and strength, metallic meteoroids are better able to survive the atmosphere entry. Such cosmic bodies are much less abundant than stone bodies. It is estimated that their share is only about 4% of the observed falls of meteoroids by number and only 2% of all meteorites that have been found and observed during the fall. However, the mass of iron meteorites is the main part of the mass of all known meteorites, since they are denser, more resistant to weathering and more easily recognized (or detected) as meteorites, especially when they are buried under the ground.

Due to the difference in density and strength, an iron cosmic body about 50 m in diameter (i.e., similar to the estimated diameter of the Tunguska asteroid) is able to create a km-sized crater on the Earth's surface. The famous Barringer Meteorite Crater, aka Meteor Crater in Arizona (1.2 km in diameter, age 50 000 years) was probably formed due to the impact of an iron asteroid 45–65 m in diameter after its breakup in the atmosphere (Artemieva and Pierazzo, 2009, 2011). Meteorites found around the crater rim were named after nearby Canyon Diablo, the closest community to the crater at the end of the nineteenth century. Recent modeling efforts demonstrated that the impact conditions that appear to best agree with the known crater morphology are consistent with the impact of a tight swarm of fragments that are not strongly decelerated in the atmosphere. During the flight through the atmosphere the projectile lost probably 30–70% of its mass, mainly because of mechanical ablation and gross fragmentation. Starting from an impactor pre-atmospheric velocity of ~18 km s^{-1}, which is close to the average value for Earth-crossing asteroids, the most probable impact velocity at Earth's surface for a tight swarm is around 15 km s^{-1} or higher (Artemieva and Pierazzo, 2009, 2011). However, Melosh and Collins (2005) proposed that the impact velocity at the surface was lower, about 12 km s^{-1}, which provides a possible explanation for why the crater does not contain large volumes of rock melted by the impact.

The impact of the Canyon Diablo asteroid was accompanied by strong thermal radiation and a shock wave, which had a significant effect on the neighboring region (Kring, 2017). In the type of environment that existed at that time, the most destructive effects of the impact event were ejected debris, the fireball and the shock wave in the air. These effects were confined to a small region. A small amount of seismic energy was generated and small amounts of climatically-active gases (e.g., CO, CO_2, SO_2 and/or SO_3, H_2O, Cl and Br) were released. Plants and animals at ground zero were annihilated. Bedrock below and around the vapor-melt zone was ejected and overturned, burying the topography and any plants and animals not already swept away by the air blast. The ensuing winds severely damaged trees in any forested area within a diameter of 32 km. Grass, small shrubs, and soil were probably stripped from the area near the crater. Throughout a circular region up to 32 km in diameter, large mammals would have been killed or wounded by the air blast (Kring, 2017).

There are twenty-five confirmed impact craters not exceeding the Meteor Crater in size in the Earth Crater database (www.passc.net/EarthImpactDatabase/Diametersort.html). Most of these (seventeen cases) were formed by an iron impactor, and two small craters (Haviland, 15 m, and Dalgaranga, 24 m) were formed by stony-iron impactors. For four craters the origin of the impactor was not determined, and only two craters were formed by chondritic cosmic objects (recent Carancas, 13.5 m in size, and Tswaing, 1.13 km). There is evidence of different entry scenarios for these impactors, some of which produce crater strewn fields with well-separated small craters (i.e., Sikhote-Alin; Nemtchinov and Popova, 1997; Svetsov, 1998), craters with overlapping rims can be formed (e.g., Henbury craters, Hodge and Wright, 1971), and the formation of a single crater due to the impact of a tight swarm of fragments is also possible (Artemieva and Pierazzo, 2009). The behavior in the atmosphere is dependent on the size and mass of an impactor, as well on its composition, strength, and entry conditions, such as angle and velocity of entry (Shuvalov and Trubetskaya, 2010; Svetsov et al., 1995).

Small (one to several meters) iron meteoroids penetrate deeper into Earth's atmosphere than do stony and especially cometary objects, so they represent an increased danger. For example, the Morasko strewn field located near Poznan (Poland) includes seven impact craters from 20 to 90 m in diameter, covering an area of about 0.4–0.5 km, and numerous iron meteorites have been recovered in the area. Recent work has constrained the impact scenario that resulted in the formation of the Morasko strewn field (Bronikowska et al., 2017). It was estimated that the entry mass is 600–1100 t (i.e., diameter 4–8 m), the velocity range is from 16 to 18 km s^{-1}, and the entry angle is 30–40°. Such entry velocities and angles are typical values for near-Earth asteroids, although the initial mass can be considered as small (Bronikowska et al., 2017). The total energy of the Morasko event was about 30 kilotons, which is more than 10 times less than the estimated total energy of the Chelyabinsk event. Hence, it is likely that the environmental effects of the Morasko impact were strongly localized (Bronikowska et al., 2017).

It turns out that some small objects can reach the ground without significant destruction in the atmosphere. For example, it is assumed that the 45-m-wide Kamilsky crater in Egypt was formed by a 40–60-t iron projectile about 2–2.4 m in size that passed through the atmosphere without disruption (Ott et al., 2014).

Among recent small events, several impacts have left footprints on the ground. There are holes about one meter in size that were formed by the fall of chondritic meteorites Kunya-Urgench (1988 June 20) and Jilin (1976 March 8). These holes are termed penetration pits with a depth-to-diameter ratio greater than 1 (Mukhamednazarov, 1999; Academia Sinica, 1977), which is typical of impacts at terminal velocity. The fall of the Sterlitamak iron meteorite (1990 May 17) was tentatively classified as a transitional morphological type between a hypervelocity meteorite crater and a terminal velocity impact pit (Petaev, 1992). The largest fragment, weighing 315 kg, was extracted from the 12 m depth of the 10-m-diameter hole. The initial mass of the impactor is estimated to be about 1.5 t.

Craters were formed due to the fall of the Sikhote-Alin iron asteroid (1947 February 12) with an estimated mass of 200–500 t (size 4–5 m) (Nemtchinov and Popova, 1997). The largest crater formed in the strewn field had a diameter of 26.5 m and a depth of 6 m (Krinov, 1971), with a depth-to-diameter ratio of ∼0.2, typical for hypervelocity impact craters. This fall is an example of a successive fragmentation event. By generalizing eyewitness reports, it was assumed that fragmentation occurred in several stages at altitudes of 58, 34, 16 and 6 km (Divari, 1959). Individual fragments weighing up to 1700 kg, decelerated and fell to the surface at speeds of up to several kilometers per second (Svetsov, 1998). The main crater field of the Sikhote-Alin fall (zone of possible damage) extends to 0.3–0.6 km.

Most meteoroids that produce fireballs begin breakup at altitudes that indicate that the mean strength of the meteoroid is about 0.1–1 MPa (Popova et al., 2011). An exception is the recent fall of the Carancas stony meteorite in Peru (2007 September 15) which appears to have been substantially stronger, survived atmospheric entry and led to the formation of a 13.5-m-wide impact crater (Borovička and Spurný, 2008; Tancredi et al., 2009). Several houses at distances of several hundred meters, up to 1300 meters from the crater, experienced broken windows. One person and one bull fell at a distance of about 100–200 m from the crater due to the action of a shock wave during the impact (Tancredi et al., 2009).

This meteoroid probably did not experience significant atmospheric fragmentation, although detailed observational data were not available. It was classified as an ordinary chondrite H4-5 (Connolly et al., 2008). The meteoroid probably had an initial size about 0.9–1.7 m (i.e., a mass of 1.3–10 t) (Borovička and Spurný, 2008). Borovička and Spurný (2008) assumed that the Carancas meteoroid could survive atmospheric loading if its strength was 20–40 MPa, and concluded that Carancas was a rare example of a monolithic meteoroid that was free of internal cracks. Kenkmann et al. (2009) obtained lower estimates of the Carancas meteoroid strength (12–18 MPa) for a vertical impact of a 2–4-t body (entry velocity $V \sim 14$ km s^{-1}). Kenkmann et al., however, preferred the possibility of a shallow impact, which allowed them to reduce the maximum loading along the trajectory to 8–12 MPa or even lower (Kenkmann et al., 2009), even though this scenario requires a much larger initial meteoroid mass of 10–50 t. The Carancas event confirms that meteoroid strength can vary significantly from case to case. However, even though 10–30 bodies of this size enter Earth's atmosphere every year (Nemtchinov and Popova, 1997; Brown et al., 2002) it should be emphasized that the formation of small craters on Earth is extremely rare due to the effective disruption of meteoroids in the atmosphere.

12.8 Action on People

As mentioned in Section 12.3, the main dangerous effects of impacts of small cosmic bodies are related to the action of the shock wave in the air and thermal radiation. Overpressure can harm people by creating a pressure differential between the body's internal pressure and ambient pressure. People can suffer from lung damage, eardrum rupture, concussion, being rendered unconscious, etc. Strong winds can throw people against objects or vice versa, resulting in cuts, bruises, bone fractures, and other internal and external injuries (Glasstone and Dolan, 1977). Thermal radiation can cause skin burns and indirect burns due to the ignition of materials in the environment. The burns can be of different natures and levels of severity, depending on the source and duration of the radiation. The effect of thermal radiation from nuclear explosions was described by Glasstone and Dolan (1977). These data are often used in radiation hazard assessments (e.g., Rumpf et al., 2016; Mathias et al., 2017). However, the spectral dependence of radiation emitted during an asteroid impact can be very different from the spectral radiation emitted by a nuclear explosion. Sunburns, retinal and conjunctival burns, temporary blinding and heat sensations can arise due to airburst radiation. All theoretical approaches to risk assessment should be verified, and the Chelyabinsk and Tunguska events provide rare opportunities.

The recent Chelyabinsk event ($\sim$ 500 kt impact, 2013 February 15) occurred over a populated area. Data on injuries were collected from government reports, from phone and Internet surveys shortly after the event and from several local hospitals more recently (Kartashova et al., 2018). The epicenter of the event was about 40 km south of the city of Chelyabinsk, which has a population of more than 1.15 million inhabitants.

It was reported that 1613 people (about 0.1% of inhabitants) sought medical assistance at hospitals; sixty-nine people were hospitalized, two in serious condition. Those in serious condition included one child with a severely injured eye and one woman with a spinal fracture, both from Kopeysk. They evacuated for treatment to Moscow (Popova et al., 2013; Kartashova et al., 2018). In the first days after the event, the most reported injuries were caused by cuts from broken glass and trauma from the shock wave (falls and blows from objects, causing brain concussions, bruises, etc.). A few days later, the causes were vegetative-emotional syndrome, reaction to stress and hypertension (Akimov et al., 2015). The majority of injuries occurred in the densely populated Chelyabinsk city, but the highest percentage of people seeking assistance was near the trajectory track in the Korkinsky district (0.16%). The percentage of people who sought medical assistance declined sharply with distance from the airburst source (Popova et al., 2013; Kartashova et al., 2018).

Telephone interviews collected in February 2013 revealed that 1% of 500 respondents were injured or had some other ailment. Another 1% reported that both the respondent and close relatives had received some injuries/ailments. The combined fraction of respondents (2%) is much higher than the 0.1% of inhabitants who went to hospital. In addition, 7% of respondents reported that members of their nearest family were injured. Based on this, several thousand people received at least minor injuries or ailments as a result of the Chelyabinsk airburst event.

An Internet survey collected the answers of 1813 residents. Most of these residents were from Chelyabinsk, but some residents were further from the meteoroid trajectory (Popova et al., 2013; Kartashova et al., 2018). A total of 386 respondents to the online survey (21%) reported some injuries. Among them, more than 50% complained of sore eyes, and about 37% reported temporal stun or brain concussion/headache. A small percentage reported sunburn, and about 9% experienced stress or panic immediately after the event (Kartashova et al. 2018). Again, the distribution of respondents in the Internet survey confirmed that the total number of people affected by this event was much larger than the $\sim$1600 who sought medical care in hospitals.

Recent official data from several hospitals confirm that the type of injuries included cuts, bruises, brain injuries, and stress (Kartashova et al., 2018). These data also confirm the occurrence of eye injuries and bone fractures, which previously were not reported (Popova et al., 2013). Injuries caused by overpressure were more severe than those caused by thermal and UV radiation. A significant number of people were panicked and stressed after the arrival of shock waves, so that post-event injuries included hypertension owing to stress.

To determine the levels of thermal radiation and overpressure created by the Chelyabinsk asteroid impact, the impact was modeled using a quasi-liquid model (Shuvalov et al., 2017). The distribution of injuries is in general agreement with the results of calculations of radiation and overpressure in the Chelyabinsk event (Shuvalov et al., 2017; Kartashova et al., 2018).

The much more powerful ($\sim$5–20 Mt) Tunguska event on 1908 June 30 occurred in an area with low population density, where the population basically consisted of local reindeer nomads (Evenks). The trading post Vanavara was located about 70 km from the epicenter. Most of the eyewitness reports were collected long after the event, in 1921–1930, 1938, 1959–1969 (Vasilyev et al., 1981). Many accounts retold the stories of other people. The locations of reported injuries are approximate. Nevertheless, some information can be extracted from the eyewitness interviews.

Eyewitness reports that contained information on injuries were extracted from the catalogue of eyewitness accounts (Vasilyev et al., 1981). Corresponding locations are shown in Figure 12.11. The locations of injuries are mainly in the region up to 300–500 km from the epicenter. At the largest distances, there were signs of stress and panic, sometimes these cases were accompanied by objects falling from high places (bench, roof, Russian stove). About 50 eyewitness reports describe events in locations closer than 130 km to the epicenter. More serious injuries occurred there. The injuries mentioned included concussion, stunning and fainting, a broken arm, burns, aphasia and blindness. Concussion and fainting were the most-often mentioned cases. People remained unconscious for up to two days. People at the Evenk meeting in 1926 reported three casualties. According to the information recorded by I. M. Suslov, these losses might have occurred at distances of 30–50 km from the epicenter (Vasilyev et al., 1981), but locations are approximate. The exact number of injured is not known, but it was probably not large, as the area close to epicenter was largely unpopulated.

There is also information on glass damage reported at distances from 65 to about 300 km from the epicenter. Despite the fact that they were collected in 1960, these data look reliable. A few reports mentioned roof damage. According to Gi et al. (2018) an approximate boundary of glass damage corresponds to 200–500 Pa. An approximate model for rapid assessment of hazardous effects (Glazachev et al., 2019) predicts the

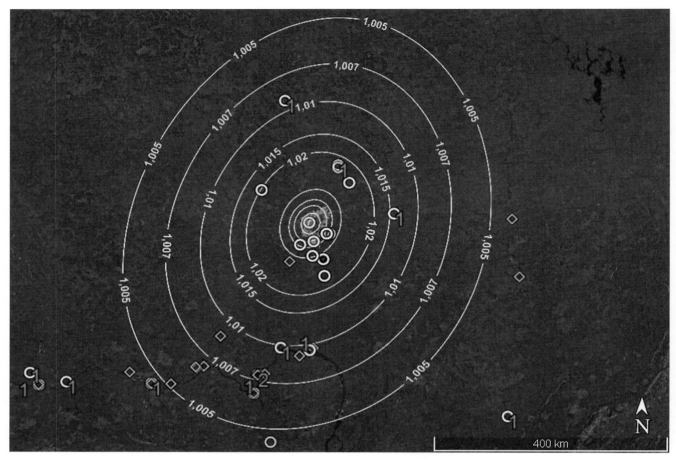

Figure 12.11. A map of approximate locations of injuries due to the Tunguska event. The area of fallen trees is shown near the epicenter (Longo et al., 2005). Locations with reported injuries are marked by circles. The numbers at the circles indicate reported stress/panic cases. Diamonds mark damaged windows and roofs. Oval contours correspond to the approximation model for rapid estimates of hazardous effects (Glazachev et al., 2019); the numbers correspond to the relative overpressure levels. (A black and white version of this figure will appear in some formats. For the colour version, please refer to the plate section.)

overpressure levels at those distances to be 500–700 Pa for a 10 Mt event, which can be considered as a reasonable agreement.

12.9 Assessment of Hazard and Fast Calculators

The Chelyabinsk airburst event was unpredicted and caused ~27M EUR damage on the ground and injured more than 1600 people (Popova et al., 2013). It was a reminder of the potential hazard of meteoroids and small asteroids, the vast majority of which are undetected, and motivates the need for accurate hazard and risk assessment.

Airbursts caused by meteoroids and small asteroids have rather localized effects; damage from even the most energetic impacts of stony asteroids up to 300 m in diameter at 20 km s^{-1} is restricted to a radius <50 km (Table 12.3). Hence, the threat from such objects depends on both the frequency and location of their occurrence. It is conventional to consider a risk, which is defined as the consequence of an occurring event – the hazard – weighted by the probability of the event occurring. It can also include a vulnerability factor, which accounts for the likelihood of survival for a given stress (Rumpf et al., 2017b).

The event probability includes the probability of a given asteroid encounter, as well as the likelihood of the event occurring at a given location with a given population and infrastructure. The probability distributions of several meteoroid parameters, such as size, speed and angle of entry into the atmosphere, are known from observations and calculations to a reasonable degree of confidence. However, other important asteroid properties, such as density and strength, remain quite uncertain. Estimating the energy and altitude of the airburst, which are the key parameters that control the damage extent, introduces several additional uncertainties, such coefficients of atmospheric drag, dispersion and ablation. The location uncertainty is of particular significance owing to the highly non-uniform distribution of the global population.

The large number of uncertain parameters implies that thorough impact risk assessment is a significant computational challenge and motivates the need for simplified approaches and fast damage calculators or the development of an authoritative scenario database using more rigorous approaches. Here we summarize the most common simplified approaches used in fast calculators (an example of a fast calculator is the Impact Earth! web program, http://impact.ese.ic.ac.uk) and the findings of recent hazard assessments that applied these methods. For meteoroids and small asteroids, such approaches typically involve two components: a method for predicting energy deposition in the atmosphere; and a method for converting this into ground damage.

12.9.1 Atmospheric Entry

Several (semi)-analytical approaches have been developed to describe the deceleration, ablation, fragmentation and disruption of meteoroids in an atmosphere (Passey and Melosh, 1980; Chyba et al., 1993; Hills and Goda, 1993; Artemieva and Shuvalov, 2001; Collins et al., 2005; Avramenko et al., 2014; Wheeler et al., 2017). Of these, the most successful approach to replicate the rate of energy deposition in the atmosphere inferred from light curves appears to be one in which the meteoroid undergoes catastrophic disruption at the point of fragmentation leading to a rapid and dramatic increase in drag and deceleration (Avramenko et al., 2014; Collins et al., 2017; Register et al., 2017; Wheeler et al., 2017). While these alternative models of meteoroid disruption and deceleration in the atmosphere differ in their prescriptions of spreading rate and when spreading stops, in general they give similar results for a specific airburst scenario (e.g., Chelyabinsk; Avramenko et al., 2014; Collins et al., 2017; Wheeler et al., 2017) depending on the choice of internal model parameters.

An important criticism of these catastrophic fragmentation models is that excessive spreading of the asteroid results in non-physical geometries for the deformed meteor (Artemieva and Pierazzo, 2009). However, the success of these models in replicating the energy deposition curves for Chelyabinsk appears to imply that the drag (effective cross-sectional area) predictions of the models are correct even if the geometrical interpretation of the "pancaking" is unphysical. Collins et al. (2017) recommended that these models should therefore be regarded in a similar manner to an empirical equation of state, and used to predict phenomena in regimes where they have been calibrated and applied only with caution in other regimes. Chelyabinsk (∼0.5 Mt) provides the only opportunity of calibration based on measured physical data; however, catastrophic fragmentation models were successfully applied to model the consequences of Tunguska (5–20 Mt), collision of comet Shoemaker-Levy 9 with Jupiter, and some bolide light curves. For strong or dense impactors that can survive to strike the ground at high speed, alternative approaches, such as the separated fragmentation model or hybrid models, are more appropriate to describe atmospheric disruption (e.g., Artemieva and Shuvalov, 2001; Wheeler et al., 2017).

Short calculation times have allowed catastrophic fragmentation models to be performed for a wide range of impactor and model parameters, revealing the sensitivity of burst altitude and energy to asteroid properties (Collins et al., 2017; Wheeler et al., 2017). Such analysis reveals that for a fixed impactor size, burst altitude is most sensitive to trajectory angle, while burst energy is most sensitive to impact speed. Strength is particularly important for smaller asteroids, such as Chelyabinsk.

12.9.2 Airburst Damage Models

With a prediction of burst altitude and energy, the next step in hazard assessment is to estimate the extent of damage from the shock wave (airblast), ensuing winds and thermal radiation. If any surviving fragments strike the ground at high speed, additional hazards include seismic shaking, ejecta dispersal and tsunami waves, if impact occurs in water. The hazards from crater-forming impacts on land or water are not considered here.

The most practical approach to estimate airblast damage is the application of scaling laws based on the results of nuclear weapons tests (Glasstone and Dolan, 1977) and numerical simulations of point explosions with a higher energy of 250 Mt (Aftosmis et al., 2018; Stokes et al., 2017). Convenient analytical fits to these data have been developed to describe the 4-psi overpressure blast radius as a function of burst altitude and energy (Hills and Goda, 1993, Eq. 2) and blast overpressure as a function of distance, burst altitude and energy (Collins et al., 2005, Eqs. (54–58)).

High winds in the aftermath of the shock wave can also be hazardous. Collins et al. (2005) provide a recipe for estimating peak wind speed as a function of distance from ground zero based on nuclear explosion data (Glasstone and Dolan, 1977), which has been used in risk assessment (Rumpf et al., 2017a; Reinhardt et al., 2016). However, given the similarity in the severity and spatial distribution of peak overpressures and peak wind speeds, the most parsimonious approach is to consider a common blast radius that combines the effects of both pressure and wind speed (Stokes et al., 2017; Mathias et al., 2017).

Thermal radiation effects have also been considered in some airburst hazard assessments (Rumpf et al., 2017a; Mathias et al., 2017). Based on nuclear explosion data, Collins et al. (2005) derived a scaling relationship between the thermal radiation exposure (energy per unit area) and distance from the fireball produced by a crater-forming impact of a given energy, which can be compared with the exposure required to cause burns and ignite combustible materials. Mathias et al. (2017) extended this relationship to be applied to airbursts, assuming the fireball was centered at the first altitude. Rumpf et al. (2017a), on the other hand, applied a scaling relationship for thermal exposure on the ground based on hydrodynamic calculations of radiation from airbursts (Nemtchinov et al., 1994). Results of comprehensive hazard assessments suggest that while thermal radiation can be the dominant hazard in rare cases, in most airburst scenarios its effects are secondary to damage from the atmospheric shock wave (Rumpf et al., 2017a; Mathias et al., 2017), which provides justification for airburst hazard assessments based solely on airblast damage (e.g., Stokes et al., 2017).

A criticism of using results of static-source nuclear explosions to approximate the consequences of impacts has been that the along-the-trajectory transfer of momentum from the asteroid to the air is neglected (Boslough and Crawford, 2008). However, comparison of simulations of explosion-generated blastwaves with a static and moving source suggest that neglecting along-the-trajectory momentum is of little consequence at ranges farther than three times the burst altitude and results in underestimation of the blast overpressure at ground zero by less than a factor of 2 (Aftosmis et al., 2016; Collins et al., 2017). Moreover, static explosion-based scaling laws are in good agreement with hydrodynamic simulations of atmospheric shocks generated by airbursts (Shuvalov et al., 2013, 2017). Hence, the static-source approach provides an adequate approximation of the azimuthally averaged airblast for probabilistic hazard assessment (Collins et al., 2017).

12.9.3 Hazard and Risk Assessment of Small Asteroids

The approaches described in the section have been used to assess the risk of small asteroid airbursts in several recent studies (National, 2010; Reinhardt et al., 2016; Mathias et al., 2017; Rumpf et al., 2017a; Shuvalov et al., 2017; Stokes et al., 2017). The probabilistic risk from asteroids up to 200 m in size is estimated at about 5 casualties per year (Mathias et al., 2017), with the largest number of casualties per year begin caused by asteroids 40–60 m in diameter (National, 2010; Shuvalov et al., 2013). According to this simple statistic, the airburst hazard is comparable to that posed by shark attacks and plane accidents. However, the average annual number of casualties is not a meaningful statistic to describe the asteroid hazard over short timescales, owing to the highly non-uniform distribution of the world's population and the frequency of fatal airbursts. The nature of the danger of asteroids lies in the fact that rare events can lead to a large number of human victims, with little or no warning. The average recurrence interval of a 50-m diameter asteroid impact is estimated as $>$ 500 years (Harris and D'Abramo, 2015). Moreover, the vast majority of airbursts will occur in unpopulated areas and be of no consequence, while only rare events that occur over dense population centers will be hugely devastating. For example, based on 2000 UN-census data of the global population distribution, Mathias et al. (2017) showed that for impacts by asteroids of 80–100 m diameter, more than 85% of scenarios affected no one; while less than 1% of events affected $>$ 100 000 people. Thus, a more nuanced metric is needed to adequately describe the low-probability/high-consequence impact hazard. Mathias et al. (2017) proposed the use of probability distributions that provide a likelihood of causing a given number of casualties. According to their ensemble risk analysis, each millennium there is a one-in-a-hundred chance of affecting a large town ($\sim$50 000 people), a one-in-a-thousand chance of affecting a large city ($\sim$1.4M people) and a one-in-ten-thousand chance of affecting one of the world's major conurbations ($\sim$ 6M people).

12.10 Summary and Future Issues

We described various effects that can be dangerous when small asteroids enter the atmosphere above the ground. Starting at a size of about 20 m, the atmospheric shock wave can cause appreciable damage and is the greatest danger. The seismic waves are less dangerous on the ground than a shock wave, but can cause the destruction of underground devices. Thermal radiation is a significant danger along with the shock wave and can cause fires if the size of the asteroid exceeds 50 m. Asteroids of this scale produce significant perturbations of the ionosphere, which can disrupt radio communication and navigation. The threat of these objects is regional. Meteoroids and asteroids of less than 20 m in size can be locally dangerous if they are strong enough to reach the surface at a high speed. Despite the fact that impacts of small hazardous asteroids are rare and that the calculated average annual number of casualties appears to be small, a dangerous event will happen sooner or later. In 2005, the United States Congress directed NASA to find at least 90% of potentially hazardous near-Earth objects that are 140 m in size or larger by the end of 2020. As we have shown, smaller asteroids are also dangerous, but their number is huge, and there is still no way, at least at the moment, to detect a substantial proportion of these bodies.

We described the current state of research on the prediction of the main effects of the impacts of asteroids with sizes ranging from 20 to about 100–200 m. In this chapter, we have paid more attention to modeling the fall of cosmic objects on the basis of the hydrodynamic equations that neglect the strength of the body. The hydrodynamic modeling of meteoroid entry into the atmosphere is challenging not only because of the high required accuracy of the numerical solutions but also because of the need for reliable tables of equations of state of different materials and absorption coefficients of radiation. Hydrodynamic simulations of entry require significant computer time, but make it possible to reproduce the flight pattern in detail and calculate propagation of shock waves and radiation fluxes emitted by the heated vapor and air. However, in these hydrodynamic models, the possibility of fragmentation into large, strong fragments is not taken into account. Existing semi-analytical models are much simpler but allow energy deposition during flight and the height of maximum energy release to be estimated at low cost. They are advantageous for explaining specific well-recorded falls by tuning the model parameters. With a known energy deposition along a trajectory, it is possible to calculate overpressure on the ground produced by the shock wave propagating in the atmosphere either from scaling relations for airbursts or, more accurately, using three-dimensional gasdynamic calculations in which the deposited energy is prescribed as an initial condition. The overpressure represents the most hazardous factor. In addition, the influence of pressure on the Earth creates earthquakes, the magnitudes of which were calculated. Using these semi-analytical approaches and hydrodynamic modeling, adequate models were constructed and the details of specific events (Tunguska, Chelyabinsk) were explained.

The unpredictability of impacts requires timely assessment of the consequences of such events. Existing fast calculators allow us to predict the effects of asteroid impacts associated with the shock wave in the air and formation of a crater. However, these predictions have large uncertainties and may be unreliable in some special cases. They should be improved in the future. Recently-calculated estimates of seismic efficiency and luminous efficiency of airbursts produced by the falls of small asteroids may improve the estimates of seismic and thermal radiation damage obtained from these calculators.

The physical and mathematical models used currently for numerical modeling should be improved in the future. Higher accuracy may be required if a dangerous object approaching the Earth is detected. In specific cases, it may be necessary to determine more precisely the pressure on the surface by taking into account terrain, water basins, location of industrial objects, buildings, and population. The model of a quasi-liquid object that does not have strength, although it showed fairly good results in application to the Chelyabinsk event, does not take into account the possibility of the fall of strong (for example, iron) meteoroids, when the body is broken up into separate large fragments and a mass of small debris. The further development of the hydrodynamic model can include taking into account the properties of real-body vapors, mechanical ablation, turbulence, mixing of vapors with air, and replacing the liquid model with a better one (for example, rubble pile, accounting for cracks and vapor seepage into them). Radiation transfer has been taken into account using certain approximations. It is desirable to perform hydrodynamic simulations with radiation transfer, solving the full equation for the radiation intensity; however, this requires considerable computer resources. Least investigated is the danger associated with perturbations of the ionosphere. Until now, only density perturbations have been determined, and plasma dynamics, calculation of currents and magnetic disturbances and their effects on various devices remain open problems.

To solve the problem of asteroid danger, in addition to searching for potentially dangerous bodies with available means, development of new systems for their detection is needed, which should include detection in those cases when the asteroid approaches the Earth from the side of the Sun, as in the case of the Chelyabinsk event. It is also necessary to continue studying the physical properties of small asteroids, of which we know very little. Despite the progress in understanding impact consequences and in the search for potentially dangerous objects, humankind remains unprepared for the impact events. In theory, there are a number of methods by which near-Earth objects can be diverted or destroyed in order to prevent the devastating effects of an impact. However, practical demonstrations of such technologies have yet to be performed and are likely to be sensitive to the internal structure and physical properties of the asteroid. NASA's Double-Asteroid Redirect Test (DART) mission, which is planned to intercept the secondary of the Didymos binary asteroid system in near-Earth orbit in October 2022, would be the first artificial deflection of an asteroid in which the magnitude of the resulting change in velocity is measured (Cheng et al., 2016). Successful demonstration of asteroid deflection by a kinetic impactor would be an important milestone for planetary defense.

Acknowledgements

The study was partially supported by the program of the Russian Academy of Sciences No. 0146-2017-0003.

The authors are grateful to Peter Brown and Donovan Mathias for thoughtful and constructive comments and suggestions.

References

Academia Sinica, Joint investigation group of the Kirin Meteorite Shower. 1977. A preliminary survey on the Kirin Meteorite Shower. *Scientia Sinica*, **4**, 502–512.

Afraimovich, E. L., Astafyeva, E. I., Demyanov, V. V. et al. 2013. A review of GPS/GLONASS studies of the ionospheric response to natural and anthropogenic processes and phenomena. *Journal of Space Weather and Space Climate*, **3**, A27.

Aftosmis, M. J., Nemec, M., Mathias, D., and Berger, M. 2016. Numerical simulation of bolide entry with ground footprint prediction. In: *54th AIAA Aerospace Sciences Meeting*, pp. 1–18. AIAA SciTech Forum. American Institute of Aeronautics and Astronautics. Available at http://arc.aiaa.org/doi/10.2514/6.2016-0998

Aftosmis, M. J., Mathias, D. L., and Tarano, A. M. 2018. Simulation-based height of burst map for asteroid airburst damage prediction. *Acta Astronautica*. doi: 10.1016/j.actaastro.2017.12.021.

Aki, K., and Richards, P. G. 1980. *Quantative Seismology: Theory and Methods*. Freeman, San Francisco.

Akimov, V. A., Glazachev, D. O., Emelyanenko, V. V. et al. 2015. *Asteroids and Comets Hazard. Strategy of Counteraction*. Moscow, FGBU VNII GOChS. In Russian.

Artemieva, N., and Pierazzo, E. 2009. The Canyon Diablo impact event: Projectile motion through the atmosphere. *Meteoritics & Planetary Science*, **44**(1), 25–42.

Artemieva, N., and Pierazzo, E. 2011. The Canyon Diablo impact event: 2. Projectile fate and target melting upon impact. *Meteoritics & Planetary Science*, **46**(6), 805–829.

Artemieva, N. A., and Shuvalov, V. V. 2001. Motion of a fragmented meteoroid through the planetary atmosphere. *Journal of Geophysical Research: Planets*, **106**(E2), 3297–3309.

Avilova, I. V., Biberman, L. M., Vorob'ev, V. S. et al. 1970. *Properties of Heated Air*. Moscow: Nauka. In Russian.

Avramenko, M. I., Glazyrin, I. V., Ionov, G. V., and Karpeev, A. V. 2014. Simulation of the airwave caused by the Chelyabinsk superbolide. *Journal of Geophysical Research: Atmospheres*, **119**(12), 2013JD021028.

Ben-Menahem, A. 1975. Source parameters of the Siberian explosion of June 30, 1908, from analysis and synthesis of seismic signals at four stations. *Physics of the Earth and Planetary Interiors*, **11**(1), 1–35.

Ben-Menahem, A., and Singh, S. J. 1981. *Seismic Waves and Sources*. New York: Springer.

Ben-Menahem, A., and Toksöz, M. N. 1964. Excitation of seismic surface waves by atmospheric nuclear explosions. *Journal of Geophysical Research*, **69**(8), 1639–1648.

Bergeot, N., Bruyninx, C., Defraigne, P. et al. 2011. Impact of the Halloween 2003 ionospheric storm on kinematic GPS positioning in Europe. *GPS Solutions*, **15**(2), 171–180.

Borovička, J., and Spurnỳ, P. 2008. The Carancas meteorite impact – Encounter with a monolithic meteoroid. *Astronomy & Astrophysics*, **485**(2), L1–L4.

Boslough, M. B. E., and Crawford, D. A. 2008. Low-altitude airbursts and the impact threat. *International Journal of Impact Engineering*, **35**(12), 1441–1448.

Boslough, M. B., Crawford, D. A., Trucano, T. G., and Robinson, A. C. 1995. Numerical modeling of Shoemaker-Levy 9 impacts as a framework for interpreting observations. *Geophysical Research Letters*, **22**(13), 1821–1824.

Bronikowska, M., Artemieva, N. A., and Wünnemann, K. 2017. Reconstruction of the Morasko meteoroid impact – Insight from numerical modeling. *Meteoritics & Planetary Science*, **52**(8), 1704–1721.

Brown, P., Spalding, R. E., ReVelle, D. O., Tagliaferri, E., and Worden, S. P. 2002. The flux of small near-Earth objects colliding with the Earth. *Nature*, **420**(6913), 294–296.

Brown, P., Wiegert, P., Clark, D., and Tagliaferri, E. 2016. Orbital and physical characteristics of meter-scale impactors from airburst observations. *Icarus*, **266**, 96–111.

Brown, P. G., Assink, J. D., Astiz, L. et al. 2013. A 500-kiloton airburst over Chelyabinsk and an enhanced hazard from small impactors. *Nature*, **503**(7475), 238–241.

Cheng, A. F., Michel, P., Jutzi, M. et al. 2016. Asteroid Impact and Deflection Assessment mission: Kinetic impactor. *Planetary and Space Science*, **121**, 27–35.

Chyba, C. F., Thomas, P. J., and Zahnle, K. J. 1993. The 1908 Tunguska explosion: Atmospheric disruption of a stony asteroid. *Nature*, **361**(6407), 40–44.

Collins, G. S., Melosh, H. J., and Marcus, R. A. 2005. Earth impact effects program: A web-based computer program for calculating the regional environmental consequences of a meteoroid impact on Earth. *Meteoritics & Planetary Science*, **40**(6), 817–840.

Collins, G. S., Lynch, E., McAdam, R., and Davison, T. M. 2017. A numerical assessment of simple airblast models of impact airbursts. *Meteoritics & Planetary Science*, **52**, 1542–1560.

Connolly, H. C., Smith, C., Benedix, G. et al. 2008. The Meteoritical Bulletin, No. 93, 2008 March. *Meteoritics & Planetary Science*, **43**(3), 571–632.

Crampin, S. 1966. Higher-mode seismic surface waves from atmospheric nuclear explosions over Novaya Zemlya. *Journal of Geophysical Research*, **71**(12), 2951–2958.

Divari, N. B. 1959. Phenomena that accompany meteorite shower and its atmospheric trajectory. In: V. G. Fesenkov (ed), *Sikhote-Alin Iron Meteorite Shower*, vol. 1. Moscow: USSR Acad. Sci., pp. 26–48. in Russian.

Dunn, T. L., Burbine, T. H., Bottke, W. F., and Clark, J. P. 2013. Mineralogies and source regions of near-Earth asteroids. *Icarus*, **222**(1), 273–282.

Edwards, W. N., Eaton, D. W., and Brown, P. G. 2008. Seismic observations of meteors: Coupling theory and observations. *Reviews of Geophysics*, **46**(4), 1–21. RG4007.

Emelyanenko, V. V., and Naroenkov, S. A. 2015. Dynamical features of hazardous near-Earth objects. *Astrophysical Bulletin*, **70**(3), 342–348.

Gi, N., Brown, P., and Aftosmis, M. 2018. The frequency of window damage caused by bolide airbursts: A quarter century case study. *Meteoritics & Planetary Science*, 53(7), 1413–1431.

Glasstone, S., and Dolan, P. J. 1977. *The Effects of Nuclear Weapons*. Washington, GPO.

Glazachev, D., Podobnaya, E., Popova, O. et al. 2019. Impact effects calculator. Scaling relations for shock wave and radiation effects applied to Chelyabinsk and Tunguska events. *Proceedings of the IMC 2018 Conference, 30 August–2 September, Pezinok-Modra, Slovakia*. In press.

Griggs, D. T., and Press, F. 1961. Probing the Earth with nuclear explosions. *Journal of Geophysical Research*, **66**(1), 237–258.

Grigorian, S. S. 1979. Motion and destruction of meteors in planetary atmospheres. *Cosmic Research*, **17**(6), 724–740.

Grigoryan, S. S., Ibodov, F. S., and Ibadov, S. I. 2013. Physical mechanism of Chelyabinsk superbolide explosion. *Solar System Research*, **47**(4), 268–274.

Halliday, I., Griffin, A. A., and Blackwell, A. T. 1996. Detailed data for 259 fireballs from the Canadian camera network and inferences concerning the influx of large meteoroids. *Meteoritics & Planetary Science*, **31**(2), 185–217.

Harkrider, D. G., Newton, C. A., and Flinn, E. A. 1974. Theoretical effect of yield and burst height of atmospheric explosions on Rayleigh wave amplitudes. *Geophysical Journal International*, **36**(1), 191–225.

Harris, A. W., and D'Abramo, G. 2015. The population of near-Earth asteroids. *Icarus*, **257**, 302–312.

Hills, J. G., and Goda, M. P. 1993. The fragmentation of small asteroids in the atmosphere. *The Astronomical Journal*, **105**, 1114–1144.

Hodge, P. W., and Wright, F. W. 1971. Meteoritic particles in the soil surrounding the Henbury Meteorite Craters. *Journal of Geophysical Research*, **76**(17), 3880–3895.

Ivanov, B. 2008. Size-frequency distribution of asteroids and impact craters: Estimates of impact rate. In: V. V. Adushkin and I. V. Nemchinov (eds), *Catastrophic Events Caused by Cosmic Objects*. Berlin: Springer, pp. 91–116.

Ivanov, B. A., and Hartmann, W. K. 2007. Exogenic dynamics, cratering and surface ages. *Treatise on Geophysics*, **10**, 207–242.

Jacobsen, K. S., and Andalsvik, Y. L. 2016. Overview of the 2015 St. Patrick's day storm and its consequences for RTK and PPP positioning in Norway. *Journal of Space Weather and Space Climate*, **6**, A9.

Kartashova, A. P., Popova, O. P., Emel'yanenko, V. V. et al. 2018. Study of injuries from the Chelyabinsk airburst event. *Planetary and Space Science*, **160**, 107–114.

Kenkmann, T., Artemieva, N. A., Wünnemann, K. et al. 2009. The Carancas meteorite impact crater, Peru: Geologic surveying and modeling of crater formation and atmospheric passage. *Meteoritics & Planetary Science*, **44**(7), 985–1000.

Korea Dynamite Explosion November 11, 1977. 1979. Summary Report. Available at: https://core.ac.uk/download/pdf/51172936.pdf.

Kosarev, I. B. 1999. Calculation of thermodynamic and optical properties of the vapors of cosmic bodies entering the earth's atmosphere. *Journal of Engineering Physics and Thermophysics*, **72**(6), 1030–1038.

Kosarev, I. B., Loseva, T. V., and Nemchinov, I. V. 1996. Vapor optical properties and ablation of large chondrite and ice bodies in the Earth's atmosphere. *Solar System Research*, **30**, 265–278.

Kring, D. A. 2017. *Guidebook to the Geology of Barringer Meteorite Crater, Arizona (aka Meteor Crater)*. Houston: Lunar and Planetary Institute.

Kring, D. A., and Boslough, M. 2014. Chelyabinsk: Portrait of an asteroid airburst. *Physics Today*, **67**(9), 32–37.

Krinov, E. L. 1971. New studies of the Sikhote-Alin iron meteorite shower. *Meteoritics*, **6**, 127–138.

Kuznetsov, N. M. 1965. *Functions and Impact Adiabates for Air under High Temperature*. Moscow: Mashinostroenie. In Russian.

Longo, G., Di Martino, M., Andreev, G. et al. 2005. A new unified catalogue and a new map of the 1908 tree fall in the site of the Tunguska Cosmic Body explosion. In: *Asteroid-Comet Hazard-2005*, pp. 222–225. Institute of Applied Astronomy of the Russian Academy of Sciences, St. Petersburg, Russia.

Luo, X., Gu, S., Lou, Y. et al. 2018. Assessing the performance of GPS precise point positioning under different geomagnetic storm conditions during solar cycle 24. *Sensors*, **18**(6), 1784.

Mannan, S. 2012. *Lees' Loss Prevention in the Process Industries*. 4th edn. Elseiver Butterworth-Heinemann.

Mathias, D. L., Wheeler, L. F., and Dotson, J. L. 2017. A probabilistic asteroid impact risk model: Assessment of sub-300 m impacts. *Icarus*, **289**, 106–119.

Melosh, H. J. 1989. *Impact Cratering: A Geologic Process*. New York: Oxford University Press.

Melosh, H. J., and Collins, G. S. 2005. Planetary science: Meteor Crater formed by low-velocity impact. *Nature*, **434**(7030), 157.

Mukhamednazarov, S. 1999. Observation of a bolide and the fall of the first large meteorite in Turkmenistan. *Pisma v Astronomicheskii Zhurnal (Astronomy Letters)*, **25**(2), 150–152.

National Academy of Sciences. 2010. *Defending Planet Earth: Near-Earth-Object Surveys and Hazard Mitigation Strategies: Final Report*. Washington, DC: The National Academies Press. Available at: http://books.nap.edu/catalog.php?record_id=12842.

Nemtchinov, I., Shuvalov, V., Kovalev, A., Kosarev, I. B., and Zetzer, Yu. 2008. Ionospheric and magnitospheric effect. In: Adushkin, V. V., and Nemchinov, I. V. (eds), *Catastrophic Events Caused by Cosmic Objects*. Berlin: Springer, pp. 313–332.

Nemtchinov, I. V., and Popova, O. P. 1997. An analysis of the 1947 Sikhote-Alin event and a comparison with the phenomenon of February 1, 1994. *Solar System Research*, **31**, 408–420.

Nemtchinov, I. V., Popova, O. P., Shuvalov, V. V., and Svetsov, V. V. 1994. Radiation emitted during the flight of asteroids and comets through the atmosphere. *Planetary and Space Science*, **42**(6), 491–506.

Ott, U., Merchel, S., Herrmann, S. et al. 2014. Cosmic ray exposure and pre-atmospheric size of the Gebel Kamil iron meteorite. *Meteoritics & Planetary Science*, **49**(8), 1365–1374.

Ovchinnikov, V. M., and Pasechnik, I. P. 1988. Earthquake caused by an explosion of Chulym bolide. *Meteoritika*, **47**, 10–20. in Russian.

Pasechnik, I. P. 1970. *Characteristics of Seismic Waves in Nuclear Explosions and Earthquakes*. Moscow: Nauka. In Russian.

Pasechnik, I. P. 1976. Evaluation of parameters of the Tunguska meteorite explosion from seismic and microbarographic data. In: Sobolev, V.S. (ed), *Kosmicheskoe Veshchestvo na Zemle*, pp. 24–54. Novosibirsk, USSR: Nauka. In Russian. Available at: www.geokniga.org/books/13015.

Passey, Q. R., and Melosh, H. J. 1980. Effects of atmospheric breakup on crater field formation. *Icarus*, **42**(2), 211–233.

Petaev, M. I. 1992. The Sterlitamak meteorite – A new crater-forming fall. *Solar System Research*, **26**, 384–398. In Russian.

Pierazzo, E., Vickery, A. M., and Melosh, H. J. 1997. A reevaluation of impact melt production. *Icarus*, **127**, 408–423.

Popova, O., Borovička, J., Hartmann, W. K. et al. 2011. Very low strengths of interplanetary meteoroids and small asteroids. *Meteoritics & Planetary Science*, **46**(10), 1525–1550.

Popova, O. P., Jenniskens, P., Emel'yanenko, V. et al. 2013. Chelyabinsk airburst, damage assessment, meteorite recovery, and characterization. *Science*, **342**(6162), 1069–1073.

Rabinowitz, D., Helin, E., Lawrence, K., and Pravdo, S. 2000. A reduced estimate of the number of kilometre-sized near-Earth asteroids. *Nature*, **403**(6766), 165–166.

Register, P. J., Mathias, D. L., and Wheeler, L. F. 2017. Asteroid fragmentation approaches for modeling atmospheric energy deposition. *Icarus*, **284**, 157–166.

Reinhardt, J. C., Xi, C., Wenhao, L., Manchev, P., and Pate-Cornell, M. E. 2016. Asteroid risk assessment: A probabilistic approach. *Risk Analysis*, **36**(2), 244–261.

Rumpf, C., Lewis, H. G., and Atkinson, P. M. 2016. On the influence of impact effect modelling for global asteroid impact risk distribution. *Acta Astronautica*, **123**, 165–170.

Rumpf, C. M., Lewis, H. G., and Atkinson, P. M. 2017a. Asteroid impact effects and their immediate hazards for human populations. *Geophysical Research Letters*, **44**(8), 3433–3440.

Rumpf, C. M., Lewis, H. G., and Atkinson, P. M. 2017b. Population vulnerability models for asteroid impact risk assessment. *Meteoritics & Planetary Science*, **52**(6), 1082–1102.

Schunová-Lilly, E., Jedicke, R., Vereš, P., Denneau, L., and Wainscoat, R. J. 2017. The size-frequency distribution of $H > 13$ NEOs and ARM target candidates detected by Pan-STARRS1. *Icarus*, **284**, 114–125.

Shorten, J. R. 1970. *The Johannesburg Saga*. Cape Town and Johannesburg: John R. Shorten (Proprietary) Ltd.

Shustov, B. M., Naroenkov, S. A., and Efremova, E. V. 2017. On population of hazardous celestial bodies in the near-Earth space. *Solar System Research*, **51**(1), 38–43.

Shuvalov, V., Svetsov, V., Popova, O., and Glazachev, D. 2017. Numerical model of the Chelyabinsk meteoroid as a strengthless object. *Planetary and Space Science*, **147**, 38–47.

Shuvalov, V. V. 1999. Multi-dimensional hydrodynamic code SOVA for interfacial flows: Application to the thermal layer effect. *Shock Waves*, **9**(6), 381–390.

Shuvalov, V. V., and Artemieva, N. A. 2002. Numerical modeling of Tunguska-like impacts. *Planetary and Space Science*, **50**(2), 181–192.

Shuvalov, V. V., and Khazins, V. A. 2018. Numerical simulation of ionospheric disturbances generated by the Chelyabinsk and Tunguska space body impacts. *Solar System Research*, **52**, 129–138.

Shuvalov, V. V., and Trubetskaya, I. A. 2007. Aerial bursts in the terrestrial atmosphere. *Solar System Research*, **41**(3), 220–230.

Shuvalov, V. V., and Trubetskaya, I. A. 2010. The influence of internal friction on the deformation of a damaged meteoroid. *Solar System Research*, **44**(2), 104–109.

Shuvalov, V. V., Svetsov, V. V., and Trubetskaya, I. A. 2013. An estimate for the size of the area of damage on the Earth's surface after impacts of 10–300-m asteroids. *Solar System Research*, **47**(4), 260–267.

Shuvalov, V. V., Popova, O. P., Svetsov, V. V., Trubetskaya, I. A., and Glazachev, D. O. 2016. Determination of the height of the "meteoric explosion". *Solar System Research*, **50**, 1–12.

Stokes, G. H., Barbee, B. W., Bottke, W. F. et al. 2017. *Update to determine the feasibility of enhancing the search and characterization of NEOs*. Report of the Near-Earth Object Science Definition Team, NASA, September, 2017. Available at: https://cneos.jpl.nasa.gov/doc/2017_neo_sdt_final_e-version.pdf.

Svetsov, V. V. 1994. Explosions in the lower and middle atmosphere: The spherically symmetrical stage. *Combustion, Explosion and Shock Waves*, **30**(5), 696–707.

Svetsov, V. V. 1998. Enigmas of the Sikhote-Alin crater field. *Solar System Research*, **32**, 67–79.

Svetsov, V. V. 2002. Comment on "Extraterrestrial impacts and wildfires". *Palaeogeography, Palaeoclimatology, Palaeoecology*, **185**(3-4), 403–405.

Svetsov, V. V. 2007. Estimates of the energy of surface waves from atmospheric explosions and the source parameters of the Tunguska event. *Izvestiya, Physics of the Solid Earth*, **43**(7), 583–591.

Svetsov, V. V., and Shuvalov, V. V. 2018. Thermal radiation and luminous efficiency of superbolides. *Earth and Planetary Science Letters*, **503**, 10–16.

Svetsov, V. V., Nemtchinov, I. V., and Teterev, A. V. 1995. Disintegration of large meteoroids in Earth's atmosphere: Theoretical models. *Icarus*, **116**(1), 131–153.

Svetsov, V. V., Artemieva, N. A., and Shuvalov, V. V. 2017. Seismic efficiency of meteor airbursts. *Doklady Earth Sciences*, **475**(2), 935–938.

Svetsov, V. V., Shuvalov, V. V., and Popova, O. P. 2018. Radiation from a Superbolide. *Solar System Research*, **52**(3), 195–205.

Tancredi, G., Ishitsuka, J., Schultz, P. H. et al. 2009. A meteorite crater on Earth formed on September 15, 2007: The Carancas hypervelocity impact. *Meteoritics & Planetary Science*, **44**(12), 1967–1984.

Tauzin, B., Debayle, E., Quantin, C., and Coltice, N. 2013. Seismoacoustic coupling induced by the breakup of the 15 February 2013 Chelyabinsk meteor. *Geophysical Research Letters*, **40**(14), 3522–3526.

Thompson, S. L., and Lauson, H. S. 1972. *Improvements in the Chart D radiation-hydrodynamic CODE III: Revised analytic equations of state*. Albuquerque, New Mexico: Sandia National Laboratory. Report SC-RR-71 0714.

Tillotson, J. H. 1962. *Metallic equations of state for hypervelocity impact*. General Atomic Report GA 3216.

Tricarico, P. 2017. The near-Earth asteroid population from two decades of observations. *Icarus*, **284**, 416–423.

Trilling, D. E., Valdes, F., Allen, L. et al. 2017. The size distribution of Near-Earth Objects larger than 10 meters. *Astronomical Journal*, **154**, 1–10.

Vasilyev, N. V. 1998. The Tunguska meteorite problem today. *Planetary and Space Science*, **46**(2-3), 129–150.

Vasilyev, N. V., Kovalevsky, A. F., Razin, S. A., and Epiktetova, L. E. 1981. *Testimony of witnesses of Tunguska fall (catalogue)*. Tomsk. in Russian. Available at: http://tunguska.tsc.ru/ru/science/1/0.

Wheeler, L. F., Register, P. J., and Mathias, D. L. 2017. A fragment-cloud model for asteroid breakup and atmospheric energy deposition. *Icarus*, **295**, 149–169.

White, J. 1994. Exploding myths: The Halifax harbor explosion in historical context. *Ground zero*, 251–274. Available at http://jaywhite.ca/wp-content/uploads/2013/03/Exploding_Myths1994.pdf.

Xiong, C., Stolle, C., and Lühr, H. 2016. The Swarm satellite loss of GPS signal and its relation to ionospheric plasma irregularities. *Space Weather*, **14**(8), 563–577.

Index

1I/'Oumuamua, 244, 248
3-body problem, 189

ablation, 9, 10, 22, 24, 30, 120, 124–126, 136, 143, 187, 201, 278, 289, 292, 293, 295
    coefficient, 13, 22–24, 30, 52, 102
    differential, 12, 68, 69, 103
    equations, 12, 13
    mass loss, 12, 14, 15
    model, 102
    model mass determination, 68
    non-thermal, 14
    spraying, 12
    sputtering, 14, 15, 70, 103, 200
abundance, 39–41, 45, 46, 53–55, 103–104, 200
    elemental, 188, 200
    solar, 188, 191, 200
achondrite, *see* meteorite
Advanced Meteor Orbit Radar, 74–77, 218, 236, 241–243, 247
afterglow, 54
airburst, 28, 31, 128, 276, 280, 281, 286, 287, 291–294
ALTAIR, *see* ARPA Long-Range Tracking and Instrumentation Radar
AMOR, *see* Advanced Meteor Orbit Radar
AMOS Video Meteor Network, 92, 218
apex of Earth, 173, 177
Arecibo radar, 69–71, 241, 243
ARPA Long-Range Tracking and Instrumentation Radar, 69, 82
asteroid, 46, 55
    density, 37
    individual
        (1) Ceres, 37, 51
        (2) Pallas, 178, 191
        (4) Vesta, 37, 39
        (21) Lutetia, 37
        (253) Mathilde, 37
        (433) Eros, 37
        (1566) Icarus, 195, 197, 198, 202
        (3200) Phaethon, 55, 121, 163, 169, 178, 191–193, 198, 202
        (5496) 1973 NA, 177
        (25143) Itokawa, 37, 46
        (101955) Bennu, 46
        (155140) 2005 UD, 178, 191, 192, 197, 202
        (162173) Ryugu, 46
        (196256) 2003 EH_1, 51, 168, 172–176, 193, 197, 202
        1983 TB, *see* (3200) Phaethon
        1994 JX, 173
        1999 LT_1, 173
        1999 YC, 178, 191, 192, 197, 202
        2000 PG_3, 173
        2002 AR_{129}, 173
        2002 EX_{12}, 194
        2002 KF_4, 173
        2002 UO_3, 173
        2002 XM_{35}, 178
        2003 QC_{10}, 178
        2003 UL_3, 178
        2003 WY_{25}, 196, 199
        2003 YS_1, 173
        2004 BZ_{74}, 173
        2004 TG_{10}, 178
        2005 UD, *see* (155140)
        2005 UR, 102
        2006 KT_{67}, 195
        2007 MK_6, 195, 197, 198, 202
        2008 TC_3, 41, 106
        2014 AA, 106
        2015 TX_{24}, 102
        2017 MB_1, 194, 202
        2018 LA, 106
    rubble-pile, 198
    types (*see also* taxonomy), 37
atmosphere
    air density, 81
    ambipolar diffusion, 73, 80, 82
    gravity wave momentum flux, 80
    gravity waves, 77, 78, 80, 83
    ionosphere, 124, 126–128
    Jupiter, 128, 129
    Mars, 119, 120, 123, 127
    metal layers, 120, 124–126, 128
    temperature, 77, 80, 83
    Venus, 119–121
    wind field, 77–79, 82, 83
atmospheric dynamics in the meteor zone
    ionospheric perturbations, 294, 295
atmospheric entry, 275–279, 281, 286–290, 292–294

β Pictoris, 242
blackbody, 151
bolide, 10, 12, 23, 24, 26, 30, 31, 55
    classification, 23, 52, 55
        A_L parameter, 52
        P_E parameter, 51
    European Network, 24
    individual
        Almahata Sitta, 28
        Antarctic 2005, 28
        Benešov, 26, 27, 29, 53–55, 104, 106
        Bunburra Rockhole, 106
        Carancas, 52

bolide (cont.)
    Čechtice, 54, 55
    Chelyabinsk (*see also* Chelyabinsk), 93
    EN311015 (Taurid), 53
    Hradec Králové, 53
    Jesenice, 53
    Kácov, 54, 55
    Karlštejn, 53, 55
    Košice, 53, 93, 106
    Križevci, 53
    Maribo, 53, 106
    Marshall Island, 28, 29
    Morávka, 52
    Neuschwanstein, 106
    Příbram, 91
    Romanian (2015 January 7), 53
    Senohraby, 54, 55
    Stubenberg, 53
    Šumava, 53–55
    Tagish Lake, 53
    Žďár nad Sázavou, 52, 53
    satellite-observed, 24, 28
    spectra, 53
        components of, 54
break-up, 196
    rotational instability, 198
    thermal disintegration, 198
breccia, 41

Ca, 50, 54, 55
    ions, 127
    neutral, 124, 130
CAI, *see* calcium- and aluminum-rich inclusion
Calcium- and aluminum-rich inclusion, 39, 44
Cameras for Allsky Meteor Surveillance, 49, 92
CAMO, *see* Canadian Automated Meteor Observatory
CAMS, *see* Cameras for Allsky Meteor Surveillance
Canadian Automated Meteor Observatory, 16, 19, 47, 92, 93, 242, 247
Canadian Meteor Orbit Radar, 69, 73–77, 83, 241, 243, 244
Canary Island Long-Baseline Observatory, 47, 218
carbonaceous chondrite, *see* meteorite
charge-to-mass ratio, 237, 238, 245
Chelyabinsk, 25, 31, 52, 53, 105, 275–277, 281–284, 286, 287, 290–295
CHON material, 43
chondrite *see* meteorite
chondrule, 39, 41, 44
CI composition, 39, 40, 42, 45, 46, 50, 51, 54, 56
CILBO, *see* Canary Island Long-Baseline Observatory
circulation, orbital parameter, 165–167, 171, 173
close encounter, 127, 163, 167, 168, 188, 242, 255
cluster analysis, 169, 214–215
CMN, *see* Croatian Meteor Network
CMOR, *see* Canadian Meteor Orbit Radar
coefficient
    accommodation, 12
    drag, 13, 22
    erosion, 30
    heat transfer, 12, 13, 22–24
    ionisation, *see* ionisation, coefficient
    luminous efficiency, *see* luminous efficiency
    shape-density, 13, 22, 30
cohesive strength, 198

color (asteroid, comet), 189, 191, 195, 197, 202
comet, 55
    crust, 51, 53
    density, 37
    dormant, 55, 199, 202
    dust trail, 46, 161, 162, 169, 227
    dust / gas ratio, 195
    Encke-type, 188
    Halley-type, 164, 168, 188
    individual
        1P/Halley, 45, 54, 121, 165, 169
        2P/Encke, 53, 121, 123, 162, 168, 171, 178, 194, 202, 235
        3D/Biela, 169, 196
        4P/Faye, 162
        5D/Brorsen, 173
        17P/Holmes, 162
        21P/Giacobini-Zinner, 169
        22P/Kopff, 162
        55P/Tempel-Tuttle, 53, 164, 169–171
        67P/Churyumov-Gerasimenko, 37, 45, 46, 54, 162
        81P/Wild 2, 37, 45
        96P/Machholz 1, 167, 168, 172, 173, 193, 197, 202
        109P/Swift-Tuttle, 152, 169
        126P/IRAS, 168
        141P/Machholz 2, 173
        169P/NEAT, 194, 197
        206P/Barnard-Boattini (1892 T1), 173
        226P/Pigott-LINEAR-Kowalski (1783 W1), 173
        289P/Blanpain (D/1819 W1), 196, 199
        C/1490 Y1, 173, 176
        C/1771 A1, 177
        C/1861 G1 (Thatcher), 169
        C/1979 Y1 (Bradfield), 177
        C/1995 O1(Hale–Bopp), 239
        C/2007 H2 (Skiff), 122, 123
        C/2013 A1 (Siding Spring), 123, 124, 126, 127, 270
        D/1819 W1, *see* 289P/Blanpain
        P/2003 T12 (SOHO), 194
    Jupiter-family, 168, 188, 192, 195, 196, 201
    Kuiper belt, 189
    long-period, 168
    Oort cloud, 72, 189, 201, 224, 226
    short-period, 199
    sungrazing, 171
    sungrazing group
        Kracht, 173
        Kreutz, 173
        Marsden, 167, 173, 176
        Meyer, 173
    Whipple model, 162
complex
    Andromedids, 196
    asteroid-meteoroid complex, 187, 189
    Capricornids, 194
    PGC, *see* Phaethon-Geminid Complex
    Phaethon-Geminid Complex, 178, 191, 197, 202
    Phoenicids, 196
    Quadrantids, 168, 172, 193
    Sekanina's (1973) Taurid-Perseids, 195, 197
    Taurid Complex, 164, 178, 188, 194, 202
cosmic spherules, 44
Croatian Meteor Network, 92, 218
Curiosity rover, 130

D'Alembert rules, 165
D-criteria, *see* orbital similarity
damage, *see also* spacecraft damage and anomalies, 275, 276, 280, 281, 287–294
data catalogues
    IAU Meteor Data Center, 99
database
    EDMOND, *see* European viDeo MeteOr Network Database
    European viDeo MeteOr Network Database, 92, 97, 101, 179, 218–219, 222, 242
    IAU Meteor Data Center, 165, 168, 179, 187, 210, 215, 217, 219, 236, 240, 241, 243
        meteor shower database, 219
    SonotaCo (Japanese video network), 92, 97, 104, 179, 218, 219, 222, 242
Desert Fireball Network, 93, 95, 96
diamond, 43
dust
    asteroidal, 46
    cometary, 44–46, 56
        composition, 45, 46, 56
        organic content, 45, 46
        polarimetry, 46
    detector, 136, 236, 244, 246–248, 266, 270
    flux, 245, 248
    interplanetary (*see also* zodiacal cloud), 243, 244
    interstellar, *see* interstellar
dust trail, *see* comet
dynamic pressure, 52

EDMOND, *see* database
effect
    general relativistic, 163, 179
    Lense-Thirring, 163, 179
    Poynting–Robertson, 45, 162, 163, 168, 177, 178, 201, 225, 237
    Yarkovsky, 162, 163, 179, 217, 237
    YORP, 162, 163, 198
efficiency
    destruction, 280
    luminous, *see* luminous efficiency
EISCAT, *see* European Incoherent Scatter Scientific Association
EISCAT 3D, 248
ejection velocity, 161, 163
electron density, 124, 125, 127, 128
elongation of the radiant from the apex, 240
Enceladus, 51
escape velocity, 142
European Fireball Network, 55, 93, 94, 96
European Incoherent Scatter Scientific Association, 47, 67–71, 244

F-parameter, 48, 49
Fe, 50
    ions, 124, 126, 127
    neutral, 120, 124
fire ignition, 287, 289–291, 293
fireball, *see* bolide
Fireball Recovery and InterPlanetary Observation Network, 92, 218
flash-heating, 42, 43
flow regime, 10, 11
    continuous, 10, 12, 24
    free-molecular, 10–12
    transition, 11, 20
flux (*see also* meteoroid, flux), 241, 245, 247

force
    Coulomb drag, 237
    gravitational, 162, 165, 237, 239, 248
    Lorentz, 238, 244–246
    magnetic field, 236, 237
    non-gravitational, 162, 163, 188, 226
        absorption, 162
        scattering of light, 162
    out-gassing, 167
    Poynting–Robertson drag, 45, 130, 162, 163, 168, 172, 177, 178, 201, 217, 225, 226, 237
    radiation pressure, 45, 162, 163, 193, 198, 199, 201, 237–239, 248
    Yarkovsky, 162–163, 179, 237
fragmentation, 9, 10, 19, 27, 52, 53, 68, 69, 71, 73, 76, 82, 102, 104, 277, 279, 289, 290, 293, 294
    indirect evidence, 47
    lateral speed, 48
    observation, 47
fragmentation model, 18, 23, 27, 31
    dustball, 18
    gross fragmentation, 18
    husking, 19
    hybrid, 29
    pancake, 28, 29, 277
    progressive, 18, 29
    quasi-continuous, 18, 19, 30
    quasi-liquid, 26, 31, 295
    semi-empirical, 29, 30
Fresnel holography, 47
FRIPON, *see* Fireball Recovery and InterPlanetary Observation Network, 92, 105, 218

GEMS, *see* glass with embedded metal and sulfides
geocentric velocity, 248
glass with embedded metal and sulfides, 42, 43
goethite, 41
gravitational focusing, 141, 256
gravitational scattering, 248

Harvard Radio Meteor Project, 74, 77, 241, 243
heliocentric velocity, 248
heliosphere, 236–238, 245, 248
HRMP, *see* Harvard Radio Meteor Project
hydrodynamic model, 10, 20, 27, 31, 277, 281, 283, 287, 293–295
hyperbolic
    excess, 235, 240–242, 244
    meteoroid, 235, 236, 239–242
    speed, 235, 239, 243
hypervelocity impacts, 257
    craters, 258
    damage equations, 259, 267, 271
    plasma, 263
    residue, 259
    severing of thin structures, 262
    spall, 258

IDP, *see* Interplanetary Dust Particle
impact
    crater, 52, 148, 153–155, 276, 277, 280, 283, 287–290, 293
        apparent diameter, 148
        rim-to-rim diameter, 148
    frequency, 275
    melt, 51, 55

impact (cont.)
    vaporisation, 129, 130
    velocity, 255
impact flashes
    cooling, 140
    energy, 139
    figure of merit, 142
    impact angle, 139
    Jupiter, 128, 129, 131
    kinetic energy threshold, 144
    lightcurve, 137
    lunar, 136–140, 145, 193
    peak magnitude, 140
    probability parameter, 142–145, 147
    spectrum, 138, 141, 155
    subradiant, 139, 143
    temperature, 138, 151
    velocity, 142
in-situ experiments
    Cosmic Dust Analyzer (Cassini), 270
    DEBIE, 270
    EURECA, EUropean REtrievable CArrier, 261
    GIADA (Rosetta), 271
    GORID, 270
    impact detectors, 270
    LDEF, 261
    NGIMS, 126
    Pioneer, 271
    PVDF detectors, 270
    PZT detectors, 270
    Stardust, 270
    Ulysses, 270
incomplete evaporation, 27, 54, 55
injuries, 291, 292
InSight lander, 130
interplanetary dust cloud, see zodiacal cloud
Interplanetary Dust Particle, 42–43, 47, 55
    chondritic porous, 42, 43, 47, 55
    cluster, 42, 43, 55
    hydrated chondritic, 42, 55
    non-chondritic, 43
    sulfide, 42–46
interstellar
    asteroid, 248
    cloud, local, 236–238, 246, 248
    dust, 236–238, 244–248
        flux, 238, 246, 247
        population, 235
    fireballs, 244
    gas-to-dust mass ratio, 236
    medium, 245, 246, 248
    meteor, 239, 241–244, 247
    meteoroid, 235, 236, 238, 239, 241–248
        flux, 242, 246, 247
    origin, 235, 236, 239
    source, 235, 242
Io, 51
ionisation/ionization, 9, 10, 15, 17, 18, 27
    coefficient, 17, 18
    probability, 17

Jicamarca radar, 66, 67, 70, 72, 81

Knudsen number, 10
    modified, 10
Kozai mechanism, see Lidov-Kozai mechanism

Laser Induced Breakdown Spectroscopy, 26, 53
Late Heavy Bombardment, 56
Lense-Thirring effect, 163, 179
libration, orbital parameter, 165, 166, 171, 176
LIBS, see Laser Induced Breakdown Spectroscopy
Lidov-Kozai mechanism, 167, 171, 179, 187, 188, 199, 202
lifetime, 168
    collisional, 167, 198, 201, 226
    dynamical, 199
LINEAR survey, 142
luminous efficiency, 15–17, 24, 25, 52, 137, 287–289, 294
    impact, 139–141, 145, 152, 155
    integral, 25, 26
lunar
    regolith bulk density, 151
    terminator, 137

MAARSY, see Middle Atmosphere Alomar Radar System
magnetic fields, 237–239
magnetite rim, 43, 44
MARS, see Meteor Automatic Radar System
Mars Exploration Rovers, 120, 130
mass determination
    ablation model, 68
    dynamical, 68
    scattering model, 68
mass influx, 56
mass loss, see ablation
measurement
    accuracy, 2, 178, 236, 240, 246, 247
    error, 235, 236, 239, 241, 242, 244
    speed, 235, 244, 247
        geocentric, 240, 242, 243, 248
        heliocentric, 235, 237, 239–242, 244
        pre-atmosphere, 240
Mercury, 129, 163, 200, 239
meteor
    anomalous, 104
    catalogues, 241
    classification, 48, 52, 55
        k_B parameter, 48
        k_C parameter, 49
    cluster, 105
    high-altitude, 69, 104
    high-velocity, 242
    light curve, 48, 104
    spectra, 27, 50, 53, 93, 94, 103, 104
        hot component, 27, 200
        intensity ratio, 50, 188, 191, 200
        main component, 27, 200
    sporadic, 66, 70, 72–77, 83, 201
    trail, 18
    train, 103
Meteor Automatic Radar System, 243
meteor observations, see observations
meteor shower, see also meteoroid stream, 14, 167, 169, 170, 172, 174, 269

ecliptic, 173
IAU definition, 210, 211
individual
    α-Cetids, 173, 176
    α-Draconids, 173
    α-Microscopiids, 177
    α-Piscids, 173
    β-Arietids, 173
    β-Taurids, 178
    δ-Aquariids, 173
    η-Aquariids, 72, 241
    η-Piscids, 173
    γ-Bootids, 177
    κ Cygnids, 54, 98
    λ-Taurids, 176
    o-Orionids, 178
    θ-Carinids, 176
    ζ-Perseids, 178
    April Lyrids, 99, 149, 168, 169, 210, 217, 220, 241
    Arietids, 76, 97, 173
    Aurigids, 101
    Camelopardalids, 76, 103
    Carinids, 173, 176
    Comae Berenicids, 102
    Daytime Sextantids, 97
    December α-Draconids, 176
    Draconids, *see* October Draconids
    Eta Aquariids, 241
    Geminids, 19, 48, 49, 51, 53, 54, 72, 75, 76, 98
    July γ Draconids, 91
    July Pegasids, 177
    Leonids, 11, 18–22, 48, 49, 51, 53, 54, 66, 82, 83, 99, 105, 121, 122, 143, 241, 269
    Lyrids, *see* April Lyrids
    Northern δ-Aquariids, 173
    Northern Taurids, 178
    November ι-Draconids, 176
    October Camelopardalids, 102
    October Draconids, 3, 48, 49, 51, 72, 76, 99–102
    Orionids, 48, 49, 72, 73, 98, 99, 241
    Perseids, 48, 49, 51, 54, 69, 72, 101, 143, 241
    Phoenicids, 101
    Puppid-Velids, 173
    Quadrantids, 51, 75, 76, 102, 173
    September ϵ-Perseids, 101, 105
    Southern δ-Aquariids, 49, 51, 173
    Southern Piscids, 178
    Southern Taurids, 102
    Taurids, 48, 49, 51, 53–55, 75
    Ursids, 102, 176
    Virginids, 153
list of established meteor showers, 222
mass distribution index, 91
number density, 91
outburst, 3, 77, 91, 98–104, 121, 146–148, 164, 170–172, 269
parent body, 217
population index, 91, 98, 141, 145
toroidal, 171, 173, 176, 224
transitory, 177
Working Group on Meteor Shower Nomenclature, 215, 219
zenithal hourly rate, 91, 143, 145, 146, 191
meteor storm, 164–166, 179
    Leonid, 66, 83, 99, 104, 121, 152, 164, 171, 269
meteorite, 9, 11, 25, 29, 38, 105

acapulcoite, 39
achondrite, 39
carbonaceous chondrite, 39, 41, 42, 53, 55
chondrite, 39
classification, 38
composition, 25–27, 40
cosmic ray exposure age, 40, 41
CRE, *see* cosmic ray exposure age
density, 40, 41
diogenite, 39, 55
enstatite chondrite, 39, 41
eucrite, 39
fall statistics, 39, 40
formation, 39
HED, 39
howardite, 39
individual
    Almahata Sitta, 41, 55, 106
    Benešov, 106
    Chelyabinsk, 28, 29, 31
    Galim, 55
    Gao-Guenie, 55
    Innisfree, 24
    Košice, 29, 42
    La Ciénega, 55
    Markovka, 55
    Tagish Lake, 39, 42
iron, 38, 41, 55
lodranite, 39
matrix, 39, 41
mesosiderite, 39
orbits, 38
ordinary chondrite, 39, 41, 42, 44, 46, 53
pallasite, 39
petrologic type, 41
physical properties, 40, 41
pore space, 41
porosity, 40, 41
shock stage, 41
stony, 39, 55
stony-iron, 39, 55
strength, 27, 40, 41, 53
strewn field, 9, 13, 29, 55, 90, 290
structure, 39
ureilite, 39
meteorite fall, *see* bolide
Meteoritical Bulletin Database, 55
meteoroid, 161
    β-meteoroid, 201, 239
    abundances, 53–55, 103–104, 126, 188, 191, 200
    bulk density, 256
    composition, 9–11, 15, 16, 19, 20, 24, 26, 27, 50, 51, 53, 54, 56, 102, 103
    Fe poor, 51
    iron, 49, 50, 55
    Na enhanced, 51
    Na free, 51, 53, 55
    Na poor, 51, 55
    Na rich, 51
    Normal, 51
cracks, 52, 55
definition, 1, 256
density, 49, 102, 210, 211
flux, 71, 73, 75, 76, 83, 244, 245, 247, 256

meteoroid (cont.)
    uncertainty, 256, 272
    forces on, 162, 163, 238, 239
    heterogeneous, 55
    interstellar, see interstellar
    mass, 68, 69, 72, 75
    orbit, 70, 72, 74, 77, 82, 169, 236, 239
    origin, 235
    population, 235, 243
    size-density diagram, 55
    speed, 70–77, 256
    sporadic, 2, 50, 54, 127, 142–146, 168, 171, 202, 222–224
    strength, 27, 29, 51–53, 55, 102
        scaling law, 28, 29
    structure, 102
    vapor cloud, 10, 11, 20–22, 27
meteoroid stream, see also meteor shower, 119, 121, 122, 130, 162, 169, 171, 187, 189, 201
    definition, 169
    detection methods, 169
        break-point method, 170, 215
        clustering, 170
        iterative method, 214
        method of indices, 169, 214
        single linkage, 215
        wavelet transform method, 170, 214
    IAU definition, 210, 211
    identification methods, see detection methods
    individual
        δ-Aquariids, 167, 173
        η-Aquariids, 169
        Andromedids, 169, 196, 202
        April Lyrids, 140, 168, 169
        Arietids, 167, 168, 173
        Capricornids, 194, 202
        Carinids, 173
        Daytime Capricornids-Sagittariids, 194
        Geminids, 136, 140, 152, 168, 169, 171, 178, 188, 191, 192, 202, 256
        July γ-Pegasids, 177
        Leonids, 140, 152, 153, 164–166, 169–171, 188, 256, 267
        October Draconids, 169
        Orionids, 165, 166, 169
        Perseids, 152, 165, 166, 168, 169, 188, 256, 270
        Phoenicids, 171, 188, 196, 202
        Quadrantids, 51, 164, 167, 168, 172, 173, 176, 188, 193, 202
        Sekanina's (1973) Taurid-Perseids, 195, 197, 202
        September ϵ-Perseids, 154
        Southern α-Pegasids, 177
        Taurids, 140, 178, 194
        Ursids, 173
    mass index, 75, 141, 144
    modelling, 161, 210, 215, 217
    orphan, 164, 167
    parent body, 167, 172, 174, 176
    toroidal, 171
meteoroid swarm, 102, 142, 162, 196
Mg, 50, 51, 54, 55
    ions, 124, 126, 127
    neutral, 120, 124, 126
micrometeorite, 43–45, 47, 55
    porosity, 44
    scoriaceous, 44
    ultracarbonaceous, 44, 47

MIDAS, see Moon Impacts Detection and Analysis System
Middle Atmosphere Alomar Radar System, 66, 67, 72, 81–83
minor meteor shower, 216
    definition, 215
    source of, 216
mission, see also spacecraft
    Cassini, 245, 246
    Galileo, 246
    Giotto, 45
    Hayabusa, 37, 46
    Hayabusa2, 46
    Helios, 246
    LADEE, 136
    OSIRIS-REx, 46
    Rosetta, 37, 46, 56
    Stardust, 37, 45, 245
    STEREO, 245
    Ulysses, 245, 246
    Vega 1 and 2, 45
    WIND, 245
model, 276, 277, 280, 283, 286–288
    ablation, 12, 119, 120, 124
    air-beam, 20
    Direct Simulation Monte Carlo, 21
    dustball, 48, 104
    engineering, 255, 268
    fragmentation, see fragmentation model
    numerical simulation, 293
    spectra, 10, 20, 26
Moon Impacts Detection and Analysis System, 138
MORP, see Meteorite Orbit and Recovery Project, 93

Na, 49–51, 54, 55, 188, 191, 200
    ions, 126, 127
    neutral, 124, 130
NASA All-Sky Fireball Network, 92
NASA Cosmic Dust Program, 42
NASA Lunar Impact Monitoring System, 138
NEA, near-Earth asteroid, see near-Earth object
Near-Earth object, 188, 198, 201, 213, 217, 223
near-Earth object (NEO), 275, 290, 294
near-Sun object, 191, 197–199, 202
NELIOTA (NEO Lunar Impacts and Optical TrAnsients), 138
NEO, see near-Earth object
nodal point, 167, 171
numerical simulation, see model

observations, 239, 248
    data analysis, 96, 97
    history, 65, 66, 74, 90, 235
    in situ, 162, 236, 244, 247, 248
    infrared, 225
    meteor camera networks, 92, 94, 95, 218
    optical, 14, 47, 90, 97, 137, 225, 239
    photographic, 48, 91, 217, 218, 235, 240
    radar, 47, 218, 235, 241–244, 248
    radio occultation, 124–127
    scattered light, 225
    spectroscopy, 26, 50, 93, 94, 104, 188, 191, 202
    thermal infrared, 225
    ultraviolet, 225
    video, 48, 49, 91, 128, 129, 218, 236, 242
    video observation networks, 218
    visual, 90, 235

orbit
    heliocentric, 242
        elements, 235, 240, 241, 243
        hyperbolic, 235–237, 239–243, 247, 248
        parabolic, 239–242
        retrograde, 239, 242, 246
orbit-to-orbit distance, 121, 123
orbital debris, 137, 259
orbital similarity, orbital difference
    D-distance functions, D-criteria, 74, 169, 187, 202, 211, 217
        similarity threshold, 214, 215
        similarity threshold determination, 215
            break-point method, 170, 215
            statistical approach, 215, 216
ordinary chondrite, *see* meteorite
overpressure, 276, 279–282, 291–294

packing effect, 201
PANSY radar, 67
PFN, *see* Polish Fireball Network
photoionization, 198
planetary perturbations, 130, 170, 193, 195, 235, 236, 239
planetary shielding, 256
Polish Fireball Network, 94, 101
porosity, 37
potentially hazardous meteoroid stream, 195
Prairie Network, 93
precession
    general relativistic, 163
    secular, 167
presolar grains, 41, 44

radar
    collecting area, 73, 75–77, 83
    critical frequency, 67
    finite velocity effect, 76
    head echo, 18, 47, 65, 70, 83
    head echo aspect sensitivity, 71
    high-power large-aperture, 18, 65, 66
    HPLA, *see* high-power large-aperture
    initial radius effect, 72, 76
    limiting mass, 70, 72, 75–77
    Middle and Upper atmosphere radar, 67, 72, 73
    MU, *see* Middle and Upper atmosphere radar
    non-specular trail echo, 65, 81, 82
    overdense scattering, 66, 68, 69, 73
    PRF, *see* pulse repetition frequency effect
    pulse repetition frequency effect, 76
    radar cross section, 68–73
    RCS, *see* radar cross section
    specular meteor radar (SMR), 65, 73, 77
    specular trail echo, 65, 73
    trail, 47
    underdense scattering, 66, 73
radiant, 72–75, 142, 211, 222, 240–242
radiation efficiency
    impact, 140, 155
radio waves, 136
regolith, 46
resonance
    exterior, 164, 165
    interior, 164, 165
    Jovian, 164, 165, 172
    mean motion, 163–166, 172

    Saturnian, 165, 166
    three-body, 166, 179
    Uranian, 164, 165
risk factor, 276, 280, 281, 291–294
rotational period, 197
rubble pile, 37, 53, 55

SAAMER, *see* Southern Argentina Agile Meteor Radar
Saha function, 27, 200
salt, 51
seismic magnitude, 283, 286, 287, 294
seismic wave, 136, 276, 283, 286, 294
selenographic coordinates, 137
shock wave, 10, 11, 13, 16, 20, 22, 23, 25, 26, 29, 31, 54, 275–277, 279–281, 283, 286, 287, 289–291, 293, 294
slingshot effect, 239
sodium, *see* Na
solar gravity, 238
solar wind, 130, 162, 200, 201, 217, 226, 238, 266
SOMN, *see* Southern Ontario Meteor Network
Southern Argentina Agile Meteor Radar, 74, 75, 77, 80, 83
Southern Ontario Meteor Network, 92, 218
space debris, *see* orbital debris
space weathering, 46
spacecraft, *see also* mission, 245–248
    Akatsuki, 131
    Cassini, 125, 239, 244
    Deep Impact, 193
    DESTINY+, 193
    Galileo, 125, 236, 243, 244, 247
    Hayabusa, 107
    Helios, 244
    LADEE, 130
    Mars Express, 125, 127, 128
    Mars Global Surveyor, 125
    Mars Reconnaissance Orbiter, 127, 128, 270
    MAVEN, 124, 126–128, 130, 131, 270
    MESSENGER, 131
    Midcourse Space eXperiment, 120
    Pioneer 10, 125
    Pioneer 11, 125
    Pioneer Venus Orbiter, 120, 124, 125
    Solar Terrestrial Relations Observatory, 192, 246
    Stardust, 106
    STEREO, *see* Solar Terrestrial Relations Observatory
    Ulysses, 236, 238, 239, 243–248
    Venus Express, 124, 125
    Voyager, 125, 128
    WIND, 246
spacecraft damage and anomalies, *see also* damage
    Chandra, 267
    Gaia, 260
    Hipparcos, 260
    HST, 261
    ISEE-3, 266
    ISS, 261
    LISA Pathfinder, 260
    Mariner IV, 266
    Olympus 1, 266
    Swift, 260
    XMM-Newton, 259
Spanish Meteor Network, 92, 95
spin-barrier, 197
SPMN, *see* Spanish Meteor Network

sporadic background, 168–171, 174
sporadic meteoroids, 267, 269
    cometary-asteroidal classification, 222, 223
    one and two parameter C-A classification, 222, 223
superbolide, *see* bolide
surface boundary exosphere, 119, 123, 129–131

Tajikistan Fireball Network, 94, 95
taxonomy
    B-type, 191, 197
    Bus, 197
    C-type, 197
    Q-type, 197
    T-type, 197
    Tholen, 197
    V-type, 197
    X-type, 197
thermal desorption, 51
thermal process, 199
thermal radiation, 276, 277, 286–289, 291, 293, 294
Tisserand parameter, 188, 189, 223
trans-Neptunian planet, 179

Trojan
    Jovian, 191, 197
Tunguska, 31, 275, 276, 283–289, 291–294

ureilite, *see* meteorite

wake, 19–21, 25, 27, 31, 48, 54
Weibull theory, 52
Whipple shield, 259, 269

YORP effect, 162, 163, 198

zodiacal cloud, 43, 45, 56, 201, 224
    cometary dust trail, 227
    dust band, 225, 227
        asteroid cluster, 227
    F-corona, 225
    Gegenschein, 225
    model, 225–227
    observation, 225, 226
    source, 226
    spatial distribution, 225, 226
    zodiacal light, 45, 169, 224